Theory of Transformations
in Steels

Theory of Transformations in Steels

Harshad K. D. H. Bhadeshia

CRC Press
Taylor & Francis Group
Boca Raton London New York

CRC Press is an imprint of the
Taylor & Francis Group, an **informa** business

First edition published 2021
by CRC Press
6000 Broken Sound Parkway NW, Suite 300, Boca Raton, FL 33487-2742

and by CRC Press
2 Park Square, Milton Park, Abingdon, Oxon, OX14 4RN

© 2021 Taylor & Francis Group, LLC

CRC Press is an imprint of Taylor & Francis Group, LLC

Library of Congress Cataloging-in-Publication Data
Names: Bhadeshia, H. K. D. H. (Harshad Kumar Dharamshi Hansraj), 1953- author. Title: Theory of transformations in steels / Harshad K.D.H. Bhadeshia. Description: Boca Raton : CRC Press, 2021. | Includes bibliographical references and indexes. | Summary: "Written by the leading authority in the field, this work is the first to provide readers with a complete discussion of the theory of transformations in steel. It offers comprehensive treatment of solid-state transformations, covering the vast number in steels. It features discussion of physical properties, thermodynamics, diffusion, and kinetics and covers ferrites, martensite, cementite, carbides, nitrides, substitutionally-alloyed precipitates, and pearlite. It also contains a comprehensive list of references as further and recommended reading. With its broad and deep coverage of the subject, this work aims at inspiring research within the field of materials science and metallurgy"-- Provided by publisher. Identifiers: LCCN 2020043503 (print) | LCCN 2020043504 (ebook) | ISBN 9780367518080 (hardback) | ISBN 9781003056782 (ebook) Subjects: LCSH: Steel alloys. | Physical metallurgy. | Phase transformations (Statistical physics) Classification: LCC TN756 .B49 2021 (print) | LCC TN756 (ebook) | DDC 669/.96142--dc23 LC record available at https://lccn.loc.gov/2020043503 LC ebook record available at https://lccn.loc.gov/2020043504

ISBN: 978-0-367-51808-0 (hbk)
ISBN: 978-1-003-05678-2 (ebk)

Typeset in Computer Modern font
by KnowledgeWorks Global Ltd.

To my late father and mother.

Contents

Preface

There are books that are enjoyable to read, study and re-read. Amongst them is Ziman's *Models of disorder* published shortly after I started my own research. Ziman considered steel and glass, earth and water to be disordered phases of condensed matter, even though the opening page of his first chapter identifies a crystal as an example of the highest degree of spatial order, other than that of the mythical perfect vacuum. But he could be forgiven for associating steel with the level of order in glass and water, because by definition, steel never is a pure crystal – if atoms other than iron or vacancies within the structure are not ordered, then the strict meaning of spatial order as taught to undergraduates is lost; there is no long-range periodicity in an unordered crystalline solution. From my perspective, Ziman's beautifully composed book was readable in parts and challenging otherwise. That, I concluded, is the essence of a good book because the challenge inspires, as it does so in sport. An absence of challenge might make the book dull to read, but if the challenge is insurmountable, then the book is abandoned. I took this as a working principle in my teaching. An important caveat: challenge does not equate with poor delivery.

Another book that I would love to take with me if I ever were to be abandoned on a deserted island is Christian's *Theory of Transformations in Metals and Alloys*, the most respected book on the subject. I spent an entire sabbatical studying the 1975 edition from cover-to-cover, now a bedraggled copy that I continue to cherish. I also have the complete and final edition, but I remain loyal to my well-used version. Christian also wrote the most amazing, concise and deep reviews on a variety of transformation-related topics, many of which formed the foundations of understanding. I got to know Professor Christian through some exhilarating discussions and we managed eventually to pen a really quite nice paper together.

Given Christian's writings, why did I find it necessary to write this book? The simple fact is that there is a great deal of theory specific to steels that is not covered in any other book. There are monographs dealing with specifics such as Nishiyama's book on martensitic transformations, chemical thermodynamics by Lupis, or my own on bainite. The monographs taken together do not cover the spectrum of transformations in steels. And there is a lot to gain from a single text that can enable the big-picture of the myriads of transformations possible in steels to be revealed and rationalised. By this I do not claim that there is a grand unification theory, but it is possible, finally, to go a fair way towards that notion.

Iron and its alloys laid the foundations of metallurgy, perhaps beginning with Sorby's revelation of the microcrystalline structure of steel, followed by many attempts to relate such observations to the behaviour of the metal. It could be argued similarly that materials science, as opposed to the basic sciences, is essentially the study of relationships between structure and properties. And metallurgy from the beginning has been interdisciplinary – Sorby began his microscopy with the study of rocks. Adolf Martens, after whom martensite is named, began as a locksmith; his sketches of the microstructure of steel are consummate works of art and precision, that reveal structure to the naked eye. The term *interdisciplinary* is a modern caricature; scholarship in the not-so-distant past was simply that, with no distinction between the humanities and sciences. It was all about natural philosophy. Mechanical engineers determined the chemical composition of cementite back in 1885. A decade before that, William Henry Johnson, the Managing Director of the 'Iron Metal and Electrical Company' discovered hydrogen embrittlement; his paper on the subject contains the essence what we know today about the pernicious role of the lightest element on the properties of the most stable element in the universe. This passage of time since the study of iron became a yearning means that there is a great deal known about iron and its alloys across the modern version of disciplines, be it astrophysics or the history and philosophy of science.

Given this long-standing learning associated with iron and its alloys, the subject today is not for the faint hearted. As Ziman stated in another context, the subject is no longer "an elementary scientific discipline, to be entered without long preparation". I hope that this book helps in establishing a marker from which to create without falling into the all too common trap of re-discovery. In this respect, I must acknowledge *Google Scholar* in helping me discover the most lonely of articles. Who would have thought, for example, that titanium carbide with its foreboding hardness, undergoes martensitic transformation? And more, that its features fit like a glove to the crystallographic theory of martensite.

Without thermodynamics in all its forms, it would not be possible to understand the most peculiar behaviour of iron, even in its purest form. It's allotropic transformation at ambient pressure from ferrite and then back into ferrite is not replicated by any other substance. The cause, revealed by an analysis of its heat capacity, is entirely a consequence of the magnetic characteristics of ferrite and austenite, rather than anything to do with the thermal component of entropy. This is why, for many decades, first principles calculations failed to reproduce the correct ground state of iron. In fact, even austenite has a variety of magnetic states – of vital importance in explaining its large thermal expansion coefficient that makes austenitic steels less useful in the construction of power plant. To deal with this, I have chosen to invest in the Weiss two-state theory because it captures the essence of the problem, rather than the first principles approach which reveals somewhat more detail but not to an extent that is useful. On the other hand, first principles calculations for terapascal pressures have revealed that at ambient temperatures, the ferrite when pressurised transforms into the hexagonal allotrope, then at higher pressures back into austenite and then into a slightly tetragonal form of ferrite; at the moment, these last two stages are not amenable to experimental validation and do not account for magnetism. Quantitative thermodynamics is now routine in steel metallurgy, thanks to the pioneering efforts that led to the creation of databases and the algorithms to implement those assessed data in the estimation of phase equilibria in multicomponent steels. I have tried to capture the many ways in which thermodynamics has revealed the secrets of steels and has placed boundaries on hypotheses such as the role of host vacancies on the diffusion of interstitials. The story on thermodynamics would not be complete without an excursion into the irreversible version, which helps in dealing with steady-state processes driven by multiple forces and fluxes.

The detail with which we understand diffusion in iron is astonishing. Diffusion is driven by a free energy gradient, that much is clear. This sometimes explains the concentration dependence of diffusivity, but not entirely in the case of carbon. There is a strong repulsive force between carbon atoms in adjacent sites, which further promotes its migration along a gradient of concentration. Hydrogen too has its foibles, that it is easily trapped during its passage through steel, making the apparent diffusion coefficient much smaller than expected. There is an anomaly in the migration of large atoms at the Curie transition in ferrite. Onsager's irreversible thermodynamics is applied generously in dealing with multicomponent or multiforce driven migration.

After dealing with thermodynamics and diffusion in some detail, I chose to focus on ferrite and martensite. This is because they represent the archetypal reconstructive and diffusionless transformations respectively. As such, they lay the foundations for broad-ranging phenomena which need not be repeated when dealing with any other transformation in steels. For example, the structure and consequences of the structure of transformation interfaces, the rate-controlling phenomena and interface response-functions, evidence-backed nucleation mechanisms, the geometric partitioning of the austenite, and so on. There are some logical anomalies that are revealed during this journey.

The theory for bainite has contributed much to the development of novel steels, a subject covered comprehensively in the three editions of *Bainite in Steels*. I present instead a decisive exposition of just the theoretical framework based on harsh observations rather than pander to hypotheses that neglect the totality of evidence. The story of displacive transformations (excluding carbides) concludes in the chapter on Widmanstätten ferrite, with an exciting mathematical rationalisation of martensite, bainite and Widmanstätten ferrite. It becomes clear why all three of these products of austenite decomposition do not necessarily occur in all steels, with an attractive reconciliation of theory and observations.

There are some hidden depths to the precipitates that occur in steels, i.e., graphite, carbides, nitrides and intermetallic compounds. In this context, a number of chapters cover their mechanisms, crystallography, where possible a quantitative explanation of orientation relationships, thermodynamic stability and physical properties. A knowledge of the crystallographic methods used is assumed because they are dealt with in the book *Geometry of crystals, polycrystals and phase transformations*. There are some titillating outcomes that explain shapes and growth directions. Many of these precipitates are metastable so they may have a fleeting existence but nevertheless leave a legacy of influence on subsequent reactions. This is where the reader will first encounter methods for dealing with serial and parallel chains of reactions.

Pearlite is perhaps better described as a clever way in which nature permits the decomposition of austenite by diffusion, but without changing its average composition, at least in the Fe-C system. But even in this simple binary alloy, theory does not yield a growth rate, but rather a Péclet number which is the product of the rate and an interlamellar spacing. Like it or not, there is no clear way of deciphering this number; only experiments can reveal whether the spacing actually obtained corresponds to a particular hypothesis, such as Zener's assumption that growth should occur at a maximum rate. When it comes to substitutionally alloyed steels, the theory is a poor predictor if local equilibrium is assumed at the transformation fronts.

Kinetic theory is introduced throughout the book, but it was necessary to add a final chapter that does not deal with individual transformations, but a more realistic scenario in which many reactions occur at the same time, albeit at different rates. There also is a limited treatment of grain growth, recrystallisation and thermomechanical processing. Although solidification *is* a transformation, I have avoided it primarily because I lack the expertise. I hope the reader can find solace elsewhere. I could have limited the title to solid-state transformations, but I thought that would be clumsy, rather like the formal and lengthy titles of some heads-of-state.

This book was written over a time period that is too painful to mention. Time seems to be the only resource that humans cannot muster at will. But I would like to finish by mentioning a few names. I recall with affection how the late Professor Robert Honeycombe, who as an external examiner for my undergraduate degree at the City of London Polytechnic, recruited me to Cambridge for research in his steels group. He and my immediate supervisor David Edmonds gave me the freedom to pursue ideas even when they believed that I was going down alleyways and therefore likely to be mugged! Professor Jack Christian of Oxford had a thrilling and lasting influence on my understanding of theory. He too was a scholar and a gentleman who treated a mere student from another University without regard for status. I was able to spend some quality time with Professor Morris Cohen and Greg Olson at MIT when we explored strange couplings in steels. I never had the opportunity to publish with Professor Colin Humphreys, but owe a lot to him as a human being, for his selfless nurture. Professor Hubert Aaronson and I were deep adversaries when it came to the theory of bainite, but strangely, I suffer when some of the work he and others did on the ledge mechanism of ferrite growth is neglected in modern publications where unrealistic hypotheses are proposed to bring a closure between ill-conceived theory and ill-designed experiments.

The unexpected endowment of my Chair by Tata Steel touched me to the core; the value of this generosity will continue in perpetuity. I have visited what seems like a vast part of the planet during my academic adventures, but South Korea was special; I was recruited by Professor Hae-Geon Lee to POSTECH and spent 3-4 months each year for ten years working with amazing people on steels and steels alone. But I have to say, of all the places in all of the world, Cambridge to me is just that bit extra special.

Harshad K. D. H. Bhadeshia
Cambridge

Author

Harshad K. D. H. Bhadeshia is the Tata Steel Professor of Metallurgy at the University of Cambridge. His main interest has been on the theory of solid-state phase transformations with emphasis on the prediction and verification of microstructural development in complex metallic alloys, particularly multicomponent steels. He has authored or co-authored some 700 research publications. His books include *Bainite in steels* (3rd edition 2015), *Steels* (4th edition 2017, with Robert Honeycombe), *Geometry of crystals, polycrystals and phase transformations* (2018) and *Innovations in everyday engineering materials* (2021, with Tarashankar DebRoy). He became a Fellow of the Royal Society in 1998, of the Royal Academy of Engineering in 2002, and a Foreign Fellow of the National Academy of Engineering (India) in 2004 and was awarded the Bessemer Gold Medal in 2006. He was knighted for 'services to science and technology' in 2015.

Acronyms etc.

bcc Body-centred cubic

bct Body-centred tetragonal

fcc Face-centred cubic

hcp Hexagonal close-packed

F Face-centred

I Body-centred

IC Interface composition contour

IV Interface velocity contour

ILS Invariant-line strain

IPS Invariant-plane strain

KS Kurdjumov-Sachs orientation relationship

NP-LE Negligible partitioning local equilibrium

NW Nishiyama-Wasserman orientation relationship

PLE Partitioning local equilibrium

P Primitive

R Trigonal

TTT Time-temperature-transformation

The use of braces, for example, $x\{y\}$ implies a functional relationship, i.e., x is a function of y.

Vector and matrix notation Bowles and MacKenzie's notation is particularly useful in avoiding confusion between frames of reference. A set of vectors $\mathbf{a}_i$ ($i = 1, 2, 3$) defines the coordinate system with a basis symbol 'A'. A vector $\mathbf{u}$ in real space, referred to this basis will have components u_i along $\mathbf{a}_i$, i.e.,

$$[A; \mathbf{u}] = [u_1 \ u_2 \ u_3]$$

where the square brackets imply a column vector; round brackets are read as row vectors. The components of $\mathbf{u}$ will be different in another basis 'B':

$$[B; \mathbf{u}] = (B \ J \ A)[A; \mathbf{u}]$$

where (B J A) is a 3×3 coordinate transformation matrix relating the two bases. Note the juxtapositioning of the basis symbols, which reduces the chances of making errors during the expression of a chain of separate operations. The inverse of (B J A) is written (A J B) and its transpose as (A J' B). A deformation referred to basis 'A' is written (A S A) and it is noted that a rigid body rotation falls in this category:

$$[A; \mathbf{v}] = (A \ S \ A)[A; \mathbf{u}], \qquad [B; \mathbf{v}] = (B \ J \ A)(A \ S \ A)[A; \mathbf{u}].$$

A plane normal is a vector in reciprocal space, generally written as a row vector $(\mathbf{h}; A^*)$ where A^* is the reciprocal basis corresponding to the basis A:

$$(\mathbf{h}; B^*) = (\mathbf{h}; A^*)(A \ J \ B)$$

Methods of dealing mathematically with crystallography, interfacial structure and homogeneous deformations are described thoroughly in the book *Geometry of crystals, polycrystals and phase transformations*, so are not reproduced here.

Nomenclature

α	Ferrite, or sometimes specifically, allotriomorphic ferrite
α''	$Fe_{16}N_2$ nitride
α'	Martensite
α_a	Acicular ferrite
α_b	Bainite
α_i	Idiomorphic ferrite
α_{lb}	Lower bainite
α_{ub}	Upper bainite
α_W	Widmanstätten ferrite
α_{id}	Parabolic rate constant, numerical subscript referring to the dimension
e_V	Volume expansion coefficient
β	Isothermal compressibility or angle
β_O, β_{Te}	Number of octahedral (O) or tetrahedral (Te) interstices per solvent atom
β_p	Parameter in the Aziz theory for solute-trapping
β_q	Quantity in quasi-chemical solution model
γ	Austenite
γ'	Fe_4N nitride
Γ	Activity coefficient
δ_b	Grain boundary thickness
δ_s	Interplanar spacing
Δ	Uniform dilatational strain
$\dot{\varepsilon}$	Strain rate
ε	Elastic strain
ε_*	Strain during transfer of interstitial between adjacent sites
ε_λ	Strain caused by transferring C atom between not-preferred to preferred sites
$\varepsilon_{C\square}$	Binding energy between carbon and iron-vacancy
ε_{uu}	Energy per carbon-carbon pair
ε_u	Energy per interstitial-vacancy/carbon pair
ε	$Fe_{2.4}C$ or $Fe_{2-3}N$ nitride
ε	Hexagonal close-packed iron
ε_p	Plastic strain
ζ	Fe_2N nitride
ζ	Dilatational strain normal to habit plane
η_i	Principal distortions, the subscript identifying the principal axis
η	Fe_2C
θ	Ratio of number of C atoms to total number of solvent atoms
θ_i	Fraction of sites occupied by interstitials on plane i
θ	Cementite
$\Theta_1, \Theta_2, \Theta_3$	Functions of z, D and t in diffusion-controlled growth theory
λ	Parameter in quasi-chemical interstitial-solution model
λ^*	Value of λ when the interstitial atoms are randomly distributed
$\overline{\lambda}$	λ yielding largest term in summation contained in partition function
μ°	Gibbs free energy per mole of pure substance
μ_i	Chemical potential per mole of element i in solution
μ_B	Bohr magneton
L	Liquid

ν	Attempt frequency in diffusion, or Poisson's ratio in elasticity
ρ	Density; ledge height, isolated or leading ledge in a train of ledges
ρ^*	Critical height of ledge for successful nucleation
$\rho_\perp$	Dislocation density
ρ_i	Height of trailing ledge divided by ρ
ρ_A	Density of atoms in a close-packed plane, moles per unit area
σ_+, σ_-	Energy per unit length of positive or negative interface edge
σ_0	Stress driving the motion of a grain boundary in a pure material
σ_d	Impurity drag-stress on grain boundary
σ_f	Stacking fault energy per unit area
σ_P	Flow stress for plastic deformation
$\sigma_{\alpha\gamma}$	Energy per unit area of α/γ interface
τ	T/T_C, or time in overall kinetics
ϕ	Parameter in quasi-chemical solution model, or electrical potential
ϕ_1, ϕ_2	Variables in solute-drag theory
ϕ_O	Fraction of carbon atoms in octahedral interstices
ϕ_{Te-Te}	Fraction of carbon atoms in tetrahedral interstices, which jump by a Te-Te route
Φ	Electrical field, or capillarity constant
Φ_i	Functions in theory for precipitate growth, martensite nucleation, etc.
χ	Hägg carbide, $Fe_{2.2-2.5}C$
ψ	Fraction of austenite region that transforms into a plate of martensite
ψ	Order parameter in phase field theory
ω	Change in binding energy per atom during mixing to form binary solution
ω_γ	Carbon-carbon interaction energy in phase γ
ω_D	Debye frequency
Ω	Partition function
a	Activity, or atomic percent of element identified in subscript
a_α	Lattice parameter of phase α
a_γ	Lattice parameter of austenite
a_i^α	Activity of i in phase α
a_m	Activity of activated complex
A	Atomic weight, or area of interface
Ac_3	Temperature at which a sample becomes fully austenitic during heating
Ae_1	Temperature separating the $\alpha+\gamma$ and α phase fields for a specific alloy
Ae_3	Temperature separating the $\alpha+\gamma$ and γ phase fields for a specific alloy
Ar_3	Temperature at which γ begins to transform to α during cooling
b	Number of lattice points per unit cell, or magnitude of Burger's vector
b_i	Empirical fitting constants
B_i	Number of sites excluded from occupation on plane i
$\bar{c}^\gamma$	Average carbon concentration in the austenite
$\bar{c}_i$	Average concentration of i in alloy, moles per unit volume
c	Molar concentration per unit volume
c_i	Concentration of component i, moles per unit volume
$c_i^{\alpha\gamma}$	Concentration of i in α in equilibrium with γ, moles per unit volume
c_i^α	Concentration of i in α at α/γ interface, moles per unit volume
$c_i^{\gamma\alpha}$	Concentration of i in γ in equilibrium with α, moles per unit volume
c_i^γ	Concentration of i in γ at γ/α interface, moles per unit volume
c_b	Concentration of solute in a stationary grain boundary
c_e^α	Electronic specific heat coefficient for a mole of phase α
c_V	Specific heat capacity per unit volume

c_V^μ	Magnetic component of specific heat capacity per unit volume
$c_{i\alpha}$	Concentration of i in homogeneous phase α, moles per unit volume
$c_{i\gamma}$	Concentration of i in homogeneous phase γ, moles per unit volume
C	Euler's constant, $0.5772\ldots$
C_1	Function relating C_P and C_V
C_e^α	Electronic specific heat coefficient for phase α
C_L	Lindemann constant associated with Debye temperature
C_P	Specific heat capacity at constant pressure
C_P^e	Electronic specific heat function (constant pressure)
$C_P^{\mu\alpha}$	Magnetic component of the specific heat capacity of phase α at constant pressure
C_V	Specific heat capacity at constant volume
C_V^e	Electronic specific heat function (constant volume)
C_V^L	Debye specific heat function
$C_V^{\mu\alpha}$	Magnetic component of the specific heat capacity of phase α, at constant volume
C_{ij}	Elastic stiffness constants
D_{11}^*	Tracer diffusion coefficient for carbon
ΔK	Fraction of kinetic energy of the migrating atom that leads to diffusion
$\overline{D}$	Interdiffusion coefficient defined relative to the laboratory frame of reference
$\overline{D}_i$	Intrinsic diffusivity of component i
D	Chemical or interdiffusion coefficient for a binary solution
D^*	Tracer diffusion coefficient
$D^{\text{Te}-\text{Te}}$	Diffusion by path indicated in superscript
D_o	Pre-exponential factor in diffusion
D_T	Thermal diffusivity
D_{ij}	Chemical or interdiffusion coefficient for a ternary solution
m	Mass of an atom
e	Coefficient of thermal expansion; fundamental charge on an electron
e_{ik}	Wagner interaction parameter
E	Young's modulus of elasticity
E'	Term in the equation for the elastic strain energy for a martensitic transformation
E_i	ith energy level in partition function
E_B	Binding energy per atom in a crystal relative to that of a free atom
E_F	Fermi energy
E_s	Shear modulus
Ei	Exponential integral function, Abramowitz and Stegan (1964)
f^*	Attempt frequency for atoms jumping across boundary
f_μ	Fraction of magnetic component of enthalpy retained beyond T_C
f_i	Fractional supersaturation of i, or dimensionless concentration
f_c	Bardeen-Herring correlation factor in diffusion theory; also, distance between the focus of parabola and its tip
f_o	Fraction of component of enthalpy retained beyond T_C during chemical ordering
F	Helmholtz free energy
F, f	Functions arising in diffusion-controlled growth theory
$F^{\mu\gamma}$	Magnetic component of the molar Helmholtz free energy of γ
F_1, F_2	Functions of concentration
g_i	Degeneracy of energy level i in partition function
$\Delta G'$	Total free energy available to drive an interface
$\Delta G'\{x,\bar{x}\}$	Molar Gibbs free energy change accompanying the transfer of a small amount of material of composition x, from γ of composition $\bar{x}$, to α of composition x in Fe-C
$\Delta G^{\alpha\gamma}$	Molar Gibbs free energy change for $\alpha \to \gamma$ transformation, $= G^\gamma - G^\alpha$

$\Delta G_3^{\gamma\alpha}$	Molar Gibbs free energy change for $\gamma \to \alpha$ transformation in pure Fe, $= G^\alpha - G^\gamma$
ΔG_F	Gibbs free energy of formation for a vacancy
ΔG_M	Change in Gibbs free energy per mole due to mixing during formation of a solution
ΔG_m	Maximum free energy change accompanying nucleation
ΔG_V	Gibbs free energy change per unit volume
$\Delta_e G$	Excess free energy per mole of solution
G	Gibbs free energy
G'	Rate of change in activation energy with carbon concentration
G^*	Activation free energy
G_M^*	Activation free energy for the migration of a vacancy
$G^{\mu\gamma}$	Magnetic contribution to the molar Gibbs free energy of γ
G^e	Strain energy per mole
G^α	Molar Gibbs free energy of phase α
G_{Debye}	Molar Gibbs free energy component due to Debye specific heat function
G_D	Free energy dissipated in diffusion
G_I	Free energy dissipated in interfacial processes
G_O^*	Activation free energy for a jump from an octahedral to tetrahedral site
G_s	Solute-boundary interaction energy
G_{Te}^*	Activation free energy for a jump from a tetrahedral to octahedral site
$G_t\{x,\bar{x}\}$	Molar Gibbs free energy corresponding to a point x on a tangent at point $\bar{x}$ on the γ free energy curve for Fe-C
G_V^e	Elastic strain energy per unit volume
h	Planck constant
H_T	Heat of transport per mole of atoms during thermomigration
H_M^*	Activation enthalpy for migration of vacancy or atom
$\Delta H^\mu\{T\}$	Excess component of magnetic enthalpy as a function of temperature
$\Delta H^{\alpha\gamma}$	Molar enthalpy change accompanying the transformation of α to γ
ΔH_F	Enthalpy of formation of a vacancy
ΔH_M	Change in enthalpy due to mixing during formation of a solution
$\Delta_e H$	Excess enthalpy per mole of solution
$\bar{H}$	Partial molar enthalpy with respect to solute atom at rest in vacuum
H	Enthalpy
H_0	Enthalpy at 0 K
H_{mag}	Magnetic field strength
i	Electrical current
i_*	Electrical current density
$\mathbf{I}$	Component array in generalised regular solution model
I_V	Nucleation rate per unit volume
$\bar{J}_i$	Flux of i relative to Kirkendall frame
J	Diffusion flux
$J_\square$	Flux of vacancies
J_i	Diffusion flux of species i
J_B	Flux defined relative to a volume-fixed frame of reference
k	Boltzmann constant
k_e	Equilibrium solute-partitioning coefficient
k_p	Solute-partitioning coefficient that deviates from equilibrium
K	Bulk modulus of elasticity
K_0	Modified Bessel function of zero order
v_ℓ	Plate or needle lengthening rate
$\bar{L}$	Mean lineal intercept defining grain size

$\bar{L}_o$	Mean lineal intercept defining initial grain-size
L	Long-range order parameter, or liquid
L_{sr}	Short-range order parameter
$L_{AB,C}$	Binary interaction parameter in two-sublattice solution model
L_{ABC}	Ternary interaction parameter in solution model
L_{AB}	Binary interaction parameter in solution model
m^+ or m^-	Transmission coefficient in diffusion theory
m_A	Number of atoms per particle of A
M	Magnetisation, or mobility in phase-field theory
M_{ij}^K	Mobility coefficient in Kirkendall frame of reference, diffusion theory
M_i, M_{ij}	Mobility coefficient in diffusion theory
M_A	Mobility of A atoms during diffusion in a chemical potential graident
M_A^e	Mobility of A atoms during migration in an electrical potential gradient
M_A^T	Mobility of A atoms during migration in a thermal gradient
M_b	Boundary mobility
M_m	Maximum magnetisation
M_S	Martensite-start temperature
M_{ik}	Coefficient relating force X_k to flux J_i in Onsager relations
n	Number of planes in a stacking fault
n_s	Number of ledges per unit length
N	General designation of total number
N^{Te}, N^O	Number of occupied tetrahedral or octahedral interstices
N_i	Number of γ atoms in level i
N_{AA}	Number of A-A bonds
N_A	Number of A atoms
N_a	Avogadro's number
N_b	Number of lattice points per unit cell
N_e	Number of free electrons in sample
N_V	Number per unit volume
N_V^o	Initial number per unit volume
O	Boundary area
p	Momentum; probability; Péclet number
p_a	Autocatalysis factor in nucleation model
P	Pressure
P	Pearlite
q	Heat transferred into a system
Q	Activation energy
r_c	Critical particle radius at which the growth rate becomes zero
r_i	Radius of an interstitial atom
r_o	Radius of largest sphere that can be accommodated without distortion in an interstice
R	Universal gas constant
$\bar{s}_a$	Mean spin imbalance per atom in an alloy
s	Shear strain parallel to habit plane
s_a	Net spin imbalance per atom
S_M^*	Activation entropy
ΔS_F	Entropy of formation of a vacancy
ΔS_M	Change in configurational entropy due to mixing during formation of a solution
$\Delta_e S$	Excess entropy per mole of solution
$\dot{S}$	Rate of entropy production
$\bar{S}^v$	Partial molar non-configurational entropy

S	Entropy
S^*	Activation entropy
S_M^*	Activation entropy for the migration of a vacancy
$S^{\mu\gamma}$	Magnetic contribution to the molar entropy of γ
S_I	Interlamellar spacing in pearlite
S_{ij}	Elastic compliance constants
T_N	Néel temperature
T	Absolute temperature
T_0	Temperature at which α and γ have identical composition and free energies
T_0'	As for T_0 but taking strain energy into account
T_a	Thermal arrest temperature during cooling
T_C	Curie temperature
T_D	Debye temperature
T_E	Eutectoid temperature
T_F	Fermi temperature
T_m	Melting temperature
T_q	Temperature to which γ is quenched, with $T_q \le M_S$
U	Internal energy
$U^{\mu\gamma}$	Magnetic component of molar internal energy of γ
U_O	Site energy for an octahedral interstice
U_C	Energy to move C atom from preferred to a less favoured octahedral site
U_{rep}	Short-range repulsive component for energy of C-C interaction
U_{Te}	Site energy for a tetrahedral interstice
$\square$	Vacancy
v	Speed with which an interface moves
v_K	Velocity of Kirkendall markers
v_s	Step velocity
$\overline{V}_i$	Partial molar volume of component i
ΔV_F	Change in volume of formation of defect in standard state
ΔV_M	Change in volume when defect reaches saddle point during migration
$\overline{V}$	Mean volume
$\overline{V}_i$	Partial molar volume of component i in an n component solution
V	Volume
V^*	Activation volume
V_0	Volume at ambient pressure
V_a	Volume per atom
V_m	Molar volume
V_V	Volume per unit volume, i.e. the volume fraction
w	Work done by a closed system
w_i	Weight percent of solute i
W	Number of octahedral interstices around a single such interstice
$\overline{x}$	Average concentration in alloy
$\overline{x}_I^\alpha$	Concentration at interface in phase α
x	Concentration, mole fraction
$x_i^{\alpha\gamma}$	Concentration of i in α in equilibrium with γ
x_i^α	Concentration of i in α at α/γ interface
$x_i^{\gamma\alpha}$	Concentration of i in γ in equilibrium with α
x_i^γ	Concentration of i in γ at γ/α interface
$x_{i\alpha}$	Concentration of i in homogeneous phase α
$x_{i\gamma}$	Concentration of i in homogeneous phase γ

X_i	Force term in the Onsager force-flux relations
y	Fraction of atoms in ferromagnetic form of γ
y'_A	Site fraction of A in two-sublattice solution model
$y_{A,s}$	Site fraction of A within specified sublattice s
y_A	Concentration expressed as site fraction of A on a lattice
$\mathbf{Y}$	Matrix of site fractions in generalised regular solution model
z	Position coordinate, or coordination number
z_ℓ	Length of plate modelled as oblate spheroid
z_t	Thickness of plate modelled as oblate spheroid
Z	Position of interface
$Z_\uparrow$	Number of nearest neighbours with spin in favoured direction
Z_k	Zener-Holloman parameter representing combined effect of temperature and strain rate
Z_v	Effective valency

1 Crystal structures and mechanisms

1.1 ALLOTROPES OF IRON

Why do metals adopt the crystal structures that they do? This no longer is a curiosity because the metallic state is so well understood that it is possible to select, from a calculation of the cohesive energies of trial structures, that which should be the most stable. Figure 1.1 shows the cohesive energy as a function of the density and crystal structure. Of all the test structures, hexagonal close-packed (hcp) iron is found to have the highest cohesion and therefore should represent the most stable form. This contradicts experience, but the calculations do not account for the ferromagnetism of body-centred cubic iron (ferrite), which would make it more stable than the hcp form. There are in fact magnetic transitions in each of the allotropes of iron, details of which are reserved for Chapter 2.

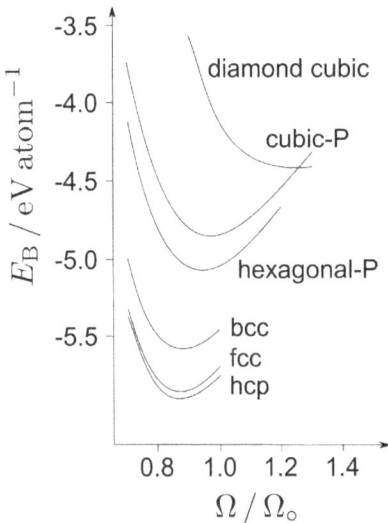

Figure 1.1 Plot of cohesive energy E_B for $0\,\mathrm{K}$ and $0\,\mathrm{Pa}$ pressure versus the normalised volume per atom for a variety of crystal structures of iron. E_B is the binding energy per atom in the crystal relative to that of a free atom. Hexagonal-P and Cubic-P are primitive structures; like the diamond cubic form, they do not exist on earth. Adapted with permission from [1]. Copyrighted by the American Physical Society.

Only three allotropes of iron occur in nature, in bulk form; Figure 1.2 shows the phase diagram for pure iron. Each point on a boundary between the phase fields represents an equilibrium state in which two phases can coexist. The *triple point* where the three boundaries intersect represents an equilibrium between all three co-existing phases. It is seen that in pure iron, the hexagonal close-packed form is stable only at high pressures, consistent with its high density.

There are two further allotropes which can be created in the form of thin films. Face-centred tetragonal iron can be prepared by coherently depositing iron as a thin film on a $\{100\}$ plane of a substrate such as copper with which the iron has a mismatch. The atoms in the first deposited layer replicate the positions of those on the substrate. A few monolayers can be forced into coherency in the plane of the substrate with a corresponding distortion normal to the substrate. This gives the deposit a face-centred tetragonal structure which in the absence of any mismatch would be face-centred cubic [2, 3]. Eventually, as the film thickens during the deposition process to beyond about ten monolayers (on copper), the structure relaxes to the low-energy bcc form, a process accompanied by the

1

formation of dislocation defects which accommodate the misfit with the substrate.

Growing iron on a misfitting $\{111\}$ surface of a face-centred cubic substrate similarly causes a distortion along the surface normal, giving trigonal iron [4].[1] Graphene is a single layer of carbon atoms that has a hexagonal structure with a lattice parameter of 0.245 nm. The crystal is seldom perfect, often containing holes. Compounds used in its manufacture provide a source of iron atoms that can attach themselves to the edges of these holes, building up into monoatomic layers that are suspended in the graphene films. These "free-standing" single-atom-thick two-dimensional arrays of iron form a square lattice with a parameter 0.265 ± 0.005 nm [5]. This particular structure has been suggested to be a consequence of lattice matching with the "armchair" configuration of carbon atoms at the graphene edges where the bonding between the iron and carbon atoms is stronger than when the iron attaches to the "zig-zag" edges of the graphene [5]. The largest stable monolayer of iron on holes in graphene is found to be about 3 nm^2 in area.

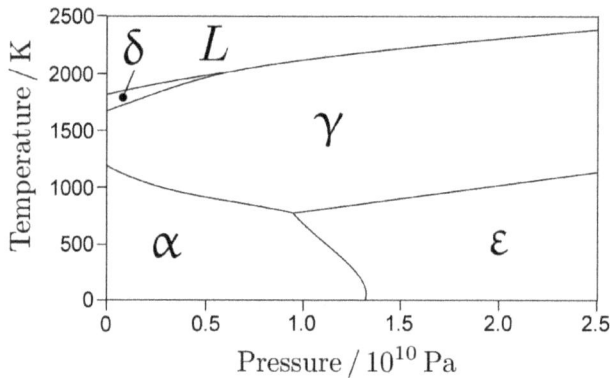

Figure 1.2 Temperature versus pressure equilibrium phase diagram for pure iron, based on assessed thermodynamic data. The triple point temperature and pressure are 490°C and 11 GPa respectively. α, γ and ε refer to ferrite, austenite and ε-iron respectively. δ is simply the higher temperature designation of α. Diagram courtesy of Shaumik Lenka, calculated using the database 'TCFE8', and ThermoCalc version 2015b.

Table 1.1 lists some transformation temperatures and thermodynamic data for the natural allotropes of iron. The free energy changes during the solid-state transformations are really quite small, even when compared with the energy associated with the magnetic disordering of ferrite. In a way, this is a reflection of the fact that their crystal structures are not all that different (Figure 1.3). For example, assuming a hard-sphere model, the crystal structure of austenite and ε-iron can be generated by stacking close-packed layers of iron atoms in such a way that each successive layer sits in the depressions of the underlying layer. There are in each layer, two types of depressions, so the close-packed layers can be stacked either in the sequence ... $ABCABCABC$... or ... $ABABABABA$.... The former sequence, which has a stacking period of three, generates the austenite structure, whereas the latter gives the hcp structure of ε-iron with a stacking period of two. The best comparison of the relative densities of the phases is made at the triple point where the allotropes are in equilibrium at an identical temperature and pressure and where the sum of all the volume changes is zero:

$$\left.\begin{array}{l} \Delta V(\text{bcc} \rightarrow \text{hcp}) = -0.34 \\ \Delta V(\text{hcp} \rightarrow \text{fcc}) = +0.13 \\ \Delta V(\text{fcc} \rightarrow \text{bcc}) = +0.21 \end{array}\right\} \quad \text{cm}^3\,\text{mol}^{-1}.$$

Ferrite is the least dense of all the allotropes under ambient conditions.

The fact that the hcp iron is denser than austenite is not expected from the hard sphere approximation, which also predicts an ideal lattice parameter ratio for hcp (c/a) of $\sqrt{8/3} \simeq 1.633$. Experimental measurements at high pressures indicate that the ratio is close to ideal, but the results do not

Table 1.1

Transformation temperatures and thermodynamic data for pure iron at ambient pressure (after Hoffman, Tauer, Paskin, Weiss, Chipman and Orr [6–9]). The transformation temperatures are consistent with the International Practical Temperature Scale, which was in 1968 modified by raising the designated melting point of palladium by 2 K. T_C^α is the Curie temperature for the transition between the ferromagnetic and paramagnetic states of ferrite; the energy and entropy quoted alongside T_C are those required to completely disorder ferromagnetic iron. The approximate Curie and Néel temperatures for the two states of austenite are also listed.

	$T/°C$	T/K	Enthalpy change $\Delta H/\mathrm{J\,mol}^{-1}$	Entropy change $\Delta S/\mathrm{J\,K^{-1}\,mol}^{-1}$
$\alpha \to \gamma$	911.5	1184.65	900	0.7548
$\gamma \to \alpha$	1394.0	1667.15	837	0.5025
$\gamma \to L$	1527.0	1800.15		
$\alpha \to L$	1538.0	1811.15	13807	7.6325
T_C^α	769.0	1042.15	8075	9.21
T_C^γ		1800		
T_N^γ		55–80		

agree with respect to its variation with pressure, Figure 1.4. Measurements made at ambient pressure on retained hcp phase generated by shock-loading Fe-14Mn wt% alloys tend to support the data of Mao et al., with a lattice parameter ratio of about 1.61 [10]. The fact that hcp iron has the highest density of all the common allotropes of iron means that it is likely to be the stable solid phase of iron at the inner core of the Earth. The crystal structure of the Fe-10Ni wt% alloy that may represent the core composition is thought to be hcp, based on X-ray diffraction data from *in-situ* measurements at 340 GPa and 4700 K [11]. Simulations using techniques such as *ab initio* molecular dynamics show that under Earth–like conditions, it is indeed the hcp phase that is stable at the inner core [12]. Exoplanets have been discovered outside of our solar system that are much larger than the Earth. This has prompted studies of the iron phase diagram under conditions of extraordinarily large pressures and temperatures, This is because like Earth, such exoplanets and even some so-called gas giants may contain iron-rich cores that determine their magnetic fields. Calculations using *ab initio*

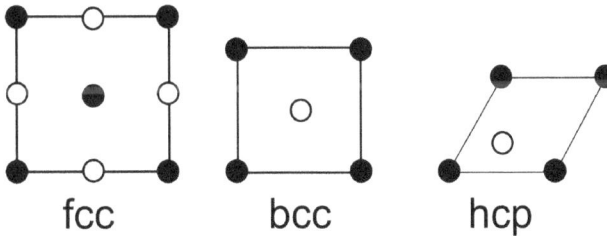

Figure 1.3 Projections of atom positions on plane normal to the *z*-axes of the crystal structures of austenite, ferrite and ε-iron, drawn to scale with respect to their observed lattice parameters. The black atoms have fractional *z*-axis coordinates 0 and 1, whereas the white atoms are at $\frac{1}{2}z$.

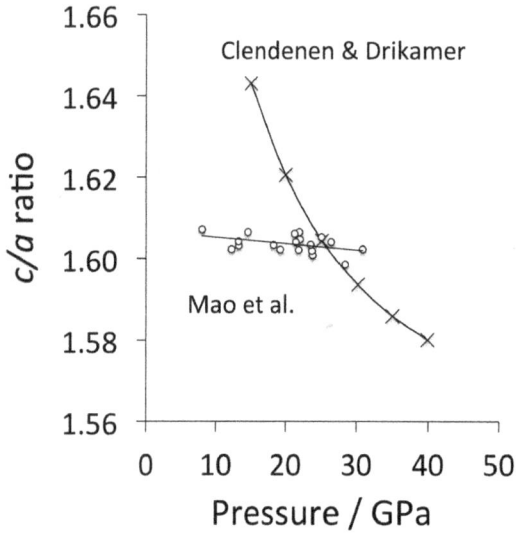

Figure 1.4 The measured lattice parameter ratio c/a for hcp iron as a function of pressure (after Clendenen and Drickamer [13] and Mao et al. [14]). The ideal value of the ratio in the hard-sphere approximation is $\sqrt{8/3} = 1.633$.

methods are shown in Figure 1.5 for temperatures up to 40,000 K and pressures reaching 100 TPa [15]. Counter to intuition, the hcp ceases to be stable at very high pressures. This is because the electronic structures of the phases change – the configuration of the nonmagnetic bcc iron atoms under normal conditions is $3d^6 4s^2$ with the core $3s$ and $3p$ bands being very narrow and at energy levels where they are tightly bound to the nucleus and hence unable to participate significantly in the metallic bonding process. Extreme pressure changes that, with the core electron bands widening, overlapping and hence participating more effectively in the bonding process leading to changes in the relative stabilities of the allotropes of iron [15, 16]. It is interesting that direct measurements using laser pulses for pressure and X-ray diffraction for crystal structure have shown that while Fe-6.5Si becomes hexagonal close-packed at pressures in the range 400–1314 GPa, Fe-15Si becomes body-centred cubic over the same pressure range [17]. The estimated temperatures involved in these experiments lie in the range 1500–3000 K.

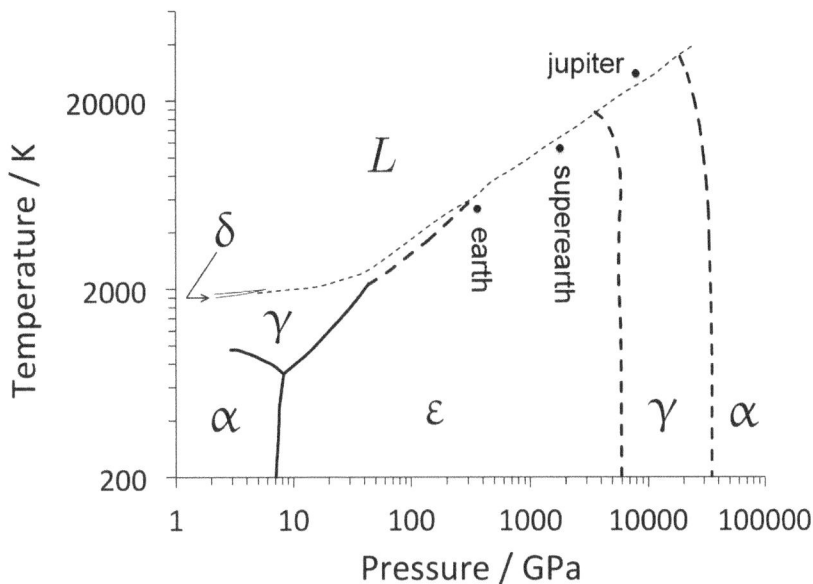

Figure 1.5 Phase diagram of iron under extreme conditions calculated using finite temperature density functional theory. The continuous curves are from experimental data whereas the calculations, that neglect magnetic contributions, are represented by dashed curves. Pressure-temperature conditions at the centre of the Earth, Jupiter and a planet five times the size of the Earth ("superearth") are also illustrated. The appearance of α at pressures in excess of 34 TPa is a body-centred tetragonal distortion ($c/a \approx 0.975$) of the bcc form [16]. Adapted with permission from Strixrude [15]. Copyrighted by the American Physical Society.

1.2 THE ILLOUSARY OMEGA PHASE

In alloys based on titanium, zirconium and hafnium, a metastable phase ω can form from the body-centred cubic β. The $\beta \rightleftharpoons \omega$ transformation is reversible and diffusionless with a coordinated motion of atoms [18]. The structure of β has a sequence ....ABCABC.... in the stacking of {111} planes which are not close-packed. The $\beta \rightleftharpoons \omega$ transformation occurs by the passage of a longitudinal displacement wave along ⟨111⟩ which causes the B and C planes to collapse into each other, leaving the A planes unaffected.[2] The stacking sequence therefore changes to ...AB'AB'AB'.... in which the B' planes have twice the density of atoms as the A planes. The ...AB'AB'AB'.... stacking is consistent with the hexagonal crystal structure of ω with a $c/a \approx 0.6$. The longitudinal displacement waves are responsible for the intense streaking observed in the electron diffraction patterns in a mixture of the matrix and ω-phase.

There are periodic reports in the literature that a metastable ω phase forms in α-iron [e.g., 19]. However, these are based on a misinterpretation of the additional reflections that occur due to double diffraction between twin-related variants of α. The characteristic streaking observed in the diffraction patterns from titanium alloys undergoing the ω-transformation is also absent in the studies of twin-free iron.

Figure 1.6 is an electron diffraction pattern taken from an internally twinned martensite plate in a Fe-4Ni-0.4C wt% steel. It contains a pair of twin-related $⟨011⟩_\alpha$ zones [20]. Reflections such as the one marked *dd* are due to double diffraction when an electron beam diffracted from one crystal passes through another one and acts as an incident beam. These reflections have frequently been misinterpreted to be due to the ω-phase. In some cases it is claimed that the reflections come from an untwinned region but this has not been established systematically.

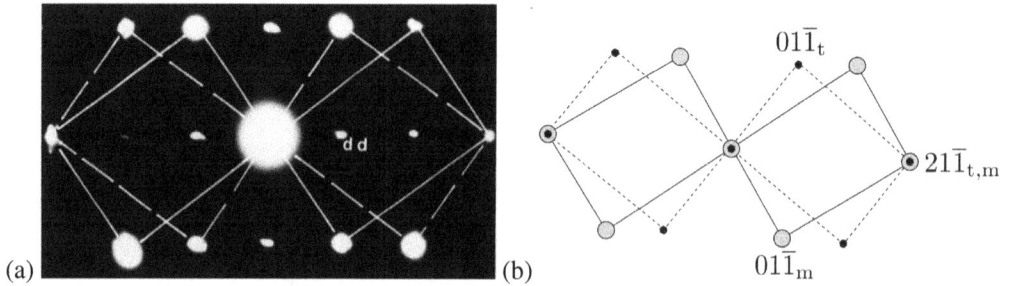

Figure 1.6 (a) Electron diffraction pattern from a martensite plate (m) and its twin (t). Spots not connected by lines (e.g. "dd") arise from double diffraction. (b) Interpretation of the diffraction pattern. After [20].

1.3 AMORPHOUS IRON

The classic work on gold-silicon alloys that led to the discovery of metallic glass [21] spawned an entire field of research that continues to thrive. Whereas iron has formed the basis of many metallic glasses, it does not seem possible to obtain sizeable sample of *pure iron* in a glassy state. Particles about 30 nm in size, of iron of about 96% purity, produced by subjecting iron pentacarbonyl to intense ultrasound have been shown to be amorphous [22]. Ultrasound of this kind leads to the formation, growth and collapse of bubbles, processes associated with momentary high temperatures and pressures. The particles are found to be ferromagnetic under ambient conditions with a magnetic moment of $1.7\mu_B$ per atom.

Molecular dynamic simulations indicate that liquid iron quenched at about $10^{13}\,\mathrm{K\,s^{-1}}$ for pressures up to 20 GPa would lead to the glassy state, but if the same experiment is conducted at greater pressures, then it tends to crystallise because the crystalline state has a smaller specific volume [23].

1.4 MECHANISMS OF TRANSFORMATION

The atomic arrangement in a crystal can be altered either by breaking all the bonds and rearranging the atoms into an alternative pattern (*reconstructive* transformation), or by homogeneously deforming the original pattern into a new crystal structure (*displacive* transformation), Figure 1.7.

In the displacive mechanism the change in crystal structure also alters the macroscopic shape of the sample when the latter is not constrained. The shape deformation during constrained transformation is accommodated by a combination of elastic and plastic strains in the surrounding matrix. The product phase grows in the form of thin plates to minimise the strains. The atoms are displaced into their new positions in a coordinated motion. Displacive transformations can therefore occur at temperatures where diffusion is inconceivable within the time scale of the experiment. Some solutes may be forced into the product phase, a phenomenon known as solute trapping. Both the trapping of atoms and the strains make displacive transformations less favourable from a thermodynamic point of view.

It is the diffusion of atoms that leads to the new crystal structure during a reconstructive transformation. The flow of matter is sufficient to avoid any shear components of the shape deformation, leaving only the effects of volume change. This is illustrated phenomenologically in Figure 1.8, where displacive transformation is followed by diffusion, which eliminates the shear. This *reconstructive diffusion* is necessary even when transformation occurs in pure iron [24]. In alloys, the diffusion process may also lead to the redistribution of solutes between the phases in a manner consistent with a reduction in the overall free energy.

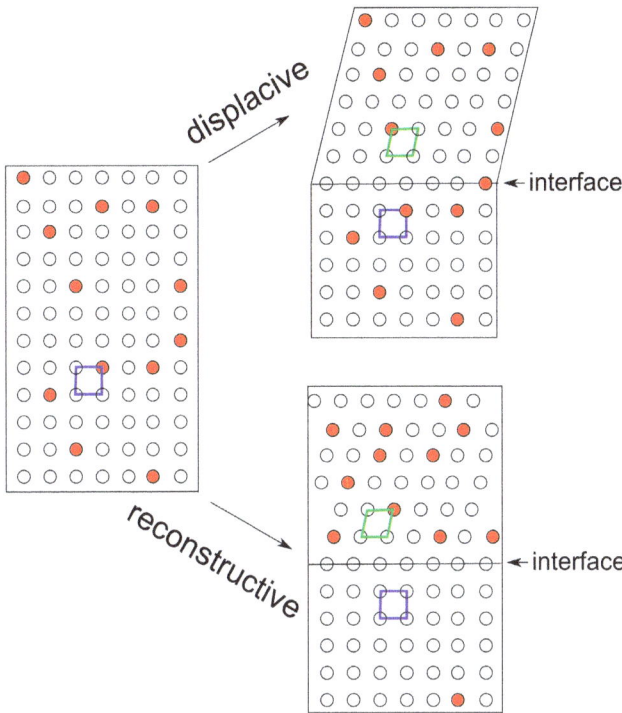

Figure 1.7 The main mechanisms of transformation. The parent crystal contains two kinds of atoms. The figures on the right represent partially transformed samples with the parent and product unit cells. The transformations are unconstrained in this illustration.

Virtually all the phase transformations in steels can be discussed in the context of these two mechanisms. There are, of course, important details which are best described during the discussion of specific microstructures.

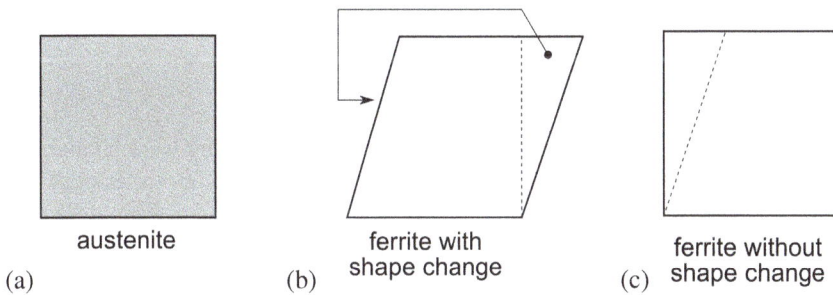

austenite   ferrite with   ferrite without
                 shape change   shape change
(a)   (b)   (c)

Figure 1.8 A phenomenological interpretation of reconstructive transformation. (a) Parent phase; (b) product phase generated by a homogeneous deformation of the parent phase. The arrow shows the mass transport that is necessary in order to eliminate the shear component of the shape deformation; (c) shape of the product phase after the reconstructive-diffusion has eliminated the shear component of the shape deformation.

Any phase which forms by reconstructive transformation is likely to be closer to the equilibrium state than that generated by displacive transformation. In the latter case, the strains and the absence of solute partitioning contribute to the deviation from the equilibrium state. The equilibrium phase diagram (Figure 1.2) shows that in pure iron, the hcp phase is stable only at pressures above 13 GPa at room temperature. Nevertheless, there are experiments where ε-iron has been observed at pressures as small as 8 GPa. These experiments were carried out using shock waves to generate the required pressure. The time scale is then so short that the mechanism is displacive with a glissile ε/α interface [10]. The defects introduced during shock deformation make it difficult for the transformation front to glide reversibly so there is a large *hysteresis*, as illustrated in Figure 1.9. At ambient temperature, ε-iron forms in increasing quantities as the pressure is increased beyond 13 GPa. But when the pressure is subsequently reduced, the ε-iron does not revert to ferrite until 8.1 GPa, the specimen only becoming fully ferritic when the pressure becomes less than 4.5 GPa. Deviations like these should be much smaller when the ε-iron transforms by a reconstructive mechanism. The defects in the ferrite then would not interfere with the uncoordinated transfer of atoms at the interface. On the contrary, they would encourage a faster rate of transformation as the defects are eliminated by the passage of the interface in a process akin to recrystallisation.

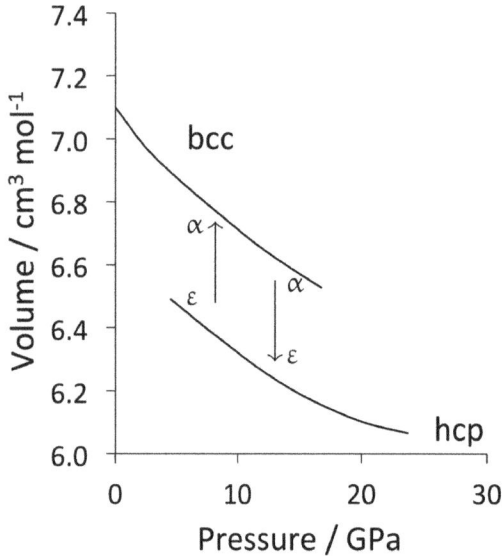

Figure 1.9 Pressure hysteresis (overshooting of the equilibrium transformation pressure of 13 GPa) of the $\alpha \leftrightarrow \varepsilon$ martensitic transformation. The vertical arrows indicate the pressures at which the $\alpha \rightarrow \varepsilon$ and $\varepsilon \rightarrow \alpha$ changes begin. Data from Giles, Longenbach and Marder [25].

1.5 CRYSTALLOGRAPHIC SIMILARITIES

It was stated earlier that the three natural allotropes of iron are in many ways similar, so the free energy change on transformation can be quite small when compared with, for example, the stored energy in a highly supersaturated solution. Although the crystal structures have different symmetries, it is easy to demonstrate similarities.

The fcc and hcp structures can be related in terms of the stacking sequences of the close-packed planes. A transformation between austenite and ε-iron can be visualised simply as a change in the stacking sequence of the close-packed planes. The unit cell projections in Figure 1.3 are drawn to scale. It is readily demonstrated that the lattice parameters of the ε and γ are related: $a_\varepsilon \simeq a_\gamma/\sqrt{2}$ and $c_\varepsilon \simeq 2a_\gamma/\sqrt{3}$.

The least dense form of iron, with the body-centred cubic crystal structure, at first sight appears to differ significantly from the other allotropes. However, Bain in 1924 [26] showed how a relatively simple deformation can generate the ferrite lattice from that of austenite (Figure 1.10). Thus, $a_\alpha \simeq 0.8a_\gamma$.

It might be argued that the orientation relationships between the forms of iron are obvious from these crystallographic similarities. This is partly true, but there are other considerations arising from the need to ensure a better fit at the interface which may not be neglected. The observed orientation relationships are almost always irrational and the reason for this will become evident in Chapter 5. The Bain deformation is one of many possible mechanisms for converting austenite into ferrite.

fcc & bct unit cells of austenite

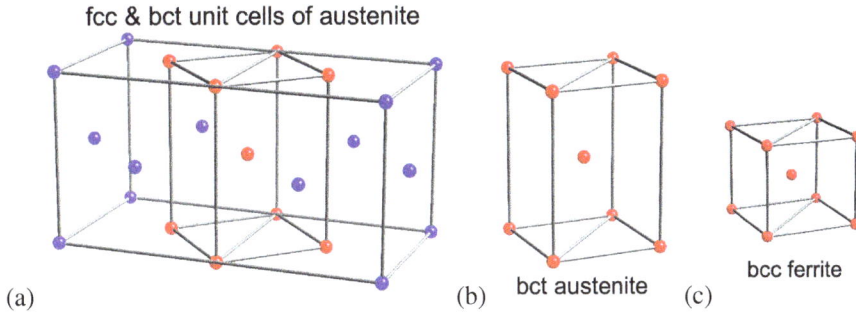

(a)   (b) bct austenite   (c)   bcc ferrite

Figure 1.10   The Bain strain. (a) A body-centred tetragonal unit cell (red) defined from the conventional face-centred cubic representation of austenite. (b) The body-centred tetragonal cell of austenite illustrated in isolation. This can be compressed along the vertical axis and expanded uniformly along the other two principal axes to give the body-centred cubic unit cell of ferrite, illustrated in (c).

The primitive unit cells of both austenite and ferrite are trigonal. The trigonal cell for austenite has its cell edges made up of $\frac{a_\gamma}{2}\langle 1\,1\,0\rangle$ of the conventional fcc cell, with the angles between the cell edges being $60°$ (Figure 1.11a). For ferrite, the trigonal cell has its edges made up of $\frac{a_\alpha}{2}\langle 1\,1\,1\rangle$ with axial angles $109°\,28'$ (Figure 1.11b). It is obvious that the austenite trigonal cell can be converted into that of ferrite by varying the characteristic angle and distorting along the $\langle 1\,1\,1\rangle$ trigonal axis. This amounts to an alternative to the Bain strain for converting fcc to bcc but it involves larger deformations and hence is not favoured.

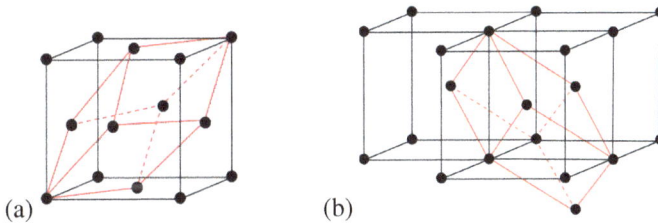

(a)   (b)

Figure 1.11   The primitive unit cells of (a) austenite and (b) ferrite, constructed within the conventional fcc and bcc cells.

1.6   IRON-CARBON PHASE DIAGRAM

Although many aspects of the Fe-C equilibrium phase diagram will be elaborated in later chapters, it is nevertheless introduced here because a steel must contain carbon and a binary steel is the simplest of the extraordinarily large numbers of phase equilibria possible in multicomponent alloys of iron. Figure 1.12 illustrates the equilibrium between the allotropic forms of iron and cementite or graphite (red lines). The solubility of carbon in ferrite that is in equilibrium with any other phase (except liquid iron) is very small at all temperatures. The high-temperature form of ferrite is for historical reasons labelled δ-ferrite but there is no difference in the crystal structure from α when the latter is at temperatures above T_C. Below that temperature, α-iron is not strictly cubic because the magnetic spins are aligned, say along the z axis, so that rotations about the x or y axes must be combined with time reversal to preserve the directions of the spins. The magnetic point group therefore becomes tetragonal. It follows that the ferrite structure below T_C is tetragonal, but the extent of tetragonality is small enough to be neglected in most experiments on steels. It does, however, manifest during magnetostriction.

The diagram shows a peritectic reaction ($L + \delta \rightarrow \gamma$), eutectic reaction ($L \rightarrow \gamma + [\text{graphite or } \theta]$) and a eutectoid ($\gamma \rightarrow \alpha + \theta$). When in equilibrium with iron, the graphite is said to be the more stable phase than cementite [8]. Long-term heat treatment can therefore cause the cementite to disintegrate into graphite and ferrite. The formation of graphite on iron nanoparticles coated with carbon is often interpreted using the Fe-C phase diagram.

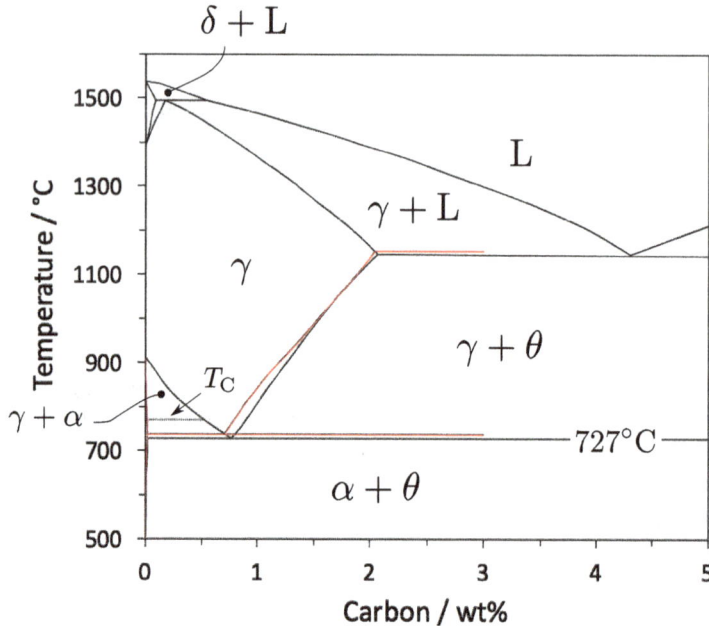

Figure 1.12   The iron-carbon equilibrium phase diagram, in part calculated using assessed thermodynamic data. T_C represents the Curie temperature of ferrite, θ the cementite and L the liquid phase. The more stable equilibrium between the allotropic forms of iron and graphite is represented by the red lines. However, graphite may not actually form due to kinetic limitations.

It has been speculated that iron nanoparticles have biomedical applications, particularly in drug delivery because in their ferritic form, they can be manipulated by an external magnetic field. Because of their reactivity, they would need to be protected using graphite coatings. After production using a variety of techniques, the particles consist of iron α, γ, θ, iron oxide and amorphous carbon. On annealing, the carbon reduces any oxide and cementite is replaced by graphite on the surface [27]. Observations like these are interpreted by appealing to the greater stability of graphite in Fe-C equilibrium phase diagram when compared with cementite. However, the rate of graphitisation during annealing is rapid (800 °C, 20 min) when compared with bulk steels. The iron-carbon phase diagram is not strictly applicable given the large curvature of the iron nanoparticles, which would tend to favour the formation of the most stable phase.[3] Experiments on carbon deposited on thin films of iron, followed by annealing, show that the graphite is formed at the Fe_3C-carbon interface as the cementite decomposes [28].

There are many allotropic forms of crystalline carbon. Hexagonal-graphite is more stable at ambient pressure than the fullerenes, carbon onions, single and multiwall tubes (closed ends or not), nanotube ropes, graphene and diamond [29]. The stability of graphite remains the highest even when nano-sized, though the ranking of the other forms may change somewhat. Given the interest in commercial diamonds, there has been some effort to construct the Fe-diamond phase diagram [30] but diamond remains metastable with respect to graphite when under ambient pressure.

1.7 CLASSIFICATION SCHEME

The microstructures that can be generated in pure iron can be increased considerably by the addition of carbon, but they can still all be classified within the general scheme of displacive and reconstructive mechanisms, as illustrated in Figure 1.13.

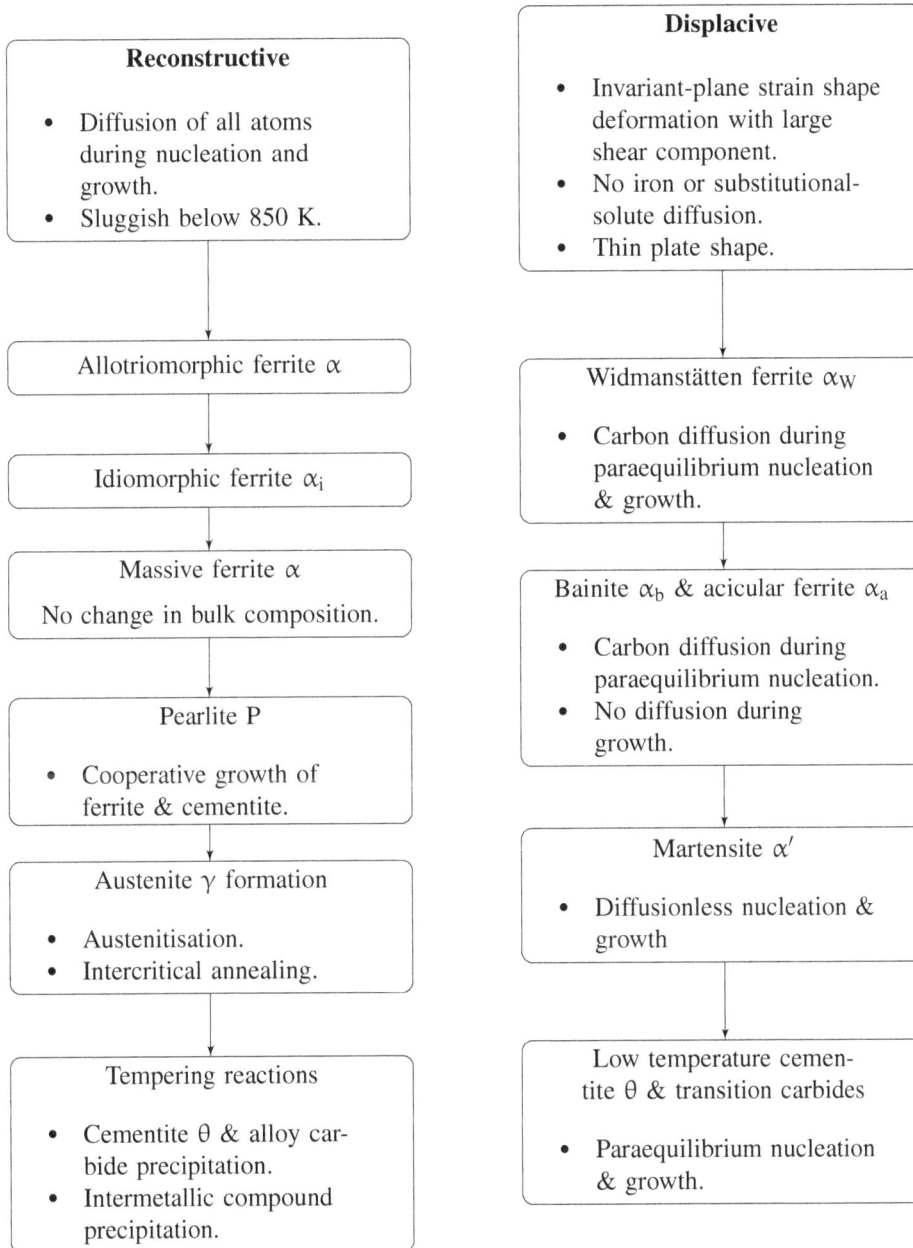

Reconstructive

- Diffusion of all atoms during nucleation and growth.
- Sluggish below 850 K.

Allotriomorphic ferrite α

Idiomorphic ferrite α_i

Massive ferrite α

No change in bulk composition.

Pearlite P

- Cooperative growth of ferrite & cementite.

Austenite γ formation

- Austenitisation.
- Intercritical annealing.

Tempering reactions

- Cementite θ & alloy carbide precipitation.
- Intermetallic compound precipitation.

Displacive

- Invariant-plane strain shape deformation with large shear component.
- No iron or substitutional-solute diffusion.
- Thin plate shape.

Widmanstätten ferrite α_W

- Carbon diffusion during paraequilibrium nucleation & growth.

Bainite α_b & acicular ferrite α_a

- Carbon diffusion during paraequilibrium nucleation.
- No diffusion during growth.

Martensite α'

- Diffusionless nucleation & growth

Low temperature cementite θ & transition carbides

- Paraequilibrium nucleation & growth.

Figure 1.13 Characteristics of transformations in steels.

All of the phases where growth occurs by a displacive mechanism evolve in the form of thin plates, because the strain associated with the displacements is minimised by that shape. The detailed nature of the shape strain is such that it leaves the habit plane (the broad face of the plate) macroscopically undistorted and unrotated. This invariant-plane strain can be factorised into a shear on the habit plane and a dilatation normal to that plane. One important consequence of this mechanism is that the transformation product is confined to the parent grain in which it grows – after all, a coordinated motion of atoms cannot be sustained across a grain boundary. The most important displacive transformations are:

1. Martensitic transformation is the simplest to understand because both nucleation and growth occur without any diffusion whatsoever. Of all the transformation products in steel, martensite represents the largest deviation from equilibrium. Any solutes in the austenite become trapped within the martensite even though they may prefer to partition during transformation. There are defects created which are intrinsic to the growth mechanism and there is a large strain energy associated with the shape change. Martensite can in some circumstances grow at speeds that are a large fraction of the speed of sound in the metal. Indeed, there are no other transformations in steel or in any other alloy system which even approach the maximum growth rates of martensite.
2. Bainite occurs at somewhat higher temperatures than martensite, where interstitial solutes are mobile but the substitutional atoms are not. The smaller driving force available at the higher temperatures means that the partitioning of carbon is essential during nucleation. However, growth appears to be diffusionless in the first instance. There is a tendency for the microstructure to temper fairly quickly as it forms, causing a redistribution of the carbon that either is precipitated as carbides or goes on to enrich the remaining austenite.
3. At even higher temperatures, displacive transformation occurs by a paraequilibrium mechanism in which the carbon atoms partition between the parent and product phases but the ratio of the substitutional solute to iron atoms remains constant everywhere. Thus, Widmanstätten ferrite nucleates and then grows in the form of coarse plates that lengthen at a rate which is controlled by the diffusion of carbon in the austenite ahead of the transformation interface.

Reconstructive transformations play an increasingly important role in the decomposition of austenite as the mobility of atoms increases with temperature. At the same time the driving force for transformation becomes smaller as the Ae_3 temperature is approached, making it difficult to sustain transformations that deviate substantially from equilibrium. The diffusion of atoms (iron, substitutional and interstitial solutes) frequently becomes the rate-controlling process. This diffusion also helps avoid the strains that would otherwise accompany the change in crystal structure. Therefore, it is an essential feature of reconstructive transformations in pure iron. This is why in similar circumstances, reconstructive transformations can be sluggish when compared against those occurring by a displacive mechanism. The main products of the reconstructive transformation of austenite are as follows:

1. Allotriomorphic ferrite is the first phase to precipitate when steel is cooled into the two phase $\alpha + \gamma$ field. It nucleates and grows most readily along the austenite grain surfaces. This gives it an irregular shape which from a macroscopic point of view appears irregular and not crystallographic. Idiomorphic ferrite, on the other hand, nucleates intragranularly and can be faceted. The growth of both of these phases leads to the partitioning of carbon, and can also involve the partitioning of substitutional solutes. Unlike displacive transformations, growth by a reconstructive transformation mechanism is not limited by austenite grain boundaries. Massive ferrite grows without the substantive redistribution of solutes; its growth rate can therefore be rapid. The few grains which nucleate first grow rapidly, across austenite grain boundaries, giving an exceptionally large ferrite grain size and hence the term "massive".
2. Pearlite consists of cementite and ferrite. The two phases grow together at a common transformation front with the austenite. This cooperative growth leads to the formation of

a colony which is a bicrystal of interpenetrating cementite and ferrite. A two-dimensional section of a pearlite colony gives the false impression that it is a stack of alternating lamellae of cementite and ferrite – all of the cementite within a colony is connected in three dimensions, as is all of the ferrite.[4] Pearlite can sometimes grow at the same temperature as bainite albeit at a much slower rate. The average composition of a pearlite bicrystal ("colony") can in plain carbon steels be the same as that of the steel as a whole, so its growth does not enrich the residual austenite and the transformation can go to completion. In two-dimensional sections, pearlite appears to consist of alternate lamellae of ferrite and cementite.

1.7.1 THERMODYNAMIC CLASSIFICATION

Phase transformations are classified traditionally as first or second order. In a first-order transformation the two phases coexist at the transition temperature, for example, ferrite and austenite in pure iron at about 910°C. In a second-order transformation the two phases do not coexist but change continuously into one another. Examples of second-order transitions are rare but the Curie point of a ferromagnet, above which the magnetic moment of a material vanishes, is one case.

The formal thermodynamic definition of the order of a transformation is due to Ehrenfest [31]. The classification is based on the successive differentiation of a thermodynamic potential (e.g. Gibbs free energy) with respect to an external variable such as temperature or pressure. The *order* of the transformation is given by the lowest derivative to exhibit a discontinuity. Thus, in a first order transformation, the partial derivative of the Gibbs free energy with respect to temperature is discontinuous at the transition temperature (Figure 1.14). There is thus a latent heat of transformation evolved at a sharp transformation interface which separates the coexisting parent and product phases. The phase change occurs at a well-defined boundary which separates perfect forms of the parent and product phases.

In a second-order transformation, the Gibbs free energy functions have precisely the same slope at the transformation temperature (Figure 1.14b) so enthalpy evolves continuously as the transformation temperature is approached. Since the free energy curves have different curvatures, it is the second derivative that is discontinuous at the transformation temperature. The specific heat capacity changes abruptly at that temperature. The two phases tend to gradually become similar as the transformation temperature is approached, and become identical at that temperature. This is why the free energy curves merge beyond T_t in Figure 1.14b.

The Ehrenfest classification is useful in the context of steels because it adds clarity to the mechanism of transformation in a way that is independent of the time scales involved. A first-order transformation always involves the nucleation and growth of the product phase.

This is in contrast to a second-order structural change, where the crystal structure of the parent phase would change gradually, through an infinite series of intermediate states, into that of the product. The parent and product phases therefore never coexist and hence there is no interface in the normal sense. The transformation occurs simultaneously at all locations.

All of the allotropic transformations in steels are of first order. The product phase is initiated by a large fluctuation in the structure of the parent phase, the two regions being separated by a well-defined interface.

Second-order transformations do not involve an interface but a gradual change where the concept of nucleation does not apply.

The martensitic transformation in steels is sometimes regarded as "instantaneous" with the implication that it does not involved nucleation, simply because it can occur rapidly. This is wrong, it is of first order with a well-defined interface and is governed by the processes of nucleation and growth. It cannot in this sense be distinguished from any other structural change in steels.

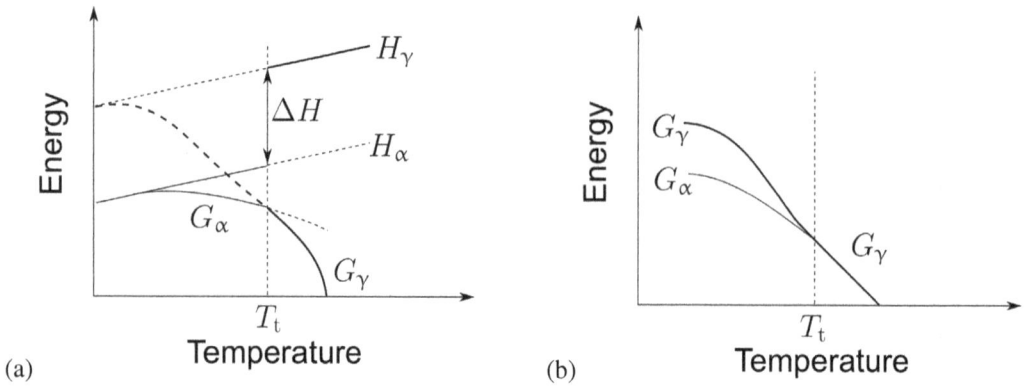

Figure 1.14 Free energy curves for (a) first-order transformation, (b) second-order transformation. T_t represents the transformation temperature.

REFERENCES

1. A. T. Paxton, M. Methfessel, and H. M. Polatoglou: 'Structural energy-volume relations in first-row transition metals', *Physical Review B*, 1990, **41**, 8127–8138.

2. S. Peng, and H. J. F. Jansen: 'Electronic structure of face-centered tetragonal iron', *Journal of Applied Physics*, 1990, **67**, 4567–4569.

3. S. S. Peng, and H. J. F. Jansen: 'Antiferromagnetism in face centered tetragonal iron', *Journal of Applied Physics*, 1991, **69**, 6132–6134.

4. S. Fox, and H. J. F. Jansen: 'Structure and properties of trigonal iron', *Physical Review B*, 1996, **53**, 5119–5122.

5. C. Zhao, W. Q. Cao, C. Zhang, Z. G. Yang, H. Dong, and Y. Q. Weng: 'Effect of annealing temperature and time on microstructure evolution of 0.2C-5Mn steel during intercritical annealing process', *Materials Science and Technology*, 2014, **30**, 791–799.

6. J. A. Hoffmann, A. Paskin, K. J. Tauer, and R. J. Weiss: 'Analysis of ferromagnetic and antiferromagnetic second-order transitions', *Journal of the Physics and Chemistry of Solids*, 1956, **1**, 45–60.

7. R. I. Orr, and J. Chipman: 'Thermodynamic functions of iron', *TMS-AIME*, 1967, **239**, 630–633.

8. J. Chipman: 'Thermodynamics and phase diagram of the Fe-C system', *Metallurgical Transactions*, 1972, **3**, 55–64.

9. R. J. Weiss: 'The origin of the 'Invar effect', *Proceedings of the Physics Society*, 1963, **82**, 281–288.

10. A. Christou, and N. Brown: 'High-pressure transition and demagnetization in shock compressed Fe-Mn alloys', *Journal of Applied Physics*, 1971, **42**, 4160–4170.

11. S. Tateno, K. Hirose, T. Komabayashi, H. Ozawa, and Y. Ohishi: 'The structure of Fe-Ni in Earth's inner core', *Geophysical Research Letters*, 2012, **39**, L12305.

12. B. K. Godwal, F. González-Cataldo, A. K. Verma, L. Strizrude, and R. Jeanloz: 'Stability of iron crystal structures at 0.3–1.5 TPa', *Earth and Planetary Science Letters*, 2015, **409**, 299–306.

13. R. L. Clendenen, and H. G. Drickamer: 'The effect of pressure on the volume and lattice parameters of ruthenium an iron', *Journal of the Physics and Chemistry of Solids*, 1964, **25**, 865–868.

14. H.-K. Mao, W. A. Bassett, and T. Takahashi: 'Effect of pressure on crystal structure and lattice parameters of iron up to 300 kbar', *Journal of Applied Physics*, 1967, **38**, 272–276.

15. L. Stixrude: 'Structure of iron to 1 Gbar and 40000 K', *Physical Review Letters*, 2012, **108**, 055505.

16. C. J. Pickard, and R. J. Needs: 'Stable phases of iron at terapascal pressures', *Journal of Physics – Condensed Matter*, 2009, **21**, 452205.

17. J. K. Wicks, R. F. Smith, D. E. Frananduono, F. Coppari, R. G. Kraus, M. G. Newman, J. R. Tygg, J. H. Eggert, and T. S. Duffy: 'Crystal structure and equation of state of Fe-Si alloys at super-Earth core conditions', *Science Advances*, 2018, **4**, eaao5864.

18. S. L. Sass: 'The structure and decomposition of Zr and Ti bcc solid solutions', *Journal of Less Common Metals*, 1972, **28**, 157–173.

19. D. H. Ping, and W. T. Geng: 'A popular metastable omega phase in body-centered cubic steels', *Materials Chemistry and Physics*, 2013, **139**, 830–835.

20. H. K. D. H. Bhadeshia: Geometry of Crystals: 2nd ed., Institute of Materials, 2001.

21. W. Klement Jun., R. H. Willens, and P. Duwez: 'Non-crystalline structure in solidified gold–silicon alloys', *Nature*, 1960, **187**, 869–870.

22. K. S. Suslick, S.-B. Choe, A. A. Cichowlas, and M. W. Grinstaff: 'Sonochemical synthesis of amorphous iron', *Nature*, 1991, **353**, 414–416.

23. J. Mo, H. Liu, Y. Zhang, M. Wang, L. Zhang, B. Liu, and W. Yang: 'Effects of pressure on structure and mechanical property in monatomic metallic glass', *Journal of Non-Crystalline Solids*, 2017, **464**, 1–4.

24. H. K. D. H. Bhadeshia: 'Diffusional formation of ferrite in iron and its alloys', *Progress in Materials Science*, 1985, **29**, 321–386.

25. P. M. Giles, M. H. Longenbach, and A. R. Marder: 'High pressure alpha to epsilon martensitic transformation in iron', *Journal of Applied Physics*, 1971, **42**, 4290–4295.

26. E. C. Bain: 'The nature of martensite', *Transactions of the AIME*, 1924, **70**, 25–46.

27. B. David, N. Pizúrová, O. Schneeweiss, P. Bezdička, I. Mojan, and R. Alexandrescu: 'Preparation of iron/graphite core–shell structured nanoparticles', *Journal of Alloys and Compounds*, 2004, **378**, 112–116.

28. M. C. Borah: 'Effect of metals on the graphitization of thin carbon films': Ph.D. thesis, Imperial College of Science and Technology, London, U.K., 1970.

29. O. A. Shenderova, V. V. Zhirnov, and D. W. Brenner: 'Carbon nanostructures', *Critical reviews in Solid State and Material Sciences*, 2002, **27**, 227–356.

30. A. A. Zukhov, and R. L. Snezhnoi: 'The iron-carbon system. new developments – III. The iron-diamond phase diagram', *Acta Metallurgica*, 1975, **23**, 1103–1110.

31. P. Ehrenfest: 'Phasenumwandlungen im ueblichen und erweiterten sinn, classifiziert nach dem entsprechenden sigularitaeten des thermodynamischen potentiales', *Verhandelingen der Koninklijke Akademie van Wetenschappen (Amsterdam)*, 1933, **36**, 153–157.

Notes

[1] The trigonal basis is characterised by three equal length vectors from the origin with the same angle separating each of them. When these angles are 90° the unit cell becomes cubic but a distortion along the body diagonal gives a trigonal structure.

[2] Also described as a periodic shift of the $(111)_\beta$ planes in the $[111]_\beta$ direction by 0, $\sqrt{3}a_\beta/6$ and $-\sqrt{3}a_\beta/6$. The space group is $P6/mmm$.

[3] This is the Gibbs-Thompson effect described later in the text, Section 7.8.

[4] Imagine a colony as a cabbage in a bucket of water. The leaves of the cabbage, which are all connected, represent the single crystal of cementite, and the water the single crystal of ferrite.

2 Thermodynamics

2.1 INTRODUCTION

Thermodynamics facilitates the linking together of "the many observable properties so that they can be seen to be a consequence of a few" [1]. It provides a firm basis for the rules that macroscopic systems follow at equilibrium. When combined with phenomena associated with the approach to equilibrium, it forms the foundations of kinetic theory. It was on this basis that Zener attempted to rationalise the transformations that occur in steels [2, 3] so that the effect of alloying elements, atomic mobility, nucleation and mechanism could all be incorporated into a single hypothesis.

After an introduction to some essential concepts, the remainder of this chapter deals with theory that is relevant particularly to iron and its solutions.

2.2 DEFINITIONS

2.2.1 INTERNAL ENERGY AND ENTHALPY

The change in the internal energy ΔU of a closed system can be written as

$$\Delta U = q - w \tag{2.1}$$

where q is the heat transferred into the system and w, the work done by the system. The sign convention is that heat added and work done by the system are positive, whereas heat given off and work done on the system are negative. Equation 2.1 may be written in differential form as

$$dU = dq - dw. \tag{2.2}$$

For the special case where the system does work against a constant atmospheric pressure, this becomes

$$dU = dq - PdV \tag{2.3}$$

where P is the pressure and V the volume.

The specific heat capacity of a material represents its ability to absorb or emit heat during a unit change in temperature. Heat changes the distribution of energy amongst the particles in the system (atoms, electrons, ...) and it is these fundamental mechanisms that control the heat capacity, defined formally as dq/dT. Since $dq = dU + PdV$, the specific heat capacity measured at constant volume is given by

$$C_V = \left(\frac{\partial U}{\partial T}\right)_V.$$

It is convenient to define a new function H, the enthalpy of the system:

$$H = U + PV.$$

A change in enthalpy accounts for both the heat absorbed at constant pressure and the work done by the $P\Delta V$ term. The specific heat capacity measured at constant pressure is therefore given by

$$C_P = \left(\frac{\partial H}{\partial T}\right)_P.$$

Heat capacity can be measured using a variety of calorimetric methods. The data can then be used to estimate enthalpy changes as a function of temperature and pressure:

$$\Delta H = \int_{T_1}^{T_2} C_P \, dT. \tag{2.4}$$

2.2.2 ENTROPY, FREE ENERGY

In the reversible Carnot cycle, a gas is placed in contact with a heat reservoir at temperature T_2, expands isothermally on absorbing a quantity of heat q_2, in the process doing work $-w_1$. The gas is then insulated and expands adiabatically, does work $-w_2$ as its temperature drops to T_1. It is then placed in contact with a heat reservoir at T_1, compressed reversibly and isothermally with work w_3 done upon it and giving up heat $-q_1$ to the reservoir. To complete the cycle, the gas is insulated, compressed reversibly and adiabatically by doing work w_4 upon it, causing its temperature to rise back to T_2 [4]. The change in internal energy $\Delta U = q_1 + q_2 + w$ on completion of the cycle is therefore zero, where $w = \Sigma_{i=1}^{4} w_i$. The work output of the engine, $-w$, is the difference between heat taken and heat returned to the reservoirs, i.e. $-w = q_2 - (-q_1)$ so the maximum efficiency is defined as the ratio of the work output to the heat absorbed:

$$\text{efficiency} = \frac{-w}{q_2} = \frac{q_2 + q_1}{q_2}.$$

Kelvin used this to define the absolute temperature,

$$\text{efficiency} = \frac{q_2 + q_1}{q_2} = \frac{T_2 - T_1}{T_2} \quad \text{so that} \quad \frac{q_2}{T_2} + \frac{q_1}{T_1} = 0. \tag{2.5}$$

By considering a cyclic process in terms of infinitesimal parts [4], it can be demonstrated that the following relationship holds for any reversible cycle,

$$\frac{dq_2}{T_2} + \frac{dq_1}{T_1} = 0 \quad \text{with} \quad \oint \frac{dq}{T} = 0$$

making the quantity dq/T a function of state S with

$$dS = \frac{dq}{T}. \tag{2.6}$$

Clausius during the 19th century named this function S as entropy; in the absence of any change in enthalpy, a reaction can occur spontaneously and irreversibly in an isolated system if it leads to an increase in entropy, i.e., $\Delta S > 0$. It is evident that in general, neither the enthalpy nor the entropy change can in isolation be assumed to reliably indicate whether a reaction can occur spontaneously. The Gibbs free energy G is therefore defined as a combination of these two terms,

$$G = H - TS. \tag{2.7}$$

The Helmholtz free energy F is the corresponding term at constant volume, when H is replaced by U in Equation 2.7. A process can occur spontaneously if it leads to a reduction in the free energy. Quantities such as H, G and S are all *functions of state*.

From an experimental perspective, a change in entropy can be measured via the heat capacity:

$$\Delta S = \int_{T_1}^{T_2} \frac{C_P}{T} \, dT.$$

2.2.3 CONFIGURATIONAL ENTROPY

Figure 2.1a shows a mixture of two kinds of atoms, with like atoms segregated with no mixing; there is only one way of achieving this arrangement. On the other hand, if they are allowed to mix ideally, then there are many more ways of configuring them, three of which are illustrated in Figure 2.1c–d. A mixing of the atoms is obviously more *probable*.

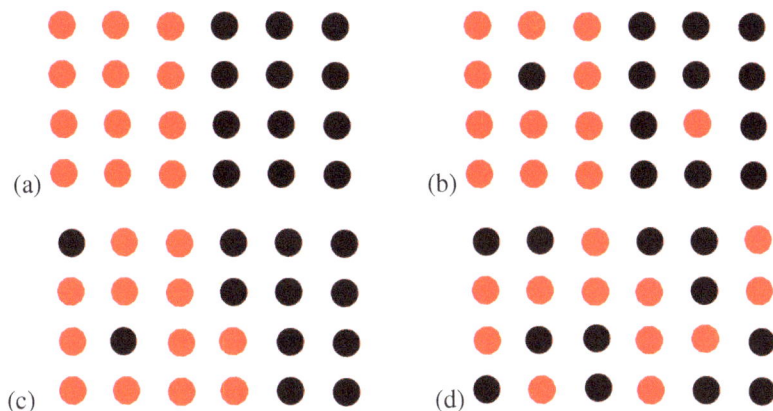

Figure 2.1 Four different configurations of a mixture of two kinds of atoms. (a) The two kinds of atoms are partitioned into their own spaces, without mixing. (b–d) If the atoms are allowed to mix, then many more arrangements are possible, here only three of the many are illustrated.

Suppose there are N sites amongst which are distributed n atoms of type A and $N - n$ of type B. The first A atom can be placed in N different ways and the second in $N - 1$ different ways. These two atoms cannot be distinguished so the number of different ways of placing the first two A atoms is $N(N-1)/2$. The number of distinguishable ways of placing all the atoms in this way, the number of distinguishable ways of placing all the A atoms is

$$\frac{N(N-1)\ldots(N-n+2)(N-n+1)}{n!} = \frac{N!}{n!(N-n)!}. \tag{2.8}$$

So if the atoms behave ideally, i.e., they do not have a preference for the type of neighbour, then the probability of a uniform distribution is much more likely than the ordered distribution.

For a real system for which the number of atoms is very large, a parameter is needed that expresses the likelihood as a function of the correspondingly large number of configurations (w_c) possible. Suppose that a term S is defined such that $S \propto \ln w_c$, where the logarithm is taken because it may be necessary to add two different kinds of disorder (after Boltzmann), then the S is identified as the configurational *entropy* $S = k \ln w_c$, where k, the proportionality constant, is known as the Boltzmann constant which for a mole of atoms is the gas constant R. The entropy is a thermodynamic function of state and it is additive. When comparing scenarios, the one that is favoured on the basis of the degree of disorder is that which has the greater entropy. In terms of solutions, entropy favours mixing over separation. On this basis, it can be shown quite simply that the change in entropy when atoms mix is given by

$$\Delta S = -R\Sigma_{i=1}^{j} x_i \ln x_i$$

where $i = 1 \ldots j$ represents the atomic species and x_i its mole fraction.

2.2.4   RELATIONSHIP BETWEEN CLAUSIUS AND BOLTZMANN ENTROPIES

The Carnot engine illustrates the spreading of energy whereas the Boltzmann approach is about mixing. The relationship between these is quite straightforward – both involve mixing. As Denbigh stated with admirable elegance, "As soon as it is accepted that matter consists of small particles which are in motion it becomes evident that every large-scale natural process is essentially a process of *mixing*" [p. 110, 5]. Energy transfer involves the motion of the atoms; it was not, at the time of Carnot, known that matter consists of atoms. It really isn't necessary to say much more.

2.3 MAXWELL RELATIONS

There are some useful relations between the thermodynamic quantities; combining Equations 2.3 and 2.6 gives:

$$dU = T\,dS - P\,dV. \tag{2.9}$$

An exact differential equation[1] such as this requires that

$$\left(\frac{\partial T}{\partial V}\right)_S = -\left(\frac{\partial P}{\partial S}\right)_V. \tag{2.10}$$

Since $H = U + PV$, it follows that

$$\begin{aligned}
dH &= dU + P\,dV + V\,dP \\
   &= T\,dS - P\,dV + P\,dV + V\,dP \\
   &= V\,dP + T\,dS.
\end{aligned}$$

$$\therefore \quad \left(\frac{\partial V}{\partial S}\right)_P = \left(\frac{\partial T}{\partial P}\right)_V.$$

Similarly, $G = H - TS$ so that

$$\begin{aligned}
dG &= dH - S\,dT - T\,dS \\
   &= V\,dP + T\,dS - S\,dT - T\,dS \\
   &= V\,dP - S\,dT.
\end{aligned}$$

$$\therefore \quad -\left(\frac{\partial V}{\partial T}\right)_P = \left(\frac{\partial S}{\partial P}\right)_T.$$

These "Maxwell relations" are embodied in Figure 2.2.

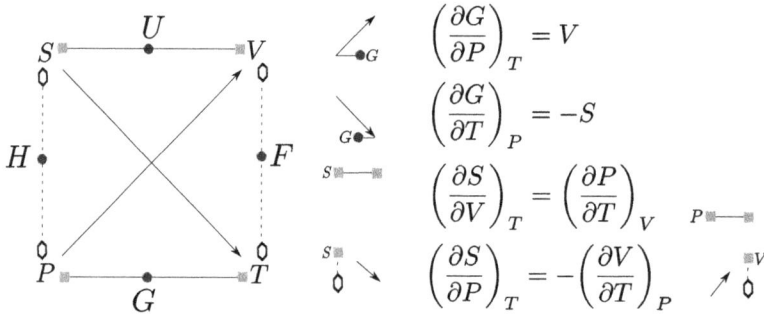

$$\left(\frac{\partial G}{\partial P}\right)_T = V$$

$$\left(\frac{\partial G}{\partial T}\right)_P = -S$$

$$\left(\frac{\partial S}{\partial V}\right)_T = \left(\frac{\partial P}{\partial T}\right)_V$$

$$\left(\frac{\partial S}{\partial P}\right)_T = -\left(\frac{\partial V}{\partial T}\right)_P$$

Figure 2.2 Aid to the use of the Maxwell relations. The starting point in the path-diagrams adjacent to the equations is at the thermodynamic quantity labelled. If a path opposes an arrow, then the partial differential takes a negative sign. When the path is along an edge, the partial differentials of the quantities on the opposite sides are set equal, unless an arrow points towards the quantity differentiated, in which case the sign is set as negative. A similar *aide memoire* is possible for magnetic systems by substituting the field strength for P and the negative of the magnetisation for V.

2.4 THERMODYNAMIC FUNCTIONS OF IRON

We have seen that heat capacity data can be related directly to the thermodynamic functions of state H, G and S. Its variation with temperature and chemical composition is therefore important in determining the relative stabilities of the phases. A number of factors can contribute independently to the ability of a material to absorb energy. It has been found useful to factorise the specific heat capacities of each phase in iron, into three components with different origins.

The bulk of the contribution comes from lattice vibrations, the electrons themselves contributing in a relatively minor way because the Pauli exclusion principle prevents all of them from participating in the energy absorption process. The third contribution, which is particularly significant for iron, comes from temperature-induced magnetic changes. The net specific heat capacity at constant pressure is therefore:

$$C_P\{T\} = C_V^L\left\{\frac{T_D}{T}\right\}C_1 + C_e T + C_P^\mu\{T\}$$

where T_D is the Debye temperature and C_V^L is the Debye specific heat with the function C_1 correcting C_V^L to a specific heat at constant pressure. C_e is the electronic specific heat coefficient and C_P^μ the component due to magnetism.

The Debye specific heat has its origins in the vibrations of atoms, which become increasingly violent as the temperature rises [6]. These elastic waves (phonons) take discrete, quantised wavelengths consistent with being bound by the periodic lattice of atoms in the solid, although the Debye model described here is a continuum model. The atoms do not all vibrate with the same frequency, so a spectrum of vibrations is considered in deriving their contribution to the internal energy U. The maximum in the spectrum is designated the Debye frequency ω_D, which is proportional to the Debye temperature T_D at which the highest frequency mode is excited:

$$T_D = \frac{h\omega_D}{2\pi k}$$

where h and k are the Planck and Boltzmann constants respectively. With the approximation that the phonon frequency is proportional inversely to the wavelength, the internal energy due to the atom vibrations is:

$$U = \frac{9NkT^4}{T_D^3}\int_0^{x_{max}}\frac{x^3}{(e^x - 1)}\,\mathrm{d}x \tag{2.11}$$

where $x = h\omega_D/(2\pi kT)$ and N is the total number of lattice points in the specimen. Since $C_V^L = \mathrm{d}U/\mathrm{d}T$, it follows that the lattice specific heat capacity at constant volume can be specified in terms of the Debye temperature and the Debye function (Equation 2.11). The theory does not provide a complete description of the lattice specific heat but has nevertheless been shown to fairly accurately represent the heat capacity data for iron [7].

At low temperatures ($T \ll T_D$), $U \to 3NkT^4\pi^4/(5T_D^3)$ so that $C_V^L \to 12\pi^4NkT^3/(5T_D^3)$ and the lattice specific heat thus follows a T^3 dependence. For $T \gg T_D$, the lattice heat capacity can similarly be shown to become temperature independent and approach a value $3Nk$, as might be expected for N classical oscillators, each with three degrees of freedom (Figure 2.3).

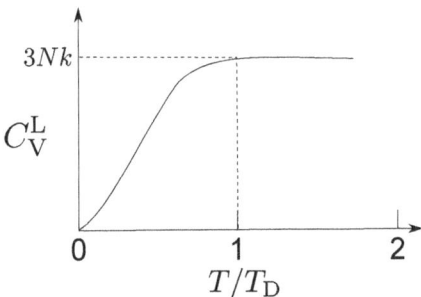

Figure 2.3 The Debye function showing how the heat capacity due to phonons varies as a function of the absolute temperature normalised by the Debye temperature.

2.4.1 RELATION BETWEEN C_P AND C_V

The Debye function is for the lattice specific heat capacity at constant volume; for solids, it is convenient to convert this to constant pressure. For a reversible change, $dq = dU + PdV = TdS$, so it follows from the definitions of the heat capacities that:

$$C_V = T\left(\frac{\partial S}{\partial T}\right)_V \qquad \text{and} \qquad C_P = T\left(\frac{\partial S}{\partial T}\right)_P. \tag{2.12}$$

$$\text{Since} \quad dS \;=\; \left(\frac{\partial S}{\partial T}\right)_V dT + \left(\frac{\partial S}{\partial V}\right)_T dV,$$

$$\therefore \frac{dS}{dT} \;=\; \left(\frac{\partial S}{\partial T}\right)_V + \left(\frac{\partial S}{\partial V}\right)_T \frac{dV}{dT}$$

$$\therefore \left(\frac{\partial S}{\partial T}\right)_P \;=\; \left(\frac{\partial S}{\partial T}\right)_V + \left(\frac{\partial S}{\partial V}\right)_T \left(\frac{\partial V}{\partial T}\right)_P.$$

On substitution into Equations 2.12, the relationship between the heat capacities is found to be

$$C_P - C_V = T\left(\frac{\partial S}{\partial V}\right)_T \left(\frac{\partial V}{\partial T}\right)_P.$$

Using the Maxwell relations (Section 2.3), and the relationships between partial derivatives,[2]

$$\left(\frac{\partial S}{\partial V}\right)_T = \left(\frac{\partial P}{\partial T}\right)_V \equiv -\left(\frac{\partial P}{\partial V}\right)_T \left(\frac{\partial V}{\partial T}\right)_P$$

so

$$C_P - C_V = -T\left(\frac{\partial P}{\partial V}\right)_T \left(\frac{\partial V}{\partial T}\right)_P^2$$

and since the volume expansion coefficient e_V and isothermal compressibility β are given by

$$e_V = \frac{1}{V}\left(\frac{\partial V}{\partial T}\right)_P, \qquad \beta = -\frac{1}{V}\left(\frac{\partial V}{\partial P}\right)_T,$$

it follows that

$$\frac{C_P}{C_V} = 1 + \frac{e_V^2 V T}{\beta C_V}. \tag{2.13}$$

2.4.2 DEBYE TEMPERATURES

The heat capacity cannot always be determined over the temperature range of interest, for example, if the phase concerned is not stable within that regime. The Debye temperature can then be estimated using the Lindemann relation [8] which is based on an assumption that melting occurs in a solid when the mean-square amplitude of vibrations of atoms about their equilibrium positions reaches a critical fraction of the interatomic spacing:

$$T_D = C_L V_a^{-\frac{1}{3}} T_m^{\frac{1}{2}} A^{-\frac{1}{2}} \tag{2.14}$$

where V_a is the volume per atom, T_m is the melting point of the phase concerned and A is the atomic weight. C_L is the Lindemann constant which when set to $16.50 \pm -0.50\,\mathrm{K}^{\frac{1}{2}}\,\mathrm{nm}\,\mathrm{g}^{\frac{1}{2}}\,\mathrm{mol}^{\frac{1}{3}}$ permits the Debye temperatures of many metals, especially those of iron and its alloys, to be estimated [9, 10].

The Lindemann relation requires a knowledge of the melting temperature. This can be problematic if the phase of interest transforms before melting but it may be possible to estimate T_M. For example, the melting point of γ-iron has been estimated to be 1800 K by using that of α iron (1811 K), the entropy of fusion of ferrite and the measured entropy change attending the $\alpha \rightarrow \gamma$ transition at 1665 K. Similarly, T_M for ε is found to be 1320 K [10, 11].

The Lindemann equation together with experimental data show that the Debye temperatures of both α and γ iron are both approximately 432 K [10], consistent with the 409 ± 9 K measured for α-iron using X-ray diffraction [12].

The Debye temperature of ε-iron has been estimated to be 375 K [11], consistent with heat capacity measurements (60-300 K) of hcp Fe-Ru alloys [13]. At a constant temperature, C_P varies linearly with the atomic fraction of Ru, enabling the Debye temperature for pure ε-iron to be obtained by extrapolation to be 385 K, in good agreement with the value obtained using the Lindemann equation. The fact that the Debye temperature of ε-iron is less than that of α-iron means that the former has a greater vibrational entropy. This goes against intuition since a loosely packed structure usually is associated with a greater vibrational entropy. Nevertheless, the relatively low melting temperature of ε-iron is in accord with its low T_D and larger vibrational entropy.

The compressibility of a material should be related to the looseness of its structure. The ε-iron crystal structure has been reported to have the same isothermal compressibility as ferrite, with a value of about 6×10^{-4} kbar^{-1} at ambient pressure (Figure 2.4) [14, 15]. This again is consistent with a large vibrational entropy associated with ε iron.

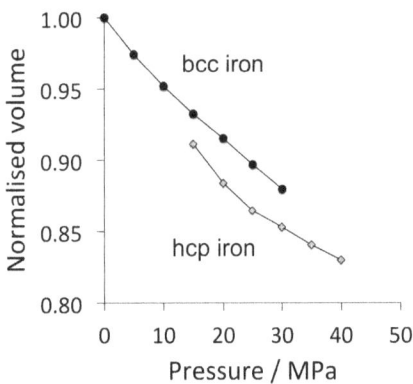

Figure 2.4 Fractional change in normalised volume as a function of pressure. The normalised volume of ferrite at zero pressure is set to be 1. After Clendenen and Drickamer [14]. The ε-iron is more dense than ferrite but nevertheless has a similar compressibility.

2.5 ELECTRONIC HEAT CAPACITY

The distribution of energy amongst a set of particles depends on the nature of the particles, specifically their spins. There are three cases:

(i) when the particles are distinguishable but with no limit to the number that can occupy any particular energy level (Maxwell-Boltzmann distribution);

(ii) when the particles cannot be distinguished and there is no limit to the number per energy state (Bose-Einstein distribution);

(iii) when the particles are indistinguishable but only selected pairs can occupy each energy state (Fermi-Dirac distribution), Figure 2.5.

When atoms in a crystal vibrate they do not all do so with the same frequency. Furthermore, the number that vibrate with a particular frequency is not restricted. The Debye function stated in the last section is based on the continuum assumption of a Maxwell-Boltzmann distribution of energy amongst the vibrating atoms. The Bose-Einstein distribution law is more appropriate since the particles are indistinguishable, but the difference between these two distributions is small at high temperatures. However, to obtain an accurate description of specific heat at low temperatures (where there is a T^3 dependence) requires the use of the Bose-Einstein distribution.

For both of these distributions, the entire distribution can change with temperature so all of the particles participate in energy absorption, leading to a large heat capacity (Figure 2.5). This is not the case when considering the ability of electrons to absorb thermal energy.

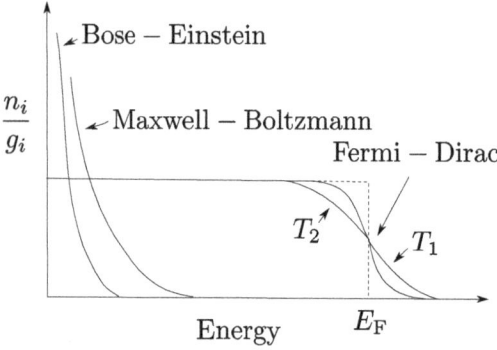

Figure 2.5 Bose-Einstein, Maxwell-Boltzmann and Fermi-Dirac distributions. The number of particles at each energy level is n_i and the number of states at each level g_i.

The metallic bond is associated with a delocalised electron gas that contributes to specific heat capacity. However, electrons are fermions so only two with opposing spins can occupy a particular energy state, the distribution of energy is given by the Fermi-Dirac function (Figure 2.5). At 0 K, the maximum energy of the states occupied is known as the Fermi energy, E_F. At higher temperatures, the exclusion principle prevents all but a few of the electrons within about kT of the Fermi energy to change their energies. The contribution of electrons to heat capacity is therefore small when compared with a classical gas at all but the lowest of temperatures [16]:

$$C_V^e \approx \frac{\pi^2 N_e kT}{2T_F}\left[1 - 0.3\pi^2\left(\frac{T}{T_F}\right)^2\right] \approx \frac{\pi^2 N_e kT}{2T_F} \tag{2.15}$$

where N_e is the number of free electrons in the sample and T_F is the Fermi temperature at which *all* the electrons can receive kinetic energy. The Fermi temperature for iron is about 1.302×10^5 K. This equation is presented alternatively as

$$C_V^e = C_e T \approx C_P^e \tag{2.16}$$

where C_e is the electronic specific heat coefficient (units: $J K^{-2}$). Writing c_e as the corresponding electronic specific heat coefficient per mole of substance, it is found that for iron [10, 11, 17, 18],

$$c_e^\gamma = c_e^\alpha = 5.021 \times 10^{-3}\,J\,mol^{-1}\,K^{-2}. \tag{2.17}$$

The electronic specific heat coefficient is somewhat larger for ε-iron, at $c_e^\varepsilon = 5.858 \times 10^{-3}\,J\,mol^{-1}\,K^{-2}$ [13]. There is some uncertainty about why the electronic specific heat coefficient measured at temperatures in the liquid helium regime tends to be greater than that measured at temperatures near the boiling point of liquid nitrogen [13, 18].

2.6 MAGNETIC SPECIFIC HEAT OF IRON

The origin of the properties described here is in the magnetic moments associated with electrons, primarily those associated with the spin of the unpaired electrons in atomic orbitals. The electron can be imagined to be a charge rotating about an axis, equivalent to a circular current flow with a corresponding magnetic moment.[3] The alignment of the spins of electrons on the arrays of atoms in a material can sometimes lead to a reduction in energy with consequent changes in the macroscopic magnetic properties. When the spins are all identically oriented, the material is said to be ferromagnetic. When neighbouring spins of equivalent magnitude point in opposite directions, the material is said to be antiferromagnetic.

Magnetic ordering has a profound influence on phase stability in iron and it alloys. The familiar allotropes of iron are ferrite (α and δ), austenite (γ) and hcp (ε). The effects associated with the Curie transition were once thought to be due to the existence of another allotrope, β-iron, which when retained by quenching made the iron extremely hard. We now know that the hardening is caused by the martensitic transformation of austenite and the other features associated with β-iron are due to the transition from a paramagnetic to ferromagnetic state – although it could be argued the β-iron should be recognised as an allotropic form of iron, given that below the Curie temperature, the ferrite is no longer cubic. This is because if the magnetic spins are aligned along a particular unit-cell axis, then rotations about either of the other two axes must be combined with time reversal to preserve the directions of the spins. The magnetic point group then becomes tetragonal, so the ferrite exhibits a small tetragonality in its lattice parameters. Aside from ferromagnetism, there are other magnetic transitions in the different allotropes, which make seminal contributions to their relative stabilities.

The spin of an electron is characterised by a spin quantum number s, which has values of $\pm\frac{1}{2}$. The unit magnetic moment is the Bohr magneton (μ_B). Because the spin can be in one of two senses, the magnetic dipole of the electron either supports or opposes an applied magnetic field. A magnetic field of strength H_{mag} (amperes per metre) may lead to an induced magnetic dipole moment per unit volume, M. The magnetic susceptibility of the material is then given by M/H_{mag}. The susceptibility is negative if the induced moment opposes the applied field.

A free electron will tend to align itself to the applied magnetic field, but in metals the vast majority of electrons are in states where the opposite spin state is already occupied. The Pauli exclusion principle permits only two electrons with opposite spins, per state. If all the electrons are spin-paired in this way, then the atom has no net magnetic moment and the material can only be *diamagnetic*.

2.6.1 DIAMAGNETISM

Diamagnetism is a weak, temperature independent negative susceptibility which causes the material to be repelled by a magnetic field. Any substance whatsoever can be diamagnetic. As discovered by Landau, an electron gas such as that associated with metallic bonding will exhibit diamagnetism because in a magnetic field, the electrons move within the metal in spirals, but in a quantised manner, about the field direction. This induced current results in a magnetic moment which, according to Lenz's law, opposes the applied field. Lenz's law states that when the flux through an electrical circuit is changed, an induced current is set up in a direction that opposes the change in flux. The real scenario may be more complex in a metal because the electron gas moves under the influence of the periodic potential associated with the ion cores within the crystal structure.

2.6.2 PARAMAGNETISM

Unlike diamagnetism, paramagnetism involves a magnetic moment that is proportional to the applied magnetic field. In a paramagnetic material, the spins associated with each atom are aligned at random except when the distribution is biased by an applied magnetic field. The effect is to cause the energies of those electrons more parallel to the applied field to decrease relative to those that oppose the field (Figure 2.6). To achieve a uniform Fermi potential, some of the electrons transfer from states with spins anti-parallel to the applied field, to those where the spin is parallel. This leaves a net imbalance in the spins. The resultant magnetisation depends on the excess number of spins. The effect is much smaller than if all electrons were able to change their spins to lie parallel with the field, but this is not permitted by the Pauli exclusion principle. Paramagnetism is therefore a weak effect which reinforces the applied field. Because diamagnetism is associated with negative susceptibility, it reduces the contribution of paramagnetism by about a third according to the Landau theory.

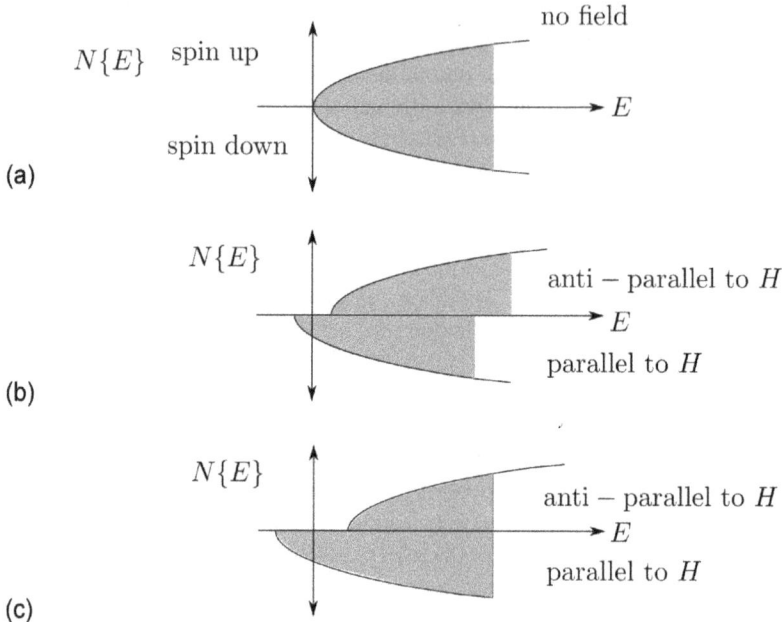

Figure 2.6 Mechanism of paramagnetism in a material with atoms containing unpaired electrons [19]. H is the magnetic field and $N\{E\}$ the number of electrons with energy E. (a) Density of states, with opposite spins separated for the purposes of illustration. The spin orientations illustrated are relative to an arbitrary direction. (b) Decrease in energy of those electrons with spins parallel with the applied field, and vice-versa. (c) The final energy distribution with uniform Fermi potential, leading to an imbalance in the pairing of electrons with opposite spins.

2.6.3 FERROMAGNETISM, ANTIFERROMAGNETISM AND FERRIMAGNETISM

Materials which are ferromagnetic, antiferromagnetic or ferrimagnetic can possess a magnetic dipole moment in the absence of any externally applied field. This is because their atoms contain electrons which are not spin-paired, making them *magnetic ions*. When these are arranged on a crystal lattice, then at low temperatures where the entropy is sufficiently small, the magnetic ions "order" spontaneously (Figure 2.7).

Ferromagnetism occurs when the dipoles from the individual atoms align parallel. If neighbouring dipoles are antiparallel then there is a zero net magnetic moment resulting in an antiferromagnetic state. In a body-centred cubic antiferromagnetic crystal, the dipoles are parallel on each of the primitive cubic sublattices, but those on one sublattice are exactly anti-parallel to those on the other, resulting in a net zero magnetic moment. In a face-centred cubic lattice, the antiferromagnetism can consist of the dipoles on adjacent planes pointing in opposite directions (Figure 2.8). In complex structures, the magnetic moments on each of the two opposing sublattices may not exactly cancel so a permanent magnetisation persists, an effect described as ferrimagnetism. The long-range ordering of magnetic ions is destroyed by thermal agitation once the Curie temperature is exceeded in a

ferromagnetic   antiferromagnetic

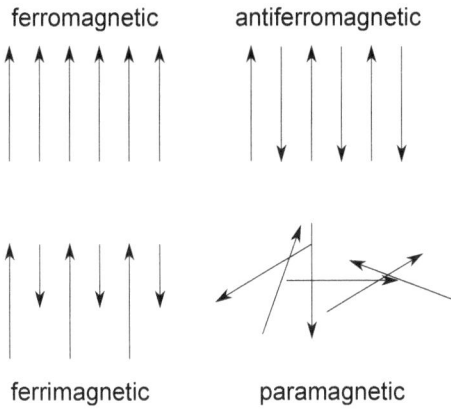

Figure 2.7 Variety of dipole alignments in materials containing atoms with un-spin-paired electrons.

ferrimagnetic   paramagnetic

structural cell parameter

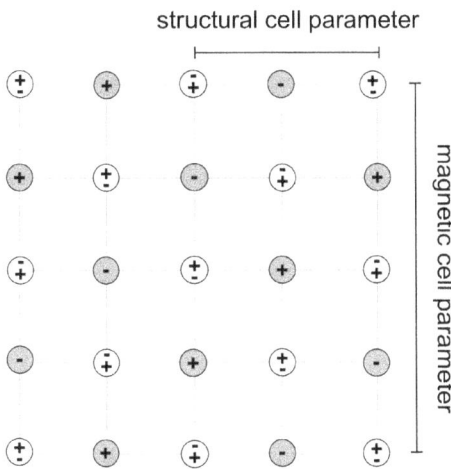

Figure 2.8 Anti-ferromagnetic ordering on a crystal with a face-centred cubic structure. The atomic arrangement is projected on to the $\{100\}$ plane. The $+$ and $-$ signs indicate that the dipole moment is either up or down. The shaded atoms are at height $\frac{1}{2}$, the unshaded ones at height 0 or 1 relative to the structural lattice parameter.

ferromagnet. The corresponding order-disorder temperature for ferrimagnetic and antiferromagnetic materials is designated the Néel temperature. The disordered state is paramagnetic.

2.6.4 HEAT CAPACITY DUE TO MAGNETISATION

Consider a ferromagnetic material containing unpaired electrons, at a temperature which is below its Curie temperature but above 0 K. The spins due to the unpaired electrons will therefore be imperfectly aligned due to thermal agitation. Magnetic disordering of this kind does not happen by the exact reversal of the aligned spins on an increasing number of sites as the temperature increases, but by correlated deviations of the spin orientation. These are called *spin waves* or magnons (Figure 2.9). The increased disorder at high temperatures excites ever shorter wavelengths and thus gives the material an ability to absorb energy. This is the origin of the magnetic component of heat capacity. Unlike phonons, magnons are not vibrational waves but involve angular displacements between adjacent spin orientations.

The ground state electron configurations for Fe, Co and Ni atoms are as follows: iron [Ar] $4s^2 3d^6$, cobalt [Ar] $4s^2 3d^7$, nickel [Ar] $4s^2 3d^8$, with [Ar] $\equiv 1s^2 2s^2 2p^6 3s^2 3p^6$. The electrons in the [Ar] core are all spin-paired, so they do not contribute to ferromagnetism; the remainder are distributed amongst the $3d$ and $4s$ in a manner that maximises spin. As a result, there will be five electrons with spin-up in the $3d$ levels, two in $4s$ and the remainder in $3d$ spin-down orbitals.

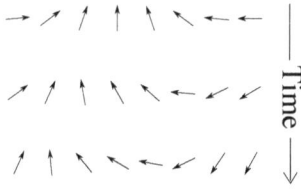

Figure 2.9 Spin wave disorder in a ferro-magnet (after Ziman [20]).

The metallic state develops when a sufficient number of atoms are in close proximity, permitting the valence electrons to become delocalised. The energy levels associated with individual atoms become energy bands, each containing N finely spaced energy levels, where N is the number of collaborating atoms in the metal. In the present context, there is an overlap in energy between the $3d$ and $4s$ bands so the delocalised electrons are shared, with on average a fractional number of atoms in each band. For iron, there are on average 7.8 and 0.2 electrons in the $3d$ and $4s$ bands respectively so the spin imbalance in the $3d$ state is 2.2, giving a magnetic moment per atom of $2.2\,\mu_B$ per atom, close to that observed for bulk iron.[4] This is in contrast to a moment of $4\,\mu_B$ for an isolated atom of iron. Small clusters of iron atoms have a magnetic moment approaching $3\,\mu_B$ per atom [21] because a large portion of the atoms are located at the free surface of the cluster, where there is a reduced number of near neighbours and hence a lesser delocalisation of the $3d$ electrons [22]. In other words, the atoms at the surface behave more like isolated atoms. For example, for clusters of iron atoms where $25 \leq N \leq 130$, the moment per atom is $3\,\mu_B$, whereas for $N = 500$, it becomes $2.2\,\mu_B$; the Curie temperatures for small clusters are also less than that for bulk iron [23]. It is worth noting that for small clusters of atoms, the average magnetic moment per atom is a function of depth [24].

2.6.5 MAGNETIC HEAT CAPACITY OF FERRITE

Ferrite in pure iron undergoes a paramagnetic to ferromagnetic transition on cooling below its Curie temperature of $T_C = 1042\,$K, which is insensitive to pressure in tests up to $9.5\,$GPa [10, 25]. The magnetic transition is not abrupt at T_C. The magnetisation M of the ferromagnetic state has a maximum value M_m below T_C when the dipoles are aligned.[5] As the temperature is raised, M_m decreases, slowly at first, but at a much more rapid rate near the Curie temperature (Figure 2.10). The thermal energy required to eliminate the magnetisation by disorienting the magnetic ions varies as $-\mathrm{d}M/\mathrm{d}T$. As a consequence, there is (ideally) a discontinuity in the slope of the specific heat curve at T_C, where the randomising of dipoles is complete as far as long-range order is concerned. The

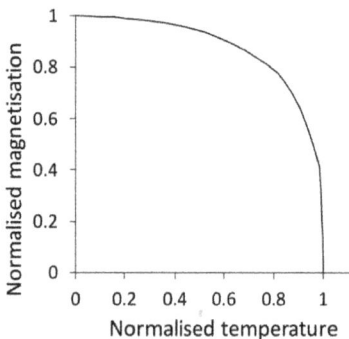

Figure 2.10 Variation in the normalised magnetisation $M_m\{T\}/M_m\{0\}$ against the normalised temperature T/T_C.

area under a plot of magnetic specific heat versus temperature gives the energy required to take the system from the state of approximately complete magnetic order at $T \ll T_C$ to one of almost complete magnetic disorder for $T \gg T_C$. The difference between the continuous and dashed curves in Figure 2.11 represents an energy of about $8078 \pm 500 \, \mathrm{J \, mol^{-1}}$ for α-iron. The entropy increase in going from the ordered to the disordered spin state has been estimated from the plot to be about $9 \pm 0.3 \, \mathrm{J \, mol^{-1} \, K^{-1}}$ [26]. Suppose that each atom has a spin imbalance $s_a = \frac{1}{2}$, then there are only two possible accessible states per atom $\uparrow$ or $\downarrow$. When $s_a = 1$, three possibilities arise, $\uparrow\uparrow$, $\downarrow\downarrow$ and $\uparrow\downarrow$. In general, the number of accessible states per ion is $2s_a + 1$ and if there are N_a ions, then the number of configurations becomes [27]

$$w_c = (2s_a + 1)^{N_a}. \tag{2.18}$$

The Boltzmann relation $S = k \ln\{w_c\}$ links the entropy to the logarithm of the number of microstates w_c that a given macroscopic state of the system can have. The change in entropy in going from a completely ordered spin state to one which is completely disordered is given therefore by

$$N_a k \ln\{2s_a + 1\} \tag{2.19}$$

for a mole of atoms.

Iron atoms in ferrite contain 7.4 electrons in the d shell, 4.8 of which oppose the spin of the remaining 2.6, leaving a net magnetic moment per atom of $2.2 \, \mu_B$ (more precise data would indicate $2.22 \, \mu_B$). The non-integral numbers of electrons are due to the overlapping of s and d states. The electrons can be thought of as spending some time in each state. Thus, the value of s_a is about 1.1 for iron atoms in the ferrite crystal structure. Consequently, the change in magnetic entropy in going from a completely magnetically ordered to a disordered state is about $9.67 \, \mathrm{J \, mol^{-1} \, K^{-1}}$, which compares well with the experimental value quoted earlier.

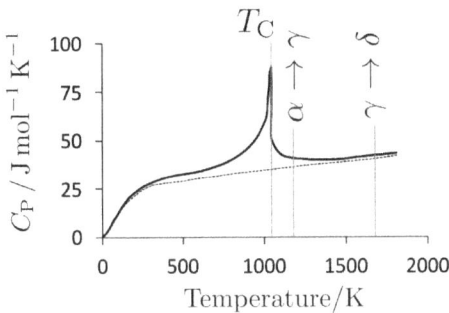

Figure 2.11 The specific heat capacities of ferrite as a function of temperature (after Kaufman [28]). The dashed curve represents the combined contributions of the phonons and electrons whereas the thicker lines also include the magnetic terms.

For $T < 300 \, \mathrm{K}$, the magnetic specific heat of α can be calculated using spin-wave theory [18]:

$$C_P^{\mu\alpha} \approx C_V^{\mu\alpha} = 4.728 \times 10^{-5} T^{\frac{3}{2}} \, \mathrm{J \, mol^{-1} K^{-1}}. \tag{2.20}$$

It has not been possible to verify this experimentally because below 300 K, $C_P^{\mu\alpha}$ is a small fraction of the total heat capacity, but the form of the equation appears valid for other metals even at temperatures where the harmonic character of the spin waves, assumed in the theory, becomes disrupted. Wallace et al. [29] measured C_P^{α} over the temperature range 298-1323 K. Using these data, and removing the Debye and electronic specific heat terms, Kaufman et al. [10] obtained $C_P^{\mu\alpha}$ for $T = 0$-1183 K, values of which are plotted in Figure 2.11. Later work has justified this factorisation of specific heat capacity, since experimental measurements for ferrite over the temperature range 60-300 K were found to be excellent agreement with calculations carried out using $T_D^{\alpha} = 432 \, \mathrm{K}$ and $c_e^{\alpha} = 5.02 \times 10^{-3} \, \mathrm{J \, mol^{-1} \, K^{-2}}$ [13].

Figure 2.11 shows that the magnetic component of specific heat persists above T_C even though long-range spin order vanishes. This is attributed to short-range magnetic order above the Curie temperature [30, 31]. For ferromagnetic and antiferromagnetic materials at temperatures above the Curie or Néel temperature respectively, the magnetic component of the heat capacity varies approximately as T^{-2} [10]:

$$C_P^{\mu\alpha} = 7.7822 \times 10^6 T^{-2} \qquad \text{for } T > 1183\,\text{K} \qquad \text{J\,mol}^{-1}\text{K}^{-1}. \qquad (2.21)$$

The work described above does not give a model for the magnetic component of the specific heat over the entire range of temperature such as would be useful in common algorithms for the estimation of phase diagrams. Semi-empirical expressions accurately describing the magnetic contribution to the specific heat of ferromagnetic metals in general, and ferritic iron in particular, have been derived by Inden [32]:

$$C_P^{\mu\alpha} = b_1 \ln\{2s_a + 1\} R \ln\left\{\frac{1 + \tau^3}{1 - \tau^3}\right\} \qquad \text{for} \qquad \tau < 1$$

$$C_P^{\mu\alpha} = b_2 \ln\{2s_a + 1\} R \ln\left\{\frac{\tau^5 + 1}{\tau^5 - 1}\right\} \qquad \text{for} \qquad \tau > 1$$

where $\tau = T/T_C$. The constants, obtained by fitting to experimental data, are $b_1 = 0.6417$ and $b_2 = 0.9180$ for ferrite [33]. The respective magnetic specific heats of ferromagnetic cobalt and nickel are much smaller at 80% and 40% that of iron below the Curie temperature, and 50% and 25% above the Curie temperature [32], emphasising the predominant role of ferromagnetism in iron. The application of this model shows the importance of short-range order above T_C. Inden found that the fraction of the total magnetic enthalpy which is absorbed above the Curie temperature is about 0.4 for bcc metals and about 0.28 for fcc metals. Table 2.1 shows the variation in the percentage of magnetic enthalpy retained at T_C as a function of the chromium concentration in Fe-Cr alloys [34]. The model is useful in conducting generalised calculations involving many variables as will become apparent when the thermodynamics of both pure elements and solutions are discussed in detail (Section 2.19.4).

Table 2.1

The percentage of magnetic enthalpy at 0 K, which is retained at T_C for ferritic iron and for ferritic Fe-Cr alloys (after Miodownik [34]).

Cr / at.%	T_C/K	μ_B per atom	% magnetic enthalpy retained
0	1042.15	2.22	35
3	1042	2.14	21
9	1025	2.04	27
11	1012	2.00	29
16	975	1.90	8

2.6.6 MAGNETIC HEAT CAPACITY OF AUSTENITE

Austenitic iron alloyed with manganese can exhibit antiferromagnetic ordering, whereas that with nickel can show ferromagnetic ordering. In the latter case, there is evidence that the Fe-Ni alloys can be magnetically *heterogeneous* if the nickel concentration is less than about 36 at.%; some regions of the sample remain paramagnetic below the Curie temperature and transform to the antiferromagnetic state at even lower temperatures. Similar observations have been reported for Fe-Cu alloys [35].

This and other features of the peculiar magnetic behaviour of iron are illustrated in Figure 2.12, which shows the magnetic phase diagram for $Fe_{50}(Ni_xMn_{1-x})_{50}$ and some data for the binary Fe-Ni and Fe-Mn alloys [36]. There is a two-phase ferromagnetic+paramagnetic phase field followed at lower temperatures by a ferromagnetic+antiferromagnetic phase field. Thus, the coexistence of two magnetic phases in an apparently single crystalline phase is a physical reality. Work on manganese oxides indicates that because the magnetic phases are determined essentially by electron correlations, the likelihood is that the magnetic-phase segregation occurs on a short length scale [37]. Long-lived antiferromagnetic and ferromagnetic clusters have been observed to coexist in manganese perovskites [38].

Precisely such a coexistence of magnetic phases has been proposed by Weiss and co-workers for austenite in pure iron, involving two different states of electron spin. This will be discussed in detail since the theory is capable of explaining a vast range of anomalous data on austenite.

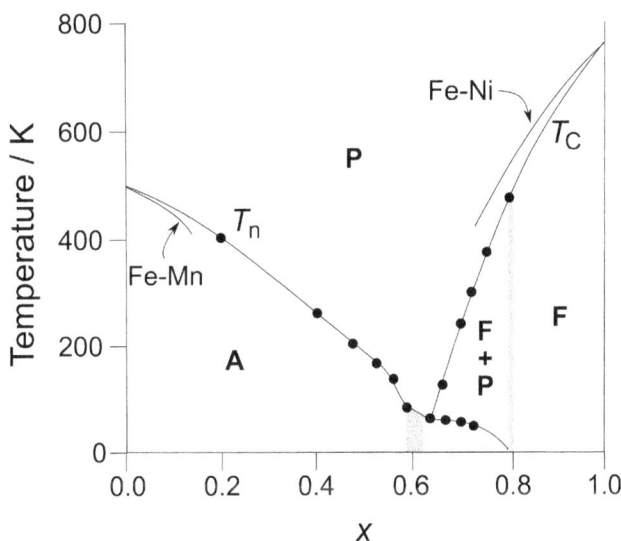

Figure 2.12 Magnetic phase diagram of the $Fe_{50}(Ni_xMn_{1-x})_{50}$ system (after Ettwig and Pepperhoff [36]). The Néel and Curie temperatures for the binary Fe-Mn and Fe-Ni systems respectively, are also shown. The shaded regions represent boundaries which are experimentally uncertain. The abbreviations AF, F and P stand for antiferromagnetic, ferromagnetic and paramagnetic states respectively.

The idea that a single crystal can entertain simultaneously, two different magnetic states, the so-called "magnetically heterogeneous" state, is not as alien as it might seem. It was the basis of a famous two-state theory of austenite by Weiss and co-workers. Whereas the real scenario may be more complex as indicated by first-principles calculations, the theory continues to serve well in representing the experimental data on austenite, far more so than the *ab initio* methods have been able to do so.

In the two-state theory, the γ-iron atoms are considered to exist in two electronic states, separated by an energy gap E_1 [10, 39]. Electrical resistivity and neutron diffraction experiments show that the average magnetic moment of γ-iron at high temperatures is about 2 Bohr magnetons per atom, similar to that of α iron [40]. Pure iron can be retained in its austenitic state to low temperatures by coherent precipitation in copper. At temperatures close to 0 K, both neutron diffraction and Möss-bauer spectroscopy suggest that fcc iron has an antiferromagnetic structure, the Néel temperature of which is 55-80 K with a magnetic moment of about 0.56-0.70 μ_B per atom [41, 42]. It appears, therefore, that γ-iron has a small magnetic moment and is antiferromagnetic at low temperatures, but has a larger moment (2.8 μ_B) and is ferromagnetic at high temperatures with a Curie temperature of about 1800 K [43].

In pure iron, the ground state (designated γ_0) at 0 K is antiferromagnetic but the higher energy ferromagnetic state (γ_1) can be thermally excited. It only requires a shift of an electron form one half of a d-band to the other half to convert from γ_0 to γ_1. Such a reversal in spin accounts for a difference of 2 Bohr magnetons between the two states, which compares well with the observed difference of about 2.3 μ_B per atom.

Consider a total number N of austenite atoms. At any finite temperature a number N_0 of the atoms are in the antiferromagnetic ground state, whereas N_1 ($= N - N_0$) atoms occupy the higher level ferromagnetic state, with an energy E_1 relative to the ground state. The fraction of atoms in the two states at a temperature T and at zero pressure is given by

$$\frac{N_0}{N} = \frac{g_0}{g_0 + g_1 \exp\left\{-\frac{E_1}{kT}\right\}}$$

$$\frac{N_1}{N} = \frac{g_1 \exp\left\{-\frac{E_1}{kT}\right\}}{g_0 + g_1 \exp\left\{-\frac{E_1}{kT}\right\}}$$

where g_i represents the degeneracy of the ith energy level. The degeneracy gives the number of states with the same energy. When the system is completely (magnetically) disordered, the degeneracies are given by $2s + 1$, although in each case the degeneracy must have a minimum value of 2 because with $s = \frac{1}{2}$, the spin can point either up or down.

In each of these equations, the term in the denominator is called the partition function Ω; in general, for a multi-level system,

$$\Omega = \sum_i g_i \exp\left\{-\frac{E_i}{kT}\right\}$$

where E_i is the energy relative to the ground state. The degeneracy ratio g_0/g_1 is found to be about 0.559 and $E_1 = 5.6971 \times 10^{-21}$ J atom^{-1} [10]. As seen from Figure 2.13, even at 1000 K, the fraction of atoms in the higher energy ferromagnetic state is only 0.5 so that long-range coupling of dipoles and macroscopic ferromagnetism is not observed for pure γ-iron. Consequently, for pure iron at least, the austenite can be regarded as magnetically disordered since the Néel temperature is so low. The degeneracy ratio g_0/g_1 is therefore expected to have the value $2/(2.8 + 1) = 0.53$, which compares well with that quoted above.[6]

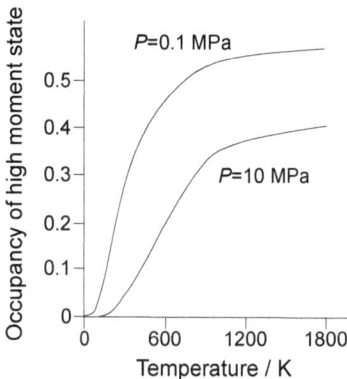

Figure 2.13 Relative population of high moment spin states in austenitic iron as a function of temperature and pressure [28].

The fraction of atoms in the higher energy state increases with temperature, with a corresponding increase in the internal energy of the system by $\Delta U^{\mu\gamma}$, which for 1 mole of atoms is given by

$$\Delta U^{\mu\gamma} = E_1 \times \frac{N_a N_1}{N} = \frac{N_a E_1}{1 + \frac{g_0}{g_1} \exp\left\{\frac{E_1}{kT}\right\}}$$

where N_a is Avogadro's number. The corresponding molar magnetic component of the specific heat capacity for austenitic iron, $C_V^{\mu\gamma}$, is obtained by differentiating this equation with respect to temperature:

$$
\begin{aligned}
C_V^{\mu\gamma} &= \left(\frac{\partial \Delta U^{\mu\gamma}}{\partial T}\right)_V \equiv \frac{\partial}{\partial T}\left(RT^2 \frac{\partial \ln \Omega}{\partial T}\right)_V \\
&= R\left(\frac{E_1}{kT}\right)^2 \frac{g_0}{g_1} \exp\left\{\frac{E_1}{kT}\right\} \Big/ \left(1 + \frac{g_0}{g_1}\exp\left\{\frac{E_1}{kT}\right\}\right)^2 .
\end{aligned}
$$

Figure 2.14 illustrates the magnetic component of the specific heat capacity of austenite; it is negligibly small at temperatures above 1185 K.

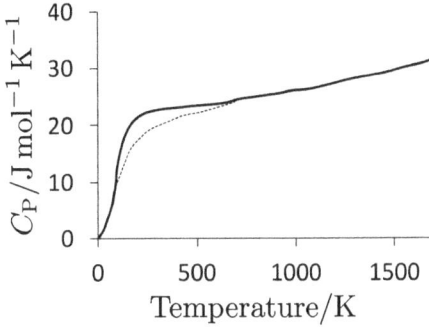

Figure 2.14 The specific heat capacities of austenite as a function of temperature [28]. The dashed curve represents the combined contributions of the phonons and electrons whereas the thicker lines also include the magnetic terms.

The Helmholtz free energy $F^{\mu\gamma}$ per mole is obtained from the partition function:

$$
\begin{aligned}
F^{\mu\gamma} &= -RT \ln\{\Omega\} \\
&= -RT \ln\left\{g_0 + g_1 \exp\left\{-\frac{E_1}{kT}\right\}\right\} \\
&= -RT \ln\{g_0\} - RT \ln\left\{1 + \frac{g_1}{g_0}\exp\left\{-\frac{E_1}{kT}\right\}\right\}.
\end{aligned}
$$

Therefore, the *change* in free energy as atoms are promoted from the ground state is

$$\Delta F^{\mu\gamma} = -RT \ln\left\{1 + \frac{g_1}{g_0}\exp\left\{-\frac{E_1}{kT}\right\}\right\}.$$

The corresponding molar Gibbs free energy change is given by

$$\Delta G^{\mu\gamma} = \Delta F^{\mu\gamma} + PV_m^\gamma.$$

Noting that $\Delta F^{\mu\gamma} = \Delta U^{\mu\gamma} - T\Delta S^{\mu\gamma}$, the change in molar entropy $\Delta S^{\mu\gamma}$ of austenite at a temperature T, due to the promotion of atoms from the ground state, is

$$
\begin{aligned}
S^{\mu\gamma} &= R\ln\left\{\frac{\Omega}{g_0}\right\} + \frac{\Delta U^{\mu\gamma}}{T} \\
&= R\ln\left\{1 + \frac{g_1}{g_0}\exp\left\{-\frac{E_1}{kT}\right\}\right\} + \frac{\Delta U^{\mu\gamma}}{T}.
\end{aligned}
$$

This entropy is a direct consequence of the magnetic properties of γ and is about $8.40\,\mathrm{J\,mol^{-1}\,K^{-1}}$ at the α/γ transition temperature 1185 K.

For atmospheric pressure, $\Delta G^{\mu\gamma} \approx \Delta F^{\mu\gamma}$ but for greater pressures a more accurate value of $C_P^{\mu\gamma}$ can be deduced by expressing the energy gap as a function of pressure:

$$E_1\{P\} = E_1 + \frac{P(V_m^{\gamma 0} - V_m^{\gamma 1})}{N_a},$$

where $V_m^{\gamma 0}$ and $V_m^{\gamma 1}$ are the molar volumes of γ_0 and γ_1 respectively. By assuming that each state of austenite has the same expansion coefficient as ferrite, Kaufman et al. [10] found that:

$$V_m^{\gamma 0} = 6.695(1 + 2.043 \times 10^{-5}T + 1.52 \times 10^{-8}T^2) \qquad \mathrm{cm^3\,mol^{-1}}$$

$$V_m^{\gamma 1} = 7.216(1 + 2.043 \times 10^{-5}T + 1.52 \times 10^{-8}T^2) \qquad \mathrm{cm^3\,mol^{-1}}$$

This difference in the molar volumes between the two electronic states of austenite explains why the magnetic heat capacity of ferrite hardly varies with pressure, but that of austenite is sensitive to pressure. The lower volume antiferromagnetic state of austenite naturally is favoured by external pressure. Experimental work provides strong support for the idea that the antiferromagnetic state should be the one with the smaller volume [44, 45]. Precipitated particles of fcc iron show antiferromagnetism, whilst thin films of iron deposited on copper substrates show ferromagnetism. This is because the particles are large enough to have lost coherency with the copper matrix and hence have a smaller lattice parameter than the film which is forced to match the larger lattice parameter of the Cu substrate. It is known from the Bethe-Slater curve that the exchange interaction is sensitive to atomic distance. The interaction can change sign with a small increase in the interatomic distance, giving the transition from antiferromagnetic to ferromagnetic ordering. Antiferromagnetic iron particles in copper become ferromagnetic if the lattice parameter of the copper is increased, for example by alloying the copper with gold [45]. Alternatively, an austenite film deposited on a hot copper substrate (with an expanded lattice parameter) is found to exhibit ferromagnetism whilst maintained at the deposition temperature [46].

These observations are reinforced by studies on $Fe_{100-x}Cu_x$ metastable solid solutions prepared by a mechanical alloying technique [47–49].[7] The magnetic moment of iron atoms as a function of the copper content in fcc $Fe_{100-x}Cu_x$ alloys is found to be close to $2.3\,\mu_B$ for $x \leq 60$ but drops rapidly to less than $1\,\mu_B$ for $x \simeq 70$. This is because the lattice constant of fcc $Fe_{100-x}Cu_x$ for $x \leq 60$ is about 0.364 nm, comparable with that of ferromagnetic austenite. For $x = 70$ the lattice constant is smaller at 0.3619 nm, corresponding to the low-moment state of austenite.

The general result for Fe-Cu alloys is that austenite becomes paramagnetic for small lattice parameters, (low-moment) antiferromagnetic for $a_\gamma \simeq 0.36\,\mathrm{nm}$ and (large moment) ferromagnetic for $a_\gamma > 0.36\,\mathrm{nm}$. The details can vary depending on the binding interactions specific to different crystallographic planes of the substrate on which the iron is deposited [46].

The fact that the Debye temperatures and electronic specific heat coefficients of ferrite and austenite are equal implies that the expansion coefficients of these phases should also be equal. However, because the densities of the two forms of austenite differ significantly, and since the relative proportion of atoms in the γ_0 and γ_1 states changes with temperature, the apparent expansion coefficient of austenite as a whole, as detected experimentally, is much larger than that of ferrite. This can be seen from the equations representing the molar volumes of ferrite and austenite as a function of temperature [28] $(300 \to 1775\,\mathrm{K})$:

$$\begin{aligned} V_m^\alpha\{T\} &= 7.061(1 + 2.043 \times 10^{-5}T + 1.52 \times 10^{-8}T^2) \quad \mathrm{cm^3\,mol^{-1}} \\ V_m^\gamma\{T\} &= (1-y)V_m^{\gamma 0}\{T\} + yV_m^{\gamma 1}\{T\} \quad \mathrm{cm^3\,mol^{-1}} \end{aligned} \qquad (2.22)$$

where y is the fraction of atoms of austenite in the γ_1 state. The thermal expansion coefficients of α and γ_0 and γ_1 are equal, even though that of austenite as a whole is higher than that of ferrite. The volumetric relations between α, γ_0, γ_1 and γ are shown in Figure 2.15.

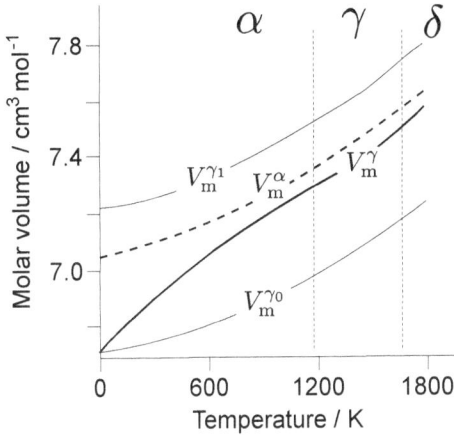

Figure 2.15 Molar volumes of the bcc and fcc forms of iron (after Kaufman [28]).

The large thermal expansion coefficient of austenitic steels makes them particularly susceptible to thermal fatigue when applied in conditions, such as in power plant, where the operating temperature fluctuates. They therefore are considered unsuitable for any application where the component has a thick cross-section, even though they can be designed for with superior creep-resistance relative to ferritic steels.

2.6.7 INVAR EFFECT AND ASSOCIATED PHENOMENA

The two-state model was proposed originally for fcc iron but it has been established for other metals such as Mn, Cr, Co and Ni. However, as pointed out by Miodownik [34], the thermodynamic consequences of the two states are likely to be appreciable only for iron because of the relatively small value of E_1, as illustrated in Figure 2.16. On the other hand, when these particular elements are added to fcc iron as alloying elements, they obviously cause a change in E_1 and therefore in the stability and physical properties of alloyed austenite. Some of these effects can be quite remarkable.

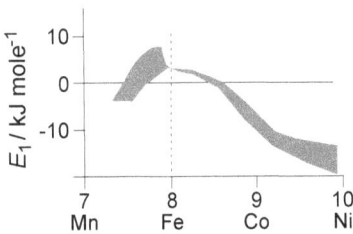

Figure 2.16 Variation in E_1 with the "electron concentration" [34]. Note that E_1 is the elevation of the ferromagnetic state relative to the antiferromagnetic ground state.

The notion that there are two electronic configurations of iron atoms in the fcc structure can be exploited to explain a variety of observations on the effect of certain alloying elements on the physical properties of iron. In particular, it is found experimentally that the expansion coefficient of austenite decreases markedly, even tending towards zero, as the concentration of elements such as nickel, platinum or palladium is increased. This is called the *Invar effect*. The apparent thermal expansion of austenite is due to two effects, changes in the proportions of γ_0 and γ_1 (Figure 2.15), and the normal thermal contraction resulting from the anharmonic shape of the force-interatomic separation curve. If the relative stability of the two states of austenite is altered by solute additions, then it is natural to expect the apparent thermal expansivity to change, with accompanying changes in the magnetic properties of the austenite.

Weiss [43] has shown that many of the unusual properties of iron-nickel alloys can be understood by assuming a specific dependence of E_1 on the nickel concentration (Figure 2.17). In pure iron, it is the antiferromagnetic condition of austenite which is the ground state. But the addition of sufficiently large concentrations of nickel causes a reversal of stability, with the ferromagnetic state eventually becoming the ground state (Figure 2.17). When that happens, the fraction of *lower density* ferromagnetic austenite increases as the temperature drops, an effect which opposes the normal anharmonic source of thermal contraction. It then becomes possible to get zero or negative expansion with rising temperature. It is emphasised that the plot in Figure 2.17 is an *assumed* variation of the kind required to explain the thermal expansion data.

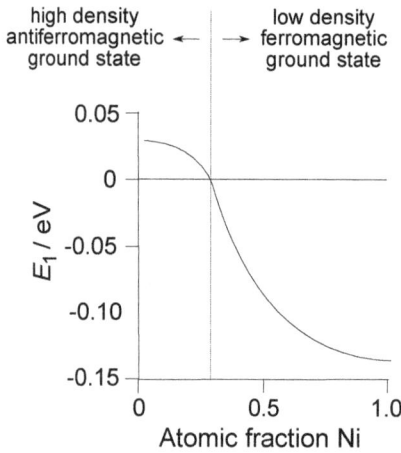

Figure 2.17 The variation in the energy gap E_1 for the two electronic states of austenite, as a function of the nickel concentration (after Weiss [43]). A negative value implies that the antiferromagnetic form of austenite is the ground state. Note that the switching of ground states occurs at a nickel fraction of about 0.29.

The lattice parameters of austenite for atom fractions x of nickel greater than and less than 0.29 are given by

$$a^\gamma = a^{\gamma_0}\left[1 - \frac{a^{\gamma_0} - a^{Ni}}{a^{\gamma_0}}x + \frac{a^{\gamma_1} - a^{\gamma_0}}{a^{\gamma_0}}\frac{N_1}{N}(1-x)\right](1 + e[T - 290]) \qquad x < 0.29 \qquad (2.23a)$$

where e is the coefficient of linear thermal expansion. The second term in the square brackets allows for the Vegard's law effect of nickel, the third term for the fraction of iron atoms which are not in the ground state.

$$a^\gamma = a^{\gamma_1}\left[1 - \frac{a^{\gamma_1} - a^{Ni}}{a^{\gamma_1}}x - \frac{a^{\gamma_1} - a^{\gamma_0}}{a^{\gamma_1}}\frac{N_0}{N}(1-x)\right](1 + e[T - 290]) \qquad x > 0.29 \qquad (2.23b)$$

where $a^{\gamma_1} = 0.354$ nm and $a^{\gamma_1} = 0.364$ nm are respectively, the lattice parameters of the antiferromagnetic (γ_0) and ferromagnetic γ_1 states of austenite, at 290 K. The ratios N_0/N and N_1/N, as usual, give the fractions of iron atoms in the γ_0 and γ_1 states respectively. In calculating these ratios, the degeneracy ratio g_0/g_1 for $x < 0.29$ is 1.79, as in ordinary steels [28]. However, for nickel concentrations greater than 0.29, the ferromagnetic state is the ground state, so that the austenite is well below its Curie temperature of about 1800 K. The spins are therefore aligned, giving a degeneracy of 2, which is identical to that of the disordered spin state. Consequently, for $x > 0.29$, $g_0/g_1 = 1$. Figure 2.18 shows the lattice parameters of nickel-alloyed austenite as a function of temperature. These calculations are due to Weiss, using Equation 2.23 and the assumed E_1 function. The invar effect is apparent for the alloy containing 34.4 at.% nickel, with an almost zero thermal expansion coefficient near ambient temperature. At that concentration of nickel, the ground state is the high volume ferromagnetic austenite, so raising the temperature (below T_C) will have the thermal expansion compensated by the promotion of the low volume antiferromagnetic austenite. The net thermal expansion can become zero; this is the so-called invar effect. In contrast, at low nickel concentrations where the ground state is the low volume antiferromagnetic austenite, raising the temperature leads to an exaggerated expansion as more of the high-volume state is promoted.

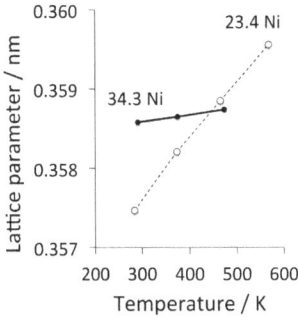

Figure 2.18 The lattice parameters of austenite as a function of the atomic percent of nickel and the temperature. The data for the 34.3 at.% nickel are all below the Curie temperature. Selected data from Weiss [43].

The Invar effect has not been studied for fcc iron-copper alloys because of the small solubility of iron in copper and the difficulty in obtaining fcc iron at low temperatures. However, copper-rich Fe-Cu solutions with an fcc structure can be created by mechanical alloying. The Curie temperature of fcc $Fe_{1-x}Cu_x$ with $50 < x < 61$ is found to be about 500 K with a magnetic moment per atom of $2.35\,\mu_B$ [Jiang:1994b]. It would be interesting to measure the thermal expansion coefficient of such alloys.

Curie temperature

The Weiss two-state model for austenite manages to predict accurately the variation in the ferromagnetic Curie temperature of austenite as a function of the nickel concentration. The prediction is based partly on an empirical equation that works rather well for many transition metals and alloys:

$$T_C/K \simeq 113.5 \mid Z_\uparrow - Z_\downarrow \mid \ln(2s_a + 1) \qquad (2.24)$$

where $Z_\uparrow$ is the number of nearest neighbours with spin aligned in the direction favoured by the *exchange integral* and $Z_\downarrow$ is the number with opposite spins [50]. When the electron wave functions of two atoms overlap, the electrons are shared and therefore they must correlate in some way, since they can *exchange* their roles. This correlation, including that of the spins, can lead to ferromagnetism or antiferromagnetism. For ferromagnetic austenite, all the dipoles point in the same direction. Since each atom has twelve nearest neighbours along $\langle 110 \rangle$ directions, $\mid Z_\uparrow - Z_\downarrow \mid = 12$.
The overall structure of antiferromagnetic austenite must contain an equal number of opposing spins. This could arise if, for example, the atoms on any given $\{100\}$ plane have identical spins, but with alternating planes having opposing spins. Given that the nearest neighbour atoms lie along $\langle 110 \rangle$ directions, each atom has four nearest neighbours with the same spin, but eight of which oppose its spin. Consequently, $\mid Z_\uparrow - Z_\downarrow \mid = 8 - 4 = 4$.
To account for nickel additions to iron, Weiss [43] assumed that the average magnetic moment per atom of the alloy would scale with the concentration:

$$[2s_a]_{average} = \mu_B^{Ni} x + \frac{\mu_B^{\gamma 1}(1-x)N_1}{N} + \frac{\mu_B^{\gamma 0}(1-x)N_0}{N} \qquad (2.25)$$

where μ_B refers to the magnetic moment per atom of the species identified by the superscript. Similarly,

$$\mid Z_\uparrow - Z_\downarrow \mid_{average} = 12x + 12(1-x)\frac{N_1}{N} + 4(1-x)\frac{N_0}{N}. \qquad (2.26)$$

Both these equations must be evaluated at the Curie temperature - the agreement with experimental data was found to be excellent (Figure 2.19).

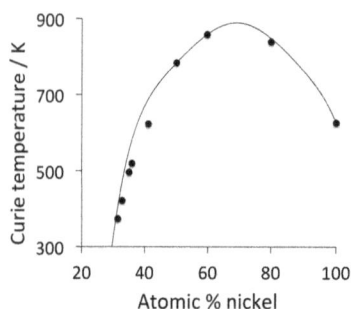

Figure 2.19 The estimated and observed values of the ferromagnetic Curie temperature of austenite in Fe-Ni alloys (after Weiss [43]).

Justification of the magnetic properties of austenite

The Weiss and Tauer two-state model for austenite is able to describe a variety of thermodynamic and magnetic phenomena in iron. There has been some unease about its physical significance which prompted many attempts at finding direct connections between the two-state hypothesis and *ab initio* calculations based on the electron theory of metals.

Quantum mechanical calculations broadly confirm that the magnetic properties of fcc iron are complicated. Consistent with the Weiss and Tauer model, Kübler [51] has shown that the ground state of γ-iron is antiferromagnetic with a relatively low volume and a magnetic moment per atom of about $0.6\,\mu_B$. However, he has found two higher energy ferromagnetic states of austenite, one of which has a low magnetic moment and the other a high magnetic moment ($< 0.6\,\mu_B$ and $> 2\,\mu_B$ per atom, respectively). This is in contrast to the single ferromagnetic state proposed by Weiss and Tauer.

With more refined calculations, Krasko [52] revealed three ferromagnetic states in austenite, two with the low spin and one with the high spin. The two low-spin states have energies similar to that of a hypothetical nonmagnetic austenite, the energies being much larger than that associated with the high-spin state. Unfortunately, Krasko did not extend the calculations to include the antiferromagnetic γ-iron.

Other quantum mechanical total-energy[8] calculations in the context of the invar effect (e.g., Mohn et al. [53]; Wassermann [54]) indicate results reminiscent of the two-state model of austenite. However, many of the studies neglect to consider antiferromagnetic austenite as an option. Where low-spin ferromagnetic states have been reported for austenite, their much greater energies may justify neglecting them, in which case the high-spin, high-volume ferromagnetic and a low-spin, low-volume antiferromagnetic ground state of austenite remain consistent with the original Weiss and Tauer model. Other results using *ab initio* calculations are discussed in the context of thin films of iron, where the crystal structure can be altered continuously between the fcc and bcc forms (Section 2.6.9).

2.6.8 MAGNETIC HEAT CAPACITY OF ε-IRON AND SUPERCONDUCTIVITY

If it is assumed that magnetic properties are a function primarily of interatomic spacing rather than the details of lattice symmetry, then hcp iron is expected to exhibit magnetic characteristics similar to those of austenite. Therefore, the ground state should be a low-volume, low-spin antiferromagnetic structure, with increasing contributions from the high-volume, high-spin ferromagnetic state as the temperature is raised. An increase in pressure should tend to favour the antiferromagnetic state.

There is only limited evidence in support of this conclusion. Whereas Mössbauer spectra indicate that at high pressures hcp iron is not ferromagnetic, it is unclear whether it is antiferromagnetic or paramagnetic [55, 56]. Iron-manganese alloys with fcc and hcp structures are known to be antiferromagnetic; the Néel temperature of a Fe-17Mn wt% hcp alloy is found to be about 240 K [30, 56]. The study of the internal fields and the Néel temperatures of the Fe-Mn alloys in both the fcc and hcp structures indicate that the magnetic properties of the iron atoms in these two environments are similar. If these data are extrapolated to pure iron, then the internal field and Néel temperature of pure hcp iron can be estimated to be similar to that of pure γ-iron [56]. The Bethe-Slater curve which describes the relationship between the exchange coupling and interatomic distance implies that fcc and hcp iron should be antiferromagnetic [57].

The mechanism of superconduction in elemental metals according to Bardeen, Cooper and Schrieffer is thoroughly embedded in the literature; a conduction electron, when it moves through the lattice, distorts the positive ions towards itself, thus creating a higher density of positive charge in its vicinity. This can attract another electron, leading to a weak electron-electron pairing. How this leads to superconduction is described in the original theory, suffice it to say that the two electrons should have opposite spins. Any metal can in principle become superconducting in its non magnetic state at a sufficiently low temperature.

Iron becomes superconducting at pressures in the range 15-30 GPa and at temperatures below 2 K [58]. The transition to the superconducting state was confirmed by both a drop in resistivity and the expulsion of magnetic flux from the iron (Meissner effect). Metallic superconductors rely on the pairing of electrons with opposite spins; the pairs are then propelled through the lattice by vibrations (phonons). Magnetic interactions change the electron spins and hence prevent them from pairing up. Ferromagnetism and superconductivity therefore compete in metallic superconductors. Ferromagnetism is eliminated in iron by studying it at a pressure and temperature where it becomes hexagonal close-packed. It is argued that this is the reason for the onset of superconduction, although it is not clear how this conclusion would be modified if the hcp iron turns out to be antiferromagnetic. It has been argued [59] that the electron pairing mechanism mediated by lattice vibrations is not the dominant mechanism representing superconduction in hcp iron, because the theory fails to explain why the superconducting state vanishes beyond 30 GPa. It is proposed instead that the superconductivity is unconventional with incommensurate antiferromagnetic spin waves inducing the electron pairing. This is based on calculations that indicate a coincidence between the onset and vanishing of magnetic order over roughly the same range pressure as the observed superconduction.

Stepakoff and Kaufman [13] have shown that C_P^ε can, at least for the temperature range 60-300 K, be represented accurately as a function of just the vibrational and electronic components of specific heat. It seems that the magnetic component of the specific heat of ε-iron can be neglected without significant loss of accuracy (Figure 2.20).

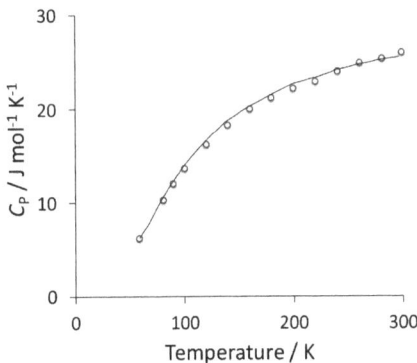

Figure 2.20 The experimentally measured specific heat capacity of ε-iron and its comparison with a curve calculated neglecting the magnetic component of the specific heat capacity (after Stepakoff and Kaufman [13]).

2.6.9   TRIGONAL AND TETRAGONAL IRON

These unusual allotropes of iron can only be created in the form of incredibly thin films. The drive to study such films of iron deposited on single-crystal substrates is allegedly from the vague possibility of applications in information technology and in the manufacture of artificial superlattices. The studies nevertheless reveal evidence for the magnetic properties of iron. The basis for the classical thermodynamic models for iron has been reinforced by the ability in thin-film experiments to study phases of iron which do not otherwise occur in nature, or to study known phases in conditions that are not natural.

Recall that coherency may be maintained with the substrate during the early stages of film deposition provided that the misfit between the iron and substrate is not so large that the elastic limit is exceeded. The deposited layers of iron may then adopt the atomic structure of the substrate surface. There is an important caveat that the clean substrate surface may reconstruct into a pattern of atoms that is different from that in the bulk crystal structure. Thus, $\{100\}_{Au}$ surface is known to reconstruct into a (5×20) structure, but the deposition of a small fraction of a monolayer of iron causes the surface to revert to the (1×1) bulk-termination structure [46].[9]

Any small in-plane misfit in coherently deposited films leads to a corresponding distortion normal to the layer, placing the film in an anisotropic state of strain. A homogeneously strained layer like this can be regarded as having either the face-centred tetragonal or trigonal structure depending on whether it is deposited to match the $\{100\}$ or $\{111\}$ substrate planes respectively.

Consider first the trigonal form of iron. The trigonal cell is defined by three basis vectors from the origin of an fcc cell, each of equal magnitude and separated by equal angles (Figure 1.11, Table 2.2). The body-cantered cubic, primitive cubic and face-centred cubic cells can all be regarded as special cases of the trigonal cell. If the primitive cubic cell is stretched along its body diagonal $[111]$ until the cell angles (α, β, γ) become $60°$, then this represents the primitive trigonal cell of the conventional fcc cell. By contrast, compression along the body diagonal to give cell angles of $109.47°$ gives the primitive trigonal cell of the conventional bcc lattice. It follows that when considering the deposition of iron on the $\{111\}$ plane of the fcc substrate, a matching atomic configuration for γ-iron is obtained when the substrate parameter (a_s) is equal to that of austenite. The distance between the atoms in the $\{111\}$ plane of the substrate is $\frac{a_s}{2}\langle 110\rangle$. The corresponding distance between the atoms on the $\{111\}$ plane of the trigonal cell for α-iron is $a_{bcc}\langle 110\rangle$. This means that α-iron is said to be deposited coherently when the substrate lattice parameter $a_s = 2a_\alpha$.

Table 2.2

Relationship between the non-primitive conventional and primitive trigonal cell parameters.

Conventional cell	Basis vectors of primitive cell relative to conventional cell				Primitive (trigonal) cell parameters	
Cubic	$a[100]$	$a[010]$	$a[001]$	$a = b = c$	$\alpha = \beta = \gamma = 90°$	
fcc	$\frac{a}{2}[101]$	$\frac{a}{2}[110]$	$\frac{a}{2}[011]$	$a = b = c$	$\alpha = \beta = \gamma = 60°$	
bcc	$\frac{a}{2}[1\bar{1}1]$	$\frac{a}{2}[11\bar{1}]$	$\frac{a}{2}[\bar{1}11]$	$a = b = c$	$\alpha = \beta = \gamma = 109.47°$	

The results of first-principle calculations using density functional theory are presented in Figure 2.21 [60]. The horizontal axis gives the substrate lattice parameter, beginning with a value which is close to that of austenite. Further increases in the substrate parameter lead to a gradual change in the trigonal cell until it becomes equivalent to the bcc form of iron. The austenite thin film is thus associated with a low magnetic moment, consistent with the fact that the calculations are for $0\,K$ although it is not clear whether the antiferromagnetic state was included in the calculations. The ferrite is correctly predicted to have a large magnetic moment per atom. The states between these natural allotropes are trigonal forms of iron. It is interesting that the magnetic moment per atom can be controlled by appropriately depositing iron in the form of thin films, although it is difficult as yet to see any practical application of this property.

Given that the energies of the trigonal forms of iron are larger than those of the fcc or bcc allotropes, they can exist only in metastable states, during forced coherency with the substrate. The excess energy relative to the equilibrium forms can be regarded as the coherency strain energy. The strain energy is clearly doubled when iron is deposited on palladium compared with when it is deposited on copper. Fox and Jansen suggest that this is why islands of bcc iron start to form on palladium when the thickness of the iron is less than a couple of atomic layers and why silver and aluminium are difficult substrates for iron.

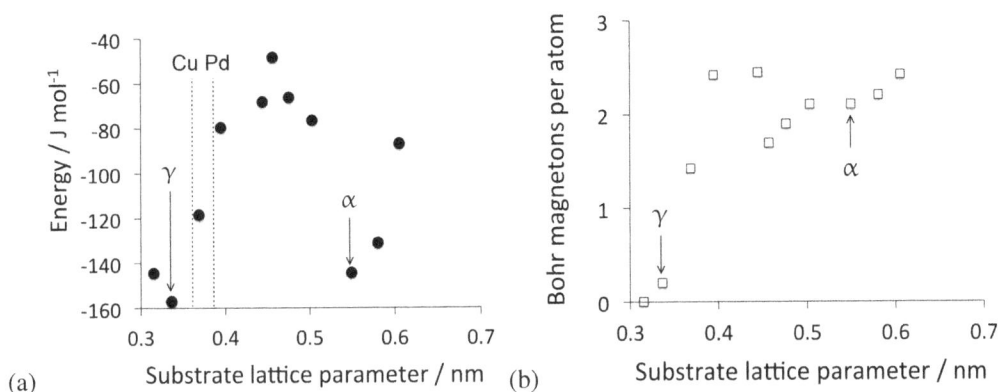

Figure 2.21 Density functional theory calculations corresponding to 0 K. (a) Plot of the minimum in the total energy of a thin film of iron deposited coherently on the {111} plane of an fcc substrate. The dashed lines indicate the positions of Cu and Pd substrates. The fcc and bcc forms of iron are also identified; it is assumed for both lattices that their {111} planes are parallel to the substrate {111} plane, and that the genesis of the three-dimensional structure corresponds to the correct plane spacing normal to those planes. The bcc is a $2 \times 2 \times 2$ supercell since the {111} plane of a single cell does not intersect the body-centering atom. (b) Corresponding plot of the calculated magnetic moment per atom. (Adapted from Fox and Jansen [60].)

When a thin film of iron is deposited coherently on a mismatching {100} fcc substrate, it assumes a face-centred tetragonal structure [61]. The tetragonal distortion gives another continuous path between the fcc and bcc structures. The famous Bain strain [62], involving a compression of the fcc cell along one of the cell edges and a uniform expansion along the other two, Figure 1.10, fits precisely within this scheme. This Bain strain is assumed throughout transformation theory to be the correct description of the pure strain which accomplishes the change in crystal structure. There are other strains which can implement the change but the Bain strain apparently has the smallest known principal deformations [63]. The trigonal deformation described above is another possibility for the lattice deformation. However, total energy calculations indicate that the barrier between the fcc and bcc structures for this deformation is about seven times larger than for the tetragonal distortion [61].

Figure 2.22 shows how the magnetic moment per atom varies as a function of the lattice parameter of a {100} fcc substrate. The variation in the ratio c/a of the in-plane to the normal lattice parameter of the iron film is also illustrated. This ratio is unity for the fcc structure but $1/\sqrt{2}$ for ferrite. This is because $a_\alpha = \sqrt{2}c_\alpha$ in order to ensure the same number of lattice points per unit cell in all the structures.

The calculations confirm that the ground sate for austenite at low temperatures is the low moment antiferromagnetic state. Antiferromagnetic ferrite has too large an energy to be stable; only ferromag-

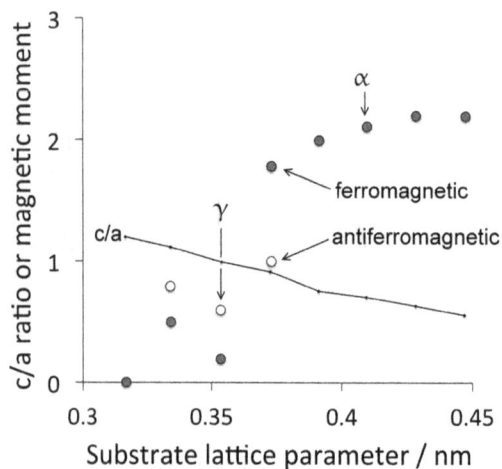

Figure 2.22 Plot of the c/a ratio and magnetic moment per atom (Bohr magnetons) for a thin film of iron deposited coherently on the $\{100\}$ plane of an fcc substrate.

netic ferrite is stable.[10] The face-centred tetragonal structures in between the fcc and bcc forms are either low moment austenite-like or high moment ferrite-like with a rather abrupt change at a substrate lattice parameter of 0.37 nm. Iron films grown coherently on $\{100\}$ on copper ($a = 0.3602$ nm) are expected to be slightly tetragonal and antiferromagnetic with a small moment. This is found experimentally to be the case when the films are about three monolayers thick. Thicker films (about eight layers) tend to be fcc in structure with defects present at the interface with the copper. It is assumed that the strain accumulated as the film thickens beyond three monolayers is sufficient to cause a breakdown of coherency and a relaxation to the fcc structure [64]. Since even this is not the equilibrium structure it is expected that further thickening or annealing must cause a transformation to bcc iron. Alternatively, when the mismatch with the substrate is large (e.g., with Pd or Ag), it becomes difficult to deposit anything but the bcc iron. In fact, there is only a 0.8% mismatch between the fcc $\{100\}$ Ag and bcc $\{100\}$ Fe surface after a $45°$ rotation so it would not be favourable to precipitate fcc iron given the larger misfit [65].

It is striking that almost all of the *ab initio* calculations, including those discussed above, fail to predict that bcc iron is more stable than fcc iron at 0 K. It appears that this is an inherent difficulty with the local density approximation [60]. To overcome this difficulty, Krasko and Olson [66] made an empirical adjustment to one of the magnetic terms (the Stoner exchange parameter) to ensure a match with experimental data. This leads to the surprising result that ferromagnetic austenite is unstable to tetragonal distortion and hence does not exist. In contrast, all the other *ab initio* calculations predict that the ferromagnetic states of fcc and bcc iron are stable, defined by energy minima in structure space.

These problems are resolved by using full-potential band-structure calculations in the density functional formalism but together with a generalised gradient approximation. The method correctly predicts that ferromagnetic bcc iron is the most stable form at low temperatures. It confirms that the ferromagnetic fcc phase has the low-moment low-spin and high-moment high-spin states. Furthermore, the high-density antiferromagnetic form of fcc iron is found to be the ground state for austenite but with increasing amounts of the low-density high-spin ferromagnetic austenite excited with rising temperature, explaining the large apparent thermal expansivity of austenite and also the invar type effects [67, 68]. This is reminiscent of the classical Weiss and Tauer two-austenite state model.

Magnetic anisotropy in thin films of iron

There are two special magnetic effects associated with thin films of iron. First, the magnetic moment per atom becomes especially large (3.1 μ_B) when compared with bulk iron (2.2 μ_B). Secondly, there exists a large magnetic-anisotropy in thin epitaxial films of iron.

The increase in the magnetic moment per atom is due to the smaller coordination number for atoms in a thin film. The atoms of the substrate used to produce the thin film do not contribute to the coordination number because there is a lack of hybridisation between the electronic states of the iron layer and the substrate [65, 69]. The d-bands in bulk ferromagnets are much broader than they would be for a single atom because of hybridisation between atoms. In reducing the number of nearest neighbours, the hybridisation is reduced so the bands become "atom-like". This squashing of the d-bands increases the density of states at the Fermi level and resolves the majority spin-up band from the minority spin-down band. A low-coordination atom therefore has more electrons in its majority spin-up band, and so has a larger moment per atom. An isolated atom has the highest moment and the bulk material the lowest. Reducing the coordination makes the material less bulk-like and more single-atom-like.

The magnetic anisotropy seen in thin films is a general feature found even in bulk iron where it is more readily magnetised along the $\langle 100 \rangle$ axis [70]. Anisotropy is caused by the coupling of the directions of the spin magnetic moments and orbital magnetic moments. For a thin film of iron, the net effect is often such that it causes the spins to align in a direction normal to its plane. Thin layers of iron separated by intervening layers of chalcogenides have been found to be highly anisotropic with the internal field perpendicular to the plane [71]. Such materials show a large change in resistance as the magnetic field is altered and could conceivably have applications in recording devices.

The details of this spin alignment normal to the plane of the film can only be accessed via precise calculations of the band structure along different crystallographic directions. However, Van Vleck [72] proposed a model that gives some intuitive feel for the problem. Quantum mechanics shows that the spin direction and spin orbit energy are coupled. As the spin direction changes, the spin orbit energy changes so the electron wave function must redistribute in order to comply. This affects the electron overlap between nearest neighbour atoms, with consequences on electrostatic interactions. The energy of the system becomes a function of the direction of the electron spin with respect to the crystalline lattice. The resulting magnetocrystalline anisotropy therefore reflects the symmetry of the crystalline lattice. In bulk iron the symmetry is cubic whereas in a monolayer it must be uniaxial so the spins align normal to the plane.

There is a further consequence of the fact that the magnetisation tends to be normal to the film of iron [73]. Bulk samples of iron generally contain a magnetic-domain structure to minimise the overall energy. But domain formation is opposed by the exchange energy which acts to make the magnetic moment of each atom line up with those of neighbouring atoms. Consequently, sufficiently small particles tend to have a single domain structure. This applies to thin films. Thus, thin films of cobalt with in-plane magnetisation are single-domain whereas iron films where the magnetisation in normal to the film plane contain a domain structure. Reducing the thickness of an iron film has no influence on its domain structure.

When thin layers of iron are deposited on slightly mismatching substrates, the spins mostly tend to align along the normal to the film. This is confirmed both experimentally [65, 74] and using total energy calculations which show that the spin-alignment is in-plane for a free-standing monolayer of iron, but perpendicular when a monolayer of iron is deposited on a monolayer of Au, Ag or Pd [69].[11] The difference with free-standing iron arises because of an interaction of the Fe $3d$ and substrate $4d$ band electrons. The interaction can induce weak magnetic moments in the adjacent substrate atoms (Au, Ag, Pd) which normally are not magnetic.

The degree of magnetic anisotropy varies with the thickness of the iron film. For austenite "clamped" to the $\langle 100 \rangle$ plane of copper, the anisotropy changes as a function of the number of monolayers deposited, becoming maximum at a 5-layer thick film. This is because the fcc iron films expand normal to the surface with the strain reaching a maximum value of 6% at 5-layer thick films. This phenomenon is attributed to subtle details of the electronic band structure that is dependent on the strain [75].

Curie temperature of monolayers of iron

Equation 2.24 indicates that the Curie temperature varies directly with the atomic coordination number. A {100} monolayer of bcc iron has only four nearest neighbours compared to eight for bulk iron. It is not surprising, therefore, that the Curie temperature of the monolayer is only about 500 K, i.e., about half that found for bulk iron [46, 75, 76].

2.7 HEAT CAPACITY OF LIQUID IRON

Most metals, at their melting temperatures, have values of constant pressure heat capacities in the range 30-40 J mol^{-1} K^{-1}. The temperature dependence of the heat capacity tends to be small [77]. It can be shown empirically (the Neumann-Kopp law) that the heat capacities of liquid alloys can be estimated by a weighted mean of the component elemental liquids. A comprehensive list of heat capacity data for liquid metals can be found in Guthrie and Iida [78].

At 1811 K pure iron, which has a bcc structure, melts to give liquid iron. Orr and Chipman [79] concluded that the thermodynamic data for liquid iron can be represented adequately by assuming a constant value $C_P^L = 46$ J mol^{-1} K^{-1} over the temperature range 1811-3200 K, within the limits of experimental error. This is consistent with data generated for the purposes of geophysics, at 47 ± 3 J mol^{-1} K^{-1} at 1811 K [80].

2.8 FREE ENERGY FUNCTIONS OF IRON

Having established the heat capacity terms for α, ε and γ, the expression for the total heat capacity, i.e.,

$$C_P\{T\} = C_V\left\{\frac{T_D}{T}\right\}C_1 + c_e T + C_P^\mu\{T\},\tag{2.27}$$

can be integrated to give the corresponding components of the Gibbs free energy of the phase concerned:

$$G\{T\} = H_0 + G_{\text{Debye}}\left\{\frac{T_D}{T}\right\} + G^\mu\{T\} + 0.5 c_e T^2\tag{2.28}$$

where the integration constant H_0 is the molar enthalpy at 0 K. This enthalpy cannot be fixed absolutely; it is for α set by convention to zero at 0 K and atmospheric pressure. This does not cause difficulties since the usual interest is in differences in the free energies of phases.

Since the electronic specific heat coefficients of α and of γ are equal, it follows that the change in molar Gibbs free energy on transforming α to γ at a pressure of 1 atmosphere is given by

$$\Delta G^{\alpha\gamma} = \Delta H^{\alpha\gamma}\{0\text{K}\} + G^{\mu\gamma}\{T\} - G^{\mu\alpha}\{T\}\tag{2.29}$$

where $\Delta H^{\alpha\gamma}\{0\text{K}, 1\text{atm}\}$ is the enthalpy change accompanying the transformation of a mole of α to γ, found to be about 5452 J mol^{-1} [10]. The function $\Delta G^{\alpha\gamma}$ is illustrated in Figure 2.23. The molar Gibbs free energy changes accompanying the $\gamma \to \varepsilon$ and $\alpha \to \varepsilon$ transformations can similarly be found, but because of the uncertain magnetic properties of ε-iron, $G^{\mu\varepsilon}$ has to be assumed to be zero:

$$\Delta G^{\alpha\varepsilon} = \Delta H^{\alpha\varepsilon}\{0\text{K}\} + G_{\text{Debye}}^\varepsilon\left\{\frac{T_D}{T}\right\} - G_{\text{Debye}}^\alpha\left\{\frac{T_D}{T}\right\} - G^{\mu\alpha}\{T\} + 0.5T^2(c_e^\varepsilon - c_e^\alpha)$$

$$\Delta G^{\varepsilon\gamma} = \Delta H^{\varepsilon\gamma}\{0\text{K}\} + G_{\text{Debye}}^\gamma\left\{\frac{T_D}{T}\right\} - G_{\text{Debye}}^\varepsilon\left\{\frac{T_D}{T}\right\} - G^{\mu\gamma}\{T\} + 0.5T^2(c_e^\gamma - c_e^\varepsilon).$$

Some thermodynamic data for iron are summarised in Table 2.3. Figure 2.23 illustrates the variation, at a constant pressure of 1 atmosphere, of the free energy changes accompanying the $\alpha \to \gamma$, $\alpha \to \varepsilon$ and $\alpha \to L$ transformations.

Table 2.3
Thermodynamic properties of α, ε and γ iron.

	α	γ	γ_0	γ_1	ε
T_m/K	1811	1800	-	-	1320
$V_m\{0\,K, 1\,atm\}/cm^3\,mol^{-1}$	7.061	-	6.695	7.216	6.731
$H_0/J\,mol^{-1}$	0	5452	-	-	4812
T_D/K	432	432	-	-	385
$c_e/J\,mol^{-1}K^{-2}$	0.00502	0.00502	-	-	0.00586
T_C/K	1042	-	-	≈ 1800	-
T_n/K	-	-	55-80	-	-
μ_B / Bohr magnetons	2.2	-	0.56-0.7	≈ 2.8	-

The work of Kaufman and co-workers explains much of the peculiar behaviour of iron. Because of its ferromagnetism, ferrite is stable relative to austenite at low temperatures, but the two-spin state electronic structure of austenite gives it an extra entropy that stabilises it above 1185 K. The reversion to ferrite above 1667 K results from its magnetic entropy which builds up rapidly above the Curie temperature and eventually overrides the two-spin state entropy of austenite. Indeed, it is this magnetic contribution to the entropy of ferrite which dominates at high-temperatures to produce a minimum in the $\Delta G^{\alpha\varepsilon}$ versus T curve, even though at low temperatures, the vibrational entropy of ε is greater than that of α. Figure 2.23 also shows that ε is more stable than γ at 0 K, but that as the temperature rises, γ, because of its magnetic entropy, becomes the more stable phase above 340 K. If pure, unconstrained γ-iron could be quenched to a temperature below 340 K, it should tend to transform into ε-iron.

Kaufman [81] pointed out that if α-iron were not to be ferromagnetic, its free energy at 0 K would rise by such a large amount that austenite would become stable even at low temperatures. Indeed, since the Debye temperatures and electronic specific heat coefficients of α and γ are equal, the stability of austenite relative to ferrite would persist at all temperatures. Below 340 K, iron would be in the hcp structure in these circumstances. Life would clearly be dull in the absence of ferrite and ferromagnetic iron!

The scenario just described actually arises in ruthenium, which is considered to be an "iron-analogue" because it lies in the same column of the periodic table. Its bcc α phase is not ferromagnetic and consequently is less stable than its fcc γ phase for all temperatures [81]. Following the earlier logic, ruthenium is found to have a hcp structure at ambient temperature.

2.8.1   LIQUID IRON

For temperatures in excess of 1667 K where ferrite is once again more stable than austenite, an approximate expression for the molar Gibbs free energy change accompanying the $\alpha \to L$ transfor-

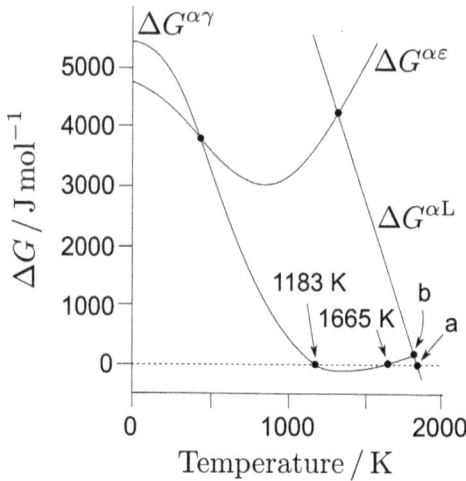

Figure 2.23 Gibbs free energy-temperature relations for the phases of pure iron, at a pressure of 1 atmosphere (after Kaufman, Clougherty and Weiss [10]; Blackburn, Kaufman and Cohen [11]; and Stepakoff and Kaufman [13]). The points "a" and "b" represent the melting temperatures of α and γ respectively.

mation can be obtained using the following experimental data due to Orr and Chipman [79]:

$$
\begin{aligned}
C_P^{\mathrm{L}} &\approx 46.0 && \mathrm{J\,mol^{-1}\,K^{-1}} \\
C_P^{\alpha} &\approx 41.8 && \mathrm{J\,mol^{-1}\,K^{-1}} \\
\Delta S^{\alpha \mathrm{L}}\{T_{\mathrm{m}}\} &= 7.6325 && \mathrm{J\,mol^{-1}\,K^{-1}} \\
\Delta H^{\alpha \mathrm{L}}\{T_{\mathrm{m}}\} &= 13807 && \mathrm{J\,mol^{-1}}.
\end{aligned}
$$

Pure metals which do not have allotropic transitions have an entropy of fusion of approximately 9.2 $\mathrm{J\,mol^{-1}\,K^{-1}}$ [82]. For those metals like iron which do undergo solid state transformations, it is the total entropy change due to all the transitions from 0 K to the liquid state that has this value. Therefore, for iron, the entropy of fusion and the entropy changes at the $\gamma \rightarrow \delta$ and $\alpha \rightarrow \gamma$ add up as follows (Table 1.1):

$$7.6325 + 0.5025 + 0.7548 = 8.89 \qquad \mathrm{J\,mol^{-1}\,K^{-1}}.$$

The data show that the change in heat capacity on solidification is approximately constant and quite small. The data can be integrated to yield the thermodynamic variables as a function of temperature:

$$\Delta H^{\alpha \mathrm{L}}\{T\} = \Delta H^{\alpha \mathrm{L}}\{T_{\mathrm{m}}\} + \int_{T_{\mathrm{m}}}^{T} (C_P^{\mathrm{L}} - C_P^{\alpha})\,\mathrm{d}T$$

$$\Delta S^{\alpha \mathrm{L}}\{T\} = \Delta S^{\alpha \mathrm{L}}\{T_{\mathrm{m}}\} + \int_{T_{\mathrm{m}}}^{T} \frac{C_P^{\mathrm{L}} - C_P^{\alpha}}{T}\,\mathrm{d}T$$

so that $\qquad \Delta G^{\alpha \mathrm{L}} = 6229 - 3.4486T - 4.184T \ln\left\{\dfrac{T}{T_{\mathrm{m}}}\right\} \qquad \mathrm{J\,mol^{-1}}.$

A more approximate equation, the form of which is consistent with the approximation that $C_P^L - C_P^\alpha = 0$, is given by Kaufman [81], valid for $T > 1667\,\text{K}$, is

$$\Delta G^{\alpha L} = 15397 - 8.4935T \qquad \text{J mol}^{-1}.$$

Kaufman also found that for the $\gamma \to L$ and $\varepsilon \to L$ metastable transformations,

$$\Delta G^{\gamma L} = 16339 - 9.0584T \qquad \text{J mol}^{-1}$$

$$\Delta G^{\varepsilon L} = 15397 - 12.522T \qquad \text{J mol}^{-1}.$$

All of these thermodynamic data are subject to approximations and likely to fail if excessive extrapolations are made into temperature regimes where one of the phases is metastable. There are, for example, no *measurements* of the heat capacity of ferrite above its melting temperature, nor of the supercooled liquid iron. The extrapolation of the properties of liquids has assumed new significance because of access to rapid quenching techniques, the aim being either to refine microstructure by forcing transformation to initiate at large undercoolings, or to induce the liquid into a glass transition. The latter occurs when the liquid is undercooled to such an extent, that its configuration becomes frozen at the glass transition temperature. Pure liquid iron is unlikely ever to be made into a glass; modelling based on molecular dynamics indicates that a quenching rate of $10^{13}\,\text{K s}^{-1}$ would be necessary to achieve amorphous iron [83]. Nevertheless, work on glass-forming alloys gives some valuable clues on the thermodynamic properties of pure metals as well.

Although there are no data for supercooled liquid iron, a number of other pure metals have been investigated at undercoolings in excess of 200 K. The experimental heat capacity data have been compared with a variety of approximations for the difference $\Delta C_P = C_P^{\text{liquid}} - C_P^{\text{solid}}$,

$$\Delta C_P = b_3 + b_4 T$$

where b_3 and b_4 are empirical constants. The approximations involved can then be described as follows (presented in order of increasing accuracy):

(a) $b_3 = b_4 = 0$ [81]. This turns out to be a reasonable assumption for pure metals that do not undergo a glass transition, where the *change* ΔC_P tends to be small when compared with the absolute value of the heat capacity of either the liquid or solid phases. The approximation would fail for a polymeric liquid where the change in heat capacity on transition to the solid state is expected to be large [84].

(b) $b_4 = 0$, so that $\Delta C_P = b_3$ [85].[12] The value of ΔC_P usually is taken to be that at the melting temperature [86].

(c) b_3 and b_4 are taken to be finite. The problem here is to obtain appropriate values for the constants in the absence of experimental data.

2.9 EFFECT OF PRESSURE

2.9.1 LIQUID IRON

It is natural that the melting temperatures of both δ and γ should be raised by increasing pressure because the solid phases are denser than the liquid (Figure 1.2). This is particularly important in geophysics where the increase of the melting temperature limits the temperature distribution in the core of the earth, which is predominantly iron [87]. δ-Iron is eliminated at the sort of pressures that exist in the earth's core, so it is the melting temperature of austenitic iron that is of greatest interest. The experimental data over the pressure range $P = 3\text{-}20\,\text{GPa}$ are represented as follows [87]:

$$T_m^\gamma / {}^\circ\text{C} = 1718 + 38.5(P - 5.2) - 1.95(P - 5.2)^2 + 6.24 \times 10^{-2}(P - 5.2)^3$$

The actual pressure at the core is of the order of 300 GPa; this equation does not extrapolate well because the melting temperature at 225 GPa is much greater at about 5100 K [88]. The outer liquid-core of the earth is not of course pure iron, so its melting temperature is likely to be lower than implied by this equation. Indeed, the density of the core liquid is smaller [89] and the bulk modulus larger, than that of pure liquid-iron [90].

2.9.2 FERRITE AND AUSTENITE

The compressibility of iron in all but its vapour state is small so the consequences on the free energy of the solid or liquid phases as a function of pressure are noticeable only at large pressures. Some data are presented in Table 2.4.

Table 2.4

Bulk modulus K, and pressure derivatives K', K'', for allotropes of iron.

Phase	Bulk modulus or pressure derivatives	Source
Liquid iron	$K = 109.7 \pm 0.7$ GPa, at 1181 K, 10^5 Pa	[80]
	$K' = 4.661 \pm 0.040$, $K'' = -0.043 \pm 15.3$ GPa^{-1}	[80]
Austenitic stainless steels	$K = 157.5\text{-}163.3$ GPa, ambient conditions	[91]
	$K' = 5.39\text{-}5.57$	[91]
γ-iron	$K = 199$ GPa	[92]
	$K' = 5.5$	[92]
ε-iron	$K = 164.8$ GPa	[93]
	$K' = 5.33$	[93]
α-iron	$K = 166.4$ GPa, ambient temperature	[94]
	$K' = 5.29$	[94]

In the case of austenite, an increase in pressure favours the low-volume antiferromagnetic state. The energy gap E_1 between the γ_0 and γ_1 states increases by $P\Delta V_m^{\gamma_0-\gamma_1}$ assuming that the molar volumes $V_m^{\gamma_0}$ and $V_m^{\gamma_1}$ are both independent of pressure, because any change in pressure can be accommodated by a change in the fraction of atoms in each of the two states. The molar Gibbs free energy of γ at zero pressure follows from Equation 2.28:

$$G^\gamma\{T\} = H_0^\gamma + G_{\text{Debye}}^\gamma\left\{\frac{T_D}{T}\right\} + G^{\mu\gamma}\{T\} + 0.5c_e^\gamma T^2.$$

If the electronic specific heat coefficient c_e remains constant, the effect of pressure is to modify this equation to:

$$G^\gamma\{T,P\} = H_0^\gamma + PV_m^\gamma\{T,P\} + G_{\text{Debye}}^\gamma\left\{\frac{T_D}{T}\right\} + G^{\mu\gamma}\{T,P\} + 0.5c_e^\gamma T^2.$$

The magnetic component $G^{\mu\gamma}$ must depend on pressure via E_1. Since the magnetic properties of ferrite are insensitive to pressure [10, 25],

$$G^\alpha\{T,P\} = H_0^\alpha + PV_m^\alpha\{T,P\} + G_D^\alpha\left\{\frac{T_D}{T}\right\} + G^{\mu\alpha}\{T\} + 0.5c_e^\alpha T^2.$$

Given that the Debye temperatures and electronic specific heat coefficients of α and γ are identical, it follows that

$$\Delta G^{\alpha\gamma}\{T,P\} = H_0^{\alpha\gamma} + P(V_m^\gamma\{T,P\} - V_m^\alpha\{T,P\}) + G^{\mu\gamma}\{T,P\} - G^{\mu\alpha}\{T\}.$$

2.9.3 HEXAGONAL CLOSE-PACKED IRON

Following the same principles, the pressure dependence of $\Delta G^{\alpha\varepsilon}$ is given by

$$\Delta G^{\alpha\varepsilon}\{T,P\} = \Delta G^{\alpha\varepsilon}\{T\} + P\Delta V_m^{\alpha\varepsilon}\{T\}.$$

Assuming that the molar volumes of α and ε, and $G^{\mu\alpha}$ are independent of pressure. For $\gamma \rightarrow \varepsilon$,

$$\Delta G^{\gamma\varepsilon}\{T,P\} = \Delta G^{\gamma\varepsilon}\{T,P\} + \Delta G^{\gamma\alpha}\{T,P\}. \tag{2.30}$$

2.10 MECHANICAL MIXTURES AND SOLUTIONS

All steels are solutions. The distinction between a compound and a solution is that the free energy of the former increases sharply with a change in its chemical composition. For a solution, the variation in free energy with composition is much more gentle so that the range of composition over which it can exist is greater. It is useful to examine the nature of a solution by considering its evolution as the components are mixed together. For when is a mixture, a solution?

2.10.1 ALLOYING BY DEFORMATION

Mechanical alloying is a process invented by Benjamin [95], in which mixtures of fine powders consisting of elemental metals or master alloys are changed into solid solutions, apparently without any melting (Figure 2.24). The powders are forced to collide with each other and with much larger, hardened steel balls whilst contained in a ball mill. The collisions are energetic, involve large contact pressures, and lead eventually to the formation of an intimate solid solution. Refractory oxides can also be introduced into the mechanically alloyed powder for dispersion strengthening. The alloyed powder is finally extruded to form full density bulk samples in rod, sheet or other useful shapes. The process has been used commercially to make iron alloys containing large amounts of aluminium for oxidation resistance, and yttria particles to guard against creep.

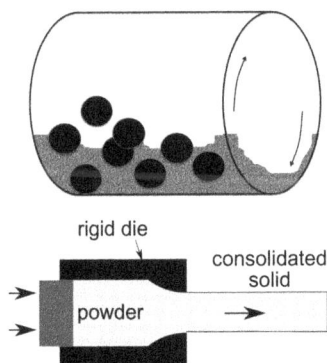

Figure 2.24 Mixture of metallic powders and compounds ball-milled together until alloying occurs. The top part shows a cylindrical drum containing a mixture of elemental powders and large steel balls. When the drum is rotated the balls collide causing the powder particles to coalesce and fragment repeatedly. The resulting powders are then canned and hot-extruded to produce solid metal.

The preparation of a binary alloy by this route can be considered in terms of the two elemental powders ('A' and 'B') which are mixed such that the mole fraction of B is x. The pure powders have the molar free energies μ_A° and μ_B° respectively, Figure 2.25. The free energy of this mechanical mixture of powders is given by

$$G\{\text{mixture}\} = (1-x)\mu_A^\circ + x\mu_B^\circ + \Delta S_M \tag{2.31}$$

where ΔS_M is the accompanying change in configurational entropy. It has been assumed here, that there is no change in enthalpy in the process, i.e., the atoms in the context of bonding are indifferent to the type of neighbouring atom. The change in configurational entropy as a consequence of

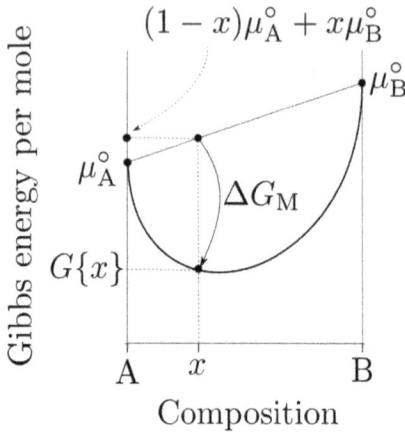

Figure 2.25 Plot of free energy versus composition, both for mechanical mixtures and a solid solution. ΔG_M is the free energy of mixing when the mechanical mixture turns into a solid solution.

mixing can be obtained using the Boltzmann equation $S = k \ln\{w_c\}$ where w_c is the number of configurations. Suppose that there are m_A atoms per powder particle of A, and m_B atoms per particle of B; the powders are then mixed in a proportion which gives an average concentration of B which is the mole fraction x.

There is only one configuration when the heaps of powders are separate. When the powders are mixed randomly, the number of possible configurations for a mole of atoms becomes [96]

$$\frac{\left(N_a([1-x]/m_A + x/m_B)\right)!}{(N_a[1-x]/m_A)! \ (N_a x/m_B)!}. \tag{2.32}$$

The numerator in Equation 2.32 is the total number of particles, and the denominator is the product of the factorials of the A and B particles respectively; N_a is Avogadro's number. Using Stirling's approximation, the molar entropy of mixing is

$$
\begin{aligned}
\frac{\Delta S_M}{kN_a} &= \frac{(1-x)m_B + xm_A}{m_A m_B} \ln\left\{ N_a \frac{(1-x)m_B + xm_A}{m_A m_B} \right\} \\
&\quad - \frac{1-x}{m_A} \ln\left\{ \frac{N_a(1-x)}{m_A} \right\} \\
&\quad - \frac{x}{m_B} \ln\left\{ \frac{N_a x}{m_B} \right\}
\end{aligned}
\tag{2.33}
$$

subject to the condition that the number of particles remains integral and non-zero.[13]

The largest reduction in free energy occurs when the particle sizes are atomic, Figure 2.26, which shows the molar free energy of mixing for a case where the average composition is equiatomic. Such a composition maximises configurational entropy. When it is considered that phase changes often occur at appreciable rates when the accompanying reduction in free energy is just $10\,\text{J}\,\text{mol}^{-1}$, Figure 2.26 shows that the entropy of mixing cannot be ignored when the particle size is less than a few hundreds of atoms. In commercial practice, powder metallurgically produced particles are typically 100 μm in size, in which case the entropy of mixing can be entirely neglected, though solution formation must be considered to be advanced when the processing reduces particle dimensions to some 10^2 atoms. These comments must be qualified due to the neglect any enthalpy change during mixing.

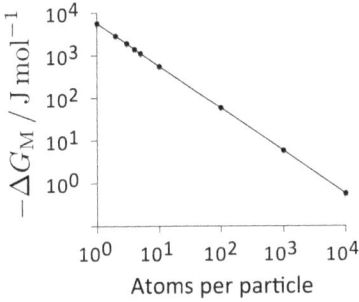

Figure 2.26 The molar Gibbs free energy of mixing, $\Delta G_M = -T\Delta S_M$, for a binary alloy, as a function of the particle size when all the particles are of uniform size in a mixture, the average composition of which is equiatomic, $T = 1000\,\mathrm{K}$.

2.11 CHEMICAL POTENTIAL

In a single-phase equilibrium diagram such as that for iron as a function of temperature and pressure, the boundaries between the phase fields represent the locus of all points along which the adjacent phases are in equilibrium, i.e., they have an identical free energy. For example, the α/γ phase boundary is defined by setting (Figure 2.27):

$$G^\alpha = G^\gamma. \qquad (2.34)$$

This is because allotropic transitions are considered here as a function of variables such as temperature and pressure, where the crystal structure changes but not the chemical composition.

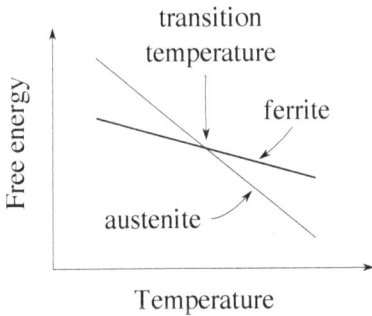

Figure 2.27 The transition temperature for an allotropic transformation.

A different approach is needed when the chemical composition is variable. Consider a single-phase alloy consisting of two components A and B. The molar free energy $G\{x\}$ of that phase will in general be a function of the mole fractions $(1-x)$ and x of A and B respectively, written as a weighted mean of the free energy contributions from each component:

$$G\{x\} = \underbrace{(1-x)\mu_A}_{\text{contribution from A atoms}} + \underbrace{x\mu_B}_{\text{contribution from B atoms}}. \qquad (2.35)$$

The terms μ_B and μ_B, known as the *chemical potentials* per mole of A and B respectively, in effect partition the free energy $G\{x\}$ into a component purely due to A atoms and another due to B atoms alone. This equation is illustrated in Figure 2.28 by the tangent at the coordinate $[G\{x\}, x]$. Consistent with Equation 2.35, the intercepts of this tangent on the vertical axes give μ_A and μ_B. Since the slope of the tangent depends on the composition, so do the chemical potentials. Note that the free energies of the pure components are written μ_A° and μ_B°.

It should be obvious from Figure 2.28 that

$$\mu_A = G\{x\} - x\frac{\partial G}{\partial x}$$

$$\text{and} \qquad \mu_B = G\{x\} + (1-x)\frac{\partial G}{\partial x}$$

where $\partial G/\partial x$ is the slope of the tangent so the product on the right-hand side of the equations simply represents the difference in μ and G. In general, for a system with n components [97, p. 57]:

$$\mu_i = G\{x_i\} + \sum_{j=2}^{n} (\delta_{ij} - x_j)\frac{\partial G}{\partial x_j} \tag{2.36}$$

where δ_{ij} is the Kronecker delta ($\delta_{ij} = 0$ for $i \neq j$ and $\delta_{ij} = 1$ for $i = j$).

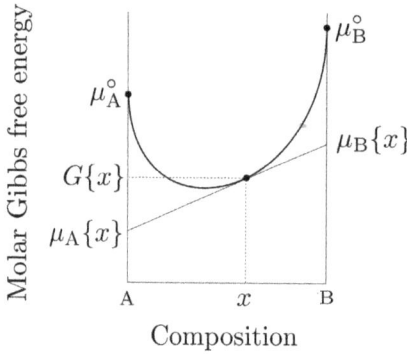

Figure 2.28 Illustration of the chemical potential μ for a binary solution, with μ° representing the free energy of the pure component.

The chemical potential $\mu\{x\}$ of a component is known also as its *partial* molar free energy, describing a part of the *integral* molar free energy $G\{x\}$. There are in fact many quantities which can be expressed using relationships of the form implied by Equation 2.35. Thus, the volume of a solution might be written in terms of the partial molar volumes of the components:

$$V_m = \overline{V}_A x_A + \overline{V}_B x_B \tag{2.37}$$

where $\overline{V}_i$ refers to the partial molar volume of component $i = $A,B.

2.12 EQUILIBRIUM BETWEEN SOLUTIONS

Consider now two phases α and γ that are placed in intimate contact in a binary steel. The phases will only be in equilibrium with each other if the carbon atoms in γ have the same chemical potential as the carbon atoms in α, and if this is true also for the Fe atoms:

$$\begin{aligned}\mu_C^\alpha &= \mu_C^\gamma \\ \mu_{Fe}^\alpha &= \mu_{Fe}^\gamma.\end{aligned} \tag{2.38}$$

In fact, in a binary solution, the chemical potentials of A and B when sharing a tangent are not independent so this last condition is redundant. This is apparent from Figure 2.28, where the two potentials are connected by the tangent.

If the atoms of a particular species have the same chemical potential in both the phases, then there can be no tendency for them to migrate across the phase boundaries. The system will be in stable equilibrium if this condition applies to all species of atoms. The way in which the free energy of a phase varies with concentration is unique to that phase, so the *concentration* of a particular species of atom need not be identical in phases which are at equilibrium. Therefore, in general,

$$
\begin{aligned}
x_C^{\alpha\gamma} &\neq x_C^{\gamma\alpha} \\
x_{Fe}^{\alpha\gamma} &\neq x_{Fe}^{\gamma\alpha}
\end{aligned}
\tag{2.39}
$$

where $x_i^{\alpha\gamma}$ describes the mole fraction of element i in phase α which is in equilibrium with phase γ etc.

The condition that the chemical potential of each species of atom must be the same in all phases at equilibrium is general. For the binary alloy, two phase case, it follows that the equilibrium compositions can be found on a plot of free energy versus composition, by constructing a tangent that is common to the two free energy curves as illustrated in Figure 2.29.

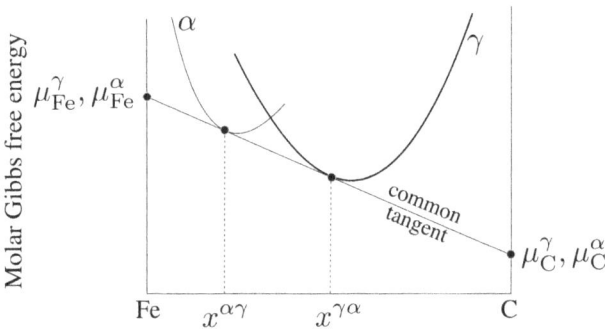

Figure 2.29 The common tangent construction giving the equilibrium compositions $x^{\alpha\gamma}$ and $x^{\gamma\alpha}$ of the two phases at a fixed temperature.

2.13 ACTIVITY

The chemical potential μ_A^α of the A atoms in the α phase may be expanded in terms of a contribution from the pure component A and a concentration dependent term as follows:

$$
\mu_A^\alpha = \mu_A^{\circ\alpha} + RT \ln a_A^\alpha
\tag{2.40}
$$

where $\mu_A^{\circ\alpha}$ is the free energy of pure A in the structure of α, and a_A is the *activity* of atom A in the solution of A and B.

The activity of an atom in a solution can be thought of as its effective concentration in that solution. For example, there will be a greater tendency for the A atoms to evaporate from solution, when compared with pure A, if the B atoms repel the A atoms. The effective concentration of A in solution will therefore be greater than implied by its atomic fraction, i.e., its activity is greater than its concentration. The opposite would be the case if the B atoms attracted the A atoms.

The atom interactions can be expressed in terms of the change in energy as an A-A and a B-B bond is broken to create 2(A-B) bonds. An ideal solution is formed when there is no change in energy in the process of forming A-B bonds. The activity is equal to the mole fraction in an ideal solution (Figure 2.30). If, on the other hand, there is a reduction in energy than the activity is less than ideal and vice versa. The activity and concentration are related via an activity coefficient Γ:

$$
a = \Gamma x.
\tag{2.41}
$$

The activity coefficient is in general a function of the chemical composition of all the elements present in the solution but tends to be constant in dilute solutions (i.e., in the Henry's law region).

In this discussion, the activity of the solute was defined with respect to a Raoultian reference state, i.e., $a = 1$ for $x = 1$. Other definitions are sometimes convenient. A common alternative for dilute solutions being that the activity tends to unity as the concentration tends to 1 wt%.

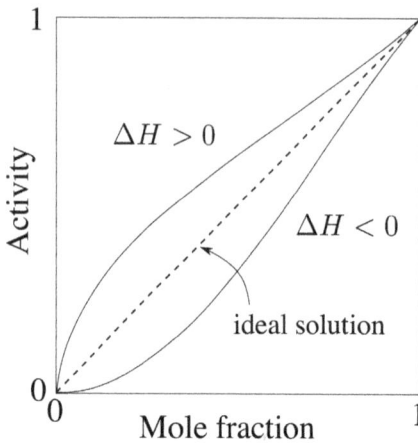

Figure 2.30 Variation in Raoultian activity as a function of its concentration in a binary solution. The ideal solution represents the case where the enthalpy of mixing is zero; the atoms are indifferent to the specific nature of their neighbours. The case where the activity is larger than the concentration is for solutions where the enthalpy of mixing is greater than zero, with like atoms preferred as near neighbours. When the activity coefficient is less than unity, unlike atoms are preferred as near neighbours, the enthalpy of mixing being negative.

Note that solutions where the enthalpy of mixing is positive tend to exhibit clustering at low temperatures whereas those with a negative enthalpy of mixing will tend to exhibit ordering at low temperatures. The effect of temperature is to mix all atoms since both clustering and ordering cause a reduction in entropy (i.e., a reduction in entropy). The product $-T\Delta S$ becomes increasingly positive at high temperatures, so much so that it eventually overcomes the enthalpy effects and causes the mixing of all atoms.

2.14   IDEAL SOLUTION

An ideal solution is one in which the atoms at equilibrium are distributed randomly; the interchange of atoms within the solution causes no change in the potential energy of the system. For a binary (A-B) solution the numbers of the different kinds of bonds can therefore be calculated using simple probability theory:

$$N_{AA} = \frac{1}{2}N(1-x)^2$$

$$N_{BB} = \frac{1}{2}Nx^2$$

$$N_{AB} = N(1-x)x$$

where N_{AB} represents both A-B and B-A bonds which cannot be distinguished. N is the total number of atoms and x the fraction of B atoms. The factor of $\frac{1}{2}$ avoids counting A-A or B-B bonds twice. For an ideal solution, the entropy of mixing is given by Equation 2.33 with $m_B = m_A = 1$. There is no enthalpy of mixing since there is no change in energy when bonds between like atoms are broken to create those between unlike atoms. This is why the atoms are randomly distributed in the solution. The molar free energy of mixing is therefore:

$$\Delta G_M = N_a kT[(1-x)\ln\{1-x\} + x\ln\{x\}]. \tag{2.42}$$

Figure 2.31 shows how the configurational entropy and the free energy of mixing vary as a function of the concentration. ΔG_M is at a minimum for the equiatomic alloy because that is when the entropy of mixing is at its largest; the curves are naturally symmetrical about $x = 0.5$. The form of the curve does not change with temperature though the magnitude at any concentration scales with the temperature. It follows that at 0 K there is no difference between a mechanical mixture and an ideal solution.

From Equation 2.40, the chemical potential per mole for a component in an ideal solution is given by:

$$\mu_A = \mu_A^\circ + N_a kT \ln\{1 - x\}$$

and there is a similar equation for B. Since $\mu_A = \mu_A^\circ + RT \ln a_A$, it follows that the activity coefficient is unity.

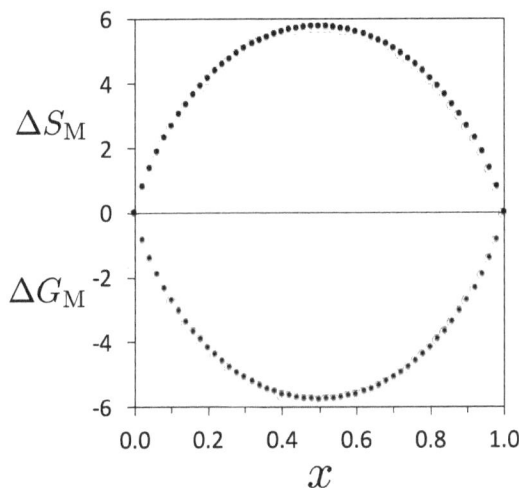

Figure 2.31 The entropy of mixing $(kJ\,mol^{-1}\,K^{-1})$ and the free energy of mixing $(kJ\,mol^{-1})$ as a function of concentration in an ideal binary solution where the atoms are distributed at random. The free energy is for a temperature of 1000 K. The data are plotted as dots rather than curves because concentration is strictly a discrete variable. So the slope at the vertical axes is not $\pm\infty$ as implied by Equation 2.42, but finite though very large.

2.15 REGULAR SOLUTIONS

There are no solutions of iron that are ideal. The iron-manganese liquid phase is close to ideal, though even that has an enthalpy of mixing which is about $-860\,J\,mol^{-1}$ for an equiatomic solution at 1000 K, which compares with the contribution from the configurational entropy of about $-5800\,J\,mol^{-1}$. The ideal solution model is nevertheless useful because it provides a reference. The free energy of mixing for a non-ideal solution often is written with an additional term, the *excess* free energy $(\Delta_e G = \Delta_e H - T\Delta_e S)$ that indicates the deviation from ideality:

$$\begin{aligned} \Delta G_M &= \Delta_e G + N_a kT [(1-x)\ln\{1-x\} + x\ln\{x\}] \\ &= \Delta_e H - T\Delta_e S + N_a kT [(1-x)\ln\{1-x\} + x\ln\{x\}] \end{aligned} \qquad (2.43)$$

One of the components of the excess enthalpy of mixing comes from the change in the energy when new kinds of bonds are created during the formation of a solution. This enthalpy is, in the *regular solution* model, estimated from the pairwise interactions between adjacent atoms. The term *regular solution* was proposed by Hildebrand [98] to describe mixtures, the properties of which when plotted varied in an aesthetically regular manner; he went on to suggest that a regular solution, although not ideal, would still contain a random distribution of the constituents.[14] Following Guggenheim [99], the term regular solution is now restricted to cover mixtures that assume an ideal entropy of mixing but have a non-zero interchange energy.

In the regular solution model, the enthalpy of mixing is obtained by counting the different kinds of near neighbour bonds when the atoms are mixed at random; this information together with the binding energies gives the required change in the enthalpy on mixing. The binding energy may be defined by considering the change in energy as the distance between a pair of atoms is decreased from infinity to an equilibrium separation (Figure 2.32). The change in energy during this process is the binding energy, which for a pair of A atoms is written $-2\varepsilon_{AA}$. It follows that when $\varepsilon_{AA} + \varepsilon_{BB} < 2\varepsilon_{AB}$, the solution will have a larger than random probability of bonds between unlike atoms. The converse is true when $\varepsilon_{AA} + \varepsilon_{BB} > 2\varepsilon_{AB}$ since atoms then prefer to be neighbours to their own kind. Notice that for an ideal solution it only is necessary for $\varepsilon_{AA} + \varepsilon_{BB} = 2\varepsilon_{AB}$, and not $\varepsilon_{AA} = \varepsilon_{BB} = \varepsilon_{AB}$ [99].

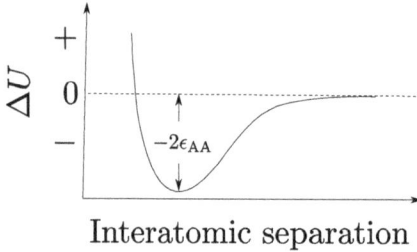

Figure 2.32 Change in energy as a function of the distance between a pair of A atoms. $-2\varepsilon_{AA}$ is the binding energy for the pair of atoms. There is a strong repulsion at close range.

Suppose now that the approximation that atoms are randomly distributed is retained, even though the enthalpy of mixing is not zero. The number of A-A, A-B and B-B bonds in a mole of solution is then $\frac{1}{2}zN_a(1-x)^2$, $\frac{1}{2}zN_ax^2$ and $zN_a(1-x)x$ respectively, where z is the coordination number. It follows that the molar enthalpy of mixing is given by

$$\Delta H_M \simeq N_a z(1-x)x\omega \qquad \text{where} \qquad \omega = \varepsilon_{AA} + \varepsilon_{BB} - 2\varepsilon_{AB}. \qquad (2.44)$$

The product $zN_a\omega$ is often called the regular solution parameter, which in practice will be temperature and composition dependent. A composition dependence also leads to an asymmetry in the enthalpy of mixing as a function of composition about $x = 0.5$. For the nearly ideal Fe-Mn liquid phase solution, the regular solution parameter is $-3950 + 0.489T \, \text{J mol}^{-1}$ if a slight composition dependence is neglected.

A positive ω favours the clustering of like atoms whereas when it is negative there is a tendency for the atoms to order. This second case is illustrated in Figure 2.33, where an ideal solution curve is presented for comparison. Like the ideal solution, the form of the curve for the case where $\Delta H_M < 0$ does not change with the temperature, but unlike the ideal solution, there is a free energy of mixing even at 0 K where the entropy term ceases to make a contribution.

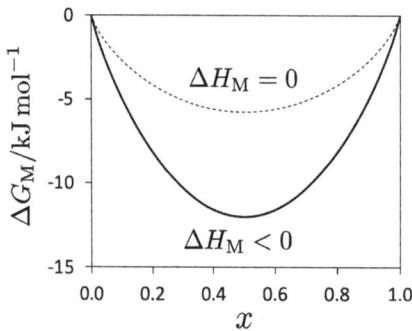

Figure 2.33 Free energy of mixing at 1000 K, as a function of concentration in a binary solution where there is a preference for unlike atoms to be near neighbours. The free energy curve for the ideal solution ($\Delta H_M = 0$) is included for reference.

The corresponding case for $\Delta H_M > 0$ is illustrated in Figure 2.34, where the form of the curve is seen to change with the temperature. The contribution from the enthalpy term can largely be neglected at high temperatures where the atoms become randomly mixed by thermal agitation so the free energy curve then has a single minimum. However, as the temperature is reduced, the opposing contribution to the free energy from the enthalpy term introduces two minima at the solute-rich and solute-poor concentrations. This is because like-neighbours are preferred. On the other hand, there is a maximum at the equiatomic composition because that gives a large number of unfavoured unlike atom bonds. Between the minima and the maximum lie points of inflexion which are of importance in spinodal decomposition, to be discussed later.

For a regular solution, Equation 2.35 shows that the chemical potential per mole is given by

$$\mu_B = \mu_A^\circ + zN_a x^2 \omega + N_a kT \ln\{1 - x\} \qquad (2.45)$$

and that the activity coefficient is $\exp\{zx^2\omega/kT\}$. Some of the properties of the different kinds of solutions are summarised in Table 2.5.

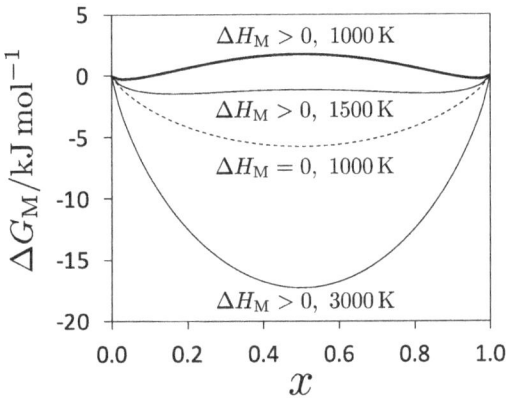

Figure 2.34 Free energy of mixing as a function of concentration and temperature in a binary solution where there is a tendency for like atoms to cluster. The free energy curve for the ideal solution ($\Delta H_M = 0$) is included for reference.

Table 2.5
Elementary thermodynamic properties of solutions

Type	ΔS_M	ΔH_M
Ideal	Random	0
Regular	Random	$\neq 0$
Quasichemical	Not random	$\neq 0$

2.16 QUASICHEMICAL SOLUTION

The regular solution model assumes a random distribution of atoms even though the enthalpy of mixing is not zero, whereas in reality a random solution is only expected at high temperatures when the entropy term overwhelms any tendency for ordering or clustering of atoms. It follows that the configurational entropy of mixing should therefore vary with the temperature. The *quasichemical* solution model has a better treatment of configurational entropy which accounts for a non-random distribution of atoms. The model is so-called because it has a mass-action equation that has similarity to chemical reactions [100]. However, the presentation below follows derivations by Christian [101] and Lupis [97].

Recalling that zN_{AB} represents the number of A-B bonds, the total energy of the assembly for a particular value of N_{AB} is $U_{N_{AB}} = -z(N_A\varepsilon_{AA} + N_B\varepsilon_{BB} - N_{AB}\omega)$ where $\omega = \varepsilon_{AA} + \varepsilon_{BB} - 2\varepsilon_{AB}$. In a non-random solution there are many values that N_{AB} can adopt; each value corresponding to one or more arrangement of atoms with an identical value of U is therefore associated with a degeneracy $g_{N_{AB}}$ which is the number of arrangements possible for a given value of U. The partition function is therefore the sum over all possible N_{AB}:

$$\Omega = \sum_{N_{AB}} g_{N_{AB}} \exp\left\{-\frac{U_{N_{AB}}}{kT}\right\}$$

$$= \sum_{N_{AB}} g_{N_{AB}} \exp\left\{\frac{z(N_A\varepsilon_{AA} + N_B\varepsilon_{BB} - N_{AB})\omega}{kT}\right\}. \tag{2.46}$$

For a given value of N_{AB}, the different non-interacting *pairs* of atoms can be arranged in the following number of ways ($N = N_A + N_B$)

$$g_{N_{AB}} \propto \frac{(\frac{1}{2}zN)!}{(\frac{1}{2}z[N_A - N_{AB}])!\,(\frac{1}{2}z[N_B - N_{AB}])!\,(\frac{1}{2}zN_{AB})!\,(\frac{1}{2}zN_{BA})!} \tag{2.47}$$

where the first and second terms in the denominator refer to the numbers of A-A and B-B bonds respectively, and the third and fourth terms the numbers of A-B and B-A pairs respectively. This is not an equality because the various pairs are not independent, as illustrated in Figure 2.35; the distribution of pairs is not random. Guggenheim addressed this difficulty by using a normalisation

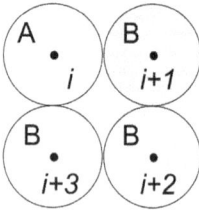

Figure 2.35 Why pairs of atoms cannot be distributed at random on lattice sites which are marked as small dots. Once the bonds connecting the coordinates $(i, i+1)$, $(i+1, i+2)$, $(i+2, i+3)$ are made as illustrated, the final bond connecting $(i, i+3)$ is necessarily occupied by a pair AB. Adapted from Lupis [97].

factor such that the summation of all possible degeneracies equals the total number of possible configurations as follows.

Suppose that the number of arrangements of pairs of atoms possible in a random solution is identified with an asterix, then from the proportionality 2.47, it is seen that

$$g^* \propto \frac{(\frac{1}{2}zN)!}{(\frac{1}{2}z[N_A - N_{AB}^*])!\,(\frac{1}{2}z[N_B - N_{AB}^*])!\,(\frac{1}{2}zN_{AB}^*)!\,(\frac{1}{2}zN_{BA}^*)!}. \tag{2.48}$$

This again will overestimate the number of possibilities (Figure 2.35), but for a random solution it is known already that

$$g^* = \frac{N!}{N_A!\,N_B!}. \tag{2.49}$$

It follows that $g_{N_{AB}}$ can be normalised as

$$g_{N_{AB}} = \frac{(\frac{1}{2}z[N_A - N_{AB}^*])!\,(\frac{1}{2}z[N_B - N_{AB}^*])!\,(\frac{1}{2}zN_{AB}^*)!\,(\frac{1}{2}zN_{BA}^*)!}{(\frac{1}{2}z[N_A - N_{AB}])!\,(\frac{1}{2}z[N_B - N_{AB}])!\,(\frac{1}{2}zN_{AB})!\,(\frac{1}{2}zN_{BA})!} \times \frac{N!}{N_A!\,N_B!}. \tag{2.50}$$

With this, the partition function Ω is defined explicitly and the problem is in principle solved. It is usual, however, to simplify first by assuming that the sum in Equation 2.46 can be replaced by its maximum value. This is because the thermodynamic properties that follow from the partition function depend on its logarithm, in which case the use of the maximum is a good approximation. The equilibrium number N^e_{AB} of A-B bonds may then be obtained by setting $\partial \ln\{\Omega\}/\partial N_{AB} = 0$ [97, 101]:

$$N^e_{AB} = \frac{2Nzx(1-x)}{\beta_q + 1} \tag{2.51}$$

with β_q being the positive root of the equation

$$\beta^2_q - (1-2x) = 4x(1-x)\exp\{2\omega/kT\}, \tag{2.52}$$

so that

$$N^e_{AB} = \frac{2Nzx(1-x)}{[1-2x+4x(1-x)\exp\{2\omega/kT\}]^{\frac{1}{2}}+1}.$$

The percentages of the different pairs are plotted in Figure 2.36. Equation 2.51 obviously corresponds to the regular solution model if $\beta_q = 1$ with a random arrangement of atoms. As expected, the number of unlike pairs is reduced when clustering is favoured, and increased when ordering is favoured.

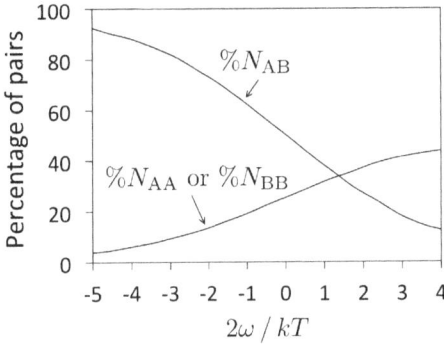

Figure 2.36 Calculated percentages of pairs for the quasichemical model with $x = (1-x) = 0.5$. The result is independent of the coordination number z.

The free energy of the assembly is

$$G = F = -kT \ln\{\Omega\} = U_{N^e_{AB}} - kT \ln g_{N^e_{AB}} \tag{2.53}$$

so that the free energy of mixing per mole becomes

$$\begin{aligned}
\Delta G_M &= zN^e_{AB}\omega - N_a kT \ln g_{N^e_{AB}} \\
&= \underbrace{\frac{2z\omega N_a x(1-x)}{\beta_q + 1}}_{\text{molar enthalpy of mixing}} - RT \ln g_{N^e_{AB}}. \tag{2.54}
\end{aligned}$$

The second term on the right-hand side has the contribution from the configurational entropy of mixing. By substituting for $g_{N^e_{AB}}$, and with considerable manipulation, Christian has shown that this can be written in terms of β_q so that the molar free energy of mixing becomes

$$\begin{aligned}
\Delta G_M &= \frac{2z\omega N_a x(1-x)}{\beta_q + 1} \\
&+ RT\left[(1-x)\ln\{1-x\} + x\ln\{x\}\right] \\
&+ \frac{1}{2}RTz\left\{(1-x)\ln\frac{\beta_q + 1 - 2x}{(1-x)(\beta_q + 1)} + x\ln\frac{\beta_q - 1 + 2x}{x(\beta_q + 1)}\right\}.
\end{aligned}$$

The second term in this equation is the usual contribution from the configurational entropy of mixing in a random solution, whereas the third term can be regarded as a quasichemical correction for the entropy of mixing because the atoms are not randomly distributed.

It is not possible to give explicit expressions for the chemical potential or activity coefficient since β_q is a function of concentration. Approximations using series expansions are possible [97] but the resulting equations are not as easy to interpret physically as the corresponding equations for the ideal or regular solution models.

The expressions in the quasichemical (or *first approximation*) clearly reduce to those of the regular solution (or *zeroth approximation*) model when $\beta_q = 1$. Although a better model has been obtained, the first approximation relies on the absence of interference between atom-pairs. However, each atom in a pair belongs to several pairs so that better approximations can be obtained by considering larger clusters of atoms in the calculation. Such calculations are known as the "cluster variation" method proposed originally by Kikuchi [102]. The improvements obtained with these higher approximations are usually rather small though there are cases where pairwise interactions simply will not do.

It is worth emphasising that although the quasichemical model has an excess entropy, this comes as a correction to the configurational entropy. The excess entropy from this model is always negative; as Lupis pointed out [97], there is more disorder in a random solution than in one that is biased. Therefore, the configurational entropy from the quasichemical model is always less than expected from an ideal solution. Thermal entropy or other terms such as magnetic or electronic are additional contributions.

The procedure in the development of the quasichemical models is illustrated in Figure 2.37.

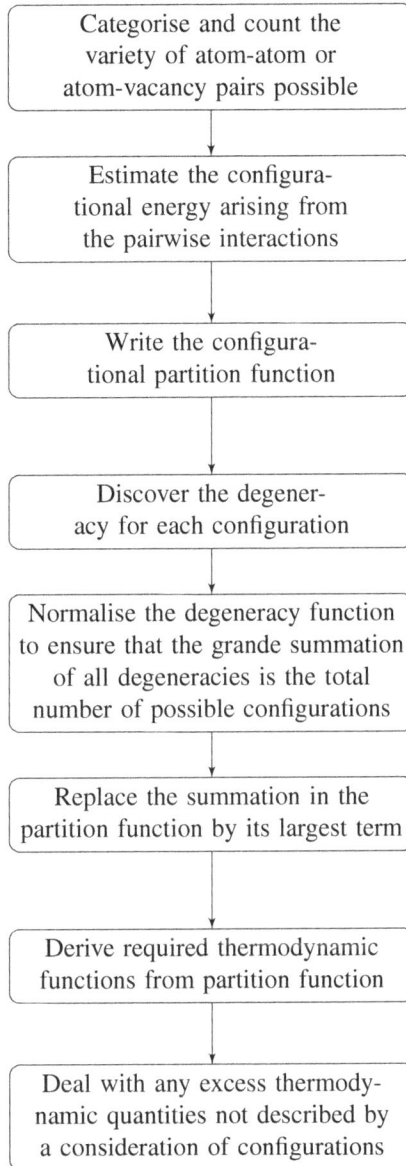

Figure 2.37 Steps in the construction of a quasichemical solution model.

2.17 QUASICHEMICAL MODEL FOR CARBON IN AUSTENITE

There is a particularly useful application of the quasichemical model to the solution of carbon in austenite [103].[15] The theory is founded on the fact that there is a repulsion between the carbon atoms which has the effect of reducing the probability by which a neighbouring interstitial site is occupied. The history of such "site exclusion models" has been reviewed by [103, 104], but the focus here is on the McLellan and Dunn model that removes many of the difficulties of earlier treatments. Furthermore, the model appears to have considerable physical significance, both in explaining fine detail in thermodynamic data and in the prediction of diffusion phenomena (Chapter 3).

The essential problem in the construction of a quasichemical model is, of course, the partition function. Carbon dissolves in the octahedral interstices between the iron atoms in austenite. The number of Fe-Fe pairs and the number of Fe-C pairs do not change for a given composition for all configurations. The partition function can therefore be described solely in terms of the carbon atoms u and the octahedral sites u_o. A u-u pair therefore refers to an adjacent carbon-carbon pair and u-u_o pair is a carbon atom next to an unoccupied interstitial site.

Consider a system with N_u carbon atoms, N_{Fe} iron atoms, and therefore, $\beta_O N_{Fe}$ octahedral sites where β_O is the number of octahedral interstices per iron atom ($\beta_O = 1$ for octahedral holes in austenite and $\beta_O = 3$ for octahedral holes in ferrite). The variety of pairs of species is listed in Table 2.6, where the number of u-u_o and u_o-u pairs is written $W\lambda$; the parameter $W = 12$ is the number of octahedral sites around a single such interstice in austenite (for ferrite, $W = 4$). Naturally, W is the same for the carbon atoms.

Table 2.6

Pair interactions in Fe(γ)-C quasichemical model (after McLellan and Dunn [103]). The energy zero is for an atom at rest in a vacuum so the energies listed are numerically negative.

Kind of pair	No. of such pairs	Energy per pair	Total energy
u_o-u_o	$\frac{1}{2}W(N_{Fe}\beta_O - N_u - \lambda)$	0	0
u_o-u and u-u_o	$W\lambda$	ε_u	$W\lambda\varepsilon_u$
u-u	$\frac{1}{2}W(N_u - \lambda)$	ε_{uu}	$\frac{1}{2}W(N_u - \lambda)\varepsilon_{uu}$
Total	$\frac{1}{2}WN_{Fe}\beta_O$		$W\left[\lambda\varepsilon_u + \frac{N_u-\lambda}{2}\varepsilon_{uu}\right]$

The configurational partition function using the data listed in Table 2.6 is, therefore,

$$\Omega = \sum_\lambda g_\lambda \exp\left\{-W\left[\lambda\varepsilon_u + \frac{N_u - \lambda}{2}\varepsilon_{uu}\right]/kT\right\}. \tag{2.55}$$

Proceeding as in Section 2.16 with the assumption of non-interacting pairs, the degeneracies g_λ are proportional to

$$g_\lambda \propto \frac{[\frac{1}{2}WN_{Fe}\beta_O]!}{\underbrace{[\frac{1}{2}W(N_{Fe}\beta_O - N_u - \lambda)]!}_{u_o-u_o} \underbrace{[\frac{1}{2}W(N_u - \lambda)]!}_{u-u} \underbrace{[\frac{1}{2}W\lambda]! [\frac{1}{2}W\lambda]!}_{u-u_o \text{ and } u_o-u}}. \tag{2.56}$$

As before, this proportionality can be converted into an equality by normalising with the degeneracy corresponding to a completely random solution, for which λ has the value λ^* given by the product of the number of solute atoms (N_u) and the chance of finding an unoccupied octahedral site:

$$\lambda^* = N_u\left(1 - \frac{N_u}{N_{Fe}\beta_O}\right). \tag{2.57}$$

$$\therefore \quad g_\lambda = \frac{[N_{Fe}\beta_O]!}{[N_{Fe}\beta_O - N_u]!\,N_u!}$$

$$\times \frac{[\frac{1}{2}W(N_{Fe}\beta_O - N_u - \lambda^*)]!\,[\frac{1}{2}W(N_u - \lambda^*)]!\,[\frac{1}{2}W\lambda^*]!\,[\frac{1}{2}W\lambda^*]!}{[\frac{1}{2}W(N_{Fe}\beta_O - N_u - \lambda)]!\,[\frac{1}{2}W(N_u - \lambda)]!\,[\frac{1}{2}W\lambda]!\,[\frac{1}{2}W\lambda]!}.$$

The partition function Ω is now solved, but to proceed and obtain some useful thermodynamic quantities, the usual approximation is made to replace the summation in Equation 2.56 by the largest term corresponding to $\lambda = \overline{\lambda}$, which can be obtained by differentiation with respect to λ:

$$\overline{\lambda} = \frac{N_{Fe}\beta_O}{2\phi}\left\{1 - \left[1 - 4\phi\frac{\theta}{\beta_O}\left(1 - \frac{\theta}{\beta_O}\right)\right]^{\frac{1}{2}}\right\}$$

where $\theta = N_u/N_{Fe}$ and

$$\phi = 1 - \exp\left\{\frac{-\omega_\gamma}{kT}\right\}$$

and $\omega_\gamma = \varepsilon_{uu} - 2\varepsilon_u$ is the *carbon-carbon interaction energy*, describing the repulsion between carbon atoms in adjacent interstitial sites.

When ω_γ is large, all octahedral sites adjacent to a carbon atoms are blocked from occupation, whereas if the distribution of carbon atoms were to be random, there would be a finite chance of carbon atoms in adjacent sites. Figure 2.38 shows how significant differences develop between the blocking and random approximations as θ becomes greater than about 0.04. The partition function

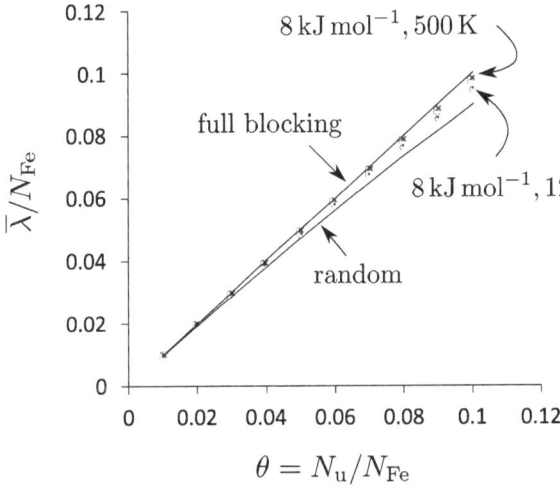

Figure 2.38 Plot of a measure of the carbon atoms that have empty adjacent octahedral interstices ($\overline{\lambda}/N_{Fe}$) values against a measure of the carbon concentration (θ). Complete blocking is represented here by $\overline{\lambda}/N_{Fe}$ for $\omega_\gamma = 800,000$ J mol^{-1} and $T = 1200$ K and random mixing by λ^*/N_{Fe}. The points correspond to $\overline{\lambda}/N_{Fe}$ values for $\omega_\gamma = 8000$ J mol^{-1} at 1200 K and 500 K. Note that $\omega_\gamma = \varepsilon_{uu} - 2\varepsilon_u$

now becomes

$$\Omega = \frac{[N_{Fe}\beta_O]!}{[N_{Fe}\beta_O - N_u]!\, N_u!}$$

$$\times \frac{[\frac{1}{2}W(N_{Fe}\beta_O - N_u - \lambda^*)]!\,[\frac{1}{2}W(N_u - \lambda^*)]!\,[\frac{1}{2}W\lambda^*]!\,[\frac{1}{2}W\lambda^*]!}{[\frac{1}{2}W(N_{Fe}\beta_O - N_u - \overline{\lambda})]!\,[\frac{1}{2}W(N_u - \overline{\lambda})]!\,[\frac{1}{2}W\overline{\lambda}]!\,[\frac{1}{2}W\overline{\lambda}]!} \tag{2.58}$$

$$\times \exp\left\{-W\left[\overline{\lambda}\varepsilon_u + \frac{N_u - \overline{\lambda}}{2}\varepsilon_{uu}\right]\Big/kT\right\} \tag{2.59}$$

which reduces to that for a random solution when $\overline{\lambda} = \lambda^*$. This can happen when the temperature is high enough to allow thermal agitation to mix up the atoms irrespective of their binding tendencies. The configurational free energy F and chemical potential μ follow:

$$F = -kT\ln\{\Omega\}$$

$$\mu = -kT\ln\left\{\frac{\partial \ln\{\Omega\}}{\partial N_u}\right\}_{N_{Fe},T}.$$

McLellan and Dunn used Stirling's approximation to obtain the configurational part of the chemical potential:

$$\frac{\partial}{\partial N_u} \ln\left\{ \frac{[N_{Fe}\beta_O]!}{[N_{Fe}\beta_O - N_u]!\, N_u!} \right\} = -\ln\left\{ \frac{\theta/\beta_O}{1 - \theta/\beta_O} \right\}$$

$$\frac{\partial}{\partial N_u} \ln\left\{ \left[\frac{1}{2} W (N_{Fe}\beta_O - N_u - \lambda) \right]! \right\} = -\frac{W}{2} \ln\left\{ \frac{W}{2} [N_{Fe}\beta_O - N_u - \lambda] \right\}$$

$$\frac{\partial}{\partial N_u} \ln\left\{ \left[\frac{1}{2} W (N_u - \lambda) \right]! \right\} = \frac{W}{2} \ln\left\{ \frac{W}{2} (N_u - \lambda) \right\}$$

$$\frac{\partial}{\partial N_u} \ln\left\{ \left[\frac{1}{2} W \lambda \right]! \right\} = 0$$

$$\frac{\partial}{\partial N_u} \left\{ -W\left(\overline{\lambda} \varepsilon_u + \frac{N_u - \overline{\lambda}}{2} \varepsilon_{uu} \right) \right\} = -\frac{W \varepsilon_{uu}}{2}. \tag{2.60}$$

Noting that $W\varepsilon_{uu}/2 = W\varepsilon_u + W\omega_\gamma/2$, the chemical potential per atom is given by

$$\mu = kT \ln\left\{ \frac{\theta/\beta_O}{1 - (\theta/\beta_O)} \right\}$$

$$-kT \ln\left\{ \left(\frac{\theta/\beta_O}{1 - (\theta/\beta_O)} \right)^W \left(\frac{1 - (\theta/\beta_O) - (\overline{\lambda}/N_{Fe}\beta_O)}{(\theta/\beta_O) - (\overline{\lambda}/N_{Fe}\beta_O)} \right)^{\frac{W}{2}} \right\} + W\varepsilon_u + W\frac{\omega_\gamma}{2}.$$

The term $W\varepsilon_u$ is interpreted as the partial energy of solution of carbon in austenite at infinite dilution, a component of μ_o in $\mu = \mu_o + RT\ln\{a\}$. The activity of carbon is, in this derivation, with respect to pure carbon with the crystal structure of austenite. It is more convenient to define the activity of carbon relative to pure graphite as the standard state. This is easily done, since

$$\frac{a_C\{graphite\}}{a_C\{austenite\}} = \exp\left\{ \frac{\mu_C^{\circ,austenite} - \mu_C^{\circ,graphite}}{kT} \right\}$$

$$= \exp\left\{ \frac{\Delta\mu_C^\circ}{kT} \right\} \tag{2.61}$$

where $\Delta\mu_C^\circ$ is the change in free energy accompanying the transfer of carbon from its standard state of graphite to a standard state based on an infinitely dilute solution as the reference state (it is also called the *relative* partial free energy of a solute atom in solution with respect to the pure solute at infinite dilution). The value of $\Delta\mu^\circ$ is determined experimentally from thermodynamic data measured at sufficiently small concentrations (Figure 2.39).[16] It follows that the activity of carbon with respect to graphite is, according to McLellan and Dunn,

$$a_u = \frac{\theta/\beta_O}{1 - (\theta/\beta_O)} \exp\left\{ \frac{\Delta\mu^\circ}{kT} \right\}$$

$$\times \left(\frac{\theta/\beta_O}{1 - (\theta/\beta_O)} \right)^{-W} \left(\frac{1 - (\theta/\beta_O) - (\overline{\lambda}/N_{Fe}\beta_O)}{(\theta/\beta_O) - (\overline{\lambda}/N_{Fe}\beta_O)} \right)^{-\frac{W}{2}} \exp\left\{ \frac{W\omega_\gamma}{2kT} \right\}.$$

The term containing ε_u is absorbed into $\Delta\mu^\circ$.

There is a difficulty in this derivation of the chemical potential that λ^* and $\overline{\lambda}$ are assumed not to be functions of N_u. This approximation was not made in a parallel treatment of the zeroth order quasichemical model [107]. A version it therefore developed here which allows for the dependence

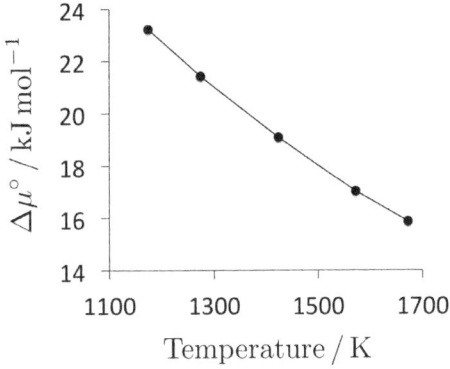

Figure 2.39 Variation of the experimentally determined relative partial free energy of carbon in solution in austenite, with respect to pure graphite at infinite dilution (after Ban-ya, Elliott and Chipman [105, 106]).

of λ on N_u:

$$\frac{\partial}{\partial N_u} \ln\left\{ \left[\frac{1}{2} W(N_{Fe}\beta_O - N_u - \lambda) \right]! \right\} = -\frac{W}{2}\left(1 + \frac{\partial \lambda}{\partial N_u}\right) \ln\left\{ \frac{W}{2}[N_{Fe}\beta_O - N_u - \lambda] \right\}$$

$$\frac{\partial}{\partial N_u} \ln\left\{ \left[\frac{1}{2} W(N_u - \lambda) \right]! \right\} = \frac{W}{2}\left(1 - \frac{\partial \lambda}{\partial N_u}\right) \ln\left\{ \frac{W}{2}(N_u - \lambda) \right\}$$

$$\frac{\partial}{\partial N_u} \ln\left\{ \left[\frac{1}{2} W\lambda \right]! \right\} = W\left(\frac{\partial \lambda}{\partial N_u} \right) \ln\left\{ \frac{W}{2}\lambda \right\}$$

$$\frac{\partial}{\partial N_u}\left\{ -W\left(\bar{\lambda}\varepsilon_u + \frac{N_u - \bar{\lambda}}{2}\varepsilon_{uu} \right) \right\} = -\left(\frac{\partial \lambda}{\partial N_u} \right) W\varepsilon_u - \frac{W\varepsilon_{uu}}{2} + \left(\frac{\partial \lambda}{\partial N_u} \right)\frac{W\varepsilon_{uu}}{2}$$

$$\dot{\lambda}^* = \frac{\partial \lambda^*}{\partial N_u} = 1 - 2(\theta/\beta_O)$$

$$\dot{\bar{\lambda}} = \frac{\partial \bar{\lambda}}{\partial N_u} = \frac{1 - 2(\theta/\beta_O)}{[1 - 4\phi(\theta/\beta_O) + 4\phi(\theta/\beta_O)^2]^{\frac{1}{2}}}.$$

The configurational chemical potential thus becomes [108]:

$$\mu = kT \ln\left\{ \frac{\theta/\beta_O}{1 - (\theta/\beta_O)} \right\}$$

$$-kTW \ln\left\{ \theta^{\frac{2\theta}{\beta_O}} \left[\frac{\theta}{\beta_O} - \left(\frac{\theta}{\beta_O} \right)^2 \right]^{1 - \frac{2\theta}{\beta_O}} \left[1 - \frac{\theta}{\beta_O} - \left(\frac{\bar{\lambda}}{N_{Fe}\beta_O} \right) \right]^{\frac{1}{2}(1+\dot{\bar{\lambda}})} \right\}$$

$$+kTW \ln\left\{ \left(\frac{\bar{\lambda}}{N_{Fe}\beta_O} \right)^{\dot{\bar{\lambda}}} \left[\frac{\theta}{\beta_O} - \frac{\bar{\lambda}}{N_{Fe}\beta_O} \right]^{\frac{1}{2}(1-\dot{\bar{\lambda}})} \left[1 - 2\frac{\theta}{\beta_O} + \left(\frac{\theta}{\beta_O} \right)^2 \right]^{1 - \frac{\theta}{\beta_O}} \right\}$$

$$+W\dot{\bar{\lambda}}\varepsilon_u + \frac{1}{2}W(1 - \dot{\bar{\lambda}})\varepsilon_{uu} \tag{2.62}$$

and the activity of carbon in austenite with respect to graphite is [108]

$$
\begin{aligned}
a_u^\gamma \;=\;& \frac{\theta/\beta_O}{1-(\theta/\beta_O)}\exp\left\{\frac{\Delta\mu^\circ}{kT}\right\} \\[2mm]
&\times\left\{\frac{\theta}{\beta_O}^{\frac{2\theta}{\beta_O}}\left[\frac{\theta}{\beta_O}-\left(\frac{\theta}{\beta_O}\right)^2\right]^{1-\frac{2\theta}{\beta_O}}\left[1-\frac{\theta}{\beta_O}-\left(\frac{\overline{\lambda}}{N_{Fe}\beta_O}\right)\right]^{\frac{1}{2}(1+\overset{.}{\overline{\lambda}})}\right\}^{-W} \\[2mm]
&\times\left\{\left[\frac{\overline{\lambda}}{N_{Fe}\beta_O}\right]^{\overset{.}{\overline{\lambda}}}\left[\frac{\theta}{\beta_O}-\frac{\overline{\lambda}}{N_{Fe}\beta_O}\right]^{\frac{1}{2}(1-\overset{.}{\overline{\lambda}})}\left[1-\frac{2\theta}{\beta_O}+\left(\frac{\theta}{\beta_O}\right)^2\right]^{1-\frac{\theta}{\beta_O}}\right\}^{W} \\[2mm]
&\times\exp\left\{\frac{(1-\overset{.}{\overline{\lambda}})W\omega_\gamma}{2kT}\right\}.
\end{aligned}
\tag{2.63}
$$

The term with $W\varepsilon_u$ is again absorbed into $\Delta\mu^\circ$.

2.17.1   DILUTE SOLUTION, $\omega_\gamma \to 0$ LIMIT

In the limit that ω_γ tends to zero, $\overline{\lambda}\to\lambda^*$ and $\overset{.}{\overline{\lambda}}\to 1-2(\theta/\beta_O)$. Noting that $\omega_\gamma=\varepsilon_{uu}-2\varepsilon_u$, the limiting form of Equation 2.62 becomes

$$
\mu\{\omega_\gamma\to 0\}=kT\ln\left\{\frac{\theta/\beta_O}{1-(\theta/\beta_O)}\right\}+W\varepsilon_u+W\omega_\gamma\frac{\theta}{\beta_O}.
\tag{2.64}
$$

Naturally, $\overline{\lambda}$ must also tend to λ^* at high temperatures so the high temperature limit of this first-order quasichemical theory will then become equivalent to the zeroth-order mixing treatment where the solute atoms are distributed at random [107].

2.17.2   INFINITE REPULSION LIMIT

In this case, as $\omega_\gamma\to\infty$, $\overline{\lambda}\to N_u$ and $\overset{.}{\overline{\lambda}}\to 1$. Therefore,

$$
\begin{aligned}
\mu\{\omega_\gamma\to\infty\}\;=\;& kT\ln\left\{\frac{\theta/\beta_O}{1-(\theta/\beta_O)}\right\} \\[2mm]
&-kT\ln\left\{\left(1-2\frac{\theta}{\beta_O}\right)\left(1-\frac{\theta}{\beta_O}\right)^{1-2\frac{\theta}{\beta_O}}\right\} \\[2mm]
&+kT\ln\left\{\left(1-\frac{\theta}{\beta_O}\right)^{2-2\frac{\theta}{\beta_O}}\right\} \\[2mm]
&+W\varepsilon_u
\end{aligned}
$$

which, at values of θ *sufficiently small* to allow the expansion $\ln\{1-\theta\}\simeq-\theta$, can be rewritten as

$$
\mu\{\omega_\gamma\to\infty\}=kT\ln\left\{\frac{\theta/\beta_O}{1-(W+1)(\theta/\beta_O)}\right\}+W\varepsilon_u.
\tag{2.65}
$$

This is consistent with all the interstitial sites adjacent to a carbon atom being blocked from occupation. Notice that the term $-W(\theta/\beta_O)\omega_\gamma$ is absent from all these complete-blocking equations because there are no carbon-carbon near neighbour pairs in that scenario.

2.17.3 ZEROTH-ORDER QUASICHEMICAL MODEL

The zeroth approximation of quasichemical theory has the solute atoms distributed at random. The partition function and all associated functions for the zeroth-order treatment can be deduced from the corresponding equations for the first-order treatment simply by setting $\overline{\lambda} = \lambda^*$. It is nevertheless interesting to compare the first and zeroth approximations as was first reported by Alex and McLellan [107], and indeed to illustrate the difference between a substitutional and interstitial solution [97, 107].

From Equation 2.55, the partition function is obtained as

$$\Omega = \frac{[N_{Fe}\beta_O]!}{[N_{Fe}\beta_O - N_u]!\, N_u!} \times \exp\left\{-W\left[\lambda^*\varepsilon_u + \frac{N_u - \lambda^*}{2}\varepsilon_{uu}\right]\bigg/ kT\right\}.$$

On substituting for λ^* (Equation 2.57) and ε_{uu} this becomes:[17]

$$\Omega = \frac{[N_{Fe}\beta_O]!}{[N_{Fe}\beta_O - N_u]!\, N_u!} \times \exp\left\{-W\left[2N_u\varepsilon_u + \frac{N_u^2}{N_{Fe}\beta_O}\omega_\gamma\right]\bigg/ 2kT\right\}. \tag{2.66}$$

The configurational chemical potential therefore becomes:

$$\mu = kT\ln\left\{\frac{\theta/\beta_O}{1 - (\theta/\beta_O)}\right\} + W\varepsilon_u + W\omega_\gamma\frac{\theta}{\beta_O}. \tag{2.67}$$

Comparison with Equation 2.64 shows that the two equations are identical.

The zeroth approximation is in reality identical to the regular solution model. For a substitutional solution consisting of A and B atoms, the chemical potential per atom of A is given by Equation 2.45:

$$\mu_A = \mu_A^\circ + zx^2\omega + kT\ln\{1 - x\} \tag{2.68}$$

which clearly has a different compositional dependence than that given by Equation 2.67.

2.17.4 REFERENCE STATE FOR CARBON

In an earlier discussion the activity of carbon defined with respect to a hypothetical pure carbon with the austenite crystal structure was converted into a reference state of pure graphite (Equation 2.61). For graphite, $\mu_C^{\circ,\text{graphite}}$ can be obtained from standard tables. The term $\mu_C^{\circ,\text{austenite}}$ may be decomposed further into a relative partial enthalpy ($\overline{H}_C$) and a nonconfigurational entropy ($\overline{S}_C^v$). The relative partial enthalpy comes from the insertion of carbon atoms from rest in a vacuum, into infinitely dilute austenite. The entropy term is entirely nonconfigurational because $\mu_C^{\circ,\text{austenite}}$ refers to pure carbon in the austenite crystal structure; this is why $\Delta\overline{S}_C^v$ has the superscript.

Although Figure 2.39 shows that $\Delta\mu_C^\circ$ varies nonlinearly with the temperature, Dunn and McLellan [104] have shown that the curvature is due to the temperature dependence of $\mu_C^{\circ,\text{graphite}}$. They find $\Delta\overline{H}_C$ and $\Delta\overline{S}_C^v$ to be independent of temperature, for carbon in austenite or in ferrite [109], Table 2.7.

The data in Table 2.7 can be used in conjunction with those for graphite to obtain $\Delta\mu^\circ$, which in general is temperature dependent. Table 2.8 contains some values of $\Delta\mu^\circ$ for transformations between a variety of different states of carbon.

2.17.5 CARBON-CARBON INTERACTION ENERGY IN AUSTENITE

The pairwise C-C interaction energy in austenite, $\omega_\gamma = \varepsilon_{uu} - 2\varepsilon_u$, features prominently in the quasichemical models for Fe-C solutions. It determines the distribution of carbon atoms in solution. Its specific value to some extent depends on the nature of the quasichemical model that is used to

Table 2.7

The partial molar enthalpy with respect to a solute atom at rest in a vacuum and the non-configurational entropy for solute in austenite and in ferrite [104, 109, 110]. These quantities are found to be temperature independent.

	$\overline{H}/\mathrm{J\,mol}^{-1}$	$\overline{S}_C^v/\mathrm{J\,mol}^{-1}\,\mathrm{K}^{-1}$
Carbon in austenite	−649691	5.78
Carbon in ferrite	−602747	6.56
Nitrogen in austenite	−342670	7.20
Nitrogen in ferrite	−310871	6.20

Table 2.8

The stability of different forms of carbon. 'L' stands for liquid and βMn and αMn are complicated crystal structures of manganese. The data are due to Kaufman and Nesor [111] for temperatures in excess of 300 K.

Transformation	$\mathrm{J\,mol}^{-1}$
$\mu_\mathrm{L}^\circ - \mu_\mathrm{bcc}^\circ$	$-32635 - 12.552T$
$\mu_\mathrm{bcc}^\circ - \mu_\mathrm{hcp}^\circ$	32635
$\mu_\mathrm{hcp}^\circ - \mu_\mathrm{fcc}^\circ$	-24267
$\mu_\mathrm{fcc}^\circ - \mu_{\beta\mathrm{Mn}}^\circ$	-7531
$\mu_{\beta\mathrm{Mn}}^\circ - \mu_{\alpha\mathrm{Mn}}^\circ$	6694
$\mu_\mathrm{L}^\circ - \mu_\mathrm{graphite}^\circ$	$114223 - 27.196T$

interpret measured activity data, but it generally is accepted that carbon atoms in austenite repel with $\omega_\gamma \simeq 8000\,\mathrm{J\,mol}^{-1}$ [104].

Mössbauer spectroscopy can be used to study more directly, the distribution of carbon atoms relative to an iron atom. The measured probabilities of different kinds of clusters of atoms can then be compared against Monte Carlo simulations to deduce interaction energies [112, 113].

The Mössbauer data show that it is necessary to consider four kinds of Fe-C configurations (Figure 2.40): iron atoms which do not have neighbouring carbon atoms; an iron atom with a single near neighbour carbon atom; an iron atom with two neighbouring carbon atoms in a 180° configuration; an iron atom with two neighbouring carbon atoms in a 90° configuration. The pair of carbon atoms in the 90° configuration are near neighbours whereas those in the 180° configuration are next-near neighbours. A surprising result is that the repulsion between near neighbour carbon pairs is smaller than that between next-near neighbours. The mechanism for this apparently strange behaviour, which is not found for nitrogen-nitrogen pairs, does not seem to have been considered.

Oda et al. [112] compared the local interaction energies derived from Mössbauer results, with the global value obtained by applying quasichemical theory to thermodynamic data, by taking a weighted average:

$$W\omega_\gamma \simeq W\omega_\gamma^{(1)} + W_2\omega_\gamma^{(2)} \tag{2.69}$$

where, as usual, $W = 12$ is the number of C-C near neighbour pairs that can form about one C atom whereas $W_2 = 6$ is the number of C-C next near neighbour pairs that can form about one C atom. $\omega_\gamma^{(1)}$ and $\omega_\gamma^{(2)}$ are the C-C interaction energies for the $90°$ and $180°$ configurations respectively. As shown in Table 2.9, the Mössbauer and quasichemical data compare well except for nitrogen where there is some uncertainty in the thermodynamic data. However, an independent study has verified the general trend that there is a much stronger N-N repulsion than the C-C interaction in the first coordination shell whereas the opposite is found for the second coordination shell [114].

The consequence of such interactions on the non-random nature of the distribution of interstitial atoms in austenite has also been measured. Both carbon and nitrogen are heterogeneously distributed on a scale of about 30 nm, the inhomogeneities being more abrupt for nitrogen than for carbon [114].

Table 2.9

Comparison of the overall interaction energy between interstitial atoms determined from the Mössbauer method (column 4) against values derived using quasichemical models (column 5). The energies are in $J\,mol^{-1}$ (after Oda et al. [112]).

	Mössbauer results			Quasichemical
	$\omega_\gamma^{(1)}$	$\omega_\gamma^{(2)}$	$\omega_\gamma \simeq (W\omega_\gamma^{(1)} + W_2\omega_\gamma^{(2)})/W$	ω_γ
C-C	3500	7333	7167	$\simeq 8000$
N-N	8167	1000	8667	$\simeq 4000$

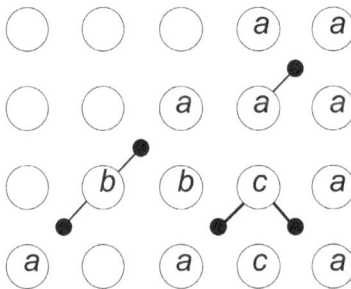

Figure 2.40 The distribution of carbon atoms (dark) on a single $\{001\}_\gamma$ plane of austenite. The symbol a represents an iron atom with a single carbon neighbour, b the case where Fe has two carbon atoms forming a $180°$ configuration, and c where Fe has two carbon atoms in a $90°$ configuration.

2.17.6 CARBON-CARBON INTERACTION ENERGY IN FERRITE

The problem for ferrite is more complicated. Following on from Equation 2.63, the activity of carbon in ferrite relative to pure graphite as the standard state is as follows [108]:

$$
a_u^\alpha = \frac{\theta/\beta_O}{1-(\theta/\beta_O)} \exp\left\{\frac{\Delta\mu^\circ}{kT}\right\}
$$

$$
\times \left\{ \frac{\theta}{\beta_O}^{\frac{2\theta}{\beta_O}} \left[\frac{\theta}{\beta_O} - \left(\frac{\theta}{\beta_O}\right)^2\right]^{1-\frac{2\theta}{\beta_O}} \left[1 - \frac{\theta}{\beta_O} - \left(\frac{\overline{\lambda}}{N_{Fe}\beta_O}\right)\right]^{\frac{1}{2}(1+\dot{\overline{\lambda}})} \right\}^{-W}
$$

$$
\times \left\{ \left[\frac{\overline{\lambda}}{N_{Fe}\beta_O}\right]^{\dot{\overline{\lambda}}} \left[\frac{\theta}{\beta_O} - \frac{\overline{\lambda}}{N_{Fe}\beta_O}\right]^{\frac{1}{2}(1-\dot{\overline{\lambda}})} \left[1 - \frac{2\theta}{\beta_O} + \left(\frac{\theta}{\beta_O}\right)^2\right]^{1-\frac{\theta}{\beta_O}} \right\}^{W}
$$

$$
\times \exp\left\{ \frac{(1-\dot{\overline{\lambda}})W\omega_\alpha}{2kT} \right\}
\tag{2.70}
$$

where $\dot{\overline{\lambda}} = \partial\overline{\lambda}/\partial N_u$.

This corrects the earlier work in which the activity function was stated as [103, 104]:

$$
a_u = \frac{\theta/\beta}{1-(\theta/\beta)} \exp\left\{\frac{\Delta\mu^\circ}{kT}\right\}
$$

$$
\times \left(\frac{\theta/\beta}{1-(\theta/\beta)}\right)^{-W} \left(\frac{1-(\theta/\beta)-(\overline{\lambda}/N_{Fe}\beta)}{(\theta/\beta)-(\overline{\lambda}/N_{Fe}\beta)}\right)^{-\frac{W}{2}}
$$

$$
\exp\left\{\frac{W\omega_\alpha}{2kT}\right\}.
\tag{2.71}
$$

On the basis of Equation 2.70, ω_α is found to be large, probably greater than $150\,\mathrm{kJ\,mol^{-1}}$ [115, 116]. However, both studies reported large and unsystematic variations in the value of ω_α as a function of different experimental data.[18]

The corrected quasichemical theory (Equation 2.70) in fact compounds the problem of deducing ω_α. Whereas in Equation 2.71 the activity (even at small concentrations) is a significant function of ω_α, this is not the case for Equation 2.70. This can be seen from the behaviour of $\overline{\lambda}$ and $\dot{\overline{\lambda}}$ at small concentrations. As $\theta \to 0$, $\overline{\lambda} \to N_{Fe}\beta$ and $\dot{\overline{\lambda}} \to 1$. Equation 2.70 therefore ceases to be a function of ω_α, but Equation 2.71 does not because the extreme right-hand exponential terms are different:

$$
\text{Equation 2.70:} \quad \exp\left\{\frac{(1-\dot{\overline{\lambda}})W\omega}{2kT}\right\} \qquad \text{Equation 2.71:} \quad \exp\left\{\frac{W\omega}{2kT}\right\}
\tag{2.72}
$$

For Equation 2.70 this exponential term tends to zero via $(1-\dot{\overline{\lambda}})$ as $\theta \to 0$ whereas this is not the case for equation 2.71.

The sad conclusion is that it is not possible, using the low carbon concentration data available for ferrite, to deduce the value of ω_α if the correct quasichemical model is used. On the other hand, this must be a reasonable common sense conclusion since the probability of finding a pair of carbon atoms in close proximity also vanishes as the concentration tends to zero. Thus, for carbon in ferrite, it does not seem relevant to worry about the magnitude of the interaction. This may not, of course, be the case for supersaturated solutions of carbon in ferrite, such as in bainitic or martensitic transformations, but no activity measurements have been conducted for those solutions (supersaturated solutions of carbon are not stable).

This topic of carbon-carbon interactions in ferrite will be revisited in Chapter 3.

2.18 ZENER ORDERING

The transformation of austenite without diffusion is achieved by a Bain deformation [62]. The carbon atoms in both the parent and product phases occupy octahedral interstices, but there are three times as many of these interstices per atom of iron in the ferrite than in austenite. The Bain strain therefore places all the carbon atoms on to one sublattice of octahedral interstices in the ferrite, resulting in a tetragonal unit cell. The Zener order [117] is therefore imposed by the action of the Bain strain. This tetragonality can have dramatic effects on the properties of the ferrite, for example, an increase in the solubility of carbon in tetragonal-ferrite that is in equilibrium with austenite [118]. Although the state of Zener order [2, 117] is forced on to the product lattice by the Bain strain, its subsequent stability can be examined using the theory of order-disorder transformations. There are three irregular-octahedral interstice sublattices in the bcc structure, each with the short axis aligned to one of the three $\langle 100 \rangle$ axes. The strain energy due to a misfitting carbon atom will be $\frac{1}{2}E_{100}\varepsilon^2$, where E_{100} is the elastic modulus along $\langle 100 \rangle$ and ε is the elastic strain. Consider ordering to occur specifically along a preferred axis [100], with N_V^p and N_V^n being the number of carbon atoms per unit volume in preferred and not-preferred sites, with the total number of carbon atoms per unit volume being $N_V = N_V^p + N_V^n$. An order parameter can be defined as follows:

$$z = \frac{3}{2}\left(\frac{N_V^p}{N_V} - \frac{1}{3}\right) \qquad \begin{cases} z = 1 & \text{if} \quad N_V^p = N_V \\ z = 0 & \text{if} \quad N_V^p = \frac{1}{3}. \end{cases}$$

The free energy change per unit volume on ordering is given by

$$\Delta G_{\text{Zener}} = -\frac{1}{2}\left(\frac{2}{3}N_V\varepsilon_\lambda z\right)^2 E_{100} + \frac{1}{3}N_V kT[2(1-z)\ln\{1-z\} + (1+2z)\ln\{1+2z\}] \qquad (2.73)$$

where ε_λ is the strain caused by transferring a carbon atom from a non-preferred site to a preferred site, in units of strain per (number of atoms per unit volume). The first term on the right represents the change in free energy when carbon atoms are transferred to the preferred sites, and the second the corresponding change in configurational entropy. The entropy change associated with ordering on sublattices is discussed in detail in Section 2.20. The factor $\frac{2}{3}$ is because from a random distribution of carbon atoms, it would only require that fraction of carbon atoms to relocate to obtain the fully ordered state. The temperature at which the ordering parameter becomes finite is the critical ordering temperature T_{Zener} that can be derived by setting the differential of Equation 2.73 to zero, which gives

$$T = \frac{2}{3k}N_V\varepsilon_\lambda^2 E_{100} \times z \left/ \ln\left\{\frac{1+2z}{1-z}\right\}\right.$$

so that

$$T_{\text{Zener}}/\text{K} \approx 28080 \times \frac{x_C}{1-x_C} \qquad (2.74)$$

where x_C is the mole fraction of carbon.[19]
Khachaturyan and Shatalov [119, 120] used the microscopic elasticity theory [121] which recognises the discrete nature of the lattice during elastic interactions between solute and solvent atoms, to derive a more elaborate model for carbon atom ordering. This gives a critical ordering temperature (still labelled T_{Zener}) [122]:

$$T_{\text{Zener}}/\text{K} \approx (11400 \quad \text{or} \quad 23400) \times \frac{x_C}{1-x_C}$$

where the choice of the numerical value depends on whether the original [120, 122] strain interaction parameter is used or another based on molecular dynamic simulation is substituted [123]; in the latter case, the result is almost identical to Equation 2.74.

2.19 COMPUTER CALCULATION OF PHASE DIAGRAMS

The thermodynamic methods described thus far in this chapter are revealing and have been applied towards the understanding and modelling the behaviour of iron and its alloys. It nevertheless is too complicated in the context of multicomponent steels where individual solute concentrations can vary over a large range. Therefore, methods have been developed for doing this in a seamless manner; these methods have been so successful that they now represent the first step in any alloy development project. The subject of the computer calculation of phase diagrams based on experimental data has been reviewed extensively, e.g., [124–129]. The focus here is on the framework for such generic calculations, which necessarily involves a degree of educated, clever empiricism. The process has also led to the systematic compilation and assessment of experimental data on a scale that is perhaps unique in science. All this was initiated by like-minded scientists long before "big data" or computer modelling became fashionable.

One possibility is to represent thermodynamic quantities by a series expansion with sufficient adjustable parameters to adequately fit the experimental data. There has to be a compromise between the accuracy of the fit and the number of terms in expansion. However, such expansions do not generalise well when dealing with complicated phase diagram calculations involving many components and phases. Experience suggests that the specific heat capacities for the pure elements are better represented by a polynomial with a form that describes most of the known experimental data:

$$C_P = b_8 + b_9 T + b_{10} T^2 + \frac{b_{11}}{T^2}. \tag{2.75}$$

If the fit with experimental data is found not to be good enough, the polynomial is applied to a range over which the fit is satisfactory, and more than one polynomial is used to represent the full dataset with care exercised to ensure continuity over the range. A standard element reference state is defined with a list of the measured enthalpies and entropies of the pure elements at 298 K and one atmosphere pressure, for the crystal structure appropriate for these conditions. With respect to this state, the Gibbs free energy is obtained by integration to be

$$G = b_{12} + b_{13} T + b_{14} T \ln\{T\} + b_{15} T^2 + b_{16} T^3 + \frac{b_{17}}{T}. \tag{2.76}$$

This free energy is defined with respect to a reference (included in b_{12}), i.e., relative to the enthalpy at 298.15 K and entropy at 0 K of the stable states of the element(s) concerned at 298.15 K. Allotropic transformations can be included if the transition temperatures, enthalpy of transformation and the C_P coefficients for all the phases are known.

Any specific contributions to C_P, such as due to magnetic transitions, are dealt with separately, as are the effects of pressure. Once again, the equations for these effects are chosen carefully in order to maintain generality.

The excess Gibbs free energy for a binary solution with components A and B is written:

$$\Delta_e G_{AB} = x_A x_B \sum_{i=0}^{j} L_{AB,i} (x_A - x_B)^i. \tag{2.77}$$

For $i = 0$ this gives a term $x_A x_B L_{AB,0}$ which is familiar in regular solution theory, where the coefficient $L_{AB,0}$ is, as usual, independent of chemical composition and to a first approximation describes the interaction between components A and B. If all other $L_{AB,i}$ are zero for $i > 0$, then the equation reduces to the regular solution model with $L_{AB,0}$ as the regular solution parameter. Further terms ($i > 0$) are added to allow for any composition dependence not described by the regular solution constant.

In the first approximation, the excess free energy of a ternary solution can be represented purely by a combination of the binary terms in Equation 2.77:

$$\Delta_e G_{ABC} = x_A x_B \sum_{i=0}^{j} L_{AB,i}(x_A - x_B)^i$$

$$+ x_B x_C \sum_{i=0}^{j} L_{BC,i}(x_B - x_C)^i + x_C x_A \sum_{i=0}^{j} L_{CA,i}(x_C - x_A)^i.$$

The advantage of the representation embodied in Equation 2.77 is clear, that for the ternary case, the relation reduces to the binary problem when one of the components is set to be identical to another, e.g., B≡C [130].

Experimental data may indicate significant ternary interactions, in which case a term $x_A x_B x_C L_{ABC,0}$ is added to the excess free energy. If this does not adequately represent the deviation from the binary summation, then it can be converted into a series which properly reduces to a binary formulation when there are only two components:

$$x_A x_B x_C \quad \left[L_{ABC,0} + \frac{1}{3}(1 + 2x_A - x_B - x_C) L_{ABC,1} \right.$$

$$\left. + \frac{1}{3}(1 + 2x_B - x_C - x_A) L_{BCA,1} + \frac{1}{3}(1 + 2x_C - x_A - x_B) L_{CAB,1} \right].$$

This method can be extended to any number of components as long as appropriate thermodynamic data are available, with the advantage that few coefficients have to be changed when the data due to one component are improved. The information necessary to derive the coefficients become sparse for systems with more than three components.

2.19.1 STOICHIOMETRIC PHASES: REGULAR SOLUTION MODEL

A compound differs from a solution in that its free energy increases rapidly when the composition deviates from a specific ratio of the components (Figure 2.41). When the compound is crystalline it usually is the case that the lattice can be considered to consist of interpenetrating sublattices. Thus, for a binary compound each sublattice would be occupied mostly by one of the components. A widely used regular solution model for multicomponent compounds, as developed by Hillert and Staffansson [131] from work on ionic compounds and melts, is presented here.

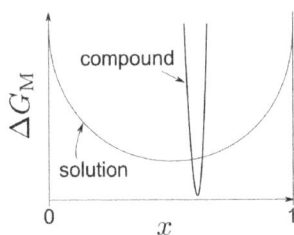

Figure 2.41 Free energy of a (binary) compound increases sharply on deviation from a stoichiometric composition, when compared with that of a solution.

Consider a stoichiometric phase $(A,B)_a (C,D)_c$ in which the components A,B and C,D are located on different sublattices, the stoichiometry being a/c with $a + c = 1$. A new composition parameter, the site fraction y' is introduced to deal with the sublattices in isolation:

$$y'_A = \frac{n_A}{n_A + n_B} = \frac{x_A}{a}$$

$$y'_C = \frac{n_C}{n_C + n_D} = \frac{x_C}{c}$$

where n_i represent the number of moles of the component identified in the subscript. If the two sublattices are treated as independent with random mixing on each, then the ideal configurational entropy of mixing for the system consisting of N atoms is [132]

$$\Delta S_M = -R\left(n_A \ln\{y'_A\} + n_B \ln\{y'_B\} + n_C \ln\{y'_C\} + n_D \ln\{y'_D\}\right) \qquad \mathrm{J\,K^{-1}}$$

so dividing both sides this equation by $n_A + n_B + n_C + n_D$ gives the molar quantity:

$$\Delta S_M = -R\left(x_A \ln\{y'_A\} + x_B \ln\{y'_B\} + x_C \ln\{y'_C\} + x_D \ln\{y'_D\}\right) \qquad \mathrm{J\,mol^{-1}\,K^{-1}}.$$

The free energy of the phase is then written relative to the four possible pure binary compounds, each of which contains just one component on each sublattice[20]

$$G = \left(y'_A y'_C \mu^\circ_{A_a C_c} + y'_A y'_D \mu^\circ_{A_a D_c} + y'_B y'_C \mu^\circ_{B_a C_c} + y'_B y'_D \mu^\circ_{B_a D_c}\right) - T\Delta S_M + \Delta_e G.$$

Referring to Figure 2.42, the bottom left hand corner corresponds to the pure compound $A_a C_c$. Thus, $y'_A = y'_C = 1$ and $y'_B = y'_D = 0$ so it is found correctly that $G = \mu^\circ_{A_a C_c}$. On the other hand, half way along the horizontal axis, $y'_A = y'_B = 0.5$, $y'_C = 1$ so that $G = 0.5\mu^\circ_{A_a C_c} + 0.5\mu^\circ_{B_a C_c} - T\Delta S_M + \Delta_e G$, where the entropy term is due solely to mixing on the AB sublattice. The excess free energy on the other hand must involve interactions between all the species present. To account for this, Hillert and Staffannson expressed the excess free energy as

$$\Delta_e G = y'_A y'_B y'_C L_{AB,C} + y'_A y'_B y'_D L_{AB,D} + y'_C y'_D y'_A L_{CD,A} + y'_C y'_D y'_B L_{CD,B}$$

so the regular solution parameters for each sublattice depend also on the occupancy of the other sublattice. The L parameters embody interactions: e.g., $L_{CD,B}$ the interaction between C and D when the other sublattice is filled completely with B. It also is possible, in deriving the partial quantities,

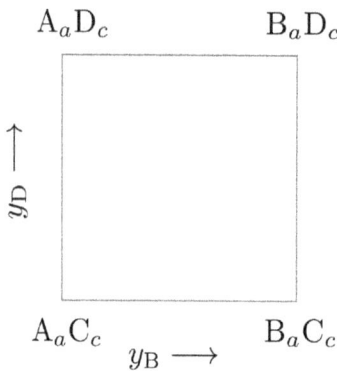

$$A_a D_c \qquad\qquad\qquad B_a D_c$$

$y_D \longrightarrow$ (vertical axis)

$$A_a C_c \qquad\qquad\qquad B_a C_c$$
$$y_B \longrightarrow$$

Figure 2.42 Representation of composition for a $(A,B)_a(C,D)_c$ compound (after Hillert and Staffansson [131]). The corners are binary compounds whereas a point along the edges has mixtures of atoms on the sublattices.

to treat the binary compound as a component in the quarternary solution. Using Equation 2.36, it is seen that

$$\mu_{A_a C_c} = G + (1 - y'_A)\frac{\partial G}{\partial y'_A} + (1 - y'_C)\frac{\partial G}{\partial y'_C}.$$

2.19.2 INTERSTITIAL SOLUTION

An interstitial solution can be accommodated in the sublattice model by treating the interstitial vacancies ($\square$) as a component in the stoichiometric phase $(A,B)_a(C,\square)_c$ and, as before, $a + c = 1$;

the ratio a/c is 1 and $\frac{1}{3}$ for carbon dissolved in austenite and ferrite respectively. Consider the Fe-Mn-C system where Fe (solvent) and Mn are atoms in the substitutional sites.[21] The site fractions for the substitutional elements are therefore:

$$y_{Fe} = x_{Fe}/(1-x_C)$$
$$y_{Mn} = x_{Mn}/(1-x_C)$$

The number of interstitial sites is the product of the number of moles of substitutional atoms ($n_{Fe} + n_{Mn}$) and the ratio c/a so the site fraction for carbon is given by:

$$y_C = \frac{n_C}{\frac{c}{a}(n_{Fe}+n_{Mn})} = \frac{ax_C}{c(1-x_C)}$$
$$y_\square = 1-y_C.$$

The configurational entropy of mixing for a system containing a total of $N = N_C + N_{Fe} + N_{Mn}$ atoms is, therefore,

$$\Delta S_M = -R(n_{Fe}\ln\{y_{Fe}\}+n_{Mn}\ln\{y_{Mn}\}+n_C\ln\{y_C\}+n_\square\ln\{y_\square\}). \tag{2.78}$$

Since the number of unoccupied interstitial vacancies is

$$n_\square = \frac{c}{a}(n_{Fe}+n_{Mn})-n_C,$$

the molar configurational entropy of mixing becomes

$$\Delta S_M = -R\left[x_{Fe}\ln\{y_{Fe}\}+x_{Mn}\ln\{y_{Mn}\}+x_C\ln\{y_C\}+\left(\frac{c}{a}-x_C\frac{a+c}{a}\right)\ln\{1-y_C\}\right].$$

In order to obtain the molar Gibbs free energy, it is necessary to define the reference states (μ°) for each of the pure components, in the context of the stoichiometric phase model. Since an $Fe_a\square_c$ compound is simply pure Fe, $\mu^\circ_{Fe_a\square_c} = \mu^\circ_{Fe}$, the free energy of a mole of pure Fe. Similarly, for manganese, $\mu^\circ_{Mn_a\square_c} = \mu^\circ_{Mn}$.

The reference state for carbon which is interstitially dissolved is a little more awkward. In an Fe-C-$\square$ system it is given by the difference between a pure vacancy-free $Fe_a - C_c$ carbide and pure Fe:

$$\mu^\circ_{Fe_aC_c} - \mu^\circ_{Fe_a\square_c} = \mu^\circ_{Fe_aC_c} - \mu^\circ_{Fe}.$$

This standard state can of course be related to pure graphite as discussed on page 64. Similarly, in a B-C-$\square$ system it is given by the difference between a pure vacancy-free Mn_aC_c carbide and pure Mn:

$$\mu^\circ_{Mn_aC_c} - \mu^\circ_{Mn_a\square_c} = \mu^\circ_{Mn_aC_c} - \mu^\circ_{Mn}. \tag{2.79}$$

It follows that the molar Gibbs free energy is given by

$$\begin{aligned}
G &= x_{Fe}\mu^\circ_{Fe}+x_{Mn}\mu^\circ_{Mn}+x_c(\mu^\circ_{Fe_aC_c}-\mu^\circ_{Fe_a\square_c}) \\
&\quad +RT\left[x_{Fe}\ln\{y_{Fe}\}+x_{Mn}\ln\{y_{Mn}\}\right.\\
&\qquad\left.+x_C\ln\{y_C\}+\left(\frac{c}{a}-x_C\frac{a+c}{a}\right)\ln\{1-y_C\}\right]\\
&\quad +\Delta_eG
\end{aligned}$$

where the excess free energy is

$$\begin{aligned}
a\times\Delta_eG &= y_Cx_{Mn}\left[(\mu^\circ_{Fe}-\mu^\circ_{FeC_c})-(\mu^\circ_{Mn}-\mu^\circ_{MnC_c})\right]\\
&\quad +x_{Fe}y_{Mn}\left[y_CL_{FeMn,C}+(1-y_C)L_{FeMn,\square}\right]\\
&\quad +y_C(1-y_C)\left[x_{Fe}L_{C\square,Fe}+x_{Mn}L_{C\square,Mn}\right].
\end{aligned}$$

The chemical potentials can be derived from this integral free energy. For austenite, $c/a = 1$ so the following equations are obtained for the chemical potentials [133]; the subscripts identifying the phase as austenite have been omitted for clarity:

$$
\begin{aligned}
\mu_{Fe} = {} & \mu_{Fe}^{\circ} + RT \left[\ln\{y_{Fe}\} + \ln\{1 - y_C\} \right] \\
& - y_{Mn} y_C \left[(\mu_{Fe}^{\circ} - \mu_{FeC}^{\circ}) - (\mu_{Mn}^{\circ} - \mu_{MnC}^{\circ}) \right] \\
& - y_{Mn} y_C \left[L_{FeMn,C} - L_{FeMn,\square} + L_{C\square,Mn} - L_{C\square,Fe} \right] \\
& + y_{Mn}^2 L_{FeMn,\square} + y_C^2 L_{C\square,Fe} \\
& + 2 y_{Mn}^2 y_C (L_{FeMn,C} - L_{FeMn,\square}) + 2 y_{Mn} y_C^2 (L_{C\square,Mn} - L_{C\square,Fe}).
\end{aligned}
\tag{2.80}
$$

$$
\begin{aligned}
\mu_{Mn} = {} & \mu_{Mn}^{\circ} + RT \left[\ln\{y_{Mn}\} + \ln\{1 - y_C\} \right] \\
& + y_{Fe} y_C \left[(\mu_{Fe}^{\circ} - \mu_{FeC}^{\circ}) - (\mu_{Mn}^{\circ} - \mu_{MnC}^{\circ}) \right] \\
& + y_{Fe} y_C \left[L_{FeMn,\square} - L_{FeMn,C} + L_{C\square,Mn} - L_{C\square,Fe} \right] \\
& + y_{Fe}^2 L_{FeMn,\square} + y_C^2 L_{C\square,Mn} \\
& + 2 y_{Fe}^2 y_C (L_{FeMn,C} - L_{FeMn,\square}) + 2 y_{Fe} y_C^2 (L_{C\square,Fe} - L_{C\square,Mn}).
\end{aligned}
$$

$$
\begin{aligned}
\mu_C = {} & \mu_{FeC}^{\circ} - \mu_{Fe}^{\circ} + RT \left[\ln\{y_C/(1 - y_C)\} \right] \\
& + y_{Mn} \left[(\mu_{Fe}^{\circ} - \mu_{FeC}^{\circ}) - (\mu_{Mn}^{\circ} - \mu_{MnC}^{\circ}) \right] \\
& + y_{Mn} \left[L_{FeMn,C} - L_{FeMn,\square} + L_{C\square,Mn} - L_{C\square,Fe} \right] \\
& + (1 - 2 y_C) L_{C\square,Fe} \\
& + 2 y_{Mn} y_C (L_{C\square,Fe} - L_{C\square,Mn}) + y_{Mn}^2 (L_{FeMn,\square} - L_{FeMn,C}).
\end{aligned}
$$

Typical values of some of the parameters for ternary Fe-Mn-C austenite are listed in Table 2.10. A model such as this can readily be extended to many more components. An application to the Fe-Mo-W-C system with numerous alloy carbide phases in addition to austenite can be found in Uhrenius and Harvig [134].

Table 2.10

Some typical thermodynamic parameters for ternary Fe-Mn-C austenite (after Hillert and Waldenström [133]). The units are all in J mol^{-1}.

Parameter	J mol^{-1}
$L_{C\square,Fe}/(1 - x_C)$	$-21{,}058 - 11.581T$
$L_{C\square,Fe} - L_{C\square,Mn}$	0
$(\mu_{Fe}^{\circ} - \mu_{FeC}^{\circ}) - (\mu_{Mn}^{\circ} - \mu_{MnC}^{\circ})$	-48500
$L_{FeMn,C} - L_{FeMn,\square} + L_{C\square,Mn} - L_{C\square,Fe}$	8500
$L_{FeMn,C} - L_{FeMn,\square}$	8668
$\mu_{FeC}^{\circ} - \mu_{Fe}^{\circ} - \mu_C^{\circ,graphite}$	$67{,}208 - 7.64T$
$\mu_{MnC}^{\circ} - \mu_{Mn}^{\circ} - \mu_C^{\circ,graphite}$	$18708 - 7.64T$

2.19.3 GENERALISED REGULAR SOLUTION MODEL

Sundman and Ågren [135] enhanced the sublattice-regular solution concept of Hillert and Staffansson to allow for an arbitrary number of components and sublattices. The familiar site fraction for a particular sublattice thus becomes:

$$y_{i,s} = N_{i,s} \bigg/ \left(N_{\square,s} + \sum_i N_{i,s} \right)$$

where $y_{i,s}$ is the site fraction of component i on sublattice s. The site fraction of vacant sites is therefore,

$$y_{\square,s} = 1 - \sum_i y_{i,s}.$$

A parameter a_s is defined to identify the number of sites on the sublattice s per mole of formula unit of the phase (this is equivalent to the ratio c/a in the Hillert and Staffanson model). The molar entropy of mixing becomes:

$$-\Delta S_M/R = \sum_s a_s \sum_i y_{i,s} \ln y_{1,s}.$$

As pointed out by Sundman and Ågren, it is the use of site fractions instead of mole fractions that leads to this simple and recognisable form of the entropy equation. The mole fractions are related to the site fractions by

$$x_i = \frac{\sum_s a_s y_{i,s}}{\sum_s a_s (1 - y_{\square,s})}.$$

The site fractions are now be arranged on a matrix with as many rows and columns as sublattices (N_s) and components (N_c) respectively:

$$\mathbf{Y} = \begin{pmatrix} y_{1,1} & y_{2,1} & y_{3,1} & \cdots & y_{N_c,1} \\ y_{1,2} & y_{2,2} & y_{3,2} & \cdots & y_{N_c,2} \\ \vdots & & & & \vdots \\ \vdots & & & & \vdots \\ y_{1,N_s} & y_{2,N_s} & y_{3,N_s} & \cdots & y_{N_c,N_s} \end{pmatrix}.$$

Quantities, such as $\mu^\circ_{B_aD_c}$, the Gibbs free energy of a mole of pure compound B_aD_c, were used to define the reference state in the Hillert-Staffansson model for stoichiometric phases. Retaining the old notation, the reference state for a four component, two-sublattice model was deduced to be

$$G^\circ = y_A y_C \mu^\circ_{A_aC_a} + y_A y_D \mu^\circ_{A_aD_c} + y_B y_C \mu^\circ_{B_aC_c} + y_B y_D \mu^\circ_{B_aD_c}.$$

With two sublattices there are only binary compounds. With four components and two sublattices, it is possible to generate 16 compounds but only four are used here since A and B are restricted to enter just one sublattice and C and D the other. The sixteen possibilities are,

$$\begin{array}{llll} y_A y_A A_a A_c & y_B y_A B_a A_c & y_C y_A C_a A_c & y_D y_A D_a A_c \\ y_A y_B A_a B_c & y_B y_B B_a B_c & y_C y_B C_a B_c & y_D y_B D_a B_c \\ y_A y_C A_a C_c & y_B y_C B_a C_c & y_C y_C C_a C_c & y_D y_C D_a C_c \\ y_A y_D A_a D_c & y_B y_D B_a D_c & y_C y_D C_a D_c & y_D y_A D_a D_c. \end{array}$$

All of these can be deduced from $\mathbf{Y}$. Each term can be identified by the components and the location of those components with respect to the sublattices. For example, when considering the compound B_aD_c the *component array* $\mathbf{I_0} = (B \leftrightarrow a, D \leftrightarrow c)$ indicates the complete information both about the compound and about its premultiplier $y_B y_D$. The corresponding $\mu^\circ_{A_a A_c}$ can be written in brief as $\mu^\circ_{I_0}$. It follows that

$$G^\circ = \sum_{I_0} \Pi_{I_0}\{Y\}\mu^\circ_{I_0}$$

where $\Pi\{Y\}$ represents the product of the site fractions from the Y matrix.

The enthalpy of mixing, which forms the excess free energy in the regular solution model for a multicomponent single-lattice phase, is given by

$$\Delta_e G = \frac{1}{2}\sum_i \sum_j x_i x_j K_{ij}$$

where K_{ij} represents the change in pairwise binding energies on mixing. In the Hillert-Staffanson two-sublattice model [131], it is necessary when considering interactions within one sublattice, to take account of the species present in the other sublattice, so

$$\Delta_e G = \frac{1}{2}\sum_s \sum_i \sum_j \sum_k y_{i,s} y_{j,s} y_{k,t} L^{sst}_{ijk}$$

where s denotes one sublattice and t the other. The location of each component is identified by comparing the corresponding terms in the subscript and superscript for L. For brevity, the scripts of L may be represented by another component array $I_1 = (i \leftrightarrow s, j \leftrightarrow s, k \leftrightarrow t)$ which this time allows more than one component in a given sublattice. The excess molar Gibbs free energy is then given by

$$\Delta_e G = \sum_{I_1} \Pi_{I_1}\{Y\}L_{I_1}.$$

Higher order interactions can be represented in an equivalent way by introducing higher order component arrays, but with the restriction that the array must not contain any component twice in the same sublattice.

2.19.4 MAGNETIC EFFECTS

Inden's theory [32, 33, 128] for the heat capacity due to magnetic effects was described briefly in Section 2.6.5:

$$C_P^{\mu,f} = b_1 \ln\{2s_a + 1\}R\ln\left\{\frac{1+\tau^3}{1-\tau^3}\right\} \qquad \text{for} \qquad \tau < 1$$

$$C_P^{\mu,p} = b_2 \ln\{2s_a + 1\}R\ln\left\{\frac{\tau^5+1}{\tau^5-1}\right\} \qquad \text{for} \qquad \tau > 1$$

where the additional superscripts "p" and "f" are intended to emphasise the paramagnetic and ferromagnetic states. $\tau = T/T_C$ and b_1 and b_2 are constants dependent on the crystal structure. These equations in principle completely define the thermodynamic properties since they can be integrated to reveal the magnetic contributions to enthalpy and entropy [128]. For $\tau < 1$,

$$\frac{H^\mu\{T\} - H^\mu\{0\}}{b_1 \ln\{2s_a + 1\}RT_C} = (1-\tau)\ln\{1-\tau\} + \tau\ln\left\{\frac{1+\tau^3}{1+\tau+\tau^2}\right\}$$

$$+ \ln\left\{\frac{1+\tau}{\sqrt{\tau^4+\tau^2+1}}\right\} + \sqrt{3}\,\text{arctg}\left\{\frac{2\tau-1}{\sqrt{3}}\right\}$$

$$- \sqrt{3}\,\text{arctg}\left\{\frac{2\tau+1}{\sqrt{3}}\right\} + 2\sqrt{3}\,\text{arctg}\left\{\frac{1}{\sqrt{3}}\right\}$$

The corresponding enthalpy for $\tau > 1$ is

$$
\begin{aligned}
\frac{H^\mu\{T\} - H^\mu\{T_C\}}{b_2 \ln\{2s_a + 1\}RT_C} &= (\tau+1)\ln\{\tau+1\} - (\tau-1)\ln\{\tau-1\} + \tau\ln\left\{\frac{1-\tau+\tau^2-\tau^3+\tau^4}{1+\tau+\tau^2+\tau^3+\tau^4}\right\} \\
&\quad - \cos\left\{\frac{\pi}{5}\right\}\ln\left\{\tau^4 - 2\tau^2\cos\left\{\frac{\pi}{5}\right\} + 1\right\} \\
&\quad - \cos\left\{\frac{\pi}{5}\right\}\ln\left\{\tau^4 + 2\tau^2\cos\left\{\frac{\pi}{5}\right\} + 1\right\} \\
&\quad + 2\sin\left\{\frac{\pi}{5}\right\}\left[\operatorname{arctg}\left\{\frac{\tau-\cos\{\pi/5\}}{\sin\{\pi/5\}}\right\} - \operatorname{arctg}\left\{\frac{\tau+\cos\{\pi/5\}}{\sin\{\pi/5\}}\right\}\right] \\
&\quad + 2\sin\left\{\frac{3\pi}{5}\right\}\left[\operatorname{arctg}\left\{\frac{\tau-\cos\{3\pi/5\}}{\sin\{3\pi/5\}}\right\} - \operatorname{arctg}\left\{\frac{\tau+\cos\{3\pi/5\}}{\sin\{3\pi/5\}}\right\}\right] \\
&\quad + \ln\left\{\frac{5}{4}\right\} + \cos\left\{\frac{\pi}{5}\right\}\ln\left\{4\sin^2\{\pi/5\}\right\} + \left\{\frac{3\pi}{5}\right\}\sin\left\{\frac{\pi}{5}\right\} \\
&\quad + \cos\left\{\frac{3\pi}{5}\right\}\ln\left\{4\sin^2\{3\pi/5\}\right\} - \left\{\frac{\pi}{5}\right\}\sin\left\{\frac{3\pi}{5}\right\}.
\end{aligned}
$$

The corresponding entropy for $\tau < 1$ is

$$
\frac{S^\mu\{T\} - S^\mu\{0\}}{b_1 \ln\{2s_a + 1\}R} = \frac{1}{3}\Phi\{\tau^3\}
$$

and for $\tau > 1$:

$$
\frac{S^\mu\{T\} - S^\mu\{T_C\}}{b_2 \ln\{2s_a + 1\}R} = \frac{1}{5}\left[\Phi\{1\} - \Phi\{\tau^{-5}\}\right]
$$

where the function $\quad \Phi\{y\} = 2\left(y + \frac{y^3}{3^2} + \frac{y^5}{5^2} + \cdots\right).$

The total entropies for long- and short-range order obtained using these equations are, respectively,

$$
S^\mu\{T_C\} - S^\mu\{0\} = \frac{b_1}{12}\ln\{2s_a + 1\}R\pi^2 \tag{2.81a}
$$

$$
S^\mu\{\infty\} - S^\mu\{T_C\} = \frac{b_2}{20}\ln\{2s_a + 1\}R\pi^2. \tag{2.81b}
$$

It is found in practice that Equation 2.81 can be applied as an approximation even when the spin is not integral [18].

The parameters b_1 and b_2

The Curie temperature, the spin per atom and the parameters b_1 and b_2 define the magnetic component of the specific heat capacity and hence all the necessary thermodynamic functions. We present here the procedure used by Inden to estimate b_1 and b_2. For a pure component with an integral number $2s_a$ of Bohr magnetons, the total magnetic entropy must, according to Equation 2.19, equal

$$
S^\mu\{\infty\} - S^\mu\{0\} = R\ln\{2s_a + 1\}.
$$

Using Equation 2.81, it follows that

$$\frac{b_1}{12}\ln\{2s_a+1\}R\pi^2 + \frac{b_2}{20}\ln\{2s_a+1\}R\pi^2 = R\ln\{2s_a+1\} \qquad \therefore \pi^2\left(\frac{b_1}{12}+\frac{b_2}{20}\right)=1$$

Having established a relationship between b_1 and b_2, the actual values of these constants can be obtained by considering the magnetic enthalpies. If the fraction of the total magnetic enthalpy that is absorbed above the Curie temperature is given by f_μ, then

$$f_\mu = \frac{H^\mu\{\infty\} - H^\mu\{T_C\}}{(H^\mu\{\infty\} - H^\mu\{T_C\})(H^\mu\{T_C\} - H^\mu\{0\})}.$$

On substituting for the various enthalpies, and using equation 2.81, Inden obtains

$$b_1 = 1.217 - \frac{0.470 f_\mu}{0.598 - 0.211 f_\mu}$$

$$b_2 = \frac{0.784 f_\mu}{0.598 - 0.211 f_\mu}.$$

The value of f_μ is reliably found to be about 0.4 for bcc metals and 0.28 for fcc metals.

The Gibbs free energy

The magnetic Gibbs free energy G^μ is defined with respect to the completely disordered state at $T = \infty$; this is consistent with the fact that any magnetic ordering produces a stabilisation effect. For temperatures above the Curie point,

$$G^\mu = \int_\infty^T C_P^{\mu,p}dT' - T\int_\infty^T \frac{C_P^{\mu,p}}{T'}dT'$$

and for $\tau < 1$

$$G^\mu = \int_\infty^{T_C} C_P^{\mu,p}dT' - T\int_\infty^{T_C} \frac{C_P^{\mu,p}}{T'}dT' + \int_{T_C}^T C_P^{\mu,f}dT' - T\int_{T_C}^T \frac{C_P^{\mu,f}}{T}dT'.$$

Figure 2.43 illustrates the accuracy with which Inden was able to represent the magnetic heat capacity of bcc iron. The curves are calculated using the following data:

$$Fe: T_C = 1043\,K \quad f_\mu = 0.40 \quad S^\mu\{\infty\} - S^\mu\{0\} = R\ln\{2.2+1\}.$$

Figure 2.43 shows also the significant influence of short-range order beyond T_C in iron.

Alloys

With the assumption that $f_\mu = 0.4$ for bcc metals and $f_\mu = 0.28$ for fcc metals, Inden's theory can be used to estimate the magnetic entropy, enthalpy and Gibbs free energy changes not only for pure metals but also for alloys if the spin imbalance per atom (s_a) and Curie temperature T_C are known.

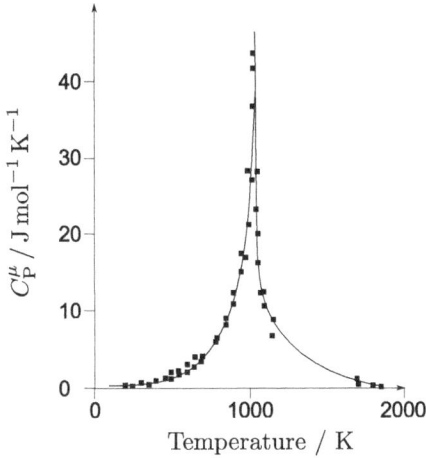

Figure 2.43 Magnetic specific heat capacity of iron. The curve is calculated whereas the points are derived by subtracting the nonmagnetic components from experimental data (after Inden [128]).

Assuming that the atoms in an alloy contribute independently, the total magnetic entropy is given by

$$S^\mu\{\infty\} - S^\mu\{0\} = R\sum_i x_i \ln\{2s_{a,i} + 1\}$$

where $s_{a,i}$ is the spin for species i with mole fraction x_i. If the individual moments are not known, then this expression can be approximated by

$$S^\mu\{\infty\} - S^\mu\{0\} \simeq R\ln\{2\bar{s}_a + 1\}$$

where $2\bar{s}_a$ is the average magnetic moment per atom. This depends on the concentration of the components:

$$\bar{s}_a = x_{Fe}s_{a,Fe} + x_X s_{a,X}$$

where X is a substitutional solute. It has been suggested that for the addition of a nonmagnetic solute to iron, $s_{a,X}$ may be set to zero if the appropriate data are not available, whilst keeping $s_{a,Fe}$ constant even for alloys [33, 136].

Figure 2.44 shows a calculated heat capacity curve for $Fe_{0.79}Cr_{0.21}$ solid solution [128], using the following data:

$$Fe_{0.79}Cr_{0.21} : T_C = 923\,K \quad f_\mu = 0.40 \quad S^\mu\{\infty\} - S^\mu\{0\} = R\ln\{1.61 + 1\}.$$

The mean value of the magnetic moment was obtained from experimental measurements of the variation in magnetisation with the chromium concentration [137]:

$$2\bar{s}_a = 2.11 - 2.36x_{Cr}.$$

The nonmagnetic component of the specific heat for $Fe_{0.79}Cr_{0.21}$ was derived simply as a weighted mean of the corresponding values for iron and chromium. The agreement between calculation and experiment is seen to be rather good.

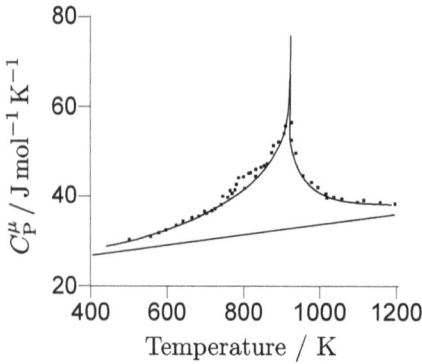

Figure 2.44 Specific heat capacity of $Fe_{0.79}Cr_{0.21}$ [128]. The straight line represents the nonmagnetic component, and the curve the sum of the nonmagnetic and calculated magnetic components of the specific heat capacity. The points are experimental data.

Simplification

Inden's integrated expressions are complicated for implementation into a general scheme for phase diagram calculations. Hillert and Jarl [33] therefore suggested that in the heat capacity expressions, the terms in τ should be expanded as a series which is truncated at an appropriate level of usefulness, prior to the integrations:

$$C_P^{\mu,f} = 2b_1 R\ln\{2s_a+1\}\left(\tau^3+\frac{\tau^9}{3}+\frac{\tau^{15}}{5}\right) \qquad \text{for} \qquad \tau < 1$$

$$C_P^{\mu,p} = 2b_2 R\ln\{2s_a+1\}\left(\tau^{-5}+\frac{\tau^{-15}}{3}+\frac{\tau^{-25}}{5}\right) \qquad \text{for} \qquad \tau > 1.$$

As expected theoretically, Inden's equations give a heat capacity which tends to infinity at $T = T_C$ whereas the series expansions tend to the finite limits $(46/15)R\ln\{2s_a + 1\}b_1$ and $(46/15)R\ln\{2s_a + 1\}b_2$ which is convenient from the point of view of computer implementation and yet does not lead to significant errors in the enthalpy or entropy terms, which are obtained by integration (Equations 2.4 and 2.6) as

$$S^\mu\{T \geq T_C\} - S^\mu\{T = T_C\} = \int_{T_C}^T \frac{C_P^{\mu,p}}{T'}dT' \tag{2.82}$$

$$= -2Rb_2\ln\{2s_a+1\}\left[\left(\frac{\tau^{-5}}{5}+\frac{\tau^{-15}}{45}+\frac{\tau^{-25}}{125}\right)-0.6\frac{518}{675}\right]$$

$$S^\mu\{T \leq T_C\} - S^\mu\{T = 0\} = \int_0^T \frac{C_P^{\mu,f}}{T'}dT'$$

$$= Rb_1\ln\{2s_a+1\}\left(\frac{\tau^3}{3}+\frac{\tau^9}{27}+\frac{\tau^{15}}{75}\right)$$

giving

$$S^\mu\{T = \infty\} - S^\mu\{T = 0\} = \frac{518}{675}R\ln\{2s_a+1\}(b_1+0.6b_2).$$

The magnetic entropy is also given by Equation 2.19 as $N_a k\ln\{2s_a+1\}$ so

$$b_1 + 0.6b_2 = \frac{675}{518}. \tag{2.83}$$

One further condition is needed to solve for the parameters b_1 and b_2, obtained by considering the enthalpy terms:

$$H^\mu\{T \geq T_C\} - H^\mu\{T = T_C\} = \int_{T_C}^{T} C_P^{\mu,p} dT'$$

$$= RT_C b_2 \ln\{2s_a + 1\}\left[\frac{79}{140} - \left(\frac{\tau^{-4}}{2} + \frac{\tau^{-14}}{21} + \frac{\tau^{-24}}{60}\right)\right]$$

$$H^\mu\{T \leq T_C\} - H^\mu\{T = 0\} = \int_{0}^{T} C_P^{\mu,f} dT'$$

$$= RT_C b_1 \ln\{2s_a + 1\}\left(\frac{\tau^4}{2} + \frac{\tau^{10}}{15} + \frac{\tau^{16}}{40}\right)$$

so that

$$H^\mu\{T = \infty\} - H^\mu\{T = T_C\} = \frac{79}{140} RT_C b_2 \ln\{2s_a + 1\}$$

$$H^\mu\{T = T_C\} - H^\mu\{T = 0\} = \frac{71}{120} RT_C b_1 \ln\{2s_a + 1\}.$$

The fraction of enthalpy retained beyond the Curie temperature, f_μ, is therefore

$$f_\mu = \left(\frac{79}{140} b_2\right) \bigg/ \left(\frac{79}{140} b_2 + \frac{71}{120} b_1\right).$$

On solving this simultaneously with Equation 2.83, it is seen that

$$b_1 = b_2\left[\frac{474}{497}\left(\frac{1 - f_\mu}{f_\mu}\right)\right] \quad \text{with} \quad b_2 = \left[\frac{518}{1125} + \frac{11692}{15975}\left(\frac{1 - f_\mu}{f_\mu}\right)\right]^{-1}.$$

Taking the magnetic contribution to the molar Gibbs free energy to be zero at infinite temperature, for $\tau > 1$, it follows that

$$G^\mu = \int_{\infty}^{T} C_P^{\mu,p} dT' - T\int_{\infty}^{T} \frac{C_P^{\mu,p}}{T'} dT'$$

$$= -b_2 \ln\{2s_a + 1\} RT_C\left(\frac{\tau^{-4}}{10} + \frac{\tau^{-14}}{315} + \frac{\tau^{-24}}{1500}\right)$$

and for $\tau < 1$

$$G^\mu = \int_{\infty}^{T_C} C_P^{\mu,p} dT' - T\int_{\infty}^{T_C} \frac{C_P^{\mu,p}}{T'} dT' + \int_{T_C}^{T} C_P^{\mu,f} dT' - T\int_{T_C}^{T} \frac{C_P^{\mu,f}}{T'} dT'$$

$$= -b_2 \ln\{2s_a + 1\} RT_C\left(\frac{79}{140} - \frac{518\tau}{1125}\right)$$

$$- b_1 RT_C\left(\frac{\tau^4}{6} + \frac{\tau^{10}}{135} + \frac{\tau^{16}}{600} + \frac{71}{120} - \frac{518\tau}{675}\right).$$

2.19.5 MAGNETISATION

It is a good approximation when the temperature is close to the Curie temperature, to take the spontaneous magnetisation to be given by the relation

$$\frac{M_{\mathrm{m}}\{T\}}{M_{\mathrm{m}}\{0\}} = \left(1 - \frac{T}{T_{\mathrm{C}}}\right)^{b_{18}}$$

where the term on the left is the reduced magnetisation, i.e., the magnetisation normalised relative to that at $0\,\mathrm{K}$. For pure ferritic iron it is found experimentally that:

$$b_{18} = 0.33 \qquad \text{for} \qquad 10^{-2} > \left(1 - \frac{T}{T_{\mathrm{C}}}\right) > 10^{-1}$$

$$b_{18} = 0.27 \qquad \text{for} \qquad 10^{-1} > \left(1 - \frac{T}{T_{\mathrm{C}}}\right) > 2 \times 10^{-1}.$$

Inden and Meyer [138] found that the values of b_{18} do not seem to be sensitive to the addition of cobalt to iron, and hence were able to use the relationship to accurately fix the Curie temperatures from magnetisation measurements at low temperature, for Fe-Co alloys where T_{C} is not accessible because the sample undergoes a phase change before becoming paramagnetic.

Summary

The Inden formalism is useful – it permits the estimation of the magnetic component of heat capacity from a knowledge of just the magnetic moment per atom (s_{a}), the Curie temperature (T_{C}) and assumed values of the partitioning of enthalpy between long and short-range order (f_{μ}). There is evidence that the method is applicable equally to chemical ordering.

2.20 ORDER PARAMETER

The phenomenon of ordering is discussed first, in the context of thermodynamics, before introducing the crystallographic aspects of typical ordering reactions in iron alloys.

The classical Bragg-Williams model contains many of the essential features for ordering. The derivation presented here is due to Christian [101] and is for an equiatomic binary alloy of components A and B when ordering changes a cubic-I lattice into one which is primitive cubic. The most popular example of this kind of an ordering reaction is CuZn, but FeAl and FeCo are relevant examples in the context of iron. The iron aluminides in particular may become important engineering materials in their own right.

The extent of order is described in terms of a long-range order parameter L which is unity for the fully ordered crystal and zero when the distribution of atoms is random in the context of a long range:

$$L = \frac{r_{\mathrm{A}} - x_{\mathrm{A}}}{1 - x_{\mathrm{A}}} = \frac{r_{\mathrm{B}} - x_{\mathrm{B}}}{1 - x_{\mathrm{B}}}$$

where r_A is the probability of an A site being rightly occupied by an A atom, equal in our example to the probability of a B site being occupied by a B atom, so that with $x_A = x_B = 0.5$,

$$L = 2r - 1.$$

When $r = 1$, the alloy is fully ordered with $L = 1$ and when $r = \frac{1}{2}$, $L = 0$ with the alloy being fully disordered.

The probability of A-A pairs is the chance of finding an A site that is occupied correctly (r) multiplied by that of finding a B site which is incorrectly occupied ($1-r$). A similar rationale for the B-B and A-B pairs gives:

$$N_{AA} = \frac{N}{2}zr(1-r)$$

$$N_{BB} = \frac{N}{2}zr(1-r)$$

$$N_{AB} = \frac{N}{2}z[r^2 + (1-r)^2]$$

since only $\frac{N}{2}$ sites can be occupied correctly by either species for the cubic-I to cubic-P ordering transition. z is a coordination number and N is the number of atoms. In deducing the number of unlike bonds, r^2 is the probability of finding two adjacent sites that are occupied correctly and the additional term $(1-r)^2$ accounts for the probability of finding two adjacent sites which are incorrectly occupied. Both of these circumstances lead to A-B bonds.

The configurational internal energy U is then given by the sum of all bond strengths,

$$\begin{aligned} U &= -Nzr(1-r)(\varepsilon_{AA} + \varepsilon_{BB}) - Nz[r^2 + (1-r)^2]\varepsilon_{AB} \\ &= -Nz\varepsilon_{AB} - Nzr(1-r)\omega \\ &= -Nz[\varepsilon_{AB} + r(1-r)\omega]. \end{aligned}$$

(2.84)

The factor of two drops out because of the way in which the binding energy is defined (Figure 2.32). The configurational Gibbs free energy is given by assuming that the enthalpy of ordering is equal to the internal energy, and adding the contribution from the change in configurational entropy. The ordered lattice can be imagined to consist of a series of sublattices to enable the entropy of any distribution of atoms on these sublattices to be estimated; for a binary alloy without vacancies, this is given by[22]

$$S = -\frac{R}{n}\sum_{i=1}^{n}(x_A^{(i)}\ln\{x_A^{(i)}\} + x_B^{(i)}\ln\{x_B^{(i)}\})$$

(2.85)

where n is the number of sublattices and $x_A^{(i)}$ the fraction of A atoms on the sublattice i. For the present example, $N = 2$ so:

$$G \approx F = U + NkT[r\ln r + (1-r)\ln(1-r)].$$

(2.86)

When $r = 1$, the configurational entropy is zero. To find the equilibrium value of r, dG/dr is set to 0, to obtain

$$\frac{1}{2r-1}\ln\frac{r}{1-r} = z\omega/kT \tag{2.87}$$

$$(2/L)\tanh^{-1}L = z\omega/kT. \tag{2.88}$$

The variation of the long range order parameter as a function of the reduced temperature T/T_C is illustrated in Figure 2.45. T_C is the temperature where $L = 0$. The increase in disorder is at first small as the temperature rises towards T_C, but this resistance to disordering decreases once the process gets hold.

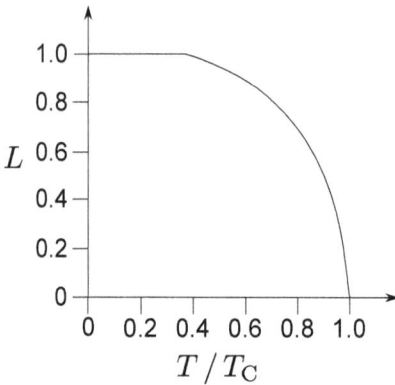

Figure 2.45 The equilibrium value of the long-range order parameter as a function of the reduced temperature, using the Bragg-Williams model.

2.20.1 SHORT-RANGE ORDER

This Bragg-Williams model for the order-disorder transition is analogous to the Weiss model for ferromagnetism by spin orientation ordering. They both assume that the atoms (or moments) are distributed at random on each sublattice of the superlattice structure and hence predict completely random mixing at all temperatures above T_C. Short-range order cannot therefore be treated with these models.[23] As with magnetism, the probability of finding A-B pairs remains larger than random when the long-range order parameter becomes zero. An excess heat capacity (i.e., relative to an ideal solution model) persists beyond T_C; important implications of this phenomena are discussed in Chapter 3 which deals with diffusion.

It is well established that models based on near neighbour interactions alone cannot explain the stability of many of the superlattices found in practice. Second, sometimes third near neighbour interactions have to be taken into account. This will become obvious when the detailed crystal structures are discussed in the next section. In some cases it becomes necessary to consider clusters of atoms rather than just pairwise interactions in order to predict the correct equilibrium state. The historical development of the theory can be described in terms of quasichemical approximations, where the Bragg-Williams model is the zeroth, and the Beth model the first order quasichemical approximation. Cluster variation methods are higher approximations of the quasichemical method. These are not described here partly because they have been discussed thoroughly by Christian [101] but also because in the context of iron, there are few data against which they can be tested. One difficulty is that many important ordering reactions occur at temperatures where it takes a long time to reach equilibrium.

A pragmatic approach is to apply Inden's heat capacity equations (Section 2.19.4) to the problem of chemical ordering. As with magnetism, all that is required is a knowledge of the critical temperature and the factor f_o which gives the ratio of the enthalpy due to short-range order to the total amount of enthalpy due to ordering. Inden [128] has suggested that in the absence of good experimental data, f_o can be taken to be identical to f_μ, i.e., 0.4 for bcc and 0.28 for fcc systems.

2.21 SUPERLATTICES

2.21.1 ORDERED CRYSTALS

Ordering leads to an increase in the volume of the primitive cell, hence the term *superlattice* is used to describe crystal structures in which the different atomic species are ordered. Some of the technologically important superlattices of iron-based intermetallic compounds are listed in Table 2.11. All of these structures lead to an increase in the number of unlike neighbours when compared with the case where the atoms are disordered, randomly dispersed.

FeAl and FeCo intermetallic compounds have a cubic-I structure in the disordered state and cubic-P when ordered. The tendency for ordering (i.e., enthalpy change on ordering) is much larger for FeAl than FeCo so that the former is often described as a *strongly ordered* compound, a description that is not a measure of the order parameter.

Ni_3Fe is a classic ordered compound which in the disordered state is cubic-F that on ordering becomes cubic-P with Ni atoms at the face centres and Fe atoms at the cube corners. Fe_3Al and Fe_3Si have a cubic-I crystal structure when disordered.

Table 2.11
Some of the ordered compounds that occur in iron alloys.

| Compound | Crystal Structure | | T_m / K | T_C / K | Density / $kg\,m^{-3}$ |
	Ordered	Disordered			
FeAl	B2, Cubic-P	Cubic-I	1523-1673	1523-1673	5560
Ni_3Fe	$L1_2$, Cubic-P	Cubic-F			
$(Fe_{22}Co_{78})_3Fe$	$L1_2$, Cubic-P	Cubic -F	1673	1223	7800
$(Fe_{60}Ni_{40})_3(V_{96}Ti_4)$	$L1_2$, Cubic-P	Cubic -F	1673	953	7600
Fe_3Al	DO_3, Cubic-F	Cubic-I	1813	813	6750
Fe_3Si	DO_3, Cubic-F	Cubic-I	1543	1543	7250

These and other ordered crystals can, for the purposes of thermodynamic analysis (Section 2.19.1), be represented conveniently by subdividing the bcc and fcc arrangements of atoms into four and eight sublattices as illustrated in Figure 2.46.

Table 2.12

Location of atoms in the sublattices of the perfectly ordered phases. The sites 1, 2 ... 8 are identified in Figure 2.46. The first two compounds are based on the four sublattices of the bcc arrangement, and the last on the eight sublattices of the fcc arrangement.

Compound	I	II	III	IV	V	VI	VII	VIII
AB, B2	A	A	B	B	-	-	-	-
A_3B, DO_3	B	A	A	A	-	-	-	-
A_3B, $L1_2$	B	B	A	A	A	A	A	A

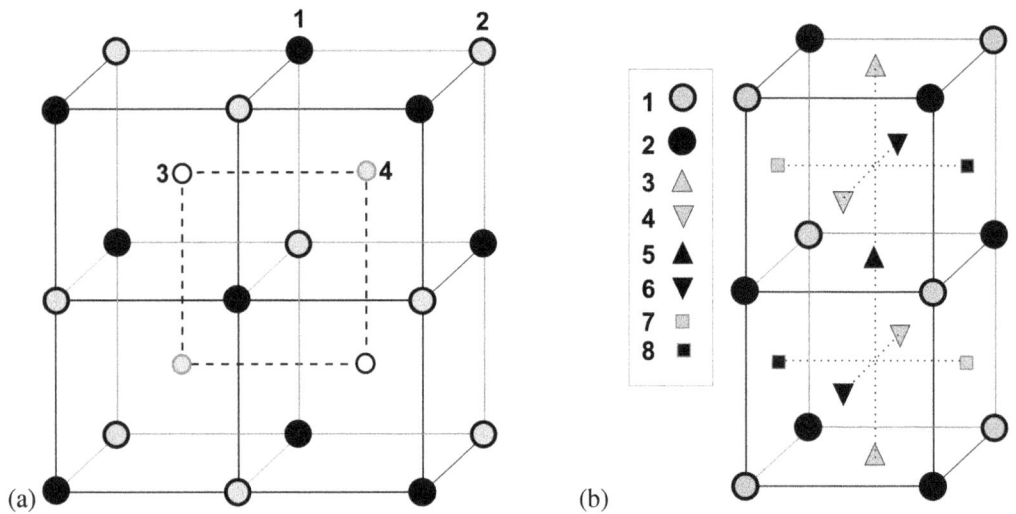

Figure 2.46 Four and eight sublattices of the bcc and fcc arrangements respectively, as an aid to illustrating the ordering of atoms (adapted from [128]).

2.22 THERMODYNAMICS OF IRREVERSIBLE PROCESSES

Thermodynamics as a subject is limited to the equilibrium state. Properties such as entropy and free energy are, on an appropriate scale, static and time-invariant during equilibrium. There is an extension of the subject to systems that are close to equilibrium so that they can be divided into subsystems where the rules of equilibrium can be applied locally [139]. Parameters not relevant to the discussion of equilibrium, such as thermal conductivity, diffusivity and viscosity, then enter the picture because they can describe a second kind of time independence, that of the steady state. For example, the concentration profile does not change during steady-state diffusion, even though energy is being dissipated during diffusion.

The thermodynamics of irreversible processes deals with systems that are not at equilibrium but are nevertheless *stationary*. The theory in effect uses thermodynamics to deal with *kinetic* phenomena. There is, nevertheless, a distinction between the thermodynamics of irreversible processes and kinetics. The former applies strictly to the steady state, whereas there is no such restriction on kinetic theory.

2.22.1 REVERSIBILITY

A process, the direction of which can be changed by an infinitesimal alteration in the external conditions is called reversible, because an exact reversal leads to no net dissipation of energy. Figure 2.47 shows the response of an ideal gas contained at uniform pressure within a cylinder, any change being achieved by the motion of the piston. For any starting point on the pressure-volume curve, the application of an infinitesimal force may cause the piston to move to an adjacent position still on the curve, while the removal of the infinitesimal force restores the system to its original state. This process is reversible because there is no net dissipation in displacing and recovering the frictionless piston.

If the motion of the piston in the cylinder entails friction, then deviations occur from the P/V curve as illustrated by the cycle in Figure 2.47. An infinitesimal force cannot move the piston because energy must be dissipated to overcome the friction; this energy is the area enclosed by the cycle on the P/V plot. A process such as this, which involves the dissipation of energy, is classified as irreversible with respect to an infinitesimal change in the external conditions. More generally,

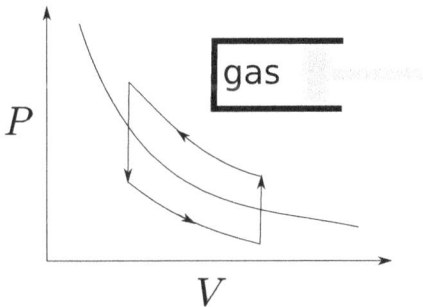

Figure 2.47 The curve represents the variation in pressure within the cylinder as the volume of the ideal gas is altered by the frictionless positioning the piston. The cycle represents the dissipation of energy when the motion of the piston causes friction.

reversibility means that it is possible to pass from one state to another without appreciable deviation from equilibrium. Real processs are not reversible so equilibrium thermodynamics can only be used approximately, though the same principles define whether or not a process can occur spontaneously without ambiguity.

For irreversible processes the *equations* of classical thermodynamics become *inequalities*. For example, at the equilibrium melting temperature, the free energies of the liquid and solid are identical ($G_{\text{liquid}} = G_{\text{solid}}$) but not so below that temperature ($G_{\text{liquid}} > G_{\text{solid}}$). Such inequalities are much more difficult to deal with though they indicate the natural direction of change. For steady-state processes however, the thermodynamic framework for irreversible processes as developed by Onsager [140] is particularly useful in obtaining relationships even though the system may not be at equilibrium.

2.22.2 LINEAR LAWS

There is no change in entropy or free energy at equilibrium. An irreversible process dissipates energy and entropy is created continuously. In the example illustrated in Figure 2.47, the dissipation was due to friction; diffusion ahead of a moving interface is dissipative. The rate at which energy is · dissipated is the product of the temperature and the rate of entropy production:

$$T\dot{S} = JX \tag{2.89}$$

where J is a generalised flux of some kind, and X a generalised force. In the case of an electrical current, the heat dissipation is the product of the current (J) and the electromotive force (X).

As long as the flux-force sets can be expressed as in Equation 2.89, the flux must naturally depend in some way on the force. It may then be written as a function $J\{X\}$ of the force X. At equilibrium, the force is zero. $J\{X\}$ can be expanded in a Taylor series about equilibrium ($X = 0$):

$$\begin{aligned}
J\{X\} &= \sum_0^\infty a_n X^n \\
&= J\{0\} + J'\{0\}\frac{X}{1!} + J''\{0\}\frac{X^2}{2!} \cdots
\end{aligned} \tag{2.90}$$

In this expansion, $J\{0\} = 0$ because there is no flux in the absence of force. If the high-order terms are neglected, then a proportionality between the force and flux is revealed:

$$J \propto X.$$

Therefore, the forces and their conjugate fluxes are linearly related whenever the dissipation can be expressed as in Equation 2.90, at least when the deviations from equilibrium are not large. This caveat is illustrated nicely by the relationship between the rate at which an interface moves and the driving force. In Chapter 4, Equation 4.64, which is limited to small driving forces, shows a linear relationship between the two quantities, whereas Equation 4.63 which is derived without limits on the magnitude of the driving force shows that the rate and driving force are *not* in general linearly related.

In another example, consider a closed system in which a quantity dH of heat is transferred in a time interval dt across an area A in a direction z normal to that area, from a region at temperature T_h to a lower temperature T_ℓ. The receiving part increases its entropy by dH/T_ℓ whereas the depleted region experiences a reduction dH/T_h, so the change in entropy is

$$dS = dH\left(\frac{1}{T_\ell} - \frac{1}{T_h}\right).$$

The rate of entropy production per unit volume is therefore

$$\dot{S} = \frac{1}{V}\frac{dS}{dt} = \frac{1}{V}\frac{dH}{dt}\left(\frac{1}{T_\ell} - \frac{1}{T_h}\right). \tag{2.91}$$

The flux of heat J is defined as $A^{-1}\mathrm{d}H/\mathrm{d}t$ so Equation 2.91 becomes

$$\dot{S} = J\frac{A}{V}\left(\frac{1}{T_\ell} - \frac{1}{T_\mathrm{h}}\right) \equiv J\left(-\frac{1}{T^2}\right)\frac{\mathrm{d}T}{\mathrm{d}z}$$

$$\text{or} \quad T\dot{S} = \underbrace{J}_{\text{flux}} \underbrace{\left(-\frac{1}{T}\right)\frac{\mathrm{d}T}{\mathrm{d}z}}_{\text{force}}$$

Some examples of forces and fluxes in their generic form are listed in Table 2.13.

Table 2.13

Examples of forces and their conjugate fluxes. z **is the distance over which the gradient exists,** ϕ **is the electrical potential and** μ **the chemical potential.**

Force	Flux
$-\frac{\partial \phi}{\partial z}$	Electrical current
$-\frac{1}{T}\frac{\partial T}{\partial z}$	Heat flux
$-\frac{\partial \mu_i}{\partial z}$	Diffusion flux
Stress	Strain rate

2.22.3 MULTIPLE IRREVERSIBLE PROCESSES

There are circumstances whereby a number of irreversible processes occur together. In a ternary Fe-Mn-C alloy, the diffusion flux of carbon depends not only on the gradient of carbon, but also on that of manganese. A uniform distribution of carbon will tend to become inhomogeneous in the presence of a manganese concentration gradient. Similarly, the flux of heat may not depend on the temperature gradient alone; heat can be driven also by an electromotive force (Peltier effect).[24] Electromigration involves diffusion that is driven by an electromotive force. When there is more than one dissipative process, the total energy dissipation rate can still be written

$$T\dot{S} = \sum_i J_i X_i. \tag{2.92}$$

In general, if there is more than one irreversible process occurring, it is found *experimentally* that each flow J_i is related not only to its conjugate force X_i, but is also linearly related to all other forces present. Thus,

$$J_i = M_{ij} X_j \tag{2.93}$$

with i, $j = 1,2,3\ldots$. Therefore, a given flux depends on all the forces causing the dissipation of energy.

2.22.4 ONSAGER RECIPROCAL RELATIONS

Equilibrium in real systems is dynamic on a microscopic scale. It seems obvious that to maintain equilibrium under these dynamic conditions, a process and its reverse must occur at the same rate on the microscopic scale. The consequence is that provided the forces and fluxes are chosen from the dissipation equation and are independent, $M_{ij} = M_{ji}$. This is known as the Onsager theorem, or the Onsager reciprocal relations. It applies to systems near equilibrium when the properties of interest have even parity, and assuming that the fluxes and their corresponding forces are independent. An exception occurs with magnetic fields in which case there is a sign difference $M_{ij} = -M_{ji}$ [141].

REFERENCES

1. A. B. Pippard: Classical Thermodynamics: Cambridge, U. K.: Cambridge University Press, 1981.
2. C. Zener: 'Kinetics of the decomposition of austenite', *Transactions of the American Institute of Mining and Metallurgical Engineers*, 1946, **167**, 550–595.
3. C. Zener: 'Equilibrium relations in medium alloy steels', *Trans. A.I.M.E.*, 1946, **167**, 513–534.
4. W. J. Moore: Physical Chemistry: 5th ed., New Jersey, USA: Prentice-Hall, 1972.
5. K. G. Denbigh: The Principles of Chemical Equilibrium: 4th ed., Cambridge, U. K.: Cambridge University Press, 2018.
6. P. Debye: 'Zur theorie der spezifischen wärmen', *Annalen der Physik*, 1912, **344**, 789–839.
7. L. Kaufman, S. V. Radcliffe, and M. Cohen: 'Thermodynamics of the bainite reaction', In: V. F. Zackay, and H. I. Aaronson, eds. *Decomposition of Austenite by Diffusional Processes.* New York, USA: Interscience, 1962:313–352.
8. F. A. Lindemann: 'Ueber die berechnung molekularer eigenfrequenzen (the calculation of molecular Eigen-frequencies)', *Physikalische Zeitschrift*, 1910, **11**, 609–612.
9. L. Kaufman: 'Estimation of the entropy of NaCl type compounds', *Transactions of the Metallurgical Society of AIME*, 1962, **224**, 1006–1009.
10. L. Kaufman, E. V. Clougherty, and R. J. Weiss: 'The lattice stability of metals – iii iron', *Acta Metallurgica*, 1963, **11**, 323–335.
11. L. D. Blackburn, L. Kaufman, and M. Cohen: 'Phase transformations in iron-ruthenium alloys under high pressure', *Acta Metallurgica*, 1965, **13**, 533–541.
12. F. H. Herbstein, and J. Smuts: 'Determination of the Debye temperature of alpha-iron by X-ray diffraction', *Philosophical Magazine*, 1963, **8**, 367–385.
13. G. L. Stepakoff, and L. Kaufman: 'Thermodynamic properties of hcp iron and iron-ruthenium alloys', *Acta Metallurgica*, 1968, **16**, 13–22.
14. R. L. Clendenen, and H. G. Drickamer: 'The effect of pressure on the volume and lattice parameters of ruthenium an iron', *Journal of the Physics and Chemistry of Solids*, 1964, **25**, 865–868.
15. H.-K. Mao, W. A. Bassett, and T. Takahashi: 'Effect of pressure on crystal structure and lattice parameters of iron up to 300 kbar', *Journal of Applied Physics*, 1967, **38**, 272–276.
16. A. H. Wilson: The Theory of Metals: Cambridge, U. K.: Cambridge University Press, 1958.

17. W. H. Keesom, and B. Kurrelmeyer: 'The atomic heat of iron from 1.1° to 20.4°K', *Physica*, 1939, **6**, 633–672.

18. J. A. Hofmann, A. Paskin, K. J. Tauer, and R. J. Weiss: 'Analysis of ferromagnetic and antiferro-magnetic second-order transitions', *Journal of Physics and Chemistry of Solids*, 1956, **1**, 45–60.

19. P. Wilkes: Solid State Theory in Metallurgy: Cambridge, U. K.: Cambridge University Press, 1973.

20. J. M. Ziman: Models of disorder: Cambridge, U. K.: Cambridge University Press, 1979.

21. I. M. K. Billas, A. Chatelain, and W. A. de Heer: 'Magnetism from the atom to the bulk in iron, cobalt, and nickel clusters', *Science*, 1994, **265**, 1682–1684.

22. A. J. Freeman, and R. Q. Wu: 'Electronic structure theory of surface, interface and thin film magnetism', *Journal of Magnetism and Magnetic Materials*, 1991, **100**, 497–514.

23. I. M. K. Billas, J. A. Baker, A. Chatelain, and W. A. de Heer: 'Magnetic moments of iron clusters with 25 to 700 atoms and their dependence on temperature', *Physical Review Letters*, 1993, **71**, 4067–4070.

24. L. C. nd X. L. Wan, and K. M. Wu: 'Three-dimensional analysis of ferrite allotrimorphs nucleated on grain boundary faces, edges and corners', *Materials Characterization*, 2010, **61**, 580–583.

25. L. Patrick: 'The change in ferromagnetic Curie points with hydrostatic pressure', *Physical Review*, 1954, **93**, 384–392.

26. J. A. Hoffmann, A. Paskin, K. J. Tauer, and R. J. Weiss: 'Analysis of ferromagnetic and antiferromagnetic second-order transitions', *Journal of the Physics and Chemistry of Solids*, 1956, **1**, 45–60.

27. C. Kittel: Introduction to Solid State Physics: 4th ed., New York, USA: John Wiley and Sons Inc., 1971.

28. L. Kaufman: 'Condensed-state reactions at high pressures', In: W. M. Mueller, ed. *Energetics in Metallurgical Phenomena*, vol. 3. New York, USA: Gordon and Breach Publishers, 1967:55–84.

29. D. C. Wallace, P. H. Sidles, and G. C. Danielson: 'Specific heat of high purity iron by a pulse heating method', *Journal of Applied Physics*, 1960, **31**, 168–176.

30. R. J. Weiss, and K. J. Tauer: 'Components of the thermodynamic functions of iron', *Physical Review*, 1956, **102**, 1490–1495.

31. H. A. Gersch, C. G. Shull, and M. K. Wilkinson: 'Critical magnetic scattering of neutrons by iron', *Physical Review*, 1956, **103**, 525–534.

32. G. Inden: 'Determination of chemical and magnetic interchange energies in bcc alloys', *Zeitschrift für Metallkunde*, 1975, **66**, 577–582.

33. M. Hillert, and M. Jarl: 'A model for alloying effects in ferromagnetic metals', *CALPHAD*, 1978, **2**, 227–238.

34. A. P. Miodownik: 'The calculation of magnetic contributions to phase stability', *CALPHAD*, 1977, **1**, 133–158.

35. H. Ino, K. Hayashi, T. Otsuka, D. Isobe, K. Tokumitsu, and K. Oda: 'Appearance of ferromagnetism in fcc solid solutions of binary and ternary Fe-Cu based systems prepared by mechanical alloying techniques', *Materials Science & Engineering A*, 2001, **304-306**, 972–974.

36. H. H. Ettwig, and W. Pepperhoff: 'On magnetism of gamma-Fe-Ni-Mn alloys', *Physica Status Solidi A*, 1974, **23**, 105–111.

37. J. M. D. Coey: 'Magnetic localization and magnetoresistance in mixed-valence manganites and related ferromagnetic oxides', *Philosophical Transactions of the Royal Society A*, 1998, **356**, 1519–1541.

38. T. G. Perring, G. Aeppli, S. M. Hayden, Y. Tokura, Y. Moritomo, J. P. Remeika, and S.-W. Cheong: 'Magnetic dynamics in colossal magnetoresistive perovskite manganites', *Philosophical Transactions of the Royal Society, London*, 1998, **356**, 1563–1575.

39. K. J. Tauer, and R. J. Weiss: 'Unusual magnetic structure of face centered cubic Fe', *Bulletin of the American Physical Society*, 1961, **6**, 125.

40. R. J. Weiss, and A. S. Marotta: 'Spin-dependence of the resistivity of magnetic metals', *Journal of the Physics and Chemistry of Solids*, 1956, **9**, 302–308.

41. S. C. Abrahams, L. Guttman, and J. S. Kasper: 'Neutron diffraction determination of antiferromagnetism in face-centered cubic (gamma) iron', *Physical Reviews*, 1962, **127**, 2052–2055.

42. U. Gonser, C. J. Meechan, A. H. Muir, and H. Wiedersich: 'Determination of the néel temperatures in fcc iron', *Journal of Applied Physics*, 1963, **34**, 2373–2378.

43. R. J. Weiss: 'The origin of the 'Invar effect'', *Proceedings of the Physics Society*, 1963, **82**, 281–288.

44. J. G. Wright: 'Ferromagnetism in epitaxial FCC iron films', *Philosophical Magazine*, 1971, **24**, 217–233.

45. U. Gonser, K. Irischel, and S. Nasu: 'Ferromagnetic ordering in fcc gamma-iron precipitates in Cu-Au alloys', *Journal of Magnetism and Magnetic Materials*, 1980, **15-18**, 1145–1146.

46. S. D. Bader, and E. R. Moog: 'Magnetic properties of novel epitaxial films', *Journal of Applied Physics*, 1987, **61**, 3729–3734.

47. J. Z. Jiang, and F. T. Chen: 'A study of the microstructure of mechanically alloyed fcc FeCu', *Journal of Physics – Condensed Matter*, 1994, **6**, L343–L348.

48. J. Z. Jiang, Q. A. Pankhurst, C. E. Johnson, C. Gente, and R. Borman: 'Magnetic properties of mechanically alloyed FeCu', *Journal of Physics – Condensed Matter*, 1994, **6**, L227–L232.

49. J. Z. Jiang, C. Gente, and R. Bormann: 'Mechanical alloying in the Fe-Cu system', *Materials Science & Engineering A*, 1998, **242**, 268–277.

50. K. J. Tauer, and R. J. Weiss: 'Magnetic second-order transitions', *Physical Reviews*, 1955, **100**, 1223–1224.

51. J. Kübler: 'Magnetic moments of ferromagnetic and antiferromagnetic bcc and fcc iron', *Physics Letters A*, 1981, **81**, 81–83.

52. G. L. Krasko: 'Metamagnetic behaviour of fcc iron', *Physical Review B*, 1987, **36**, 8565–8569.

53. P. Mohn, K. Schwarz, and D. Wagner: 'Magnetoelastic anomalies in Fe-Ni invar alloys', *Physical Review B*, 1991, **43**, 3318–3324.

54. E. F. Wassermann: 'The invar problem', *Journal of Magnetism and Magnetic Materials*, 1991, **100**, 346–362.

55. M. Nicol, and G. Jura: 'Mössbauer spectrum of iron-57 in iron metal at very high pressure', *Science*, 1963, **141**, 1035–1038.

56. D. N. Pipkorn, C. K. Edge, P. Debrunner, G. De Pasquali, H. G. Drickamer, and H. Frauenfelder: 'Mössbauer effect in iron under very high pressure', *Physical Review A*, 1964, **135**, 1604–1612.

57. J. S. Kouvel: 'Magnetic properties of solids under pressure', In: W. Paul, and D. M. Warschauer, eds. *Solids under Pressure*. New York, USA: McGraw Hill, 1963:277–301.

58. K. Shimizu, T. Kimura, S. Furomoto, K. Takeda, K. Kontani, Y. Onuki, and K. Amaya: 'Superconductivity in the non-magnetic state of iron under pressure', *Nature*, 2001, **412**, 316–318.

59. V. Thakor, J. B. Staunton, J. Poulter, S. Ostanin, B. Ginatempo, and E. Bruno: 'Ab initio calculations of incommensurate antiferromagnetic spin fluctuations in hcp iron under pressure', *Physical Review B*, 2003, **67**, 180405.

60. S. Fox, and H. J. F. Jansen: 'Structure and properties of trigonal iron', *Physical Review B*, 1996, **53**, 5119–5122.

61. S. Peng, and H. J. F. Jansen: 'Electronic structure of face-centered tetragonal iron', *Journal of Applied Physics*, 1990, **67**, 4567–4569.

62. E. C. Bain: 'The nature of martensite', *Transactions of the AIME*, 1924, **70**, 25–46.

63. C. M. Wayman: 'The growth of martensite since E C Bain (1924) - some milestones', *Materials Forum*, 1990, **56-58**, 1–32.

64. S. S. Peng, and H. J. F. Jansen: 'Structural and magnetic properties of tetragonally distorted iron', *Ultramicroscopy*, 1992, **47**, 361–366.

65. B. T. Jonker, K. H. Walker, E. Kisker, G. A. Prinz, and C. Carbone: 'Spin-polarized photoemission study of epitaxial Fe(001) films on Ag(001)', *Physical Review Letters*, 1986, **57**, 142–143.

66. G. L. Krasko, and G. B. Olson: 'Ferromagnetism and crystal lattice stability of bcc and fcc iron', *Journal of Applied Physics*, 1990, **67**, 4570–4572.

67. H. C. Herper, E. Hoffmann, and P. Entel: 'Ab initio investigations of iron-based martensitic stystems', *Le Journal de Physique IV*, 1997, **7**, C5-71–C5-76.

68. H. C. Herper, E. Hoffmann, and P. Entel: 'Ab initio full-potential study of the structural and magnetic phase stability of iron', *Physical Review B*, 1999, **60**, 3839–3848.

69. C. Li, A. J. Freeman, H. J. F. Jansen, and C. L. Fu: 'Magnetic anisotropy in low-dimensional ferromagnetic systems: Fe monolayers on Ag (001), Au(001) and Pd(001) substrates', *Physical Review B*, 1990, **42**, 5433–5422.

70. H. J. Haugan, B. D. McCombe, and P. G. Mattocks: 'Structural and magnetic properties of thin epitaxial Fe films on (110) GaAs prepared by metalorganic chemical vapor deposition', *Journal of Magnetism and Magnetic Materials*, 2002, **247**, 296–304.

71. A. S. Edelstein, J. S. Murday, and B. B. Rath: 'Challenges in nanomaterials design', *Progress in Materials Science*, 1997, **42**, 5–21.

72. J. H. van Vleck: 'On the anisotropy of cubic ferromagnetic crystals', *Physical Review*, 1937, **52**, 1178–1198.

73. C. Stamm, F. Marty, A. Vaterlaus, V. Weich, S. Egger, U. Maier, U. Ramsperger, H. Fuhrmann, and D. Pescia: 'Two-dimensional magnetic particles', *Science*, 1998, **282**, 449–451.

74. S. T. Purcell, B. Heinrich, and A. S. Arrott: 'Perpendicular anisotropy at the (001) surface of bulk iron single crystals', *Journal of Applied Physics*, 1988, **64**, 5337–5339.

75. R. F. Willis, J. A. C. Bland, and W. Schwarzacher: 'Ferromagnetism in ultrathin metastable films of fcc Fe, Co, Ni', *Journal of Applied Physics*, 1988, **63**, 4051–4056.

76. S. D. Bader, E. R. Moog, and P. Grunberg: 'Magnetic hysteresis of epitaxially deposited iron in the monolayer range: a Kerr effect experiment in surface magnetism', *Journal of Magnetism and Magnetic Materials*, 1986, **53**, L295–L298.

77. T. Baykara, R. H. Hauge, N. Norem, P. Lee, and J. L. Margrave: 'A review of containerless thermophysical property measurements for liquid metals and alloys', *High Temperature and Materials Science*, 1994, **32**, 113–154.

78. R. I. L. Guthrie, and T. Iida: 'Thermodynamic properties of liquid metals', *Materials Science & Engineering A*, 1994, **178**, 35–41.

79. R. I. Orr, and J. Chipman: 'Thermodynamic functions of iron', *TMS-AIME*, 1967, **239**, 630–633.

80. W. W. Anderson, and T. J. Ahrens: 'An equation of state for liquid iron and implications for the Earth's core', *Journal of Geophysical Research*, 1994, **99**, 4273–4284.

81. L. Kaufman: 'Thermodynamics of martensitic fcc⇌bcc and fcc⇌hcp transformations in the iron-ruthenium system', In: *Physical Properties of Martensite and Bainite, Special Report 93*. London, U.K.: Iron and Steel Institute, 1965:49–52.

82. G. P. Tiwari: 'Modification of Richard's rule and correlation between entropy of fusion and allotropic behaviour', *Metal Science*, 1978, **12**, 317–320.

83. J. Mo, H. Liu, Y. Zhang, M. Wang, L. Zhang, B. Liu, and W. Yang: 'Effects of pressure on structure and mechanical property in monatomic metallic glass', *Journal of Non-Crystalline Solids*, 2017, **464**, 1–4.

84. C. V. Thompson, and F. Spaepen: 'On the approximation of the free energy change on crystallisation', *Acta Metallurgica*, 1979, **27**, 1855–1859.

85. J. D. Hoffman: 'Thermodynamic driving force in nucleation and growth processes', *Journal of Chemical Physics*, 1958, **29**, 1192–1193.

86. D. R. H. Jones, and G. A. Chadwick: 'An expression for the free energy of fusion in the homogeneous nucleation of solid from pure melts', *Philosophical Magazine*, 1971, **190**, 995–998.

87. L.-G. Liu, and W. A. Bassett: 'The melting of iron up to 200 kbar', *Journal of Geophysical Research*, 1975, **80**, 3777–3782.

88. J. H. Nguyen, and N. C. Holmes: 'Melting of iron at the physical conditions of the Earth's core', *Nature*, 2004, **427**, 339–342.

89. O. L. Anderson, and D. G. Isaak: 'Another look at the core density deficit of earth's outer core', *Physics of the Earth and Planetary Interiors*, 2002, **131**, 19–27.

90. H. Ichikawa, T. Tsuchiya, and Y. Tange: 'The P-V-T equation of state and thermodynamic properties of liquid iron', *Journal of Geophysical Research*, 2014, **119**, 240–252.

91. D. Gerlich, and S. Hart: 'Pressure dependence of the elastic moduli of three austenitic stainless steels', *Journal of Applied Physics*, 1984, **55**, 880–884.

92. M. Müller, P. Erhart, and A. Karsten: 'Analytic bond-order potential for bcc and fcc iron— comparison with established embedded-atom method potentials', *Journal of Physics – Condensed Matter*, 2007, **19**, 3262220.

93. H. K. Mao, Y. Wu, L. C. Chen, and J. F. Shu: 'Static compression of iron to 300 GPa and $Fe_{0.8}Ni_{0.2}$ alloy to 260 GPa: implications for composition of the core', *Journal of Geophysical Research*, 1990, **95**, 21737–21742.

94. M. W. Guinan, and D. N. Beshers: 'Pressure derivatives of the elastic constants of α-iron to 10 kb', *Journal of the Physics and Chemistry of Solids*, 1968, **29**, 541–549.

95. J. S. Benjamin: 'Oxide dispersion strengthened (ODS) superalloys directional recrystallisation', *Metallurgical Transactions*, 1970, **1**, 2943–2951.

96. A. Y. Badmos, and H. K. D. H. Bhadeshia: 'Evolution of solutions', *Metallurgical & Materials Transactions A*, 2000, **28**, 2189–2193.

97. C. H. P. Lupis: Chemical Thermodynamics of Materials: New York, USA: North-Holland, 1983.

98. J. H. Hildebrand: 'Solubility. xii. regular solutions', *Journal of the American Chemical Society*, 1929, **51**, 66–80.

99. E. A. Guggenheim: 'The statistical mechanics of regular solutions', *Proceedings of the Royal Society A*, 1935, **148**, 304–312.

100. E. A. Guggenheim: Mixtures: Oxford, U. K.: Oxford University Press, 1952.

101. J. W. Christian: Theory of Transformations in Metals and Alloys, Part I: 2 ed., Oxford, U. K.: Pergamon Press, 1975.

102. R. Kikuchi: 'A theory of cooperative phenomena', *Physical Review*, 1951, **81**, 988–1003.

103. R. B. McLellan, and W. W. Dunn: 'A quasichemical treatment of interstitial solid solutions: its application to carbon in austenite', *Journal of Physics and Chemistry of Solids*, 1969, **30**, 2631–2637.

104. W. W. Dunn, and R. B. McLellan: 'The application of a quasichemical solid solution model to carbon in austenite', *Metallurgical Transactions*, 1970, **1**, 1263–1265.

105. S. Ban-Ya, J. F. Elliott, and J. Chipman: 'Activity of C in Fe-C alloys at 1150 C', *Transactions of the Metallurgical Society of AIME*, 1969, **245**, 1199–1206.

106. S. Ban-Ya, J. F. Elliott, and J. Chipman: 'Thermodynamics of austenitic Fe-C alloys', *Metallurgical Transactions*, 1970, **1**, 1313–1320.

107. K. Alex, and R. B. McLellan: 'A zeroth order mixing treatment of interstitial solid solutions', *Journal of the Physics and Chemistry of Solids*, 1970, **31**, 2751–2753.

108. H. K. D. H. Bhadeshia: 'Quasichemical model for interstitial solutions', *Materials Science and Technology*, 1998, **14**, 273–276.

109. W. W. Dunn, and R. B. McLellan: 'The thermodynamic properties of carbon in body-centered cubic iron', *Metallurgical Transactions*, 1971, **2**, 1079–1086.

110. R. B. McLellan: 'The thermodynamics of dilute interstitial solid solutions', In: P. S. Rudman, and J. Stringer, eds. *Phase Stability in Metals and Alloys*. New York, USA: McGraw Hill, 1967:393–417.

111. L. Kaufman, and H. Nesor: 'Coupled phase diagrams and thermochemical data for transition metal binary systems - IV', *CALPHAD*, 1978, **2**, 295–318.

112. K. Oda, H. Fujimura, and H. Ino: 'Local interactions in carbon-carbon and carbon-M (M: Al, Mn, Ni) atomic pairs in fcc gamma-iron', *Journal of Physics – Condensed Matter*, 1994, **6**, 679–692.

113. A. L. Sozinov, A. G. Balanyuk, and V. G. Gavriljuk: 'C-C interaction in iron-base austenite and interpretation of Mössbauer spectra', *Acta Materialia*, 1997, **45**, 225–232.

114. V. M. Nadutov, L. A. Bulvain, and V. M. Garamus: 'Analysis of small-angle neutron scattering from Fe–18Cr–10Mn–16Ni–0.5N and Fe–21Cr–10Mn–17Ni–0.5C austenites', *Materials Science & Engineering A*, 1999, **264**, 286–290.

115. H. K. D. H. Bhadeshia: 'Application of first-order quasichemical theory to transformations in steels', *Metal Science*, 1982, **16**, 167–169.

116. Y. Mou, and H. I. Aaronson: 'The carbon carbon interaction energy in alpha Fe-C alloys', *Acta Metallurgica*, 1989, **37**, 757–765.

117. J. C. Fisher, J. H. Hollomon, and D. Turnbull: 'Kinetics of the austenite to martensite transformation', *Metals Transactions*, 1949, **185**, 691–700.

118. J. H. Jang, H. K. D. H. Bhadeshia, and D. W. Suh: 'Solubility of carbon in tetragonal ferrite in equilibrium with austenite', *Scripta Materialia*, 2012, **68**, 195–198.

119. A. G. Khachaturyan, and G. A. Shatalov: 'Elastic-interaction potential of defects in a crystal', *Soviet Physics - Solid State*, 1969, **11**, 118–123.

120. A. G. Khachaturyan, and G. A. Shatalov: 'Theory of ordering of carbon-atoms in a martensite crystal', *Physics of Metals and Metallography*, 1971, **32**, 5–13.

121. A. G. Khachaturyan: Theory of Phase Transformations and Structure of Solid Solutions: New York, USA: John Wiley & Sons, Inc., 1983.

122. D. A. Mirzayev, A. Mirzoev, and P. V. Chirkov: 'Ordering of carbon atoms in free martensite crystal and when enclosed in elastic matrix', *Metallurgical & Materials Transactions A*, 2016, **47**, 637–640.

123. P. V. Chirkov, A. A. Mirzoev, and D. A. Mirzaev: 'Role of stresses and temperature in the Z ordering of carbon atoms in the martensite lattice', *Physics of Metals and Metallography*, 2016, **117**, 1138–1143.

124. L. Kaufman: 'The stability of metallic phases', *Progress in Materials Science*, 1969, **14**, 57–96.

125. T. G. Chart, J. F. Counsell, G. P. Jones, W. Slough, and P. J. Spencer: 'Provision and use of thermodynamic data for the solution of high-temperature practical problems', *International Metals Reviews*, 1975, **20**, 57–82.

126. M. Hillert: 'Prediction of iron base phase diagrams', In: D. V. Doane, and J. S. Krikaldy, eds. *Hardenability Concepts with Applications to Steels*. Warrendale, Pennsylvania, USA: The Metallurgical Society of AIME, 1977:5–27.

127. I. Ansara: 'Comparison of methods for the thermodynamica calculation of phase diagrams', *International Metals Reviews*, 1979, **1**, 20–53.

128. G. Inden: 'The role of magnetism in the calculation of phase diagrams', *Physica B*, 1981, **103**, 82–100.

129. K. Hack, ed.: The SGTE Casebook: Thermodynamics at Work: The Institute of Materials, London, 1996.

130. M. Hillert: 'Empirical methods of predicting and representing thermodynamic properties of ternary solution phases': Tech. Rep. 0143, Royal Institute of Technology, Stockholm, Sweden, 1979.

131. M. Hillert, and L. I. Staffansson: 'The regular solution model for stoichiometric phases and ionic melt', *Acta Chimica Scandinavica*, 1970, **24**, 3616–3626.

132. M. Temkin: 'Mixtures of fused salts as ionic solutions', *Acta Phsicochimica URSS*, 1945, **20**, 411–417.

133. M. H. M. Waldenström: 'A thermodynamic analysis of the Fe-Mn-C system', *Metallurgical & Materials Transactions A*, 1977, **8**, 5–13.

134. B. Uhrenius, and H. Harvig: 'A thermodynamic evaluation of carbide solubilities in the Fe-Mo-C, Fe-W-C, and Fe-Mo-W-C systems at 1000 C', *Metal Science*, 1975, **9**, 67–82.

135. B. Sundman, and J. Ågren: 'A regular solution model for phases with several components and sublattices, suitable for computer applications', *Journal of the Physics and Chemistry of Solids*, 1981, **42**, 297–301.

136. L. F. Bates: Modern Magnetism: Cambridge, U. K.: Cambridge University Press, 1963.

137. M. V. Nevitt, and A. T. Aldred: 'Ferromagnetism in V-Fe and Cr-Fe alloys', *Journal of Applied Physics*, 1963, **34**, 463–468.

138. G. Inden, and W. O. Meyer: 'Approximate determination of the Curie temperatures of bcc Fe-Co alloys', *Zietschrift für Metallkunde*, 1975, **66**, 725–727.

139. K. G. Denbigh: The Thermodynamics of the Steady State: New York, USA: John Wiley & Sons, Inc., 1955.

140. L. Onsager: 'Reciprocal relations in irreversible processes – i', *Physical Review*, 1931, **37**, 405–426.

141. D. G. Miller: 'Thermodynamics of irreversible processes: The experimental verification of the Onsager reciprocal relations.', *Chemical Reviews*, 1960, **60**, 15–37.

142. J. Z. Jiang, U. Gonser, C. Gente, and R. Bormann: 'Mössbauer investigations of mechanical alloying in the Fe-Cu system', *Applied Physics Letters*, 1993, **63**, 2768–2770.

143. J. Z. Jiang, U. Gonser, C. Gente, and R. Bormann: 'Thermal stability of the unstable fcc-FeCu phase prepared by mechanical alloysing', *Applied Physics Letters*, 1993, **63**, 1056–1058.

144. J. G. Gay, and R. Richter: 'Spin anisotropy of ferromagnetic films', *Physical Review Letters*, 1986, **56**, 2728–2731.

145. J. G. Gay, and R. Richter: 'Spin anisotropy of ferromagnetic slabs and overlayers', *Journal of Applied Physics*, 1987, **61**, 3362–3365.

146. J. H. Perepezko, and J. S. Paik: 'Thermodynamic properties of undercooled liquid metals', *Journal of Non-crystalline Solids*, 1984, **61**, 113–118.

147. G. J. Shiflet, J. R. Bradley, and H. I. Aaronson: 'A re-examination of the thermodynamics of the proeutectoid ferrite transformation in steels', *TMS-AIME*, 1978, **9A**, 999–1008.

148. A. H. Cottrell: 'Constitutional vacancies in NiAl', *Intermetallics*, 1995, **3**, 341–345.

Notes

[1]The necessary and sufficient condition that the differential equation

$$M\{x,y\}\mathrm{d}x + N\{x,y\}\mathrm{d}y = 0 \tag{2.94}$$

is exact requires that

$$\left(\frac{\partial M}{\partial y}\right)_x = \left(\frac{\partial N}{\partial x}\right)_y \tag{2.95}$$

[2]This follows from the relationships between partial derivatives of variables $x, y z$. The total differentials are

$$\mathrm{d}x = \left(\frac{\partial x}{\partial y}\right)_z \mathrm{d}y + \left(\frac{\partial x}{\partial z}\right)_y \mathrm{d}z \quad \text{and} \quad \mathrm{d}y = \left(\frac{\partial y}{\partial x}\right)_z \mathrm{d}x + \left(\frac{\partial y}{\partial z}\right)_x \mathrm{d}z$$

Substituting for $\mathrm{d}y$ in the equation for $\mathrm{d}x$ yields:

$$\left[\left(\frac{\partial x}{\partial y}\right)_z \left(\frac{\partial y}{\partial x}\right)_z - 1\right]\mathrm{d}x + \left[\left(\frac{\partial x}{\partial y}\right)_z \left(\frac{\partial y}{\partial z}\right)_x + \left(\frac{\partial x}{\partial z}\right)_y\right]\mathrm{d}z = 0$$

Since each expression in the square brackets must vanish independently, if follows that

$$\left(\frac{\partial z}{\partial x}\right)_y \left(\frac{\partial x}{\partial y}\right)_z \left(\frac{\partial y}{\partial z}\right)_x = -1.$$

[3]This picture is convenient but inaccurate because in a classical calculation, the velocity at which the electron would have to spin to produce the observed magnetic moment would be greater than that of light. The term "spin" really refers to a quantised rotation that has no counterpart in classical mechanics.

[4]There is a rule that says that electrons are distributed in a level to maximise spin. Therefore, in $3d$ there will be five electrons with spin-up, and 2.8 with spin-down, giving a net moment of 2.2.

[5]M is the magnetic moment per unit volume, induced by a magnetic field of strength H_{mag}. It is sometimes called the intensity of magnetisation or simply magnetisation. The ratio of M/H_{mag} is the magnetic susceptibility.

[6]It is assumed here that g_0 has the minimum value of 2, and that the spin per atom for the γ_1 state is 1.4, so that $2s + 1 = 2.8 + 1$.

[7]When bcc iron and copper powders are heavily deformed together in a ball mill, the powders naturally become finer but there also is an increase in contact between iron and copper particles. This eventually causes the iron particles to transform into austenite as they are forced into coherency with the copper. Continued milling leads to interdiffusion and finally, the formation of a true Fe-Cu fcc solid solution [49, 142, 143].

[8]The term "Total energy" describes first-principles calculations of the ground-state energy of an interacting electron gas. It is the sum of the many-body kinetic energy, the many-body Coulomb energy and the external energy due to the atomic nuclei and other external fields. The energy is really an enthalpy since it refers to the ground state at 0 K.

[9] If the two-dimensional unit cell of the unrelaxed surface has parameters a and b, then a 2×1 reconstruction refers to the relaxed pattern having a larger unit cell with parameters $2a$ and b.

[10] The antiferromagnetic structure used in these calculations involves the continuous deformation of the simplest bcc case where the corner and body-centering atoms have opposite spins.

[11] There are two difficulties. Experiments on austenite monolayers deposited on $\{100\}_{Au}$ show results which appear to contradict these calculations that the magnetisation is in the plane of the film [46]. Secondly, calculations of spin anisotropy by Gay and Richter [144, 145] predicted magnetisation perpendicular to the surface of a free-standing iron monolayer; this is believed to be a consequence of computational difficulties [69].

[12] Some difficulties with the use of this approximation are described in [84, 146].

[13] This equation reduces to the familiar

$$\Delta S_M = -kN_a[(1-x)\ln\{1-x\}+x\ln\{x\}] \tag{2.96}$$

when $m_A = m_B = 1$.

[14] Hildebrand's definition: "A regular solution is one involving no entropy change when a small amount of one of its components is transferred to it from an ideal solution of the same composition, the total volume remaining unchanged" [98].

[15] It is straightforward to apply this also to carbon dissolved in ferrite. The parameters β_O and W to be described later have to be set to different values consistent with the geometry of the body-centred cubic lattice.

[16] The relationship illustrated in Figure 2.39 can be described by the following equation which is due to Ban-ya, Elliott and Chipman:

$$\Delta\mu° = 4.184(17250 + 12.44T\log\{T\} - 48.15T) \quad \text{J mol}^{-1}. \tag{2.97}$$

[17] It is worth emphasising again that this substitution for λ^* recognises implicitly that λ^* is a function of N_u. This is in contrast to the first order quasichemical theory by Dunn and McLellan, where during the differentiation of the partition function with respect to N_u, λ was treated as a constant. The comparison reported by Alex and McLellan is therefore between models based on different assumptions. The comparison here is with the corrected first-order quasichemical theory.

[18] It has been believed that ω_α may be negative, i.e., that carbon atoms in ferrite attract each other [147]. However, the analysis used an incorrect coordination number [115].

[19] A similar ordering effect has been found for the cubic compound NiAl, where there are strong distortions around vacancies with atomic displacements parallel to $\langle 111 \rangle$ [148]. The vacancies, when they are close to each other, are therefore expected to align parallel to the $\langle 111 \rangle$ axes to reduce the strain energy. This is said to be responsible for the transformation of the cubic NiAl structure into that of Al_3Ni_2 which has a trigonal crystal structure.

[20] The equation that follows should be compared with the usual one for a multicomponent solution, i.e.,

$$G = \sum_i x_i\mu_i° - T\Delta S_M + \Delta_e G$$

[21] There is a variation in terminology in the published literature. The reader should note particularly the definition of the terms a and c to avoid confusion

[22] Assuming a random distribution of the atoms on each sublattice

[23] Short-range order is usually discussed in terms of the Bethe parameter given by

$$L_{sr} = \frac{N_{AB} - N_{AB}^{random}}{N_{AB}^{ordered} - N_{AB}^{random}}$$

where the numerator is the difference between the number of A-B bonds and the number expected in a random alloy. The denominator is the corresponding difference relative to a fully ordered alloy. Clearly, $L_{sr} = 0$ when the solution is random and $L_{sr} = 1$ when it is fully ordered. In this respect it is similar to the long range order parameter L, but the definition is with respect to nearest neighbours only rather than segregation into different lattices.

[24] In the Peltier effect, the two junctions of a thermocouple are kept at the same temperature but the passage of an electrical current causes one of the junctions to absorb heat and the other to liberate the same quantity of heat. This Peltier heat is found to be proportional to the current.

3 Diffusion

3.1 INTRODUCTION

There probably are no examples of useful steels that are even close to equilibrium. But the rate at which they may approach equilibrium determines the evolution of microstructures during manufacture and the stability of the steel during service. This rate depends on the mobility of atoms in the solid-state, i.e., on diffusion and on the forces that drive diffusion. It will be seen that iron has some distinctive features when it comes to diffusion. Magnetic transitions cause marked anomalies in diffusion data. There are huge disparities in the rates at which interstitial and substitutional solutes migrate. On occasions, the concentration dependence of the diffusion coefficients cannot be explained by thermodynamics alone. And there is much more detail that is important to the development of phase transformation theory for steels.

3.2 FICK'S LAW AND DIFFUSION COEFFICIENTS

Fick's first law for diffusion in a binary system along a coordinate z is based on the intuitively reasonable premise that the solute flux J is related directly to the concentration gradient via a proportionality constant which is the diffusion coefficient D:

$$J = -D\frac{dc}{dz}. \tag{3.1}$$

The negative sign is because the flux is along $+z$ whereas the concentration increases along $-z$ and it is assumed that diffusion occurs down a concentration gradient. The hypothesis was based on his work on diffusion in liquids, at the Department of Anatomy in Zürich. To quote Fick: "According to this law, the transfer of salt and water occurring in a unit of time, between two elements of space filled with differently concentrated solutions of the same salt, must be, *cæteris paribus*, directly proportional to the difference of concentration, and inversely proportional to the distance of the elements from one another" [1].

The flux is the rate of transfer of the diffusing substance through a unit of area and the concentration gradient is measured normal to this area [1]. A large number of measurements have been made within the framework of this law. Deviations from Fick's law are therefore treated by making the diffusion coefficient concentration dependent in order to retain the basic proportionality. The justification for this will be explored later in this chapter.

Some elements diffuse faster in a solution than others; each component of the solution is associated with its characteristic *intrinsic* diffusion coefficient $\overline{D}_A$. This represents the flux $\overline{J}_A$ of component "A" in a binary A-B substitutional solution containing a concentration gradient of A (and hence of B).

When the two species in an interdiffusion experiment have unequal intrinsic mobilities in a system where diffusion occurs by a vacancy mechanism, there will be a net flow of matter in the direction of the more mobile element. If a diffusion couple is constructed with inert markers at the interface, fixed to the laboratory bench, then the specimen will move relative to the markers during diffusion (Figure 3.1), a phenomenon known as the Kirkendall effect [2, 3]. This bulk flow of matter contributes to composition change at any point but is not accounted for in the definition of intrinsic coefficients. The substitution of intrinsic diffusion coefficients into Fick's first law is therefore said to define fluxes $\bar{J}_i$ relative to the Kirkendall (or lattice fixed) frame of reference in which $\sum_i \bar{J}_i = -J_\square$, where $J_\square$ is the flux of vacancies. The Kirkendall frame moves with the velocity of the inert markers relative to the laboratory frame to which the specimen is fixed.

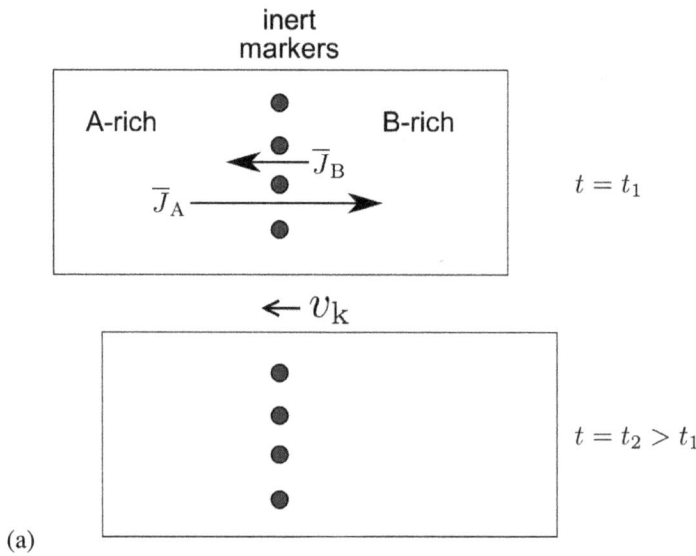

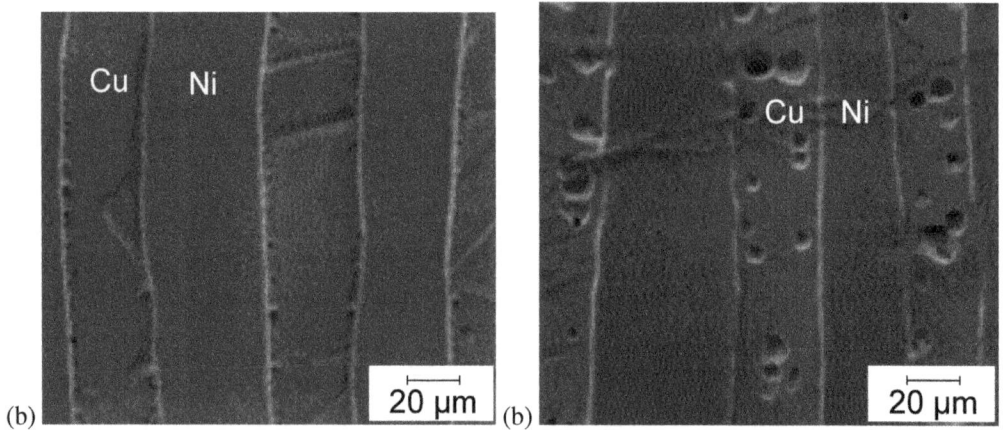

(a)

(b)

(b)

Figure 3.1   (a) The Kirkendall effect during diffusion by a vacancy mechanism. $\bar{J}$ represents the fluxes of the two elements A and B in the diffusion couple and t is the time. $\bar{D}_A > \bar{D}_B$. The inert markers are fixed to the laboratory frame. Unequal fluxes result in a net transport of matter relative to the markers. (b) A Cu-Ni multilayer heat treated at 1273 K for 1 min. (c) The same after holding at 1273 K for 15 min. Cu diffuses faster than Ni in the Cu-Ni solid solution, leading to a net flow of vacancies into the copper-rich side, vacancies that condense eventually to form pores [4]. Micrographs courtesy of David Matlock.

An observer in the laboratory frame sees the change in composition at a point in the sample due to both intrinsic diffusion and the Kirkendall flow of matter. The *chemical* or *interdiffusion* coefficient D is defined relative to the laboratory frame in order to take both of these effects into account, representing the rate at which mixing or unmixing occurs. It is the interdiffusion coefficient that is used most frequently in practical applications, for example when dealing with transformation kinetics, homogenisation or tempering reactions. In circumstances where $\overline{D}_A = \overline{D}_B$, there is no Kirkendall effect and $\overline{D}_A = \overline{D}_B = D_{A,B}$; in other words, the laboratory and Kirkendall frames of reference become identical.

It may be convenient to define the interdiffusion coefficient D relative to other frames of reference. In the volume-fixed frame, the fluxes J_i are measured across a section such that the total volume on either side of the section remains constant during diffusion. The origin of the coordinate system is at the instantaneous centre of volume. In this case,

$$\sum_i^n J_i \overline{V}_i = 0,$$

where $\overline{V}_i$ is the partial molar volume of component i in an n component solution. A coordinate system such as this can be defined only if the partial molar volumes are constant.

The laboratory frame is also known as the number-fixed frame. In this case, the fluxes occur across a section defined such that the total number of atoms on either side of the section remains constant:

$$\sum_i^n J_i = 0.$$

If the integral molar volume remains constant, then the volume-fixed frame coincides with the number-fixed frame. This may be true approximately in dilute solutions where any changes in the integral molar volume as a function of chemical composition will be small.

The frames of reference are summarised in Table 3.1; Miller [5] states that with few exceptions, most experimental determinations of interdiffusion coefficients assume constant volume and hence refer to the volume-fixed frame. If the further assumption is made that the molar volume remains constant, then these coefficients are valid for the laboratory frame of reference.

Table 3.1
Frames of reference for diffusion. $J_\square$ is the flux of vacancies.

Reference frame	Definition
Kirkendall (lattice-fixed)	$\sum_i^n \overline{J}_i = -J_\square$
Laboratory or number-fixed	$\sum_i^n J_i = 0$
Volume-fixed	$\sum_i^n J_i \overline{V}_i = 0$

The flux described in the laboratory frame includes a component $\overline{J}_i$ due to intrinsic diffusion, and another due to the bulk flow of matter (Kirkendall effect), i.e., $c_i v_K$ where v_K is the velocity of the Kirkendall markers and c is the local molar concentration per unit volume. It follows that [6]

$$J_A = \overline{J}_A + c_A v_K \qquad \text{and} \qquad J_B = \overline{J}_B + c_B v_K. \qquad (3.2)$$

The velocity of the markers depends on the volume of matter passing through a unit area in a unit of time. The net flux across the markers is $\overline{J}_A + \overline{J}_B$ (moles m^{-2} s^{-1}). Noting that $1/(c_A + c_B)$ is simply the volume per mole,

$$v_K = -\frac{\text{volume past markers}}{\text{area} \times \text{time}} = -\frac{\overline{J}_A + \overline{J}_B}{c_A + c_B}. \qquad (3.3)$$

The negative sign indicates that the markers move in a direction opposite to the net flux. Since $c_A/(c_A + c_B)$ is the mole fraction x_A, substitution for the Kirkendall velocity into Equation 3.2 yields

$$J_A = x_B \overline{J}_A - x_A \overline{J}_B. \qquad (3.4)$$

The relationship between the chemical diffusion coefficient and intrinsic coefficients therefore becomes

$$D = x_B \overline{D}_A + x_A \overline{D}_B. \qquad (3.5)$$

This description of interdiffusion in a binary alloy requires just one diffusion coefficient, though some information is omitted unless the Kirkendall velocity also is stated.

The *tracer* diffusion coefficient, D^*, describes the motion of radioactively labelled isotopes in an otherwise chemically homogeneous solution. When the tracer atoms are of the same species as the non-tracer atoms, the coefficient becomes the self-diffusion coefficient. Tracer and self-diffusion coefficients describe purely random motion, not intended to represent diffusion in concentration gradients. In self-diffusion measurements there is a mass difference between the radioactive isotope and the solvent, which causes the two isotopes to vibrate with different frequencies in the lattice. The vibration frequency is proportional to the inverse square root of the mass m of the atom. For two different isotopes of the same species, it follows that

$$\frac{D_1^* - D_2^*}{D_2^*} \propto \frac{m_1^{-\frac{1}{2}} - m_2^{-\frac{1}{2}}}{m_2^{-\frac{1}{2}}} \qquad (3.6)$$

with the proportionality constant $f_c \Delta K$, where the Bardeen-Herring correlation factor f_c accounts for the non-random path of a defect such as a vacancy around a "solute" because some of its jumps are reversed. It is possible that a number of adjacent atoms might participate in the vibration mode which causes a particular atom to migrate; ΔK is the fraction of the total kinetic energy attributed to the vibration mode, which resides in the migrating atom [7]. ΔK is unity when the migrating atom is not coupled with its neighbours.

In contrast with self-diffusion, an additional virtual force acts on the diffusing species in a chemical potential gradient:

$$D = x_B \overline{D}_A + x_A \overline{D}_B = (x_B D_A^* + x_A D_B^*) \times \text{thermodynamic factor} \qquad (3.7)$$

where the nature of the thermodynamic factor will be described later (Section 3.16). When the thermodynamic factor is close to unity or if the solute concentration $x_B \ll 1$, then $D \approx \overline{D}_B \approx D_B^*$.

3.2.1 REACTION RATE EXPRESSION

The diffusion coefficient for transport by a vacancy mechanism must depend on the probability of finding a vacancy in a neighbouring site, on the barrier to the migration of that vacancy and on the frequency with which diffusion jumps are attempted along the appropriate direction:

$$D = va^2 f_c \exp\left\{-\frac{\Delta G_F + G_M^*}{kT}\right\}$$ (3.8)

where ΔG_F is the free energy of formation of a vacancy with $\exp\{-\Delta G_F/kT\}$ representing the probability of forming a vacancy. G_M^* is the activation free energy for the migration of the vacancy, v is the frequency with which diffusion is attempted in the direction of the overall flux, a is the lattice parameter and f_c the Bardeen-Herring correlation factor. This is a rather general description of the use of reaction rate theory in understanding diffusion. Much more detail about the individual factors can be found elsewhere [8]; the important issue in the present context is the form of the equation. In particular, the free energies can be factorised into their entropic S and enthalpic H components:

$$
\begin{aligned}
D &= va^2 f_c \exp\left\{-\frac{\Delta S_F + S_M^*}{k}\right\} \exp\left\{-\frac{\Delta H_F + H_M^*}{kT}\right\} \\
&\equiv D_0 \exp\left\{-\frac{Q}{kT}\right\}
\end{aligned}
$$ (3.9)

where D_0 is the pre-exponential factor and Q is the activation enthalpy for diffusion. In the case of interstitial diffusion, there is in general no shortage of vacant interstices so the activation enthalpy for interstitial diffusion is much smaller because $\Delta H_F = 0$. Interstitial diffusion is therefore much faster than substitutional diffusion by a host-atom vacancy mechanism.

3.3 DIFFUSION OF CARBON IN FERRITE

A carbon atom is about 60% of the size of an iron atom so consistent with the Hume-Rothery rules, occupies the interstices within the crystal structure. Because the concentration of carbon tends to be small, the chances of finding a vacant interstice adjacent to an atoms of carbon are large. This is unlike substitutional solutes, so carbon is much more mobile in solid iron; it can, for example, rearrange into a different location at ambient temperature leading to changes in macroscopically observable properties, such as the yield point effect or Snoek damping. At high temperatures, its solid-state diffusivity compares with the self-diffusivity of iron in its liquid state at about $10^{-10}\,\mathrm{m^2 s^{-1}}$.

3.3.1 INTERSTITIAL SITES IN FERRITE

There are two kinds of interstitial sites in iron capable of accommodating small atoms such as carbon, nitrogen or hydrogen. These are the tetrahedral and octahedral sites, illustrated in Figure 3.2.

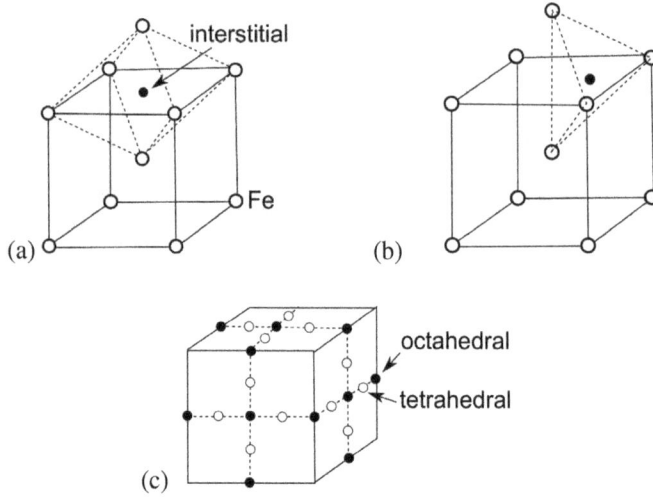

Figure 3.2 The main interstices in the body-centred cubic structure of ferrite. (a) An irregular octahedral-interstice; (b) a tetrahedral interstice; (c) location of both kinds of interstices.

The octahedral interstices in ferrite are not regular. One of the axes has a length a_α, the lattice parameter of ferrite, whereas the other two are along the $\langle 1\ 1\ 0 \rangle$ directions with length $\sqrt{2}a_\alpha$. The shortest axis has four-fold symmetry so a vacant octahedral site in ferrite has tetragonal symmetry $\frac{4}{m}\frac{2}{m}\frac{2}{m}$. Adjacent octahedra have their axes inclined at $90°$ to each other. Since the close-packed direction is $\langle 1\ 1\ 1 \rangle$, the radius of an iron atom in ferrite is $a_\alpha\sqrt{3}/4$. The smallest axis of the octahedron is parallel to the cell edge so the radius of the largest sphere which can fit in the octahedron without any distortion is $a_\alpha(1/2 - \sqrt{3}/4) = 0.067a_\alpha$. This compares with the space available along the $\langle 1\ 1\ 0 \rangle$ where the corresponding radius is much larger, at $a_\alpha(1/\sqrt{2} - \sqrt{3}/4) = 0.274a_\alpha$.

The tetrahedral interstice in ferrite is also irregular (Figure 3.2). Two of its edges lie parallel to $\langle 1\ 0\ 0 \rangle$ directions, the remaining four being along $\langle 1\ 1\ 1 \rangle$ directions, with lengths a_α and $a_\alpha\sqrt{3}/2$ respectively. The lines joining the corners of the tetrahedron to its centre are of the form $\langle 2\ 1\ 0 \rangle$. These lines each have a length $a_\alpha\sqrt{5}/4$; given that an iron atom in ferrite has a radius $a_\alpha\sqrt{3}/4$, the radius of the largest atom which will fit without distortion is $a_\alpha(\sqrt{5}/4 - \sqrt{3}/4) = 0.126a_\alpha$. This is much larger than is the case for the octahedral hole. It follows that a sphere which fits into an octahedral hole without distortion has free passage through the crystal since the intervening tetrahedral holes are larger. No interstitial atom, with the possible exception of hydrogen, satisfies this condition.

Calculations of the distortions caused by carbon in ferrite are summarised in Table 3.2 [9], where the position vectors originate at the carbon atom and are in units of $a_\alpha/2$. The distortion falls off rapidly with distance (cf. the iron atoms at 1,0,0 and 3,0,0 for the octahedral hole). The huge displacements at 1,0,0 exclude the octahedral interstice located at distance a_α along the tetragonal axis from occupation [10], as verified experimentally using a diffuse X-ray scattering technique [11].

The radius of an interstitial atom such as carbon depends on its environment. There is evidence

Table 3.2
Positions of iron atoms relative to a carbon atom located at the origin, for octahedral and tetrahedral interstices. The position vectors are in units of $a_\alpha/2$ and the last column describes the number of equivalent iron neighbours. The data are from Johnson, Dienes and Damask [9] where a more comprehensive listing can be found.

OCTAHEDRAL

Perfect position	Displaced position			Number
1 0 0	1.225	0.000	0.000	2
3 0 0	3.046	0.000	0.000	2
0 1 1	0.000	0.958	0.958	4
0 3 3	0.000	2.955	2.955	4
1 0 2	1.017	0.000	1.986	8

TETRAHEDRAL

$0\,\frac{1}{2}\,1$	0.000	0.529	1.116	4
$1\,\frac{3}{2}\,0$	0.983	1.459	0.000	4
$2\,\frac{1}{2}\,1$	2.011	0.506	1.015	8

that both carbon and nitrogen in austenite are positively charged [12–15]. A good estimate can be made by measuring the change in the lattice parameter of untempered, high-carbon martensite as a function of the carbon concentration. Based on such data, Beshers [16] concluded that both carbon and nitrogen in ferrite have a radius of 0.082 nm ($\approx 0.286a_\alpha$).

Carbon mainly occupies the octahedral interstices in ferrite and martensite. It is large enough to push all six of the neighbouring iron atoms apart. The octahedral hole becomes nearly spherical, occupying some of the space that is in the undistorted lattice assigned to the neighbouring four tetrahedral interstices (Figure 3.2c). The activation volume V^* in general has the components:

$$V^* = \Delta V_F + \Delta V_M \tag{3.10}$$

where ΔV_F is the change in volume upon the formation of one (interstitial) vacancy, in this case zero since the vacancies are always there; ΔV_M is the difference in volume between the stable configuration, and the configuration at the saddle point during the jump between adjacent sites.

The activation volume is related to the activation free energy G^* as follows:

$$V^* = \left(\frac{\partial G^*}{\partial P}\right)_T \tag{3.11}$$

and if $D = D_o \exp\{-G^*/kT\}$ then if follows that

$$V^* = -kT\left(\frac{\partial \ln\{D/D_o\}}{\partial P}\right)_T. \tag{3.12}$$

The activation volumes for the diffusion of carbon and nitrogen in ferrite determined by relaxation methods at temperatures in the range 238-253 K and 300 MPa pressure is close to zero [17–19], i.e., the respective diffusion coefficients are insensitive to pressure. However, measurements of carbon diffusion at elevated temperatures (889-1000 K) reveal larger activation volumes that vary with temperature, with significant sensitivity to pressure [20]. The mechanism for the observations is not entirely established but it is argued that there may be a change in the mechanism of diffusion.

Beshers [16] has estimated the strain and activation energies for each interstitial site and diffusion path using continuum elasticity theory. The strain energy was assumed to be proportional to the square of the displacement and to the Young's modulus in the direction of motion. Any movement at right angles to the direction of motion were discounted; parallel displacements were added before squaring. Neglecting Poisson's effects, the strain energy associated with the occupation of an octahedral (U_O) and tetrahedral (U_{Te}) site is given by

$$U_O \propto 2(r_i - r_o)^2 E_{100}, \qquad U_{Te} \propto 4(r_i - r_o)^2 E_{210} \tag{3.13}$$

where r_i is the interstitial-atom radius, r_o is the radius of the undistorted hole and E the Young's modulus along the crystallographic direction identified in the subscript. For $r = 0.08$ nm, $U_{Te}/U_O = 1.38$. On this basis, the octahedral site is favoured, even though the tetrahedral interstice can accommodate somewhat larger interstitials. Some calculations suggest that there is a difference in the chemical energy of solution between the octahedral and tetrahedral sites that also makes the former interstice the favoured location for carbon atoms [21].

The activation energies for diffusion given $r_i = 0.08$ nm were found to be $Q_{O \to O}/U_O = 1$, $Q_{O \to T}/U_O = 0.69$, $Q_{T \to O}/U_O = 0.31$ and $Q_{T \to T}/U_O = 0.91$. The favoured diffusion path is therefore $O \to T \to O$; direct jumps between octahedral sites are unlikely.

An interesting consequence of the misfit between the interstitial and interstice is the Snoek effect in which ordering occurs under the influence of a stress. The strain field about the (irregular) octahedral hole has tetragonal symmetry. The application of a stress makes some of the sites less tetragonal than others and hence favours their occupation. Beshers demonstrated that the asymmetry of the strain field about the tetrahedral sites can in principle lead to a similar ordering of interstitials [16].

3.3.2 DUAL-SITE OCCUPANCY

Because of its limited solubility in ferrite, the diffusion coefficient of carbon often is measured using internal friction experiments, where the damping of imposed vibrations yields data about the diffusion of interstitials between adjacent sites. Traditional mass flow methods can be used to determine diffusivity at high temperatures. The combined data from these different techniques cover some fourteen orders of magnitude in the diffusion coefficient. The data do not strictly follow an Arrhenius relationship over this range of temperatures, with perceptible deviations from the straight line on a plot of $\ln\{D\}$ versus T^{-1}, Figure 3.3. They can, over this range, be described empirically by the equation [22]:

$$\ln D = -11.297 - 1.197\left(\frac{10^4}{T}\right) + 0.0037\left(\frac{10^4}{T}\right)^2 \qquad \mathrm{m^2\,s^{-1}}. \tag{3.14}$$

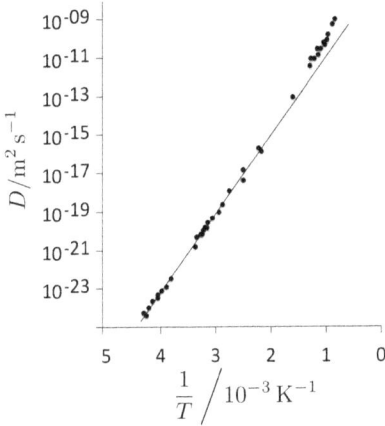

Figure 3.3 Diffusion data for carbon in ferrite, showing deviations at high temperatures from the Arrhenius behaviour. Data from compilation by McLellan, Rudee and Ishibachi [23].

One explanation for this nonlinear behaviour relies on the two kinds of interstitial sites available in ferrite, the octahedral (O) and tetrahedral (Te) sites [23]. Although the fraction of carbon atoms that reside in the tetrahedral sites is small (about one atom in a thousand at high temperatures), the deviation from linearity is nevertheless significant because the activation energy for jumps between adjacent tetrahedral sites is relatively small.

Let the activation free energies for the O→Te and Te→O jumps be G_O^* and G_{Te}^* respectively. The jump probability for O→Te is proportional to $\beta_{Te} \exp\{-G_O^*/RT\}$, where β_{Te} is the number of tetrahedral sites per solvent atom, into which a carbon atom originally in the octahedral site can jump. The probable ratio of the number of occupied tetrahedral sites (N_{Te}) to occupied octahedral sites is given by

$$\frac{N_{Te}}{N_O} = \frac{\beta_O}{\beta_{Te}} \exp\left\{ -\frac{G_{Te}^* - G_O^*}{RT} \right\}. \tag{3.15}$$

Since $\beta_O = 3$ and $\beta_{Te} = 6$, the fraction ϕ_O of carbon atoms in the octahedral sites is

$$\phi_O = 1 - \left(\frac{3}{6} \exp\left\{ -\frac{G_{Te}^* - G_O^*}{RT} \right\} + 1 \right)^{-1}. \tag{3.16}$$

Given the distribution of carbon atoms amongst the octahedral and tetrahedral sites, it becomes possible to consider the combined effect of the diffusion paths between near neighbour interstices:

(i)   An atom in a tetrahedral site can jump either into an adjacent Te site, or an O site.
(ii) Given that direct jumps between neighbouring octahedral sites must involve passage through a tetrahedral site, a third diffusion path is O-Te-O.

The overall diffusion coefficient D is then

$$\begin{aligned}
D = {} & \phi_O D_{O-Te-O} + (1 - \phi_O)\phi_{Te-Te} D_{Te-Te} \\
& + (1 - \phi_O)(1 - \phi_{Te-Te}) D_{Te-O-Te}
\end{aligned} \tag{3.17}$$

where ϕ_{Te-Te} is the fraction of carbon atoms in the tetrahedral sites which jump by the Te→Te route. By fitting this equation to the experimental data, McLellan et al. were able to explain the nonlinear behaviour in the Arrhenius plot, assuming the parameter values given in Table 3.3.

Table 3.3

Parameters describing the diffusion of carbon in ferrite. The first set is due to McLellan, Rudee and Ishibachi [23], derived by fitting the dual-occupancy model to experimental data. $D_{o,Te-Te}$ was corrected slightly by Condit and Beshers [24]. The second set is due to Beshers [16]. D_o is the pre-exponential in the equation $D = D_o \exp\{-Q/RT\}$, where Q is an activation enthalpy. G^* the activation free energy. U_O is the energy of a carbon atom in an octahedral site.

Dual occupancy model			Calculations using elasticity theory	
$D_{o,O-Te-O}$	3.3×10^{-7}	$m^2 s^{-1}$	Q_{O-O}/U_O	1
$D_{o,Te-Te}$	2.6×10^{-4}	$m^2 s^{-1}$	Q_{O-Te}/U_O	0.69
Q_{O-Te-O}	80.8	$kJ\,mol^{-1}$	Q_{Te-O}/U_O	0.31
Q_{Te-Te}	61.5	$kJ\,mol^{-1}$	Q_{Te-Te}/U_O	0.91
$G_{Te}^* - G_O^*$	$30.1 + 4.4RT$	$kJ\,mol^{-1}$		
ϕ_{Te-Te}	0.86			

The data in Table 3.3 show that the two models give a different trend for the activation energy for jumps between tetrahedral sites when compared with the O-Te-O route. Beshers predicts the latter route to have the smallest activation energy, whereas McLellan et al. attribute a smaller value for the Te-Te jumps. The diffusion data for carbon in ferrite have been measured using a variety of techniques including internal friction and mass flow experiments. In internal friction the carbon atoms jump to octahedral sites that are favoured by the applied stress so the appropriate activation energy is Q_{O-Te-O}. Atoms located in different kinds of sites can be distinguished by the temperature or frequency corresponding to the maximum in the damping of vibrations so there is less ambiguity in the diffusion path. In contrast, mass flow experiments access all possible diffusion paths. It is difficult to reconcile the dual occupancy model with the fact that the two kinds of experiments give identical diffusion coefficients in a regime that shows deviations from the strict Arrhenius law [22]. Only a single stage diffusion mechanism would give identical diffusivities with internal friction and mass flow experiments.

This discussion will be developed further when considering the influence of the magnetic transition in ferrite on the diffusion process.

3.4   DIFFUSION OF CARBON IN MARTENSITE

The irregular octahedral interstices with axes of magnitude a_α, $\sqrt{2}a_\alpha$ and $\sqrt{2}a_\alpha$ in the body-centred cubic lattice of ferrite are all crystallographically equivalent. There are three such interstices per iron atom. The anisotropic distortion that accumulates when carbon atoms can locate preferentially on one of the three sublattices of interstitial sites causes the lattice to become tetragonal and those particular interstices are then slightly expanded along the c-axis of the bcc lattice relative to the other two unoccupied sites, which are contracted relative to the cubic lattice, Figure 3.4a. During diffusion it may be necessary for the carbon atom to jump into the less favoured contracted holes. The diffusion of carbon in tetragonal martensite is therefore expected to be slower than in ferrite [25].

For the body-centred tetragonal lattice, Zener [10, 26, 27] has estimated the energy U_C to move a single carbon atom from a preferred to an unfavourable adjacent octahedral site. Assuming that the tetragonality in the first instance is caused by the application of a stress along [100] in a single-crystal sample of the random solution, the resulting strain ε causes an increase in the internal energy by $\frac{1}{2}E_{100}\varepsilon^2$, where E_{100} is the appropriate elastic modulus. If the random solution is now induced to order completely, then the total change in energy is

$$\frac{1}{2}E_{100}\varepsilon^2 - \frac{2}{3}N_V^C U_C \tag{3.18}$$

since only two-thirds of the carbon atoms need to redistribute in the transition from the random to the ordered state. The term N_V^C is the number of carbon atoms per unit volume. When the work done by the applied stress is equal to the reduction in energy on the ordering of the carbon atoms, the external stress is not needed to maintain the strain ε. This condition is realised by setting the derivative of the total energy with respect to the strain to zero,

$$E_{100}\varepsilon - \frac{2}{3}N_V^C \frac{dU_C}{d\varepsilon} = 0. \tag{3.19}$$

Zener assumed that the energy U_C, which is the difference in the energy of a carbon atom in a preferred site and the corresponding term in the unfavourable site, is proportional to the strain, i.e., $U_C \propto -E_{100}\varepsilon$. This is then converted into an equality by multiplying by the strain ε_* introduced when one carbon atom per unit volume is transferred into a preferred site:

$$U_C = E_{100}\,\varepsilon_*\varepsilon \tag{3.20}$$

For a sample in which randomly distributed carbon atoms change into a fully ordered configuration, the strain $\varepsilon_* = \varepsilon/(\frac{2}{3}N_V^C)$, so that

$$U_C = \frac{2}{3}E_{100}\,\varepsilon_*^2 N_V^C. \tag{3.21}$$

Hillert has estimated the $U_C = 4.6543 \times 10^{-20}\,w_C$ J mol^{-1}, where w_C is the weight percent of carbon. Hillert [28] argued that the effect of introducing tetragonality to the cubic lattice is to lower the energy of the preferred site by $\frac{1}{2}U_C$ and to raise that of the contracted site by the same amount (Figure 3.4b). In this way, the height of the barrier from the preferred site is increased by $\frac{1}{2}U_C$, so that the jump probability from that site is correspondingly reduced by a factor $\exp\{-N_aU_C/2RT\}$. Assuming that long-range diffusion is dominated by the largest barrier, he concluded that

$$D_{\alpha'} \approx D_\alpha \exp\{-N_aU_C/2RT\} \tag{3.22}$$

where α (bcc) and α' (bct) refer to ferrite and martensite respectively. The diffusivity in martensite should decrease exponentially with its carbon concentration assuming that the martensite is tetragonal at the temperature and concentration of interest.

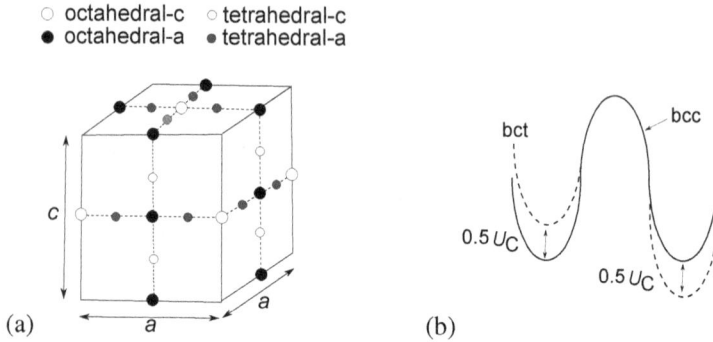

Figure 3.4 Octahedral and tetrahedral sites on the surface of a body-centred tetragonal cell [29]. The c/a ratio has been exaggerated for the purposes of illustration. Hollow circles represent expanded holes, whereas filled circles represent those that are contracted, the deformations being defined with respect to the cubic lattice. (b) The influence of tetragonality on the barrier to diffusion between adjacent octahedral sites.

3.5 INTERACTIONS BETWEEN CARBON ATOMS

3.5.1 REPULSION BETWEEN CARBON ATOMS

Carbon atoms in adjacent interstitial sites in austenite or in ferrite repel each other. The resulting interaction energy ω is defined in thermodynamic models as follows (p. 66):

$$\omega = \varepsilon_{uu} - 2\varepsilon_u \tag{3.23}$$

where ε_u is the energy per interstitial-vacancy/carbon pair and ε_{uu} is the energy per C-C pair. The properties of graphite suggest that the short-range repulsive component of the C-C interaction can be written [30, 31]:

$$U_{rep} = 1856 \exp(-35.75r) \qquad \text{eV} \tag{3.24}$$

per C-C interaction at distance r in nm.

It is found from the analysis of thermodynamic data that $\omega_\gamma = 8250\,\mathrm{J\,mol^{-1}}$ or 0.086 eV per C-C pair [32] and $\omega_\alpha \approx 150,000\,\mathrm{J\,mol^{-1}}$ or 1.54 eV per C-C pair [33]. The closest approach distance of carbon atoms in austenite is 0.252 nm, giving $U_{rep} = 0.23$ eV per pair. Similarly, the closest approach distance of carbon atoms in ferrite is 0.143 nm, giving $U_{rep} = 11.0\,\mathrm{eV}$ per pair. The values of U_{rep} and ω are not for direct comparison since the latter is a net interaction energy. But U_{rep} does explain why $\omega_\alpha \gg \omega_\gamma$; there is a much shorter approach distance for carbon atoms in ferrite.

Equation 3.24 is derived for graphite, where the C-C spacing between planes is large (0.335 nm) and it refers to the residual repulsion when the bonding molecular orbitals of the carbon atoms are filled. The equation should not be structure dependent provided that the carbon orbitals are saturated, as they probably are for carbon in transition metals. However, the expression strictly applies at large

C-C distances. While in principle, there could be a close distance of approach in ferrite (0.143 nm), in practice the carbon atoms are never likely to get as close as that, preferring to occupy second near neighbour sites. They are a good deal apart even in cementite (Chapter 8). On the other hand, carbon atoms get as close as 0.127 nm in certain transition metal carbides [34, pp. 66-67]. This happens when there is direct carbon-carbon covalent bonding, similar to that in diamond and within the basal planes of graphite, where the bond length is just 0.13 nm. In the unlikely event that the carbon atoms get as close as this in ferrite, then the expression for U_{rep} would have to be used alongside a strong molecular bonding term. There is much evidence against the existence of covalently bonded carbon pairs in ferrite, for example from the high mobility of carbon in ferrite.

The repulsion between carbon atoms complicates the analysis of diffusion. Experimental measurements usually assume that the diffusion coefficient does not depend on the magnitude of the *gradient* of concentration. It will become apparent later that the repulsive forces between the carbon atoms affect the migration of the carbon in a differential manner with respect to motion up or down the gradient. This problem is important in austenite because large gradients can arise, but not in ferrite where the low solubility of carbon limits the concentration gradients that can develop.

3.5.2   CLUSTERING OF INTERSTITIAL ATOMS

The probability in a random, ideal solution that interstitial atoms occupy adjacent sites is proportional to x^2, where x is the mole fraction of the solute. However, the adjacent sites are so close in ferrite ($\frac{1}{2}a_\alpha$) that electrostatic, thermodynamic and elastic strain considerations completely exclude them from simultaneous occupation. In austenite the nearest approach distance for a pair of interstitials is larger, but the probability that adjacent sites are simultaneously occupied is much less than expected at random because of the repulsion between carbon atoms.

In ferrite, the interstices located at $0, \frac{1}{2}, \frac{1}{2}$ and $0, 0, \frac{1}{2}$ have tetragonal symmetry; the long axis is in each case parallel to an edge of the body-centred cubic unit cell. Carbon atoms that at first are randomly dispersed amongst the three tetragonal orientations can order under the influence of stress. The favoured orientation is that where the distortion caused by the stress makes it easier to accommodate the carbon. This stress-induced ordering is responsible for the Snoek effect in which the damping of vibrations by the migration of atoms between sites can be used to estimate the mobility and concentration of dissolved carbon. A similar effect does not exist in austenite where all the octahedral interstices (which have cubic symmetry) are equivalent even under the influence of an external load. Consequently, the only information available about the distribution of carbon in austenite comes from thermodynamic analysis, that the carbon atoms in adjacent sites repel.

The clustering of nitrogen and carbon in ferrite has been assessed on the basis of the associated strain energy. The methods range from macroscopic elasticity [10] to those in which the strain energies are estimated using interatomic potentials. The focus here is on the latter which ought to be more reliable given that the local distortions may exceed the limits of linear elasticity theory. As emphasised in Section 3.5.1, clusters involving near-neighbour pairs are impossible in ferrite; the calculations therefore refer to more distant pairs of atoms as illustrated in Figure 3.5. The results rely on strain interactions alone without accounting for electrostatic effects; there is evidence to suggest that the interstitial atoms in iron are positively charged [12–15].

Strain interactions undoubtedly support a tendency for both carbon and nitrogen to cluster in iron [9, 29, 35, 36]. The di-carbon, tri-carbon and tetra-carbon binding energies have been estimated to be about 0.13, 0.36 and 0.66 eV respectively, with the binding energies for even larger clusters increasing by about 0.31 eV per additional atom. These energies are insensitive to the carbon concentration. The clusters are predicted to form as thin plates on $\{0\,0\,1\}$ planes when the carbon atoms are in octahedral sites. They can migrate by the motion of peripheral atoms, although the activation energy for a concerted motion is large.

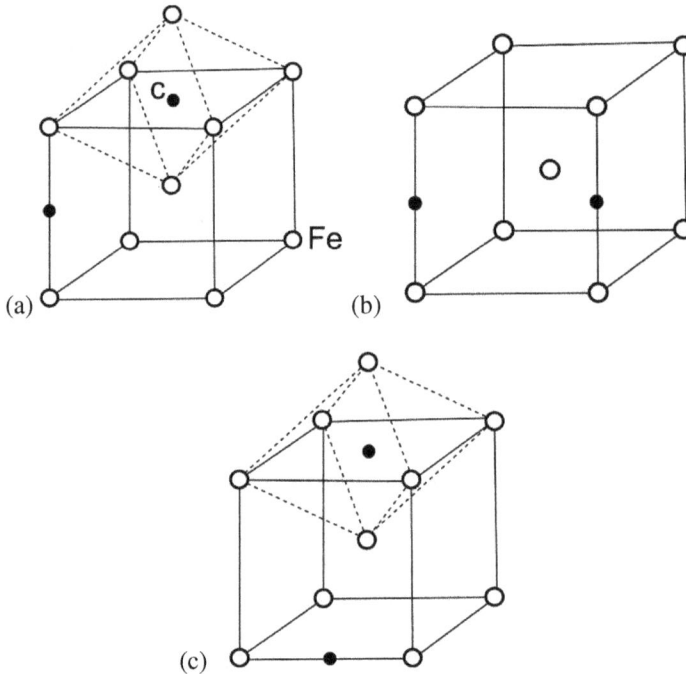

Figure 3.5 Three possible di-carbon clusters, which on the basis of strain energy alone are stable with binding energies of 0.13 eV, 0.11 eV and 0.08 eV for (a), (b) and (c) respectively. Note that in all cases the carbon atoms do not occupy near-neighbour interstitial sites. After Johnson, Dienes and Damask [35].

There is experimental evidence for the existence of carbon atom clusters in ferrite, and even for their diffusion through the lattice. Keefer and Wert [37, 38] in their studies of stress-induced ordering found that at low temperatures ($\approx 250\,\mathrm{K}$), the vast majority of nitrogen or carbon atoms in ferrite are isolated from each other. However, about 5% associate as pairs or triplets. The binding energies for C-C and N-N pairs were estimated to be about 0.08 eV and 0.065 eV respectively, consistent with the calculations by Johnson et al. [35]. For triplets, the corresponding binding energies were found to be 0.26 eV and 0.22 eV for carbon and nitrogen respectively. The mobility of clusters must naturally be smaller than that of individual atoms since their diffusion requires either the coordinated movement of the constituent atoms, or the breaking and reforming of the groups.

Keefer and Wert's experiments were based on the damping of vibrations so it was not possible to obtain a precise definition of the clusters. Given the large repulsion between carbon atoms in adjacent sites in ferrite, the clusters are presumably formed of next near-neighbour interstitial atoms separated by a_α.

The strain fields of the atoms that constitute the carbon pairs illustrated in Figure 3.5a,b are parallel; each pair can translate to the right by $a_\alpha/2$ [35]. The resulting reorientation of the strain fields can be detected in an internal friction experiment. For the case illustrated in Figure 3.5c, the two carbon atoms have their tetragonal axes perpendicular to each other and this would remain the case after diffusion; such motion cannot therefore be detected in internal friction experiments such as those of Keefer and Wert.

3.5.3 ASSOCIATION OF CARBON WITH DEFECTS

Much of the information on the migration of carbon atoms associated with defects such as iron-vacancies and iron-interstitials comes from irradiation experiments simulations of defect-clusters and their mechanisms of migration using interatomic potentials. The defects can trap solutes and hence alter the nature of precipitation reactions in irradiated iron.

Interstitial type complexes

The stable Fe-interstitial in α-iron has a "split" configuration in which two atoms are symmetrically disposed in a $\langle 110 \rangle$ direction about a normal site which is vacant. In Figure 3.6a, the carbon atom is lower than its normal position by $0.025a_\alpha$; the centre of the split interstitial is similarly displaced downwards by $0.06a_\alpha$, the distance between the atoms of the split interstitial being $0.75a_\alpha$ [9]. An interstitial such as this has a binding energy with the carbon atom of about $0.5\,\mathrm{eV}$. The migration energy for the isolated iron interstitial is $0.3\,\mathrm{eV}$. A complex may be formed when a migrating iron interstitial is trapped by association with carbon. For the iron-interstitial to escape from the carbon will necessitate overcoming an energy barrier greater than $0.8\,\mathrm{eV}$. The alternative proposal that the carbon can escape from the cluster is not relevant because the migration energy of carbon is larger ($0.86\,\mathrm{eV}$), meaning that it would need $1.36\,\mathrm{eV}$ to escape. It goes against intuition to think that the carbon is less mobile than the Fe-interstitial, but because the split interstitial is a high-energy configuration, the Fe can migrate freely above approximately $80\,\mathrm{K}$, whereas carbon becomes mobile in this context, at much higher temperatures ($\approx 200\,\mathrm{K}$).

An iron di-interstitial consisting of two parallel split interstitials is also possible and has a binding energy with the Fe-interstitial/carbon complex of about $1\,\mathrm{eV}$. This grande complex therefore dissociates at higher temperatures.

Vacancy type complexes

An iron vacancy in ferrite has a binding energy with an atom of carbon of about $\varepsilon_{C\square} = 0.41 - 0.48\,\mathrm{eV}$ [39]. A carbon atom encountering a vacancy does not fall into it, but moves towards it along $\langle 100 \rangle$ as illustrated in Figure 3.6b, remaining attached to five of the adjacent iron atoms. Positron annihilation studies indicate that the carbon atom is about $0.365a_\alpha$ from the vacancy centre, forming a dumbbell configuration with the vacancy [40]. In irradiated samples, precipitation causes

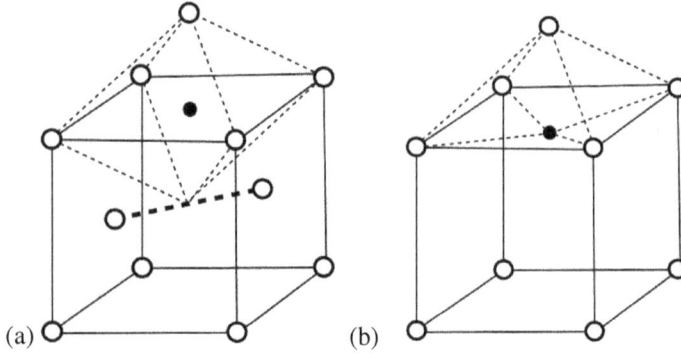

Figure 3.6 The most stable position of a carbon atom (black) in a defective unit cell of ferrite [9, 29]. (a) There is a split interstitial of a pair of iron atoms (heavy dashed line along $\langle 110 \rangle$) causing the carbon atom to relax downwards by 2.5% of the lattice parameter. (b) The effect of an iron vacancy on the position of carbon.

the carbon-vacancy complexes to dissociate at approximately 250 °C [41]. Notice that this does not mean, as is sometimes argued, that under equilibrium conditions the complex decomposes above 250 °C, since at that temperature the binding energy is almost an order of magnitude greater than the thermal energy kT. The problem is important because the association of a carbon atom with a host vacancy may influence its mobility. There is theory to handle this scenario, based on a generic treatment of the thermodynamics of interstitial-vacancy interactions in solid solutions [42], that has been applied to these interactions in ferrite [43]. The most probable complexes involve the pairing of a single host vacancy with a single carbon atom. The fraction $y_{C\square}$ of carbon atoms complexing in this way is given by [43]:

$$y_{C\square} = \frac{z x_{\square} \exp\{\varepsilon_{C\square}/kT\}}{\beta_0 (1 - f\{\theta, \varepsilon_{C\square}, x_{\square}\})}$$

$$f\{\theta, \varepsilon_{C\square}, x_{\square}\} = x_{\square}(1 + 4\theta \exp\{\varepsilon_{C\square}/kT\}) \tag{3.25}$$

where the equilibrium concentration of vacancies, $x_{\square} = \exp\{-\Delta H_F^{\square}/kT\}$, where $\Delta H_F^{\square} \approx 1.5\,\text{eV}$ is the enthalpy of formation of a vacancy.[1] Using an exaggerated value of the binding energy $\varepsilon_{C\square} = 0.8\,\text{eV}$ gives the fraction of carbon atoms that are present as carbon-vacancy pairs as $\approx 10^{-7}$. The logical conclusion is that this cannot perceptibly affect either the thermodynamic function of carbon in iron nor the diffusivity of carbon or iron [43]. This remains the case even if, for example, the concentration of vacancies is increased a hundred fold. Consistent with this, first principles calculations combined with Monte Carlo simulations have failed to reveal any significant effect of interstitial-vacancy complexes on equilibria between ferrite and nitride or carbide in Fe-N and Fe-C systems respectively [44]; these calculations reveal $\varepsilon_{X\square}$ for X=O, N or C as $1.43, 0.73$ and $0.41\,\text{eV}$. An iron vacancy in ferrite could in principle accommodate a pair of carbon atoms. However, kinetic data suggest that such di-carbon/vacancy clusters do not form in practice, presumably because of electrostatic repulsion between the carbon atoms [29].

It is suggested that the presence of carbon in austenite, by association, reduces the vacancy formation energy and thereby increases the self-diffusion coefficient of iron. However, these claims do not seem to be well-founded, first because of the small fraction of carbon atoms that would associate with vacancies, and second because the space within the host vacancy to associate with the carbon will be reduced and hence make self-diffusion more difficult.

There is a strong repulsion ($-7\,eV$) between a carbon atom and an iron vacancy in austenite [39]. The carbon should therefore remain in the octahedral holes even in the presence of adjacent vacancies. This is because a carbon atom in an iron vacancy would hardly interact with its now more distant iron atom neighbours.

3.6 DIFFUSION OF CARBON IN AUSTENITE

The diffusion coefficient of carbon in austenite increases by a factor of five as the concentration changes from 0 to 0.05 mole fraction [45, 46]. This cannot be explained in terms of the thermodynamic factors that are used to convert diffusion based solely on concentration gradients to the more logical free-energy gradients, as discussed in Section 3.16.

Carbon occupies octahedral interstices in austenite (Figure 3.7). Unlike ferrite, the octahedra are regular with all three axes equal in size. As a consequence, the distortion caused by carbon is spherically symmetrical and essentially interacts only with the hydrostatic component of the stress field of the dislocation. Carbon is therefore much less effective as a solid solution strengthening agent in austenite than in ferrite. Since the distortion it causes is symmetrical, there is no tendency for the carbon atoms in austenite to order.

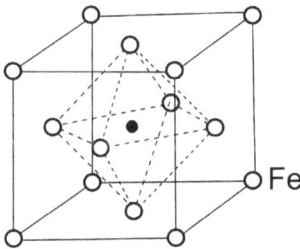

Figure 3.7 The octahedral interstices in austenite. Unlike the case for ferrite, each of the three axes of the octahedron is equivalent. There is one octahedral interstice per iron atom in austenite.

3.6.1 DEPENDENCE OF DIFFUSIVITY ON COMPOSITION

The sensitivity of the diffusivity of carbon in austenite has been investigated [47] using Eyring's absolute reaction rate theory. During diffusion, an atom which is in a potential well must overcome a barrier in order to translate into an adjacent equilibrium position. The rate theory postulates that an activated complex forms at the peak of the barrier, the energy of which differs from that in the potential well by the activation free energy G^*. The ratio of the activity of the activated complex to that of the reactant in the ground state is given by

$$\frac{a_m}{a} = \exp\left\{-\frac{G^*}{kT}\right\}. \tag{3.26}$$

Since the activity is related to the atom fraction by an activity coefficient Γ, the fraction of solute atoms that are in the activated state is given by

$$\frac{x_m}{x} = \frac{\Gamma}{\Gamma_m} \exp\left\{ -\frac{G^*}{kT} \right\} \tag{3.27}$$

where Γ_m is the activity coefficient of the solute atoms at the saddle point, assumed constant given that the concentration of activated complexes will always be small.

If the number of solute atoms per unit volume is c, then the number per unit area of a plane with spacing δ_s is $c\delta_s$; of these, $(x_m/x)c\delta_s$ will be in the activated state. The number of jumps in a unit of time across a unit area of the $\{002\}$ planes designated 2 and 3 in Figure 3.8 is proportional to the number of activated complexes and the attempt frequency kT_D/h where T_D is the Debye temperature.

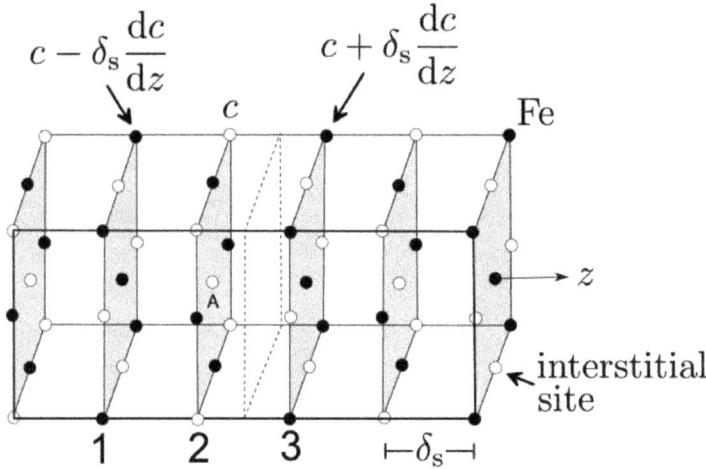

Figure 3.8 Illustration of the concentration gradient along the z direction of the austenite lattice. The iron atoms and interstitial sites are illustrated. The $\{002\}$ austenite planes are shaded. Diffusion is considered across the imaginary (dashed) plane sandwiched between planes 2 and 3.

This leads to the flux in the positive z direction:

$$\begin{aligned}
J^+ &= c\delta_s \frac{kT_D}{h} \exp\left\{ -\frac{G^*}{kT} \right\} \frac{\Gamma}{\Gamma_m} \frac{1}{m^+} \\
&= c\delta_s K_o \frac{\Gamma}{\Gamma_m} \frac{1}{m^+}
\end{aligned} \tag{3.28}$$

where the activity of solute on plane 2 has been replaced by the product of its concentration and activity coefficient. The aggregation term K_o is therefore a constant at a fixed temperature. The transmission coefficient $\frac{1}{m^+}$ represents the fraction of all possible jumps which are in the forward direction. The flux in the reverse direction (i.e., from plane 3 to 2) is

$$J^- = \frac{K_o \delta_s}{\Gamma_m} \left(c + \delta_s \frac{dc}{dz} \right) \left(\Gamma + \delta_s \frac{d\Gamma}{dz} \right) \frac{1}{m^-} \tag{3.29}$$

since the concentration and activity of solute changes by $\delta_s \frac{dc}{dz}$ and $\delta_s \frac{d\Gamma}{dz}$ respectively on going from plane 2 to plane 3. If it is assumed that $m^+ = m^- \equiv m$ and when terms higher than δ_s^2 are neglected, the net diffusion flux is found to be

$$J = J^+ - J^-$$
$$= -\frac{K_o \delta_s^2}{m \Gamma_m} \underbrace{\Gamma \left(\frac{d \ln \Gamma}{d \ln c} + 1 \right)}_{\text{thermodynamic factor}} \frac{dc}{dz}. \tag{3.30}$$

Figure 3.9 shows how the variation in the diffusion coefficient with the carbon concentration as measured experimentally is more pronounced that that calculated using Equation 3.30. Evidently, there are other effects which cause diffusion to be faster than just the thermodynamic factor. The term $\frac{d \ln \Gamma}{d \ln c} + 1$ is alternatively written $c \frac{d \ln \Gamma}{dc} + 1$ or $\frac{d \ln a}{d \ln c}$.

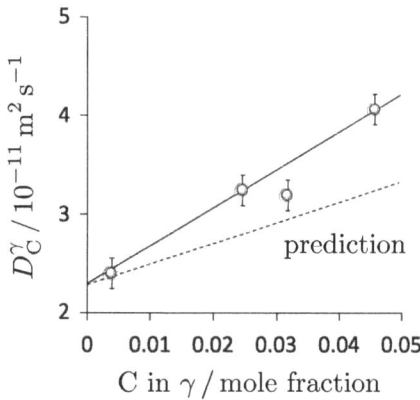

Figure 3.9 Comparison of the measured diffusivity of carbon in austenite, against that calculated allowing for the thermodynamic factor. Data from Kirkaldy [48].

3.6.2 DEPENDENCE OF DIFFUSIVITY ON C-C INTERACTIONS

Smith [46], and later Kirkaldy [48], demonstrated that neither the thermodynamic factor nor the proportionality of mobility to the activity coefficient can account for the composition dependence of the diffusivity of carbon in austenite.

There is another effect peculiar to the strong repulsion between carbon atoms in solution that resolves the puzzle [49, 50]. The *gradient* of carbon concentration then influences migration because the probability of interstitial site occupation in the vicinity of an occupied site depends on the repulsion. In such a gradient, a carbon atom attempting random motion therefore sees an exaggerated difference in the number of available sites in the forward and reverse directions, making diffusion down the gradient particularly favourable. A carbon atom is affected more by the repulsion when it jumps into a carbon-rich region up a concentration gradient then when it enters a lower concentration regime down that gradient.

These repulsive interactions contribute to the concentration dependence of the diffusion coefficient; the description that follows is due to Siller and McLellan [49, 50], who incorporated the repulsion explicitly via the transmission coefficients $1/m^+$ or $1/m^-$. Referring to Figure 3.8, an atom on the

ith plane can jump on to planes $(i-1)$, (i) or $(i+1)$; of all these possibilities, just a few contribute to flow in the intended direction. It is this fraction of correctly oriented jumps that the transmission coefficient represents. Figure 3.8 shows also the twelve possible near-neighbour sites that a carbon atom located at position "A" can jump into, four on its own plane and four each on the two adjacent planes. Geometric considerations alone therefore reduce the chance of a jump in the forward direction by a factor of three; there is a further reduction by a factor of 2 because the activated complex can relax with equal probability in the forward and backward directions [51]. Therefore, the transmission coefficient $\frac{1}{m^+}$ is, for diffusion jumps from plane 2 to 3, given by

$$\frac{1}{m^+} = \frac{1}{2} \times \frac{s_3}{s_1 + s_2 + s_3} \tag{3.31}$$

where the subscripts give the location of $\{0\,0\,2\}$ planes in the sequence shown in Figure 3.8 and the terms s_i represent the number of available sites for diffusion on the ith plane. If $s_i = 1$ for all i, then the transmission coefficient is $\frac{1}{6}$ since four of the twelve available sites are on the forward plane. Similarly, for jumps in the reverse direction from plane 3 to 2, the transmission coefficient is

$$\frac{1}{m^-} = \frac{1}{2} \times \frac{s_2}{s_2 + s_3 + s_4}. \tag{3.32}$$

It remains to estimate the terms s_i. On each plane there will be some sites already occupied by carbon atoms, and others that are vacant but excluded from occupation by the repulsive forces between adjacent carbon atoms. Noting that in austenite there is one octahedral interstitial site per iron atom,

$$s_i = N_P^{Fe}\left\{1 - \theta_i - \frac{B_i}{N_P^{Fe}}\right\} \tag{3.33}$$

where N_P^{Fe} is the number of iron atoms per plane, θ_i accounts for the fraction of sites that are occupied by carbon, and B_i is the number of sites on plane i which are excluded from occupation because of the repulsion from carbon atoms in adjacent sites. Thus,

$$B_i = \frac{1}{3} \sum_{j=i-1}^{j=i+1} N_j^{e\text{-}u} p_j \tag{3.34}$$

where $N_j^{e\text{-}u}$ is the number of unoccupied interstitial sites which are nearest neighbours to carbon atoms on plane j (the "e" representing an empty site, and "u" a site occupied by a carbon atom); p is the probability that one of these sites is excluded from occupation by a neighbouring carbon atom. The factor of $\frac{1}{3}$ is to avoid counting the same exclusion three times when considering separately, each plane in the set of three that are included in Equation 3.34.

Using the appropriate thermodynamic model for the solution of carbon in austenite, McLellan and Dunn (pp. 61-63) showed that the equilibrium number of e-u pairs is

$$N_{e\text{-}u} = \frac{W\bar{\lambda}}{2} \tag{3.35}$$

$$= \frac{W}{2}\frac{N_P^{Fe}}{2\phi}[1 - (1 - 4\phi\theta(1-\theta))^{1/2}] \tag{3.36}$$

where $\theta = x_1/(1-x_1)$ is the ratio of the number of carbon atoms to the total number of solvent atoms, W is the number of octahedral interstices around a single such interstice; $W = 12$ for octahedral sites in austenite, $\phi = 1 - \exp(-\omega_\gamma/kT)$; ω_γ is the nearest neighbour carbon-carbon interaction energy in austenite. It remains to solve for the probability p, which is

$$p = 1 - \frac{N_{\text{u-u}}}{N_{\text{u}}} \tag{3.37}$$

where N_{u} is the number of carbon atoms and $N_{\text{u-u}}$ is the number of carbon-carbon pairs. The latter is the difference between the number of carbon atoms and the number of carbon-vacancy pairs, i.e.,

$$N_{\text{u-u}} = \frac{1}{2}W(N_{\text{u}} - \bar{\lambda}) \tag{3.38}$$

All the parameters necessary to solve for the transmission coefficients are therefore determined. As before, the coefficients can be used to calculate the forward and reverse fluxes across a plane and hence the net flux (and ignoring terms higher than δ_s^2):

$$
\begin{aligned}
J &= J^+ - J^- \\
&= -\frac{K_o \delta_s^2}{6\Gamma_m}\Gamma\left[\frac{Wc}{1 - (\frac{W}{2}+1)\theta + \frac{W}{2}(\frac{W}{2}+1)(1-\phi)\theta^2}\frac{d\theta}{dc} + \frac{d\ln\Gamma}{d\ln c} + 1\right]\frac{dc}{dz}.
\end{aligned}
\tag{3.39}
$$

By comparing the net flux with Fick's first law, the diffusion coefficient of carbon (D_{11}) that defines its flux relative to the gradient in carbon concentration is given by[2]

$$D_{11}\{x_1, T\} = \frac{kT_D}{h}\left(\frac{\delta_s^2}{6\Gamma_m}\right)\exp\left\{-\frac{G^*}{kT}\right\}\eta\{\theta\} \tag{3.40}$$

with

$$
\begin{aligned}
\frac{\eta\{\theta\}}{a_1^\gamma} &= 1 + \left[\frac{W(1+\theta)}{1 - (0.5W+1)\theta + (0.25W^2 + 0.5W)(1-\phi)\theta^2}\right] \\
&\quad + (1+\theta)\frac{1}{a_1^\gamma}\frac{da_1^\gamma}{d\theta}
\end{aligned}
\tag{3.41}
$$

where G^* is, in the notation of absolute reaction rate theory, the free energy difference between the activated complex and the reactants when each is in its standard state at the temperature of the reaction, defined here to be independent of the temperature and composition of the austenite; this is comparable to the activation free energy of the standard Arrhenius equation. Bhadeshia [52] found $\delta_s G^*/k = 21,230\,\text{K}$ and $\ln\{6\Gamma_m/\delta_s^2\} = 31.84$. Table 3.4 shows typically how the diffusivity depends on the C-C interaction energy.

3.6.3 DILATATION EFFECTS

Carbon atoms are larger than the interstices in which they reside in iron. The resulting distortions can be categorised into two types. At constant pressure, there is a mean distortion of the crystal as a

Table 3.4

Diffusion coefficient of carbon in Fe-C austenite at 1273 K, calculated using Siller and McLellan's method [50] for a variety of carbon-carbon interaction energies.

C / wt%	C-C interaction energy / $J\,mol^{-1}$	Diffusivity / $10^{-10}\,m^2\,s^{-1}$
0.2	0	0.198
0.2	8250	0.221
0.2	16500	0.233
0.2	24750	0.239
0.9	0	0.321
0.9	8250	0.509
0.9	16500	0.626
0.9	24750	0.693

whole, the magnitude of which is approximately a linear function of the concentration (Vegard's law, [53]). There are the local distortions around each atom of carbon, defined relative to the positions of the atoms in the homogeneously distorted crystal [54]. In the context of diffusion, the homogeneous distortions lead to a variation in the specific volume along any concentration gradient. The activation free energy for diffusion then becomes a function of the position along the gradient. Its dependence on the gradient of concentration can be approximated as a linear function of concentration [55, 56]:

$$G^* = G_o^* + \theta\,\frac{dG^*}{d\theta} \tag{3.42}$$

where $\theta = x_1/(1 - x_1)$ is as usual, the ratio of the number of carbon atoms to the total number of solvent atoms and G_o^* is the characteristic activation free energy for an isolated solute atom. The term $dG^*/d\theta$ has a value of about $-40\,kJ\,mol^{-1}$ when compared with $G_o^* \approx 150\,kJ\,mol^{-1}$, so G^* decreases by about $2.5\,kJ\,mol^{-1}$ as θ changes from 0 to 0.06.

$dG^*/d\theta$ is the product of the activation volume V^*, and the rate at which the ghost pressure required to restore the austenite to its undilated state changes with carbon concentration [57]:

$$\frac{dG^*}{d\theta} = -K \times \left[\frac{3}{4}\,\frac{N_a}{V_m}\,a_\gamma^2\,\frac{da_\gamma}{d\theta}\right] \times V^* \tag{3.43}$$

where a_γ and V_m are the lattice parameter and molar volume of austenite, respectively; K is the bulk elastic modulus. The terms containing a_γ represent the volume strain per unit concentration of carbon so its product with the bulk modulus gives the ghost pressure. After making appropriate substitutions it is found that the activation volume is about $0.6 \times 10^{-6}\,m^3\,mol^{-1}$ [58], which is small, and compares against a measurement for carbon in ferrite by Bass and Lasarus [19] of almost zero at $(0.0 \pm 0.1 \times 10^{-6}\,m^3\,mol^{-1})$; it may as well be assumed to be zero [59]. Substitutional atoms have large activation volumes in the solid-state (almost equal to the molar volume) because their migration requires both the formation and motion of vacancies. The formation part is absent for carbon and its motion seems to cause little change in volume, perhaps because it is partly ionised

[59]. The conclusion is that pressure does not have a significant influence on the mobility of carbon atoms over the range studied.

These results emphasise that the diffusion behaviour of a carbon atom in a concentration gradient is expected to be especially different from that in a homogeneous solution. Parris and McLellan [60] demonstrated this by measuring the mobility of carbon in a diffusion couple, one half of which contained ^{12}C atoms and the other half the same concentration of ^{14}C atoms. The *interstitial tracer diffusivity* of carbon was found to be smaller in the couple than when diffusion occurred in a finite concentration gradient (Figure 3.10). Notice that in the tracer experiments, there nevertheless is a dependence of D on θ, but the differential effects of the gradient are eliminated.

For tracer diffusion, the transmission coefficients in the forward and reverse directions are equal, i.e., $1/m^+ = 1/m^- = 1/m$, since there is no concentration gradient. But the transmission coefficient still depends on concentration since some of the possible jump sites may already be occupied or excluded by the C-C repulsion. From Equations 3.33 and 3.36,

$$
\begin{aligned}
\frac{6}{m} &= N_P\left\{1 - \theta - \frac{B\{\theta\}}{N_P}\right\} \\
&= 1 - \theta - \frac{3\chi}{\phi}\left[1 - \frac{1}{\theta}\left(1 - \frac{\chi}{2\phi\theta}\right)\right]
\end{aligned}
\tag{3.44}
$$

where

$$
\chi = [1 - (1 - 4\phi\theta(1 - \theta))^{1/2}].
\tag{3.45}
$$

There also is the concentration dependence via the thermodynamic factors (Equation 3.30):

$$
\Gamma\left(\frac{d\ln\Gamma}{d\ln c} + 1\right) = a + (1 + \theta)\frac{da}{d\theta}
\tag{3.46}
$$

so the interstitial tracer diffusion coefficient D_C^* becomes

$$
D_C^* = \frac{K_o\delta_s^2}{6\Gamma_m}\left\{1 - \theta - \frac{3\chi}{\phi}\left[1 - \frac{1}{\theta}\left(1 - \frac{\chi}{2\phi\theta}\right)\right]\right\} \times \left\{a + (1 + \theta)\frac{da}{d\theta}\right\}.
\tag{3.47}
$$

In these tracer experiments, although $d\theta/dz = 0$, $da/d\theta$ is not zero, as assumed by Liu et al. [61]. There are other treatments [62] of the tracer diffusivity which appear to neglect all but the thermodynamic factors and therefore are unable to predict correctly at large concentrations.

3.6.4 ÅGREN'S METHOD

In this practically useful method [63], the concentration dependence of diffusivity of carbon in austenite in Fe-C alloys is as usual a function of the thermodynamic factor, together with an empirical concentration dependence of the activation free energy of diffusion:

$$
\begin{aligned}
D_C^\gamma &= 4.53 \times 10^{-7}\left[1 + \theta(1 - \theta)\frac{8339.9}{T}\right] \times \\
&\quad \exp\left\{-\left(\frac{1}{T} - 2.221 \times 10^{-4}\right)(17767 - 26436\theta)\right\} \ m^2 s^{-1}
\end{aligned}
\tag{3.48}
$$

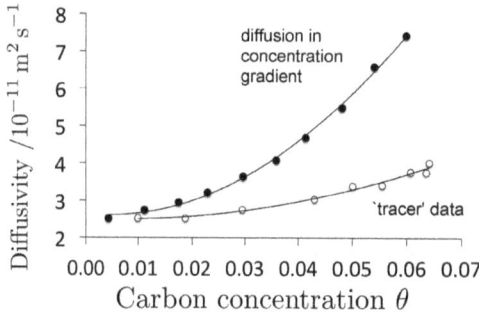

Figure 3.10 The interstitial tracer-diffusivity at 1273 K of carbon as a function of θ, compared with corresponding data for diffusion in a finite concentration gradient. Data from Parris and McLellan [60].

3.7 DIFFUSION OF NITROGEN IN FERRITE

Nitrogen causes a slightly smaller distortion of the iron lattice than carbon:

$$\Delta a_\alpha = (0.84 \pm 0.08) \times 10^{-3} \quad \text{nm per at.\% C}$$

$$\Delta a_\alpha = (0.79 \pm 0.07) \times 10^{-3} \quad \text{nm per at.\% N} \tag{3.49}$$

where Δa_α is the change in the lattice parameter of ferrite. As a consequence, its activation energy for diffusion is smaller than that for carbon (Table 3.5).

Table 3.5

The diffusion parameters for carbon and nitrogen in ferrite, derived using low temperature ($< 75°$C) data only [22].

Interstitial	$D_o / \text{m}^2 \text{s}^{-1}$	$Q / \text{kJ mol}^{-1}$
Carbon	1.67×10^{-7}	78.1
Nitrogen	1.26×10^{-7}	73.4

As for carbon, nitrogen in ferrite shows deviations from the Arrhenius law at high temperatures. To a high level of accuracy, the diffusion data for nitrogen in ferrite can be described by the empirical equation [22]:

$$\ln D_N^\alpha = -13.695 - 0.9979 \left(\frac{10^4}{T} \right) + 0.0014 \left(\frac{10^4}{T} \right)^2 \quad \text{m}^2 \text{s}^{-1} \tag{3.50}$$

The third coefficient in this equation represents the deviation from the Arrhenius plot; its magnitude is about $2\frac{1}{2}$ times smaller than the corresponding term for carbon in ferrite (0.0037, Equation 3.14). Data such as those presented in Table 3.5 should therefore be treated with caution because the activation energy as interpreted in diffusion theory is temperature dependent. This will be discussed further when considering the effect of the ferromagnetic to paramagnetic transition in ferrite on the diffusion process.

The activation volume for the diffusion of nitrogen in ferrite is $(3.7 \pm 3.3) \times 10^{-8}\,\mathrm{m^3\,mol^{-1}}$, which is only about 4% of the volume of a mole of nitrogen in solid solution [17]. Hydrostatic pressure therefore has only a minor influence on diffusivity.

3.8   DIFFUSION OF NITROGEN IN AUSTENITE

By monitoring the outgassing of a thin, nitrided sample of austenitic stainless steel (20Cr-25Ni wt%) containing 0.5 wt% of diffusible nitrogen over the temperature range 1223-1323 K, the diffusion coefficient of nitrogen in austenite has been determined to be [64]

$$D_N^\gamma = 2.18 \times 10^8 \exp\left\{-\frac{3.16 \pm 0.05 \times 10^5}{RT}\right\} \quad \mathrm{m^2\,s^{-1}}, \qquad (3.51)$$

obeying an Arrhenius relationship over the range of the data investigated. The activation energy is about half that for nitrogen diffusion in ferrite.

Very large concentrations of nitrogen (38 at.%) can be introduced into the surfaces of austenitic steels either by diffusion or using techniques such as ion implantation [65, 66]. The lattice parameter of the austenite then expands dramatically by 10%. There is sharp transition observed between the lattice parameters of the expanded austenite and that of the unaffected substrate. The diffusion coefficients for nitrogen in the expanded austenite in a stainless steel then become concentration dependent, as illustrated for 693 K in Figure 3.11; the data are compared against a calculation done using Equation 3.51. It is evident that at low and high nitrogen concentrations, there isn't much of a difference with the value extrapolated from the high-temperature data, but at intermediate concentrations, the nitrogen diffuses much faster in the expanded austenite. At first, the increase in nitrogen leads to a greater dilation of the austenite, making it easier for it to diffuse. At very large concentrations, the diffusivity decreases again because of the blocking action of interstitial sites that already are occupied [67].

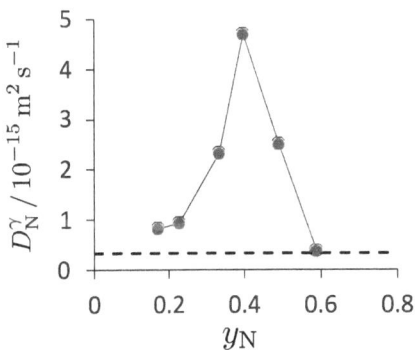

Figure 3.11   The points represent a selected data for the diffusion of nitrogen at 693 K in expanded austenite (304 stainless steel enriched with nitrogen) [67]. y_N represents the fraction of octahedral sites occupied by nitrogen. The dashed line is a calculation done by extrapolating the high-temperature data for ordinary austenite [64], using Equation 3.51.

3.9   DIFFUSION OF C AND N IN CEMENTITE AND HÄGG CARBIDE

The diffusivity of carbon in cementite has been deduced by measuring the rate of cementite formation on iron powders in CH_4/H_2 gas [68]. At 723 K the diffusivity is found to be about 10^{-19}-

$10^{-20}\,\text{m}^2\,\text{s}^{-1}$, which is many orders of magnitude smaller than for carbon in ferrite. There may be a simple explanation for the much reduced mobility in cementite; the number of vacant interstitial sites in ferrite is much greater. The diffusion coefficient increases with the activity of carbon in the gas, consistent with the fact that the cementite is not strictly stoichiometric (Chapter 8) with diffusion occurring by an interstitial or interstitialcy mechanism.

Recent measurements using similar techniques give the self-diffusion coefficients of carbon in cementite and Hägg carbide at 773 K as [69]

$$
\begin{aligned}
&\text{Cementite (Fe}_3\text{C)} &&6.0 \times 10^{-18}\,\text{m}^2\,\text{s}^{-1} \\
&\text{Hägg carbide (Fe}_5\text{C}_2) &&8.5 \times 10^{-18}\,\text{m}^2\,\text{s}^{-1}.
\end{aligned}
$$

The approximate temperature dependence of the self-diffusion coefficient of carbon in cementite, obtained by combining published data, is expressed as [69]

$$
D_C^\theta \approx 1.8 \times 10^{-6} \exp\left\{-\frac{172.8\,\text{kJ}\,\text{mol}^{-1}}{RT}\right\}\,\text{m}^2\,\text{s}^{-1}. \tag{3.52}
$$

The carbon atoms in cementite occupy four energetically preferred prismatic interstitial sites (Section 8.1) and there are four unoccupied but smaller octahedral sites within the unit cell. Molecular dynamics simulations indicate that the mechanism involves carbon diffusion jumps to the favoured sites via the unoccupied interstices (Figure 3.12), although the calculated activation energy for this process is somewhat smaller than indicated in Equation 3.52, at about $125\,\text{kJ}\,\text{mol}^{-1}$ [70, 71]. First principle calculations for the same mechanism indicate a diffusion barrier for carbon of $221\,\text{kJ}\,\text{mol}^{-1}$ and that the barrier is independent of the jump direction implying that carbon diffusion in cementite is likely to be isotropic in spite of its orthorhombic symmetry [72].

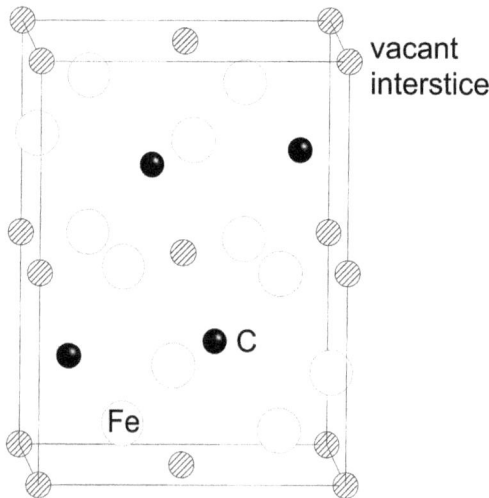

Figure 3.12 Showing the location of vacant octahedral interstitial sites within the cementite crystal structure. There are four such sites per unit cell at the equivalent positions (Wyckoff 4a) listed in Table 8.1.

The activation energy for the tracer-diffusion of nitrogen for dilute solutions in cementite is far smaller than that of carbon, at just $65 \pm 2\,\text{kJ}\,\text{mol}^{-1}$ [73].

3.10   MIGRATION OF POINT DEFECTS

The equilibrium concentration of iron-interstitials will be extremely small, but not so when the metal is irradiated with neutrons or other high energy particles. Some of the steel in a typical fission reactor may experience 20-30 displacements per atom over its service life. The stability and mobility of the resulting defects has been studied both experimentally and using theory.

3.10.1   IRON INTERSTITIALS

Many kinds of iron interstitials have been investigated; of these, six configurations are found to be at metastable equilibrium because of their symmetry (Figure 3.13 [74]). The first three cases illustrated are designated *split* because the pair of atoms in the interstitial configuration are placed symmetrically about a vacant iron-site. The spacing between each such pair is about $0.75a_\alpha$.

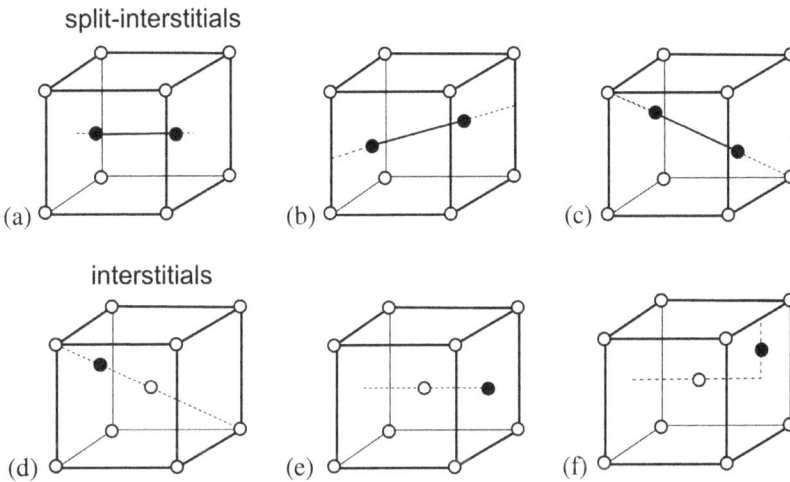

Figure 3.13 Computed structures of iron interstitials in ferrite [9]. The dark circles represent the atoms in interstitial configuration. (a) $\langle 100 \rangle$ split interstitial; (b) $\langle 110 \rangle$ split interstitial; (c) $\langle 111 \rangle$ split interstitial or crowdion; (d) activated crowdion; (e) octahedral interstitial; (f) tetrahedral interstitial.

The crowdion configuration illustrated in Figure 3.13c is so-called because it has the distortion relaxed along the close-packed row of atoms, constraining it to move along that direction. This particular crowdion is barely stable with a small local minimum in energy. Its migration along $\langle 111 \rangle$ would take it into the arrangement illustrated in Figure 3.13d before it recovers its original configuration; since the path involved is $c \to d \to c$, the interstitial of Figure 3.13d is said to be an activated crowdion, the activation energy for the motion being only 0.04 eV.

The calculated energies of the variety of interstitials are given in Table 3.6, from which it is apparent that the $\langle 110 \rangle$ split interstitial is the minimum energy configuration. The motion of such an interstitial is illustrated in Figure 3.14; the activation energy for the sequence is about 0.33 eV. Johnson also found many cases where two split-interstitials occupy adjacent positions with a reduction in energy, leading to the formation of a *di-interstitial*. For example, two $\langle 110 \rangle$ split interstitials located on adjacent lattice sites with their axes perpendicular to the line joining them have a binding energy relative to the separated pair of 1.08 eV. The migration of a di-interstitial complex occurs by dissociation and regeneration. It is worth noting that the relative stabilities of defects in ferrite are different from those in other bcc transition metals. This is because the magnetic moments, particularly on incorrectly located atoms and near neighbouring atoms, change significantly relative to the defect-free lattice [75].

Table 3.6
The energy of each of the interstitial configuration illustrated in Figure 3.13, less that of the stable configuration, Figure 3.13b. Ω' is the volume expansion relative to the atomic volume of iron. Configuration (g) is not illustrated in Figure 3.13 because it is the saddle point configuration during the motion of the $\langle 110 \rangle$ split interstitial and is difficult to describe. After Johnson, [74]. The values in brackets are from first principles calculations due to Derlet et al. [76].

Configuration	Relative energy / eV		Ω'
a	1.29	(1.12)	1.7
b	0.00	(0.00)	1.6
c	0.32	(0.68)	1.7
d	0.36	(0.71)	1.7
e	1.12	(1.28)	1.4
f	0.85	(0.39)	1.5
g	0.33		1.7

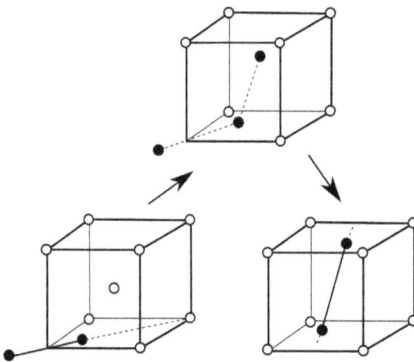

Figure 3.14 The migration sequence of a $\langle 110 \rangle$ split interstitial in ferrite. The cube at the top of the diagram is the saddle point configuration. Adapted from Johnson [74].

3.10.2 IRON DEFECTS IN AUSTENITE

There do not seem to have been any calculations of the iron-interstitial configurations in austenite, but work on copper [77, 78] indicates the stable forms illustrated in Figure 3.15. In the dumb-bell the insertion of the additional atom is accommodated by sharing with an adjacent atom with most of the distortion along $\langle 100 \rangle$. As might be expected, the body-centred interstitial causes more isotropic distortion. Figure 3.15c shows the $\langle 110 \rangle$ crowdion, an arrangement which allows the distortions to be spread over a large distance along a straight line. It is estimated that the crowdion energy is much larger than that of the other two stable configurations. Di-interstitials are also possible, particularly pairs of dumb-bells.

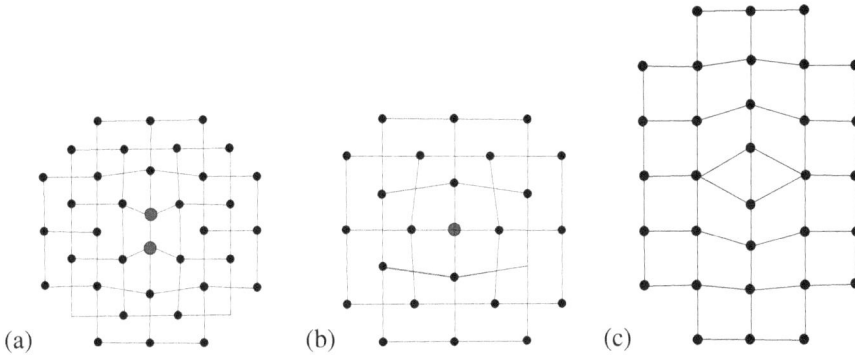

Figure 3.15 Iron-interstitial defects in austenite. (a) $\langle 100 \rangle$ dumb-bell configuration; (b) body-centred interstitial; (c) crowdion. Based on calculations by Johnson and Brown [77] on copper, with diagrams adapted from Thompson [78].

3.10.3 VACANCIES

A vacancy is an iron atom that is absent from its normal site in the lattice; its motion is simple, a place exchange with a neighbouring atom. For ferrite this has a negligible activation volume [74]. The energy barrier to vacancy motion has a slight local minimum at the intermediate position (Figure 3.16), of about 0.04 eV which is small and probably of no practical importance. In Figure 3.16, the computed curve does not extend exactly to the [000] or [111] positions because the migrating atom relaxes towards the vacancy.

Di-vacancies are simply pairs of vacancies in close proximity, with a binding energy given relative to the two isolated vacancies. It is found that the binding energies vary from 0.13 eV, 0.20 eV and 0.05 eV for the nearest, second-nearest and fourth-nearest neighbour pairs respective. All other combinations do not give appreciable binding energies. The second-nearest configuration is the most stable of these; there is a reduction of 0.1 atomic volume when the two vacancies merge. Di-vacancies migrate in a step-wise process in two possible ways: $2 \rightarrow 1 \rightarrow 2$ and $2 \rightarrow 3 \rightarrow 2$, where the numbers 1, 2 and 3 refer to the first, second and third nearest neighbour configurations.

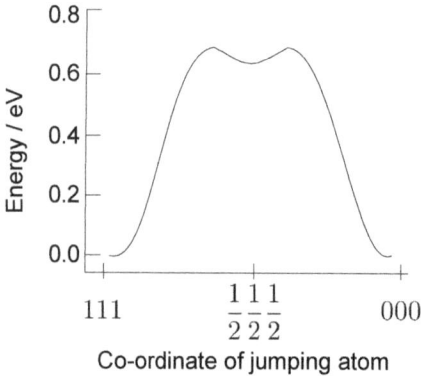

Figure 3.16 The barrier to the migration of a vacancy in ferrite. The plot is of the change in energy as a function of the movement of the atom which swaps positions with the vacancy, beginning at a cell corner and ending up at the opposite cell corner along a body diagonal [74].

3.11 MIGRATION OF HYDROGEN AND DEUTERIUM

3.11.1 DIFFUSION IN FERRITE

Hydrogen is notorious in its ability to embrittle ferritic iron in particular, even though it has an equilibrium solubility which is only a few tens of atoms per billion at 300 K. The concentrations are so small that it is only when the hydrogen can diffuse and concentrate at the stress field of a crack that it embrittles. At 300 K, its diffusivity in ferrite is comparable with the self-diffusion coefficient of water [79]:

$$D_H^\alpha = \left(1.6^{+0.94}_{-0.59}\right) \times 10^{-7} \exp\left\{-\frac{Q}{RT}\right\} \quad m^2 s^{-1}$$

$$\text{with} \quad Q = 7076 \pm 126 \, J \, mol^{-1} \text{for the temperature range: 322-799 K} \qquad (3.53)$$

The diffusion of H or ^{2}H is in practice complicated by interactions with defects such as dislocations and other traps within the microstructure [80, 81]. The effect of such traps is to reduce the apparent diffusivity. There are two kinds of traps, those which can become saturated (e.g., dislocations, grain boundaries) and others (e.g., pores) that establish an equilibrium with the dissolved hydrogen. In the latter case, the hydrogen probably becomes permanently trapped in its molecular form.

The temperature dependence of the diffusion coefficient of H or ^{2}H is consistent with Arrhenius behaviour with an activation energy that is identical for both isotopes. The pre-exponential factor, which contains the attempt frequency, is smaller for deuterium than for hydrogen, by a factor which changes from 1.8 at 322 K to 1.3 at 639 K [80]. On the basis of the isotope effect alone (Equation 3.6), the ratio would be expected to be $\sqrt{2}$ because deuterium has twice the mass of hydrogen; the natural frequency of vibration scales with mass$^{-1/2}$. It is possible that quantum mechanical effects cannot be neglected for these light particles [82]. It is considered that classical behaviour dominates when kT is much greater than the zero-point energy. Heller [83] has estimated that for a particle with the mass of a proton in the ferrite lattice, kT becomes equal to the zero-point energy at about 480 K, indicating that classical diffusion theory is not strictly appropriate for hydrogen.

Quantum tunnelling becomes important when the particle momentum is small, because the de Broglie wavelength then becomes large. The wavelengths for hydrogen and deuterium are 0.98-0.18 nm and 0.69-0.13 nm over the temperature range 10-300 K, which are greater than or similar to the typical widths of activation barriers. According to calculations, tunnelling leads to a noticeable upturn in the Arrhenius plot for the diffusion of hydrogen in ferrite at temperatures below about 250 K [82]. This is illustrated in Figure 3.17 where the calculated curve clearly is not linear – the calculations have been shown to be consistent with a range of experimental data in the domain where they exist. This kind of positive curvature in the plot of the log rate versus reciprocal temperature has been predicted for some time by applying quantum statistics with tunnelling to formulate a rate theory for light particles [84].

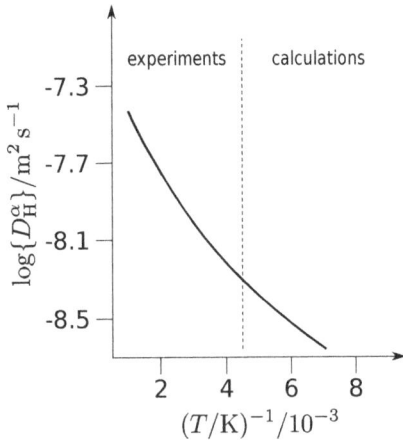

Figure 3.17 Curve showing the diffusion coefficient of hydrogen in ferrite as a function of the temperature. The dashed line shows the boundary between the domain in which experimental data exist, with the remainder based on calculations alone. Adapted from [82].

At higher temperatures, the isotopes of hydrogen all show Arrhenius behaviour, with the pre-exponential factors exhibiting the expected inverse square-root mass dependence. An explanation of this requires a rate theory that includes quantum statistics applied to a particle trapped in a potential well, with the particle described as a time-dependent Gaussian wave packet. The theory is a generalisation of the classical equilibrium statistical approach to yield an effective activation energy as follows [84, 85]:

$$Q_{\text{effective}} = Q_\text{o} + \frac{1}{4}\frac{h\omega}{2\pi} \tag{3.54}$$

where Q_o is the classical activation energy and ω is the oscillation frequency which varies with the inverse square root of the particle mass. Therefore, the activation energies for diffusion increase in the order $H > {}^2H > {}^3H$. There is experimental evidence to confirm such a trend [85–88].

3.11.2 DIFFUSION OF HYDROGEN IN AUSTENITE

The diffusivity of hydrogen in austenite is known to be many orders of magnitude slower than in ferrite, Figure 3.18 [89]:

$$D_\text{H}^\gamma = 2.5 \times 10^{-6} \exp\left\{-\frac{Q}{RT}\right\} \text{m}^2\text{s}^{-1} \quad \text{with} \quad Q = 59300\,\text{J mol}^{-1} \tag{3.55}$$

The difference between austenite and ferrite is because the hydrogen occupies relatively larger octahedral interstices in austenite but the smaller tetrahedral interstices in ferrite. The solubility of hydrogen in austenite is therefore expected intuitively to be greater. At the same time, the number of octahedral interstices in austenite is just one per iron atom, when compared with six of the tetrahedral variety per iron atom in ferrite. This may explain why the diffusion coefficient for hydrogen in ferrite is so much larger than in austenite.

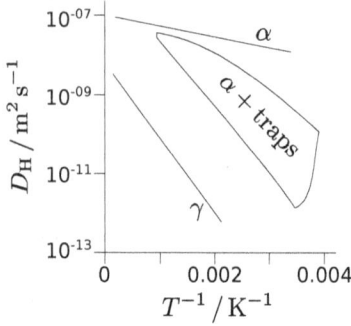

Figure 3.18 A comparison of the diffusion rates of hydrogen in ferrite and in austenite. The shaded region represents the apparent diffusion rate when hydrogen is trapped at defects in the ferrite [81].

3.11.3 HYDROGEN-VACANCY INTERACTIONS

The presence of interstitially dissolved hydrogen in metals such as nickel, palladium, platinum and austenitic iron can increase the concentration of vacancies (Figure 3.19 [90]). This is entirely due to the binding energy between dissolved hydrogen and a host-vacancy. One consequence would be a reduction of density, and another an increase in the diffusivity of atoms in substitutional sites. Both of these effects have been observed experimentally.

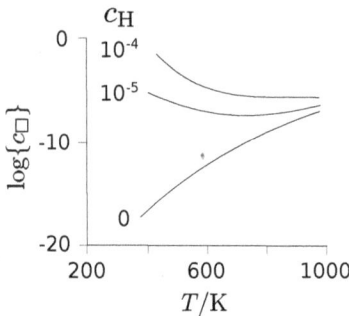

Figure 3.19 Estimate of the host-vacancy concentration $c_\square$ in hydrogenated austenitic iron. The hydrogen concentration is c_H indicated alongside each curve (after McLellan and Xu [90]).

3.12 SELF-DIFFUSION IN IRON

Self-diffusion in austenitic iron follows Arrhenius behaviour over the temperature range 910-1400°C for which experimental data exist, Figure 3.20a. A comprehensive analysis of the experimental data by Oikawa [91–93] gave the following estimates for the diffusion parameters:

$$D_{Fe}^{\gamma} = (0.89^{+0.40}_{-0.28}) \times 10^{-4} \exp\{-Q/RT\}\, m^2\, s^{-1}$$

with $Q = 291.3 \pm 4.5\, kJ\, mole^{-1}.$ (3.56)

The activation energy for diffusion is larger than the corresponding value for ferrite, even though the melting temperature of austenite is slightly lower than that of ferrite (Table 1.1). This presumably is because austenite is a more densely packed phase than ferrite.

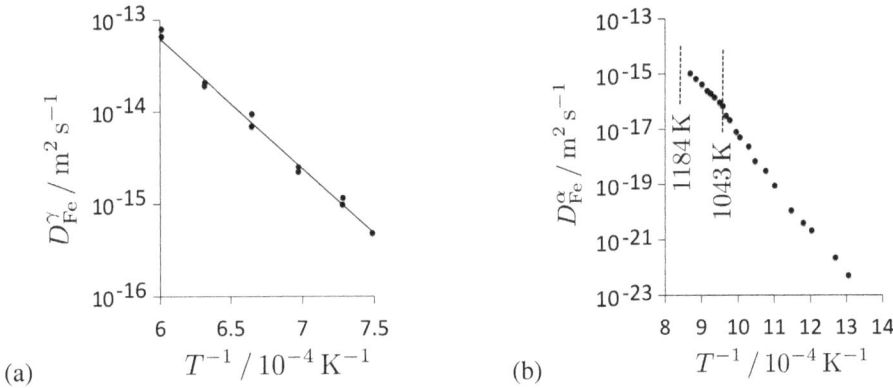

Figure 3.20   (a) Self-diffusion in austenite (data from Buffington, Hirano and Cohen [94]). (b) Self-diffusion coefficient of iron in α-ferrite [95]. The Curie temperature is 1043 K and the formation of austenite begins at 1184 K.

In contrast, the self-diffusion data for ferrite show considerable anomalies, both in the region of the paramagnetic to ferromagnetic transition and in the δ-ferrite temperature range, Figure 3.20b [94, 96]. The anomaly has been observed also for the diffusion of nickel in ferrite [97] and in a vast range of other measurements including mechanical properties such as creep at elevated temperatures.

Both the vacancy formation energy and its ability to migrate are changed in the ferromagnetic state [96]. This can be explained by magnetic interactions [98, 99]; in the fully ordered ferromagnetic state, the magnetic interaction energy of an atom which has z_c nearest neighbours is $W = -z_c \omega_0$ where $-\omega_0$ is the interaction energy per pair of atoms. This assumes that only near neighbour terms contribute significantly to W. The diffusion of iron or substitutional atoms occurs by a vacancy mechanism. In a process where a vacancy is created by removing an atom, z_c pairs of contacts are broken, half of which are recreated when the discarded atom is replaced at the free surface. The enthalpy of formation of a vacancy is therefore increased by $\frac{1}{2} z_c \omega_0$. A similar argument gives a further contribution to the migration energy as $(z_v - z_*) \omega_0$ where z_v is the number of nearest neighbours to an atom paired with a vacancy, and z_* the number of nearest neighbours to an atom in its activated state, neglecting relaxation effects. The value of ω_0 is proportional to the square of the saturation magnetisation $M_m\{0\}$ at absolute zero. Bearing this is mind, the activation enthalpy for self-diffusion becomes

$$Q = Q_0 \left[1 + b_1 \left(\frac{M_m\{T\}}{M_m\{0\}} \right)^2 \right] \tag{3.57}$$

where Q_0 is the activation energy when there is no long range magnetic order (i.e., for paramagnetic ferrite). The constant b_1 is obtained by fitting to experimental data; it can in principle be partitioned into two components, one representing the increment due to magnetic spin order in the vacancy formation energy and the other in the migration energy. This is difficult to do in practice because the calculations require a knowledge of the saddle point configuration during the migration over the activation barrier. The important point is that Equation 3.57 gives a temperature dependence to the activation energy, making it possible to interpret the deviations from the Arrhenius relationship. Precise measurements by Iijima et al. [95] confirm the relation over a wide temperature range (766-1148 K), with $Q_o = 250.6 \pm 3.8 \, \mathrm{kJ \, mol^{-1}}$, $b_1 = 0.156 \pm 0.003$ and the pre-exponential factor in the Arrhenius equation given by $D_o = (2.76^{+1.42}_{-1.04}) \times 10^{-4} \, \mathrm{m^2 \, s^{-1}}$ (Figure 3.20b). The temperature dependence of the self-diffusion coefficient apparently shows the normal Arrhenius relationship above the Curie temperature. Magnetisation data are necessary to use this diffusion model; these have been reported by Potter [100] and Crangle and Goodman [101] as illustrated in Figure 2.10. There are two further issues to consider. Early research indicated that the diffusion anomaly extends to temperatures beyond the Curie temperature, which would be inconsistent with a theory that relies on just the long-range magnetic-order parameter since the magnetisation terms then tend to zero towards T_C. Modern high-precision measurements on well prepared samples do not reveal anomalous effects beyond T_C.

Secondly, measurements for δ-ferrite at high temperatures do not seem to fit the extrapolated data for paramagnetic α-ferrite (Figure 3.21). The activation energy of $Q = 296 \, \mathrm{kJ \, mol^{-1}}$ for δ-ferrite is larger and $D_o = 9.21 \times 10^{-3} \, \mathrm{m^2 \, s^{-1}}$. It is possible that at high temperatures diffusion by a divacancy mechanism adds to the usual flux from the monovacancy mechanism [95, 102]. The estimated parameters for self-diffusion by the divacancy mechanism in ferrite are

$$D_{2\square} = 5.2 \exp\left\{ -\frac{406,000 \, \mathrm{J \, mol^{-1}}}{RT} \right\} \qquad \mathrm{m^2 \, s^{-1}}. \tag{3.58}$$

This makes $D_{2\square}$ about half $D_{\square}$ at the melting temperature.[3]

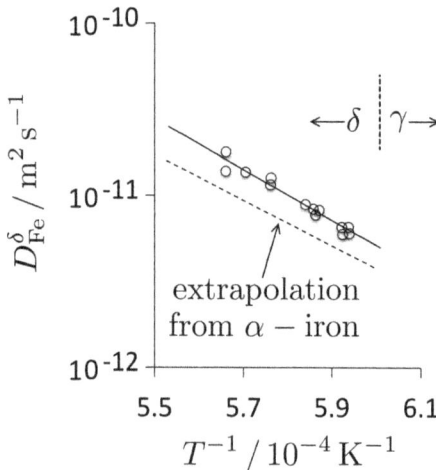

Figure 3.21 Self-diffusion coefficient of iron in δ-ferrite. The $\gamma \to \delta$ transformation occurs at 1665 K with melting occurring at 1811 K. Data from Iijima, Kimura and Hirano [95].

Short-range magnetic order persists to temperatures well in excess of the Curie temperature, Figure 3.22. A number of models have been proposed to account for short-range order effects and at the same time allow δ-ferrite and paramagnetic α-ferrite to be treated identically. The models use a number of fitting parameters and do not seem to explain why the diffusion anomaly vanishes at $T/T_C \approx 1.16$, whereas the short range order persists to even greater temperatures (Figure 3.21).

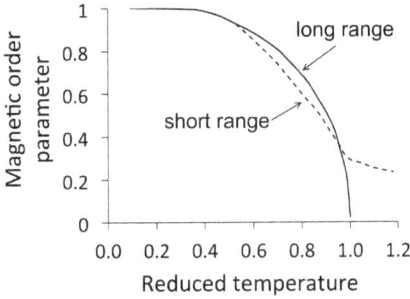

Figure 3.22 The calculated long-range and short-range magnetic order parameters for bcc iron (Kučera [103]). The reduced temperature is the ratio of the absolute temperature to the Curie temperature.

To cope with this difficulty, Kučera [103] defined a fictitious paramagnetic temperature ($T_p > T_C$) beyond which there is no measurable effect on diffusion of (short-range) magnetisation with

$$Q = Q_o(1 + q\{T\})$$ (3.59)

where $q\{T\}Q_o$ expresses the excess activation enthalpy associated with local demagnetisation arising from vacancy formation and migrations, defined such that contributions from long-range order go to zero at T_C and those from short-range order become zero at T_p. Zener [104] argued for a temperature-dependent activation free energy because an atom during its course of migration strains the surrounding lattice; the ability of the lattice to resist deformation changes with temperature. The temperature coefficient of the ratio of the activation free energy $G^*\{T\}/G^*\{T = 0\}$ should therefore be the same as that of the shear moduli $E_s\{T\}/E_s\{T = 0\}$. Any temperature dependence of the activation free energy can be incorporated into the pre-exponential term in the diffusion coefficient via the activation entropy, the change in which becomes

$$\Delta S^* \approx \frac{b_2}{T_m} \frac{d(E_s\{T\}/E_s\{T = 0\})}{d(T/T_m)}$$ (3.60)

where T_m is the melting temperature and b_2 is a fitted constant. Since the shear modulus has a dependence on the magnetic state, Kučera incorporated the function $q\{T\}$ into the pre-exponential factor:

$$D = D_o \exp\left\{\frac{0.35q\{T\}Q_o}{RT_m}\right\} \exp\left\{-\frac{Q}{RT}\right\}.$$ (3.61)

Other models have involved empirical modifications of the Kučera approach, for example by the substitution of the excess enthalpy attributed to the magnetic effect for the function $q\{T\}$ whilst retaining the same form as Equation 3.61 [105], or by explicitly including the shear moduli in the calculations [106, 107]. Some models include two additional terms in the activation energy function: the variation in the long range order parameter and in the elastic modulus [107]. Jönsson [108] pointed out that these models are difficult to distinguish as empirical representations because the excess enthalpy and elastic modulus vary in a similar manner with the temperature. Neither the excess enthalpy nor the difference in shear moduli between the ferromagnetic and paramagnetic states reach zero at the temperature where the diffusion anomalies vanish.

Bearing in mind that enthalpy data are more frequent than modulus data for alloys[4] Jönsson adapted Braun and Feller-Kniepmeier's method where the diffusion anomaly is expressed purely as a function of the excess enthalpy, so in Equation 3.61,

$$q\{T\} = b_3 \, \Delta H^{\mu}\{T\} \tag{3.62}$$

where b_3 is a fitting constant and $\Delta H^{\mu}\{T\}$ is the excess magnetic enthalpy at the temperature T. Equation 3.61 contains the melting temperature whose meaning in the context of diffusion is not clear given that an alloy melts over a range of temperatures. However, there is a well-known empirical relationship between the activation energy for diffusion and the melting temperature of a pure metal,

$$Q_{\mathrm{o}} = 140 T_{\mathrm{m}} \qquad \mathrm{J\,mol^{-1}} \tag{3.63}$$

so T_{m} can be eliminated from Equation 3.61. By fitting the equation to ferritic iron, Jönsson demonstrated that with appropriate thermodynamic data for ΔH^{μ}, it was possible to estimate the tracer diffusion of elements such as cobalt, nickel in ferrite both in the ferromagnetic and paramagnetic regions. The activation energy and D_{o} for the paramagnetic state is derived by fitting in each case and b_3 is assumed to depend only on the lattice geometry.

The method used for incorporating magnetic spin order into the diffusion equation has been extended to the representation of long-range chemical order [109]. The magnetic order parameter in Equation 3.57 is replaced directly by the long-range configurational order parameter. There is, nevertheless, an essential difference between configurational order and magnetic spin order. In a configurationally ordered system, those diffusion paths which minimise the disruption of order are favoured; the degree of order can be influenced by the redistribution of atoms during diffusion.

3.12.1 THE ISOTOPE EFFECT

The isotopes ^{55}Fe and ^{59}Fe diffuse at different rates related by the parameter ΔK which describes the fraction the kinetic energy associated with the mode of vibration that leads to diffusion, that resides in the migrating atom (Equation 3.6). A large value of ΔK implies that the neighbouring atoms do not participate much in the vibrations that lead to the successful jump of the migrating atom. The vibrations can only be decoupled in this sense if the activation volume is large, i.e., there is little relaxation of the surrounding atoms when a vacancy is created. Figure 3.23a shows that ΔK is smallest for paramagnetic iron indicating that the relaxation of the lattice around a vacancy is larger than in the ferromagnetic state. It is obvious from Figure 3.23b that many more atoms would need to participate in the successful migration of an atom in the paramagnetic ferrite when compared with the ferromagnetic state.

The particularly small values of ΔK associated with δ-ferrite may be a reflection of diffusion by both the vacancy and divacancy mechanisms [110]. The measured diffusion coefficient D is then a sum of two diffusivities, $D_{\square}$ and $D_{2\square}$, for migration by the mono and di-vacancy mechanisms respectively. The effective value of ΔK is then given by

$$f_c \Delta K = f_{c,\square} \Delta K_\square \frac{D_\square}{D} + f_{c,2\square} \Delta K_{2\square} \frac{D_{2\square}}{D}. \tag{3.64}$$

Since $f_{c,2\square}$ and $\Delta K_{2\square}$ are both much less than the corresponding terms for monovacancy diffusion, the operation of both diffusion mechanisms leads to an effective ΔK which is smaller than for the monovacancy mechanism alone.

The form of the variation illustrated in Figure 3.23, with a large drop in $f_c \Delta K$ at elevated temperatures, is typical of many metals, even some of which do not show magnetic spin ordering. This adds weight to the argument that there is a significant di-vacancy contribution to the diffusion flux at temperatures close to the melting point.

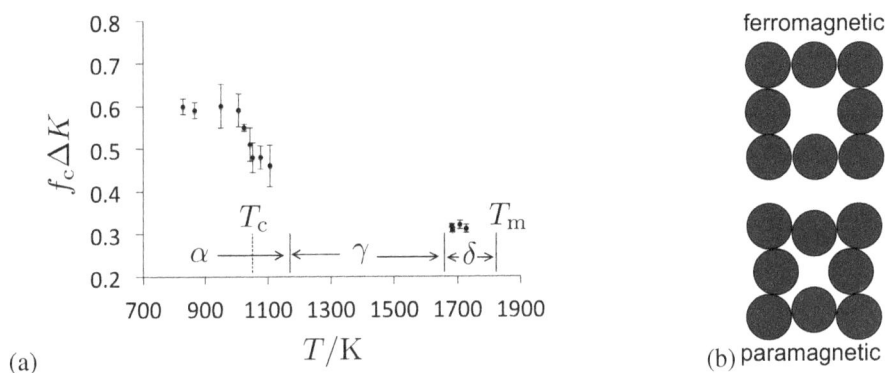

Figure 3.23   (a) Temperature dependence of ΔK in the isotope effect for diffusion of ^{55}Fe and ^{59}Fe in α and δ iron [95]. (b) Schematic illustration of the greater relaxation of atoms around a vacancy in paramagnetic ferrite. The open circles are vacancies and the relaxation is exaggerated.

3.13   MAGNETISM AND INTERSTITIAL DIFFUSION IN FERRITE

It was argued in the context of substitutional diffusion in ferromagnetic ferrite that the creation of a vacancy reduces the number of pairwise magnetic interactions and consequently leads to an increase in the energy for vacancy formation. This substantially explains the diffusion anomaly although other interpretations exist, for example Borg's proposal that the activation strain energy is changed as the elastic moduli depend on the magnetic order parameters. Whatever the true mechanism, there is evidently a powerful influence on diffusion of the magnetic transition.

That the diffusion of carbon in bcc iron cannot be described by a single Arrhenius equation is clear from the experimental data; this also is true for nitrogen in bcc iron [111]. The possibility of the magnetic transition influencing interstitial diffusion must therefore be considered. However, vacancy formation is not really an issue in the context interstitial diffusion. Any anomalous diffusion has to be attributed to effects such as the change in elastic moduli.

The diffusion data for carbon and nitrogen have been fitted to include the effect of magnetism [112]. The listing in Table 3.7 can be used in combination with Equation 3.61 to provide a good representation of experiments; for nitrogen the fit extends to δ-ferrite though similar data are not available to test the case for carbon in δ-ferrite. This approach obviates the need to consider a dual-site occupancy model, but it suffers from the difficulties encountered in treating the self-diffusion of iron, that short-range order effects beyond T_C are not taken into account. The physical basis of the effect of magnetism on interstitial diffusion is not clear because if the anomalies are attributed to changes in elastic moduli, then the pre-exponential factor must also contain a term dependent on temperature (Equation 3.61).

The data in Table 3.7 confirm earlier work that the diffusion anomaly is smaller for nitrogen than for carbon in ferrite. However, the values of the constant b_1 are about twice as large as that for self-diffusion in ferrite. This indicates that the activation energy for the migration of the interstitial is influenced more by magnetic spin ordering, but the associated reasoning is not clear.

Table 3.7

The diffusion parameters for carbon and nitrogen in ferrite. The coefficient b_1 determines the temperature dependence of the activation energy (Equation 3.57) and Q_o can be regarded as the activation energy for diffusion in paramagnetic ferrite. After Iijima [112].

Interstitial	$D_o / m^2 s^{-1}$	$Q_o / kJ\,mol^{-1}$	b_1	Applicability
Carbon	2.72×10^{-7}	59.6	0.337	230-1170 K
Nitrogen	2.42×10^{-7}	59.7	0.266	220-1742 K

The interstitial diffusion anomalies observed in iron exist also in cobalt which has a Curie temperature of 1396 K. The anomaly in cobalt is confined to $T \leq T_C$, where the activation energy depends on the extent of magnetic spin ordering. The Arrhenius law is obeyed for $T > T_C$. This is found to be the case for both substitutional and interstitial solutes in cobalt. Although the results cast doubt on the hypothesis that the diffusion anomaly must extend beyond T_C where short-range order persists, there is no physical explanation of why it should be confined to below T_C.

3.14 SUBSTITUTIONAL SOLUTES

Substitutional solutes diffuse at rates not dissimilar to the self-diffusion of iron. This is illustrated in Figure 3.24 which shows empirically how the ratio of the substitutional-solute/iron diffusivities varies with the atomic number for the first series of transition metals in iron; the same kind of variation is assumed for the remaining series where data are sparse. The relations mostly represent diffusion at 1000 °C [113]. There is no obvious explanation for the variation illustrated but the observed patterns are useful in estimating unknown diffusivities. For example, the diffusion coefficient of zinc in iron was not known at the time of the compilation, but was later measured at 998 K to be 0.5-$2 \times 10^{-17}\,m^2\,s^{-1}$ [114], which is remarkably consistent with [113] at $1.4 \times 10^{-17}\,m^2\,s^{-1}$. It is interesting that heavy elements such as molybdenum, niobium or tungsten diffuse faster in iron than iron itself. For niobium, an explanation offered is that the larger atom causes a distortion in the surrounding lattice which results in an attractive interaction with vacancies, thereby reducing the enthalpy of formation of a vacancy, making diffusion easier [115]. Ruthenium and osmium are iron analogues in that they have similar valance electron shells so it may be logical that the ratios of their diffusivities relative to iron are unity.

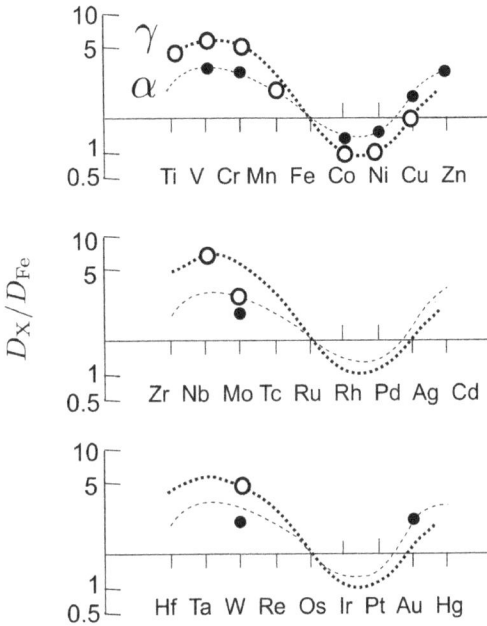

Figure 3.24 Ratio of the diffusivity of element (D_X) to the self-diffusivity of iron (D_{Fe}) in austenite or ferrite, after Fridberg, Törndahl and Hillert [113]. An anomalously low value for the diffusivity of Hf in austenite has been left out of this diagram.

A similar rationalisation, i.e., keeping the activation energies as for the self-diffusion of iron, for the light elements Al, Si, P and S is presented in Figure 3.25. They all diffuse more rapidly than iron. Some of the important difference between the diffusion of solutes in austenite and ferrite is, of course, the seminal influence of the magnetic transition in the latter. Austenite can also be ferromagnetic, for example in nickel-rich alloys, but there does not appear to have been any systematic search for diffusion anomalies in that context. Although expanded austenite can be ferromagnetic, the large interstitial concentrations cause considerable dilation and site blocking so it is difficult to interpret the role of magnetic effects per se.

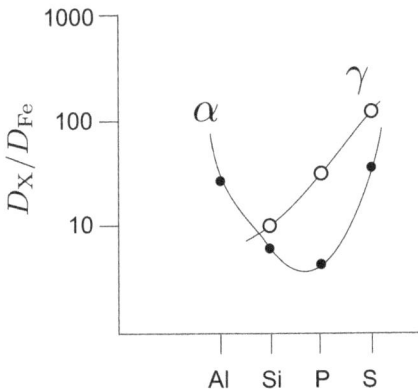

Figure 3.25 Light elements: ratio of the diffusivity of element to the self-diffusivity of iron in austenite or ferrite, after Fridberg, Törndahl and Hillert [113].

The strict application of the theory for the effect of the Curie transition on solute diffusion in ferrite requires a knowledge of the spontaneous magnetisation and Curie temperature as a function of alloy chemical composition. Such data usually are not available, but Jönsson's approach, where the diffusion anomaly is expressed in terms of the excess magnetic enthalpy alone, is an adequate practical approach to the problem of substitutional-solute diffusion in ferromagnetic ferrite. The following discussion covers a few interesting diffusion effects associated with substitutional solutes in iron; there is a useful review by Iijima and Hirano [110].

Cobalt enhances and chromium diminishes the magnetic moment of an iron-alloy relative to pure iron [116, 117]. It is not surprising, therefore, that the temperature dependence of the activation energy for the diffusion of cobalt is smaller and that of chromium larger, when compared with self diffusion in ferromagnetic ferrite (Table 3.8). The quantity b_1 in Table 3.8 is described in Equation 3.57; it scales with the temperature dependence of the activation energy in the ferromagnetic state. The results in the vicinity of the Curie temperature in pure iron are difficult to interpret given inevitable changes in T_C caused by the alloying.

Table 3.8

Diffusion parameters. The quantity b_1 scales with the temperature dependence of the activation energy in the ferromagnetic state. After Iijima and Hirano [117].

	$D_o/\mathrm{m^2\,s^{-1}}$	$Q_o/\mathrm{kJ\,mol^{-1}}$	b_1
Self-diffusion in α-iron	2.76×10^{-4}	250.6	0.156
Cobalt in α-iron	2.76×10^{-4}	251	0.23
Chromium in α-iron	37.3×10^{-4}	267.4	0.133

3.15   GRAIN BOUNDARY DIFFUSION

Grain boundaries are less dense than the perfect lattice; atoms are therefore more mobile during transport along the boundary plane. It is difficult experimentally to separate D_o and the boundary thickness δ_b so the pre-exponential term is expressed as a product of these two quantities. In their assessment of diffusion data, Fridberg et al. [113] concluded that all grain boundary diffusion data for austenite and ferrite, including substitutional solutes, can be approximated adequately by

$$D_o \times \delta_b = 5.4 \times 10^{-15}\,\mathrm{m^3\,s^{-1}} \qquad Q = 155{,}000\,\mathrm{J\,mol^{-1}}. \tag{3.65}$$

The parameters here are somewhat similar to those for diffusion in liquid iron (p. 148) indicating that the grain boundaries considered must contain a large free volume and would not be representative of the more coherent interfaces that occur during transformations or in textured steels.

Figure 3.26 shows tracer diffusion data for a variety of austenitic steels, with a comparison between the diffusion parameters at grain boundaries and lattice diffusion. To facilitate the comparison, it is assumed that the boundary has a thickness 0.5 nm. The data refer to migration along the grain boundary plane rather than across it. Boundary diffusivity is necessarily anisotropic depending on the many degrees of freedom that determine its atomic structure [118].

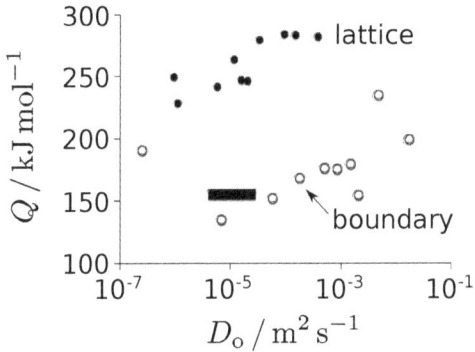

Figure 3.26 Comparison of diffusion parameters for flow in the grain boundary plane and that in the defect free lattice. The lattice data [110] refer to the isotopes ^{59}Fe, ^{51}Cr, ^{63}Ni and ^{54}Mn diffusion in austenite [110]. The horizontal bar represents an overall value for grain boundary diffusion in ferrite or austenite assuming $\delta_b = 5 \times 10^{-10}$ m [119].

3.16 PHENOMENOLOGICAL TREATMENT OF BINARY DIFFUSION

Darken's experiment [120, 121] proved that diffusion can occur against a concentration gradient; carbon initially distributed uniformly migrated across a diffusion couple from a silicon-rich to a silicon-poor steel. In spinodal decomposition, a chemically uniform solution can become spontaneously inhomogeneous as diffusion enhances concentration gradients. Artificially created multilayered structures also exhibit this *uphill diffusion* [122]. Consequently, the forces that drive diffusion are best described in terms of gradients of free energy rather than concentration [123, 124].

Consider substitutional diffusion in a binary system consisting of components "A" and "B". The fluxes are referred to the volume-fixed frame of reference because this best represents the usual experimental data. Diffusion can then be described in terms of just one (inter-)diffusion coefficient and one mobility coefficient. The theory is developed in terms of the component B but an equivalent description can be obtained by selecting A. The process of diffusion dissipates energy (Section 2.22); it would not otherwise occur at all:

$$T\dot{S} = J_A\left(-\frac{\partial \mu_A}{\partial z}\right) + J_B\left(-\frac{\partial \mu_B}{\partial z}\right) \tag{3.66}$$

where $\dot{S}$ the rate of entropy production per unit volume. An equation such as this is useful only when the forces and fluxes in it are independent. This is not the case for the chosen reference frame since conservation requires that $\sum J_i \overline{V}_i = 0$. Furthermore, the chemical potentials of the two components are related by the Gibbs-Duhem equation. To ensure that the dissipation equation contains independent forces and fluxes, the force is redefined by substituting $J_A = -(\overline{V}_B/\overline{V}_A)J_B$:

$$T\dot{S} = -J_B\left(\frac{\partial \mu_B}{\partial z} - \frac{\overline{V}_B}{\overline{V}_A}\frac{\partial \mu_A}{\partial z}\right)$$

so that $\quad X_B = -\frac{\partial}{\partial z}\left(\mu_B - \frac{\overline{V}_B}{\overline{V}_A}\mu_A\right).$ $\tag{3.67}$

There now is only one independent force driving diffusion. The dissipation Equation 3.66 is therefore correctly formulated in the context of Onsager's theory.[5] That the entropy change must be positive for an irreversible process, $J_B X_B > 0$ ensuring that the mobility coefficient M will always exceed zero.

Recalling now Equation 2.93 and Fick's law,

$$J_B \;=\; M X_B \qquad \text{and} \qquad J_B = -D \nabla c_B$$

$$\therefore \quad D \;=\; M \frac{\partial}{\partial c_B} \left(\mu_B - \frac{\overline{V}_B}{\overline{V}_A} \mu_A \right). \tag{3.68}$$

Noting that $c_B = x_B / V_m$, and that $\partial \mu_A / \partial x_B = 0$, we get

$$
\begin{aligned}
\frac{\partial \mu_B}{\partial c_B} \;&=\; V_m \frac{\partial \mu_B}{\partial x_B} \\
&=\; \frac{R T V_m}{x_B} \left(\frac{d \ln \Gamma_B}{d \ln x_B} + 1 \right) \\
&=\; \frac{R T}{c_B} \left(\frac{d \ln \Gamma_B}{d \ln x_B} + 1 \right).
\end{aligned}
\tag{3.69}
$$

Substitution into Fick's equation gives

$$J_B = - \underbrace{\frac{R T M}{c_B} \left(\frac{d \ln \Gamma_B}{d \ln x_B} + 1 \right)}_{D} \frac{\partial c_B}{\partial z}. \tag{3.70}$$

Both the mobility M and the thermodynamic factor in brackets are expected to be dependent on the chemical composition, but note that the derivation neglects any dependence of the partial molar volumes on concentration.

The chemical potentials of isotopes depend only on configurational entropy so from Equation 3.69, the diffusivity of a radioactive tracer of B in the absence of any chemical concentration gradient is simply:

$$D_B^* = M_B V_m R T. \tag{3.71}$$

Tracer diffusion data are therefore more convenient for the determination of the mobility M_B than are chemical diffusion data. For the diffusion of an interstitial element, the right hand side of Equation 3.71 must also be multiplied by the fraction of the interstitial sites that are available for diffusion (Equations 3.32-3.34).

3.17   DIFFUSION IN MULTICOMPONENT SYSTEMS

For a system of n components where the flux of i may also depend on the concentration gradients of other components, Fick's law may be generalised as follows [125, 126]:

$$J_i = - \sum_{k=1}^{n-1} D_{ik} \frac{\partial c_k}{\partial z} \tag{3.72}$$

where J_i are the $n-1$ independent fluxes, defined with respect to the volume-fixed frame of reference; D_{ik} form a matrix of empirical chemical diffusion coefficients. The virtual force X_i acting on a diffusing species really should depend on the negative gradient of its chemical potential; in a multicomponent system, this force is also a function of the chemical potential gradients of the other species.

For diffusion in a multicomponent system each J_i is not only linearly related to its conjugate X_i, but also to all the other forces which lead to entropy production (Section 2.22.3). For diffusion in an isotropic medium referred to a volume-fixed frame, and for an n-component system, this may be expressed as follows:

$$J_i = \sum_{k=1}^{n-1} M_{ik} X_k \tag{3.73}$$

where the M_{ik} are phenomenological coefficients of the various linear force-flux relations; the coefficients form a symmetric matrix of dimensions $(n-1) \times (n-1)$ and satisfy the Onsager reciprocities. If the forces are taken to be potential gradients then because of the Gibbs-Duhem relation ($\sum_i x_i \nabla \mu_i = 0$), there only are $n-1$ independent forces, given by [126]

$$X_k = -\frac{\partial}{\partial z}[\mu_k - (\overline{V}_k/\overline{V}_n)\mu_n]. \tag{3.74}$$

The particular component that is designated n in this equation can be selected arbitrarily from the set of components, but in the discussion that follows it represents the host iron atoms. In doing this, the force on the iron atoms becomes zero:

$$X_n = -\frac{\partial}{\partial z}[\mu_n - (\overline{V}_n/\overline{V}_n)\mu_n] = 0. \tag{3.75}$$

Comparison of the flux equation with Fick's law gives [126]:

$$D_{ik} = \sum_j M_{ij} \underbrace{\frac{\partial}{\partial c_k}\left(\mu_j - \frac{\overline{V}_j}{\overline{V}_n}\mu_n\right)}_{\text{thermodynamic}}. \tag{3.76}$$

The second term in this equation is entirely thermodynamic in origin.[6] The mobility may not be a purely kinetic component in this equation; it is possible to imagine circumstances where it can depend in the chemical composition. For example, when the activation energy for diffusion is affected by the composition independently of the thermodynamic effect, due to lattice dilation (Section 3.6.3). The on-diagonal diffusion coefficients have always been found to be positive, although Kirkaldy has shown that this need not be the case.

3.17.1 DIFFUSION IN TERNARY Fe-X-C ALLOYS

Consider a ternary Fe-X-C alloy, where X is a substitutional solute identified in the diffusion equations by the subscript 2, with carbon and iron identified by the subscripts 1 and 3 respectively. The rate of entropy production is

$$T\dot{S} = J_1\left(-\frac{\partial\mu_1}{\partial z}\right) + J_2\left(-\frac{\partial\mu_2}{\partial z}\right) + J_3\left(-\frac{\partial\mu_3}{\partial z}\right) \tag{3.77}$$

referred to a volume-fixed frame of reference so that

$$J_1\overline{V}_1 + J_2\overline{V}_2 + J_3\overline{V}_3 = 0 \tag{3.78}$$

with the forces related by the Gibbs-Duhem equation

$$c_1\partial\mu_1 + c_2\partial\mu_2 + c_3\partial\mu_3 = 0 \tag{3.79}$$

Equation 3.77 is valid only for independent forces and fluxes. Equations 3.78 and 3.79 can be used to eliminate the solvent iron by writing

$$\partial\mu_3 = -\frac{c_1}{c_3}\partial\mu_1 - \frac{c_2}{c_3}\partial\mu_2 \quad \text{and}$$

$$T\dot{S} = J_1\left(-\frac{\partial\mu_1}{\partial z}\right) + J_2\left(-\frac{\partial\mu_2}{\partial z} + \frac{\partial\mu_3}{\partial z}\right)$$

Assuming now that $\overline{V}_1/\overline{V}_3 = 0$ and $\overline{V}_2/\overline{V}_3 = 1$, it follows that for diffusion in one dimension along a coordinate z,[7]

$$X_1 = -\frac{\partial\mu_1}{\partial z}$$

$$X_2 = -\left(1+\frac{x_2}{x_3}\right)\frac{\partial\mu_2}{\partial z} - \frac{x_1}{x_3}\frac{\partial\mu_1}{\partial z}. \tag{3.80}$$

The solute fluxes follow:

$$J_1 = M_{11}X_1 + M_{12}X_2$$

$$J_2 = M_{22}X_2 + M_{21}X_1 \tag{3.81}$$

again, referred to a volume-fixed frame of reference.

The assumptions concerning the partial molar volumes are considered reasonable as long as the volume change of mixing is insufficient to influence the diffusion profiles significantly. By substitution of the forces in Equation 3.80 into the flux Equations 3.81, together with a comparison against Fick's law, it follows that [127]:

$$D_{11} = M_{11}\frac{\partial\mu_1}{\partial c_1} + M_{12}\frac{x_1}{x_3}\frac{\partial\mu_1}{\partial c_1} + M_{12}\left(1+\frac{x_2}{x_3}\right)\frac{\partial\mu_2}{\partial c_1} \tag{3.82a}$$

$$D_{12} = M_{11}\frac{\partial\mu_1}{\partial c_2} + M_{12}\left(1+\frac{x_2}{x_3}\right)\frac{\partial\mu_2}{\partial c_2} + M_{12}\frac{x_1}{x_3}\frac{\partial\mu_1}{\partial c_2} \tag{3.82b}$$

$$D_{22} = M_{21}\frac{\partial\mu_1}{\partial c_2} + M_{22}\frac{x_1}{x_3}\frac{\partial\mu_1}{\partial c_2} + M_{22}\left(1+\frac{x_2}{x_3}\right)\frac{\partial\mu_2}{\partial c_2} \tag{3.82c}$$

$$D_{21} = M_{21}\frac{\partial\mu_1}{\partial c_1} + M_{22}\frac{x_1}{x_3}\frac{\partial\mu_1}{\partial c_1} + M_{22}\left(1 + \frac{x_2}{x_3}\right)\frac{\partial\mu_2}{\partial c_1}. \tag{3.82d}$$

Values of D_{11} and D_{22} are readily available so these equations can be used to obtain the ratios D_{12}/D_{11} and D_{21}/D_{22}; with certain approximations, these ratios can be expressed in terms of easily accessible thermodynamic parameters [126–128]. To illustrate the approximations, it is necessary to relate the coefficients M_{ij} defined in the laboratory frame and the corresponding Onsager coefficients M_{ij}^K defined in the Kirkendall frame. This can be done using the velocity of the Kirkendall markers. Assuming that interstitials do not contribute to the Kirkendall effect, the velocity of the Kirkendall markers (v_K) is given by the flow of just the substitutional solute and iron atoms. Equation 3.3 shows that

$$
\begin{aligned}
v_K &= -\frac{\bar{J}_2 + \bar{J}_3}{c_2 + c_3} \\
&= \frac{1}{c_2 + c_3}\left(\bar{D}_{22}\frac{\partial c_2}{\partial z} + \bar{D}_{33}\frac{\partial c_3}{\partial z}\right) \\
&= \frac{1}{c_2 + c_3}\left(M_{22}^K\frac{\partial\mu_2}{\partial c_2}\frac{\partial c_2}{\partial z} + M_{33}^K\frac{\partial\mu_3}{\partial c_3}\frac{\partial c_3}{\partial z}\right)
\end{aligned}
\tag{3.83}
$$

where cross-effects have been neglected. Given the dilute nature of the solution, $\partial\mu_2/\partial c_2 \approx RT\partial\ln x_2/\partial c_2$ and since $x_3 \approx 1$, $\partial\mu_3/\partial c_3 \approx -RT(\partial x_2/\partial c_3 + \partial x_1/\partial c_3)$ so that

$$v_K = \frac{RT}{c_2 + c_3}\left[\left(\frac{M_{22}^K}{x_2} - M_{33}^K\right)\frac{\partial x_2}{\partial z} - M_{33}^K\frac{\partial x_1}{\partial z}\right]. \tag{3.84}$$

If $\partial x_1/\partial z$ is small, then the Kirkendall effect vanishes when $M_{22}^K = x_2 M_{33}^K$. To make effective use of this equation, it is useful to relate it to the M_{ii} coefficients.

The flux J_1 in the volume-fixed frame of reference is given by

$$
\begin{aligned}
J_1 &= -M_{11}\frac{\partial\mu_1}{\partial z} - M_{12}\left(1 + \frac{x_2}{x_3}\right)\frac{\partial\mu_2}{\partial z} - M_{12}\frac{x_1}{x_3}\frac{\partial\mu_1}{\partial z} \\
&= \bar{J}_1 + c_1 v_K \\
&= -M_{11}^K\frac{\partial\mu_1}{\partial z} + \frac{c_1 RT}{c_2 + c_3}\left[\left(\frac{M_{22}^K}{x_2} - M_{33}^K\right)\frac{\partial x_2}{\partial z} - M_{33}^K\frac{\partial x_1}{\partial z}\right].
\end{aligned}
\tag{3.85}
$$

It is assumed here that in the Kirkendall frame of reference the components all diffuse independently when the solution is dilute.

Within this set of equations, a comparison of the multipliers for $\partial x_2/\partial z$ in the first and third lines gives

$$M_{12} = x_1 x_2 M_{33}^K - x_1 M_{22}^K. \tag{3.86}$$

It follows that when $M_{22}^K = x_2 M_{33}^K$, $M_{12} = 0$ so that the Kirkendall effect vanishes when $M_{12} = 0$. In this limit,

$$\frac{D_{12}}{D_{11}} = \frac{\partial\mu_1}{\partial c_2} \Big/ \frac{\partial\mu_1}{\partial c_1} \tag{3.87}$$

$$\frac{D_{21}}{D_{22}} = \frac{M_{21}\frac{\partial \mu_1}{\partial c_1} + M_{22}\frac{x_1}{x_3}\frac{\partial \mu_1}{\partial c_1} + M_{22}(1+\frac{x_2}{x_3})\frac{\partial \mu_2}{\partial c_1}}{M_{21}\frac{\partial \mu_1}{\partial c_2} + M_{22}\frac{x_1}{x_3}\frac{\partial \mu_1}{\partial c_2} + M_{22}(1+\frac{x_2}{x_3})\frac{\partial \mu_2}{\partial c_2}}. \tag{3.88}$$

If it is assumed that $x_1, x_2 \ll x_3$ and $M_{21} \ll M_{22}$, then Equation 3.88 becomes

$$\frac{D_{21}}{D_{22}} \approx \frac{\partial \mu_2}{\partial c_1} \bigg/ \frac{\partial \mu_2}{\partial c_2}$$

For dilute solutions, these equations may be expressed in terms of the Wagner interaction parameters e_{ik} [129]:

$$e_{ik} = \frac{\partial \ln \Gamma_i}{\partial x_k} = e_{ki} \tag{3.89}$$

where Γ_i is the activity coefficient of i relative to the standard state at infinite dilution. Equations 3.87 and 3.88 can be simplified as follows:

$$\frac{D_{12}}{D_{11}} = \frac{e_{12}x_1}{1+e_{11}x_1} \quad \text{and} \quad \frac{D_{21}}{D_{22}} = \frac{e_{12}x_2}{1+e_{22}x_2} \tag{3.90}$$

Brown and Kirkaldy [127] have verified the validity of Equation 3.90 for ternary steels containing Mn, Co, Cr, Ni or Si as the substitutional solutes.

Kirkaldy [126] has shown that for dilute ternary iron alloys containing two substitutional solutes, similar simplification of the diffusion matrix can only be made if the diffusivity of one of the solutes is relatively large; since this is usually not the case, Kirkaldy suggests that the direct measurement of the chemical diffusion matrix is the easiest approach.

3.18 DIFFUSION IN LIQUID IRON

Measurements of the diffusion coefficients in liquid iron are of importance in the processing of steel. The self-diffusivity of iron within its liquid state has been measured to be [130]

$$D = 9.48 \times 10^{-4} \exp\{-145000/RT\} \quad \text{m}^2\,\text{s}^{-1}. \tag{3.91}$$

The measurements are over the range 1823-1923 K. This gives a value of about $5 \times 10^{-8}\,\text{m}^2\,\text{s}^{-1}$ at the melting temperature. The activation energy and pre-exponential factor are similar to that for the data assessed by Fridberg et al. [113] for diffusion parallel to grain boundaries in steel, Figure 3.26. A comprehensive compilation of diffusion data for liquid iron has been published by Kubíček and Peřřica [131], which includes a summary of the methods used to measure the diffusion coefficients. The diffusivity of interstitials in molten iron is comparable with that of many of the substitutional solutes.

The pressure dependence of diffusivity in liquid iron is expected to be small because the activation volume consists only of the changes caused during the motion of "vacancies", which presumably may be regarded as available in profusion so the formation volume can be neglected (Nachtrieb, quoted in [17]).

3.19   STRESS-INDUCED MIGRATION

Pressure is expected to influence diffusion because it alters the ability of the lattice to accommodate transient changes in volume associated with the migration of atoms from their equilibrium positions. Different interactions are possible with other forms of stress (e.g., uniaxial tension) when the distortion associated with a misfitting solute atom has a symmetry that is lower than that of the host lattice. This is because crystallographically equivalent sites which have identical energy at zero stress, separate into differing energy states under the influence of the stress. Solute atoms that happen to be lodged in the less favoured sites then tend to migrate into sites that comply best with the applied stress. For bcc-iron there are three sets of octahedral sites which can be occupied by carbon atoms. These octahedra are irregular with the shortest principal axis aligned to one of the $\langle 100 \rangle$ directions. A stress-induced elongation of the unit-cell along one of the $\langle 1\,0\,0 \rangle$ cell edges expands the set of octahedral holes with this short axis parallel to that edge. A redistribution of the carbon atoms along these preferred edges would therefore lead to a reduction in enthalpy. Stress-induced migration must obviously be dependent on the crystallographic orientation, as verified experimentally using single-crystals or crystallographically-textured samples [132].

Stress-induced migration introduces a time-dependence in the evolution of strain. This *anelasticity* is responsible for the well-known phenomenon of internal friction. The capacity of a steel to damp vibrations depends on internal friction since work is done as the atoms switch positions during the alternating stress. The effect must obviously be frequency dependent with the damping being maximised when the jump frequency and vibration frequencies are in resonance. This characteristic frequency, which will also depend on temperature, can be used to identify the migrating species and the activation energy for diffusion. It is interesting to note that although internal friction experiments monitor macroscopic effects with quite simple equipment, the measurements can in fact deal with single-atom diffusion jumps.

The experiments usually are carried out using a torsional pendulum where the angle δ_i by which the strain lags behind the stress is a measure of anelasticity. The ratio of successive amplitudes, i.e., "the logarithmic decrement" Q_i^{-1} has a maximum value proportional to the concentration of the solute atoms. Internal friction experiments can therefore be used to characterise extremely low-carbon concentrations in ferritic steels.

It is possible for interactions between the mobile interstitial atoms and immobile substitutional-solutes to influence the stress-induced migration of the former. Figure 3.27 shows that the Snoek peak has fine structure due to different interstitial/substitutional-solute pairwise interactions. It has been possible therefore to measure the interaction energies (Table 3.9). Although the chances of finding the interstitial next to a substitutional atom are small in dilute solutions, interstitial atoms are mobile even at room temperature, enabling them to migrate into positions close to the substitutional solute atoms if this leads to a greater reduction in energy [133].

The profile of the Snoek relaxation depends on the specific solutes involved. When nitrogen interacts with chromium or vanadium, there is a discernible additional peak in the relaxation profile (Figure 3.27). This is not the case for carbon in manganese-containing ferrite, where there is no extra peak but the peak strength is markedly reduced. This is because the formation of a Mn-C pair

Table 3.9

The binding energies for interstitial-substitutional solute pairs in ferrite, as measured using internal friction [133].

Substitutional solute	Binding energy / eV	
	N	C
V	0.18	
Cr	0.16–0.18	
Mn	0.16–0.20	0.12–0.16

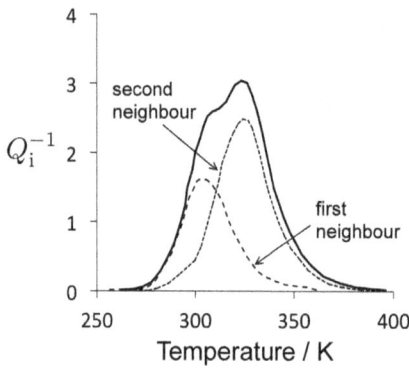

Figure 3.27 Internal friction peak in a Fe-0.1Cr-0.23N wt% ferritic steel. The continuous curve is measured whereas the other curves are calculated contributions from nitrogen atoms that are bound to a near or second-near neighbour Cr atom. After Numakura and Koiwa [133].

reduces the distortion around a carbon atom relative to an isolated carbon atom [134]. It is the elastic interaction between Mn and C which is dominant in this effect, rather than a chemical affinity. Thus, phosphorus and carbon do not chemically interact but the presence of phosphorus leads to a reduction in the Snoek peak which is even larger than that associated with manganese. This is explained by the larger elastic distortion caused by phosphorus in solution [135].

There are other traps for interstitial atoms which can affect their migration. Ke [136] showed that in cold-worked ferrite, a Snoek peak lying at about 225°C involves the stress-induced migration of atoms lying near dislocations.

3.20   ELECTROMIGRATION

Atoms can migrate in the solid-state under the influence of an electrical field, a phenomenon known as *electromigration*. Whereas the driving force for ordinary diffusion is the gradient in the chemical potential, that for electromigration is the gradient in the electrical potential ϕ. This can be demonstrated by considering the flow of current i under the influence of the potential difference $\Delta\phi$ [137]. The rate at which work is done by the current is $i\Delta\phi$; if all of this is dissipated as thermal energy in the surroundings, then it leads to the creation of entropy so

$$T\dot{S} = i\Delta\phi.\tag{3.92}$$

Using current density i_* instead of current, converts this equation into one familiar in the context of irreversible thermodynamics:

$$T\dot{S} = -i_* \frac{d\phi}{dz} \tag{3.93}$$

where the negative sign comes from the fact that the direction of current flow is opposite to that in which the electrical potential increases. It follows from the thermodynamics of irreversible processes, that the force that drives electromigration may be written as

$$X = -\frac{\partial \phi}{\partial z} = \Phi \tag{3.94}$$

where Φ is the electrical field, X is the force per atom. The current density due to this force will depend on the charge on the atom. It will be proportional to the true valency of the migrating atom, but for metals there is an additional contribution due to the effect of the electrons (and/or holes) which also migrate when a field is applied. The moving electrons interact with the diffusing atoms, imparting additional momentum in the direction of flow, the process being described as an *electron wind*. The wind effect dominates in some metals in which it opposes the flow of the positive ions and leads to an effective valency that is negative. Consequently, the true valency of the diffusing species may be masked in electromigration experiments, making it necessary to use an effective valency Z_v; this term is a dimensionless number so it must be multiplied by the magnitude of the charge per electron, $|e|$ to obtain the effective charge per atom.

The velocity v_A of the diffusing species A is related to the force by

$$v_A = M_A^e X \tag{3.95}$$

M_A^e ($m^2\,s^{-1}\,volt^{-1}$) is the electrical-mobility under the action of a unit force. The simplest explanation of this electric-mobility can be obtained by comparison with the tracer diffusion process. From Equation 3.71 the diffusion mobility M_A is given by

$$M_A = D_A^*/kT \qquad m^2\,s^{-1}\,J^{-1} \tag{3.96}$$

whereas the electric mobility has the units $m^2\,s^{-1}\,volt^{-1}$. It follows that $M_A = M_A^e/|e|Z_v$ giving

$$D_A^* = M_A^e \frac{kT}{|e|Z_v} \qquad \text{or} \qquad M_A^e = D_A^* \frac{|e|Z_v}{kT}. \tag{3.97}$$

This in essence is the Nernst equation derived for electrolytes as long ago as 1888 [138]. The flux is simply the product of the velocity and the concentration:

$$
\begin{aligned}
J_A &= v_A(c_A N_A) \\
&= M_A^e \Phi c_A N_A \\
&= \frac{D_A^* |e|Z_v}{kT} (\Phi c_A N_A) \tag{3.98}
\end{aligned}
$$

where the multiplier $c_A N_A$ takes account of the fact that X was defined originally as the force per atom. The flux in electromigration is found to respond as expected – a reversal of the field reverses the direction of flow and an alternating current has no net effect [139].

The electromigration of carbon in liquid iron and in austenite has been known ever since it has been argued that the carbon has a positive charge when dissolved in iron. For migration in the solid-state, the early research involved the passage of direct current through steel wires maintained at high temperatures with typical current densities of $30\,\mathrm{A\,mm^{-2}}$ and for time periods up to $100\,\mathrm{h}$ [12, 140–142]. Changes in carbon concentration were followed metallographically. Experimental data due to Dayal and Darken [14] are presented in Figure 3.28 in order to illustrate the magnitude of the changes involved. The data were generated by passing direct current through steel samples maintained in the austenitic state for periods of up to ten hours. Data for ferritic iron are presented in Table 3.10.

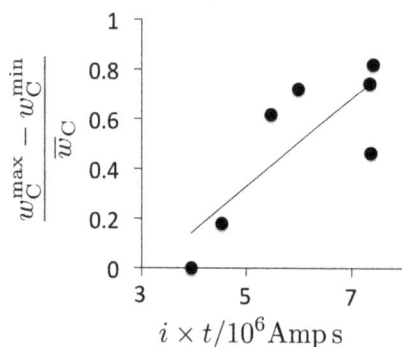

Figure 3.28 Data illustrating the electromigration of carbon in austenite [14]. The carbon concentration difference induced by the passage of electrical current is normalised with respect to the mean value at the central plane between the electrodes. The horizontal axis represents the product of the applied current and the duration of the current.

Well-designed experiments by Okabe and Guy [143] have demonstrated that $Z_v = 3.99 \pm 0.08$ for carbon in austenite, independent of the carbon concentration, the temperature or the current density. Although Z_v coincides with the known valency of carbon in its tetrahedral configuration, such a comparison may be misleading. There is evidence to indicate that the value and sign of Z_v is determined more by the electron wind effect than by the intrinsic valency of carbon. Indeed, it may be the case that the carbon, if it is in an ionised state, has its charge screened by Fermi electrons [144]. It would not then feel any direct force from the applied field, the screening charge being able to polarise in response to the field.

When data for many metals are examined together, it is found that different interstitials in a given metal all tend to have the same sign of the effective valency [145]. This might be taken to indicate the predominance of the electron wind effect.

Table 3.10
The effective valency of various elements in ferritic and austenitic iron. The unreferenced data are obtained from a compilation by Bocquet et al. [146] and Pratt and Sellors [147].

Element	Z_v	Reference
H in α	0.24	[139]
^{2}H in α	0.39	[139]
C in α	4.3	
N in α	5.7	
C in γ	3.99 ± 0.08	[143]

3.21 THERMOMIGRATION

Thermomigration is the transport of matter in a gradient of temperature; the phenomenon is also known as the Ludwig-Soret effect. The force responsible for thermomigration can once again be derived using the theory of irreversible processes [137].

Consider the steady-state flow of heat through a bridge between two bodies, one at a uniform temperature $T + \Delta T$ and the other at the uniform temperature T. When the latter body absorbs a quantity of heat q, its entropy increases by $\Delta S_1 = q/T$, whereas that of the higher temperature body has $\Delta S_2 = -q/(T + \Delta T)$, giving a net increase in entropy which is

$$\Delta S = \Delta S_1 + \Delta S_2 = \frac{q}{T} - \frac{q}{T + \Delta T} \doteq \frac{q \Delta T}{T^2} \tag{3.99}$$

The rate of entropy production during this irreversible process is then

$$\dot{S} = \frac{\mathrm{d}}{\mathrm{d}t} \left(\frac{q \Delta T}{T^2} \right) = \frac{\mathrm{d}q}{\mathrm{d}t} \frac{\Delta T}{T^2}, \tag{3.100}$$

the final equality being justified because of the steady-state nature of the process. The heat flows through the bridge where the entropy is created; the bridge has a volume V, a cross-sectional area A and a length Δz, the temperature difference between its ends being ΔT. The rate of entropy production per unit volume is therefore:

$$\frac{\dot{S}}{V} = \frac{1}{T^2} \frac{\Delta T}{\Delta z} \frac{1}{A} \frac{\mathrm{d}q}{\mathrm{d}t} \tag{3.101}$$

which in the limit of small Δ's becomes

$$T\dot{S} = \frac{1}{A} \frac{\mathrm{d}q}{\mathrm{d}t} \left(-\frac{1}{T} \frac{\mathrm{d}T}{\mathrm{d}z} \right) = J \left(-\frac{1}{T} \frac{\mathrm{d}T}{\mathrm{d}z} \right) \tag{3.102}$$

with the negative sign appearing because the flow of heat is in a direction opposite to that in which the temperature increases. J is the flux of heat. The last term on the right-hand side is the force that drives the diffusion of heat.

During thermomigration, the atoms move in a direction that reduces the temperature gradient. They must therefore carry with them a unit of enthalpy, known as the heat of transport H_T (per mole of atoms). The force responsible for thermomigration is

$$X = -\frac{1}{T}\frac{dT}{dz} \tag{3.103}$$

and the velocity of the atom is then

$$v_A = M_A^T X \tag{3.104}$$

where M_A^T is the thermal mobility with units $m^2 s^{-1}$ which by comparison with the units of the ordinary diffusion mobility gives $M_A^T = M_A/H_T$ and since $J_A = v_A c_A$ it follows that the flux in a temperature gradient is

$$J_A = -c_A H_T D_A^* \frac{1}{kT^2}\frac{dT}{dz}. \tag{3.105}$$

Table 3.11
The heats of transport for various elements in iron. The range indicated for hydrogen and deuterium in ferrite is because H_T is temperature dependent [148, 149].

Element	$H_T/\mathrm{kJ\,mol}^{-1}$	Reference
H in α	-34 to -23	[150]
^{2}H in α	-33 to -22	[150]
C in α	-100	[151]
N in α	-100	[152]
C in γ	-8	[151]

It is of course possible to have cross-effects of the kind associated with the Onsager theory. Thus, diffusion which is driven by a chemical potential gradient can lead to the development of a temperature gradient and vice-versa.

3.22 ELECTROPULSING

High frequency pulses of electrical current have long been known to induce dramatic changes in the structure and properties of materials [153], including, for example, the healing of internal cracks in microseconds [154] and electrically induced plasticity as an aid to shape forming [155]. If the temperature rise associated with the pulsing is sufficient, then the dislocation density can be affected even if the pulse duration is just a few millionths of a second [156]. The process can lead to quite dramatic changes in the microstructure of steel, as illustrated in Figure 3.29. It is believed that the observations are not related to electromigration which increases atomic mobility, because electropulsing has been found in stainless steels to *suppress* precipitation reactions [157]. It is claimed

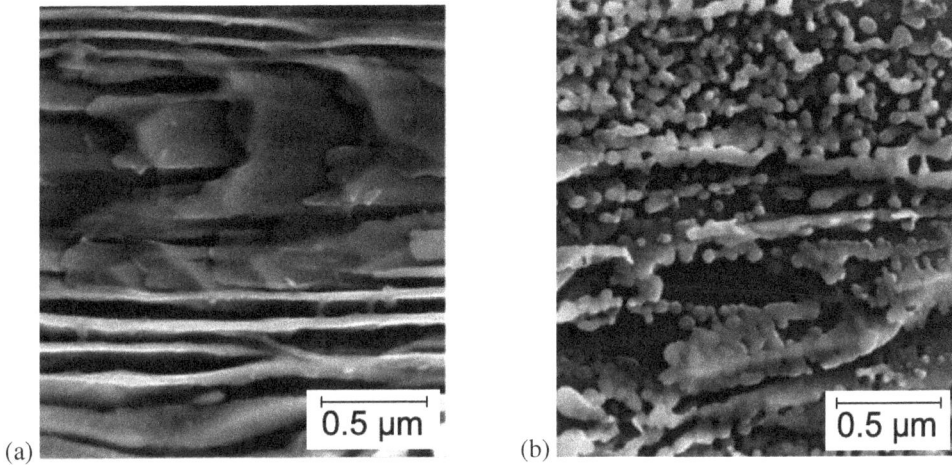

Figure 3.29 Cold-drawn pearlitic steel wire subjected to a peak current density of $9 \times 10^9 \mathrm{A m}^{-2}$ with a pulse duration of $150\,\mu s$. (a) In the cold drawn state. (b) The same following the application of electropulsing. Reproduced from Qin et al. [157] under the CCBY license.

instead that the pulsing changes the thermodynamic free energy of the system but this is not established, for example by including a free energy term into phase diagram calculations to predict the effect of electropulsing, in a manner similar to calculations for the effects of pressure and magnetic fields on phase stability. In some cases the role of joule heating is considered unimportant [158], whereas in others, it clearly is significant [159]. Other factors include stress effects caused by thermal changes, and heterogeneities due to the skin effect whereby the electrical current density varies as a function of depth from the surface of the sample.

Whereas this subject has vast quantities of experimental data to demonstrate changes due to electropulsing, it remains the case that there is no quantitative closure between any of the proposed theoretical frameworks and data.

REFERENCES

1. A. Fick: 'On liquid diffusion', *The London, Edinburgh, and Dublin Philosophical Magazine and Journal of Science*, 1855, **10**, 30–39.

2. E. O. Kirkendall: 'Diffusion of zinc in alpha brass', *Trans. A.I.M.E.*, 1942, **147**, 104–110.

3. A. D. Smigelskas, and E. O. Kirkendall: 'Zinc diffusion in alpha brass', *Transactions of the AIME*, 1947, **171**, 130–142.

4. I. D. Choi, D. K. Matlock, and D. L. Olson: 'An analysis of diffusion-induced porosity in cu-ni laminates', *Materials Science & Engineering*, 1990, **124**, L15–L18.

5. D. G. Miller: 'Thermodynamics of irreversible processes: The experimental verification of the Onsager reciprocal relations', *Chemical Reviews*, 1960, **60**, 15–37.

6. L. S. Darken: 'Diffusion, mobility and their interrelation though free energy in binary metallic systems', *Transactions of the AIME*, 1948, **175**, 184–194.

7. A. D. Le Claire: 'Some comments on the mass effect of diffusion', *Philosophical Magazine*, 1966, **14**, 1271–1284.

8. J. W. Christian: Theory of Transformations in Metals and Alloys, Part I: 2 ed., Oxford, U. K.: Pergamon Press, 1975.

9. R. A. Johnson, and A. C. Damask: 'Point defect configurations in irradiated iron-carbon alloys', *Acta Metallurgica*, 1964, **12**, 443–445.

10. C. Zener: 'Theory of strain interaction of solute atoms', *Physical Review*, 1948, **74**, 639–647.

11. M. Hayakawa, and M. Oka: 'Calculation of Fe-atom displacement around an octahedral and tetrahedral interstitial carbon atom in bcc lattice and expected X-ray diffuse intensities', *Materials Science Forum*, 1990, **56-58**, 185–190.

12. W. Seith, and O. Kubaschewski: 'Die elektrolytische überführung von kohlenstoff in festen stahl', *Zeitschrift für Elektrochemie und angewandte physikalische Chemie*, 1935, **41**, 1935.

13. V. W. Seith, and T. Daur: 'Elektrische uberfuhrung in festen metallegierungen', *Zeitschrift für Elektrochemie*, 1938, **44**, 256–260.

14. P. Dayal, and L. S. Darken: 'Migration of carbon in steel under the influence of direct current', *Transactions of the AIME*, 1950, **188**, 1156–1158.

15. M. Morinaga, N. Yukawa, H. Adachi, and T. Mura: 'Electronic state of interstitial C, N, O in FCC iron', *Journal of Physics F: Metal Physics*, 1988, **18**, 923–934.

16. D. N. Beshers: 'An investigation of interstitial sites in the BCC lattice', *Journal of Applied Physics*, 1965, **36**, 290–300.

17. J. A. Bosman, P. E. Brommer, and G. W. Rathenau: 'Influence of pressure on the mean time of stay of interstitial nitrogen in iron', *Physica*, 1957, **23**, 1001–1006.

18. A. J. Bosman, P. E. Brommer, L. C. H. Eijkelenboom, C. J. Schinkel, and G. W. Rathenau: 'The influence of pressure on the mean time of stay of interstitial carbon in iron', *Physica*, 1960, **26**, 533–538.

19. J. Bass, and D. Lazarus: 'Effect of pressure on mobility of interstitial carbon in iron', *Journal of Physics and Chemistry of Solids*, 1962, **23**, 1820–1821.

20. J. F. Cox, and C. G. Homan: 'Pressure effect on the diffusion of carbon in α-iron', *Physical Review B*, 1972, **5**, 4755–4761.

21. D. H. R. Fors, and G. Wahnström: 'Nature of boron solution and diffusion in α-iron', *Physical Review B*, 2008, **77**, 132102.

22. J. R. G. da Silva, and R. B. McLellan: 'Diffusion of carbon and nitrogen in bcc iron', *Materials Science & Engineering*, 1976, **26**, 83–87.

23. R. B. McLellan, M. L. Rudee, and T. Ishibachi: 'The thermodynamics of dilute interstitial solid solutions with dual-site occupancy and its application to the diffusion of carbon in α-iron', *Transactions of the AIME*, 1965, **233**, 1939–1943.

24. R. H. Condit, and D. N. Beshers: 'Interstitial diffusion in the bcc lattice', *Transactions of the Metallurgical Society of AIME*, 1967, **239**, 680–683.

25. B. S. Lement, and M. Cohen: 'A dislocation-attraction model for the first stage of tempering', *Acta Metallurgica*, 1956, **4**, 469–476.

26. C. Zener: 'Kinetics of the decomposition of austenite', *Transactions of the American Institute of Mining and Metallurgical Engineers*, 1946, **167**, 550–595.

27. C. Zener: Elasticity and anelasticity of metals: Illinois, USA: Chicago University Press, 1948.

28. M. Hillert: 'The kinetics of the first stage of tempering', *Acta Metallurgica*, 1959, **7**, 653–658.

29. R. A. Johnson: 'Calculation of the energy and migration characteristics of carbon in martensite', *Acta Metallurgica*, 1965, **13**, 1259–1262.

30. A. D. Cromwell: 'Potential energy functions for graphite', *Journal of Chemical Physics*, 1958, **29**, 446–447.

31. A. H. Cottrell: 'Transition metal carbides with NaCl structure', *Materials Science and Technology*, 1994, **10**, 788–792.

32. W. W. Dunn, and R. B. McLellan: 'The application of a quasichemical solid solution model to carbon in austenite', *Metallurgical Transactions*, 1970, **1**, 1263–1265.

33. H. K. D. H. Bhadeshia: 'Application of first-order quasichemical theory to transformations in steels', *Metal Science*, 1982, **16**, 167–169.

34. A. H. Cottrell: Chemical Bonding in Transition Metal Carbides: London, U.K.: Institute of Materials, 1996.

35. R. A. Johnson, G. J. Dienes, and A. C. Damask: 'Calculations of the energy and migration characteristics of carbon and nitrogen in alpha-iron and vanadium', *Acta Metallurgica*, 1964, **12**, 1215–1224.

36. R. A. Johnson: 'Clustering of C atoms in alpha-iron', *Acta Metallurgica*, 1967, **15**, 513–517.

37. D. Keefer, and C. Wert: 'The clustering of N in bcc iron', *Acta Metllurgica*, 1963, **11**, 489–497.

38. D. Keefer, and C. Wert: 'Clustering of C and N in iron', *Journal of the Physical Society of Japan*, 1963, **18**, 110–114.

39. V. Rosato: 'Comparative behaviour of carbon in b.c.c. and f.c.c. iron', *Acta Metallurgica*, 1989, **37**, 2759–2763.

40. P. Hautojärvi, J. Johansson, A. Vehanen, J. Yli-Kappila, and P. Moser: 'Vacancy-carbon interaction in iron', *Physical Review Letters*, 1980, **44**, 1326.

41. F. E. Fujita, and A. C. Damask: 'Kinetics of carbon precipitation in irradiated iron - II electrical resistivity measurements', *Acta Metallurgica*, 1964, **12**, 331–339.

42. R. B. McLellan: 'The thermodynamics of interstitial-vacancy interactions in solid solutions', *Journal of Physics and Chemistry of Solids*, 1988, **49**, 1213–1217.

43. R. B. McLellan, and M. L. Wasz: 'Carbon-vacancy interactions in B.C.C. iron', *Physica Status Solidi A*, 1988, **110**, 421–427.

44. C. Barouh, T. Schuler, C. C. Fu, and M. Nastar: 'Interaction between vacancies and interstitial solutes (C, N, and O) in α-Fe: From electronic structure to thermodynamics', *Physical Review B*, 2014, **90**, 054112.

45. C. Wells, W. Batz, and R. F. Mehl: 'Diffusion coefficient of carbon in austenite', *Transactions of the AIME Journal of Metals*, 1950, **188**, 553–560.

46. R. P. Smith: 'The diffusivity of carbon in iron by the steady state method', *Acta Metallurgica*, 1953, **1**, 579–587.

47. J. C. Fisher, J. H. Hollomon, and D. Turnbull: 'Absolute reaction rate theory for diffusion in metals', *Transactions of the AIME*, 1948, **175**, 202–215.

48. J. S. Kirkaldy: 'Crystallisation in the condensed state', In: W. M. Mueller, ed. *Energetics in Metallurgical Phenomena*, vol. 4. Gordon and Breach Publishers, 1968:197–371.

49. R. H. Siller, and R. B. McLellan: 'The variation with composition of the diffusivity of carbon in austenite', *TMS-AIME*, 1969, **245**, 697–700.

50. R. H. Siller, and R. B. McLellan: 'Application of first order mixing statistics to the variation of the diffusivity of carbon in austenite', *Metallurgical Transactions*, 1970, **1**, 985–988.

51. J. S. Kirkaldy, and D. J. Young: Diffusion in the Condensed State: London, U.K.: Institute of Metals, 1985.

52. H. K. D. H. Bhadeshia: 'Diffusion of carbon in austenite', *Metal Science*, 1981, **15**, 477–479.

53. L. Vegard: 'Die konstitution der mischkristalle und die raumfüllung der atome', *Zietschrift für Physik*, 1921, **5**, 17–26.

54. D. W. Hoffman: 'Concerning the elastic free energy of dilute interstitial alloys', *Acta Materialia*, 1970, **18**, 819–833.

55. R. B. McLellan: 'Chemical diffusion in interstitial solid solutions', *Journal of Physics and Chemistry of Solids*, 1977, **38**, 933–936.

56. R. B. McLellan: 'Cell models for interstitial solid solutions', *Acta Metallurgica*, 1982, **30**, 317–322.

57. R. B. McLellan, and T. Ishikawa: 'The diffusion motion volume of hydrogen in palladium', *Scripta Metallurgica*, 1986, **20**, 99–101.

58. R. B. McLellan, and C. Ko: 'The diffusion of carbon in austenite', *Acta Metallurgica*, 1988, **36**, 531–537.

59. R. B. McLellan: 'The diffusion of heavy interstitial atoms in absence of a particle density gradient', 1995: Personal Communication to H. K. D. H. Bhadeshia.

60. D. C. Parris, and R. B. McLellan: 'The diffusivity of carbon in austenite', *Acta Metallurgica*, 1976, **24**, 523–528.

61. W. J. Liu, J. K. Brimacombe, and E. B. Hawbolt: 'Influence of composition on the diffusivity of C in steels Part I - non alloyed austenite', *Acta Metallurgica and Materialia*, 1991, **39**, 2373–2380.

62. G. E. Murch: 'A relation between the activation energy for self diffusion and the partial molar energy in interstitial solid solutions', *Acta Metallurgica*, 1979, **27**, 1701–1704.

63. J. Ågren: 'A revised expression for the difusivity of carbon in binary Fe-C austenite', *Scripta Metallurgica*, 1986, **20**, 1507–1510.

64. R. Hales, and A. C. Hill: 'The diffusion of nitrogen in an austenitic stainless steel', *Metal Science*, 1977, **11**, 241–244.

65. H. Dong: 'S-phase surface engineering of Fe-Cr, Co-Cr and Ni-Cr alloys', *International Materials Reviews*, 2010, **55**, 65–98.

66. D. Wu, H. Kahn, J. C. Dalton, G. M. Michal, F. Ernst, and A. H. Heuer: 'Orientation dependence of nitrogen supersaturation in austenitic stainless steel during low-temperature gas-phase nitriding', *Acta Materialia*, 2014, **79**, 339–350.

67. T. L. Christiansen, and M. A. J. Somers: 'Determination of the concentration dependent diffusion coefficient of nitrogen in expanded austenite', *International Journal of Materials Research*, 2008, **99**, 999–1005.

68. B. Ozturk, V. L. Fearing, J. A. Ruth, and G. Simkovich: 'The diffusion coefficient of carbon in cementite Fe_3C at 450°C', *Solid State Ionics*, 1984, **12**, 145–151.

69. A. Schneider, and G. Inden: 'Carbon diffusion in cementite (Fe_3C) and Hägg carbide (Fe_5C_2)', *Computer Coupling of Phase Diagrams and Thermochemistry*, 2007, **31**, 141–147.

70. A. V. Evteev, E. V. Levchenko, I. V. Belova, and G. E. Murch: 'Atomic mechanism of carbon diffusion in cementite', *Defect and Diffusion Forum*, 2008, **277**, 101–106.

71. E. V. Levchenko, A. V. Evteev, I. V. Belova, and G. E. Murch: 'Molecular dynamics simulation and theoretical analysis of carbon diffusion in cementite', *Acta Materialia*, 2009, **57**, 846–853.

72. C. Jiang, S. G. Srinivasan, A. Caro, and S. A. Maoly: 'Structural, elastic, and electronic properties of Fe_3C from first principles', *Journal of Applied Physics*, 2008, **103**, 043502.

73. M. Nikolussi, A. Leineweber, and E. J. Mittemeijer: 'Nitrogen diffusion through cementite layers', *Philosophical Magazine*, 2010, **90**, 1105–1122.

74. R. A. Johnson: 'Interstitials and vacancies in alpha iron', *Physical Review*, 1964, **134**, A1329–A1336.

75. D. Nguyen-Manh, and S. L. Dudarev: 'Multi-scale modelling of defect behavior in bcc transition metals and iron alloys for future fusion power plants', *Materials Science & Engineering A*, 2006, **423**, 74–78.

76. P. M. Derlet, D. Nguyen-Manh, and S. L. Dudarev: 'Multiscale modeling of crowdion and vacancy defects in body-centered-cubic transition metals', *Physical Review B*, 2007, **76**, 054107.

77. R. A. Johnson, and E. Brown: 'Point defects in copper', *Physical Review*, 1962, **127**, 446–454.

78. M. W. Thompson: Defects and Radiation Damage in Metals: Cambridge, U. K.: Cambridge University Press, 1969.

79. N. R. Quick, and H. H. Johnson: 'Hydrogen and deuterium in iron 49-506°C', *Acta Metallurgica*, 1978, **26**, 903–907.

80. H. H. Johnson: 'Hydrogen in iron', *Metallurgical and Materials Transactions A*, 1988, **19**, 691–707.

81. H. K. D. H. Bhadeshia: 'Prevention of hydrogen embrittlement in steels', *ISIJ International*, 2016, **56**, 24–36.

82. A. T. Paxton: 'From quantum mechanics to physical metallurgy of steels', *Materials Science and Technology*, 2014, **30**, 1063–1070.

83. W. R. Heller: 'Quantum effects in diffusion: internal friction due to hydrogen and deuterium dissolved in alpha-iron', *Acta Metallurgica*, 1961, **9**, 600–631.

84. J. H. Weiner, and Y. Partom: 'Quantum rate theory for solids. One dimensional tunneling effects', *Physical Review*, 1969, **187**, 1134–1146.

85. J. H. Weiner: 'Anomalous isotope effect for hydrogen diffusion in fcc metals', *Physical Review B*, 1976, **14**, 4741–4743.

86. S. Prakash: 'Proton diffusion in noble metals', *Physical Review B*, 1978, **18**, 3980–3987.

87. S. Prakash, J. E. Bonnet, and P. Lucasson: 'Proton movement in fcc, bcc and hcp metals', *Physical Review B*, 1979, **19**, 1976–1981.

88. R. Kaur, and S. Prakash: 'Isotope effect for hydrogen diffusion in metals', *Journal of Physics F: Metal Physics*, 1982, **12**, 1381–1386.

89. F. R. Coe: 'Welding steels without hydrogen cracking': Technical Report, The Welding Institute, Abingdon, U. K., 1973.

90. R. B. McLellan, and Z. R. Xu: 'Hydrogen induced vacancies in the iron lattice', *Scripta Materialia*, 1997, **36**, 1201–1205.

91. H. Oikawa: 'Lattice self-diffusion in solid iron – critical review', *Technology Reports of Tohoku University*, 1982, **47**, 67–77.

92. H. Oikawa: 'Review on lattice diffusion of substitutional impurities in iron. A summary report', *Technology Reports of Tohoku University*, 1982, **47**, 215–224.

93. H. Oikawa: 'Lattice diffusion of substitutional elements in iron and iron-base solid solutions. A critical review', *Technology Reports of Tohoku University*, 1983, **48**, 7–77.

94. F. S. Buffington, K. Hirano, and M. Cohen: 'Self diffusion in iron', *Acta Metallurgica*, 1961, **9**, 434–439.

95. Y. Iijima, K. Kimura, and K. Hirano: 'Self diffusion and isotope effect in alpha iron', *Acta Metallurgica*, 1988, **36**, 2811–2820.

96. R. J. Borg, and C. E. Birchenall: 'Self diffusion in alpha-iron', *Transactions of the AIME*, 1960, **218**, 980–984.

97. K. Hirano, M. Cohen, and B. L. Averbach: 'Diffusion of nickel into iron', *Acta Metallurgica*, 1961, **9**, 440–445.

98. L. A. Girifalco: 'Activation energy for diffusion in ferromagnetics', *Journal of the Physics and Chemistry of Solids*, 1962, **23**, 1171–1173.

99. L. Ruch, D. R. Sain, H. L. Yeh, and L. A. Girifalco: 'Analysis of diffusion in ferromagnets', *Journal of the Physics and Chemistry of Solids*, 1976, **37**, 649–653.

100. H. H. Potter: 'The magneto-caloric effect and other magnetic phenomena in iron', *Proceedings of the Royal Society of London*, 1934, **146**, 362–387.

101. J. Crangle, and G. M. Goodman: 'The magnetization of pure iron and nickel', *Proceedings of the Royal Society A*, 1971, **321**, 477–491.

102. C.-G. Lee, Y. Iijima, and K. Hirano: 'Self-diffusion and isotope effect in face-centered cubic cobalt', *Defect and Diffusion Forum*, 1993, **95-98**, 723–728.

103. J. K. cera: 'Analysis of magnetic anomaly in bcc iron', *Czechoslovak Journal of Physics B*, 1979, **29**, 797–809.

104. C. Zener: 'Theory of diffusion', In: W. Shockley, J. H. Holloman, R. Maurer, and F. Seitz, eds. *Imperfections in Nearly Perfect Crystals*. New York, USA: John Wiley & Sons, Inc., 1952:289–314.

105. R. Braun, and M. Feller-Kniepmeier: 'On the magnetic anomalies of self-and heterodiffusion in BCC iron', *Scripta Metallurgica*, 1986, **20**, 7–11.

106. H. V. M. Mirani, R. Harthoorn, T. J. Zuurendonk, S. J. Helmerhorst, and G. de Vries: 'Influence of ferromagnetic transition on self-diffusion in bcc FeSi', *Physica Status Solidi A*, 1975, **29**, 115–127.

107. J. Cermák, M. Lubbenhusen, and H. Mehrer: 'Influence of the magnetic phase transformation on the heterodiffusion of Ni in alpha iron', *Zeitschrift für Metallkunde*, 1989, **80**, 213–219.

108. B. Jönsson: 'On ferromagnetic ordering and lattice diffusion - a simple model', *Zietschrift für Metallkunde*, 1992, **83**, 349–355.

109. Y. Iijima, and C. G. Lee: 'Self-diffusion in b.c.c. and ordered phases of an equiatomic Fe-Co alloy', *Acta Metallurgica and Materialia*, 1995, **43**, 1183–1188.

110. Y. Iijima, and K. Hirano: 'Diffusion in iron, steels and some nickel alloys', In: R. P. Agarwala, ed. *Diffusion processes in nuclear materials*. Netherlands: Elsevier Science Publishers B. V., 1992:169–200.

111. M. L. Wasz, and R. B. McLellan: 'Nitrogen diffusion in iron', *Scripta Metallurgica et Materialia*, 1993, **28**, 1461–1463.

112. Y. Iijima: 'Magnetic effect on diffusion of carbon and nitrogen in iron', *Journal of Alloys and Compounds*, 1996, **234**, 290–294.

113. J. Fridberg, L.-E. Torndähl, and M. Hillert: 'Diffusion in iron', *Jernkontorets Annaler*, 1969, **153**, 263–276.

114. J. S. Dohie: 'Rapid diffusion of zinc in iron': Master's thesis, University of Manitoba, Winnipeg, Canada, 2005.

115. N. Oono, H. Nitta, and Y. Iijima: 'Diffusion of niobium in alpha-iron', *Materials Transactions*, 2003, **44**, 2078–2083.

116. C.-G. L. Y. Iijima, T. Hiratani, and K. Hirano: 'Diffusion of chromium in alpha-iron', *Materials Transactions, JIM*, 1990, **31**, 255–261.

117. Y. Iijima, K. Kimura, C.-G. Lee, and K. Hirano: 'Impurity diffusion and isotope effect of cobalt in alpha-iron', *Materials Transactions, JIM*, 1993, **34**, 20–26.

118. A. P. Sutton, and R. W. Balluffi: Interfaces in crystalline solids: Oxford, U. K.: Clarendon Press, 1995.

119. J. Friedberg, L. E. Tomdahl, and M. Hillert: 'Diffusion in iron', *Jernkontorets Annaler*, 1969, **153**, 263–276.

120. L. S. Darken: 'Diffusion of carbon in austenite with a discontinuity of composition', *Transactions of the AIME*, 1949, **180**, 430–438.

121. H. K. D. H. Bhadeshia: 'A personal commentary on "transformation of austenite at constant subcritical temperatures"', *Metallurgical and Materials Transactions A*, 2010, **41**, 1351–1390.

122. H. E. Cook, and J. E. Hilliard: 'Effect of gradient energy on diffusion in gold-silver alloys', *Journal of Applied Physics*, 1969, **40**, 2191–2198.

123. A. Einstein: 'Ist die trägheit eines körpers von seinem energieinhalt abhängig?', *Annalen der Physik*, 1905, **18**, 639–641.

124. G. S. Hartley: 'XLI. theory of the velocity of diffusion of strong electrolytes in dilute solution', *The London, Edinburgh, and Dublin Philosophical Magazine and Journal of Science*, 1931, **12**, 463–488.

125. L. Onsager: 'Theories and problems of liquid diffusion', *Annals of the New York Academy of Sciences*, 1945, **46**, 241–265.

126. J. S. Kirkaldy: 'Isothermal diffusion in multicomponent systems', *Advances in Materials Research*, 1970, **4**, 55–100.

127. L. C. Brown, and J. S. Kirkaldy: 'Carbon diffusion in dilute ternary austenites', *Transactions of the Metallurgical Society of AIME*, 1964, **230**, 223–226.

128. J. S. Kirkaldy, and G. R. Purdy: 'The thermodynamics of dilute ternary austenite solutions', *Canadian Journal of Physics*, 1962, **40**, 202–207.

129. C. Wagner: Thermodynamics of Alloys: New York, USA: Addison-Wesley Publication Company, 1952.

130. M. Kawakami, S. Yokoyama, K. Takagi, M. Nishimura, and J. S. Kim: 'Effect of aluminium and oxygen content on diffusivity of aluminium in molten iron', *ISIJ International*, 1997, **37**, 425–431.

131. P. K. cek, and T. P. rica: 'Diffusion in molten metals and melts: application to diffusion', *International Metals Reviews*, 1983, **28**, 131–157.

132. J. C. Swartz: 'Effects of microstructure on the Snoek relaxation', *Acta Metallurgica*, 1969, **17**, 1511–1515.

133. H. Numakura, and M. Koiwa: 'Atomic interaction between interstitial and substitutional solutes in alpha iron - a message from experiments', In: S. Nishijima, and H. Onodera, eds. *Modelling and Simulation for Materials Design*. Tsukuba, Japan: National Research Institute for Metals, 1996:142–147.

134. H. Numakura, G. Yotsui, and M. Koiwa: 'Calculation of the strength of Snoek relaxation in dilute ternary bcc alloys', *Acta Metallurgica et Materialia*, 1995, **43**, 705–714.

135. H. Saitoh, and K. Ushioda: 'Influences of manganese on internal friction and carbon solubility determined by combination of infrared absorption in ferrite of low carbon steels', *ISIJ International*, 1989, **29**, 960–965.

136. T. Ke: 'Anelastic properties of iron', *Trans. A.I.M.E.*, 1948, **176**, 448–476.

137. K. G. Denbigh: The Thermodynamics of the Steady State: New York, USA: John Wiley & Sons, Inc., 1955.

138. W. H. Nernst: 'Zur kinetik der lösung befindlichen körper: Theorie der diffusion', *Zeitschrift für Physikalische Chemie*, 1888, **3**, 613–637.

139. R. A. Oriani, and O. D. Gonzalez: 'Electromigration of hydrogen isotopes dissolved in alpha iron and in nickel', *Transactions of the Metallurgical Society of AIME*, 1967, **239**, 1041–1046.

140. V. I. Prosvirin: 'Concerning ionic diffusion in metals', *Herald of Metal Industry*, 1937, **No. 12**, 13.

141. T. A. Lebedev, and V. M. Gutterman: 'Referred to in Dayal and Darken (1950)', *Doklady Akademii Nauk, SSSR*, 1948, **60**, 1201.

142. M. A. Rabkin: 'Electrolytic transfer of carbon in liquid iron-carbon alloys (translation from Russian', *Zhurnal Prikladnoi Khimii (ZhPKh)*, 1957, **30**, 832–835 (translation page numbers).

143. T. Okabe, and A. G. Guy: 'Steady state electrotransport of carbon in iron', *Metallurgical Transactions*, 1970, **1**, 2705–2713.

144. C. Bosvieux, and J. Friedel: 'Sur l'electrolyse des alliages metalliques', *Journal of the Physics and Chemistry of Solids*, 1962, **23**, 123–136.

145. H. B. Huntintdon: 'Electromigration and thermomigration in metals', In: H. I. Aaronson, ed. *Diffusion*. Metals Park, Ohio, USA: American Society for Metals, 1973:155–184.

146. J. L. Bocquet, G. Brebec, and Y. Limoge: 'Diffusion in metals and alloys', In: R. W. Cahn, and P. Haasen, eds. *Physical Metallurgy*, vol. 1. Netherlands: Elsevier Science Publishers B. V., 1996:535–666.

147. J. N. Pratt, and R. G. R. Sellors: 'Electrotransport in metals and alloys', In: Y. Adda, A. D. Le Claire, L. M. Slifkin, and F. H. Wöhlbier, eds. *Diffusion and Defect Monograph Series*. Riehen, Switzerland: Trans Tech Publications, 1973:K42–K43.

148. O. D. Gonzalez, and R. A. Oriani: 'Thermal diffusion of dissolved hydrogen isotopes in iron and nickel', *Transactions of the Metallurgical Society of AIME*, 1965, **233**, 1878–1886.

149. R. A. Oriani: 'Thermomigration in solid metals', *Journal of Physics and Chemistry of Solids*, 1969, **30**, 339–351.

150. O. D. Gonzalez, and R. A. Oriani: 'Thermal diffusion of dissolved hydrogen isotopes in iron and nickel', *Transactions of the Metallurgical Society of AIME*, 1975, **233**, 1878–1886.

151. P. Shewmon: 'The thermal diffusion of carbon in α and γ iron', *Acta Metallurgica*, 1960, **8**, 605–611.

152. L. S. Darken, and R. A. Oriani: 'Thermal diffusion in solid alloys', *Acta Metallurgica*, 1954, **2**, 841–847.

153. O. A. Troitskii: 'Electromechanical effect in metals', *JETP Letters (English version)*, 1969, **10**, 11–14.

154. Y. Zhou, J. Guo, M. Gao, and G. He: 'Crack healing in a steel by using electropulsing technique', *Materials Letters*, 2004, **58**, 1732–1736.

155. H. Conrad: 'Electroplasticity in metals and ceramics', *Materials Science & Engineering A*, 2000, **287**, 276–287.

156. W. H. Cao, J. L. Zhang, and C. H. Shek: 'Effects of electropulsing treatment on mechanical properties in Ti rich TiNi shape memory alloy', *Materials Science and Technology*, 2013, **29**, 1135–1138.

157. R. S. Qin, A. Rahnama, W. J. Lu, X. F. Zhang, and B. Eliott-Bowman: 'Electropulsed steels', *Materials Science and Technology*, 2014, **30**, 1040–1044.

158. X. R. Chu, S. X. Lin, Z. M. Yue, J. Gao, and C. S. Zhang: 'Research of initial dynamic recrystallisation for AZ31 alloy with pulse current', *Materials Science and Technology*, 2015, **31**, 1601–1606.

159. D. Guo, X. L. Wang, and W. B. Dai: 'Microstructure evolution in metals induced by high density electric current pulses', *Materials Science and Technology*, 2015, **31**, 1545–1554.

160. G. Ghosh, and G. B. Olson: 'The isotropic shear modulus of multicomponent Fe-based solid solutions', *Acta Materialia*, 2002, **50**, 2655–2675.

161. J. W. Christian: Theory of Transformations in Metals and Alloys, Part I: 3 ed., Oxford, U. K.: Pergamon Press, 2003.

Notes

[1] Other data indicate the formation and migration energies of a vacancy in iron to be 16-2.2 eV and 0.64-0.67 eV, respectively. Compilation in [76] and [82].

[2] The equation at this stage does not change if the concentration c is replaced with the mole fraction x. The following identities are useful: $x_1 = \theta/(1+\theta)$ and $\Gamma = a_1(1+\theta)/\theta$.

[3] An alternative interpretation is that the neglect of short-range order effects near T_C causes the discrepancy observed in the Arrhenius plot, but inconsistent with the fact that precise measurements made by Iijima et al. cast doubt on whether a diffusion anomaly actually exists for $T > T_C$.

[4] It now is possible to estimate the isotropic shear modulus of ferrite and austenite as a function of chemical composition [160].

[5] This assumes that there is no vacancy current [161].

[6] Onsager's theory also states that as long as the J_i and X_i which contribute to dissipation are independent, then for diffusion processes, $M_{ij} = M_{ji}$.

[7] And substituting mole fraction ratios for concentration-per-unit-volume ratios.

4 Ferrite by reconstructive transformation

4.1 INTRODUCTION

Allotriomorphic ferrite is, without a doubt, the most prolific phase in the vast majority the 1.6 billion tonnes of steel consumed annually. It is the first phase to precipitate when austenite is cooled slowly below its equilibrium transformation temperature [1–3]. The term *allotriomorphic* in this context implies that the limiting surfaces of the crystal are irregular and do not display the symmetry of its internal structure [4]. The irregular form is a consequence of the ferrite nucleating at austenite grain surfaces so subsequent diffusion-aided growth occurs more rapidly along the boundaries than in any other direction. Its shape is therefore influenced by the presence of the boundary rather than its crystal structure (Figure 4.1). Allotriomorphic ferrite need not always form at the austenite grain boundaries, but it invariably does so because of the dearth of heterogeneous nucleation sites in the austenite; modern steels are impressive in their cleanliness, with few unintended inclusions that may facilitate the intragranular nucleation of ferrite.

The term *idiomorphic* implies that the precipitate has faces belonging to its crystalline form. In steels, idiomorphic ferrite is taken to be that which has a facetted shape (Figure 4.1). Idiomorphic ferrite nucleates on inclusions or other heterogeneous nucleation sites within the austenite grains.

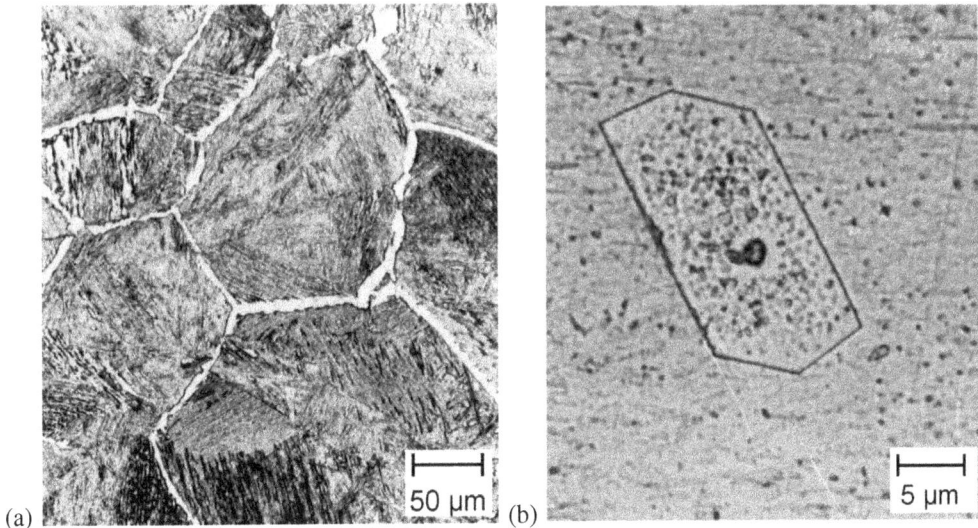

(a) 50 μm   (b) 5 μm

Figure 4.1 (a) Allotriomorphic ferrite in Fe-2.03Si-2.96Mn-0.12C wt% steel, transformed partially, at 720°C. (b) Idiomorphic ferrite in Fe-5.9W-1.95Si-0.36C wt% steel, partially transformed at 850°C (micrograph courtesy of Sunil Sahay).

These descriptions of allotriomorphs and idiomorphs are relevant when the precipitate is able to grow freely without impedance: hard impingement involves physical contact between adjacent particles whereas soft impingement describes the case where the solute-diffusion or temperature fields of nearby particles begin to overlap.[1] Impingement has the effect of constraining the microstructure. Steels, when they transform completely into allotriomorphic ferrite, have a microstructure of space-filling equiaxed ferrite grains, the shape of which bears little resemblance to the early stages of transformation.

The morphological definitions are to be interpreted loosely; they refer to a scale of observation that is typical of optical microscopy. Allotriomorphic ferrite can be crystallographically facetted when examined at a greater resolution. Similarly, it is not established whether the shape idiomorphic ferrite reflects the symmetry of the body-centred cubic crystal structure. It is more likely to comply with the combined symmetry of the γ/α bicrystal [5]. Indeed, the ability of any growing precipitate to form facets depends not just on the orientation dependence of interfacial energy but also on the driving force for transformation which determines the mechanism of interface displacement, whether the motion is stepped or continuous [6]. In principle, allotriomorphic ferrite formed at high temperatures (low supersaturations) could appear facetted while that formed at lower temperatures may not. This is because a step mechanism relies on the ability of a proportion of the interface to be pinned in a deep energy well, which is one in a periodic array of wells with a spacing normal to the interface that is equal to the step height. At large driving forces the pinned interface can escape from the energy wells and hence will tend to move continuously with every element of the interface moving in unison.

Since both idiomorphic and allotriomorphic ferrite grow by a reconstructive transformation mechanism, their growth is not limited by austenite grain boundaries. The extent of penetration into particular grains may vary given that the mobility of the transformation interface can change with the α/γ orientation relationship. Massive ferrite, which also grows by a reconstructive transformation mechanism, has the distinction that it inherits the composition of the parent austenite. The ability to cross parent austenite grain boundaries is particularly pronounced during massive transformation; the final ferrite grain size can be larger than the initial grain size of the austenite. The lack of a composition change allows the transformation to proceed until all of the austenite is consumed. These factors combine to give a single-phase microstructure of large grains of ferrite which have an approximately equiaxed morphology due to impingement between neighbouring grains. The transformation may begin with the growth of idiomorphs or allotriomorphs, so massive ferrite cannot strictly be regarded as a separate morphology in the classification scheme.

Austenite can only transform without a composition change if the temperature is below T_0, at which α and γ of identical composition have equal free energy (Figure 4.2). The T_0 temperature lies between the Ae_3 and Ae_1 temperatures which in turn define the upper and lower limits respectively of the two-phase $\alpha + \gamma$ field. For the range $Ae_3 \rightarrow T_0$, austenite can only transform to ferrite with a different chemical composition. Between T_0 and Ae_1, the growth of ferrite of equilibrium or unchanged composition is in principle possible. Below the Ae_1 temperature, massive transformation should predominate if the transformation mechanism is reconstructive.

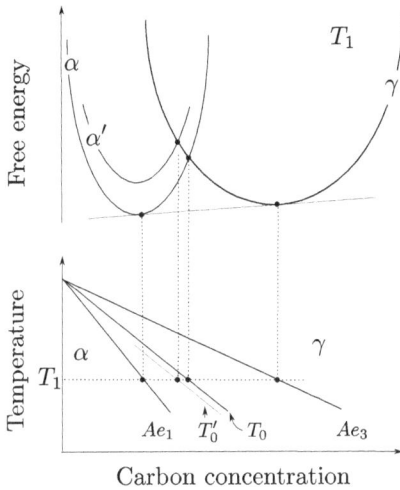

Figure 4.2 Construction of the phase boundaries using free energy curves. The tangent common to the free energy curves defines the equilibrium compositions of the austenite and ferrite at the temperature T_1; the locus of these compositions as a function of the temperature gives the Ae_1 and Ae_3 phase boundaries. The T_0 curve is the locus of all points for which austenite and ferrite of identical composition also have the same free energy.

4.2 INTERFACES

4.2.1 COHERENCY

The Ehrenfest classification (Section 1.7.1) is based on the successive differentiation of a thermodynamic potential with respect to an external variable such as temperature or pressure. The order of the transformation is defined by the lowest derivative to exhibit a discontinuity. In a first-order transformation, the partial derivative of the Gibbs free energy with respect to temperature is discontinuous at the transition temperature. There is thus a latent heat evolved at a sharp interface that separates the coexisting parent and product phases. This interface delineates perfect forms of the parent and product phases. The structure of the interface will be related to the mechanism of transformation in a manner consistent with both the nucleation and growth events.

In a second-order transformation the parent and product phases do not coexist; when such a transformation involves a lattice change, the change occurs continuously throughout the parent phase until its lattice is altered gradually into that of the product. There is no identifiable interface nor a nucleation event in the conventional sense.

All of the transformations associated with the decomposition of austenite are thermodynamically of first order. They occur by the propagation of well-defined interfaces that can be coherent, semi-coherent or incoherent. Coherency can of course be *forced* when the particle is small with consequent strain energy; stress-free coherence refers to the case where there is no breakdown of coherency as the particle grows.

A pair of crystals can be joined by a stress-free coherent interface if one of them can be generated from the other by a homogeneous deformation which is an invariant-plane strain (IPS). Such a deformation leaves one plane invariant, i.e., undistorted and unrotated, to form the coherent interface (Figure 4.3). It will be demonstrated in Chapter 5 that austenite cannot be deformed into ferrite by a strain which is an invariant-plane strain. It follows that α/γ interfaces must be semi-coherent or incoherent, except at the nucleation stage where the ferrite might be forced into coherence. For larger areas of contact when the coherency strains become intolerably large, defects are introduced into the interface that reduce the range of the strains. The structure of the interface will then consist of coherent patches separated periodically by discontinuities that prevent the misfit from accumulating over large distances.

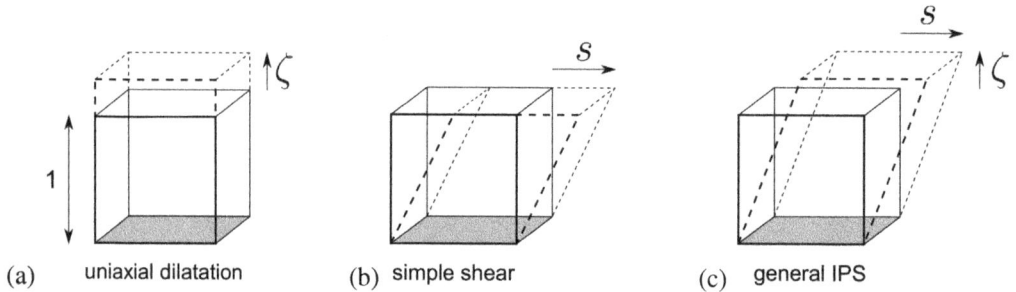

Figure 4.3 Three kinds of invariant-plane strains. The invariant-plane, shaded grey, is unaffected by any of the deformations. The terms s and ζ refer to the shear and dilatational strains respectively.

4.2.2 GLISSILE SEMI-COHERENT INTERFACES

There are two kinds of semi-coherency. If the discontinuities described above consist of a single set of screw dislocations, or dislocations with Burgers vectors that do not lie in the interface plane, then the semi-coherent interface is said to be *glissile* [4, 7, 8]. It also is necessary that the glide planes (of the misfit dislocations) associated with the ferrite lattice meet the corresponding glide planes in the austenite lattice edge to edge in the interface, along the dislocation lines [9]. A glissile interface can move conservatively, i.e., without any diffusion. When it does so, its motion causes deformation (Figure 4.4) that generates the new crystal structure, but at a rate that would be limited only by the mobility of the interface. The same interface can move slowly to produce reconstructive transformation only if its motion is accompanied by reconstructive diffusion (Figure 1.8).

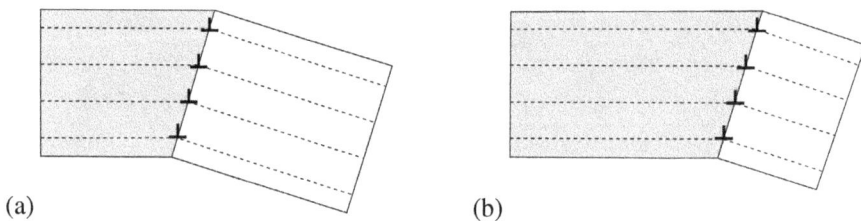

Figure 4.4 The conservative motion of a glissile interface causes deformation as it transforms the crystal structure. The dashed lines are traces of the glide planes of the interfacial dislocations, meeting edge-to-edge in the interface.

4.2.3 SESSILE SEMI-COHERENT INTERFACES

If the interfacial dislocations have Burgers vectors which lie in the interface plane, not parallel to the dislocation line, then the boundary is said to be "epitaxially semi-coherent" (Figure 4.5). Its normal displacement cannot occur without the thermally activated climb of the misfit dislocations. The rate of interface motion is therefore limited by diffusion and is expected to be sluggish at low temperatures.

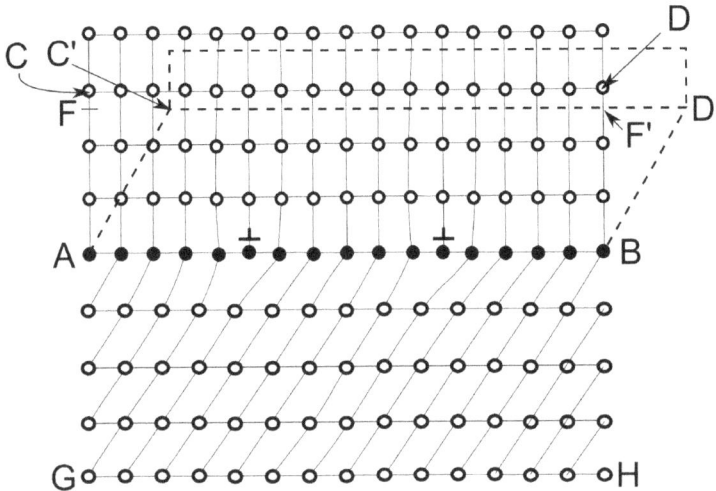

Figure 4.5 Nature of the shape change accompanying the movement of an epitaxially semi-coherent interphase interface [10].

The upwards motion of the boundary designated AB (Figure 4.5) to the new position C'D' should change the shape of a region ABCD of the parent crystal to AC'D'B of the product phase. Because of the dislocation climb implicit in the process, the total number of atoms in regions ACDB and AC'D'B will not be equal, the difference being removed by diffusion normal to the interface plane. Atom movements are therefore needed over a distance at least equal to that moved by the boundary. If this is the only diffusion flux, then the shear component of the shape change will not be eliminated, the transformation exhibiting characteristics associated with the displacive mechanism. The mobility of the interface will, of course, be limited by the dislocation climb process. This orderly removal of atoms as the interface migrates, involving the removal of the extra half-planes of the misfit dislocations, leaves only a partial atomic correspondence between the parent and product phases.[2]
However, if atoms have to migrate over large distances when an epitaxially semi-coherent interface moves, they should also be able to produce a net flow parallel to the interface, thus eliminating the shear component of the shape change, and its associated strain energy [7]. Referring to Figure 4.5, this would involve the diffusion of matter contained in the region BF'D' to region AFC', in a direction parallel to the interface (cf. Figure 1.8). It may therefore be improbable that atomic correspondence can be maintained during non-conservative interface motion.

4.2.4   INCOHERENT INTERFACES

The dislocations in the connecting interface become more closely spaced as the misfit between adjacent crystals increases. They eventually coalesce giving a boundary consisting of closely spaced vacancies or dislocation cores. Such a boundary is said to be incoherent; there is little correlation of atomic positions across the boundary. The activation energy for diffusion along such a boundary is similar to that for liquid iron (p. 142). The motion of an incoherent boundary can lead only to reconstructive transformation. The free volume and diffusivity within the boundary may be sufficiently large to confine the reconstructive processes to the proximity of the boundary itself, unlike the diffusion associated with the motion of semi-coherent interfaces.

Incoherent, coherent and semi-coherent boundaries can coexist around a particle which has grown by a reconstructive transformation mechanism. However, only coherent and semi-coherent boundaries that are glissile can enclose a particle that is generated by displacive transformation. This is because the deformation associated with the transformation is homogeneous. If an atomic correspondence exists across a particular interface of a particle, then it necessarily does so across any other interfacial orientations [4, 11].

4.3   CRYSTALLOGRAPHY

Crystallography has its origins in the description of patterns in single-crystals, but it is the relationship between crystals that is emphasised in transformations. There are at least five quantities which need to be specified in order to describe a pair of crystals which are connected at an interface. The orientation relationship, that is independent of the plane of the interface, can be described by an axis-angle pair. Since the axis can be specified with two independent direction cosines, the orientation relationship uses up three degrees of freedom (the two cosines and the angle of rotation about the axis). The interface plane requires a further two degrees of freedom since it is the normal to the plane which describes the orientation of the plane; two direction cosines fix the normal. Thus, there are five degrees of freedom necessary to describe the bicrystal (Figure 4.6). Each of these can be altered independently to obtain physically distinct bicrystals.

4.3.1   ORIENTATION RELATIONSHIPS

It is observed routinely that the orientation relationships that develop during phase transformations in the solid-state are not random [8, 12, 13]. The frequency of occurrence of certain orientation relationships usually exceeds the probability of obtaining it by arbitrarily joining two crystals. This perhaps is because the special dispositions allow the best fit at the transformation front [14, 15]. A good fit is conducive to a small value of the interfacial energy, making it easier for the transformation to nucleate. Embryos that happen by chance to be orientated for optimum fit would find it easier to grow into successful nuclei, giving rise to the non-random distribution seen experimentally. When nucleation occurs heterogeneously at austenite grain boundaries, there will also be a dependence on the relative orientations of the adjacent austenite grains.

An alternative interpretation is that nucleation occurs by the homogeneous deformation of a small

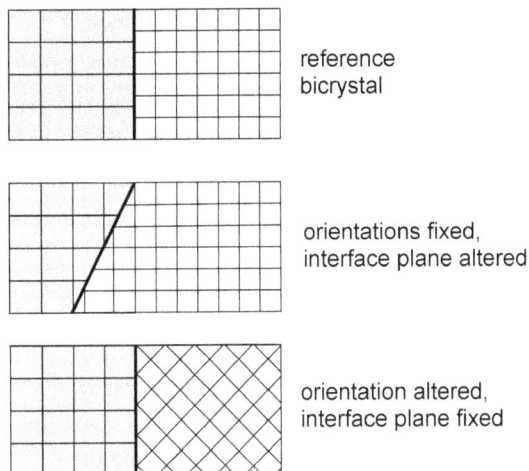

reference
bicrystal

orientations fixed,
interface plane altered

orientation altered,
interface plane fixed

Figure 4.6 Degrees of freedom in the specification of a bicrystal. The orientation of the interface can be changed without altering the orientation relationship between the crystals. Alternatively, the orientation relationship can be changed without changing the orientation of the interface.

region of the parent lattice [12]. This deformation would have to be of the kind that minimises strain energy. Of all possible ways of accomplishing the lattice change by homogeneous deformation, only a few might satisfy the minimum strain energy criterion, explaining the occurrence of preferred orientation relationships.

It has in practice been difficult to determine which of these factors control the existence of reproducible orientation relations. Nevertheless, it is found that when the α and γ have a low-energy orientation relationship, Widmanstätten ferrite which grows by a displacive mechanism can initiate from the α with an almost, but not exactly the same orientation [16]. This indicates that the nucleation mechanism of α_W is different (Chapter 7) and that of allotriomorphic ferrite is controlled by the minimisation of interfacial energy. Additional evidence for this is described in Section 4.4.1 which describes the nucleation of iron deposited from the vapour phase.

In the case of ferrite and reconstructive transformations in general, the growth process is not limited to the grain in which nucleation occurred. The experimental determination of the orientation relationship then becomes uncertain if the austenite grain which initiated the ferrite cannot be identified. The homogeneous deformation which might create the nucleus from the parent phase is not unique and various criteria have be used to select the most favoured deformation. Ryder and Pitsch [12] proposed specifically that a coherent nucleus forms as a small, thin platelet such that the plate plane and one direction within that plane are unrotated, though not necessarily undistorted during the deformation. A pair of corresponding planes from the two lattices and corresponding directions within those planes then remain parallel in the interface during nucleation. Such deformations can be described as the combination of the Bain strain with a small rigid body rotation (Chapter 5). It was found in this way that all γ/α orientation relationships should lie within about $11°$ of the Bain orientation[3] The classical Kurdjumov-Sachs [17] orientation relationship (i.e., $\{111\}_\gamma \parallel \{011\}_\alpha$ and $\langle 01\bar{1}\rangle_\gamma \parallel \langle 11\bar{1}\rangle_\alpha$) and that due to Nishiyama-Wasserman [18, 19] (i.e., $\{111\}_\gamma \parallel \{011\}_\alpha$ and $\langle 01\bar{1}\rangle_\gamma$ about $5.3°$ from $\langle 11\bar{1}\rangle_\alpha$ towards $\langle 1\bar{1}1\rangle_\alpha$) both lie within the $11°$ region about the Bain orientation.

Ryder and Pitsch went on to suggest that for precipitates which nucleate homogeneously in the austenite, the unrotated planes and directions of the two lattices should be the most closely packed planes and directions respectively. For heterogeneous nucleation at a grain boundary, the unrotated plane should correspond to the plane of the grain boundary and need not be a close-packed plane. There is a possibility that strains induced during nucleation make this displacive mechanism unlikely at low supersaturations [20], although this relies on various assumptions about the nucleus shape, composition and stored energy.

The crystallography of grain boundary nucleated ferrite indicates that the ferrite has a good-fit orientation relationship with at least one of the adjacent austenite grains [21]. The good-fit orientation is close to the classical Kurdjumov-Sachs (KS) or Nishiyama-Wasserman (NW) relationships. In fact, the range of orientations detected is more restricted than the $11°$ Bain region of Ryder and Pitsch. The orientations tend, within the limits of experimental error, to cluster at and around the KS and to a lesser extent the NW relations. It was always possible to find a close-packed plane of austenite which was within $2°$ of a close-packed plane from ferrite.[4] Within these planes, a close-packed direction of the austenite could always be found within about $8°$ of a similar direction in the allotriomorphic ferrite. There is now a vast literature on the crystallography of allotriomorphic ferrite that essentially confirms these observations, although there are violations of the general principle if the ferrite is nucleated at an intragranularly located inclusion that has the domineering influence at the nucleation stage [22].

King and Bell also found that over half of the ferrite allotriomorphs examined had a KS/NW type orientation relationship with both the adjacent austenite grains. This is unexpected since for a random population of austenite grain boundaries, about one in three might allow a ferrite orientation to be chosen which is within the Bain regions of both the adjacent matrix grains. The higher observed proportion is because of the austenite is crystallographic textured, as is inevitable in processed steels. The most likely scenario for the ferrite fitting well with both adjacent austenite grains if they have a $\Sigma = 3$ or $\Sigma = 11$ coincidence site lattice [23], although if there is some flexibility in the deviation from KS/NW, then the chances of a dual orientation developing increase monotonically with the magnitude of the "tolerable" deviation [22].

The early research used an indirect determination of the austenite orientation, based on assumed indices for the habit plane of Widmanstätten ferrite plates. It then became possible to retain the austenite. These studies have revealed a different result that the fraction of allotriomorphs exhibiting a reproducible orientation relationship with the austenite is quite small, Figure 4.7a [23, 24]. This is because of the greater mobility of more incoherent α/γ boundaries. Smith [25] proposed that when a ferrite allotriomorph nucleates at an austenite grain boundary, it has a good-fit orientation with one of the austenite grains and hence advances by the stepped motion of the α/γ_1 interface, and a random orientation with the other grain (Figure 4.7b,c). Growth, however, occurs into both of the adjacent grains, but at a more rapid pace into the austenite with which the ferrite is randomly oriented. Indeed, the ferrite can continue growing along other boundaries which had little or no influence in determining its orientation during nucleation.

The probability of finding ferrite that is well related to the adjacent austenite therefore depends

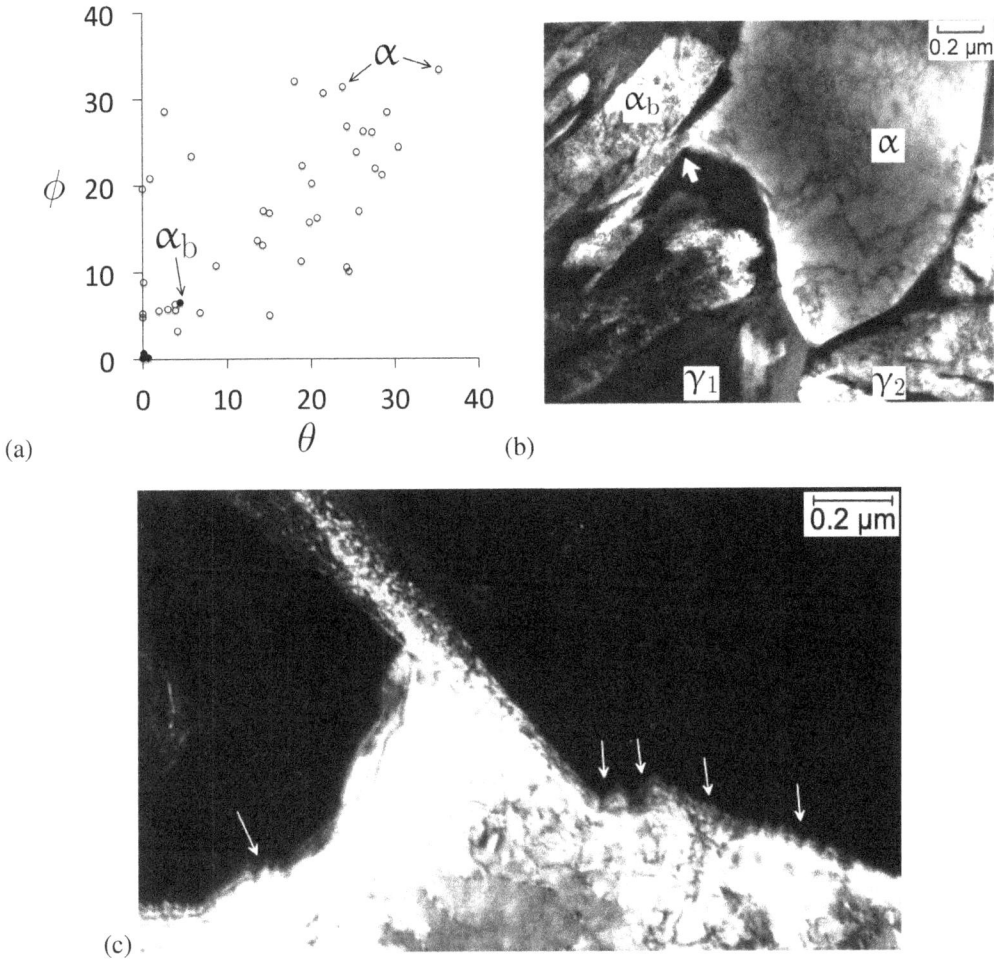

Figure 4.7 (a) The α-γ orientation relationship, where θ is the angle between $\{111\}_\gamma$ and $\{011\}_\alpha$ and ϕ is that between $\langle 10\bar{1}\rangle_\gamma$ and $\langle 11\bar{1}\rangle_\alpha$, with an accuracy of about $\pm 5°$. The orientations for the bainite are well within the Bain region whereas those for allotriomorphic ferrite are not necessarily so. (b) An allotriomorph of ferrite at an austenite grain boundary. The allotriomorph is related to γ_1 by an orientation relationship which is close to KS, but is randomly orientated with respect to the lower grain. Bainite has been able to nucleate from the allotriomorph only on the side where the orientation is suitable. (c) Enlarged version of (b) to show the steps at the α/γ_1 interface [24].

on the growth stage and this explains the small fraction of allotriomorphs that are reproducibly orientated with the austenite in which they grow. The nucleation mechanism cannot be deduced unambiguously from the observed orientations. Both interfacial energy minimisation and strain energy minimisation could lead to similar non-random distributions of orientation relations. Experiments on the nucleation of fcc α-brass rods from bcc brass showed that rods nucleated near free surfaces always had their invariant-lines parallel to the free surface [26]. This indicates an attempt to minimise the strain energy, consistent with the existence of coordinated atomic displacements during nucleation. If nucleation is controlled by the minimisation of interfacial energy, then other surface nucleation events are possible but were not found.

4.3.2 THE γ/α INTERFACE

High-resolution transmission electron microscopy of allotriomorphic ferrite/austenite interfaces has revealed regularly spaced linear discontinuities in the interface [27]. The spacing of these discontinuities varied as a function of the orientation of the interface plane.

Other studies on duplex stainless steel containing a mixture of δ-ferrite and austenite [28] where the orientation relationship was measured to be

$$(1\ 1\ 1)_\gamma \parallel (1\ 1\ 0)_\delta \qquad [0\ 1\ \bar{1}]_\gamma \parallel [\bar{1}\ 1\ 1]_\delta$$

indicated that an interface parallel to $(1\ 0\ 2)_\gamma$ contained at least two sets of intrinsic dislocations, with Burgers vectors parallel to $[1\ 1\ 0]_\gamma$ and $[1\ 0\ \bar{1}]_\gamma$ respectively, making the semi-coherent interface sessile.

Both sets of observations are consistent with a dislocation model of a partially coherent interface.

4.4 NUCLEATION OF ALLOTRIOMORPHIC AND IDIOMORPHIC FERRITE

Classical nucleation theory and the associated atomic nuances have been documented in considerable depth [4, 29]. The focus here is on the nucleation of ferrite and the difficulties of applying it in practice. There has never been an observation of the homogeneous nucleation of ferrite so that will only be discussed in order to provide context. Nucleation theory is fraught with difficulties when it comes to actual application. In the solid state, a nucleus is an enclosed particle, surrounded by a variety of interfacial orientations; most applications use a single γ/α interfacial energy, which in any case is a function of variables such as chemical composition, impurity content, crystallography and perhaps, size. The number density of nucleation sites cannot be predicted; even if measured data are used, it still is necessary to fix the fraction of the site that actually participates in the process. During heterogeneous nucleation it often is assumed that defects can be characterised by a single parameter, such as the boundary area per unit volume or a dislocation density. In reality, it is well established that there will be a spectrum of grain boundary energies, and that only a small fraction of γ/γ boundaries participate in the nucleation of ferrite [30]; similarly dislocation characteristics and configurations, and the non-uniform nature of plastic strain, are not single-valued. This level of complexity is challenging if not impossible to handle in practice. This pessimistic view can be substituted by pragmatism in the hope that a generic set of parameters can be determined by fitting to a large variety of steels and accepting that models will be approximate or indicative.

Consider the nucleation of α from the parent phase γ. The transformation does not occur spontaneously when on undercooling because the excess energy ΔG_σ associated with the creation of the interface must be accounted for. This energy scales with the surface area of the new particle, which will be some function of the square of the particle size (r), Figure 4.8. The driving force for transformation, $\Delta G_V^{\alpha\gamma}$, scales with the volume of the particle, i.e., with the cube of its size. The combined effect is in an *increase* in free energy when an embryo of the product phase forms because at small sizes, the total energy due to the interface ($\propto r^2$) overwhelms the reduction in free energy ($\propto r^3$). At some critical size r^*, the volume-dependent term begins to dominate and the net free energy change ΔG begins to decrease monotonically towards zero and then becomes negative.

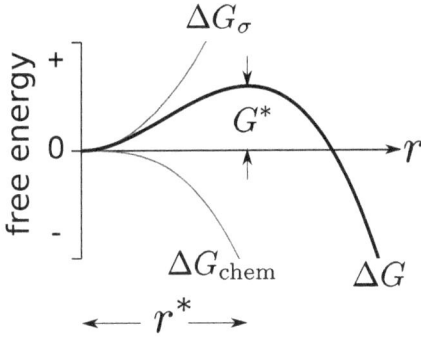

Figure 4.8 Change in free energy accompanying the nucleation of a transformation.

There is therefore an activation barrier G^* to overcome before an embryo can begin to grow; prior to reaching r^* it exists only when during thermally-induced agitation, a region of the parent by chance momentarily adopts the appropriate structure and composition in order for it to be classified as α. Such a fluctuation will not be stable when $r < r^*$. Suppose that the embryo is spherical with radius r, then

$$\Delta G = \underbrace{4\sigma_{\alpha\gamma}\pi r^2}_{\Delta G_\sigma} + \underbrace{\frac{4}{3}\pi r^3 \Delta G_V^{\gamma\alpha}}_{\Delta G_{\text{chem}}} \tag{4.1}$$

where $\sigma_{\alpha\gamma}$ is the constant interfacial energy per unit area and $\Delta G_V^{\gamma\alpha} = G^\alpha - G^\gamma$ is the chemical free energy change per unit volume. The activation energy and critical size can be obtained by differentiating this equation for the maximum in ΔG, giving

$$r^* = -\frac{2\sigma_{\alpha\gamma}}{\Delta G_V^{\gamma\alpha}} \qquad \text{and} \qquad G^* = \frac{16}{3}\frac{\pi\sigma_{\alpha\gamma}^3}{(\Delta G_V^{\gamma\alpha})^2}. \tag{4.2}$$

For transformations in the solid state, the formation of a new phase causes elastic strains with an associated strain energy ΔG^{m} which is a cost to be accounted for in the nucleation process. Any component of ΔG^{e} scales with the interfacial energy can simply be added to the latter; that which scales with the volume of the embryo can be added to $\Delta G_V^{\gamma\alpha}$, having the effect of reducing the driving force. There is an additional constant activation barrier Q for the transfer of atoms across the interface, which must be added to G^*. There also is a probability that an embryo that makes it to the top of the barrier may dissolve; this involves the so-called Zeldovich factor, neglected here given the pragmatic approach. A spherical shape was assumed in deriving G^*, but it would not be difficult to do so for other shapes although the strain energy terms may differ significantly [4]. The steady-state nucleation rate must depend on the attempt frequency ν which often is written as kT_{D}/h but this atomic vibration frequency may not be representative given that the fluctuations considered are for clusters of atoms attempting to surmount barriers. The steady-state nucleation rate per unit time and volume is then given by

$$I_V = N_V \nu \exp\left\{-\frac{G^* + Q}{RT}\right\} \tag{4.3}$$

where the units of the activation energies must be consistent with those of RT and N_V is the number per unit volume of potential nucleation sites. Q often is related to some large fraction of the activation energy for self-diffusion in iron, given that the nucleus will have a high level of forced-coherency with the matrix and that transport occurs across the interface rather than along it in order to achieve the structural change (p. 142).

Heterogeneous nucleation on defects such as the austenite grain surfaces and dislocations is much easier than when nucleation occurs homogeneously, because there is an energy gain when a part of the defect is destroyed in the process. In grain boundary nucleation, the surface per unit volume is $S_V = 2/\bar{L}_\gamma$, where the quantity in the denominator is the mean lineal intercept defining the austenite grain size. Therefore, the number density N_A per unit area of grain-surface nucleation sites will be proportional to S_V. The fraction of atoms located at the boundary surfaces of thickness δ_b is obtained as $\delta_b \times S_V$. The number density grain boundary nucleation sites per unit volume is therefore

$$N_V^{\text{grain faces}} = N_V \frac{2\delta_b}{\bar{L}_\gamma} \tag{4.4}$$

A good representation of an equiaxed grain structure in three dimensions is the Kelvin truncated-octahedron (Figure 4.9) [31]. It has 14 faces, 36 edges each of length a and 24 corners, which in an array that fills space would be shared by 2, 3 and 4 truncated octahedra respectively. The volume per grain $V_{\text{grain}} = 11.314a^3$, and since $\bar{L} = 1.69a$, the volume can be written $2.639(\bar{L}_\gamma)^3$; similar relationships exist for other shapes [32–35].

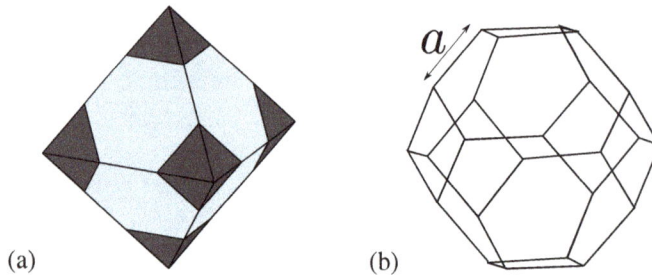

(a)   (b)

Figure 4.9 (a) A regular octahedron consisting of eight triangles. The dark regions represent the bits that have to be cut off to create the truncated octahedron consisting of eight hexagons and six squares. Adapted from [36]. (b) Truncated octahedron as a representation of an equiaxed grain. These objects can be stacked to fill space.

Assuming that grain edges have a cross-section δ_b^2 and the fact that each of the edges is shared between two of the truncated octahedra, the fraction of sites at grain edges is calculated by noting that each of the 36 edges is shared between 3 grains, so the edge length per grain is $36a/3 = 7.1\bar{L}_\gamma$ and the volume per grain is $2.639(\bar{L}_\gamma)^3$, so the fraction of sites at edges, per unit volume is

$$N_V^{\text{grain edges}} = N_V \frac{\delta_b^2 \times 7.1\bar{L}_\gamma}{2.639(\bar{L}_\gamma)^3} = 2.69 N_V \left(\frac{\delta_b}{\bar{L}_\gamma}\right)^2. \tag{4.5}$$

There are 24 corners per grain but each is shared with four other grains. The fraction of sites at corners, assuming that δ_b^3 is the volume of each corner, is therefore $(24/4)\delta_b^3/2.639(\overline{L}_\gamma)^3$ so

$$N_V^{\text{grain corners}} = \frac{6\delta_b^3}{2.639(\overline{L}_\gamma)^3} = 2.274\left(\frac{\delta_b}{\overline{L}_\gamma}\right)^3. \tag{4.6}$$

Because there is more potential for disorder at locations where large numbers of grains meet, the activation energies for nucleation increase on progressing from corner, edge, face and homogeneous sites. However, the number density of nucleation sites also increases in that order. The greatest contribution to nucleation at low driving forces ($|\Delta G^{\gamma\alpha}|$) will be at corners, but as the driving force increases, the other sites will also be active and may overwhelm the corner sites because of their greater number densities.

Equation 4.3 is said to represent the steady-state nucleation rate. Assuming that the supercooled austenite does not contain ferrite-like clusters of atoms, it would take time to establish the distribution of embryos of ferrite up to the critical size. This time represents an incubation period τ during which the nucleation rate increases towards the steady state. This transient can be represented by multiplying the steady-state rate by $\exp\{-\tau/t\}$ [37]. Russell developed a theory for the incubation time in the context of a variety of scenarios dealing with grain boundary nucleation kinetics [38]. The result can be expressed approximately as [39]

$$\tau \propto \frac{T}{(|\Delta G|)^p D} \tag{4.7}$$

where p is an exponent depending on the state of coherency of the embryo and D is an effective diffusion coefficient. This method has been used empirically to model the initiation of transformation in time-temperature-transformation diagrams [39]. ΔG here is the driving force for nucleation, written previously as $\Delta G^{\gamma\alpha} = G^\alpha - G^\gamma$. While this is correct when there is no composition change during transformation, different considerations apply for an alloy. Consider a steel containing only carbon; when the austenite of composition $\bar{x}$ is supercooled into the $\alpha + \gamma$ field and transformation allowed to proceed to equilibrium, there will exist a mixture of the two phases with a concomitant reduction in free energy relative to the fully austenitic state of $\Delta G^{\gamma \to \alpha + \gamma'}$, where γ' refers to austenite of an enriched equilibrium composition $x^{\gamma\alpha}$ with the corresponding ferrite composition at equilibrium given by $x^{\alpha\gamma}$. The equilibrium fraction of ferrite is given by $(x^{\gamma\alpha} - \bar{x})/(x^{\gamma\alpha} - x^{\alpha\gamma})$. So the free energy change $\Delta G^{\gamma \to \alpha + \gamma'}$ divided by this fraction of ferrite is ΔG_3, Figure 4.10a.

However, the formation of a nucleus cannot substantially alter the composition of the austenite which remains essentially at $\bar{x}$. In Figure 4.10a, as the fraction of ferrite is reduced, the composition of the austenite changes from $x^{\gamma\alpha}$ towards $\bar{x}$. In the limit there is no change, the line 'ab' becomes a tangent to the austenite curve at $\bar{x}$, Figure 4.10b. Any vertical arrow from that line to the ferrite curve gives the free energy reduction on forming a very small quantity of ferrite. ΔG_3 is the reduction if the terminal point of the arrow touches the ferrite free energy curve at its minimum. The maximum reduction, ΔG_m, corresponds to the ferrite composition x_m^α defined by a parallel tangent to the ferrite curve. This can be used in the nucleation rate equations.

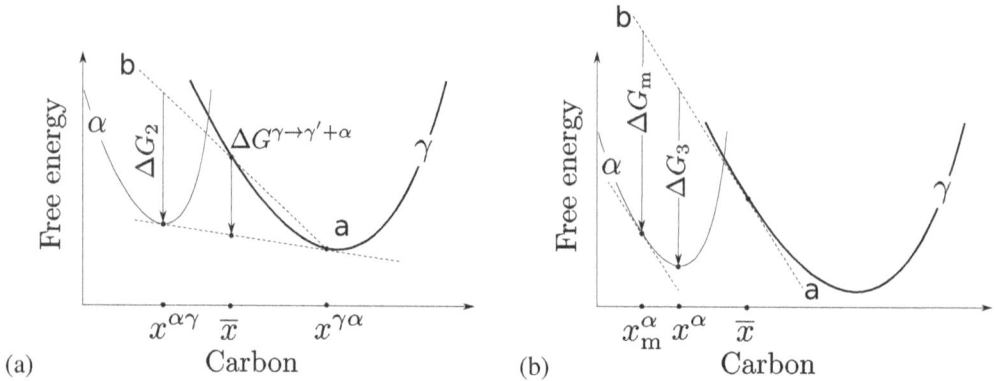

Figure 4.10 Plots of free energy versus carbon concentration at a constant temperature. (a) The equilibrium compositions of the ferrite and austenite are given by the common tangent and the arrows indicate the overall free energy change $\Delta G^{\gamma \to \alpha + \gamma'}$ when austenite of composition $\bar{x}$ decomposes into a mixture of ferrite and enriched-austenite. The second arrow indicates the overall free energy change divided by the equilibrium fraction of ferrite. (b) Similar diagram but showing how the parallel tangent construction leads to the biggest reduction in free energy when a ferrite nucleus forms without substantially altering the composition of the remaining austenite.

4.4.1 HETEROGENEOUS NUCLEATION FROM VAPOUR

There are obvious reasons why it has not yet been possible to demonstrate subtle details about the early stages of nucleation. Nucleation is dependent on random fluctuations in space and time. The dimensions of a nucleus are, for all practical purposes, minute, requiring the use of high-resolution techniques. This in turn limits the amount of material that can be studied, thereby greatly reducing the probability of capturing a nucleation event. Work on the deposition of thin films of iron from its vapour phase is interesting because it demonstrates the overriding influence of crystallography and lattice matching in heterogeneous nucleation, while at the same time revealing properties not obvious in bulk samples.[5] When depositing on to a single-crystal substrate, all orientations of iron can in principle nucleate from the vapour, but only those with the smallest activation energy will tend to do so, particularly when there is a two-dimensional match with the substrate.

The accumulated data are remarkable, obtained using techniques such as Auger electron spectroscopy, reflection high-energy electron diffraction, low-energy electron diffraction, surface X-ray diffraction etc. They include the time-dependence of the crystallography of the substrate and deposit, the chemical composition of the layers of atoms as they form, together with a myriad of revelations of the physical properties of the films of iron.

The main substrates for the deposition of iron have been Cu, Pd, Ag, Al and Ru, all of which are fcc with the exception of ruthenium which has a hcp crystal structure. The deposits are made on specific crystallographic planes, usually $\{100\}_{\text{fcc}}$, $\{111\}_{\text{fcc}}$ and $\{0001\}_{\text{hcp}}$. The deposition process has sufficient control to enable fractions of monolayers to be deposited.

Table 4.1 shows the lattice parameters of a variety of substrates used during single-crystal thin-film preparation. The deposit tends to be epitaxial when the iron is evaporated onto a low-misfit substrate such as copper. Since the structure of copper is fcc, the iron is deposited as austenite as long as the thickness of the deposit is less than a few monolayers. Thicker films tend to become ferritic. This structural transition occurs at a smaller thickness as the misfit with the substrate becomes larger. In fact, the mechanism of deposition changes at large misfits, from layer-by-layer growth to island growth.

Other results come from the deposition of iron on high-misfit substrates such as Pd, Ag and Al. On $\{111\}_{Pd}$ the first monolayer of iron deposits pseudomorphically with the Pd but there is an intermixing of Fe and Pd for experiments at both room temperature and 200°C [40].[6] Beyond a few layer equivalents, the iron morphs into large bcc $\{110\}$ islands which have a Kurdjumov-Sachs orientation relation with the Pd substrate. The misfit is so large when the substrate is silver, that the iron always deposits bcc islands, apparently in a Nishiyama-Wasserman orientation with the substrate. The islands tend to be elongated along $\langle 110 \rangle_{Ag}$ directions.

The mismatch of iron with aluminium is similar to that with silver, but the observed effect is quite different that the iron at first has an amorphous structure. It has been speculated that this is because the iron atoms rapidly diffuse into the aluminium to form an amorphous surface-alloy. An alternative possibility is that because aluminium has such an incredibly strong affinity for oxygen, it may be difficult to ensure a pure aluminium oxide-free surface prior to the deposition of iron. It would therefore be interesting to see whether similar effects are observed when iron is deposited on an appropriately oriented alumina surface.

The growth of iron on ruthenium (0001) occurs at first by the formation of a single pseudomorphic layer, with some interdiffusion of Ru into the deposited layer. All subsequent growth is by the formation of three-dimensional bcc Fe $\{110\}$ domains which are in Kurdjumov-Sachs orientation with the pseudomorphous layer of iron. These bcc islands never form directly on the ruthenium even though the pseudomorphic iron has the same atomic pattern as the substrate ruthenium [41].

It is worth emphasising that the pattern in which the atoms of a clean substrate are arranged may reconstruct into one of a series of possible surface structures; it cannot be assumed that the surface structure reflects that of parallel planes within the bulk of the sample. Thus, $\{100\}_{Au}$ surface is known to reconstruct into a (5×20) structure.[7] This must affect the nucleation of iron; it has been shown that the deposition of iron leads to a change in the surface structure of, for example, gold [42].

It is possible to reach some general conclusions about heterogeneous nucleation from these thin-film experiments:

(i)   The atomic structure of the nucleus can evolve as the particle grows. The initial arrangement might comply more with the substrate than the thermodynamically most stable form of the iron.

(ii) The atomic structure of the substrate itself may be altered during the nucleation event.

(iii) There may be an intermixing of atoms when the substrate and nucleus are chemically different.

(iv) Contamination of the substrate surface with, for example, the segregation of impurities or reaction with impurities may have a major influence on the early stages of nucleus formation.

(v) Any substrate/iron-deposit orientation is in principle possible when condensing from the vapour phase, but the observed relationships are specific and identical to those generated during transformation in bulk steels. Therefore, the crystallography during bulk-transformation is controlled by the need to minimise interfacial and strain energies at the nucleation stage, irrespective of whether the mechanism is reconstructive or displacive.

Table 4.1

Lattice parameters (a, c) of iron and of substrates used during the deposition of thin films.

	a/nm		a/nm	c/nm
Austenite	0.3567	Ferrite	0.2866	
Nickel	0.3524	Copper	0.3615	
Palladium	0.3890	Platinum	0.3923	
Gold	0.4078	Silver	0.4086	
Aluminium	0.4050	Ruthenium	0.2705	0.4281
GaAs	0.5653			

4.4.2 HETEROGENEOUS NUCLEATION ON INCLUSIONS

Much of the work on the inclusion-stimulated nucleation of ferrite comes from research on welding alloys where the structure can be refined by the intragranular nucleation of ferrite plates. Although these are not allotriomorphs, the discussion of inclusion effects remains relevant and hence will be included here.

Figure 4.11a shows that if a truncated sphere of ferrite forms on a spherical, unfaceted inclusion, the activation energy for nucleation is greater than if it nucleates at austenite grain surfaces, assuming that the inclusion has the same interfacial energy with austenite and ferrite, and that $\sigma_{\alpha\gamma} = 0.75\,\mathrm{J\,m^{-2}}$ [43]. The grain surface result will depend on the assumed values of $\sigma_{\alpha\gamma}$ and the interfacial energy between the inclusion and γ or α. The cosine of the contact angle for the geometries assumed is $\cos\theta = (\sigma_{I\gamma} - \sigma_{I\alpha})/\sigma_{\gamma\alpha}$. Figure 4.11b illustrates the potency of vanadium nitride as the substrate for ferrite nucleation, as a function of these parameters [44]. Whereas it is difficult to verify most of the quantities needed for these calculations, the size at which inclusions become more effective than austenite grain boundaries can be measured and used to fit the parameters.

In some cases, inclusions may become the most favoured site at a sufficiently large size. However, large inclusions are avoided in steel design because they compromise toughness. This model obviously has limits in its assumptions because there exist mechanisms described below that render inclusions particularly effective in stimulating ferrite.

Ferrite may deposit epitaxially on a faceted crystalline-inclusion within the austenite, but that would require some sort of a dimensional match between the two lattices at the plane of contact.[8] Given that

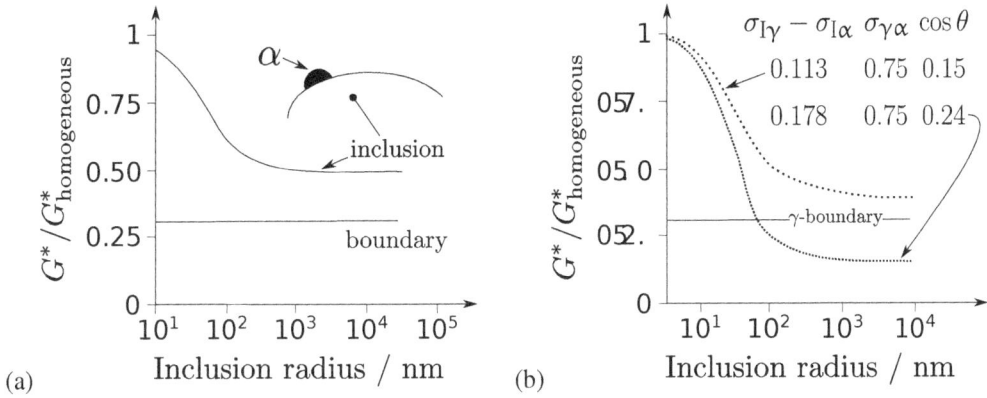

Figure 4.11 (a) Activation energy for nucleation, normalised relative to that for homogeneous nucleation, as a function of the radius of spherical, unfaceted inclusions. Also plotted is the activation energy for nucleation on an austenite grain boundary. Adapted from Ricks et al. [43]. (b) Similar calculations for ferrite nucleation on vanadium nitride, for two different sets of interfacial energy parameters. The interfacial energies are in units of Jm^{-2}. The cosine of the ferrite-inclusion contact angle is also different for the two curves. Adapted from Mu et al. [44].

the nucleation occurs in the solid state, the match must be good in order to avoid the accumulation of strains. An important consequence is that there will be reproducible, oriented growth of the ferrite with respect to the inclusion lattice. This immediately raises a problem that is not considered in most treatments of inclusion stimulated nucleation, that the ferrite orientation may be constrained by the surrounding austenite. If this is the case then the probability of simultaneously satisfying a low-energy orientation with the inclusion and austenite is likely to be small unless the inclusion precipitates in the austenite and is not inherited from the liquid state. In addition, the chemical elements and crystal structures of the non-metallic inclusions are more complex than of iron; the topography, atomic constitution and charge states of inclusion surfaces may not be consistent with the rational planes of ferrite. Such topographic matching has not been considered in any treatment of inclusion-stimulated nucleation in steel.

Dimensional matching is expressed as a mean percentage planar misfit κ_m [45]. Suppose that the inclusion is faceted on $(hkl)_I$ and that the ferrite deposits epitaxially with $(hkl)_\alpha \parallel (hkl)_I$, with a pair of corresponding rational directions $[uvw]_I$ and $[uvw]_\alpha$ inclined at an angle ϕ to each other. The interatomic spacings d along three such directions within the plane of epitaxy are examined to obtain

$$\kappa_m = \frac{100}{3} \sum_{j=1}^{3} \frac{|d_j^I \cos\phi - d_j^\alpha|}{d_j^\alpha}. \tag{4.8}$$

Table 4.2 lists some misfit data estimated assuming that epitaxy would be confined to planes of low crystallographic indices: {001}, {011} and {111}. The assumed orientation relationships include the Bain orientation which implies $\{100\}_\alpha \parallel \{100\}_I$ and $\langle 100 \rangle_\alpha \parallel \langle 011 \rangle_I$. The cube orientation occurs when the cell edges of the two crystals are parallel.

Other mechanisms of inclusion-assisted nucleation include stimulation by thermal strains and chemical heterogeneities in the vicinity of the inclusion/matrix interface; alternatively, they may simply

Table 4.2

Ambient temperature dimensional match on particular crystallographic planes of substrates and ferrite. The data are from a more detailed set published by [46, 47] for cases where the misfit is found to be less than 5%. The inclusions all have a cubic-F lattice; the ferrite is body-centred cubic. The data for MnS are from [48] with the misfit values calculated for 727°C and for the experimentally observed orientation relationship. The data for MgO are from [49].

Inclusion	Orientation	Plane of epitaxy	Misfit %
TiO	Bain	$\{100\}_\alpha$	3.0
TiO	$\{111\}_{TiO} \parallel \{110\}_\alpha$ $\{1\bar{1}0\}_{TiO} \parallel \{1\bar{1}1\}_\alpha$	$\{110\}_\alpha$	13.0
TiN	Bain	$\{100\}_\alpha$	4.6
γ-alumina	Bain	$\{100\}_\alpha$	3.2
MgO	Bain	$\{100\}_\alpha$	3.8
MgO	$\{111\}_{MgO} \parallel \{110\}_\alpha$ $\{1\bar{1}0\}_{MgO} \parallel \{1\bar{1}1\}_\alpha$	$\{110\}_\alpha$	13.8
Galaxite	Bain	$\{100\}_\alpha$	1.8
$Cu_{1.8}S$	Cube	$\{111\}_\alpha$	2.8
α-MnS	$\{4\bar{2}0\}_{MnS} \parallel \{1\bar{1}0\}_\alpha$ $\langle 1\bar{1}0 \rangle_{MnS} \parallel \langle 002 \rangle_\alpha$	$\{110\}_\alpha$	5.48
β-MnS	$\{4\bar{2}0\}_{MnS} \parallel \{1\bar{1}0\}_\alpha$ $\langle 1\bar{1}0 \rangle_{MnS} \parallel \langle 002 \rangle_\alpha$	$\{110\}_\alpha$	2.25
BN	$\{110\}_{BN} \parallel \{111\}_\alpha$	$\{111\}_\alpha$	6.5

be inert sites for heterogeneous nucleation. Pressure bonded ceramic-steel composites have been studied to reveal the potency of pure ceramic phases in stimulating the nucleation of bainite, Table 4.3 [50–53]. A simple interpretation emerges from these experiments, that the ceramics that interact chemically with steel are most effective in nucleating bainite. A significant exception is TiO, which remains inert and yet enhances bainite formation.

Table 4.3

Chemically active ceramics in experiments designed to test for ferrite nucleation at ceramic/steel bonds.

Chemically Active	Chemically Inactive
TiO_2	TiO, Ti_2O_3, TiC, TiB_2, TiN
$Al_2Si_2O_7$	Al_2O_3
MnO_2	MnO
SiC, Si	Si_3N_4, SiO_2
CoO, V_2O_5	ZrO_2, FeS, Y_2O_3

The bond experiments show that some minerals act as sources of oxygen that cause the adjacent steel to decarburise, thereby stimulating ferrite. One such mineral is TiO_2; structural and behavioural analogues of TiO_2 (SnO_2, MnO_2 and PbO_2) show similar features. TiO_2 and related minerals tend to form oxygen vacancy defects at elevated temperatures, thus releasing oxygen, which then penetrates into the steel. Therefore, all oxygen producing minerals would be expected to stimulate ferrite nucleation, while non-oxygen producing minerals would not. *Normal* perovskites (ABO_3 type) are structurally similar to *defect* perovskites (BO_3 type) but the ability of defect perovskites to produce oxygen is much greater. Therefore, WO_3, which is a defect perovskite, is an effective nucleant whereas the normal perovskite $CaTiO_3$ is not. Indeed, any oxygen source, for example, KNO_3, stimulates the nucleation of ferrite.

Neither Ti_2O_3 nor TiO is an oxygen source but both are effective nucleants. Ti_2O_3 does this by absorbing manganese; the resulting local depletion of Mn increases the driving force for transformation [51, 52, 54–56]. Therefore, Ti_2O_3 is ineffective as a nucleation site when introduced in manganese-free steel [57]. Steel deoxidised with a combination of Mn, Ti and Si can contain Mn-Ti oxides which are enveloped by manganese depletion zones which stimulate the intragranular nucleation of ferrite. The depletion is a result of the growth of the $MnTi_2O_4$ phase rather than the absorption of manganese into the oxide [58].

TiO remains chemically inert but it has a better fit with ferrite than TiN or MgO, neither of which are as effective as TiO in stimulating nucleation. The orientation relationship between TiO and the matrix is important in determining its efficacy. When thin, vapour deposited amorphous layers of TiO on steel are heat treated, they crystallise into a variety of textures. Ferrite forms readily when the texture is such that $\{1\,0\,0\}_{TiO}$ planes are parallel to the steel surface, with a much reduced effect when the texture similarly is dominated by $\{1\,1\,1\}_{TiO}$ [49]. The misfit data in Table 4.2 show that this is expected on the basis of coherency with the ferrite.

The nucleation mechanisms are summarised in Table 4.4.

Table 4.4
Mineral classification according bainite or ferrite nucleation potency.

Effective: oxygen sources	Effective: other mechanisms	Ineffective
TiO_2, SnO_2	Ti_2O_3	TiN, $CaTiO_3$
MnO_2, PbO_2	TiO	$SrTiO_3$, α-Al_2O_3
WO_3, MoO_3	MgO	γ-Al_2O_3, $MnAl_2O_4$
KNO_3		NbC

4.5 INTERFACE MOTION: RATE-CONTROLLING PROCESSES

An electrical current i flowing through a resistor will dissipate energy in the form of heat. When the current passes through a pair of resistors in series, the dissipations are $i\Delta V_1$ and $i\Delta V_2$ where ΔV_1 and ΔV_2 are the voltage drops across the respective resistors, Figure 4.12a. The total potential difference across the circuit is $\Delta V_1 + \Delta V_2$. The current that flows is identical through both of the resistors in series. Because the resistors have different $i - V$ characteristics, the potential drop across each resistor has to be such that they both yield an identical current, i.e., $i_{\circ} = f\{\Delta V_1, R_1\} = g\{\Delta V_2, R_2\}$ through each resistor; the functions f and g are as illustrated in Figure 4.12b. If $\Delta V_1 \gg \Delta V_2$ then the resistor R_1 is said to *control* the current. If, on the other hand, $\Delta V_1 \approx \Delta V_2$ then the current is said to be under *mixed* control.

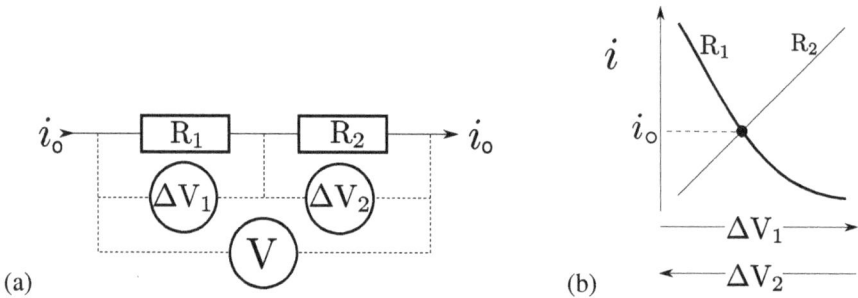

Figure 4.12 Electrical analogy for interface response functions. (a) The current flowing through each resistance is identical though the potential drop is not. (b) The two resistors have different characteristics when operating in isolation. One has a linear variation in current with voltage whereas the other is nonlinear. The actual current that passes through the circuit will be $i_{\circ}$ where the two curves intersect.

This electrical circuit is an excellent analogy to the motion of an interface, with the interface velocity and driving force (free energy change) relating to the current and applied potential difference respectively. The resistors are then the processes that impede the motion of the interface, such as diffusion or the barrier to the transfer of atoms across the interface. When most of the driving force is dissipated in diffusion, the interface is said to move at a rate *controlled* by diffusion

More formally, the rate at which an interface moves depends both on its intrinsic mobility and on the ease with which any alloying elements partitioned during transformation diffuse ahead of the moving interface.[9] It is important to realise that these dissipative processes are in series so the interface velocity as calculated from the interface mobility must equal that calculated from the diffusion of solute ahead of the transformation front. If the free energy $\Delta G'$ available to drive the interface is used primarily in driving the diffusion, growth is said to be diffusion-controlled. Interface-controlled growth occurs when most of $\Delta G'$ is dissipated in the process of transferring atoms across the interface.

Consider the growth of ferrite from supersaturated austenite in a Fe-C alloy transformed isothermally in the $\alpha + \gamma$ phase field. Figure 4.13 shows carbon concentration profiles across the α/γ interface, for a variety of scenarios; $x^{\alpha\gamma}$ is the concentration in α in equilibrium with γ and a similar interpretation applies to $x^{\gamma\alpha}$. $\bar{x}$ refers to the average concentration in the alloy, assumed here to represent a constant far-field concentration in the austenite.

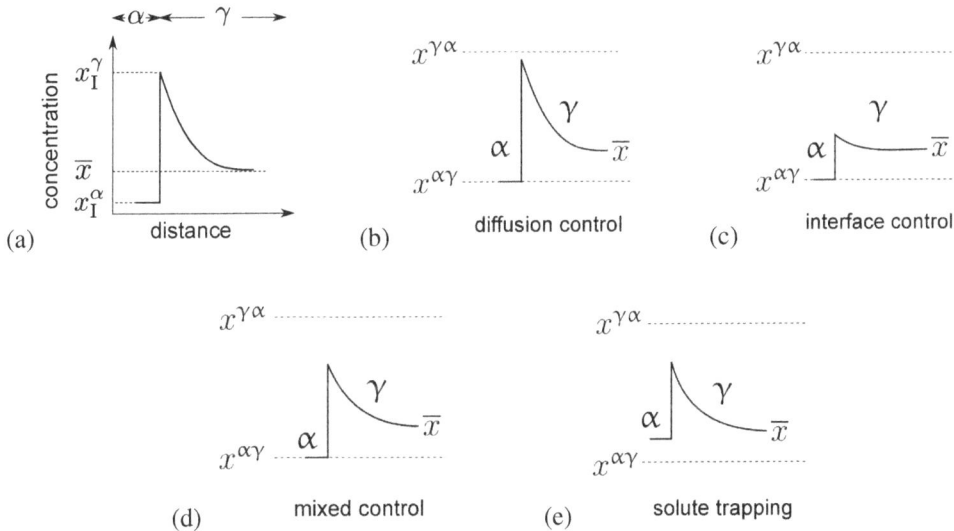

Figure 4.13   (a) The form of the concentration profile that develops in the vicinity of the moving α/γ interface when the solute-solubility α in contact with γ is less than in the γ. (b) Diffusion-control, (c) interface-control, (d) mixed interface and diffusion-control and (e) solute trapping.

Figure 4.13a describes the composition profile that develops at the α/γ interface during growth in which the ferrite is depleted and the austenite enriched in solute. The ferrite and austenite at the interface have the compositions $x_I^\alpha = x^{\alpha\gamma}$ and x_I^γ respectively. Assuming that the interface is flat, the total composition difference between the ferrite and austenite remote from the interface is

$$\Delta x = \Delta x_\infty + \Delta x_I + \Delta x_D \tag{4.9}$$

where $\Delta x_\infty = x^{\gamma\alpha} - x^{\alpha\gamma}, \qquad \Delta x_I = x_I^\gamma - x^{\gamma\alpha}, \qquad \Delta x_D = \bar{x} - x_I^\gamma.$

Δx_I and Δx_D are related to the free energies G_I and G_D dissipated in the interface and diffusion processes respectively, such that $G_D = 0$ when $\Delta x_D = 0$ and $G_I = 0$ when $\Delta x_I = 0$. Similarly, $\Delta G'$ is related to $(\Delta x_\infty - \Delta x)$ and is zero when $x^{\gamma\alpha} = \bar{x}$.

In practice the rate of interface motion always is under mixed control since the processes are in series but it is reasonable to designate it as diffusion-controlled if $|\Delta x_D| \gg |\Delta x_I|$, in which case variations in the parameters defining interface mobility have virtually no effect on velocity (Figure 4.13b). It is common, for example, in phase field modelling, to assume an infinite interface mobility for simulations involving the partitioning of solute during transformation in the solid state. Similarly, interface-control implies that $|\Delta x_D| \ll |\Delta x_I|$ and variations in diffusion parameters then have a negligible effect (Figure 4.13c). True mixed control implies that Δx_D and Δx_I are of comparable magnitude (Figure 4.13d).

The reconstructive growth of ferrite when austenite is supercooled into the $\alpha + \gamma$ phase field requires the partitioning of carbon. Consider a case where the interface motion is under mixed diffusion- and interface-control. A possible carbon concentration profile developed at the α/γ interface is illustrated in Figure 4.13d, where $x_I^\gamma \neq x^{\gamma\alpha}$ and x_I^α need not equal $x^{\alpha\gamma}$.

Displacement of a flat α/γ interface to produce ferrite is equivalent to taking a small amount of material of composition x_C^α from austenite of composition $\bar{x}_C$ and after transformation at constant composition, adding it to ferrite of composition x_C^α. The molar Gibbs free energy change $\Delta G'\{x_C^\alpha, \bar{x}_C\}$ for this process is the driving force for interface motion; from the definition of the chemical potential it follows that:

$$
\begin{aligned}
\Delta G'\{x_C^\alpha, \bar{x}_C\} &= x_{Fe}^\alpha[\mu_{Fe}^\alpha\{x_{Fe}^\alpha\} - \mu_{Fe}^\gamma\{\bar{x}_{Fe}\}] + x_C^\alpha[\mu_C^\alpha\{x_C^\alpha\} - \mu_C^\gamma\{\bar{x}_C\}] \\
&= x_{Fe}^\alpha\left(\Delta G_{Fe}^{\gamma\alpha} + RT\ln\frac{a_{Fe}^\alpha\{x_{Fe}^\alpha\}}{a_{Fe}^\gamma\{\bar{x}_{Fe}\}}\right) + x_C^\alpha RT\ln\frac{a_C^\alpha\{x_C^\alpha\}}{a_C^\gamma\{\bar{x}_C\}}
\end{aligned} \tag{4.10}
$$

where μ_i^γ and a_i^γ represent the chemical potential and activity respectively, of component i in the austenite. The activities a_{Fe}^α and a_{Fe}^γ are defined with respect to pure α-iron and pure γ-iron as the respective standard states while that of carbon is relative to pure graphite as the standard state. $\Delta G_3^{\gamma\alpha}$ is the molar Gibbs free energy for the $\gamma \to \alpha$ transformation in pure iron.

$\Delta G'$ is dissipated irreversibly in driving the diffusion of carbon ahead of the interface and in the process of transferring atoms across the interface. If G_D and G_I are the dissipations due to diffusion and interface processes respectively, then

$$
\begin{aligned}
G_I &= \Delta G' - G_D \\
&= x_{Fe}^\alpha\left(\Delta G_3^{\gamma\alpha} + RT\ln\frac{a_{Fe}^\alpha\{x_{Fe}^\alpha\}}{a_{Fe}^\gamma\{x_{Fe}^\gamma\}}\right) + x_C^\alpha RT\ln\frac{a_C^\alpha\{x_C^\alpha\}}{a_C^\gamma\{x_C^\gamma\}}.
\end{aligned} \tag{4.11}
$$

When $x_C^\gamma \simeq \bar{x}_C$ (so $|\Delta x_D| \gg |\Delta x_I|$), most of $\Delta G'$ is dissipated in interface processes and the reaction is interface-controlled; if $x_C^\gamma = x_C^{\gamma\alpha}$ then all of $\Delta G'$ is dissipated in driving the diffusion of carbon in γ ahead of the interface which moves at a rate controlled by this diffusion.

A reasonable approximation for diffusion-controlled growth, given that Δx_I is then small, is that the compositions of the phases in contact at the interface are in equilibrium, i.e., local equilibrium is said to exist at the interface. The implication is that the concentration profile can be divided into a large number of thin slices (subsystems), each of which has a definite concentration. Each of the subsystems can be considered be in local equilibrium with its neighbours, even though free energy is being dissipated in the diffusion process. These assumptions are valid if perturbations from equilibrium are not "too" large (p. 90). Equilibrium thermodynamics can then be applied locally, but not if the concentration gradients are very large [59].

An interesting consequence of the assumption of local equilibrium is that the diffusion-controlled growth and dissolution of a precipitate must occur at the same rate because the forward rate along any a reaction path must equal the reverse rate along that path [4, 60]; this is known as the principle of detailed balance.

Figure 4.13b–d represents cases where the ferrite grows with its equilibrium composition. In contrast, the ferrite in Figure 4.13e has an excess concentration of solute ($x^\alpha > x^{\alpha\gamma}$), which is said to be *trapped* because its chemical potential is increased on transfer across the interface. When the concentration of the solute is smaller than expected from equilibrium, it is the solvent that is trapped. Non-equilibrium transformations and unrealistically large concentration gradients are discussed later in the text.

4.6 DIFFUSION-CONTROLLED GROWTH

In what follows, the subscripts 1, 2, 3 will designate C, substitutional solute and iron respectively, when applied to concentration terms or diffusion coefficients.

4.6.1 GROWTH IN Fe-C ALLOYS

Consider the one-dimensional advance of a planar α/γ interface. If the rate is controlled by the diffusion of carbon in the austenite ahead of the moving interface and assuming that local equilibrium persists at the interface,[10] then $c_1^\gamma = c_1^{\gamma\alpha}$ and $c_1^\alpha = c_1^{\alpha\gamma}$. It is assumed in addition that the concentration of carbon in the austenite remote from the interface remains $\bar{c}_1$, equivalent to making the austenite semi-infinite extent along the growth direction.

Since $c_1^\alpha < c_1^\gamma$, carbon is partitioned as the ferrite grows with the excess accumulating in the γ ahead of the interface. Given the local equilibrium constraint, the maximum concentration in the austenite at the interface is limited to $c_1^{\gamma\alpha}$. The extent of the diffusion field must then increase with the fraction of ferrite so its growth rate decreases with time. From dimensional arguments[11] it can be demonstrated that the thickness Z of the ferrite is related to time t:

$$Z = \alpha_{1d}\sqrt{t} \tag{4.12}$$

where $Z = 0$ at $t = 0$ and Z defines the position of the interface along the coordinate z which is normal to the interface (and is positive in the austenite). α_{1d} is called the parabolic-thickening rate constant for one-dimensional growth. The concentration c_1 in the austenite satisfies the equation:

$$\frac{\partial c_1}{\partial t} = \frac{\partial}{\partial z}\left(D_{11}\{c_1\}\frac{\partial c_1}{\partial z}\right) \tag{4.13}$$

subject to the boundary conditions

$$\left.\begin{array}{ll} c_1 = c_1^{\gamma\alpha} & z = Z\{t\}; \quad t > 0 \\ c_1 = \bar{c}_1 & t = 0 \end{array}\right\}.$$

A further boundary condition relates the interface velocity to the concentration gradient at the interface. The rate at which solute is partitioned must equal that at which it is carried away by diffusion if $c_1^{\gamma\alpha}$ is to remain constant:

$$(c_1^{\alpha\gamma} - c_1^{\gamma\alpha})\frac{\alpha_{1d}}{2t^{\frac{1}{2}}} = D_{11}\{c_1^{\gamma\alpha}\}\frac{\partial c_1}{\partial z}\bigg|_{z=Z} \tag{4.14}$$

A general solution to Equation 4.13 in which the diffusion coefficient is concentration dependent exists, albeit for special forms of the function $D_{11}\{c_1\}$. Atkinson [61], following a method due to Philip [62, 63], has shown that when one-dimensional diffusion-controlled growth is parabolic with respect to time with a concentration-dependent diffusion coefficient, it is possible to obtain exact solutions to Equation 4.13 subject to the associated boundary conditions:

$$\frac{z}{t^{\frac{1}{2}}} = F\left\{\frac{c_1 - \bar{c}_1}{c_1^{\gamma\alpha} - \bar{c}_1}f\{\alpha_{1d}\}\right\} \tag{4.15}$$

where F is any single-valued function of c_1 such that

$$D_{11}\{c_1\} = \frac{1}{2}\frac{dF}{dc_1}\int_0^{c_1} F\,dc_1$$

and

$$\frac{c_1 - \bar{c}_1}{c_1^{\gamma\alpha} - \bar{c}_1} = f\left\{\frac{z}{t^{\frac{1}{2}}}\right\}\bigg/f\left\{\frac{Z}{t^{\frac{1}{2}}}\right\} \tag{4.16}$$

where $F = f^{-1}$. The form of Equation 4.15 is reasonable since when $c_1 = c_1^{\gamma\alpha}$, $z = Z$ and $F\{f\{\alpha_{1d}\}\} = \alpha_{1d}$ so the parabolic relation $Z = \alpha_{1d}t^{\frac{1}{2}}$ is recovered.
Differentiation of Equation 4.15 gives

$$\frac{\partial c_1}{\partial z}\bigg|_{z=Z} = \frac{c_1^{\gamma\alpha} - \bar{c}_1}{f\{\alpha_{1d}\}F'\{f\{\alpha_{1d}\}\}}\frac{1}{t^{\frac{1}{2}}} \tag{4.17}$$

where the prime denotes differentiation with respect to z. This equation is the generalised analogy of Equation 4.14. On combining it with the parabolic growth law, and the definition that $F\{f\{\alpha_{1d}\}\} = \alpha_{1d}$, the generalised equation for the parabolic rate constant is obtained:

$$f\{\alpha_{1d}\}F\{f\{\alpha_{1d}\}\}F'\{f\{\alpha_{1d}\}\} = \frac{2D_{11}(c_1^{\gamma\alpha} - \bar{c}_1)}{c_1^{\alpha\gamma} - c_1^{\gamma\alpha}}. \tag{4.18}$$

Consider again the case where the diffusion coefficient is constant [1, 61, 64]. The variation of concentration in the austenite ahead of the interface is expected to show an error function profile. This can be seen from the nature of an error function. When a thin layer of solute is sandwiched between semi-infinite bars of pure material, the solute diffuses and the resulting concentration profile decays in exponentially with distance and time. Now suppose that two semi-infinite bars of different materials form a diffusion couple, then the solute-rich bar can be considered as a set of thin exponential sources distributed over the extent of the bar. The diffusion profile is then the summation of all these exponential sources, i.e., the error function. Since in our case the concentration at the interface remains constant, the error function should describe the variation of concentration so the function f in Equation 4.16 may be written

$$f\left\{\frac{z}{t^{\frac{1}{2}}}\right\} = \text{erfc}\left\{\frac{z}{(4D_{11}t)^{\frac{1}{2}}}\right\}. \tag{4.19}$$

It follows that the concentration profile is

$$\frac{c_1 - \bar{c}_1}{c_1^{\gamma\alpha} - \bar{c}_1} = \Theta_1\{z, D_{11}\} \tag{4.20}$$

where

$$\Theta_1\{z, t, D_{ii}\} = \text{erfc}\frac{z}{(4D_{ii}t)^{\frac{1}{2}}} \Bigg/ \text{erfc}\frac{Z}{(4D_{ii}t)^{\frac{1}{2}}}. \tag{4.21}$$

Given that $F = f^{-1}$,

$$F\{y\} = 2D_{11}^{\frac{1}{2}}\text{inverfc}\{y\} \qquad \text{and} \qquad F'\{y\} = -D_{11}^{\frac{1}{2}}\pi^{\frac{1}{2}}\exp\{[\text{inverfc}\{y\}]^2\}$$

where inverfc is the inverse function of the complimentary error function erfc. These functions can be substituted into Equation 4.18, to give an implicit relation for α_{1d} as follows:

$$f_1 = \Phi_1\{D_{11}\} \tag{4.22}$$

$$\text{where} \qquad \Phi_1\{D_{ii}\} = \left(\frac{\pi}{4D_{ii}}\right)^{\frac{1}{2}}\alpha_{1d}\text{erfc}\left\{\frac{\alpha_{1d}}{2D_{ii}^{\frac{1}{2}}}\right\}\exp\left\{\frac{\alpha_{1d}^2}{4D_{ii}}\right\}$$

and f_1 is a fractional supersaturation,

$$f_1 = \frac{\bar{c}_1 - c_1^{\gamma\alpha}}{c_1^{\alpha\gamma} - c_1^{\gamma\alpha}}.$$

The case for concentration-dependent diffusivity is more complex. As pointed out previously, exact solutions such as these can only be obtained if corresponding pairs of the functions D_{11} and F can be found that satisfy Equation 4.15. Whilst mathematicians do not find this difficult to do [65], the functions D_{11} obtained are convoluted and ill-adapted to fitting experimental data on D_{11} [66]. Nevertheless, when, as for carbon in austenite, D_{11} increases with the concentration, it has shown that the following equation can be used to approximate the parabolic rate constant α_{1d} when the growth is diffusion-controlled [61]:

$$f_1 = \Phi_1\{D_{11}\{\bar{c}_1\}\}\frac{D_{11}\{\bar{c}_1\}}{D_{11}\{c_1^{\gamma\alpha}\}}.$$

It is interesting that the growth rate is found to be more sensitive to the maximum and minimum values of the diffusivity than the detailed form of the variation along the concentration profile in the austenite.

The problem of defining matching pairs of F and D functions has been circumvented using a numerical method for solving Equation 4.13 [67]. The method is expected to work for any given diffusion coefficient represented as a function, graph or table. Grain boundary allotriomorphs sometimes have an initial shape that approximates an oblate ellipsoid; Atkinson [68] has developed a similar numerical method for this morphology, assuming shape preservation during growth. An analytical solution to the same problem, for a constant diffusion coefficient, has been given by Horvay and Cahn [69].

4.6.2 LENGTHENING OF FERRITE ALLOTRIOMORPHS

Grain boundary ferrite allotriomorphs acquire their characteristic shape because the transformation rate along the γ/γ boundary is faster than that in the direction normal to the boundary plane. The initial shape of the allotriomorph is approximately that of an oblate ellipsoid [70]. The lengthening rate of allotriomorphs was at first thought to be constant, the process being described in terms of the plate lengthening theory [71, 72] but later work [73, 74] has shown that it too is parabolic with respect to time, Figure 4.14. The rate constant describing the lengthening process is simply α_{1d} divided by the thickness to length ratio (the aspect ratio) of the allotriomorph.

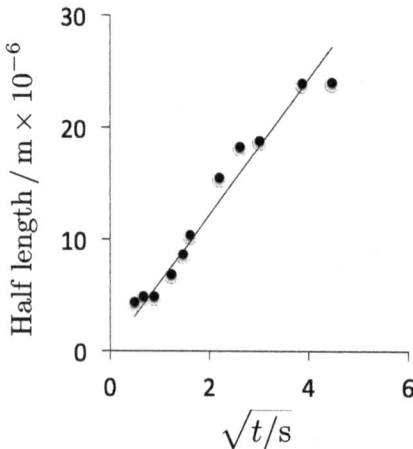

Figure 4.14 Measured parabolic lengthening-kinetics of allotriomorphic ferrite in Fe-0.11C wt% at 800°C. Selected data from Atkinson et al. [74].

The aspect ratio is found to be about $1/3$, independent of time at transformation temperature [70]. This is consistent with Horvay and Cahn's [69] theory for the diffusion-controlled growth of oblate ellipsoids. The observed aspect ratio is smaller than that expected from the balancing of interfacial tensions at the allotriomorph edges. It is suggested that this inconsistency probably arises because the shape of the allotriomorph during growth is not controlled by interfacial energy considerations alone [70]. The oblate ellipsoid is an idealisation whereas in reality the structure of the γ/γ grain boundary and the inevitable difference in the orientation relationship of the allotriomorph with each of its adjacent γ-grains must complicate the actual shape.

4.6.3 SOFT IMPINGEMENT

Many of the solutions to the diffusion equations are derived assuming that the concentration of solute far from the interface remains constant. The austenite is therefore assumed to extend to infinity beyond the α/γ interface so the continued partitioning of solute as α grows does not alter the far-field concentration. The boundary conditions for isothermal diffusion-controlled growth therefore remain constant, as does the composition of the ferrite. A different implication relevant to ternary or higher order alloys is that there is no diffusion within the ferrite since $c_i^{\alpha\gamma}$ do not change as the transformation progresses.

In reality, the diffusion fields of particles growing from different locations must eventually interfere. Even if transformation involves the growth of just one precipitate, its diffusion field may be limited by a variety of surfaces. This interference is called "soft impingement" and leads to boundary conditions that change with time. Soft-impingement is actually a difficult problem to treat exactly because it is necessary to have a prior knowledge of the distribution of all the particles. This often is the aim of the calculations rather than an input. All methods therefore use one of two approximations: (i) that the microstructure evolves from a assumed distribution of nuclei, with no subsequent nucleation; (ii) that the effect of soft impingement can be described by distributing and solute enrichment or depletion over the whole of the remaining matrix phase. The latter procedure often is called the mean field approximation in physics.

In one analytical method [75] the microstructure is specified by considering a sphere of austenite of an appropriate size. This is decorated on its surface by an infinitely thin, continuous layer of ferrite (there is no nucleation in this model). Using a solution due to Crank [66] for the radial diffusion of solute into a sphere on the surface of which, the concentration is maintained constant, the radial thickness of the ferrite at any time t is given by $r_1 - r$, where r_1 and r are the initial and instantaneous radii, respectively, of the remaining sphere of austenite. The time required to achieve a given thickness of ferrite is taken to be that necessary to partition the excess carbon from the specified thickness of ferrite, into the remaining austenite, bearing in mind that the concentration in the austenite at the transformation front is restricted to the value $c_1^{\gamma\alpha}$. From Crank's solution,

$$\frac{M_t}{M_\infty} = \Theta_0$$

where

$$\Theta_0 = 6\left(\frac{\overline{D}_{11}t}{r^2}\right)^{\frac{1}{2}}\left[\pi^{-\frac{1}{2}} + 2\sum_n \text{ierfc}\frac{nr}{(\overline{D}_{11}t)^{\frac{1}{2}}}\right] - 3\frac{\overline{D}_{11}t}{r^2}$$

and M_t represents mass of carbon which enters the stationary sphere of residual austenite and M_∞ is the maximum possible amount of carbon that the austenite can accommodate. It follows that:

$$M_t = \frac{4}{3}\pi(r_1^3 - r^3)(\bar{c}_1 - c_1^{\alpha\gamma})$$

and

$$M_\infty = \frac{4}{3}\pi r^3(c_1^{\gamma\alpha} - \bar{c}_1).$$

On combining these equations, it is evident that

$$\frac{r^3}{r_I^3} = \frac{1 - f_1}{1 - f_1 + f_1 \Theta_0}.$$

The derivation can account for any difference in the molar volumes V_m^α and V_m^γ of the ferrite and austenite respectively:

$$\frac{r^3}{r_I^3} = \frac{1 - f_1}{1 - f_1 + (V_m^\alpha / V_m^\gamma) f_1 \Theta_0}. \tag{4.23}$$

From a series of calculations of the times required to achieve the level of carbon in the austenite for each specified value of ferrite thickness, it is possible to generate the time-thickness dataset for the thickening kinetics, while accounting for soft impingement effects. The effect of soft impingement is always to retard the rate of reaction (Figure 4.15).

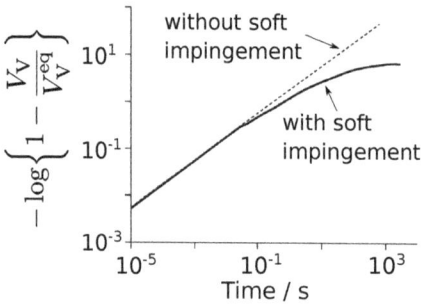

Figure 4.15 Comparison of the evolution of calculated volume fraction of allotriomorphic ferrite for cases where soft-impingement has or has not been taken into account. The volume fraction V_V is normalised to the equilibrium value. Isothermal transformation at 900 K for a low carbon steel (after Vandermeer, Vold and King [75]).

4.6.4 PHASE FIELDS

The phase field method is in principle able to deal with soft-impingement, rate-control and other phenomena associated with the kinetics of transformation, some of which have been described in the preceding section. The method is introduced here but there exist many excellent reviews and articles on the subject [76–79]. Consider the growth of a precipitate that is isolated from the matrix by an interface, the motion of which is controlled according to the boundary conditions consistent with the mechanism of transformation. The interface in this mathematical description is simply a two-dimensional surface with no width or structure; it is said to be a *sharp interface*. There are, therefore, three discrete quantities to deal with, the precipitate, interface and parent phase.

In the phase-field method, the state of the entire microstructure is represented continuously by a single variable known as the *order parameter* ψ. For example, $\psi = 1$, $\psi = 0$ and $0 < \psi < 1$ represent the precipitate, matrix and interface respectively. The latter is therefore located by the region over which ψ changes from its precipitate-value to its matrix-value (Figure 4.16). The range over which it changes is the width of the interface. The set of values of the order parameter over the whole microstructure is the phase field.

The evolution of the microstructure with time is assumed to be proportional to the variation of the

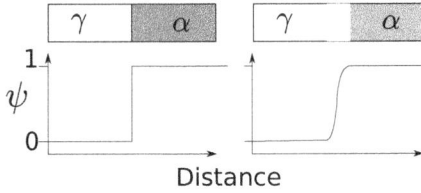

Figure 4.16 An illustration of the difference between a sharp and a diffuse interface in the context of phase field modelling. The diagram on the left shows a sharp interface between the parent γ ($\psi = 0$) and precipitate α ($\psi = 1$), whereas that on the right has a diffuse and wide interface with $\psi = 0.5$ at its centre.

free energy functional with respect to the order parameter:

$$\frac{\partial \psi}{\partial t} = M \frac{\partial g}{\partial \psi} \tag{4.24}$$

where M is a mobility. The term g describes how the free energy varies as a function of the order parameter; at constant T and P, this takes the typical form (p. 205):

$$g = \int_V \left[g_0\{\psi, T, x\} + \varepsilon(\nabla \psi)^2 \right] dV \tag{4.25}$$

where the integral is over the volume V. The second term in the brackets depends on the gradient of ψ and hence is non-zero only in the interfacial region, with $\nabla \psi \neq 0$; it is a description therefore of the interfacial energy. The term g_0 is a function of the free energies of the precipitate and matrix, the temperature and the composition x; it can contain a term describing the activation barrier across the interface. The homogeneous part of the free energy can be formulated as

$$g_0 = h g^\alpha\{x^\alpha, T\} + (1 - h) g^\gamma\{x^\gamma, T\} + Qf \tag{4.26}$$

where g^α and g^γ refer to the free energies of the phase identified in the superscript, h is a function to interpolate the free energy densities of the two phases and Q is the height of the activation barrier at the interface, with f defined as a double-well potential

$$f = \psi^2(1 - \psi)^2 \qquad \text{and} \qquad \bar{x} = \psi x^\alpha + (1 - \psi)x^\gamma.$$

The function h for the iron-carbon system is often set as [80]

$$h = \psi^3(6\psi^2 - 15\psi + 10)$$

although it can be fitted into alternative forms using thermodynamic databases [81]. Notice that the term $h g^\alpha + Qf$ vanishes when $\psi = 0$ (i.e., only γ is present), and similarly, $(1 - h)g^\gamma + Qf$ vanishes when $\psi = 1$ (i.e., only α is present). As expected, it is only when both phases are present that Qf becomes non-zero. The time dependence of the phase field according to Equation 4.24 then defines microstructural evolution provided parameter such as Q, ε and M are measured or derived assuming some mechanism of transformation.

The obvious advantage of the phase field method is that it is routinely able to deal with complexity with the numerical solution of a few equations, much fewer than the number of particles in the system. And there are many adaptations of the method to deal with the combined effects of numerous variables. It would be nice if the method were to be used to make *predictions* followed by verification experiments. Otherwise, the work to date on steels falls into the category of "the perfect computation [which] simply reproduces Nature, does not explain her" [82]. Further discussion on these particular issues is available in [83, 84].

4.6.5 FERRITE GROWTH IN Fe-X-C ALLOYS: LOCAL EQUILIBRIUM

The diffusion-controlled growth of ferrite in Fe-X-C alloys is complicated by the fact that both interstitial and substitutional solutes are mobile during transformation. There is a large disparity in the diffusivities of substitutional and interstitial species (Figure 4.17), leading to difficulties in maintaining equilibrium at the interface if the tie-line defining the interface compositions is not properly chosen [85–88].

The compositions of the phases at the interface must be connected by a tie-line of the $\alpha + \gamma$ phase field in the equilibrium Fe-X-C phase diagram if the two phases are to be in local equilibrium at the transformation front. For the discussion that follows, $X \equiv Mn$ (identified in concentration terms by the subscript $i = 2$); manganese is an austenite stabilising element but the concepts discussed are general to all substitutional alloying elements that dissolve in austenite.

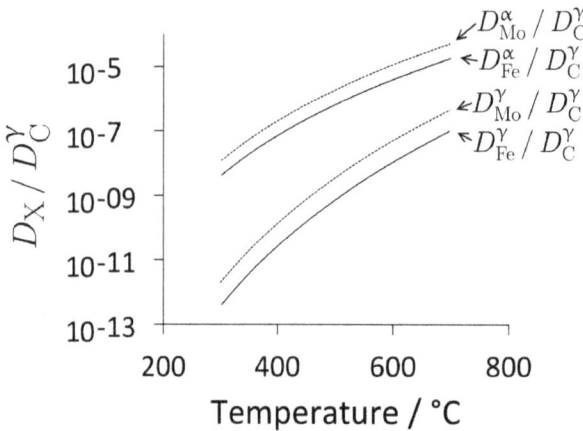

Figure 4.17 A comparison of the diffusivities of iron and substitutional solutes relative to that of carbon. Bearing in mind that the diffusivity of carbon in austenite is strongly concentration dependent, the data for austenite refer to a specific carbon concentration, 0.4C wt% (selected data from Fridberg, Torndhäl and Hillert [89]).

It might seem from our knowledge of binary alloys, that the tie-line defining interface compositions should pass through the point $\overline{c}_1, \overline{c}_2$, i.e., the average composition of the alloy. However, there is a large discrepancy in the rates at which substitutional solutes and carbon diffuse, which complicates matters if the local equilibrium condition is to be retained. Conservation of mass at a planar interface moving with a speed v in the direction z (normal to the interface plane) requires that

$$(c_1^{\gamma\alpha} - c_1^{\alpha\gamma})v = -D_{11}\frac{\partial c_1}{\partial z} - D_{12}\frac{\partial c_2}{\partial z}$$

$$(c_2^{\gamma\alpha} - c_2^{\alpha\gamma})v = -D_{22}\frac{\partial c_2}{\partial z} - D_{21}\frac{\partial c_1}{\partial z}$$

(4.27)

where all the concentration gradients are in the austenite and are evaluated at the position of the interface ($z = Z$). Since D_{12} and D_{21} are relatively small, these equations may be reduced to

$$(c_1^{\gamma\alpha} - c_1^{\alpha\gamma})v = -D_{11}\frac{\partial c_1}{\partial z}$$

$$(c_2^{\gamma\alpha} - c_2^{\alpha\gamma})v = -D_{22}\frac{\partial c_2}{\partial z}.$$

(4.28)

Because $D_{11} \gg D_{22}$, these equations cannot in general be simultaneously satisfied for the tie-line passing through the average composition $\bar{c}_1, \bar{c}_2$. Referring to Figure 4.18, it is possible to select other tie-lines that satisfy Equation 4.28, given that the entire $\alpha + \gamma$ phase field is defined by tie-lines at a constant temperature in a ternary alloy. If the tie-line is such that $c_1^{\gamma\alpha} \approx \bar{c}_1$ (e.g. line cd for alloy A of Figure 4.18a), then $\partial c_1/\partial z$ will become small that the driving force for carbon diffusion is reduced; the consequent reduced flux of carbon becomes consistent with the diffusion of manganese. Ferrite forming by this mechanism is said to grow by a "partitioning, local equilibrium" (PLE) mechanism, in recognition of the fact that $c_2^{\alpha\gamma}$ can differ significantly from $\bar{c}_2$, giving considerable partitioning and long-range diffusion of manganese into the austenite [85].

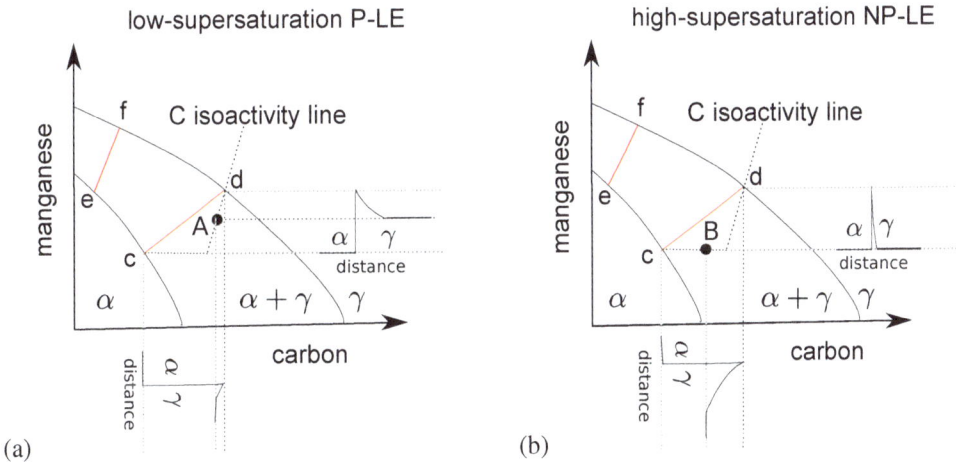

Figure 4.18 Schematic isothermal sections of the Fe-Mn-C system, illustrating ferrite growth occurring with local equilibrium at the α/γ interface. Points A and B represent $\bar{c}_1, \bar{c}_2$, designated by the symbol • in each case. (a) Growth at low supersaturations (P-LE) with bulk redistribution of manganese, (b) growth at high supersaturations (NP-LE) with negligible partitioning of manganese during transformation.

An alternative choice of tie-line could allow $c_2^{\alpha\gamma} \to \bar{c}_2$ (e.g. line cd for alloy B of Figure 4.18b), so $\partial c_2/\partial z$ is increased because of the reduced partitioning of Mn into the austenite. The flux of manganese atoms at the interface therefore increases, allowing it to keep pace with that of carbon, while satisfying Equation 4.28. The growth of ferrite in this manner is said to occur by a "negligible partitioning, local equilibrium" (or NP-LE) mechanism, in recognition of the fact that the manganese content of the ferrite approximately equals $\bar{c}_2$, so only the quantity of manganese needed to maintain local equilibrium at the interface partitions into austenite [85].

The exact choice of tie-line will be discussed quantitatively at a later stage; consider first some general points about tie-line choice. In a Fe-Mn-C alloy, both carbon and manganese are austenite stabilisers, tending to partition into austenite. It follows that $c_1^{\gamma\alpha}$ and $c_2^{\gamma\alpha}$ must always be greater than or equal to $\bar{c}_1$ and $\bar{c}_2$ respectively. $c_1^{\alpha\gamma}$ and $c_2^{\alpha\gamma}$ will always be less than or equal to $\bar{c}_1$ and $\bar{c}_2$ respectively. Tie-lines such as *ef* (Figure 4.18) are therefore inappropriate for the average alloy compositions illustrated. This leads to the division of the $\gamma + \alpha$ phase field into regimes where either the PLE or NP-LE mechanism can operate (Figure 4.19), in a mutually exclusive manner. It easily is demonstrated that these regions are exclusive; in a given domain, only one will lead to physically satisfactory concentration profiles at the transformation interface.

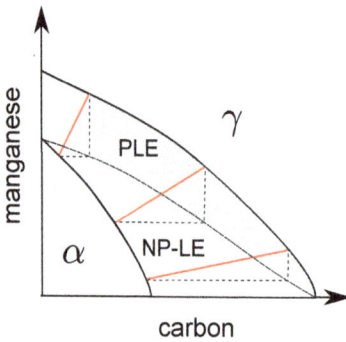

Figure 4.19 Division of the $\alpha + \gamma$ phase field into domains where either the P-LE (shaded) or the NP-LE mechanisms can operate.

When austenite is supercooled into a region close to the $\gamma/\gamma + \alpha$ phase boundary (i.e., transforming at a low supersaturation), $c_2^{\alpha\gamma}$ will be significantly smaller than $\bar{c}_2$. Therefore, it can only transform by the PLE mechanism. Similarly, for austenite quenched into the region close to the $\alpha/\alpha + \gamma$ phase boundary (i.e., transforming at a high degree of supersaturation), $c_1^{\gamma\alpha}$ will be much greater than $\bar{c}_1$, making only the NP-LE mechanism feasible.

In summary, assuming that the growth of ferrite from ternary austenite is constrained to occur at a rate which is diffusion-controlled, with local equilibrium at the moving interface, the tie-line of the $\alpha + \gamma$ phase field that defines the interface compositions does not in general pass through the point in the $\alpha + \gamma$ phase field which identifies the alloy composition. This is because the diffusivities of interstitial and substitutional alloying elements in the austenite are so different. The appropriate tie-line must be chosen to satisfy mass conservation conditions at the moving interface and must be consistent with the partitioning behaviour of the alloying elements. For this reason, the tie-line for an alloy transforming at a low supersaturation is such that there is considerable partitioning and long-range diffusion of substitutional alloying element, while the driving force for carbon diffusion is reduced to a level which allows the substitutional element flux to keep up with the carbon flux at the interface. This is the PLE mode of transformation.

At greater supersaturations, the determining tie-line is that which causes negligible partitioning of substitutional solute between the α and γ lattices, so the gradient of the X element in the austenite near the interface is large. This increases the driving force for X diffusion in austenite and allows the flux of X to keep pace with the long-range diffusion of C in austenite. This is the NP-LE mode of transformation, where the diffusion of X in austenite is short-range, being confined to the immediate vicinity of the interface.

Both of these modes of transformation involve local equilibrium at the interface and are therefore equally favoured on thermodynamic considerations alone. Both C and X diffuse during growth and their fluxes satisfy the equations for conservation of mass at the interface; it follows that the velocity calculated from the diffusion of carbon in austenite will be identical to that calculated from the diffusion of manganese in austenite. Both elements control the growth rate and neither can be said to restrict the interface motion on its own.

Quantitative determination of the tie-line and interface velocity

It may be assumed, for Fe-X-C alloys, that the distribution of carbon in austenite has a negligible effect on the diffusion of X in austenite [90, 91]. This is because $D_{11} \gg D_{22}$, so $|\partial c_2/\partial z| \gg |\partial c_1/\partial z|$. This justifies neglect of the cross diffusion coefficient D_{21}, giving:

$$
\begin{aligned}
J_1 &= -D_{11}\frac{\partial c_1}{\partial z} - D_{12}\frac{\partial c_2}{\partial z} \\
J_2 &= -D_{22}\frac{\partial c_2}{\partial z}.
\end{aligned}
\tag{4.29}
$$

Using this together with Fick's second law of diffusion, assuming one dimensional growth along a coordinate z, the differential equations for the matrix are

$$
\begin{aligned}
\frac{\partial c_1}{\partial t} &= D_{11}\frac{\partial^2 c_1}{\partial z^2} + D_{12}\frac{\partial^2 c_2}{\partial z^2} \\
\frac{\partial c_2}{\partial t} &= D_{22}\frac{\partial^2 c_2}{\partial t^2}
\end{aligned}
\tag{4.30}
$$

as long as the diffusion coefficients are concentration independent.

Kirkaldy [87] has given solutions for multicomponent diffusion during the growth of linear, cylindrical and spherical precipitates (1, 2 and 3 dimensional growth, respectively) of uniform composition in an infinite medium, for composition independent diffusion coefficients. For the boundary conditions corresponding to the one-dimensional, diffusion-controlled growth of ferrite of uniform composition along the coordinate z, Kirkaldy's solutions show that provided D_{11} does not equal

D_{22}, the concentrations in the matrix as a function of time and distance are given by [91]

$$c_1\{z,t\} = \bar{c}_1 + \frac{D_{12}(c_2^{\gamma\alpha} - \bar{c}_2)\Theta_1\{z,D_{22}\}}{(D_{22} - D_{11})}$$

$$+ \left[(c_1^{\gamma\alpha} - \bar{c}_1) - \frac{D_{12}(c_2^{\gamma\alpha} - \bar{c}_2)}{(D_{22} - D_{11})}\right]\Theta_1\{z,D_{11}\} \qquad (4.31)$$

$$\text{and} \qquad c_2\{z,t\} = \bar{c}_2 + (c_2^{\gamma\alpha} - \bar{c}_2)\Theta_1\{z,D_{22}\} \qquad (4.32)$$

$$\text{with} \qquad \Theta_1\{z,t,D_{ii}\} = \text{erfc}\frac{z}{(4D_{ii}t)^{\frac{1}{2}}} \Big/ \text{erfc}\frac{Z}{(4D_{ii}t)^{\frac{1}{2}}}$$

where $z = Z$ defines the position of the interface, so

$$Z = \alpha_{1d}t^{\frac{1}{2}} = \eta_1(D_{11}t)^{\frac{1}{2}} = \eta_2(D_{22}t)^{\frac{1}{2}}. \qquad (4.33)$$

η_i are growth constants, related to α_{1d}, the parabolic rate constant for one-dimensional growth. The equation for carbon (c_1) can be applied to Fe-C alloys by setting D_{12} to zero, in which case it reduces to Equation 4.20. By combining Equations 4.31-4.33 with the flux conditions (Equation 4.28 with $D_{21} = 0$), Coates [90] showed that

$$f_1 = \Phi_1\{D_{11}\} - \frac{c_2^{\gamma\alpha} - c_2^{\alpha\gamma}}{c_1^{\gamma\alpha} - c_1^{\alpha\gamma}}\frac{D_{12}}{D_{11} - D_{22}}[\Phi_1\{D_{22}\} - \Phi_1\{D_{11}\}] \qquad (4.34)$$

$$f_2 = \Phi_1\{D_{22}\}.$$

The function Φ_1 is defined on p. 189. When $D_{12} = 0$, then Equation 4.34 becomes identical with Equation 4.22 for an unalloyed steel.

For transformations occurring under conditions of local equilibrium at the interface, only one of $c_1^{\alpha\gamma}, c_1^{\gamma\alpha}, c_2^{\alpha\gamma}, c_2^{\gamma\alpha}$ is independent, since they are all linked by a tie-line of the $\alpha + \gamma$ phase field of the equilibrium phase diagram. The kinetic equations therefore contain only two unknowns and can be simultaneously solved to determine the growth velocity and the tie-line governing interface compositions during growth.

Two- and three-dimensional growth with local equilibrium

Coates [92], using the general solutions of Kirkaldy [87], determined the diffusion equations for two-dimensional and three-dimensional growth (radial growth of cylinders, and growth of spheres, respectively) involving local equilibrium at the γ/α interface in Fe-X-C alloys. The assumptions involved are the same as those used in the analysis of one-dimensional growth, with the additional approximation that capillarity effects may be neglected. Coates found that for two-dimensional growth,

$$f_1 = \Phi_2\{D_{11}\} - \frac{c_2^{\gamma\alpha} - c_2^{\alpha\gamma}}{c_1^{\gamma\alpha} - c_1^{\alpha\gamma}} \frac{D_{12}}{D_{11} - D_{22}}[\Phi_2\{D_{22}\} - \Phi_2\{D_{11}\}]$$

$$f_2 = \Phi_2\{D_{22}\} \tag{4.35}$$

where $\quad \Phi_2\{D_{ii}\} = \dfrac{\alpha_{2d}^2}{4D_{ii}} \exp\dfrac{\alpha_{2d}^2}{4D_{ii}} Ei\left\{\dfrac{\alpha_{2d}^2}{4D_{ii}}\right\}.$

Ei is the (tabulated) exponential integral function [93].

For three-dimensional growth,

$$f_1 = \frac{\alpha_{3d}^2}{2D_{11}}[1 - \Phi_1\{D_{11}\}]$$
$$- \frac{c_2^{\gamma\alpha} - c_2^{\alpha\gamma}}{c_1^{\gamma\alpha} - c_1^{\alpha\gamma}} \frac{D_{12}}{2(D_{11} - D_{22})} \alpha_{3d}^2 \left(\frac{1 - \Phi_1\{D_{22}\}}{D_{22}} - \frac{1 - \Phi_1\{D_{11}\}}{D_{11}}\right)$$

$$f_2 = \frac{\alpha_{3d}^2}{2D_{22}[1 - \Phi_1\{D_{22}\}]}. \tag{4.36}$$

α_{2d} and α_{3d} are the parabolic rate constants for two- and three-dimensional growth respectively. The concentration distributions in austenite during two- and three-dimensional growth can be obtained by substituting the functions Θ_2 or Θ_3 (respectively) for Θ_1 into Equations 4.31 and 4.32, where

$$\Theta_2\{z,t,D_{ii}\} = \frac{Ei\{z^2/(4D_{ii}t)\}}{Ei\{Z^2/(4D_{ii}t)\}}$$

and $\quad \Theta_3\{z,t,D_{ii}\} = \dfrac{\frac{1}{z}\exp\frac{z^2}{4D_{ii}t} - \sqrt{4\pi}\,\mathrm{erfc}\frac{z}{\sqrt{4D_{ii}t}}}{\frac{1}{Z}\exp\frac{Z^2}{\sqrt{4D_{ii}t}} - \sqrt{4\pi}\,\mathrm{erfc}\frac{Z}{\sqrt{4D_{ii}t}}}.$

These ternary equations for the concentration distributions in the matrix can be used to calculate two- and three-dimensional growth rates in binary alloys (e.g. Fe-C), if D_{12} is set to zero. The equations for binary growth were first obtained by Zener [64] and Frank [94].

Concentration-dependent diffusion coefficients

The above theory for diffusion-controlled growth is based on the assumption that the diffusion coefficients are concentration independent, and this is recognised to be an unrealistic assumption. The concentration dependence of D_{11} can be taken into account by substituting the weighted average diffusivity [95]:

$$\overline{D}_{11} = \int_{x_1^\gamma}^{\bar{x}} D_{11}\{x,T\} \frac{\mathrm{d}x}{\bar{x} - x_1^\gamma} \tag{4.37}$$

for D_{11}, even though this equation is strictly valid only for steady-state growth situations.

The ratio D_{12}/D_{11} is also concentration dependent, but the numerical calculations of Bolze et al. [96] suggest that the use of a constant D_{12}/D_{11}, evaluated at the composition $(c_1^{\gamma\alpha}, c_2^{\gamma\alpha})$, gives an adequate approximation to the problem [85].

4.6.6 INTERFACE-COMPOSITION CONTOURS

For growth constrained by local equilibrium, the compositions of the ferrite and austenite at the transformation interface, $(c_1^{\alpha\gamma}, c_2^{\alpha\gamma}, c_1^{\gamma\alpha}, c_2^{\gamma\alpha})$, for a ternary alloy of mean composition $(\bar{c}_1, \bar{c}_2)$, are obtained from a tie-line of the equilibrium phase diagram. When $D_{11} = D_{22}$, all alloys that lie on a given tie-line transform at different rates, but with identical compositions at the moving interface. If an "interface-composition contour" [90] is defined as a curve straddling the two phase $\alpha + \gamma$ phase field, identifying all alloys which transform with the same compositions at the interface, then the tie-line corresponds to an interface-composition (IC) contour for all binary alloys and also for ternary alloys where $D_{11} = D_{12}$. When D_{11} does not equal D_{22}, the tie-line determining the interface compositions does not pass through the point defining the average alloy composition and IC contours no longer coincide with tie-lines. The derivation of an IC contour depends on the ratio D_{11}/D_{22} and on the nature of interactions between components 1 and 2.

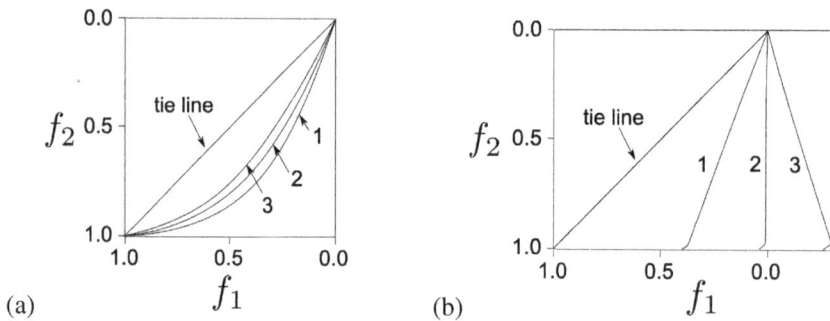

Figure 4.20 Typical IC contours for a substitutionally alloyed steel, given that the tie-line slope $c_2^{\gamma\alpha} - c_2^{\alpha\gamma}/c_1^{\gamma\alpha} - c_1^{\alpha\gamma}$ is unity [85]. (a) IC contours for a system where $D_{11}/D_{22} = 10, D_{12}/D_{11} = 0.1, 0, -0.1$ for curves 1, 2 and 3 respectively; b) IC contours for a system where $D_{11}/D_{22} = 10^6, D_{12}/D_1 = -0.3, 0, 0.3$ for curves 1, 2 and 3 respectively.

The set of alloys that for a given tie line satisfies the flux balance conditions (Equation 4.28) defines the IC contour. The composition $\bar{c}_1$ that ensures conservation for component 1 will depend on $\eta_1 D_{11}^{\frac{1}{2}}$ and similarly for component 2 it is the term $\eta_2 D_{22}^{\frac{1}{2}}$ that matters. Noting that $\eta_1 D_{11}^{\frac{1}{2}} = \eta_2 D_{22}^{\frac{1}{2}}$, both $\bar{c}_1$ and $\bar{c}_2$ which satisfy their respective mass conservation conditions can be written as a function of just η_1:

$$\bar{c}_1 = \bar{c}_1\{\eta_1\}$$

$$\bar{c}_2 = \bar{c}_2\{\eta_1\}.$$

Elimination of η_1 gives the composition $\bar{c}_2, \bar{c}_1$ which jointly satisfy mass conservation at the interface for the specified tie-line:

$$\bar{c}_2 = \bar{c}_2\{\bar{c}_1\}. \tag{4.38}$$

This defines an IC contour joining the points $(c_1^{\alpha\gamma}, c_2^{\alpha\gamma})$ and $(c_1^{\gamma\alpha}, c_2^{\gamma\alpha})$; the straight line joining these points is the tie-line defining interface compositions during the local equilibrium growth of ferrite from austenite the composition of which lies on the IC contour. With these interface compositions, the conservation conditions at the interface (Equation 4.28) are satisfied automatically for all alloys on the IC contour. All alloys on this IC contour transform with the same compositions at the interface.

The functional relation of Equation 4.38 will in general be complex but can be resolved numerically to yield the IC contours. Figure 4.20 illustrates some IC contours for a variety of D_{11}/D_{22} ratios and $f_3 D_{12}/D_{11}$, for a given tie-line. As (D_{11}/D_{22}) becomes large, the IC contour consists essentially of two straight segments which, together with the tie-line, form a triangle within the two-phase field. In the horizontal segment, $f_2 = 1$ whereas in the near vertical segment, $f_1 = -f_3 D_{12}/D_{11}$ with $f_2 = 0$. For all alloys lying on the vertical segment, the fast diffuser has its driving force for diffusion reduced to nearly zero, down to the pace of the slow diffuser; there is also considerable partitioning of component 2, giving PLE growth.

On the other hand, for all alloy compositions falling on the horizontal segment, the amount of partitioning of component 2 is reduced to minute levels so growth occurs by the NP-LE mode. The transition between these two modes of growth is logically taken to be the point $(f_2 = 1, f_1 = -f_3 D_{12}/D_{11})$, the vertex of the triangle formed by the tie-line and the two segments of the IC curve. This clearly is an approximation, because the vertex is not sharp. By joining all such points on the phase diagram, the $\alpha + \gamma$ phase field can be divided into regions, one involving growth with negligible partitioning of the substitutional-solute (NP-LE), and the other in which the fast diffuser is forced to keep pace with the slow diffuser (PLE).

Given that D_{12}/D_{11} is evaluated at $c_2 = c_2^{\gamma\alpha}$, and assuming that Equation 3.87 which gives this ratio as a function of thermodynamic parameters applies, Coates [90] has shown that the line $f_1 = -f_3 D_{12}/D_{11}$ corresponds to the carbon isoactivity line passing through $(c_1^{\gamma\alpha}, c_2^{\gamma\alpha})$. This was first deduced qualitatively by Hillert [86]. The model of Kirkaldy and Coates is quite general and is not restricted to cases where $D_{11} \gg D_{22}$, as is that of Hillert.

4.6.7 INTERFACE-VELOCITY (IV) CONTOURS

Each point on an IV contour defines a bulk composition for which precipitate grows at the same rate. The $\alpha/\alpha + \gamma$ and $\gamma/\gamma + \alpha$ phase boundaries represent two such contours since the interface velocity will be infinite and zero respectively for all alloys falling on these boundaries, assuming that interface motion remains diffusion-controlled even for large velocities.

For large D_{11}/D_{22}, the boundary between the PLE and NP-LE regions is also an IV contour. All other contours in the $\alpha + \gamma$ phase field can be derived using Equation 4.34. The velocity is in this case a function of time. The contours really represent lines with a constant parabolic rate constant α_{1d}, from which the velocity may be derived for any given value of time.

The contours are in general curves, which for $D_{11} = D_{22}$, follow roughly parallel paths to the two phase boundaries. As the ratio D_{11}/D_{22} increases, there is a progressively increased tendency for the IV contours to radiate from the point $f_1 = 0$ on the c_1 axis in the PLE region, and from the point $f_2 = 1$ on the c_2 axis in the NP-LE region. Typical IV contours are illustrated in Figure 4.21.

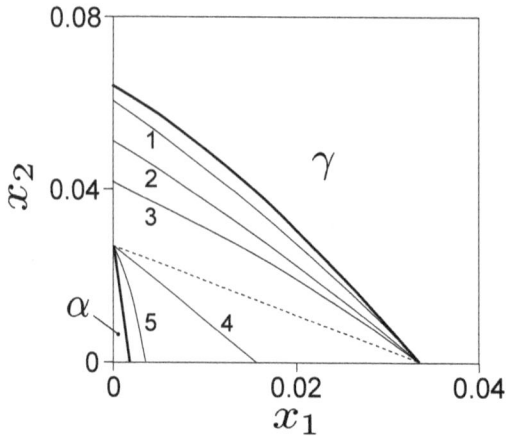

Figure 4.21 Typical IV contours for the Fe-Mn-C system (after Coates [85]). Curves 1, 2, 3, 4 and 5 correspond to $100\eta_1 = 0.01$, 0.05, 0.1, 100 and 500 respectively. The region shaded corresponds to PLE growth.

4.6.8 TIE-LINE SHIFTING DUE TO SOFT IMPINGEMENT

It is clear that in substitutionally alloyed steels, the tie-line representing interface compositions during growth constrained by local equilibrium at the transformation front does not pass through the average alloy composition. For such alloys, the effect of soft impingement is to shift the operative tie-line closer to the average composition, until eventually it does pass through the latter, causing reaction to cease as the $\alpha + \gamma$ mixture reaches equilibrium. During this process, the composition of the ferrite at the interface changes with time so the resultant gradients extend into the ferrite from the interface in the $-z$ direction, driving diffusion within the ferrite. A case such as this is treated by setting up diffusion equations for both ferrite and austenite. The mass conservation conditions must then take into account of the additional flux at $z = Z$, due to the diffusion in ferrite [97–99].

Goldstein and Randich [99] applied numerical methods to study tie-line shifting during isothermal transformation as a function of precipitate size in Fe-Ni-P alloys (P≡1, Ni≡2, Fe≡3). In this alloy, $D_{11} = 100D_{22}$. The major part of the precipitate growth was found to occur during the initial stages of transformation, prior to significant soft-impingement of the more mobile phosphorus. During this period, the interface tie-line remains constant without changing much even after the onset of P impingement. Furthermore, precipitate growth remains parabolic with respect to time, prior to the beginning of nickel impingement. It then slows down considerably when the nickel diffusion fields begin to overlap. The operative tie-line shifts towards that passing through the bulk composition. The precipitate hardly grows during this stage but the concentration gradients within the individual phases tend to homogenise by diffusion. The transformation ceases completely when the interface tie-line actually passes through the bulk composition. Goldstein and Randich point out that in a real system, where impingement distances may vary, it should not be surprising to find precipitates that have different compositions in the same alloy for the same growth time; this implies lack of equilibrium, but equilibrium in this sense may take too long to establish.

The numerical method becomes time consuming to implement when $D_{11} \gg 100D_{22}$. Gilmour et al. [100] have presented an approximate but analytical treatment (for ferrite formation by the NP-LE mechanism) of the soft impingement problem in Fe-X-C alloys. They consider the movement of a planar α/γ interface in a direction z which is normal to the interface plane; z is positive in the austenite, and $z = Z$ defines the position of the interface at any time t, with $Z = 0$ at $t = 0$. The austenite is taken to have a finite size L in the z direction and the one-dimensional growth of ferrite is assumed to begin by the NP-LE mechanism. Any diffusion fields in the austenite should strictly have the form of an error function but the profile can be approximated by assuming that the concentration varies linearly with z. The concentration profile is therefore approximated as a straight line whose end point lies a distance z_{id} ahead of the interface (Figure 4.22). The concentration gradient in austenite is uniform in this approximation, given by $(c_i^{\gamma\alpha} - \bar{c}_i)/z_{id}$, where z_{id} is an effective diffusion distance for component i, defining the region of the austenite in which the concentration differs from $\bar{c}_i$. This is the Zener [64] linearised gradient approximation and implies that the diffusion field (of element i) extends only a finite distance z_{id} into the austenite, instead of the infinite distance implied by an error function.

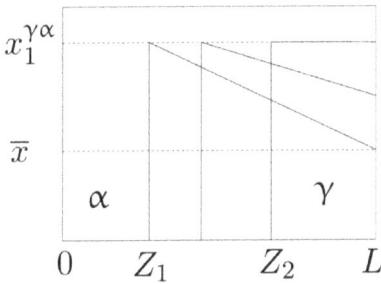

Figure 4.22 The soft-impingement process described by Gilmour, Purdy and Kirkaldy [100]. The carbon concentration of the ferrite is assumed to be zero, and $x_1^{\gamma\alpha}$ refers to the paraequilibrium carbon concentration of the γ, since Mn does not redistribute during the first two stages of soft-impingement. The first stage is completed when $Z = Z_1$, and the second when $Z = Z_2$.

The soft impingement process can be considered to occur in the three fairly distinct stages illustrated in Figure 4.22. In the first stage, the interface moves as if it is in a semi-infinite matrix, with $z_{1d} < (L - Z)$ and $c_i\{L\} = \bar{c}_i$, and Z varying parabolically with time (Equation 4.12).

During the second stage, the carbon diffusion field impinges with the boundary at $z = L$ since $z_{1d} > (L - Z)$. The position of the interface at this point is defined by $Z = Z_1$, and the next stage begins at $t = t_1$. During this stage, the carbon concentration rises everywhere in the austenite ahead of the transformation front and $c_1\{L\} > \bar{c}_1$, although $c_2\{L\} = \bar{c}_2$ since impingement of component 2 has yet to occur. When the carbon achieves uniform activity (i.e., when $Z = Z_2$, $t = t_2$) in the austenite, $c_1\{L\} \simeq c_1^{\gamma\alpha}$; notice that this latter equation is approximate because of the existence of the concentration gradient of component 2. By balancing the amount of carbon enrichment of the austenite against the carbon depletion of the ferrite (assuming $c_1^{\alpha\gamma} = 0$), it can be demonstrated that for $Z_2 \gg Z_1$,

$$c_1\{L\} = \frac{2L\bar{c}_1 - c_1^{\gamma\alpha}(L - Z)}{(L - Z)}.$$

When this is combined the condition for mass conservation at the interface (Equation 4.28), integration of the resulting differential equation gives the interface position as a function of time, during stage 2:

$$\frac{t - t_1}{D_{11}} \simeq (Z_1^2 - Z^2) + (4\ell + 2Z_2) - 2\ell^2 \ln \frac{Z_2 - Z}{Z_2 - Z_1} \qquad (4.39)$$

where $\ell = L\bar{c}_1/c_1^{\gamma\alpha}$ and $Z > Z_1$. This is not a parabolic relation between Z and t and hence is inconsistent with the more rigorous finite element method of Goldstein and Randich [99]. The time t_1 can be similarly derived:

$$t_1 \simeq \frac{L^2\bar{c}_1 c_1^{\gamma\alpha}}{D_{11}(c_1^{\gamma\alpha}-\bar{c}_1)^2}. \tag{4.40}$$

For both of these time-equations, $D_{11} = D_{11}\{c_1^{\gamma\alpha}\}$. The period t_2 can be obtained by substituting say $Z = 0.99Z_2$ into Equation 4.40.

For a Fe-Mn-C alloy, $t_1 + t_2$ turns out to be small compared with the time required for subsequent manganese diffusion and hence suggested that the tie-line governing interface compositions does not shift at all during the first two stages.

At the end of the second stage, the chemical potential of the carbon may be approximately uniform throughout; the third stage involves the slow partitioning of X from ferrite to austenite. Gilmour et al. dealt specifically with the Fe-Mn-C system; because Mn diffuses much faster in α then in γ, the concentration gradients in α were assumed to be negligible. During the third stage, the interface tie-line shifts such that the rapidly adjusting carbon distribution remains close to the conditions of uniform activity. Using methods similar to those used in deriving Equation 4.39, it can be demonstrated that the time $(t - t_2)$ taken for $c_2^{\alpha\gamma}$ to change from $c_2^{\alpha\gamma}\{t_2\}$ to any subsequent value $c_2^{\alpha\gamma}\{t\}$, is given by

$$t - t_2 = \frac{L^2}{D_{22}} \int_{F_1\{c_2^{\alpha\gamma}\{t_2\}\}}^{F_2\{c_2^{\alpha\gamma}\{t\}\}} \frac{\mathrm{d}\left[F_1\{c_2^{\alpha\gamma}\{t\}\}\right]^2}{F_2\{c_2^{\alpha\gamma}\{t_2\}\}} \tag{4.41}$$

$$\text{where} \quad F_1\{c_2^{\alpha\gamma}\} = (c_1^{\gamma\alpha}-\bar{c}_1)/c_1^{\gamma\alpha}$$

$$\text{and} \quad F_2\{c_2^{\alpha\gamma}\} = \frac{(c_2^{\gamma\alpha}-\bar{c}_2)^2}{(\bar{c}_2 - c_2^{\alpha\gamma})(c_2^{\gamma\alpha}-c_2^{\alpha\gamma})}.$$

F_1 and F_2 can be defined to be functions of just $c_2^{\alpha\gamma}$ because the other three interface compositions are not independent, being specified by a tie-line. The maximum time for the third stage depends on the equilibrium X content of the α, as determined by the equilibrium tie-line passing through the bulk composition. The model is in reasonable agreement with experimental evidence on Fe-Mn-C alloys. The method can be used to treat soft impingement in Fe-C alloys after making appropriate substitutions for the various concentration terms.

4.6.9 INVALIDITY OF THE NP-LE CONCEPT

The negligible-partitioning local-equilibrium mode becomes unphysical when transformation occurs at large undercoolings. The extent of the diffusion field for the substitutional solute decreases so much, that it becomes a mathematical formality [86, 92, 101]. There actually are no experimental data that confirm the existence of the sharp concentration spikes predicted theoretically in domains where the transformation is supposed to occur by the NP-LE mechanism [e.g., 102, 103].[12]

Coates estimated that for one-dimensional growth with local equilibrium at the interface, the extent of the substitutional-solute diffusion field in the austenite is given approximately by

$$z_{Mn} \approx 2D_{Mn}/v. \tag{4.42}$$

Coates appreciated that with steep concentration gradients, the dependence of the diffusion coefficient on the concentration gradient itself should become important. During spinodal decomposition, a homogeneous solution can develop spontaneous composition-waves that grow in amplitude. Small wavelengths are not favoured because there is a cost in creating large concentration gradients. Spinodal decomposition has been reviewed by Hilliard [104], whose treatment is followed here to estimate the energy cost of the substitutional solute spike in the austenite.

The free energy of a heterogeneous solution can be expressed by a multivariable Taylor expansion [e.g. 104, 105]:

$$
\begin{aligned}
g\{a,b,\ldots\} &= g\{\bar{c}\} + a\frac{\partial g}{\partial a} + b\frac{\partial g}{\partial b} + \ldots \\
&\quad + \frac{1}{2}\left[a^2\frac{\partial^2 g}{\partial a^2} + b^2\frac{\partial^2 g}{\partial b^2} + 2ab\frac{\partial^2 g}{\partial a\,\partial b} + \ldots\right] + \ldots
\end{aligned}
\tag{4.43}
$$

in which the variables, $a,b,\ldots$ are the spatial composition derivatives (dc/dz, d^2c/dz^2, etc.). For the free energy of a small volume element containing a one-dimensional composition variation (and neglecting third and high-order terms), this gives

$$g = g\{\bar{c}\} + \kappa_1\frac{dc}{dz} + \kappa_2\frac{d^2c}{dz^2} + \kappa_3\left(\frac{dc}{dz}\right)^2 \tag{4.44}$$

where $\bar{c}$ is the average composition and

$$\kappa_1 = \frac{\partial g}{\partial(dc/dz)} \tag{4.45}$$

$$\kappa_2 = \frac{\partial g}{\partial(d^2c/dz^2)} \tag{4.46}$$

$$\kappa_3 = \frac{1}{2}\frac{\partial^2 g}{\partial(dc/dz)^2}. \tag{4.47}$$

In this, κ_1 is zero for a centrosymmetric crystal since the free energy must be invariant to a change in the sign of the coordinate z.

The total free energy per atom, g_{ih} for the inhomogeneous solution, is obtained by integrating over the volume V:

$$g_{ih} = \int_V\left[g\{\bar{c}\} + \kappa_2\frac{d^2c}{dz^2} + \kappa_3\left(\frac{dc}{dz}\right)^2\right]. \tag{4.48}$$

On integrating the third term in this equation by parts:

$$\int\kappa_2\frac{d^2c}{dz^2} = \kappa_2\frac{dc}{dz} - \int\frac{d\kappa_2}{dc}\left(\frac{dc}{dz}\right)^2 dz. \tag{4.49}$$

As before, the first term on the right is zero, so an equation of the form below is obtained for the free energy per atom of a heterogeneous system:

$$g_{ih} = \int \left[g\{\bar{c}\} + v_a^3 \kappa \left(\frac{dc}{dz} \right)^2 \right] dV \tag{4.50}$$

where v_a is the volume per atom and κ is known as the gradient energy coefficient. The term $g\{\bar{c}\}$ is the free energy of a homogeneous solution with the average concentration $\bar{c}$. The interpretation of this equation is that the creation of gradients of concentration lead to an increase in the free energy, so steep gradients will reduce or even overcome the driving force for diffusion. There is an additional term due to strain caused by the variation of the lattice parameter with concentration. The quantity η is defined as $d\ln a / dx$, where a is the lattice parameter and x the atomic fraction of concentration, then the strain energy per atom has two components. The first is approximately $\eta^2 v_a^3 E (c - \bar{c})^2 / (1 - v)$ where E is the Young's modulus and v is the Poisson's ratio, and the second contribution has a dependence on $(dc/dz)^2$ and hence can be incorporated into the gradient energy coefficient [104, 106].

The strain energy term turns out to be rather small when considering the NP-LE mode [103], so the focus instead is on the gradient energy component. Figure 4.23 shows the increase in free energy of the solution as a consequence of the substitutional solute concentration spike in the austenite at the γ/α interface during negligible partitioning local equilibrium growth. The diffusion distance is intended to represent the width of the spike. The calculations assume a gradient energy coefficient of $3.85 \times 10^{-10}\,\text{J}\,\text{m}^{-1}$ based on the Fe-Cr system [107, 108] and that $(x - \bar{x}) = 0.03$.

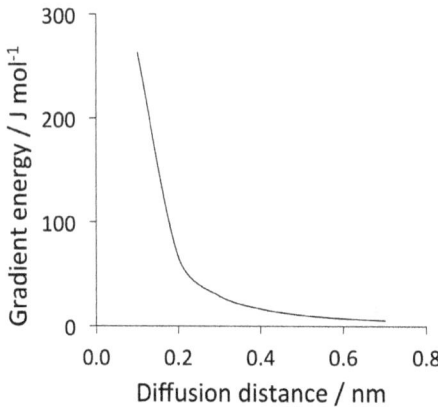

Figure 4.23 Estimate of the penalty on free energy due to the gradient of concentration in the austenite ahead of the α/γ interface [103].

Allotriomorphic ferrite growth rate calculations by Zhang et al. [109], who also conducted rate measurements over the temperature range where a transition from P-LE to NP-LE is expected, can be used to illustrate the problem. Using a value for the manganese diffusion coefficient ($1.05 \times 10^{-5} \exp(-286000/RT)\,\text{m}^2\,\text{s}^{-1}$, $R = 8.3143\,\text{J}\,\text{K}^{-1}\,\text{mol}^{-1}$) from [89], the parabolic rate constant ($5.17 \times 10^{-7}\,\text{m}\,\text{s}^{-0.5}$) for 775°C, the diffusion distance can be estimated using Equation 4.42 to be just 0.03 nm. This clearly is physically unrealistic and associated with an intolerable penalty from the gradient energy term. There are many other data in the literature that claim consistency with the

NP-LE mode based on unphysical gradients and without verifying the existence of the appropriate concentration spike.

This means that the NP-LE calculations as implemented currently are wrong because they fail to account for the gradient energy term. If such a term is incorporated, the concentration spike would be moderated to much larger widths, making the extent of partitioning greater and the growth rate slower. This is precisely analogous to spinodal decomposition, where there is a wavenumber that receives maximum amplification [Figure 4, 104], so the actual wavelength observed is of the order of 10-20 nm [104].

In conclusion, if the constraint of local equilibrium at the α/γ interface is retained, then it is necessary to account for gradient energy terms in dealing with sharp concentration profiles. The penalty due to gradient energy will stop concentration profiles from becoming unrealistically narrow–judging from work on spinodal decomposition, a diffusion distance of the order 10–20 nm should be minimum although the actual number will depend on the magnitude of the free energy change available for transformation. Naturally, the partitioning of solute would not then be negligible, making the NP-LE concept redundant.

4.7 BREAKDOWN OF LOCAL EQUILIBRIUM

4.7.1 PARAEQUILIBRIUM

Kinetic phenomena often prevent transformations from occurring under equilibrium conditions [59, 110–112]. Steels are special in that they contain atoms with strikingly different mobilities, so it becomes possible for some atoms to attain a uniform chemical potential across the phases while others cannot do so. In particular, it is possible to imagine circumstances where the sluggish substitutional solutes are unable to partition during transformation while carbon can do so [113–116]. Hultgren introduced the term *paraequilibrium* to describe the constrained equilibrium between two phases that are forced to have the same substitutional-solute to iron atom ratio, but which (subject to this constraint) achieve equilibrium with respect to carbon.

In a Fe-X-C alloy, equilibrium between austenite and ferrite of homogeneous compositions is said to exist when

$$\mu_i^\alpha\{x_{1\alpha}, x_{2\alpha}, x_{3\alpha}\} = \mu_i^\gamma\{x_{1\gamma}, x_{2\gamma}, x_{3\gamma}\} \tag{4.51}$$

$i=1, 2$ or 3. This is illustrated, for a fixed temperature, in Figure 4.24 where the equilibrium condition is defined by a common tangent-plane touching the free energy surfaces of ferrite and austenite. The intercepts on the pure-component axes represent the values of their respective chemical potentials. When Equation 4.51 is satisfied, $x_{i\alpha} = x_i^{\alpha\gamma}$ and $x_{i\gamma} = x_i^{\gamma\alpha}$, because there are no gradients of chemical potential in a system at equilibrium, either within a given phase or across phase boundaries. Austenite and ferrite are said to be in paraequilibrium when

$$\left.\begin{array}{l} x_{2\alpha}/x_{3\alpha} = x_{2\gamma}/x_{3\gamma} = \bar{x}_2/\bar{x}_3 \\ \mu_1^\alpha = \mu_1^\gamma \end{array}\right\}. \tag{4.52}$$

The Gibbs free energy change ΔG per mole reacted for reactions in a closed system when an infinitesimal amount of material of composition $x_{i\alpha}$ is transferred from austenite of composition $x_{i\gamma}$

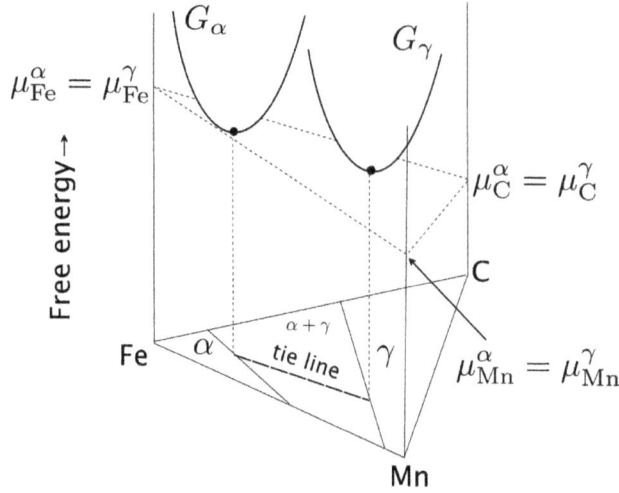

Figure 4.24 Extension of the common-tangent construction to ternary alloys. The contact points of the tangent plane to both the γ and α free energy surfaces define a tie-line on an isothermal section of the Fe-C-Mn phase diagram. The plane can be rocked whilst maintaining contact with both free energy surfaces, to generate the set of tie lines that define the $\alpha + \gamma$ equilibrium phase field. The intercepts of the tangent plane with the vertical axes give the chemical potentials of each component.

to ferrite of composition $x_{i\alpha}$ is given by

$$\Delta G = x_{1\alpha}(\mu_1^\alpha - \mu_1^\gamma) + x_{2\alpha}(\mu_2^\alpha - \mu_2^\gamma) + x_{3\alpha}(\mu_3^\alpha - \mu_3^\gamma) \qquad (4.53)$$

where the chemical potentials in the austenite and ferrite are evaluated at the compositions $x_{i\gamma}$ and $x_{i\alpha}$ respectively.

ΔG becomes zero when ferrite and austenite are in equilibrium because all elements then have uniform chemical potentials across all phases. The paraequilibrium state can also be specified by setting $\Delta G = 0$ (subject to Equation 4.52); combining Equations 4.52 and 4.53 gives [117]

$$\mu_2^\gamma - \mu_2^\alpha + (\mu_3^\gamma - \mu_3^\alpha)\frac{\overline{x}_3}{\overline{x}_2} = 0. \qquad (4.54)$$

The state of paraequilibrium is illustrated in Figure 4.25 using tangent planes and the free energy surfaces of austenite and ferrite. There is now no *common* tangent plane, but instead two planes, each of which is tangential only to one phase, the austenite or ferrite. However, these two tangent planes have a common origin on the carbon axis, because the chemical potential of carbon must be identical in both phases during paraequilibrium. The chemical potentials of the other two elements are clearly not uniform across the phases since the tangent planes do not have common intersections on either the iron or Mn axes. For the case illustrated, the chemical potential of Mn is raised on transformation whereas that of Fe is reduced, in such a way that the two effects cancel to give a zero net free energy difference between austenite and ferrite. Equation 4.54 simply expresses this in a mathematical form.

Another notable feature from Figure 4.25 is that the two tangent planes share a line of intersection, which when projected onto the isothermal section of the phase diagram, gives the locus of all points along which there is a constant ratio of Fe/Mn, as required by the definition of paraequilibrium. Consistent with Equation 4.54, it is seen from the geometry in Figure 4.25 that

$$(\mu_{Mn}^{\gamma} - \mu_{Mn}^{\alpha})\bar{x}_{Mn} = (\mu_{Fe}^{\alpha} - \mu_{Fe}^{\gamma})\bar{x}_{Fe}. \tag{4.55}$$

Equation 4.55 can be generalised to define the paraequilibrium condition for a multicomponent system in which there are n mobile components and m immobile components [118]. By analogy with the ternary case, there is a uniform chemical potential for all the mobile components:

$$\mu_i^{\alpha} = \mu_i^{\gamma} \qquad \text{for } i = 1 \text{ to } n \tag{4.56}$$

but for the immobile components,

$$\sum_{j=1}^{m} \frac{\bar{x}_j}{\bar{x}_1}(\mu_j^{\gamma} - \mu_j^{\alpha}) = 0 \qquad \text{for } j = 1 \text{ to } m. \tag{4.57}$$

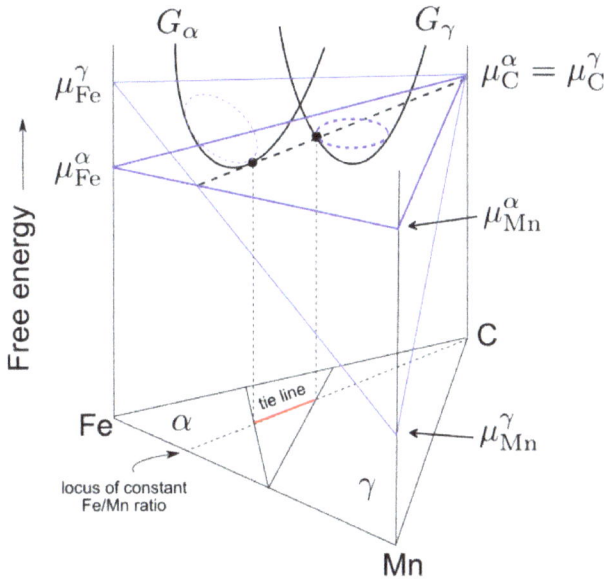

Figure 4.25 The paraequilibrium state illustrated using the austenite and ferrite free energy surfaces for a substitutionally alloyed steel at a fixed temperature. There are two tangent planes, each plane being tangential to just one of the phases whilst cutting the free energy surface of the other phase. It is only the chemical potential of carbon that is identical in austenite and ferrite during paraequilibrium. Note also that the paraequilibrium tie line is parallel to a construction line radiating from the carbon corner of the isothermal section of the phase diagram. Unlike the case for equilibrium, the $\alpha + \gamma$ two-phase field constricts to a point as the carbon concentration tends to zero.

The equilibrium and paraequilibrium concentrations of carbon in ferrite and in austenite will in general be different because the chemical potential of carbon is a function of all elements in solution. The substitutional solute concentrations are not identical for the two cases. The paraequilibrium phase diagram is constructed on the basis of Equations 4.52-4.54 rather than Equation 4.51. Hillert

[114] has shown that the paraequilibrium phase boundaries lie within the $\alpha + \gamma$ phase field of the equilibrium phase diagram. Furthermore, the tie-lines of the paraequilibrium $\alpha + \gamma$ phase field all satisfy Equation 4.52 and hence represent lines along which the substitutional to iron atom ratio is constant.[13] Typical paraequilibrium and equilibrium Fe-Mn-C diagrams are illustrated in Figure 4.26. For any given alloy, a degree of undercooling below the equilibrium transformation temperature is necessary before paraequilibrium transformation becomes feasible. This is a reflection of the lower free energy change accompanying the formation of ferrite which is forced to accept a non-equilibrium substitutional alloy content.

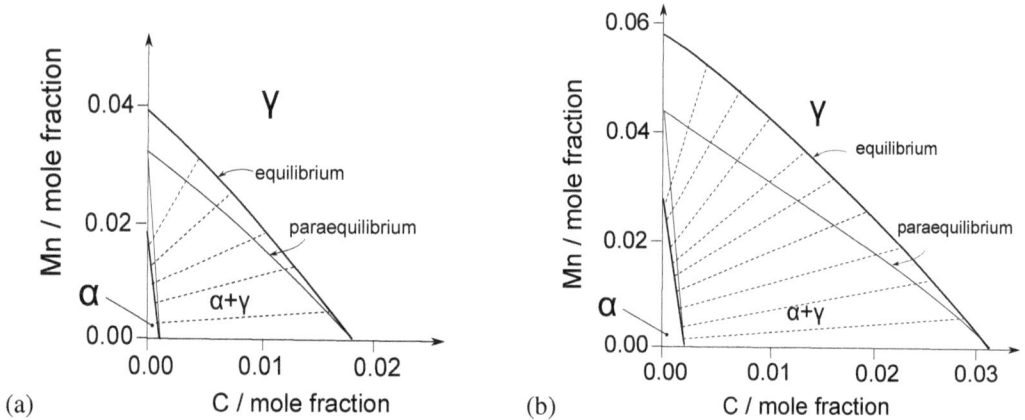

(a)   (b)

Figure 4.26 Typical calculated isothermal sections of the equilibrium and paraequilibrium phase diagrams of the Fe-Mn-C system: (a) 1053 K, (b) 1013 K (after Bhadeshia [119]). The tie-lines for the paraequilibrium diagram should be almost horizontal because the Fe/Mn ratio is constant everywhere and the carbon concentration is relatively small.

Given that only carbon partitions, its diffusion will be rate limiting, which then represents the major dissipation of the driving force, assuming that interfacial kinetics are relatively rapid. The substitutional solutes influence kinetics only through their effect on the limiting carbon concentrations at the interface; they alter the thermodynamics of the $\gamma \rightarrow \alpha$ transformation, i.e., the phase diagram. There may be a smaller effect on the diffusion coefficient of carbon through the influence of X elements on the activity of carbon in austenite (Chapter 2). Having established the paraequilibrium phase diagram, the diffusion-controlled growth rate for ferrite can be calculated using the theory relevant for Fe-C alloys, after substituting the paraequilibrium carbon concentrations for $x_1^{\gamma\alpha}$ and $x_1^{\alpha\gamma}$. This carries the assumption that paraequilibrium exists locally at the interface.

4.7.2 SOLUTE AND SOLVENT TRAPPING

During the paraequilibrium growth of ferrite in Fe-Mn-C, the chemical potential of manganese is *increased* on transfer from the austenite. The manganese normally would partition preferentially into austenite, it is, after all, the classic austenite stabilising element. Manganese is therefore passively transferred into the ferrite by the advancing α/γ interface. That its free energy is increased on transfer means that it is "solute-trapped" [120], rather like when excess carbon is inherited by martensite but would rather reside in the austenite.

If a Fe-Si-C alloy undergoes paraequilibrium transformation to α, then since silicon is a ferrite stabiliser its chemical potential must decrease on entering the α lattice. To compensate for this, the chemical potential of the major component iron must increase; this is known "solvent trapping". The change in chemical potential during the transfer of a species i is

$$
\begin{aligned}
\Delta\mu_i \;&=\; \mu_i^{\alpha} - \mu_i^{\gamma} \\
&=\; RT\ln\frac{x_{i\alpha}x_i^{\gamma\alpha}}{x_{i\gamma}x_i^{\alpha\gamma}}.
\end{aligned}
\tag{4.58}
$$

Component i is said to be trapped when $\Delta\mu_i$ is positive - i.e., when $(x_{i\alpha}x_i^{\gamma\alpha})/(x_{i\gamma}x_i^{\alpha\gamma}) > 1$.

Although ferrite growth is considered usually to occur either by paraequilibrium or equilibrium transformation, an infinite number of other possibilities exist. All compositions of ferrite which allow ΔG in Equation 4.53 to be zero or negative constitute possible α compositions that can in principle grow from austenite. For a binary alloy, this means that α formation can involve: 1) equilibrium transformation, 2) transformation in which one of the species is trapped (with $\Delta\mu = 0$ for the other element) and 3) non-equilibrium transformation in which neither solute nor solvent is trapped (with $\Delta\mu < 0$ for both elements). For a ternary alloy, the following possibilities arise:

- equilibrium transformation;
- transformation in which one of the species is trapped, with another species having equal chemical potential in both phases (e.g. paraequilibrium) as long as there is a net reduction in free energy;
- transformation in which one of the species is trapped, with no species having equal chemical potential in both phases;
- transformation in which two components are trapped; and
- non-equilibrium transformation.

Baker and Cahn [111] have pointed out that in circumstances where some components are trapped, the transfer of components across the interface cannot be independent if there is to be a net reduction in free energy.

There is an important contradiction in applying the paraequilibrium concept to a reconstructive transformation. The substitutional lattice is configurational frozen, but the phase transformation requires reconstructive diffusion (p. 168). It is not clear why the reconstructive diffusion is not accompanied by solute partitioning. There are no experimental data that verify directly the paraequilibrium state for allotriomorphic ferrite [121]. It is telling that the phase field method has never been able to predict a transition from PLE into the NP-LE or the paraequilibrium growth of ferrite; it probably is not possible to do this even in principle given the way in which the interface is modelled. It is likely that the paraequilibrium mechanism applies only to a carbon-diffusion controlled displacive transformation (Chapter 7). The items discussed in the next section should be considered along with these caveats.

Transition from local-equilibrium to paraequilibrium

When $D_{11} \gg D_{22}$, ferrite growth with local equilibrium at the interface has been considered to occur in two fairly distinct ways. At low supersaturations, there is the bulk partitioning of the slow diffuser, the activity gradient of the fast diffuser (in γ) being reduced reduced almost to zero. At large supersaturations, there is little partitioning of the slow diffuser, so its activity gradient in the austenite becomes large enough to allow it to keep pace with the faster diffusing solute; this particular mode has fundamental difficulties as described in previous sections, but there is some value in discussing it further.

Paraequilibrium transformation has zero partitioning of substitutional solutes during transformation, the ratio x_2/x_3 being constant across the interface, on the finest conceivable scale.

Hillert [101] and Coates [90] have considered the conditions leading to the onset of paraequilibrium transformation. Coates has shown that for one-dimensional growth occurring at high supersaturations, with local equilibrium at the interface, the approximate extent (z_{2d}) of the diffusion field of component 2 in the austenite ahead of the interface is given by:

$$z_{2d} = 2D_{22}/v \qquad (4.59)$$

z_{2d} is therefore a function of temperature, driving force and particle size Z. As z_{2d} decreases, either due to an increase in v or a decrease in D_{22}, the composition perturbation in the γ ahead of the interface becomes smaller in extent until z_{2d} becomes small compared with atomic dimensions and loses physical significance. Indeed, Coates [90] has suggested that the perturbation then becomes a part of the interface, since the dependence of the diffusivity on concentration gradient becomes significant, giving rise to a gradient-energy term as in spinodal decomposition. These considerations led him to suggest that growth switches from local equilibrium to the paraequilibrium mode when $z_{2d} \simeq 1\,\text{nm}$ [14].

Since z_{2d} increases with particle size, it must be the case that a particle may begin isothermal growth by a paraequilibrium mechanism but then slip into growth by the local equilibrium mechanism.

Hillert [101] has applied similar reasoning to the problem and has concluded that with increasing undercooling below the equilibrium transformation temperature, local equilibrium growth gives way to paraequilibrium but this depends on the unlikely assumption that interface velocity increases monotonically with decreasing temperature.

Whereas it seems intuitively reasonable that deviations from local equilibrium must arise when z_{2d} becomes comparable to atomic dimensions, Hillert [122] has pointed out that this reasoning fails in the case of pearlite and massive transformations. In fact, pearlite never exhibits paraequilibrium transformation, nor any mechanism which would lead the phases involved to inherit the Fe/X ratio of the austenite (Figure 12.12). A large quantity of data on the effect of alloying elements on pearlite growth show the partitioning of substitutional solutes even where the calculated values of z_{2d} are in most cases less than 0.1 nm. It may not even be reasonable to consider that a reconstructive transformation can transition into paraequilibrium.

Aziz model for solute trapping

This is a model which allows the estimation of solute trapping given an interface velocity; solute-trapping is a feature of paraequilibrium so it is relevant to understanding the onset of transformations that occur without the partitioning of substitutional solutes in a Fe-C-X system.

Figure 4.27 illustrates a transformation front between the shaded and unshaded crystals, in a binary alloy containing "A" (solvent) and "B" (solute) atoms. The smaller solute atoms in this image have a preference for the parent phase (γ). The atoms in the central layer have to move along the vectors indicated in order to transform into the product phase (α). δ_s is a typical diffusion jump distance for the solute atom; the motions required for the atoms in the interfacial layer to adjust to the new crystal structure are rather smaller. The model does not contain information about the structure of the interface.

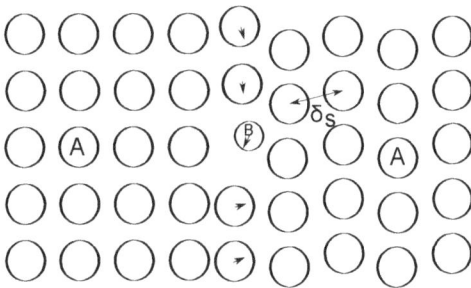

Figure 4.27 Choreography of solute trapping, adapted from Aziz [123]. The solvent is labelled A, solute B and the product phase is shaded dark. The transformation front is advancing towards the right.

Solute is trapped if the interface velocity v is larger than that at which solute atoms can diffuse away into its desired phase. The maximum diffusion speed is approximately D/δ_s since δ_s is the minimum diffusion distance, so trapping occurs when $v > D/\delta_s$. There will be no accumulation or depletion of solute so a steady state is maintained. Writing $J\{z\}$ and $c\{z\}$ as the diffusive flux across the interface and concentration respectively, both of which are functions of distance z, a steady state is obtained when

$$\frac{\partial}{\partial z}(J - vc) = 0$$

since diffusive flux must balance the rate of change in concentration caused by the partitioning of solute between the two phases as the interface moves (cf. Equation 4.14). The thickness of the interface is about the same as the minimum diffusion distance δ_s; in addition, $\partial J/\partial z = (J^{\alpha\gamma} - J^{\gamma\alpha})/\delta_s$ and $\partial c/\partial z = (c^{\gamma\alpha} - c^{\alpha\gamma})/\delta_s$, the steady-state requirement simplifies to:

$$v(c^{\gamma\alpha} - c^{\alpha\gamma}) = J^{\alpha\gamma} - J^{\gamma\alpha}. \tag{4.60}$$

Referring to Figure 4.28 and using the theory and terminology outlined in Section 3.6.1, it is seen that

$$J^{\alpha\gamma} = \frac{kT_D}{h} \frac{\delta_s \Gamma_\alpha c^{\alpha\gamma}}{\Gamma_m} \frac{1}{m^+} \exp\left\{-\frac{G^*}{RT}\right\}$$

$$J^{\gamma\alpha} = \frac{kT_D}{h} \frac{\delta_s \Gamma_\gamma c^{\gamma\alpha}}{\Gamma_m} \frac{1}{m^-} \exp\left\{-\frac{G^* + \Delta\mu}{RT}\right\}.$$

It follows that

$$
\begin{aligned}
v(c^{\gamma\alpha} - c^{\alpha\gamma}) &= \frac{kT_D}{h}\frac{\delta_s}{\Gamma_m}\frac{1}{m^+}\exp\left\{-\frac{G^*}{RT}\right\}\left(\Gamma_\alpha c^{\alpha\gamma} - \Gamma_\gamma c^{\gamma\alpha}\exp\left\{-\frac{\Delta\mu}{RT}\right\}\right) \\
&= \frac{D}{\delta_s}(c^{\alpha\gamma} - k_e c^{\gamma\alpha})
\end{aligned}
\tag{4.61}
$$

where D is the diffusion coefficient, the substitution being made according to the theory in Section 3.6.1, and k_e is the equilibrium partition coefficient obtained by setting $\mu_B^\gamma = \mu_B^\alpha$, so

$$
k_e = \frac{c^{\alpha\gamma}}{c^{\gamma\alpha}} = \frac{\Gamma_\gamma}{\Gamma_\alpha}\exp\left\{-\frac{\mu_B^\alpha - \mu_B^\gamma}{RT}\right\}
\tag{4.62}
$$

The solid that forms need not have a composition consistent with equilibrium since solute trapping is a possibility. Therefore, on replacing $c^{\alpha\gamma}$ by c^α to emphasise this point, on adapting Equation 4.62 accordingly, the actual partitioning coefficient $k_p = c^\alpha/c^{\gamma\alpha}$ is given by

$$
k_p = \frac{c^\alpha}{c^{\gamma\alpha}} = \frac{\beta_p + k_e}{\beta_p + 1} \qquad \text{where} \qquad \beta_p = \frac{v\delta_s}{D}.
$$

This equation enables the composition of a growing phase to be estimated even when it deviates from equilibrium, as long as the velocity of the interface is known. It can be used to interpret experimental data on the mechanism by which allotriomorphic ferrite grows.

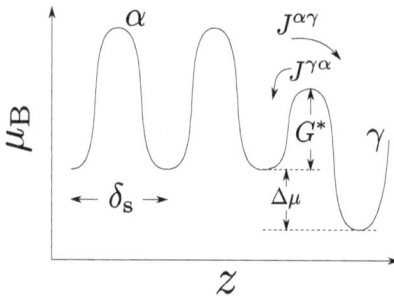

Figure 4.28 The activation energy barrier G^* for diffusion across the interface, on a plot of the chemical potential of the solute versus distance. The solute potential is lowest in the parent phase γ. The change in the solute chemical potential is $\Delta\mu = \mu^\alpha - \mu^\gamma$ on transfer from γ to α.

Consider two experiments on the growth of allotriomorphic ferrite in Fe-Mn-Si-C alloys [24]. In one case (a), the alloy was transformed isothermally at a high temperature where paraequilibrium transformation is impossible and in the second instance (b), ferrite was induced at a lower temperature where paraequilibrium transformation is thermodynamically possible. Microanalysis in each case indicated long-range solute redistribution during transformation, ruling out the paraequilibrium mechanism. Paraequilibrium would have been expected on the basis of the calculated diffusion distances in the austenite at the transformation interface (Equation 4.59). Table 4.5 shows that for case (b) these distances are very small indeed. These results are consistent with Hillert's [122] evidence that pearlite grows with partitioning of substitutional solutes even when Equation 4.59 indicates incredibly small diffusion distances. It is not reasonable therefore to model the kinetic transition on the basis of the diffusion distance alone. On the other hand, the Aziz equation correctly predicts partitioning for case (b) so it is a more rigorous approach.

Table 4.5

Calculated diffusion distances (z_d) and partition coefficients k_p for Fe-Mn-Si-C alloys transformed isothermally [24].

	Case (a)	Case (b)
z_{Mn}	18 nm	0.08 nm
z_{Si}	4 nm	0.14 nm
k_p^{Mn}		0.7
k_p^{Si}		1.7

4.8   INTERFACE-CONTROLLED GROWTH

4.8.1   PURE IRON

In pure iron, the growth of ferrite must be limited by the transfer of atoms across the interface. Consider first the case where a flat interface moves as a whole rather than by a step mechanism. It can be deduced from Figure 4.28 that the speed at which the interface moves during the $\gamma \to \alpha$ transformation is given by

$$v = \delta_b f^* \exp\left\{-\frac{G^*}{RT}\right\}\left[1 - \exp\left\{-\frac{\Delta G_3^{\alpha\gamma}}{RT}\right\}\right] \tag{4.63}$$

where δ_b is the thickness of the interface, f^* is an attempt frequency for atomic jumps across the interface and the chemical potential difference $\Delta\mu$ for pure iron becomes the free energy difference $\Delta G_3^{\alpha\gamma} = G^\gamma - G^\alpha$; consistent with earlier notation, the subscript "3" identifies iron. When the undercooling below the equilibrium transformation temperature is small, Equation 4.63 simplifies to

$$\begin{aligned} v &= \delta_b f^* \exp\left\{-\frac{G^*}{RT}\right\}\frac{\Delta G_3^{\alpha\gamma}}{RT} \\ &= M_b \frac{\Delta G_3^{\alpha\gamma}}{V_m} \end{aligned} \tag{4.64}$$

where V_m is the molar volume and the term M_b is a grain boundary mobility. The interface velocity thus becomes proportional to $\Delta G_3^{\alpha\gamma}$. The available free energy is dissipated entirely in interfacial processes (Equation 4.11) with $G_I = \Delta G_3^{\alpha\gamma}$. If the interface is curved then the net free energy change accompanying its motion is reduced in proportion to the increase in interface area due to the growth process, leading to a reduction in the growth rate, but this effect will be significant only for particles with a large surface to volume ratio. On the other hand, grain curvature is the motivation for coarsening in the theory of grain growth (Chapter 13).

The mobility $M_b = M_0 \exp\{-G^*/RT\}$ of the α/γ interface is difficult to measure experimentally, usually determined from analysis of overall transformation kinetics. Furthermore, any measurement on polycrystalline samples will tend to be some global average (Table 4.6). Studies of grain growth in single-phase ferritic samples have sometimes been used to represent the α/γ interface, but it has been shown that the mobility in the latter case is much smaller than for α/α boundaries [124]. Molecular dynamics simulations of the α/γ interfaces have the advantage that the mobility can be studied with the full specification of the crystallography involved. Such a study on the reconstructive transformation in pure iron covered an interphase boundary orientation $\{110\}_\alpha \parallel \{776\}_\gamma$, $\langle 001 \rangle_\alpha \parallel \langle \bar{1}10 \rangle_\gamma$. The orientation relationship between the austenite and ferrite was significantly tilted away from those normally observed in order to ensure that the $\{776\}_\gamma$ plane is parallel to $\{110\}_\alpha$. This is because the simulations apparently are not able to deal with flat interfaces, which do not seem to move over the typical timescales of the molecular dynamics simulations.[15] The imposition of the tilt introduced steps on the austenite side of the interface, presumably making it more mobile and introducing misfit dislocations with Burgers vectors within the plane of the boundary. A large driving force, approximately $1350\,\mathrm{J\,mol^1}$ was applied to move the interface; this would require the austenite in the pure iron to be undercooled by about $410\,^\circ\mathrm{C}$ with the simulation covering less than $20\,\mathrm{ns}$. The resulting mobility was found to be extraordinarily small (Table 4.6). Possible reasons for this have been proposed but are not established. It is obvious that the procedure taken to increase the mobility of the boundary by introducing one-sided steps that also increase the free volume associated with the boundary has consequences. This procedure was necessary to complete the simulation because that the interface otherwise does not move within the simulation timescale. The high mobility may therefore be an artefact of the simulation method.

Table 4.6
Mobility data for the α/γ interface. Note that $\Delta\mu_{\mathrm{Fe}}$ does not fully represent the driving force, hence the approximation symbols.

$M_0 / \mathrm{m\,mol\,J^{-1}\,s^{-1}}$	$G^* / \mathrm{kJ\,mol^{-1}}$	Comments
5.8×10^{-2}	140	Fe-Mn alloy, analysis of overall transformation kinetics, $\Delta G_3^{\alpha\gamma} \approx \Delta\mu_{\mathrm{Fe}}$ [125].
0.1-15	140	Fe-Cu, Fe-Co, Fe-Si, Fe-Al, FeCr alloys, analysis of overall transformation kinetics, G^* from [125–127], $\Delta G_3^{\alpha\gamma} \approx \Delta\mu_{\mathrm{Fe}}$, M_0 found to be composition dependent [128]
$7.8 \pm 1.8 \times 10^{-3}$	16.5 ± 5.3	Molecular dynamics simulation, very large driving force, pure iron [129]

4.8.2 IRON ALLOYS

There is nothing especially different about alloys when it comes to applying Equation 4.63 other than replacing the term $\Delta G_3^{\alpha\gamma}$ by the overall free energy change $\Delta G'$ (p. 184):

$$
\begin{aligned}
v &= \delta_b f^* \exp\left\{-\frac{\Delta G^*}{RT}\right\}\left[1 - \exp\left\{-\frac{\Delta G'}{RT}\right\}\right] \\
&\simeq M_b \frac{\Delta G'}{V_m}.
\end{aligned}
\tag{4.65}
$$

This assumes that almost all the free energy available is dissipated in the interface processes ($G_I \simeq \Delta G'$). Unlike diffusion-controlled growth, the composition of the matrix at the interface deviates so much from equilibrium that there is only a shallow concentration gradient ahead of the interface (Figure 4.13c). Consequently, only a small fraction of the free energy is dissipated in diffusion.

The practice of substituting $\Delta\mu_{Fe}$ for $\Delta G'$ is not justified because all elements participate in the transformation. This is easy to see by considering composition-invariant $\gamma \to \alpha$ transformation in an Fe-Si alloy. Transformation occurs even though the chemical potential of iron is *increased* as it transfers into ferrite.

In the absence of local equilibrium at the interface, the chemical potential of one or more of the components must obviously be discontinuous at the phase boundary. Neglecting any cross-effects, the flux J_i of solute across the boundary must then be some function of the chemical potential discontinuity $\Delta\mu_i$ [130]:

$$
J_i \propto \mu_i^\alpha\{c_i^\alpha\} - \mu_i^\gamma\{c_i^\gamma\}.
$$

If it is assumed that the product phase α grows with its equilibrium composition, then the following proportionalities hold approximately:

$$
\begin{aligned}
J_i &\propto \mu_i^\alpha\{c_i^{\alpha\gamma}\} - \mu_i^\gamma\{c_i^\gamma\} \\
&\propto \mu_i^\gamma\{c_i^{\gamma\alpha}\} - \mu_i^\gamma\{c_i^\gamma\} \\
&\propto c_i^{\gamma\alpha} - c_i^\gamma.
\end{aligned}
$$

If the interface velocity v is now taken to be related linearly to J_i, then

$$
v \propto (c_i^{\gamma\alpha} - c_i^\gamma).
\tag{4.66}
$$

This is interesting because it shows that the velocity, for interface-controlled growth, increases as the composition of the parent γ deviates more from its equilibrium composition. This is expected since as $c_i^\gamma \to \bar{c}_i$, less of the total free energy is dissipated in diffusion.

4.9 GROWTH WITH MIXED CONTROL

As discussed in Section 4.5, a variety of processes that affect the motion of an interface occur in series. Each of these processes has an associated *interface response function*, the simultaneous solution of which yields a single interfacial velocity and determines the free energy dissipated in each process. The process that dissipates the majority of the available free energy is said to be rate controlling. If the dissipations are similar then the interface is said to be under mixed-control.

The central problem in the calculation of the interface velocity when under mixed-control is the partitioning of the available free energy into the mobility and diffusion dissipations. Figure 4.29 shows for a Fe-C alloy how G_I and G_D depend on the concentration profile at the transformation front, which for a given set of compositions at the interface, can be calculated using Equations 4.10 and 4.11. G_D becomes an increasing fraction of $\Delta G'$ as the concentration in the austenite at the interface approaches the equilibrium concentration $x^{\gamma\alpha}$, since the rate-controlling process then becomes diffusion. Similarly, as $x_I^\gamma \to \bar{x}$, the motion becomes interface-controlled.

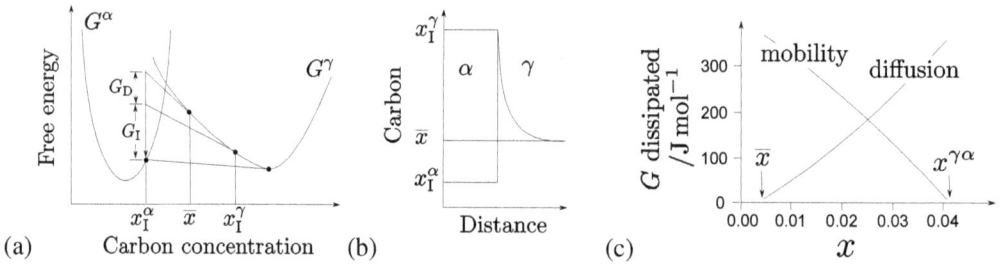

(a) Carbon concentration (b) (c)

Figure 4.29 (a) An illustration of how the available free energy is dissipated in diffusion (G_D) or in the transfer of atoms across the boundary (G_I), as a function of the concentrations $\bar{x}$, x^α and x_I^γ. (b) The resulting concentration profile at the transformation front. The concentration x_I will in general differ from the equilibrium value $x^{\gamma\alpha}$. (c) Typical calculations for Fe-0.1C wt% steel at 700°C. A greater proportion of the available free energy is dissipated in diffusion as x_I^γ approaches $x^{\gamma\alpha}$.

The actual velocity is that given by the intersection of the diffusion and mobility functions. Typical examples are plotted for the growth of allotriomorphic ferrite in Figure 4.30, in which the horizontal scale represents G_D; the dissipation G_I increases in the opposite direction since $G_I = \Delta G' - G_D$. There are three values of interface mobility used for the sake of illustration; very low, low and high corresponding to decreasing values of G^*. Diffusion-controlled growth does not occur at a constant rate:

$$\frac{\partial Z}{\partial t} = \frac{1}{2}\frac{\alpha_{1d}}{t^{\frac{1}{2}}}.$$

Because of this time dependence, the diffusion-controlled growth velocity is plotted for two stages, after 1 s and 100 s – the velocity in the latter case is ten times slower than the former.

Figure 4.30a shows a case where the transformation temperature is so high that in spite of the low interface mobility, virtually all of $\Delta G'$ is dissipated in diffusion, giving diffusion-controlled growth. The response functions for diffusion and mobility intersect almost perpendicularly so the velocities given by the marked intersections are nearly identical to those expected from a consideration of just the diffusion functions. The mobility response function can be neglected altogether and growth will occur parabolically with time and $x_I^\gamma \simeq x^{\gamma\alpha}$.

In order for growth to remain diffusion-controlled at a lower transformation temperature (500 °C), the interface would have to have a greater mobility in which case the time dependent velocities would be given by the intersections a and b in Figure 4.30b. But if the mobility is limited, then growth clearly occurs under mixed-control with the velocity remaining time dependent (intersections c and d). Even though the growth process begins with mixed control, it tends towards diffusion-control as time proceeds, as can be seen from a comparison of the dissipations at c and d. This is because at first $x_I^\gamma = \bar{x}$ but increases with the build-up of solute in front of the interface; note that this is not a soft-impingement effect since the far-field composition remains $\bar{x}$ in the calculations illustrated.

Figure 4.30c shows a case where the growth is interface-controlled so the velocity is less sensitive to time.

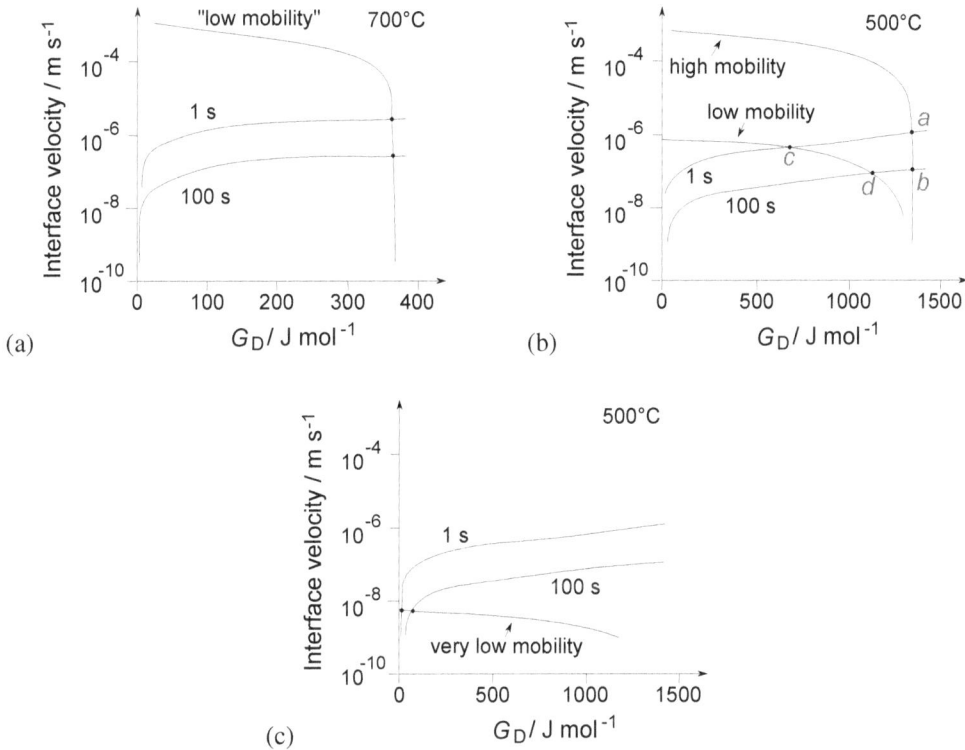

(a)

(b)

(c)

Figure 4.30 The mobility and diffusion interface-response functions for Fe-0.1C wt%, assuming that the far-field concentration remains at $\bar{x}$. Because $\Delta G' = G_D + G_I$, the horizontal scales on each of these plots also represent G_I increasing from right to left. The curves labelled 1 s and 100 s represent the carbon diffusion-controlled growth velocities for those time periods. The actual velocity is given by the intersection of the diffusion and mobility functions. (a) Diffusion-controlled growth at 700°C; (b) mixed control when the interface mobility is limited, and diffusion-controlled when the mobility is high, for transformation at 500°C; (c) interface-controlled growth.

The effect of soft-impingement on all of these examples would be to make interface-controlled growth time-dependent and diffusion-controlled growth a different function of time. A calculated example of a case where soft-impingement is taken into consideration is given in Figure 4.31a. Growth in this case begins at a rate which is interface-controlled, changes to mixed control and then evolves towards diffusion control. The concentration x_I^γ remains below equilibrium during a substantial portion of the thickening process. Figure 4.31b shows that as expected, the thickness does not scale with the square root of time, the retardation at long times being due to soft-impingement.

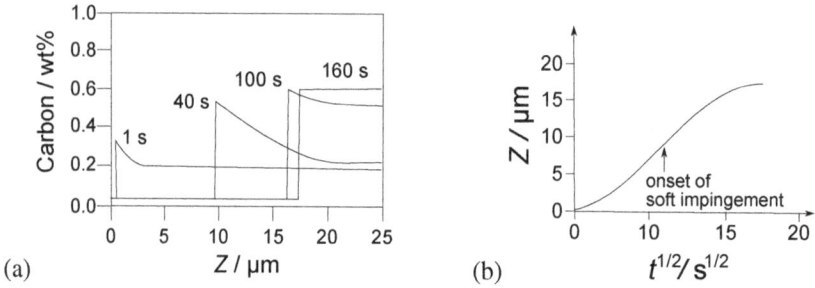

(a)

(b)

Figure 4.31 Calculations of the thickening of allotriomorphic ferrite for Fe-0.2C wt% steel transformed isothermally at 1025 K (after Krielaart [131]). (a) The development of the carbon concentration profile as a function of time; the extent of the austenite on the horizontal axis is 25 μm. (b) Corresponding plot of the ferrite thickness as a function of the square root of time.

4.10 STEP MECHANISM OF INTERFACIAL MOTION

4.10.1 BOUNDARY TOPOLOGY, LEDGE NUCLEATION

A boundary can translate in two ways: atoms may cross all parts of the interface causing it to propagate as a whole. They may alternatively attach themselves to the product phase only at favourable sites such as steps, in which case the translation of the interface as a whole requires that only the steps move. The normal displacement of the stationary part of the boundary then occurs by the passage of the steps parallel to the boundary, the magnitude of displacement depending on the step height and spacing. The term "step" is a general term, but the term "ledge" refers to a step that is linear.

When the orientation of the interface corresponds to a sharp minimum in interfacial free energy, it will tend to move by a step mechanism, rather than by the continuous displacement of every element of the interface [132–134]. Interfaces like these are said to be singular. Stepped growth depends on the existence of periodic, minimum-energy configurations, the spacing of which determines the height of the steps, but there is a dependence on the driving force for transformation [6]. Stepped growth becomes less likely at large undercoolings. Figure 4.32 shows how the free energy of the system varies as a function of the position of the boundary and of the magnitude of the driving force. Continuous motion therefore becomes possible at large driving forces. The obvious physical origin of the periodic equilibrium configurations lies in the spacing of planes within the crystal structure, when the interface is parallel to a low-index lattice plane. The step height would then equal the spacing of those lattice planes, i.e., the steps would have atomic height. This is not generally what is meant by steps in the context of allotriomorphic ferrite. The steps usually are large enough to be visible using optical microscopy or conventional transmission electron microscopy.

One possibility is that these *supersteps* occur because of the difficulty of step nucleation [135]. Figure 4.33 illustrates a ferrite allotriomorph that has grown along the prior austenite grain boundary; the form of the allotriomorph is assumed not to change in the direction normal to the plane

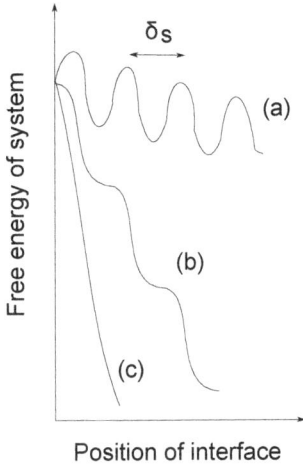

Figure 4.32 The free energy of the system as a function of interface position. The driving force increases in the order (a), (b), (c). δ_s is the distance between adjacent equilibrium positions of the interface.

of the diagram. One of the interfaces is assumed to be in a singular orientation. An idealised ledge *abcd* nucleates preferentially at the corner *a* since nucleation on a plane front would be energetically more costly. Nucleation of a ledge with cross-sectional area A and unit length in the normal direction becomes possible when

$$|A\Delta G^{\gamma\alpha}/V_m| \geq \sigma_+ + \sigma_- \tag{4.67}$$

where σ_+ and σ_- are the energies per unit length of the positive and negative corners of the ledge and V_m is the molar volume. Since there are no restrictions in this equation on the ratio ab/bc, the height of the ledge is free to vary without violating this criterion. This model is therefore unable to predict any of the geometrical features of the ledge such as the height ρ, whereas in practice the ledge height is known to vary systematically with temperature and alloy chemical-composition.

The difficulty for the step-shape illustrated in Figure 4.33a is that there is no net increase in the interfacial area when the step forms. In reality, the corner *a* should act as a *pivot* during nucleation when the interface *ab* attempts forward motion. Furthermore, the dissipation of partitioned solute will be most difficult at the corner *a* when compared with the positive edge at *c*. The process of step nucleation is therefore better represented by Figure 4.33b, a scenario verified experimentally Figure 4.33c [135, 136]. The free energy change per unit length of the step illustrated in Figure 4.33b is given by

$$\Delta G = \sigma_{\gamma\alpha}z + \sigma_+ + \frac{\Delta G^{\gamma\alpha}z\rho}{2V_m} \qquad \text{where} \qquad \underbrace{\Delta G^{\gamma\alpha}}_{-\text{ve}} = G_\alpha - G_\gamma = -\underbrace{\Delta G^{\alpha\gamma}}_{+\text{ve}}, \tag{4.68}$$

and $\sigma_{\gamma\alpha}$ is the interfacial energy per unit area of the low-energy facet plane. The ledge cross-sectional area is $A \simeq z\rho/2$. Differentiating Equation 4.68 with respect to the volume of the nucleus

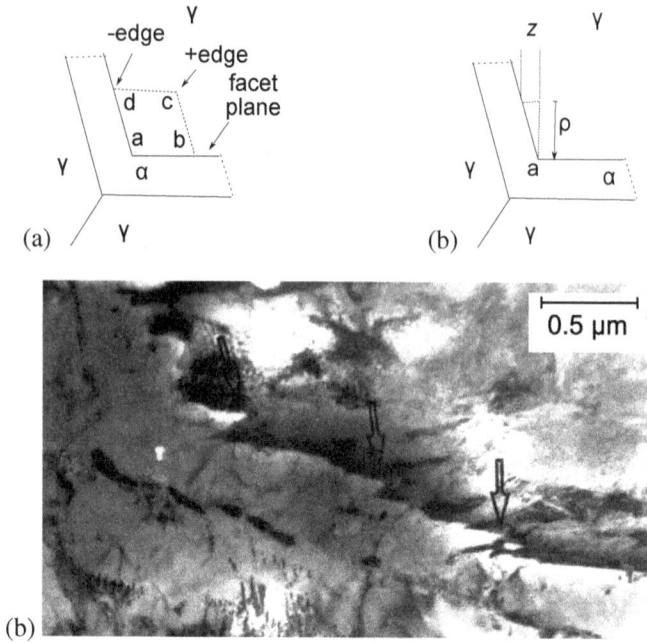

Figure 4.33 (a) Nucleation of a ledge at an α/γ interface junction. (b) Nucleation of an inclined ledge. (c) Inclined growth ledges (arrowed) on allotriomorphic ferrite in a Fe-0.39C-4.08Ni-2.05Si wt% steel transformed isothermally at 514°C for 120 s followed by quenching to ambient temperature [135].

gives

$$\frac{\mathrm{d}(\Delta G)}{\mathrm{d}A} = \frac{\mathrm{d}\rho}{\mathrm{d}A}\frac{\partial(\Delta G)}{\partial\rho} + \frac{\mathrm{d}z}{\mathrm{d}A}\frac{\partial(\Delta G)}{\partial z}$$
$$= \frac{2\sigma_{\gamma\alpha}}{\rho} - 2\frac{\Delta G^{\alpha\gamma}}{V_{\mathrm{m}}}.$$

From classical nucleation theory, the growth of an embryo into a successful nucleus becomes possible when the ledge height exceeds a critical value

$$\rho^* = \frac{\sigma_{\gamma\alpha}V_{\mathrm{m}}}{\Delta G^{\alpha\gamma}} \tag{4.69}$$

where ρ^* represents the lower limit to the ledge height for a given value of the driving force. This result, and the inverse relationship with the driving force, has been observed experimentally for the growth of allotriomorphic ferrite in steels (Figure 4.34).

It is worth considering further the geometry of a ledged interface with respect to the ratio b_{s} of the ledge height to the spacing between adjacent ledges. A ledged interface consists of a low-energy, low-mobility sessile-facet and high-energy, high-mobility ledge that translates in order to accomplish the advance of the interface as a whole. The motion of the ledges with a velocity v_{s} is equivalent to the motion of the facets with a velocity v_{l} which is related to the detailed geometry of the interface

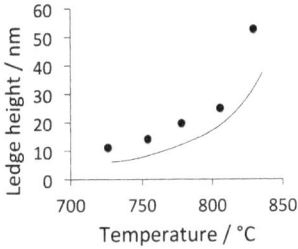

Figure 4.34 Variation in the ledge height at allotriomorphic ferrite/austenite interfaces as a function of the isothermal transformation temperature. The data are from Batte and Honeycombe [137]; the curve represents ρ^*, the calculated lower limit for the ledge height.

Figure 4.35 illustrates how the interface with $b_s > 1$ or < 1 can be displaced by the motion of the ledges or for comparison, of facets. Although the same displacement can be achieved by moving either component of the interface, it is not reasonable to consider cases where the ledge height is much smaller than the ledge spacing, as is sometimes considered in calculations of the type discussed later.

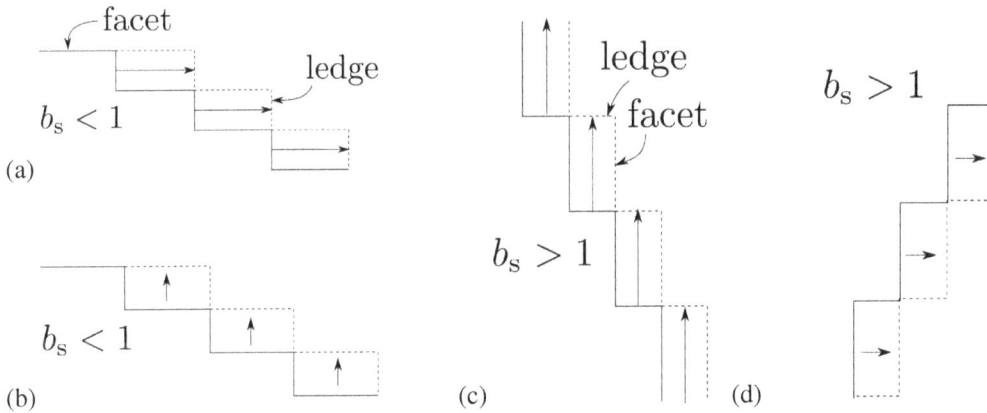

Figure 4.35 Motion of a stepped interface. The continuous lines represent the original position of the interface, the dashed lines the final position, and the arrows are velocity vectors. (a) Case where the ledge height is smaller than the ledge spacing. (b) Illustrates that the same interface displacement can be achieved as in (a) by moving the facets at a smaller velocity. (c) Case where the ledge height is smaller than the spacing. (d) The same displacement as (c) can be achieved but by moving the facets at a greater velocity.

4.10.2 MOTION OF ISOLATED LEDGE

Consider now the rate of motion of an isolated step, noting that theory has only been formulated for binary diffusion. The method is therefore restricted to Fe-C or Fe-X alloys and also Fe-X-C alloys that transform by the paraequilibrium mechanism.

For a series of ledges each of height ρ, the velocity v with which the stationary part of the interface is normally displaced (by the motion of ledges) is given by [138]

$$v = \rho n_s v_s \tag{4.70}$$

where n_s is the number of ledges per unit length and v_s is the mean ledge velocity; it is assumed that there is no interference between the diffusion fields of neighbouring ledges.

Jones and Trivedi [139] first considered the problem of determining v_s for solid-state transformations. They studied the steady-state motion of a square ended ledge moving in the direction y, the normal to the interface being in the direction z, Figure 4.36.

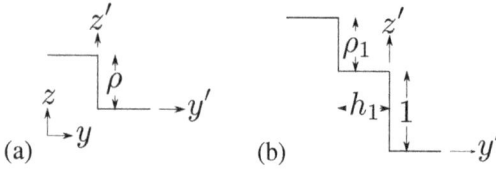

Figure 4.36 Diagram illustrating the geometry of a ledge; (a) single ledge, (b) two-ledge train.

The concentration gradient in the matrix at the ledge face is assumed constant. This is inconsistent with the steady-state growth assumption because both the concentration and its gradient should vary along the ledge face, but the approximation is not too severe [140].

For solid-state transformations, step motion can be controlled by the diffusion of atoms to the step or by the transfer of atoms across the step or by some combination of the two processes. Since it is assumed that the volume diffusion of solute in the matrix adjacent to the riser controls the growth rate, the dependence of v_s on diffusion in the parent phase can be obtained by performing a flux balance across the ledge face (cf. Equation 4.14):

$$v_s = \frac{D_{11}}{c_1^\alpha - c_1^\gamma} \frac{\partial c_1}{\partial y}\bigg|_{\text{ledge}}. \tag{4.71}$$

If growth occurs under interfacial-control, then $c_1^\gamma \simeq c_1^{\gamma\alpha}$ and assuming that v_s is proportional to the deviation of c_1^γ from equilibrium (Equation 4.66), then

$$v_s = M_b(c_1^{\gamma\alpha} - c_1^\gamma) \tag{4.72}$$

where M_b is as usual, a mobility.

Rosenthal [141] developed a number of mathematical solutions for the problem of welding in which there is a heat source that moves with a constant speed. These are steady-state solutions in the sense that an observer moving with the coordinates of the heat source does not notice any change in the environmental variables.[16] In the present context, it is the ledge that is assumed to move at a constant speed, the concentration distribution $c_1\{y', z'\}$ with respect to a moving coordinate system $(y', z',$ Figure 4.36) attached to the ledge cannot change with time. $c_1\{y', z'\}$ must therefore obey a time-independent diffusion equation in two-dimensions. By analogy with Rosenthal's work on the temperature distribution around a moving heat source, the normalised diffusion equation is given by

$$\nabla \cdot (D_{11}\nabla c_1'\{y', z'\}) + 2p\frac{\partial c_1'}{\partial y'} = 0 \tag{4.73}$$

where p is the Péclet number (a dimensionless velocity),

$$p = v_s\rho/2D_{11}$$

and the normalised, moving coordinates are defined by

$$
\begin{aligned}
y'' &= (y - v_s t)/\rho \\
z'' &= z/\rho.
\end{aligned}
$$
(4.74)

c_1' is a dimensionless solute concentration in the matrix:

$$
c_1' = \frac{c_1\{y', z'\} - \bar{c}_1}{c_1^\gamma - \bar{c}_1}.
$$
(4.75)

Using Equation 4.73, a function $\Phi_3\{p\}$ can be defined as follows:

$$
\begin{aligned}
\Phi_3\{p\} &= -\left(\frac{\partial c_1'}{\partial y'}\right)^{-1}_{\text{ledge}} \\
&= \frac{\bar{c}_1 - c_1^\gamma}{\rho(\partial c_1/\partial y)_{\text{ledge}}}.
\end{aligned}
$$
(4.76)

By combining Equations 4.71, 4.73 and 4.76, Jones and Trivedi [139] derived the step velocity to be

$$
v_s = \frac{M_b(\bar{c}_1 - c_1^{\gamma\alpha})}{1 + M_b(c_1^{\alpha\gamma} - c_1^{\gamma\alpha})(\rho/D_{11})\Phi_3\{p\} - 2p\Phi_3\{p\}}.
$$
(4.77)

The relationship takes account of both matrix diffusion and interface kinetics effects. It remains to obtain the function Φ_3 (which is the negative of the reciprocal of the concentration gradient at the ledge), by solving the differential Equation 4.73 subject to boundary conditions which allow the concentration gradient at the step face to be constant, and which include the condition that the step progresses without change of shape. The boundary conditions are [139, 140]

$$
\begin{aligned}
\partial c_1'/\partial z'' &= 0 & &\text{on } z' = 1,\ y' < 0 \\
\partial c_1'/\partial z'' &= 0 & &\text{on } z' = 0,\ y'\ 0 \\
\partial c_1'/\partial z'' &= -(\Phi_3)^{-1} & &\text{on } y' = 0,\ 0 < z' < 1.
\end{aligned}
$$

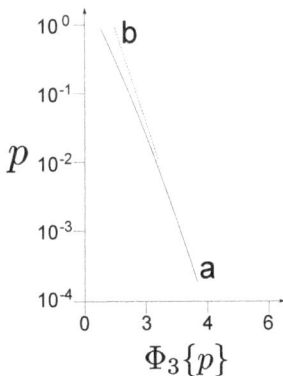

Figure 4.37 The functions $\Phi_3\{p\}$; curves (a) and (b) represent a numerical solution and an approximate solution using singular perturbation theory respectively (after Atkinson [140]).

The first two boundary conditions ensure that the isoconcentration contours around the ledge (in the matrix) are always normal to the stationary parts of the interface. The last condition specifies a constant concentration gradient at the step face. Atkinson used two methods to solve for Φ_3: the more rigorous method was numerical and another used a singular perturbation technique (valid for $p \ll 1$) in which an "outer" solution is obtained for the region where step geometry is less important and an "inner" solution in the vicinity of the step where details of its geometry are important. The two solutions are then matched to provide a complete solution. Atkinson's solutions for Φ_3 are shown in Figure 4.37, but for $p < 0.01$, singular perturbation theory gives

$$\Phi_3\{p\} = \frac{1}{\pi}[1 - C - \ln\{4\pi/p\}] \tag{4.78}$$

where C is the Euler constant equal approximately to 0.5772.

Both Atkinson, and Jones and Trivedi found the concentration distribution around the ledge to be asymmetrical in y', the type of asymmetry differing in detail, since in Jones and Trivedi's treatment c_1' becomes zero at some finite distance from the ledge. According to Atkinson, c_1' should only be zero at infinity, the concentration distribution (in the "outer" region) being defined by

$$c'\{y',z'\} = \frac{1}{\Phi_3\pi} \exp\{-py'\}K_0\{p[(y')^2 + (z')^2]^{\frac{1}{2}}\} \tag{4.79}$$

where K_0 is a modified Bessel function of zero order.

The diffusion profile decays exponentially to zero in front of the step, but decays as $p[(y')^2 + (z')^2]^{0.5}$ behind the step. The fact that the extent of the field is larger behind the ledge contradicts the results of Jones and Trivedi. Equation 4.79 is obtained using the singular perturbation method and is valid for $p < 0.01$. The diffusion field of an isolated step is shown in Figure 4.38.

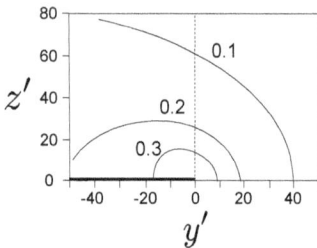

Figure 4.38 The diffusion field about an isolated ledge of height $z' = 1$, $p = 0.01$ for a variety of normalized concentrations c_1'. Adapted from Enomoto, Aaronson, Avila and Atkinson [142].

A useful form of Equation 4.71, relating f_1 to p, is as follows [139]:

$$f_1 = 2p\Phi_3 + \frac{p}{q}[1 - 2p\Phi_3] \tag{4.80}$$

$$\text{where} \quad q = \frac{M(c_1^{\alpha\gamma} - c_1^{\gamma\alpha})}{2D_{11}/\rho}. \tag{4.81}$$

As interface mobility tends to infinity, Equation 4.80 reduces to

$$f_1 = 2p\Phi_3. \tag{4.82}$$

Since a given value of the supersaturation f_1 is associated with a unique value of the Péclet number $p = \rho v_s/2D_{11}$, it follows that the ledge velocity is related inversely to the ledge height. This is physically expected since the distance that the carbon has to diffuse in order to be left in the wake of the ledge must increase with the ledge height.

As the interface mobility $M_b \to 0$, so the velocity $v_s \to M_b(c_1^{\gamma\alpha} - \bar{c}_1)$.

4.10.3 MULTILEDGE INTERACTIONS

An analysis of the growth characteristics of trains of ledges was first considered by Jones and Trivedi [143], Atkinson [144] re-analysed the problem with a corrected Φ_3 for the case where volume diffusion in the parent phase is assumed to be rate-controlling.

If the ledges in the trains all are of equal height they cannot have identical speeds. In a two-ledge train, the leading ledge always advances into fresh parent phase, whereas that which is trailing is influenced by the back diffusion field of the leading step. Therefore, for steady-state motion, the trailing ledge must have a smaller height in order to keep up with the leading ledge [144].

For the purposes of kinetic theory, a two-ledge train can be characterised in terms of dimensions normalised relative to the height ρ of the leading ledge. The normalised height of the leading ledge is then unity. The height of the trailing ledge divided by ρ is its normalised height ρ_1 and the normalised separation of the ledges is written h_1 (Figure 4.36b). When $h_1 \gg 1$, the diffusion field of a two-ledge train, in the "outer" region can be approximated by considering the ledges as line sources located at $(y' = -h_1, z' = 0)$ and $(y' = 0, z' = 0)$. For diffusion-controlled growth, Equations 4.82 and 4.79 give

$$c'\{y', z'\} = [2p/(\pi f_1)][\exp\{-py'\}K_0\{p[(y')^2 + (z')^2]^{\frac{1}{2}}\}$$
$$+ \rho_1 \exp\{-py' + ph_1\}K_0\{p[(y' + h_1)^2 + (z' - 1)^2]^{\frac{1}{2}}\}]. \tag{4.83}$$

$$\tag{4.84}$$

For a specified separation h_1, a value of ρ_1 and p consistent with the two-ledge train moving with a constant speed can be obtained by the simultaneous solution of

$$\rho_1 \exp\{-ph_1\}K_0\{-ph_1\} + [1 - C - \ln\{p/4\pi\}] = f_1\pi/2p$$
$$\exp\{ph_1\}K_0\{-ph_1\} + \rho_1[1 - C - \ln\{p\rho_1/4\pi\}] = f_1\pi/2p. \tag{4.85}$$

These equations show that for a pair of widely spaced ledges moving at the same speed, the size of the trailing ledge is smaller than that of the leading step for a given value of p. For $h_1 \gg 1$, ρ_1 decreases as h_1 becomes smaller. Extension of the analysis to include closely spaced ledges shows that ρ_1 goes through a minimum as h_1 decreases; the value of ρ_1 at the minimum decreases as p increases (Figure 4.39). The form of the curves in Figure 4.39 implies that for values of ρ_1 above the minimum, there are two spacings for which the steps travel at the same speed, the larger spacing corresponding to the stable configuration of the train. If ρ_1 becomes smaller than the minimum value, the train becomes unstable, the trailing ledge catches up and merges with the leading ledge.

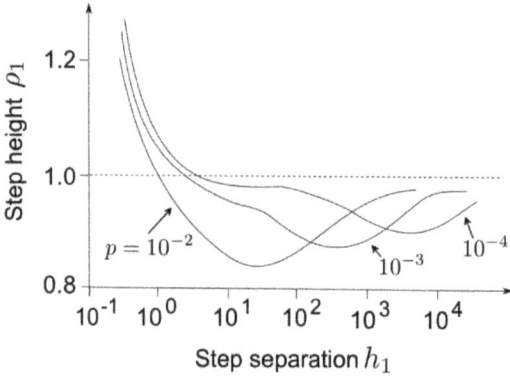

Figure 4.39 Plots of normalised height (ρ_1) of trailing ledge versus normalised ledge separation (h_1) for a two-ledge train in which the normalised height of the leading ledge is unity (after Atkinson [144]). Each curve gives the locus of conditions for which a two-step train can move at a constant speed.

Equation 4.85 can be used to show that if the heights of the two ledges are forced to be equal, then the concentration gradient at the trailing ledge will be reduced due to the dumping of solute from the diffusion field of the leading ledge. The trailing ledge will then have a smaller velocity.

Atkinson has generalised these results to trains containing many ledges, and in all cases where $h_1 \gg 1$, it is found that v_s is reduced as the number of steps in the train increases. Other results can be summarised as follows:

(i) For closely spaced pairs of ledges travelling with equal speed, ρ_1 increases as h_1 decreases (Figure 4.39).

(ii) As the separation of ledges moving with equal speed decreases, so does their velocity.

(iii) Ledges of equal height tend to separate when h_1 is large, but tend to coalesce when h_1 is small, the changeover occurring at a smaller value of h_1 as p increases.

4.10.4 LEDGE MOTION IN FINITE MEDIUM

Atkinson [145] also investigated the problem of steady-state ledge motion during solid-state diffusion-controlled transformation in a phase the extent of which is limited to a finite length L in the z' direction. The concentration gradient at an isolated step moving at a constant speed is in such circumstances a function of L and for $p \ll 1, L/\rho \gg 1$, is given by

$$-\left(\frac{\partial c_1'}{\partial y'}\right)^{-1} = \Phi_4\{p\} = \frac{1}{\pi}\left[1 - C + \Phi_5 + \ln\left\{\frac{4\pi}{p}\right\}\right] \tag{4.86}$$

where Φ_5 is a complicated function of L/ρ and p such that $\Phi_5 \to 0$ as $L/\rho \to \infty$. Φ_5 is positive as L/ρ becomes finite and for $pL/\rho \ll 1$, is given approximately by

$$\Phi_5 = \pi\rho/2pL.$$

If follows that $\Phi_4 \to \Phi_3$ as $L/\rho \to \infty$ (cf. Equations 4.86 and 4.78). The concentration gradient (and velocity) at a ledge moving in a finite medium is, as expected, smaller when compared with one that is in an infinite medium. The concentration distribution around such a ledge, in the "outer" region, is given by

$$c'\{y',z'\} = \frac{1}{\Phi_4\pi}\exp\{-py'\}[K_0\{p[(y')^2+(z')^2]^{\frac{1}{2}}\}+\Phi_5].$$

The concentration at any point in the matrix is therefore higher relative to the case where the medium is infinite. Atkinson showed that as L/ρ increases, so does v_s, but that for $p > 0.1$, $L/\rho > 10$, v_s becomes insensitive to further increases in L/ρ.

A surprising result is that $v_s = 0$ if $f_1 = \rho/L$, because the movement of a ledge in these circumstances would force the concentration in the matrix behind the ledge to a level higher than $c_1^{\gamma\alpha}$. This is an artefact of the steady-state approximation; if fresh steps nucleate then they can propagate until the steady-state diffusion field is established [146].

4.10.5 RELATIVE KINETICS OF STEPPED AND CONTINUOUS GROWTH

The velocity at which the sessile component of a regularly stepped interface is displaced by a ledge mechanism is [138]

$$v_l = b_s v_s$$

where b_s is the assumed ratio of the ledge height to the spacing between ledges. The corresponding velocity for the one-dimensional diffusion-controlled continuous motion of a planar interface is

$$v_d = \frac{1}{2}\alpha_{1d}t^{-\frac{1}{2}}.$$

Because atoms are attached only to the product phase at a fraction of the boundary, the piecewise displacement of the boundary by a step mechanism must initially be slower than continuous growth, in which every element of the interface is displaced simultaneously. However, for diffusion-controlled growth in the absence of soft-impingement effects, the rate of continuous growth decreases with time whereas the step velocity v_s is apparently independent of time. It has been argued that for this reason, ledged motion must eventually overtake continuous growth [142, 147]. On the other hand, the movement of a step across an interface causes an accumulation of solute along the interface so the boundary conditions governing the motion of any succeeding step must be altered (i.e., the supersaturation reduced). There should therefore be a progressive change in the boundary conditions of successive steps, and hence a progressive reduction in their velocity [4]. In these circumstances, stepped growth may never give a larger growth rate than that for continuous growth, as predicted by the respective linear and parabolic growth laws.

All the early analytical solutions for the isolated ledge have been for steady-state motion. This assumption might seem reasonable since the isolated ledge advances always into fresh austenite. However, Enomoto [148, 149], using a numerical analysis to solve the diffusion equation in the matrix, demonstrated that the step velocity is time dependent. The ledge velocity initially is very large (infinite at zero time, Figure 4.40) but then rapidly tends to a terminal value which is in good agreement with Atkinson's analysis of the steady motion of ledges. The steady-state analysis must therefore always underestimate the average velocity given the more rapid transients in the initial stages, although it must be recognised that there will be a similar transient in the motion of a flat interface.

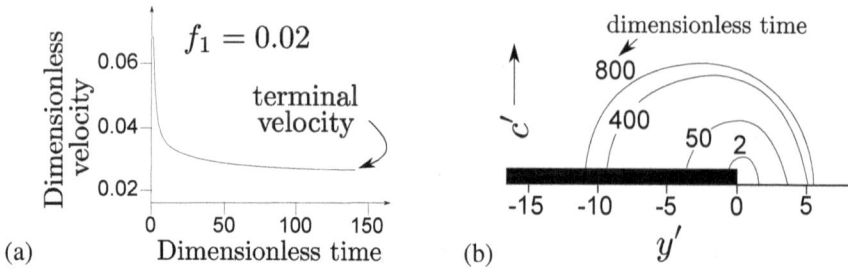

(a)

(b)

Figure 4.40 (a) Calculated variation in the dimensionless ledge velocity ($\rho v_s/D$) with dimensionless time (Dt/ρ^2), the velocity eventually tending towards a terminal value as the steady-state condition is approached. (b) An illustration of how a particular dimensionless-concentration ($[c - \bar{c}]/[c^{\gamma\alpha} - \bar{c}]$) contour changes with time as the steady-state condition is approached. After Enomoto [148].

The isolated ledge model can in principle be used to study infinite trains of ledges by using periodic boundary conditions in the finite difference model. This artificially constrains the interface to maintain its shape with the ledges equally spaced and to move at a constant speed even though there may not be a steady state in these circumstances. It appears that as $b_s \to 1$, the ratio v_l/v_d can indeed exceed unity, implying that the facet plane can in principle migrate more rapidly by a ledge mechanism (Figure 4.41). Enomoto's [149] numerical model for a finite set of ledges is more rigorous, allowing ledge spacings to vary as the trains move; it is found that when precipitate plates thicken by the motion of a small number of ledges it again is possible for the plates to thicken at a rate that is greater than v_d.

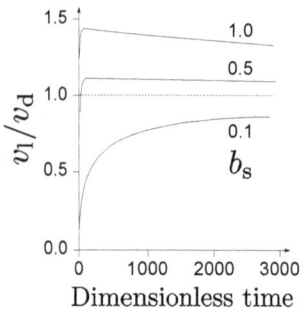

Figure 4.41 The ratio of the normal velocity of the sessile component of ledged interface to the corresponding velocity if a plane parallel to the sessile component were to move as a whole. The calculations are for an infinite train of ledges constrained to preserve the interface shape. b_s is the ratio of the ledge height to ledge spacing. Adapted from Enomoto [148].

An analytical treatment of the transient motion of ledges when $p \ll 1$ has been formulated as a generalisation of the steady-state theory [146]. One interesting outcome for a continuous distribution of steps is that the far-field solutions indicate parabolic growth kinetics even though on a finer scale the interface propagates by a step mechanism, although the parabolic rate constant is not identical to that for a flat interface.

It is worth emphasising that the theory for the growth of a particle by a ledge mechanism is worryingly incomplete. Its application must begin with an *assumed* interface shape in terms of the step heights and spacings. It is often argued that the ledges themselves move by the propagation of kinks (steps on the ledges) and there is theoretical work in this respect [146]. However, this level of sophistication hardly seems justified given that there is no method for calculating the interface topology.

4.11 SOLUTE DRAG: GRAIN BOUNDARIES

Small concentrations of solute can lead to dramatic changes in the mobility of grain boundaries in recrystallisation experiments. This is because the solute atoms associate with the moving grain boundary, causing a retarding effect known as *solute-drag*. There are no known circumstances in which solute/boundary interactions can enhance the mobility of the boundary. The first theory describing solute-drag was by Lücke and Detert [150], developed subsequently by Cahn [151] and others. Although the interest here is primarily with interphase boundaries, the theory is better established for grain boundaries which are consequently considered first.

The effect of the solute can be described in terms of a solute-boundary molar interaction energy G_s which can be negative or positive, depending on whether there is adsorption or desorption (respectively) of the impurity at the boundary:

$$c_b\{z\} = \bar{c}\,\exp\left\{-\frac{G_s\{z\}}{RT}\right\} \tag{4.87}$$

where c_b is the concentration (moles per unit volume) in the stationary boundary at a distance z from its central plane and $\bar{c}$ the average concentration. The continuum treatment that follows describes the essential details of the solute-drag phenomenon and is due to Cahn [151]. There are a number of other useful papers which reveal additional information; Lücke and Stüwe [152] proposed an approximate atomistic equivalent of the continuum model; Shirley [153] has extended the theory to concentrated solutions.

Cahn's model treats the boundary as a finite though inhomogeneous phase in which the equilibrium solute concentration varies with position. Given that drag effects are normally associated with dilute solutions, the activity may be replaced by concentration giving the chemical potential as

$$\mu = RT\ln\{c\{z\}V_m\} + G_s\{z\} + \text{constant} \tag{4.88}$$

where the constant includes μ° and is chosen such that the interaction energy is zero as $z \to \infty$; V_m is the molar volume. From Equation 2.93 and assuming that for a dilute solution the activity coefficient does not vary with concentration, the flux of solute atoms is

$$
\begin{aligned}
-J &= \frac{Dc}{RT}\frac{\partial\mu}{\partial z} \\
&= D\frac{\partial c}{\partial z} + \frac{Dc}{RT}\frac{\partial G_s}{\partial z}.
\end{aligned}
$$

On substituting this into Fick's second law ($\partial c/\partial t = -\partial J/\partial z$) and noting that the diffusion coefficient is also a function of position z:

$$
\begin{aligned}
\frac{\partial c}{\partial t} &= D\frac{\partial^2 c}{\partial z^2} + \frac{\partial D}{\partial z}\frac{\partial c}{\partial z} \\
&\quad + \left[\frac{D}{RT}\frac{\partial G_s}{\partial z}\right]\frac{\partial c}{\partial z} + \frac{c}{RT}\left[\frac{\partial D}{\partial z}\frac{\partial G_s}{\partial z} + D\frac{\partial^2 G_s}{\partial z^2}\right].
\end{aligned}
$$

If the concentration profile is to reach a steady state then this change in composition due to diffusion must be balanced by the change in concentration due to the motion of the interface at a velocity v (the usual conservation condition at the interface):

$$\frac{\partial c}{\partial t} = -v\frac{\partial c}{\partial z} \qquad v > 0$$

so for the steady state,

$$
\begin{aligned}
0 = \; & D\frac{\partial^2 c}{\partial z^2} + \left[\frac{\partial D}{\partial z} + \frac{D}{RT}\frac{\partial G_s}{\partial z} + v\right]\frac{\partial c}{\partial z} \\
& + \frac{c}{RT}\left[\frac{\partial D}{\partial z}\frac{\partial G_s}{\partial z} + D\frac{\partial^2 G_s}{\partial z^2}\right]
\end{aligned}
\tag{4.89}
$$

which describes the concentration profile for arbitrary G_s and D and has the solution [151]

$$
\begin{aligned}
c = \; & \bar{c}v\exp\left\{-\frac{G_s\{z\}}{RT} - v\int_{z_0}^{z}\frac{dz'}{D\{z'\}}\right\} \\
& \times \int_{-\infty}^{z}\exp\left\{\frac{G_s\{z''\}}{RT} + v\int_{z_0}^{z}\frac{dz'}{D\{z'\}}\right\}\frac{dz''}{D\{z'\}}
\end{aligned}
\tag{4.90}
$$

The form of this equation is such that the composition behind the moving interface is always $\bar{c}$ as if the interface had never traversed that region [151]. This must be so otherwise the motion of the grain boundary would lead to a non-equilibrium long-range redistribution of solute.

The force exerted by each excess solute atom on the boundary is $-N_a^{-1}(dG_s/dz)$, giving a total drag-stress

$$\sigma_d = -N_a\int_{-\infty}^{+\infty}(c - \bar{c})\frac{dG_s}{dz}dz.\tag{4.91}$$

The sense of the stress depends on that of the gradient dG_s/dz. If the concentration profile in the boundary is symmetrical, then the forces due to each side of the boundary cancel exactly giving zero drag. Cahn has obtained exact solutions Equation 4.90 for the segregation profiles illustrated in Figure 4.42, but has derived for the same profiles, a reasonable and simpler approximation to the drag stress is

$$\sigma_d = \frac{\phi_1 v\bar{c}}{1 + \phi_2^2 v^2}\tag{4.92}$$

$$\text{where}\quad \phi_1 = \frac{(RT)^2}{G_s\{0\}D}\left(\sinh\frac{G_s\{0\}}{RT} - \frac{G_s\{0\}}{RT}\right)$$

$$\phi_2^2 = \frac{\phi_1 RT\delta_b}{4G_s\{0\}^2 D}$$

where δ_b is the thickness of the boundary and $G_s\{0\}$ is the solute-boundary interaction energy at the centre of the boundary.

The drag theories either predict [151] or are designed [101, 154] so the solute concentration behind the interface during steady-state motion is always equal to the bulk value $\bar{c}_i$, as if the boundary did not exist (Figure 4.42). The solute concentration at $z - Z \ll -0.5\delta_b$ is thus always $\bar{c}_i$; as usual, Z defines the location of the boundary along z. For a stationary boundary, the concentration of solute differs from $\bar{c}_i$ within the region $(-\delta_b/2) < z - Z \ll \delta_b/2)$. For a moving boundary, the composition differs from $\bar{c}_i$ not only within the boundary, but also in front of it, irrespective of whether G_s is less than or greater than zero. The extent of penetration into the region beyond $z - Z = (\delta_b/2)$ depends on interface velocity amongst other factors.

The function $G_s\{z\}$ is quite general but for simplicity, it is assumed it to be a symmetrical wedge-shape with constant dG_s/dz on either side, changing sign at the minimum in the middle of the boundary. For such a function, Figure 4.42a shows how the concentration distribution changes with the velocity. It is the asymmetry of the solute distribution at the boundary that gives a drag stress. For the stationary boundary, the concentration profile is the expected symmetrical distribution, but desorption occurs everywhere, particularly at the leading edge, with some depletion in the region in front of the boundary. The drag due to these effects is shown in Figure 4.42b; at first it increases as the velocity tends towards $v\phi_2 = \frac{1}{2}$ because of the increasing deviation of the distribution of solute from the equilibrium state. It reaches a maximum when $v\phi_2 = 1$ and then falls gradually as the solute atoms fail to keep pace with the moving boundary.

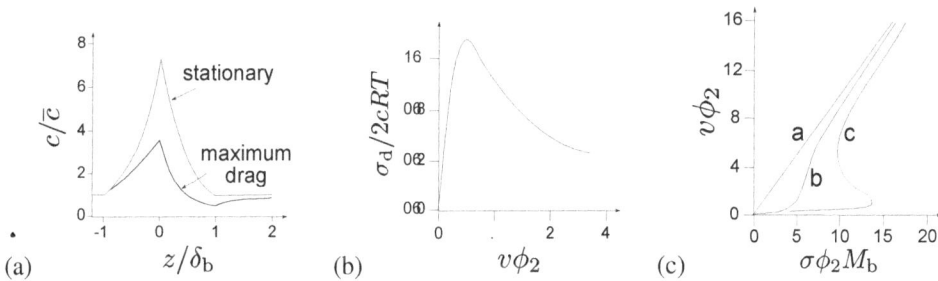

Figure 4.42 (a) Composition profiles associated with a stationary and moving boundary when there is a tendency for solute to segregate into the boundary. (b) The impurity drag as a function of velocity. (c) Plot of dimensionless velocity versus dimensionless driving force. (a) Linear relation for a pure material. (b) Corresponding material containing a small concentration of solute. There is a region where the velocity increases sharply with a small change in the concentration. (c) Corresponding material with a solute concentration which is large enough to cause the formation of an unstable branch which is shaded (after Cahn [151]).

For a pure material the grain boundary velocity is $v = M_b\Delta G$ where M_b is the grain boundary mobility and ΔG is the driving force (Section 4.8). This driving force could be written as a driving stress $\sigma_0 = V_m\Delta G$ so the net stress on a boundary moving in an impure material is $\sigma = \sigma_0 + \sigma_d$, which with Equation 4.92 becomes

$$\sigma = \frac{vV_m}{M_b} + \frac{\phi_1\bar{c}v}{1 + \phi_s^2 v^2}$$

Referring to Figure 4.42c, the velocity and stress are related linearly since the former is proportional to the driving force (curve a). The effect of the impurity is to cause a severe reduction in the velocity for a given driving force (curve b), though as expected, the effect diminishes at large velocities. There is a region of instability illustrated for curve c, where the boundary is able to break away and accelerate from the solute atmosphere.

4.11.1 SOLUTE DRAG AND DIFFUSION COEFFICIENTS

Solute-drag requires a segregation or desegregation of solute atoms to the interface - i.e., the concentration in the boundary differs from that in the bulk material. The interface itself is assumed to have a finite width δ_b, defined usually as the distance normal to the interface plane over which the solute-interface interaction free energy G_s is substantially non-zero for a stationary interface. The drag force on the boundary is obviously zero when segregation does not occur or when the composition profile due to the segregation is symmetrical with respect to the central plane of the interface. For a moving boundary, drag occurs when it becomes necessary to diffuse the solute atoms in the direction of boundary motion. One of the major difficulties in applying solute-drag theory to real problems is the suitable choice of a diffusion coefficient describing this process. A large diffusivity leads to a symmetrical distribution of solute in the boundary and consequently a zero drag force. Cahn's model assumed the diffusion coefficient to be a function of the distance from the centre of the boundary, approaching the value of the bulk diffusivity at sufficiently large distances from the interface, and a grain boundary value as the centre of the interface is approached. The diffusivity could therefore increase typically by a factor of 10^6 towards the core of the boundary [155], but this may exaggerate the real situation because grain boundary diffusion coefficients refer to transport along the boundary rather than across it. The mobility of atoms within an interface must in general be anisotropic, reflecting the nature of its defect structure. These problems are compounded because the theories generally treat the boundary width δ_b to be several interatomic distances. This is necessary because G_s may be non-zero over that range, but the diffusivity is expected to rapidly reach that of the defect-free lattice. It has also been proposed that the diffusion coefficient of a moving boundary may be larger than that of one which is stationary [156, 157]. It is not obvious how these complications can be resolved in practice.

4.11.2 INTERACTION FREE ENERGY

While both D and G_s can be expressed as functions of the distance $(z - Z)$, it usually is necessary to make simplifying assumptions about the forms of these functions. The way in which the drag force might vary with G_s has been considered by Hillert and Sundman [154]. For cases where G_s changes gently from zero (at $z - Z = \pm -0.5\delta_b$) to some other value within the boundary, the drag force goes through a maximum as the interface velocity increases. However, if G_s changes discontinuously from a constant value within the boundary to zero at $z - Z = \pm -0.5\delta_b$, then the drag stress never decreases with increasing velocity. As Hillert [101] pointed out, the former choice of G_s is probably more realistic, especially when the discrete nature of lattices is taken into account. Nevertheless, it is recognised that in the absence of detailed knowledge on solute/interface interactions, the choice of G_s must be somewhat uncertain. The form of G_s also determines the region of the boundary from which the main component of the drag force originates [151, 154]. The value of diffusivity in those particular regions would then control the drag effect. This complication may be minimised if D was always close to bulk diffusivity as might be the case for semi-coherent interfaces.

4.12 SOLUTE DRAG: INTERPHASE BOUNDARIES

The discussion of rate-controlling processes during interfacial motion was based on the dissipation of the free energy that is available to drive the interface (Section 4.5). Each of the dissipative processes is associated with a response function relating the dissipation to the velocity. Solute-drag clearly involves diffusion which dissipates free energy and hence can be treated within the same framework of rate controlling phenomena. This is the basis of Hillert's [101, 158] generalised model for solute-drag at boundaries and interphase interfaces.

The rate at which free energy is dissipated per unit volume is given by the product of the diffusion flux and the force driving the diffusion (Equation 2.89):

$$\frac{\mathrm{d}G}{\mathrm{d}t} = -V_{\mathrm{m}} \int_V J_{\mathrm{B}} \left(\frac{\partial \mu_{\mathrm{B}}}{\partial z} - \frac{\overline{V}_{\mathrm{B}}}{\overline{V}_{\mathrm{A}}} \frac{\partial \mu_{\mathrm{A}}}{\partial z} \right) \mathrm{d}V \qquad (4.93)$$

where $\overline{V}_i$ is the partial molar volume of component i. V is the volume in which the solute B is diffusing in a binary A-B alloy. With the usual notation, J_{B} is the flux defined with respect to the volume-fixed frame of reference and $\partial \mu_{\mathrm{A}}/\partial z$ represents the gradient of the chemical potential of the species A. It follows that the dissipation due to solute-drag as a unit area of interface sweeps a mole of material at a constant velocity, is:

$$G_{\mathrm{SD}} = -\frac{V_{\mathrm{m}}}{v} \int_{-\infty}^{+\infty} J_{\mathrm{B}} \left(\frac{\partial \mu_{\mathrm{B}}}{\partial z} - \frac{\overline{V}_{\mathrm{B}}}{\overline{V}_{\mathrm{A}}} \frac{\partial \mu_{\mathrm{A}}}{\partial z} \right) \mathrm{d}z.$$

This equation is for a constant interface velocity so the flux can be written as $J_{\mathrm{B}} = (x^{\gamma\alpha} - x^{\alpha\gamma}) v / V_{\mathrm{m}}$ when the theory is applied to a phase boundary between γ and α. The compositions are those in the matrix on either side of the interface and immediately adjacent to the interface.

There is a variety of other approximations to this equation. Notable amongst these is one in which the drag dissipation is taken to be that due to diffusion within the boundary only [159]. The dissipation itself will depend on specific assumptions about the diffusion coefficient and the interaction free energy function G_s. Hillert and Sundman [154] have shown that Cahn's results can be reproduced by relating the dissipation to the drag stress. The dissipation due to solute-drag must be accounted for when estimating terms such as G_1, the dissipation in the process of atom transfer across boundaries.

4.12.1 SEGREGATION TO α/γ INTERFACES

Interfaces are defects so it is expected that misfitting solute atoms may tend to segregate there. There are significant difficulties in characterising solute partitioning into the allotriomorphic ferrite-austenite boundary, particularly with respect to spatial resolution. Scanning transmission electron microscopy (STEM) data have been reported using a foil thicknesses of 200 nm, in which case the spatial resolution should not be less than this, making the results doubtful in terms of interface analysis. In some cases, deconvolution methods are used for data obtained using a particular STEM raster window scanning technique, but the extraction of the information requires a variety of assumptions that cloud the interpretation. Similarly, atom-probe data themselves have problems with spatial resolution, including small movements of atoms as they field evaporate, differences in the evaporation fields of adjacent phases leading to magnification and trajectory aberrations, larger atoms being pushed out preferentially just prior to evaporation, and surface roughness contributing to the spread [160, 161]. The data discussed here should be considered in view of these caveats.

Manganese segregation has been observed using STEM analysis at allotriomorphic-ferrite/austenite interfaces in a Fe-0.37C-3Mn-1.9Si wt% steel [162]. In some cases the monolayer coverage of the boundaries was very high, in the range 0.4-0.9, whereas in others it was much smaller at 0.07-0.27. It was speculated that in the latter case the boundaries were more coherent and less mobile, but there is no study relating the nature of the boundary to the extent of segregation. Similar data were obtained using the STEM raster window technique and simulation [163].

In atom-probe tomography of Fe-Mn-C and Fe-Mn-N, manganese was found to α/γ interfaces in the steel but not in the nitrogen alloy [164]. The thickness of the segregated region was of the order of 8 nm which is much greater than the expected width of the interfacial region. The concentration profile of the manganese about the deduced centre of the interface was symmetrical indicating a more or less stationary interface. All of these studies and others [165] address "incoherent interfaces" obtained in decarburisation or denitriding experiments. Incoherent boundaries in particular should not have long-range strain fields so it is difficult to explain the large widths of the concentration profiles that are supposed to represent segregation. When segregation is not detected it is sometimes attributed to the movement of the interface during quenching to room temperature.

4.13 MASSIVE FERRITE

A massive transformation is reconstructive in nature but with the product phase growing with the same bulk chemical composition as the parent [166]. Because of the diffusion inherent in the mechanism of transformation, the product phase is not limited by the grain boundaries of the parent phase. This ability of the product crystals to cross boundaries seems particularly pronounced in massive transformations which results in coarse α-grains the size of which may exceed that of the parent γ-phase (Figure 4.43).

As with normal allotriomorphic ferrite, there is an orientation relation with at least one of the parent crystals in contact with the product phase [168, 169]; the transformation interface is believed to be incoherent [170], consistent with the reconstructive transformation mechanism and with the fact that there is no shape change of the type expected with a displacive transformation. There is no overall change in the chemical composition but the transformation cannot occur without diffusion. Local variations in composition in the vicinity of the transformation interface have not been characterised. The reaction occurs at a rate that is interface-controlled, consistent with the observed constant growth rate [171]. The interface sometimes moves continuously, adopting a characteristic ragged contour (Figure 4.43), but there is evidence that it can migrate by a step mechanism [172].

In binary alloys, precipitation can occur without a composition change only if the temperature is below T_0 where the parent and product phases of the same chemical composition have identical free energies. The curve representing the T_0 temperature as a function of solute concentration lies within the two-phase field where the parent and product phases are in equilibrium. However, it is found experimentally that massive transformation seems to occur only when the parent phase is transformed

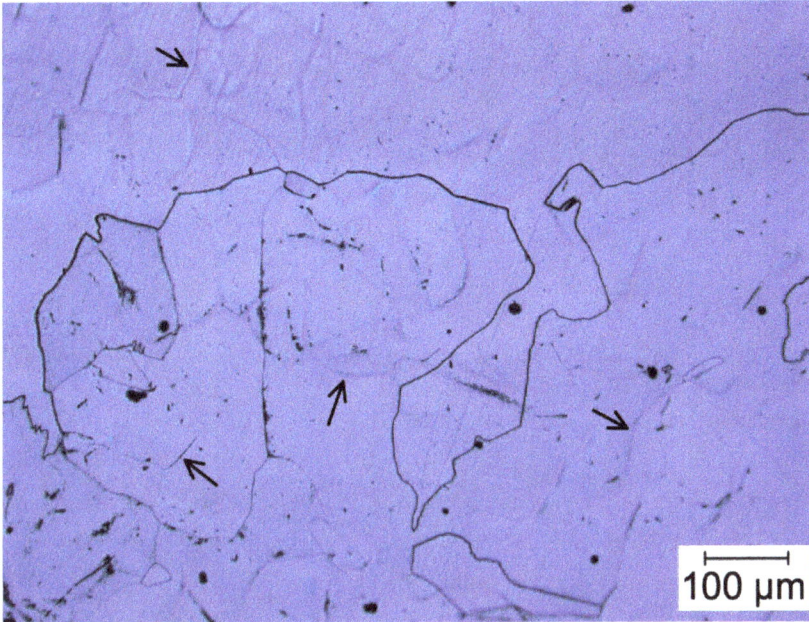

Figure 4.43   Massive ferrite in a Fe-P alloy [167]. The dark-etching lines are the massive ferrite grain boundaries whereas the thermal grooves represent the finer austenite grains (arrowed). Picture courtesy of Jeong In Kim.

at a temperature within the single-phase field where only the product phase is thermodynamically stable. This is because at small undercoolings below T_0, the massive transformation initiates at nuclei, the composition of which differs from that of the matrix. As a result, the nuclei become surrounded by a solute enriched or depleted zone. For the nucleus to develop into a massive phase, it has to be able to consume the excess solute and accelerate to the steady-state massive growth rate, a sequence that is feasible only when the parent phase is supercooled into the single-phase field representing the product [171].

Much of the work on the massive transformation in iron ($\gamma \rightarrow \alpha$) and its substitutional alloys is based on continuous cooling experiments because the rate of reaction tends to be too rapid to permit isothermal measurements. The cooling curve exhibits a thermal arrest at a temperature T_a due to the enthalpy of transformation.

In pure iron, massive ferrite can form as soon as it can nucleate under the conditions of the experiment. Gilbert and Owen [173] found that the reconstructive formation of ferrite is suppressed beyond a cooling rate of about $5500^\circ\text{C s}^{-1}$ the austenite eventually transforming into martensite at 545°C (although martensite formation was not verified by testing for surface relief effects). Some of their results are presented in Figure 4.44 which shows that T_a decreases sharply at first and then levels out at higher cooling rates. The values of T_a at a zero cooling rate represent transformation temperatures obtained from the relevant binary phase diagrams. It is evident that the influence of alloying elements on T_a depends on their corresponding effect on the equilibrium transformation temperature (i.e., on $\Delta G^{\alpha\gamma}$). Indeed, Gilbert and Owen found $\Delta G^{\alpha\gamma}$ at the plateau temperature to

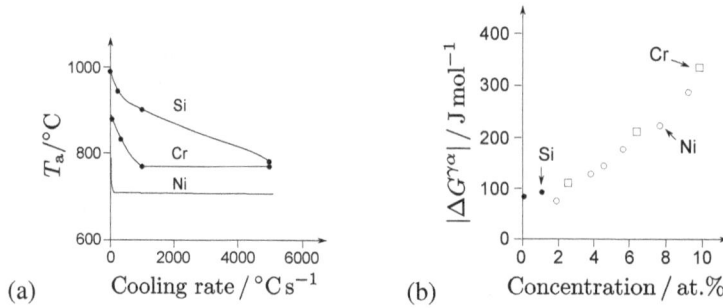

(a)   Cooling rate / $^\circ$C s^{-1} (b)   Concentration / at.%

Figure 4.44 (a) Results of continuous cooling experiments on Fe-X alloys (the Ni, Si and Cr alloys contain 2.7, 2.7 and 2.6 at.% of solute respectively); (b) Magnitude of the driving force $|\Delta G^{\gamma\alpha}|$ for massive reaction, evaluated at the plateau temperature (after Gilbert and Owen [173]).

be independent of the type of alloying element used (Figure 4.44b). This is not the whole explanation since $\Delta G^{\alpha\gamma}$ at the plateau temperature increases gently with alloy concentration (though not with alloy type).

The driving force required to initiate massive transformation is always much less than that necessary to induce martensitic reaction. In pure iron it has been demonstrated that the growth of martensite causes an invariant-plane strain shape deformation with a large shear component, which massive ferrite does not [174]. Nevertheless, there is a volume change associated with the transformation which is not entirely relieved by diffusion due to the rapid rate at which massive ferrite grows. As a result, the massive ferrite is left with a dislocation density that is less than that typically associated with martensite in low-alloy steels, but greater than that in allotriomorphic ferrite, Figure 4.45 [175].

Figure 4.45 Dislocation density in massive ferrite introduced in interstitial-free iron containing a small concentration of phosphorus, by the effect of laser welding where the cooling rates are large. The dislocation density measured to be $(1.2 \pm 0.4) \times 10^{14}\,\mathrm{m}^{-2}$; that in allotriomorphic ferrite in the same alloy contained an order of magnitude smaller dislocation density. Reproduced from open access article by Liu et al. [175].

4.14 INTERPHASE PRECIPITATION

Sometimes, a third minority-phase precipitates at the α/γ interface as the ferrite grows. The phase may be cementite, alloy carbides, Laves phase [176] or other particles such as those of copper [177] or gold [178] that have limited solubility in α-iron. There is no suggestion that the ferrite and the precipitate grow cooperatively as in the pearlite reaction. An example of interphase precipitation is illustrated in Figure 4.46, which shows $Cr_{23}C_6$ particles nucleated at the sessile part of the γ/α interface during ferrite growth by a step mechanism. The carbide particles seem to grow whilst in contact with the austenite, growth terminating after the carbides become enclosed in ferrite following the passage of a trailing step. The small concentration of carbon in the ferrite then prevents further carbide growth.

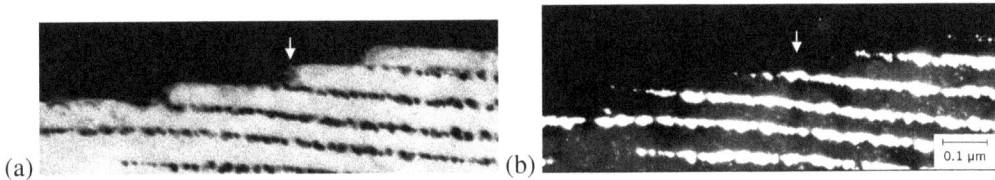

(a)   (b)

Figure 4.46 Bright-field and $Cr_{23}C_6$ corresponding dark-field transmission electron micrographs showing interphase carbide precipitation during the stepped growth of ferrite from austenite [179].

Interphase precipitation was recognised first in alloy steels containing ternary additions of strong carbide-forming elements [180, 181]. Fine dispersions of alloy carbides were observed as regular rows of particles, all of which usually have the same crystallographic orientation in any given ferrite grain. Electron microscopy of partially transformed specimens revealed that the rows of carbides are actually parts of sheets of carbides in three dimensions, the carbides nucleating at γ/α interfaces during transformation [182]. The rows only become apparent during transmission electron microscopy (using thin foil specimens) if the planes on which the carbides precipitate are virtually parallel to the beam direction [182], particularly if the sheet spacings are small compared with the foil thickness.

Interphase precipitation is for the most part associated with the step mechanism of α/γ interface motion when the carbides precipitate on the stationary component of the interface [179]. The steps themselves move too rapidly to allow successful precipitate nuclei to develop [183, 184].

Because the precipitates nucleate in contact with both the ferrite and austenite lattices, they tend to adopt a crystallographic orientation which allows good lattice matching with either phase. This restricts the number of crystallographic variants of carbide that can form [185].

Interphase precipitation is also found in cases where the α/γ boundary does not move by a rigid-step mechanism, but is displaced continuously [186, 187]. The precipitates pin the α/γ interface. The resulting precipitate dispersions in the ferrite can be random, or in the form of regular, non-planar sheets of carbides. Where random dispersions of carbides are formed, the α/γ boundary migrates by bowing in between coarsely spaced carbide particles. When the precipitation at the α/γ interface is so copious that interface bowing becomes difficult, a "quasi-ledge" mechanism operates as illustrated in Figure 4.47. The curved α/γ interface becomes pinned by the finely spaced particles, but at some position where the particle spacing is locally large, an interface bulge develops and subsequently becomes pinned but is able to spread laterally, giving in effect a ledge mechanism even though the interface energy may not be orientation dependent.

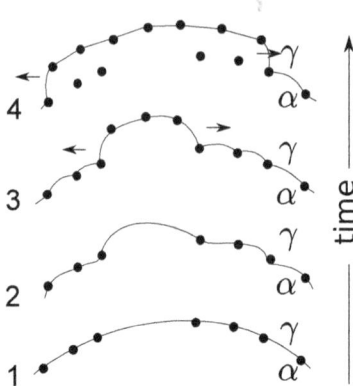

Figure 4.47 Schematic diagram showing the operation of a quasi-ledge mechanism resulting in the production of curved rows of interphase precipitates. Non-uniformly spaced particles first pin the boundary. The boundary then develops a bulge where particle spacing is locally large, only to be pinned again by further precipitation, so the bulge is only free to move laterally, giving a quasi-ledge mechanism of growth. Adapted from Ricks and Howell [187].

4.14.1 INTERPHASE PRECIPITATION: THEORY

The primary goal of any theory of interphase precipitation would be to estimate the precipitate size (r), precipitate spacing on the plane parallel to the sessile component of the ledge d, and the ledge height ρ which also is the spacing between sheets of particles. Given a volume fraction V_V^p of precipitates assumed to be spherical, it follows that [188]

$$V_V^p \propto \frac{r^3}{\rho d^2} \qquad \text{so that} \qquad r \propto (V_V^p \rho d^2)^{\frac{1}{3}}. \tag{4.94}$$

This relationship simply states that there is a fixed volume fraction of precipitate so its size must scale appropriately with the segment of ferrite associated with one precipitate particle, with considerable experimental data that support this trend [188]. It could also be argued that if the nucleation rate of precipitates at the γ/α interface is large then the precipitates will necessarily be small.

Interphase precipitation can occur only when austenite is supercooled into the equilibrium phase-domain where both ferrite and the precipitate can be stable. In typical steels designed for this microstructure, the process must begin with the formation of the majority phase ferrite. The associated local enrichment of the adjacent austenite with carbon induces the precipitate to nucleate.

There are three possible factors that control ledge height during interphase precipitation: the driving force for transformation, diffusion coefficients, particle pinning of the interface and processes during ledge nucleation.

Figure 4.48 shows that ρ increases with temperature, the increase being at a greater rate at higher temperatures. The steel containing the lower carbon and vanadium concentrations has relatively large values of the ledge height even when a comparison is made at the same transformation temperature. Both of these trends indicate that a larger magnitude of the driving force for ferrite formation leads to coarser ledge heights at a constant temperature. So in cases where the stepped motion of the interface is associated with precipitation, it is likely that particle pinning limits the size of steps;

in the comparison of the steels in Figure 4.48, the alloy with the greater V and C concentrations has a much finer microstructure.

Nucleation control of ρ can be ruled out because Equation 4.69 indicates a smaller critical size ρ^* with increasing $|\Delta G^{\alpha\gamma}|$. Diffusion during growth must play a role but given the differences observed at a constant temperature it cannot explain the data consistently, and the dependence of diffusion on temperature cannot explain why the two steels transformed at the same temperature have a different ρ. It is likely therefore that $|\Delta G^{\alpha\gamma}|$ and the tendency for particle pinning determines ledge height during interphase precipitation. In the latter context, Davenport and Honeycombe [182] first proposed that precipitates pin the γ/α interfaces with local breakaway leading to the formation of the migrating steps. A number of mathematical models have been expressed for interphase precipitation

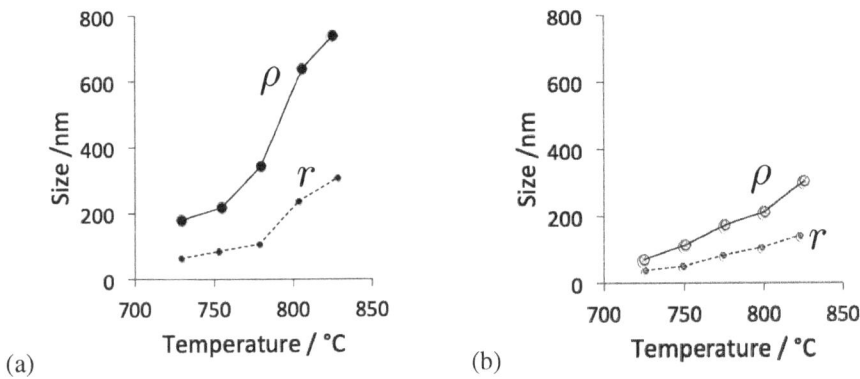

(a)   (b)

Figure 4.48   Selected data for interphase precipitation in vanadium-containing steels, as a function of the isothermal transformation temperature [189]. ρ represents the experimentally determined sheet spacing (ledge height) and r the vanadium carbide precipitate size. (a) Fe-0.09C-0.48V-0.016Nb wt%. (b) Fe-0.20C-1.04V-0.023Nb wt%.

[190–193].

The microstructure consists of arrays of particles, each of size r with intervening ferrite, separated by precipitate-free layers of ferrite of thickness ρ. The first model due to Todd and co-workers, treated the arrays as plates of average composition $x^p V_V^p + x^\alpha (1 - V_V^p)$ so the nucleation of individual particles is neglected [190, 191]; V_V^p is the equilibrium fraction of precipitate. The plates and intervening ferrite then grow together into the austenite at a common front, rather like a eutectoid reaction in a binary system where the far-field composition of the austenite is maintained constant. The growth velocity of the composite is therefore constant with the theory formulated on the basis of the diffusion of the substitutional solute alone (e.g., vanadium in the case of vanadium carbide) and so does not satisfy local equilibrium at the interface. Rios has formulated a somewhat simpler model but following similar assumptions [192].

Another effort treats the condition for a carbide particle to form at the γ/α interface as a critical driving force determined experimentally, which then is substituted into the classical nucleation rate equation scaled by a constant critical concentration for nucleation [194]. Diffusion through the γ/α interface is included in the analysis and an interesting part of the model is that ledges are generated by bowing of the interface between carbide particles as illustrated in Figure 4.47 – this implies that it is not the orientation dependence of interfacial energy that determines the step mechanism since the facet between the steps must be mobile. This model was further adapted to account for solute segregation at the interface and the resulting solute drag [193].

All of these models require fitting parameters, but none is capable of predicting the interface topology in terms of the combination of critical parameters: ledge height and ledge spacing, without appealing to experimental data. It is emphasised that a stepped mechanism of growth can occur between austenite and ferrite in the absence of interphase precipitation, when the interfacial energy is sufficiently orientation dependent.

REFERENCES

1. C. A. Dubé: 'The decomposition of austentite in hypoeutectoid steels': Ph.D. thesis, Carnegie Institute of Technology, Pittsburgh, USA, 1947.

2. H. I. Aaronson: 'Effects of nucleation site upon precipitate morphology', In: *The Mechanism of Phase Transformations in Metals*. London, U.K.: Institute of Metals, 1955:47–56.

3. C. A. Dubé, H. I. Aaronson, and R. F. Mehl: 'La formation de la ferrite proeutectoide dans les aciers au carbonne', *Revue de Metallurgie*, 1958, **55**, 201–210.

4. J. W. Christian: Theory of Transformations in Metals and Alloys, Part I: 2 ed., Oxford, U. K.: Pergamon Press, 1975.

5. J. W. Cahn, and G. Kalonji: 'Symmetry in solid state transformation morphologies', In: H. I. Aaronson, D. E. Laughlin, R. F. Sekerka, and C. M. Wayman, eds. *Solid-Solid Phase Transformations*. Pennsylvania, USA: The Metallurgical Society of AIME, 1981:3–14.

6. J. W. Cahn: 'Theory of crystal growth and interface motion in crystalline materials', *Acta Metallurgica*, 1960, **8**, 554–562.

7. J. W. Christian: 'Military transformations: an introductory survey', In: *Physical Properties of Martensite and Bainite, Special Report 93*. London, U.K.: Iron and Steel Institute, 1965:1–19.

8. J. W. Christian: 'Martensitic transformations: a current assessment', In: *The Mechanism of Phase Transformations in Crystalline Solids, Monography 33*. London, U.K.: Institute of Metals, 1969:129–142.

9. J. W. Christian, and A. G. Crocker: Dislocations in Solids, ed. F. R. N. Nabarro, vol. 3, chap. 11: Amsterdam, Holland: North Holland, 1980:165–252.

10. H. K. D. H. Bhadeshia: 'Bainite: Mobility of the transformation interface', *Journal de Physique: Colloque C4*, 1982, **43**, 449–454.

11. J. W. Christian, and D. V. Edmonds: 'The bainite transformation', In: A. R. Marder, and J. I. Goldstein, eds. *Phase Transformations in Ferrous Alloys*. Warrendale, Pennsylvania, USA: TMS-AIME, 1984:293–327.

12. P. L. Ryder, and W. Pitsch: 'The crystallographic analysis of grain boundary precipitation', *Acta Metallurgica*, 1966, **14**, 1437–1448.

13. P. L. Ryder, W. Pitsch, and R. F. Mehl: 'Crystallography of the precipitation of ferrite on austenite grain boundaries in a Co-20%Fe alloy', *Acta Metallurgica*, 1967, **15**, 1431–1440.

14. W. C. Johnson, C. L. White, P. E. Marth, P. K. Ruf, S. M. Tuominen, K. D. Wade, K. C. Russell, and H. I. Aaronson: 'Influence of crystallography on aspects of solid-solid nucleation theory', *Metallurgical & Materials Transactions A*, 1975, **6**, 911–919.

15. W. Pitsch, and P. L. Ryder: 'Generation of an orientation relationship during nucleation in precipitation processes', *Scripta Metallurgica*, 1977, **11**, 431–434.

16. D. Phelan, and R. Dippenaar: 'Widmanstätten ferrite plate formation in low-carbon steels', *Metallurgical & Materials Transactions A*, 2004, **35**, 3701–3706.

17. G. V. Kurdjumov, and G. Sachs: 'Über den mechanismus der stahlhärtung', *Zietschrift für Physik*, 1930, **64**, 325–343.

18. Z. Nishiyama: 'X-ray investigation of the mechanism of the transformation from face centered cubic lattice to body centered cubic', *Science Reports of Tohoku Imperial University*, 1934, **23**, 637–634.

19. G. Wassermann: 'Einflusse der α-γ-umwandlung eines irreversiblen nickelstahls auf kristallorientierung und zugfestigkeit', *Archiv fur das Eisenhüttenwesen*, 1933, **6**, 347–351.

20. H. I. Aaronson, D. W. Dooley, and K. C. Russell: 'Examination of a mechanism for the nucleation by shear of grain boundary allotriomorphs', *Scripta Metallurgica*, 1976, **10**, 607–610.

21. A. D. King, and T. Bell: 'Crystallography of grain boundary proeutectoid ferrite', *Metallurgical Transactions A*, 1975, **6**, 1419–1429.

22. Y. T. Chi, Y. T. Tsai, B. M. Huang, and J. R. Yang: 'Investigation of idiomorphic ferrite and allotriomorphic ferrite using electron backscatter diffraction technique', *Materials Science and Technology*, 2017, **33**, 537–545.

23. D. W. Kim, D. W. Suh, R. S. Qin, and H. K. D. H. Bhadeshia: 'Dual orientation and variant selection during diffusional transformation of austenite to allotriomorphic ferrite', *Journal of Material Science*, 2010, **45**, 4126–4132.

24. S. S. Babu, and H. K. D. H. Bhadeshia: 'Direct study of grain boundary allotriomorphic ferrite crystallography', *Materials Science & Engineering A*, 1991, **142**, 209–220.

25. C. S. Smith: 'Grains, phases and interfaces: an interpretation of microstructure', *Metals Technology*, 1948, **15**, 1–37.

26. A. Crosky, P. G. McDougall, and J. S. Bowles: 'The crystallography of the precipitation of alpha rods from beta Cu Zn alloys', *Acta Metallurgica*, 1980, **28**, 1495–1504.

27. J. M. Rigsbee, and H. I. Aaronson: 'Interfacial structure of the broad faces of ferrite plates', *Acta Metallurgica*, 1979, **27**, 365–376.

28. P. R. Howell, P. D. Southwick, R. C. Ecob, and R. A. Ricks: 'The observation of dislocation arrays in partially coherent f.c.c./b.c.c. interfaces in duplex stainless steel', In: H. I. Aaronson, D. E. Laughlin, R. F. Sekerka, and C. M. Wayman, eds. *Solid-Solid Phase Transformations*. Pennsylvania, USA: TMS-AIME, 1981:591–598.

29. K. F. Kelton, and A. L. Greer: Nucleation in Condensed Matter: Applications in Materials and Biology: Oxford, U. K.: Pergamon Press, 2010.

30. W. F. Lange III, M. Enomoto, and H. I. Aaronson: 'Kinetics of ferrite nucleation at austenite grain boundaries in Fe-C alloys', *Metallurgical Transactions A*, 1988, **19**, 427–440.

31. W. Thomson: 'On the division of space with minimum partitional area', *Acta mathematica*, 1887, **11**, 121–134.

32. E. E. Underwood: Quantitative Stereology, chap. 4: Addison-Wesley Publication Company, 1970:90–93.

33. E. J. Myers: 'Section of polyhedrons', In: *First International Congress for Stereology*. 1963:118.

34. S. B. Singh, and H. K. D. H. Bhadeshia: 'Topology of grain deformation', *Materials Science and Technology*, 1998, **15**, 832–834.

35. J. Y. Chae, R. S. Qin, and H. K. D. H. Bhadeshia: 'Topology of the deformation of a non-uniform grain structure', *ISIJ International*, 2009, **49**, 115–118.

36. Inductiveload - Own work, Public Domain: 'Creation of a truncated octahedron': 2020: URL https://commons.wikimedia.org/w/index.php?curid=4010118.

37. D. Turnbull: 'Phase changes', *Solid State Physics*, 1956, **3**, 225–306.

38. K. C. Russell: 'Grain boundary nucleation kinetics', *Acta Metallurgica*, 1969, **17**, 1123–1132.

39. H. K. D. H. Bhadeshia: 'A thermodynamic analysis of isothermal transformation diagrams', *Metal Science*, 1982, **16**, 159–165.

40. A. M. Begley, D. Tian, F. Jona, and P. M. Marcus: 'Study of ultrathin Fe films on Pd{111}, Ag{111} and Al{111}', *Surface Science*, 1993, **2880**, 289–297.

41. D. Tian, H. Li, F. Jona, and P. M. Marcus: 'Study of the growth of Fe on Ru(0001) by low-energy electron diffraction', *Solid-State Communications*, 1991, **80**, 783–787.

42. S. D. Bader, and E. R. Moog: 'Magnetic properties of novel epitaxial films', *Journal of Applied Physics*, 1987, **61**, 3729–3734.

43. R. A. Ricks, G. S. Barritte, and P. R. Howell: 'The influence of second phase particles on diffusional phase transformations in steels', In: H. I. Aaronson, D. E. Laughlin, M. Sekerka, and C. M. Wayman, eds. *Solid→Solid Phase Transformations*. Warrendale, Pennsylvania, USA: TMS-AIME, 1982:463–468.

44. W. Mu, P. G. Jönsson, and K. Nakajima: 'Prediction of intragranular ferrite nucleation from TiO, TiN, and VN inclusion', *Journal of Materials Science*, 2016, **51**, 2168–2180.

45. B. L. Bramfitt: 'The effect of carbide and nitride additions on the heterogeneous nucleation behaviour of liquid iron', *Metallurgical Transactions*, 1970, **1**, 1987–1995.

46. A. R. Mills, G. Thewlis, and J. A. Whiteman: 'Nature of inclusions in steel weld metals and their influence on the formation of acicular ferrite', *Materials Science and Technology*, 1987, **3**, 1051–1061.

47. T. Koseki, and G. Thewlis: 'Inclusions assisted microstructure control in C-Mn and low alloy steel welds', *Materials Science and Technology*, 2005, **21**, 867–879.

48. C. van der Eijk, J. Walmsley, and Ø. Grong: 'Effects of titanium containing oxide inclusions on steel weldability', In: S. A. David, J. Vitek, J. Lippold, and H. Smartt, eds. *Trends in Welding Research*. Materials Park, Ohio, USA: ASM International, 2002:730–735.

49. M. Tsutsumi, H. Kato, and T. Koseki: 'Investigation of factors affecting ferrite transformation from steel-oxide interface', In: S. A. David, T. DebRoy, J. C. Lippold, H. B. Smartt, and

J. M. Vitek, eds. *Trends in Welding Research*. Materials Park, Ohio, USA: ASM International, 2005:987–992.

50. M. Strangwood, and H. K. D. H. Bhadeshia: 'Nucleation of ferrite at ceramic/steel interfaces', In: G. W. Lorimer, ed. *Phase Transformations '87*. London, U.K.: Institute of Metals, 1988:471–474.

51. M. Gregg, and H. K. D. H. Bhadeshia: 'Bainite nucleation from mineral surfaces', *Acta Metallurgica et Materialia*, 1994, **43**, 3321–3330.

52. M. Gregg, and H. K. D. H. Bhadeshia: 'Titanium-rich mineral phases and the nucleation of bainite', *Metallurgical & Materials Transactions A*, 1994, **25**, 1603–1610.

53. C. H. Lee, S. Nambu, J. Inoue, and T. Koseki: 'Ferrite formation behavior from non-metallic compounds in steels', In: S. Babu, H. K. D. H. Bhadeshia, C. E. Cross, S. A. David, T. DebRoy, J. N. DuPont, T. Koseki, and S. Liu, eds. *Trends in Welding Research*. Ohio, USA: ASM International, 2013:43–47.

54. J. H. Shim, Y. J. Oh, J. Y. Suh, Y. W. Cho, J. D. Shim, J. S. Byun, and D. N. Lee: 'Ferrite nucleation potency of non-metallic inclusions in medium carbon steels', *Acta Materialia*, 2001, **49**, 2115–2122.

55. Y. Kang, S. Jeong, J. H. Kang, and C. Lee: 'Factors affecting the inclusion potency for acicular ferrite nucleation in high-strength steel welds', *Metallurgical & Materials Transactions A*, 2016, **46**, 2842–2854.

56. B. Wang, X. Liu, and G. Wang: 'Inclusion characteristics and acicular ferrite nucleation in Ti-containing weld metals of X80 pipeline steel', *Metallurgical & Materials Transactions A*, 2018, **49**, Metallurgical & Materials Transactions A, 49, 2124–2138.

57. J. H. Shim, Y. J. Oh, J. Y. Suh, Y. W. Cho, J. D. Shim, J. S. Byun, and D. N. Lee: 'Hot deformation and acicular ferrite microstructure in C-Mn steel containing Ti_2O_3 inclusions', *ISIJ International*, 2000, **40**, 819–823.

58. X. Zhuo, Z. Wang, W. Wang, and H. G. Lee: 'Thermodynamic calculations and experiments on inclusions to be nucleation sites for intragranular ferrite in Si-Mn-Ti deoxidized steel', *Journal of the Univesity of Science and Technology Beijing*, 2007, **14**, 14–21.

59. L. S. Darken, and R. W. Gurry: Physical Chemistry of Metals: New York, USA: McGraw-Hill, 1953.

60. G. R. Purdy, and J. S. Kirkaldy: 'Kinetics of the proeutectoid ferrite reaction at an incoherent interface, as determined by a diffusion couple', *TMS-AIME*, 1963, **227**, 1255–1256.

61. C. Atkinson: 'Concentration dependence of diffusivity in the growth or dissolution of precipitates', *Acta Metallurgica*, 1967, **15**, 1207–1211.

62. J. R. Philip: 'General method of exact solution of the concentration-dependent diffusion equation', *Australian Journal of Physics*, 1960, **13**, 1–12.

63. J. R. Philip: 'The function inverfc theta', *Australian Journal of Physics*, 1960, **13**, 13–20.

64. C. Zener: 'Theory of growth of spherical precipitates from solid solution', *Journal of Applied Physics*, 1949, **20**, 950–953.

65. J. R. Philip: 'Theory of infiltration', *Advances in Hydroscience*, 1969, **5**, 215–305.

66. J. Crank: The Mathematics of Diffusion: 2nd ed., Oxford, U. K.: Clarendon Press, 1975.

67. C. Atkinson: 'The numerical solution of problems of planar growth where the diffusion coefficient is concentration dependent', *Acta Metallurgica*, 1968, **16**, 1019–1022.

68. C. Atkinson: 'A numerical method to describe the diffusion-controlled growth of particles when the diffusion coefficient is composition-dependent', *Transactions of the Metallurgical Society of AIME*, 1969, **245**, 801–806.

69. G. Horvay, and J. W. Cahn: 'Dendritic and spheroidal growth', *Acta Metallurgica*, 1961, **9**, 695–705.

70. J. R. Bradley, J. M. Rigsbee, and H. I. Aaronson: 'Growth kinetics of grain boundary ferrite allot. in Fe-C alloys', *Metallurgical Transactions A*, 1977, **8**, 323–333.

71. R. F. Mehl, and C. A. Dubé: 'The eutectoid reaction', In: J. E. Mayer, R. Smouluschowski, and W. A. Weyl, eds. *Phase Transformations in Solids*. New York, USA: John Wiley & Sons, Inc., 1951:545–582.

72. H. I. Aaronson, C. Laird, and K. R. Kinsman: 'Mechanisms of diffusional growth of precipitate crystals', In: *Phase Transformations*. Metals Park, Ohio, USA: ASM, 1970:313–396.

73. K. R. Kinsman, and H. I. Aaronson: 'Influence of Al, Co, and Si upon the kinetics of the proeutectoid ferrite reaction', *Metallurgical and Materials Transactions B*, 1973, **4**, 959–967.

74. C. Atkinson, H. B. Aaron, K. R. Kinsman, and H. I. Aaronson: 'On the growth kinetics of grain boundary ferrite allotriomorphs', *Metallurgical Transactions*, 1973, **4**, 783–792.

75. R. A. Vandermeer, C. L. Vold, and W. E. King Jr.: 'Modelling reconstructive growth during austenite decomposition to ferrite in polycrystalline Fe-C alloys', In: J. Vitek, and S. A. David, eds. *Advances in the Science & Technology of Welding*. Ohio, USA: ASM International, 1989:223–227.

76. R. Kobayashi: 'Modeling and numerical simulations of dendritic crystal growth', *Physica D*, 1993, **63**, 410–423.

77. M. Ode, S. G. Kim, and T. Suzuki: 'Recent advances in the phase-field model for solidification', *ISIJ International*, 2001, **4**, 1076–1082.

78. L.-Q. Chen: 'Phase-field models for microstructure evolution', *Annual Reviews in Materials Science*, 2002, **32**, 113–140.

79. W. J. Boettinger, J. A. Warren, C. Beckermann, and A. Karma: 'Phase-field simulation of solidification', *Annual Review of Materials Research*, 2002, **32**, 163–194.

80. D. H. Yeon, P. R. Cha, and J. K. Yoon: 'A phase field study for ferrite-austenite transitions under para-equilibrium', *Scripta Materialia*, 2001, **45**, 661–668.

81. A. Bhattacharya, C. S. Upadhyay, and S. Sangal: 'Phase-field model for mixed-mode of growth applied to austenite to ferrite transformation', *Metallurgical & Materials Transactions A*, 2015, **46**, 926–936.

82. P. W. Anderson: 'Local moments and localized states': Nobel Lecture, Bell Telephone Laboratories, Princeton University, 1977.

83. R. S. Qin, and H. K. D. H. Bhadeshia: 'Phase field method', *Materials Science and Technology*, 2010, **26**, 803–811.

84. H. K. D. H. Bhadeshia: Bainite in Steels: Theory and Practice: 3rd ed., Leeds, U.K.: Maney Publishing, 2015.

85. D. E. Coates: 'Diffusional growth limitation and hardenability', *Metallurgical Transactions*, 1973, **4**, 2313–2325.

86. M. Hillert: 'Paraequilibrium': Technical Report, Swedish Institute for Metals Research, Stockholm, Sweden, 1953.

87. J. S. Kirkaldy: 'Diffusion in multicomponent metallic systems I – phenomenological theory for substitutional solid solution alloys', *Canadian Journal of Physics*, 1958, **36**, 899–925.

88. G. R. Purdy, D. H. Weichert, and J. S. Kirkaldy: 'The growth of proeutectoid ferrite in ternary Fe-C-Mn austenites', *Trans. TMS-AIME*, 1964, **230**, 1025–1034.

89. J. Fridberg, L.-E. Torndähl, and M. Hillert: 'Diffusion in iron', *Jernkontorets Annaler*, 1969, **153**, 263–276.

90. D. E. Coates: 'Diffusion controlled precipitate growth in ternary systems II', *Metallurgical Transactions*, 1973, **4**, 1077–1086.

91. D. E. Coates: 'Precipitate growth kinetics for Fe-C-X alloys', *Metallurgical Transactions*, 1973, **4**, 395–396.

92. D. E. Coates: 'Diffusion controlled precipitate growth in ternary systems I', *Metallurgical Transactions*, 1972, **3**, 1203–1212.

93. M. Abramowitz, and A. B. Stegan, eds: Handbook of Mathematical Functions: Washington USA: National Bureau of Standards, 1964.

94. F. C. Frank: 'Radially symmetric phase growth controlled by diffusion', *Proceedings of the Royal Society of London A*, 1950, **201**, 586–589.

95. R. Trivedi, and G. M. Pound: 'Effect of concentration dependent diffusion coefficient on the migration of interphase boundaries', *Journal of Applied Physics*, 1967, **38**, 3569–3576.

96. G. Bolze, D. E. Coates, and J. S. Kirkaldy: 'Some solutions to the ternary diffusion equations with variable off-diagonal diffusion coefficients', *Transactions ASM Quarterly*, 1969, **62**, 794–803.

97. R. A. Tanzilli, and R. W. Heckel: 'Numerical solutions to finite diffusion-controlled 2-phase moving-interface problem (with planar cylindrical and spherical interfaces)', *Transactions of the Metallurgical Society of AIME*, 1968, **242**, 2313–2321.

98. E. Randich, and J. I. Goldstein: 'Non-isothermal finite diffusion-controlled growth in ternary systems', *Metallurgical Transactions A*, 1975, **6**, 1553–1560.

99. J. I. Goldstein, and E. Randich: 'Variation of interface compositions during diffusion controlled precipitate growth in ternary systems', *Metallurgical Transactions A*, 1977, **8**, 105–109.

100. J. B. Gilmour, G. R. Purdy, and J. S. Kirkaldy: 'Partition of manganese during the proeutectoid ferrite transformation in steel', *Metallurgical Transactions*, 1972, **3**, 3213–3222.

101. M. Hillert: 'The role of interfaces in phase transformations', In: *Mechanism of Phase Transformations in Crystalline Solids*. Monograph and Report Series No. 33, London, U.K.: Institute of Metals, 1970:231–247.

102. M. Gouné, F. Danoix, J. Ågren, Y. Bréchet, C. R. Hutchinson, M. Militzer, and H. Zurob: 'Overview of the current issues in austenite to ferrite transformation and the role of migrating interfaces therein for low alloyed steels', *Materials Science & Engineering R*, 2015, **92**, 1–38.

103. H. K. D. H. Bhadeshia: 'Some difficulties in the theory of diffusion-controlled growth in substitutionally alloyed steels', *Current Opinion in Solid State and Materials Science*, 2016, **20**, 396–400.

104. J. E. Hilliard: 'Spinodal decomposition', In: V. F. Zackay, and H. I. Aaronson, eds. *Phase Transformations*. Metals Park, Ohio, USA: ASM International, 1970:497–560.

105. J. W. Cahn: 'Spinodal decomposition', *Transactions of the Metallurgical Society of AIME*, 1968, **242**, 166–179.

106. H. E. Cook, and D. de Fontaine: 'On the elastic free energy of solid solutions - I - microscopic theory', *Acta Metallurgica*, 1969, **17**, 915–924.

107. T. Ujihara, and K. Osamura: 'Kinetic analysis of spinodal decomposition process in Fe–Cr alloys by small angle neutron scattering', *Acta Materialia*, 2000, **48**, 1629–1637.

108. T. Ujihara, and K. Osamura: 'Assessing composition gradient energy effects due to spin interaction on the spinodal decomposition of Fe–Cr', *Materials Science & Engineering A*, 2001, **312**, 128–135.

109. G. Zhang, R. Wei, M. Enomoto, and D. W. Suh: 'Growth kinetics of proeutectoid ferrite in Fe-0.1C-1.5Mn-1Si quaternary and Fe-0.1C-1.5Si-0.2Al quinary alloys', *Metallurgical & Materials Transactions A*, 2012, **43**, 833–842.

110. L. S. Darken: 'Diffusion of carbon in austenite with a discontinuity of composition', *Transactions of the AIME*, 1949, **180**, 430–438.

111. J. C. Baker, and J. W. Cahn: 'Thermodynamics of solidification', In: *Solidification*. Metals Park, Ohio, USA: ASM, 1971:23–58.

112. J. W. Cahn: 'Obtaining inferences about relative stability and metastable phase sequences from phase diagrams', *Bulletin of Alloy Phase Diagrams*, 1980, **1**, 27–33.

113. A. Hultgren: 'Isotherm omvandling av austenit', *Jernkontorets Annaler*, 1951, **135**, 403–494.

114. M. Hillert: 'Använding av isoaktiva linjer i ternära till-stånds-diagram', *Jernkontorets Annaler*, 1952, **136**, 25–37.

115. E. Rudberg: 'Avbildning i rymden av fria energien för ternära system', *Jernkontorets Annaler*, 1952, **136**, 91–112.

116. H. I. Aaronson, H. A. Domian, and G. M. Pound: 'Thermodynamics of the austenite – proeutectoid ferrite transformation II, Fe-C-X alloys', *TMS-AIME*, 1966, **236**, 768–780.

117. J. B. Gilmour, G. R. Purdy, and J. S. Kirkaldy: 'Thermodynamics controlling the proeutectoid ferrite transformations in Fe-C-Mn alloys', *Metallurgical Transactions*, 1972, **3**, 1455–1469.

118. M. Enomoto, and H. I. Aaronson: 'Derivation of general conditions for paraequilibrium in multicomponent systems', *Scripta Metallurgica*, 1985, **19**, 1–3.

119. H. K. D. H. Bhadeshia: 'Ferrite formation in heterogeneous dual-phase steels', *Scripta Metallurgica*, 1983, **17**, 857–860.

120. J. C. Baker, and J. W. Cahn: 'Solute trapping by rapid solidification', *Acta Metallurgica*, 1969, **17**, 575–578.

121. H. K. D. H. Bhadeshia: 'Some unresolved issues in phase transformation theory: the role of microanalysis', *Journal de Physique Colloque*, 1989, **50**, C8–389–394.

122. M. Hillert: 'An analysis of the effect of alloying elements on the pearlite reaction', In: M. S. H. I. Aaronson, D. E. Laughlin:, and C. M. Wayman, eds. *Solid-Solid Phase Transformations.* Materials Park, Ohio, USA: TMS-AIME, 1981:789–806.

123. M. J. Aziz: 'Model for solute redistribution during rapid solidification', *Journal of Applied Physics*, 1982, **53**, 1150–1168.

124. M. Hillert, and L. Höglund: 'Mobility of α/γ phase interfaces in fe alloys', *Scripta Materialia*, 2006, **54**, 1259–1263.

125. G. P. Krielaart, and S. van der Zwaag: 'Kinetics of $\gamma \rightarrow \alpha$ phase transformations in Fe-Mn alloys containing low manganese', *Materials Science and Technology*, 1998, **14**, 10–18.

126. J. J. Wits, T. A. Kop, Y. van Leeuwen, J. Sietsma, and S. van der Zwaag: 'A study on the austenite-to-ferrite phase transformation in binary substitutional iron alloys', *Materials Science & Engineering A*, 2000, **283**, 234–241.

127. T. A. Kop, J. Sietsma, and S. van der Zwaag: 'The influence of nitrogen on the austenite/ferrite interface mobility in Fe-1at.%Si', *Materials Science & Engineering A*, 2002, **323**, 403–408.

128. S. I. Vooijs, Y. van Leeuwen, J. Sietsma, and S. van der Zwaag: 'On the mobility of the austenite-ferrite interface in Fe-Co and Fe-Cu', *Metallurgical & Materials Transactions A*, 2000, **31**, 379–385.

129. H. Song, and J. J. Hoyt: 'A molecular dynamics simulation study of the velocities, mobility and activation energy of an austenite–ferrite interface in pure Fe', *Acta Materialia*, 2012, **60**, 4328–4335.

130. F. V. Nolfi, Jr., P. G. Shewmon, and J. S. Foster: 'The dissolution and growth kinetics of spherical particles', *Transactions of the Metallurgical Society of AIME*, 1969, **245**, 1427–1433.

131. G. Krielaart: 'Primary ferrite formation from supersaturated austenite': Ph.D. Thesis, Delft University of Technology, Delft, The Netherlands, 1995.

132. J. W. Gibbs: The Scientific Papers, vol. 1: New York, USA: Dover Publications, 1961.

133. W.-K. Burton, N. Cabrera, and F. C. Frank: 'The growth of crystals and the equilibrium structure of their surfaces', *Philosophical Transactions of the Royal Society, London A*, 1951, **243**, 299–358.

134. H. I. Aaronson: 'The proeutectoid ferrite and the proeutectoid cementite reactions', In: V. F. Zackay, and H. I. Aaronson, eds. *Decomposition of Austenite by Diffusional Processes.* New York, USA: Interscience, 1962:387–548.

135. H. K. D. H. Bhadeshia: 'Diffusional transformations: the nucleation of superledges', *Physica Status Solidi A*, 1982, **69**, 745–750.

136. D. V. Edmonds, and R. W. K. Honeycombe: 'Photoemission electron microscopy of growth of grain boundary ferrite allotriomorphs in chromium steel', *Metal Science*, 1978, **12**, 399–405.

137. A. D. Batte, and R. W. K. Honeycombe: 'Strengthening of ferrite by vanadium carbide precipitation', *Metal Science*, 1973, **7**, 160–168.

138. J. W. Cahn, W. B. Hillig, and G. W. Sears: 'The molecular mechanism of solidification', *Acta Metallurgica*, 1964, **12**, 1421–1439.

139. G. J. Jones, and R. K. Trivedi: 'Lateral growth in solid-solid phase transformations', *Journal of Applied Physics*, 1971, **42**, 4299–4304.

140. C. Atkinson: 'The growth kinetics of individual ledges during solid-solid phase transformations', *Proceedings of the Royal Society of London A*, 1981, **378**, 351–368.

141. D. Rosenthal: 'Theory of moving sources of heat and its application to metal treatments', *Transactions of the ASME*, 1946, **b11**, 849–866.

142. M. Enomoto, H. I. Aaronson, J. Avila, and C. Atkinson: 'Influence of diffusional interactions among ledges on the growth kinetics of interphase boundaries', In: M. S. H. I. Aaronson, D. E. Laughlin:, ed. *Solid-Solid Phase Transformations*. Warrendale, Pennsylvania, USA: TMS-AIME, 1981:567–571.

143. G. J. Jones, and R. K. Trivedi: 'The kinetics of lateral growth', *Journal of Crystal Growth*, 1975, **29**, 155–166.

144. C. Atkinson: 'The growth kinetics of ledges during phase transformations multistep interactions', *Proceedings of the Royal Society of London A*, 1982, **384**, 107–133.

145. C. Atkinson: 'Diffusion controlled ledge growth in a medium of finite extent', *Journal of Applied Physics*, 1982, **53**, 5689–5696.

146. C. Atkinson, and P. Wilmott: 'The growth kinetics of ledged interphase boundaries: transient motion, an analytical treatment', *Proceedings of the Royal Society of London A*, 1989, **426**, 377–389.

147. C. Atkinson, K. R. Kinsman, and H. I. Aaronson: 'Relative growth kinetics of ledged and disordered interphase boundaries', *Scripta Metallurgica*, 1973, **7**, 1105–1114.

148. M. Enomoto: 'Computer modeling of the growth kinetics of ledged interphase boundaries - I. single step and infinite train of steps', *Acta Metallurgica*, 1987, **35**, 935–945.

149. M. Enomoto: 'Computer modeling of the growth kinetics of ledged interphase boundaries – II. Finite train of steps', *Acta Metallurgica*, 1987, **35**, 947–956.

150. K. Lücke, and K. Detert: 'A quantitative theory of grain-boundary motion and recrystallization in metals in the presence of impurities', *Acta Metallurgica*, 1957, **5**, 628–637.

151. J. W. Cahn: 'The impurity drag effect in grain boundary motion', *Acta Metallurgica*, 1962, **10**, 789–798.

152. K. Lücke, and H. P. Stüwe: 'On the theory of impurity controlled grain boundary motion', *Acta Metallurgica*, 1971, **19**, 1087–1099.

153. C. G. Shirley: 'Grain boundary drag in alloys', *Acta Metallurgica*, 1978, **26**, 391–404.

154. M. Hillert, and B. Sundman: 'A treatment of the solute drag on moving grain boundaries and phase interfaces in binary alloys', *Acta Metallurgica*, 1976, **24**, 731–743.

155. D. Turnbull, and R. E. Hoffman: 'Effect of relative crystal and boundary orientations on grain boundary diffusion rates', *Acta Metallurgica*, 1954, **2**, 419–426.

156. K. Smidoda, W. Gottschalk, and H. Gleiter: 'Diffusion in migrating interfaces', *Acta Metallurgica*, 1978, **26**, 1833–1836.

157. K. Smidoda, C. Gottschalk, and H. Gleiter: 'Grain boundary diffusion in migrating boundaries', *Metal Science*, 1979, **13**, 146–148.

158. M. Hillert: 'Diffusion and interface control of reactions in alloys', *Metallurgical Transactions A*, 1975, **6**, 5–19.

159. M. J. Aziz: 'Dissipation theory treatment of the transition from diffusion-controlled to diffusionless solidification', *Applied Physics Letters*, 1983, **43**, 552–554.

160. F. G. Caballero, and M. K. Miller: 'Personal communication', 2014: Interfacial width in atom probe tomography.

161. H. K. D. H. Bhadeshia: 'Anomalies in carbon concentration determinations from nanostructured bainite', *Materials Science and Technology*, 2015, **31**, 758–763.

162. H. Guo, and G. R. Purdy: 'Scanning transmission electron microscopy study of interfacial segregation of mn during the formation of partitioned grain boundary ferrite in a Fe-C-Mn-Si alloy', *Metallurgical & Materials Transactions A*, 2008, **39**, 950–953.

163. H. Guo, S. W. Yang, C. J. Shang, X. M. Wang, and X. L. He: 'A quantitative analysis of Mn segregation at partitioned ferrite/austenite interface in a Fe-C-Mn-Si alloy', *Journal of Materials Science and Technology*, 2009, **25**, 383–388.

164. H. P. van Landeghemand B. Langelier, D. Panahi, G. R. Purdy, C. R. Hutchinson, G. A. Botton, and H. S. Zurob: 'Solute segregation during ferrite growth: Solute/interphase and substitutional/interstitial interactions', *JOM*, 2016, **68**(2016), 1329–1334.

165. H. P. van Landeghemand B. Langelier, G. Gault, D. Panahi, A. Korinek, G. R. Purdy, and H. S. Zurob: 'Investigation of solute/interphase interaction during ferrite growth', *Acta Materialia*, 2017, **124**, 536–543.

166. T. B. Massalski: 'Massive transformations', In: *Phase Transformations*. Ohio, USA: ASM, 1970:433–495.

167. J. I. Kim, J. H. Pak, K. S. Park, J. H. Jang, D. W. Suh, and H. K. D. H. Bhadeshia: 'Segregation of phosphorus to ferrite grain boundaries during transformation in Fe-P alloy', *International Journal of Materials Research*, 2014, **105**, 1166–1172.

168. M. R. Plichta, J. H. Perepezko, H. I. Aaronson, and W. F. Lange: 'Part I: Nucleation kinetics of the $\beta \to \alpha_m$ transformation in Ti-Ag and Ti-Au alloys', *Acta Metallurgica*, 1980, **28**, 1031–1040.

169. M. R. Plichta, and H. I. Aaronson: 'Crystallography and morphology of the $\beta \to \zeta$ massive transformation in Ag-26 a/o Al', *Acta Metallurgica*, 1980, **28**, 1041–1057.

170. T. B. Massalski: 'The mode and morphology of massive transformations in Cu-Ga, Cu-Zn, Cu-Zn-Ga and Cu-Ga-Ge alloys', *Acta Metallurgica*, 1958, **6**, 243–253.

171. J. W. Cahn, M. Cohen, and D. A. Karlyn: 'The massive transformation in copper-zinc alloys', *Transactions of the Metallurgical Society AIME*, 1969, **245**, 197–207.

172. H. I. Aaronson, C. Laird, and K. R. Kinsman: 'Application of a theory of precipitate morphology to the massive transformation', *Scripta Metallurgica*, 1968, **2**, 259–264.

173. A. Gilbert, and W. S. Owen: 'Diffusionless transformation in iron-nickel, iron-chromium and iron-silicon alloys', *Acta Metallurgica*, 1962, **10**, 45–54.

174. M. J. Bibby, and J. G. Parr: 'The martensitic transformation of pure iron', *Journal of the Iron and Steel Institute*, 1964, **202**, 100–104.

175. B. P. Liu, T. F. Chung, J. R. Yang, J. Fu, C. Y. Chen, S. H. Wang, M. C. Tsai, and C. Y. Huang: 'Microstructure characterization of massive ferrite in laser-weldments of interstitial-free steels', *Metals*, 2020, **10**, 898.

176. S. Kobayashi: 'Formation of the Fe_2Hf Laves phase through eutectoid type reaction of $\delta \rightarrow \gamma + Fe_2Hf$ in ferritic heat resistant steels', *MRS Proceedings*, 2015, **1760**, 2–8.

177. R. A. Ricks, P. R. Howell, and R. W. K. Honeycombe: 'Formation of supersaturated ferrite during decomposition of austenite in iron-copper and iron-copper-nickel alloys', *Metal Science*, 1980, **14**, 562–568.

178. R. A. Ricks: 'A comparative study of precipitation at interphase boundaries in Fe-Cu-Ni and Fe-Au-Ni alloys', *Journal of Materials Science*, 1981, **16**, 3006–3012.

179. K. Campbell, and R. W. K. Honeycombe: 'The isothermal decomposition of austenite in simple chromium steels', *Metal Science*, 1974, **8**, 197–203.

180. J. M. Gray, and R. B. G. Yeo: 'Niobium carbonitride precipitation in low-alloy steels with particular emphasis on precipitate-row formation', *Trans. ASM*, 1968, **61**, 255–269.

181. A. T. Davenport, F. G. Berry, and R. W. K. Honeycombe: 'Interphase precipitation in iron alloys', *Metal Science*, 1968, **2**, 104–106.

182. A. T. Davenport, and R. W. K. Honeycombe: 'Precipitation of carbides at γ-α boundaries in alloy steels', *Proceedings of the Royal Society of London A*, 1971, **322**, 191–205.

183. R. W. K. Honeycombe: 'Transformation from austenite in alloy steels', *Metallurgical Transactions A*, 1976, **7**, 915–936.

184. H. I. Aaronson, M. R.Plichta, G. W. Franti, and K. C. Russell: 'Precipitation at interphase boundaries', *Metallurgical Transactions A*, 1978, **9**, 363–371.

185. P. R. Howell, J. V. Bee, and R. W. K. Honeycombe: 'The crystallography of the austenite-ferrite/carbide transformation in Fe-Cr-C alloys', *Metallurgical Transactions A*, 1979, **10**, 1213–1222.

186. R. A. Ricks, and P. R. Howell: 'Bowing mechanism for interphase boundary migration in alloy steels', *Metal Science*, 1982, **16**, 317–322.

187. R. A. Ricks, and P. R. Howell: 'The formation of discrete precipitate dispersions on mobile interphase boundaries in iron-base alloys', *Acta Metallurgica*, 1983, **31**, 853-861.

188. P. R. Rios: 'Morphology of interphase precipitation in microalloyed steels', *Journal of Material Science Letters*, 1991, **10**, 981–983.

189. A. D. Batte, and R. W. K. Honeycombe: 'Precipitation of vanadium carbide in ferrite', *Journal of the Iron and Steel Institute*, 1973, **211**, 284–289.

190. J. A. Todd, P. Li, and S. M. Copley: 'A new model for precipitation at moving interphase boundaries', *Metallurgical and Materials Transactions A*, 1988, **19**, 2133–2138.

191. J. A. Todd, and Y. J. Su: 'A mass transport theory for interphase precipitation with application to vanadium steels', *Metallurgical Transactions A*, 1989, **20**, 1647–1655.

192. P. R. Rios: 'A model for interphase precipitation in stoichiometrically balanced vanadium steels', *Journal of Materials Science*, 1995, **30**, 1872–1878.

193. R. Okamoto, and J. Ågren: 'A model for interphase precipitation based on finite interface solute drag theory', *Acta Materialia*, 2010, **58**, 4791–4803.

194. R. Lagneborg, and S. Zajac: 'Model for interphase precipitation of V-microalloyed structural steels', *Metallurgical & Materials Transactions A*, 2001, **32**, 39–50.

195. R. L. Patterson, and C. M. Wayman: 'The crystallography and growth of partially-twinned martensite plates in Fe-Ni alloys', *Acta Metallurgica*, 1966, **14**, 347–369.

196. H. K. D. H. Bhadeshia: Geometry of Crystals, Polycrystals, and Phase Transformations: Florida, USA: CRC press, ISBN 9781138070783, freely downloadable from www.phase-trans.msm.cam.ac.uk, 2017.

197. R. Kannan, Y. Wang, J. Poplawsky, S. S. Babu, and L. Li: 'Cascading phase transformations in high carbon steel resulting in the formation of inverse bainite: An atomic scale investigation', *Scientific Reports*, 2019, **9**, 5597.

198. H. K. D. H. Bhadeshia: 'Modelling of recrystallisation in mechanically alloyed materials', *Materials Science & Engineering A*, 1997, **A223**, 91–98.

Notes

[1]For solid-state transformations in steels, the temperature fields are in general minimal because of the relatively slow rates of reaction. Some martensitic transformations can occur at very high speeds, sufficient to produce microstructural effects that are due to recalescence, e.g., [195].

[2]In a diffusionless transformation, labelled rows of atoms in the parent crystal remain in the correct sequence in the product lattice. It is therefore possible to identify that a particular atom in the product must have originated from a corresponding particular atom in the parent crystal. A formal way of expressing this property is to say that there exists an *atomic correspondence* between the parent and product lattices.

[3]The Bain orientation (Figure 1.10) follows from the nature of the Bain strain:

$$\langle 100 \rangle_\gamma \parallel \langle 110 \rangle_\alpha, \langle 001 \rangle_\gamma \parallel \langle 001 \rangle_\alpha.$$

The Bain strain alone does not rotate any plane or direction by more than about $11°$, so any set of corresponding planes and directions can be made parallel after this strain by a rotation of not more than $11°$ [26].

[4]Ferrite does not have a close-packed plane but planes of the form $\{110\}$ are the most densely packed planes. To avoid clumsy terminology we shall refer to these as close-packed.

[5]Some of the properties of such films have already been discussed in Chapter 2 where original references can be found.

[6]A pseudomorphic deposit describes an epitaxial layer that is strained to match the substrate.

[7]A 1×1 surface structure means that the atoms have the same configuration at the surface as the bulk. For the same surface, a 2×1 reconstructed surface structure means that the unit cell of the surface has its edges parallel to the 1×1 cell, but is twice the length along one of the edges.

[8]The term *lattice matching* often is used but the analyses of inclusion potency are confined to two-dimensional matching at particular contact planes between the partners. Lattice matching has a more general meaning, covered for example by coincidence site or O-lattice theories [196].

[9]There may be other dissipative processes, such as solute drag and the emission of sound or heat

[10]It was emphasised earlier that all interfaces strictly move under mixed-control, but that $|\Delta x_I| \ll |\Delta x_D|$ for diffusion-controlled growth so the assumption of local equilibrium is reasonable. Unless otherwise stated, in all subsequent treatments

we assume the existence of local equilibrium at the interface during diffusion-controlled growth, bearing in mind that this is an approximation since Δx_{I} is never zero.

[11] The flux of solute from the interface must equal the rate at which solute is partitioned:

$$\underbrace{(c^{\gamma\alpha} - c^{\alpha\gamma})\frac{\partial Z}{\partial t}}_{\text{rate solute partitioned}} = \underbrace{-D\frac{\partial c}{\partial z}}_{\text{diffusion flux from interface}} \simeq -D\frac{\overline{c} - c^{\gamma\alpha}}{z_{\mathrm{d}}}$$

where z_{d} is a diffusion distance assuming that the concentration in the austenite varies linearly with distance. From overall conservation of mass:

$$(c^{\alpha\gamma} - \overline{c})Z = \frac{1}{2}(\overline{c} - c^{\gamma\alpha})z_{\mathrm{d}} \tag{4.95}$$

$$\therefore \frac{\partial Z}{\partial t} = \frac{D(\overline{c} - c^{\gamma\alpha})^2}{2Z(c^{\alpha\gamma} - c^{\gamma\alpha})(c^{\alpha\gamma} - \overline{c})} \quad \text{and} \quad Z \propto \sqrt{Dt}. \tag{4.96}$$

[12] A recent claim [197] that NP-LE is observed with cementite growth in ferrite is not justified. The atom probe composition data are not of a sufficient resolution because the concentration distributions at the interface are too wide and there has been no effort to compare the experimental data against theory to demonstrate the existence of local equilibrium.

[13] It is stated sometimes, that the *concentration* of X does not change during paraequilibrium transformation. This is not correct since some change is inevitable given that the concentration of carbon changes. However, the ratio of the iron to substitutional solute atoms is fixed during paraequilibrium transformation

[14] Karlyn et al. [171] used a similar criterion to define the onset of the massive transformation in binary alloys.

[15] If the simulation dealt with an enclosed particle, the surrounding interfaces would in any case show curvature from junction pinning [198].

[16] The differential equation for heat, in two dimensions, for the stationary rectangular coordinates y, z referred to a fixed origin is

$$(\partial^2 T/\partial y^2) + (\partial^2 T/\partial z^2) = 2D_{\mathrm{T}}(\partial T/\partial t)$$

where D_{T} is the thermal diffusivity. This equation simply states that the change in temperature with time depends on the divergence of the heat flux. For a moving heat source which has reached steady state, the equation referred to moving coordinates $y' = y - vt$, $z' = z$ located at the heat source the equation becomes

$$(\partial^2 T/\partial y'^2) + (\partial^2 T/\partial z'^2) = -2D_{\mathrm{T}}v(\partial T/\partial y') + 2D_{\mathrm{T}}(\partial T/\partial t).$$

However, the derivative with respect to time is zero since the equation describes the steady state relative to the moving coordinates located at the heat source.

5 Martensite

5.1 DIFFUSIONLESS TRANSFORMATIONS

The periodic pattern in which atoms are arranged within a crystal can in principle be changed without the need for atoms to diffuse, in which case an atomic correspondence is maintained between the parent and product crystal structures. The structure of martensite is generated in this way by a deformation, that has important consequences on the properties of the material. In steels this leads to certain combinations of strength, toughness and fitness for purpose that are unmatched by any other engineering material [1].

There are other diffusionless transformations in which the lattice deformation does not place all the atoms into their final positions; some of the atoms therefore have to undergo additional movements over distances that are fractions of the interatomic spacing. These movements are called *shuffles* to distinguish them from diffusion which involves the uncoordinated transport of atoms between lattice sites. Diffusionless transformations can therefore be categorised into those dominated by the shuffling of atoms and others in which the larger proportion of atoms is displaced into their final locations by the homogeneous deformation of the parent lattice [2, 3].

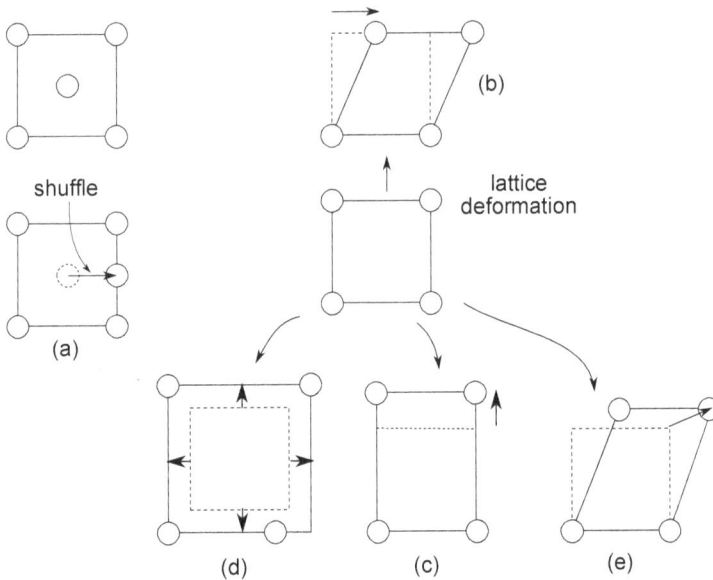

Figure 5.1 Shuffle dominant and lattice-deformation dominant mechanisms of displacive transformations (adapted from Cohen, Olson and Clapp, [2]). There are no distortions normal to the plane of the diagram.

A shuffle involves the correlated displacement of a fraction of the atoms within a unit cell, Figure 5.1a. Omega phase formation is an example common in bcc titanium alloys. The $\{111\}_{bcc}$ planes of titanium have a stacking sequence with a repeat period of three: ...**ABCABCABC**....

During the ω-transformation, the atoms move in such a way that the **A** planes remain unchanged whereas the **B** and **C** planes move symmetrically towards each other to form a single **B′** plane, Figure 5.2. This results in a new stacking sequence ...**AB′AB′AB′**... corresponding to a structure with the hexagonal symmetry of the ω-phase [4]. The **B′** planes no longer are equivalent to the **A** planes because of their doubled number density of atoms. Although a shuffle transformation such as this is diffusionless, the change in atomic positions is not dominated by macroscopic strain, so the ω-transformation is not considered to be martensitic [2].

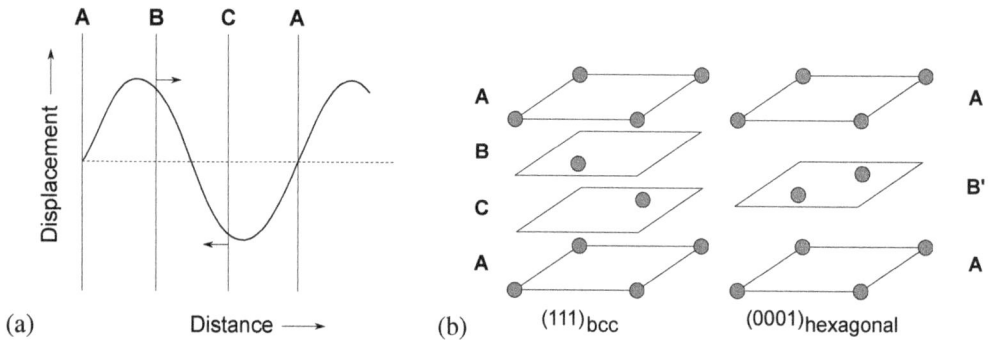

Figure 5.2 (a) Displacement wave associated with the β (bcc) to ω transformation. The **A** atoms are unaffected because they lie on the nodes of the wave. (b) How the stacking sequence and density of atoms in each plane changes on transformation (adapted from Sikka et al. [5]).

The mechanisms shown in Figure 5.1b-d all involve a homogeneous deformation of the lattice. Such deformations can be detected macroscopically because they alter the shape of the transformed region. The uniform dilatation shown in Figure 5.1d does not leave any vector undistorted, for example, when the lattice parameter of cerium contracts by 16% on cooling below 96 K [6]. The change is associated with an electronic transition that reduces the size of the ion, so the process is not considered to be martensitic even though it is diffusionless.

It is accepted that martensitic transformation must lead to a product that is in the form of a thin plate, because this shape minimises strain energy. It will be demonstrated later (page 271) that this requires at least one line to remain invariant to the total transformation strain in order to ensure a glissile interface that can move without diffusion.

Both of the cases illustrated in Figure 5.1b,c fall into the category of martensitic transformation because they leave at least one line undistorted. Each of the deformations illustrated leaves a plane invariant to the transformation strain, the normal to which is parallel to the displacement vector in Figure 5.1c but perpendicular to the shear direction in Figure 5.1b. These strains are both described as *invariant-plane strains*, the general form of which consists of both shear and dilatational components parallel and normal (respectively) to the invariant-plane (Figure 5.1e). This is representative of martensitic transformation in ordinary steels, where the shear component is of the order of 0.25 and the dilatational strain is about 0.03 (cf. typical magnitude of an elastic strain is about 10^{-3}).

The particle shape that minimises elastic strain energy due to the invariant-plane strain shape change is a thin lenticular plate on the invariant-plane [7] Martensite is therefore always in the form of thin plates whenever the transformation is constrained. The growth of a plate is arrested following a collision with other plates, austenite grain surfaces or austenite twin boundaries. Boundaries in general pose formidable crystallographic discontinuities to the coordinated movement of atoms. A plate that has been halted by collision with a hard obstacle may still continue thickening until the accumulation of strain energy can no longer be tolerated, Figure 5.3.

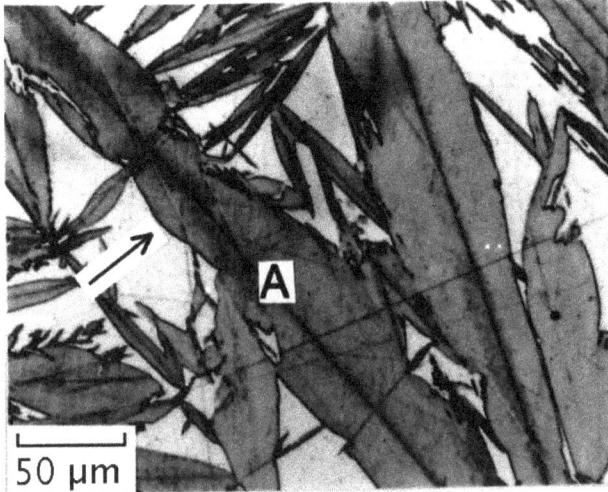

Figure 5.3 A martensite plate marked 'A' thickening by bowing between obstacles to achieve elastic equilibrium, even though its length is fixed. The steel has the chemical composition Fe-30.1Ni-0.3C wt%. The same phenomenon applies to plates of bainite which can continue thickening after their longitudinal growth is halted [8].

The fact that plates of martensite stop growing when obstructed also means that there always will be some austenite that is left untransformed, even though its chemical composition is not altered. Some austenite persists even when the alloy is cooled to cryogenic temperatures so there is no theoretically justified martensite-finish temperature; a practical value can be defined as the temperature where the fraction of austenite remaining is, for example, 0.05.

5.2 OTHER CHARACTERISTICS OF MARTENSITE

The term "martensite" was coined in honour of Adolf Martens (1850-1914) to describe a hard phase found in steels. It is now known that martensite need not be hard and that it can occur in many materials such as nonferrous alloys, pure metals, ceramics, minerals, inorganic compounds, solidified gases and polymers, even biological entities (Table 5.1). It nevertheless remains the case that the transformation is of the greatest technological importance in steels.

An obvious indicator that martensitic transformation is diffusionless is that it can occur at temperatures that are incredibly low (Table 5.1). The mobility of atoms must then be negligible over the time scale of a typical transformation experiment. That it not to say that all martensitic transformations occur at low temperatures. Indeed, many start at very high temperatures, but these too do not require the diffusion of atoms to accomplish the change in crystal structure.

Table 5.1

The temperature M_S at which martensite is first detected to form on cooling. Where available, the approximate Vickers hardness is quoted for the stated alloy or of another of similar composition.

Composition	M_S / K	Hardness HV	Reference
ZrO_2	1320	≈ 900	[9]
Fe-31Ni-0.23C wt%	83	300	
Fe-34Ni-0.22C wt%	< 4	250	
Fe-3Mn-2Si-0.4C wt%	493	600	
Cu-13Al wt%	512	240	[10]
Ytterbium	260		[11]
Ar-40N_2	39		[12]
In-25Tl at%	198		[13]
In-29Tl at%	86		[13]
Gd-35Ce at%	763		[14]
Liquid crystal MLC-2412	315		[15]

In a Fe-30Ni wt% alloy, Bunshah and Mehl recorded plates of martensite to grow at speeds in excess of $1100\,\mathrm{m\,s^{-1}}$, at temperatures as low as 78 K [16]. This compares with the terminal speed of a ballistic missile at $5000\,\mathrm{m\,s^{-1}}$. Each transformation event takes place in just 50–500 ns. Furthermore, the growth rate appears to be insensitive to temperature, indicating that there is a small activation barrier to the propagation of the martensite/austenite interface. In contrast, two of the fastest solidification front velocities ever reported are for pure-nickel at about $70\,\mathrm{m\,s^{-1}}$ [17], and for an Ni-36Fe at.% alloy at about $88\,\mathrm{m\,s^{-1}}$ [18]. For a given undercooling a pure metal is likely to solidify more rapidly but an alloy may achieve a greater undercooling before solidification begins.

The very large rate at which martensite can grow is inconsistent with any diffusion of atoms and shows that the transformation interface is exceptionally mobile. Any discontinuities in the interface must be able to move conservatively, i.e. without diffusion. The speed of the boundary is limited by that of the interfacial dislocations. The slow movement of dislocations (as in most deformation experiments) does not require any special consideration of dynamic effects. However, the energy of a dislocation increases dramatically as its velocity approaches that of transverse sound waves in the metal [19]. The velocity of sound in the metal sets an upper limit for that of dislocations. Since the martensite interface has a dislocation structure, the upper limit to its speed should be about $3000\,\mathrm{m\,s^{-1}}$ in steel.

It is known that deformation by mechanical twinning leads to the emission of sound in the form of audible clicks [20, 21]. The so-called "tin cry" on "indium cry" is due to the rapid propagation and subsequent arrest of mechanical twins. A similar effect is found for martensitic transformation, where the rapid formation of a plate causes acoustic emissions which can be used to study the nucleation and growth rates. With this method, Takashima et al. [22] determined the growth rate of martensite in austenitic stainless steel to be about 110-$200\,\mathrm{m\,s^{-1}}$ at 138 K. The growth rate has been measured to be about $10^{-5}\,\mathrm{m\,s^{-1}}$ for isothermal martensite in Fe-21Ni-4Mn wt% [23]. These values are less than the rates reported by Bunshah and Mehl, emphasising that although martensite can grow very rapidly, it need not do so as long as interfacial velocity is greater than the ratio of the diffusion coefficient to interatomic spacing, where the ratio defines the speed of atomic diffusion across the interface. The actual velocity will depend on alloying additions, the driving force for transformation and on many other factors. In the case of thermoelastic or single-interface transformation experiments, the interface motion can be followed visually by small changes in the temperature or stress, i.e., the chemical or mechanical driving forces [24].

Martensite first forms when the austenite is cooled below a critical temperature known as the martensite-start or M_S temperature. The transformation is in most cases so rapid that the amount of martensite obtained can be expressed simply as a function of the undercooling below the M_S temperature rather than the time at that temperature. This characteristic *athermal* behaviour is illustrated in Figure 5.4 where it is seen that time does not feature in the plot, where $M_S - T_q$ is the undercooling below M_S.

It is a common misconception that martensite can only be obtained by quenching rapidly from the temperature at which the austenite is stable. The rate required to achieve martensitic transformation depends strictly on the kinetics of the reactions that precede martensite. The required cooling rate can be very slow indeed if the steel has a high hardenability.

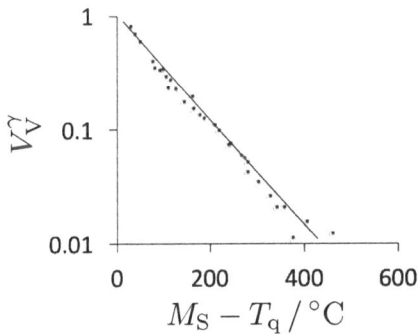

Figure 5.4 Logarithm of the volume fraction V_V^γ of austenite remaining untransformed, as a function of the undercooling below the martensite-start temperature (after Koistinen and Marburger [25]). T_q is the temperature to which the sample is cooled below the M_S temperature.

The interface between the martensite and its parent must have a structure that can translate without diffusion although there may still exist a small activation barrier akin to the Peierls barrier for dislocation motion in a periodic lattice. Such a *glissile* interface must be coherent or semi-coherent, depending upon the crystallography of the particular material undergoing transformation.

For the $\gamma \rightarrow \alpha'$ martensitic transformation, the interface will be semi-coherent because the two lattices can only be forced into coherency over small regions of the boundary surface. The misfit, as it accumulates, is relieved periodically by features such as dislocations that leave the interface glissile. Stress-free coherency is only possible for large interfaces when the parent crystal can be converted into the product phase by a homogeneous deformation which is an invariant-plane strain. This is true for the $\gamma \rightarrow \varepsilon$ martensitic transformation where the habit plane is fully coherent. The growth of the martensite then causes a change of shape which mimics the strain that converts $\gamma \rightarrow \varepsilon$. In contrast, the shape change caused by the movement of a semi-coherent martensite interface is related in a more complex way to the strain which converts the parent into the product lattice. More discussion of interfaces will follow later, but for now it is emphasised that both semi-coherent and coherent interfaces must be glissile for martensitic transformation. It follows that the motion of the interface can in principle and in practice be reversed with a consequent reversal of the shape deformation and lattice change [26].

Martensitic transformations are achieved by the coordinated movement of atoms. The crystallography of the transformation is therefore well-defined and hence reveals a great deal about the mechanism of transformation. An elastically accommodated single plate of α' martensite has the following entirely reproducible features:

(a)   A habit plane with indices that are irrational, meaning that its crystallographic indices cannot be expressed as ratios of integers. Decimal expansions of these indices do not terminate or become periodic.

(b)   An α'/γ orientation relationship which in general is also irrational.

(c)   The shape deformation consists of a shear on the habit plane and in the case of steels a smaller dilatation normal to the habit plane.

(d)   The crystallographic variant of the habit plane, shape deformation and orientation relationship are related uniquely. For example, it is not possible to have a different shape deformation for the same habit plane.

5.2.1   HABIT PLANE

The largest interface plane between the austenite and martensite is designated the "habit plane" (Figure 5.5). The macroscopic fit between the austenite and martensite is optimum on this plane but the shape is not determined by interfacial energy minimisation, rather by strain energy minimisation. The habit plane is flat when the transformation occurs without constraint. In the more usual case where the transformation is constrained by the surrounding material, the martensite grows in the form of a thin lenticular plate or a lath. The definition of a habit plane is less clear for constrained transformation because of the curvature in the shape of the plate. However, it is found experimentally that the average plane containing the major circumference of the lens corresponds closely to that expected from crystallographic theory and to that determined from unconstrained transformation. The plate aspect ratio, i.e. its thickness divided by its length, is small, typically at about 0.05, so that the average plane of the plate is a good representation of the habit plane. Some examples of the crystallographic indices of habit planes are given in Table 5.2.

5.2.2   ORIENTATION RELATIONSHIPS

The absence of diffusion ensures without exception, that there is a reproducible crystallographic relationship between the martensite and the austenite in which it grows. This is described in terms of a pair of closely parallel corresponding planes from each phase and a pair of related directions within those planes. For martensite in steels it is the most densely packed planes that are very nearly parallel together with the close-packed directions from these planes. Examples of orientation

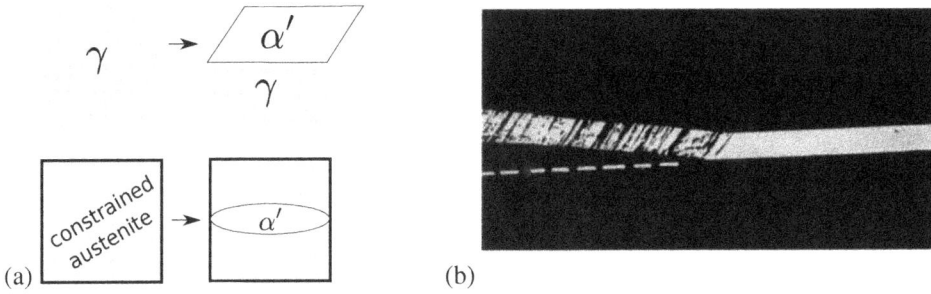

Figure 5.5 (a) The habit plane of martensite during constrained and unconstrained transformation of a single crystal of austenite. When unconstrained, the single interface moves to divide the crystal into γ and α'. When any change in the shape of the austenite is resisted by its surroundings, the α' adopts a lenticular shape, with the average crystallographic indices of its habit plane being about the same as that of the single-interface. (b) Single-interface martensitic transformation of $\alpha \rightarrow \gamma$ in a single-crystal of pure iron, by the translation of a flat interface along its length. The originally straight whisker, which has a width of about 50 μm, is kinked by the shape deformation accompanying diffusionless transformation. Reproduced from Zerwekh and Wayman [27] with permission from Elsevier.

Table 5.2
Some indices of the habit plane normal p. The stated indices are an approximation because p in fact has irrational indices. Experimental data may show some real scatter due to local changes in the strain field around the plate being measured, due for example to the formation of other plates. Results obtained using trace analysis of fully transformed samples are excluded as the boundaries between laths are not representative of the habit plane with respect to austenite.

Composition / wt%	Approximate indices	Measured indices	Reference
Fe-20Ni-5Mn, low alloy steels	$\{5\,7\,5\}_\gamma$	scattered around $\{5\,7\,5\}_\gamma$	[28, 29]
Fe-53.13Pt	$\{2\,4\,3\}_\gamma$	$\{0.1910\,0.7599\,0.6360\}_\gamma$	[30]
Fe-1.19C	$\{2\,2\,5\}_\gamma$	$\{0.3576\,0.4037\,0.8421\}_\gamma$	[31]
Fe-7.9Cr-1.11C	$\{2\,2\,5\}_\gamma$	$\{0.3582\,0.3788\,0.8533\}_\gamma$	[32]
Fe-6.14Mn-0.95C	$\{2\,2\,5\}_\gamma$	$\{0.3892\,0.4110\,0.8243\}_\gamma$	[31]
Fe-21.89Ni-0.82C	$\{2\,5\,9\}_\gamma$	$\{0.1593\,0.5596\,0.8132\}_\gamma$	[31]
Fe-53.1Pt, disordered	$\{3\,15\,10\}_\gamma$	$\{0.1920\,0.7599\,0.6214\}_\gamma$	[33]
ε-martensite in 18/8 stainless steel	$\{1\,1\,1\}_\gamma$	Exact	[34]

relations as found in steels are given below, stated in a way that emphasises the close-packed planes and directions, as illustrated in Figure 5.6:

Kurdjumov-Sachs orientation relationship [35]:

$$\{1\,1\,1\}_\gamma \| \ \{0\,1\,1\}_{\alpha'},$$
$$\langle 1\,0\,\bar{1}\rangle_\gamma \| \ \langle 1\,1\,\bar{1}\rangle_{\alpha'},$$

Nishiyama-Wasserman orientation relationship [36, 37]:

$$\{1\,1\,1\}_\gamma \parallel \{0\,1\,1\}_{\alpha'},$$

$$\langle 1\,0\,\bar{1}\rangle_\gamma \quad \text{about } 5.3° \text{ from} \langle 1\,1\,\bar{1}\rangle_{\alpha'} \text{ towards } \langle \bar{1}\,1\,\bar{1}\rangle_{\alpha'},$$

Greninger-Troiano orientation relationship [38, 39]:

$$\{1\,1\,1\}_\gamma \quad \text{about } 0.2° \text{ from } \{0\,1\,1\}_{\alpha'},$$

$$\langle 1\,0\,\bar{1}\rangle_\gamma \quad \text{about } 2.7° \text{ from } \langle 1\,1\,\bar{1}\rangle_{\alpha'} \text{ towards } \langle \bar{1}\,1\,\bar{1}\rangle_{\alpha'}.$$

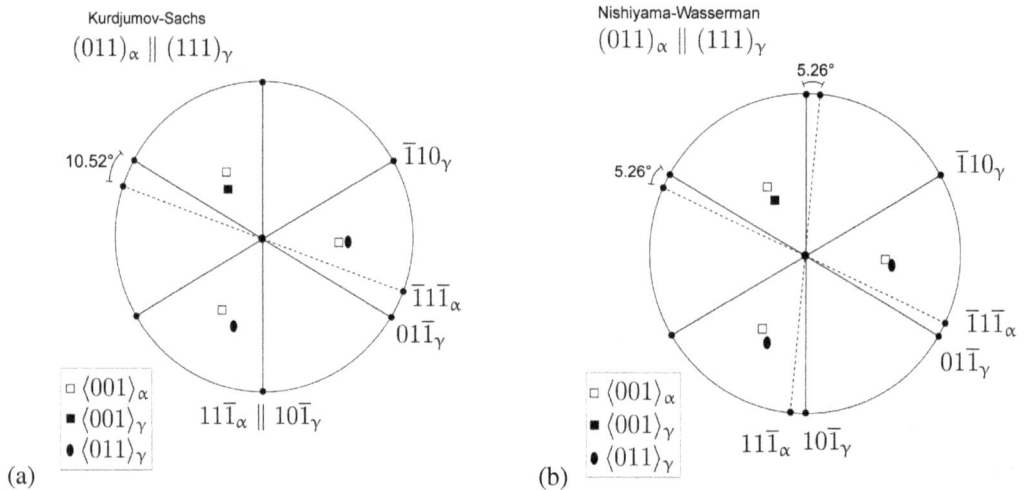

Figure 5.6 Stereographic representation of the Kurdjumov-Sachs (KS) and Nishiyama-Wasserman (NW) orientation relationships. The stereograms are both centred on $(1\,1\,1)_\gamma \parallel (0\,1\,1)_{\alpha'}$. It is seen that the NW orientation can be generated from KS by an appropriate small rotation (5.25°) about $[0\,1\,1]_{\alpha'}$. Only a few of the poles are marked to allow a comparison with the Bain orientation relationship. The neighbouring pairs of poles would superpose exactly for the Bain orientation in which $[001]_\gamma \parallel [001]_{\alpha'}$, $[100]_\gamma \parallel [110]_{\alpha'}$ and $[010]_\gamma \parallel [\bar{1}10]'_\alpha$.

These are the classic orientation relationships of which the first two are reported frequently in the literature. The vast majority of citations are on based on measurements that are not accurate enough to establish the stated exact parallelism of close-packed planes. Indeed, it is expected theoretically that they should not be exactly parallel if an invariant-line is to be obtained in the interface between the austenite and martensite [40]. Precise measurements, both on γ/α bicrystals or on hundreds of thousands of bicrystals, confirm that the exact orientations stated above do not occur in practice [41, 42]. This is illustrated in Figure 5.6, where the angle θ gives the necessary rotation about $[1\,1\,1]_\gamma$, relative to the NW orientation, that yields an invariant-line, assuming that the close-packed planes are exactly parallel; $\theta = 0°$ gives the NW relation and $\theta = 5.25°$ the KS orientation. For lattice parameters typical of steel, Figure 5.7 shows that there is no invariant-line for the exact KS or NW orientation relations.

The martensite structure can be generated by applying the Bain deformation to the austenite, Figure 1.10. This involves a compression along $[0\,0\,1]_\gamma$ and expansions along $[1\,1\,0]_\gamma$ and $[1\,\bar{1}\,0]_\gamma$. The

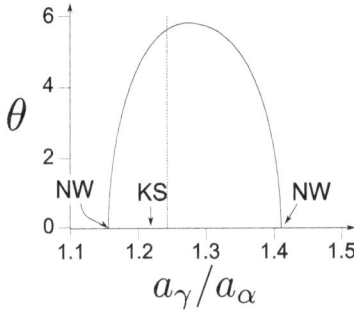

Figure 5.7 The orientation relationship required to ensure an invariant-line as a function of the austenite/martensite lattice parameter ratio. The orientation relation is represented by the angle θ, which is the rotation from the exact Nishiyama-Wasserman orientation relation assuming that the close-packed planes are exactly parallel. The dashed line gives the lattice parameter ratio typical of steel. Adapted from Knowles, Smith and Clark [40].

deformation naturally implies the (Bain) orientation relationship:

$$[0\,0\,1]_\gamma \quad \| \ [0\,0\,1]_{\alpha'}$$
$$[1\,\bar{1}\,0]_\gamma \quad \| \ [1\,0\,0]_{\alpha'}$$
$$[1\,1\,0]_\gamma \quad \| \ [0\,1\,0]_{\alpha'}$$

Figure 5.6 has these six poles plotted, the filled symbols representing the martensite and the unfilled ones the austenite. Although it is clear that the exact Bain orientation is not observed experimentally, the measured KS and NW orientations are not far from Bain. It would not take a large rotation to covert from Bain to KS or NW. *But why is the Bain orientation not observed experimentally?*

5.2.3 STRUCTURE OF THE INTERFACE

It is useful to consider the state of interfaces around an *enclosed* particle rather than focus on a particular interfacial orientation [43]. If all the interfaces around the particle are incoherent, i.e., without continuity of planes or vectors across the boundaries, then there is no correlation of atomic positions across the interface. Such boundaries are displaced only by individual atom migrations and growth can continue across the grain boundaries of the parent phase. If this migration permits long-range diffusion to remove any volume change, the strain energy becomes zero. This is the mechanism of reconstructive transformation. The interface can be semi-coherent for a reconstructive transformation, but with a structure that requires diffusion in order for it to move. The shape of the product will only be anisotropic if there is a strong variation in the interfacial energy as a function of the orientation of the interface.

Suppose now that a particle is generated by a displacive mechanism through a lattice deformation **BR** which in general consists of a pure strain and a rigid body rotation. This deformation defines the relationships between *small* lattice vectors in the parent and product. If **BR** = **P**$_1$ where **P**$_1$ is the shape deformation observed macroscopically, then the interface is fully coherent, an invariant plane.[1] It will be demonstrated later that a stress-free coherent interface is not possible during the $\alpha \leftrightarrow \gamma$ transformations. In such a case there will be additional structure in the interface, a set of glissile interface dislocations that introduce a lattice-invariant shear that permits the macroscopic shape deformation to be an invariant-plane strain. The number of atoms is conserved as the interface advances, eliminating the need for diffusion. An enclosed particle like this can only have coherent or semi-coherent interfaces; its shape will be in the form of a thin plate that minimises the strain energy due to the shape deformation.

The criteria that ensure a glissile interface that is able to move without the creation or destruction of lattice sites are as follows [44]:

(a) The interfacial dislocations glide on planes which intersect the interface plane. These glide planes will in general be differently oriented in the parent and product phases but they must meet edge-to-edge in the interface, along the dislocation lines.

(b) If more than one set of intrinsic dislocations exist, then these should either have the same line vector in the interface or their respective Burgers vectors must be parallel. This condition ensures that the interface can move as an integral unit. It implies that the deformation caused by the interface dislocations as the interface moves can be described as a simple shear caused by a resultant intrinsic dislocation that is a combination of all the intrinsic dislocations. This resultant shear would be on a plane that makes a finite angle relative to the interface and intersects it along the line vector of the resultant intrinsic dislocation.

If the glissile interface has just a single set of parallel dislocations, or a set of different dislocations that can be summed to give a single glissile intrinsic dislocation, then it follows that there must exist in the interface, a line which is parallel to the resultant intrinsic dislocation line vector, along which there is zero distortion and is unrotated by the transformation, i.e., an *invariant-line*. Therefore, for an interface to be glissile, the transformation strain relating the two lattices must at the very least leave a line in the interface plane invariant; such a strain is called an invariant-line strain (ILS). An invariant-plane strain also qualifies because all the lines within that plane are invariant.

The pure strain that can transform the austenite into the martensite lattice is known as the Bain strain ($\mathbf{B}$) [45]. There are other possibilities but the deformations involved are larger. The Bain correspondence has also been established experimentally in Fe_3Pt where the austenite is ordered with the iron atoms at the face-centres of the cubic cell and the platinum atoms at the corners [46]. On martensitic transformation, the disposition of the Fe and Pt atoms in the unit cell of martensite is entirely as expected from the Bain strain.

Consider now whether the Bain Strain ($\mathbf{B}$, Figure 5.8) is consistent with the existence of an invariant-line in the interface between austenite and martensite. There is a compression along $[001]_\gamma$ and expansions along $[1\overline{1}0]_\gamma$ and $[110]_\gamma$ such that all vectors in $(001)_\gamma$ are uniformly expanded. The Bain strain is a *pure* deformation because it leaves three mutually perpendicular directions unrotated, though distorted. The principal distortions η_i along these unrotated axes are given by the ratios of the final to the initial lengths. The Bain strain is also a homogeneous deformation which completes the lattice change, so any additional deformations associated with the formation of martensite serve other purposes.

Figure 5.9a,b shows the austenite represented as a sphere, where $\mathbf{a}_i$ with $i = 1, 2, 3$ are the basis vectors of the unit cell. The Bain strain changes the sphere into an ellipsoid of revolution about $\mathbf{a}_1$. There are no lines in the $(001)_\gamma$ plane that are undistorted. It is possible to find lines such as wx and yz that are undistorted by the deformation, but are rotated to the new positions $w'x'$ and $y'z'$ so they

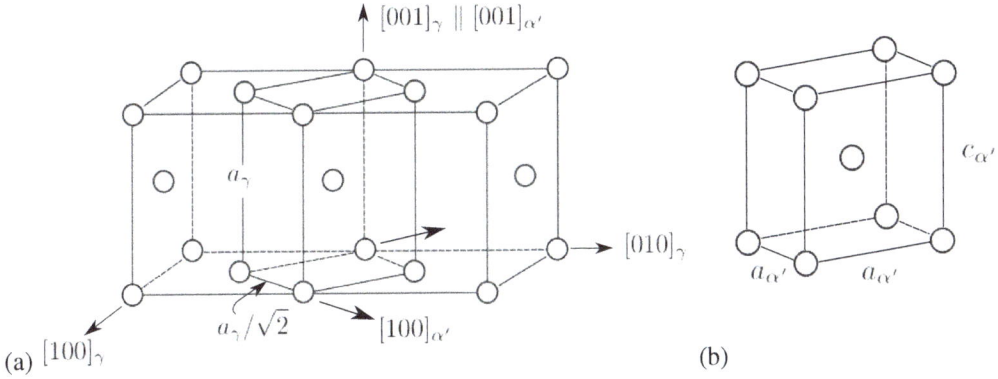

Figure 5.8 The Bain correspondence and deformation. (a) A pair of cubic austenite unit cells, with the face-centred tetragonal (fct) unit cell incorporated, showing also the orientation relationship expected with the martensite generated by the Bain strain. (b) The martensite unit cell following the Bain strain.

are not invariant. The Bain strain clearly is not an invariant-line strain, but can be converted into one by adding an appropriate rigid body rotation (**R**) that reorients the α' without affecting its crystal structure (Figure 5.9c), to enable one of the original undistorted lines (in this case yz) to be invariant. A combination **BR** is indeed an invariant-line strain. This is why the observed irrational orientation relationship differs from that implied by the Bain strain. The rotation required to generate convert **B** into an ILS yields the observed orientation of the martensite relative to the austenite.

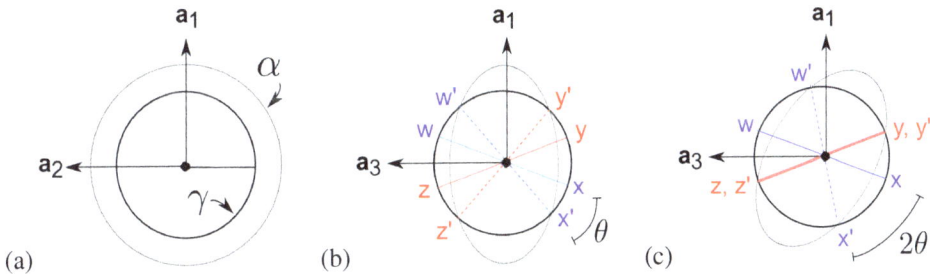

Figure 5.9 (a) and (b) show the effect of the Bain strain on austenite, which when undeformed is represented as a sphere of diameter $wx = yz$ in three-dimensions. The strain transforms it into an ellipsoid of revolution. (c) Shows the invariant-line strain obtained by combining the Bain strain with a rigid body rotation.

Figure 5.9c shows that there is no rotation that can convert **B** into an invariant-plane strain because there is no rotation capable of making two of the non-parallel undistorted lines into invariant-lines. It is impossible to convert γ into α' by a deformation that is an invariant-plane strain. The two crystals cannot ever be joined at an interface which is coherent and stress-free.

The principal distortions of **B** are such that $\eta_1 = \eta_2 > 1$ and $\eta_3 < 1$. In order to obtain an invariant-plane strain, the pure component of the total strain must be such that $\eta_1 > 1$, $\eta_2 = 1$ and $\eta_3 < 1$. Figure 5.10a shows that the pure strain itself is an invariant-line strain, because uv is undistorted and unrotated. A rotation about uv produces another non-parallel invariant-line $yz = y'z'$, making the net deformation an invariant-plane strain. This diagram is representative of the $\gamma \to \varepsilon$ fcc to

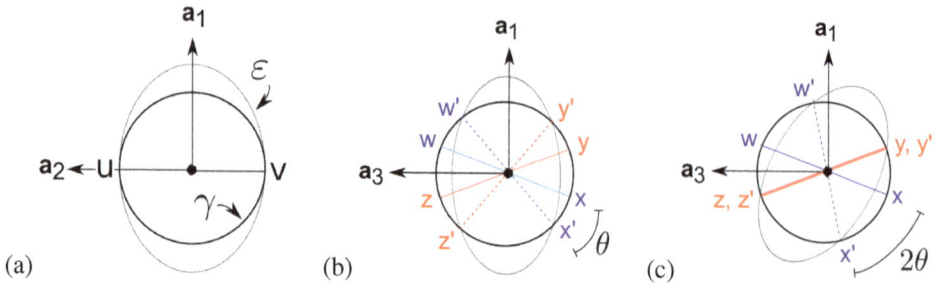

Figure 5.10 Illustration of a case where the pure strain has one of its principal distortions equal to unity, with the other two being greater or less than unity. When the pure strain is combined with a rigid body rotation the total strain is an invariant-plane strain.

hcp transformation. A fully coherent γ/ε interface is therefore possible if the principal distortions associated with the pure strain $\eta_1, \eta_3 > 1$ and $\eta_2 = 1$.

It is safe to conclude that the minimum condition for martensitic transformation to be possible at all (in any material) is that the deformation that carries the parent lattice into that of the product must as a minimum be an invariant-line strain in order to ensure a glissile interface. The caveat to this statement is that very small particles, whether constrained or not, may undergo such transformation even if they violate this condition given that they can be forced into coherency. Small, spherical iron particles precipitated in copper assume a fcc structure and can be induced to transform as a whole, into internally-twinned martensite [47]. In the Olson and Cohen theory for α' nucleation of assumes a somewhat different crystallography for the embryo which then evolves into the final macroscopically observed orientation and habit plane (Section 5.7).

The discussion thus far has been limited to the necessary condition for martensitic transformation, but the dislocations described are intrinsic with Burgers vectors that are such that they implement lattice-invariant deformation rather than phase transformation. There are other features, atom-sized steps, in the irrational interface that translate to accomplish the lattice change without diffusion. These are the coherency dislocations [48, 49], illustrated in Figure 5.11, with the step highlighted by the multi-shaded atoms at the interface. The atomic-height step has a dislocation character but there is no extra half-plane, just a tolerable distortion, which means that it can both glide and if necessary, climb without requiring diffusion. This is important because there usually is a change in density when one crystal structure is transformed into another. The coherency dislocations accomplish the lattice change but are associated with long-range strain fields surrounding the transformed particle. These fields are mitigated by introducing the anti-coherency (intrinsic) dislocations described earlier, that deform the lattice without changing its structure.   The Burgers vector of the steps is determined by the closure failure of a Burgers circuit spanning the interface [50, 51]. The Burgers vector content crossing a vector **p** in the interface (Figure 5.12, OP) can be determined from the closure failure of a Burgers circuit. The closed circuit AOBP spans the two different crystals. If a deformation (A S A) is now applied to convert one of the crystals into the other (reference crystal), then the closure failure P'P represents the required Burgers vector content $\mathbf{h}_t$ which can then be de-convoluted into individual dislocations to define the possible structure of the interface [40].

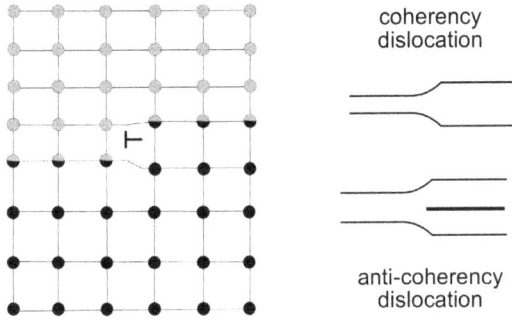

coherency
dislocation

anti-coherency
dislocation

Figure 5.11 The concept of a coherency dislocation, where the lattices are joined at a stepped interface. The dislocation represented by the step can move without creating or destroying lattice sites, i.e. it can glide and climb conservatively.

These operations are expressed formally as

$$[A; \mathbf{b_t}] = \{\mathbf{I} - (A\,S\,A)^{-1}\}[A; \mathbf{p}]. \tag{5.1}$$

In general therefore, a martensite-austenite interface will contain coherent steps of small height which for an irrational interface are arranged aperiodically, and intrinsic dislocations to accomplish the lattice invariant deformation that reduces the extent of the strain field due to the former. High-resolution imaging of martensite-austenite interfaces in steels has confirmed that the significant strain-field associated with its structure extends only 0.2-0.8 nm [52].The interface plane is then the average plane containing both components, as verified experimentally for lath martensite [28].

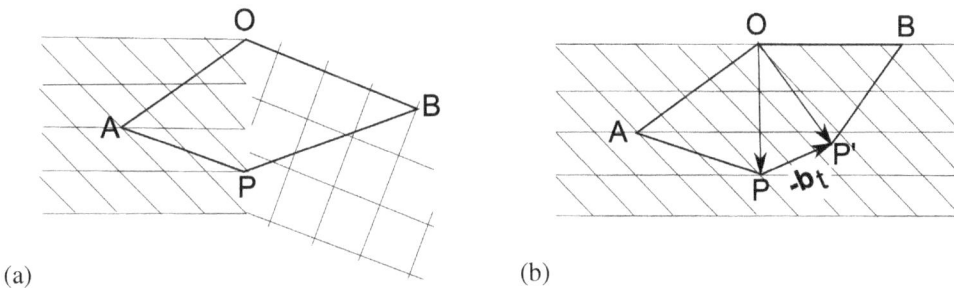

(a)   (b)

Figure 5.12 (a) Burgers circuit drawn across an interface between two different crystal structures [53]. (b) One of the structures is deformed into the other to discover the closure failure that defines the Burgers vector content crossing a vector (OP) in the interface. The homogeneous deformation (A S A) converts the assembly of crystals into a single reference-crystal with basis symbol 'A' and $\mathbf{b_t} = P'P$ is the total Burgers vector content crossing a vector $\mathbf{p} = OP$ in the interface. $\mathbf{b_t}$ can then be attributed to individual defects in the interface [40].

It is possible that in some martensitic transformations there are two lattice-invariant shears. As pointed out by Ross and Crocker, if these are independent shears then the interfacial dislocations would interact to render the interface sessile, although this might be avoided if growth involves the translation of two parallel interfaces, each containing only one set of intrinsic dislocations [54]. It has been argued that the region between the two interfaces must have a different crystallography so therefore the interfaces may not be parallel [55] but there may nevertheless be other factors such as a fault energy that could attract them. There have been no direct observations of the double-interface but molecular dynamics simulations indicate the possibility of paired interfaces that maintain a gap in-between as the martensite grows [56].[2]

5.2.4 SHAPE DEFORMATION

The shape of the pattern in which the atoms in austenite are arranged changes on transformation into martensite. Given that the atoms do not diffuse, there must be a corresponding change in the macroscopic shape of the transformed region. This deformation when observed on an initially flat surface shows that the latter becomes tilted uniformly about the line formed by the intersection of the habit plane with the surface (Figure 5.13a). Any scratch traversing the transformed region is similarly deflected while remaining connected at the α'/γ interface. These observations, and others, confirm that the measured shape deformation is an invariant-plane strain with a large shear component ($\simeq 0.22$) and a small dilatational strain ($\simeq 0.03$) directed normal to the habit plane.

The observation of this specific shape deformation obviously contradicts the fact that austenite cannot be transformed into α'-martensite by a strain that is an invariant-plane strain (Figure 5.9). The Bain strain, when combined with a rigid body rotation, can at best produce just one invariant-line. Consequently, if the observed shape deformation is applied to the austenite then it generates the wrong crystal structure.

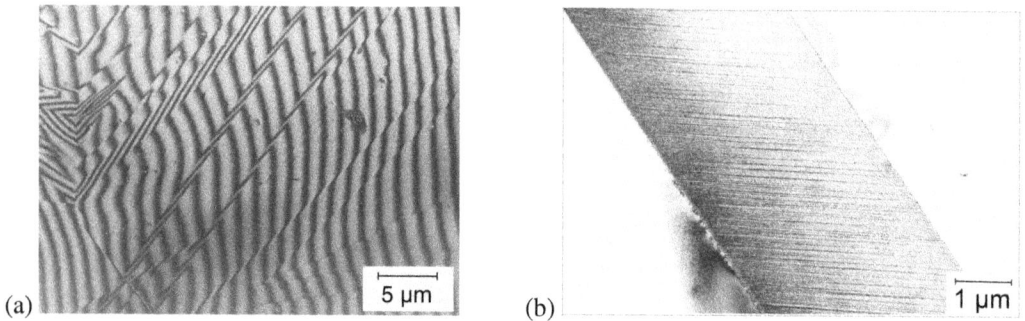

Figure 5.13 Mixture of martensite and retained austenite. (a) Fe-31Ni-10Co-Ti wt%, $M_S = 133$ K. Surface relief of thin plates of martensite embedded in austenite, revealed using Tolansky interference microscopy. Displacements caused by the formation of a plate of martensite at the surface of austenite that was polished flat before transformation. (b) Fe-30Ni-0.3C wt%, $M_S = 83$ K. Transmission electron micrograph showing finely spaced transformation-twins inside a plate of martensite. Micrographs courtesy of T. Maki.

5.2.5 MICROSTRUCTURE

Martensite is in some cases found to contain finely spaced twins when observed using transmission electron microscopy (Figure 5.13b). These twins are an intrinsic feature of the martensite in the sense that they are created during the formation of the plate as the γ/α' interface translates. Furthermore, both the austenite and the martensite regions in this thin-plate martensite are free from dislocation contrast. Although martensite is traditionally associated with a large dislocation density in the range 10^{14}-10^{16} m^{-2} [57–59], this need not be the case when its shape deformation is accommodated elastically.

The shape deformation has a strain field. The strain energy due to this can in elasticity theory be partitioned into that which resides in the precipitate and surrounding matrix, in our case the austenite [p. 467, 53]. For a shear dominated transformation, this ratio is about twice the aspect ratio at approximately $2z_t/z_\ell$ (known as the accommodation factor) where z_t is the plate thickness and z_ℓ the plate length. It is clear that for typical aspect ratios of martensite plates the majority of strain energy would be located in the austenite.

The variety of martensite morphologies that have been observed experimentally are summarised in Figure 5.14 [60]. Although focused on the Fe-Ni-C system, the mapping has generic significance. The thin plate martensite is elastically accommodated with no deformation visible of the adjacent austenite as verified using precision diffraction and imaging techniques [61]. Given the absence of dislocation debris, the interface between the thin plate and austenite is found to be reversible, with a behaviour consistent with a shape memory effect [62, 63]. A thin plate has the smallest of aspect ratios of all the common morphologies of martensite so it is not surprising that it is accommodated elastically.

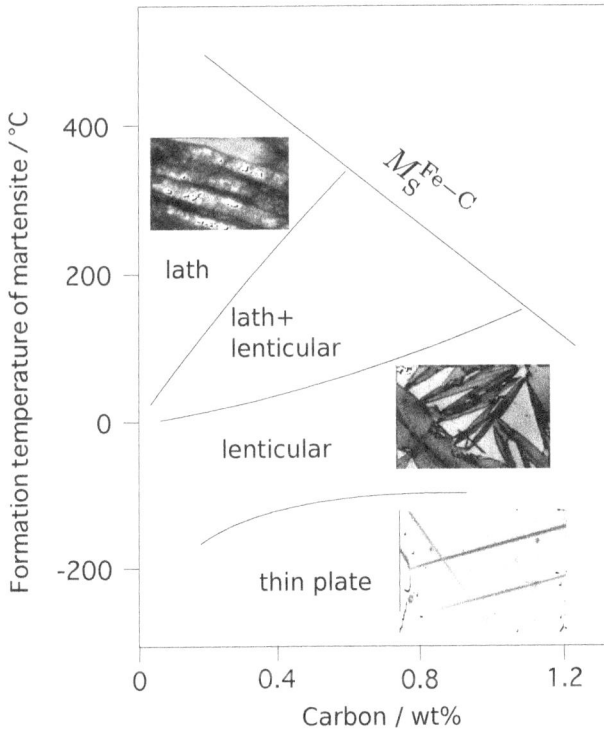

Figure 5.14 Lath, lenticular and thin-plate martensite morphologies mapped according to the transformation temperature and carbon concentration. Adapted from Maki et al. [60]. The lath martensite typically shows a great deal of dislocation contrast. The characteristic lens-like martensite occurs when the transformation temperature is depressed, giving way eventually to the thin-plate morphology.

The elastic model is less useful when considering martensite that forms at elevated temperatures. The strength of austenite decreases with temperature so it becomes prone to relax by plastic deformation when subjected to the shape deformation during transformation. But the martensite plate itself does not relax in this manner because of the low accommodation factor. Any dislocations within the lath martensite are therefore mostly inherited from the deformed austenite into which it grows [64]. The reversal of the interface then becomes more difficult, with interface velocities recorded at just 10-230 nm s^{-1} during heating of partially martensitic Fe-Ni-C alloys [65]. Dislocation debris in the austenite can limit the size of martensite laths by blocking the glissile transformation-interface, resulting in interrupted growth akin to the sub-unit mechanism of the bainite transformation (Figure 6.4).

There are other defects caused by the accommodation of the shape deformation due to martensitic transformation. Mechanical twinning is induced when differently oriented plates collide Figure 5.15; the collisions can cause microscopic cracking at the contact zones if the martensite is brittle [66].

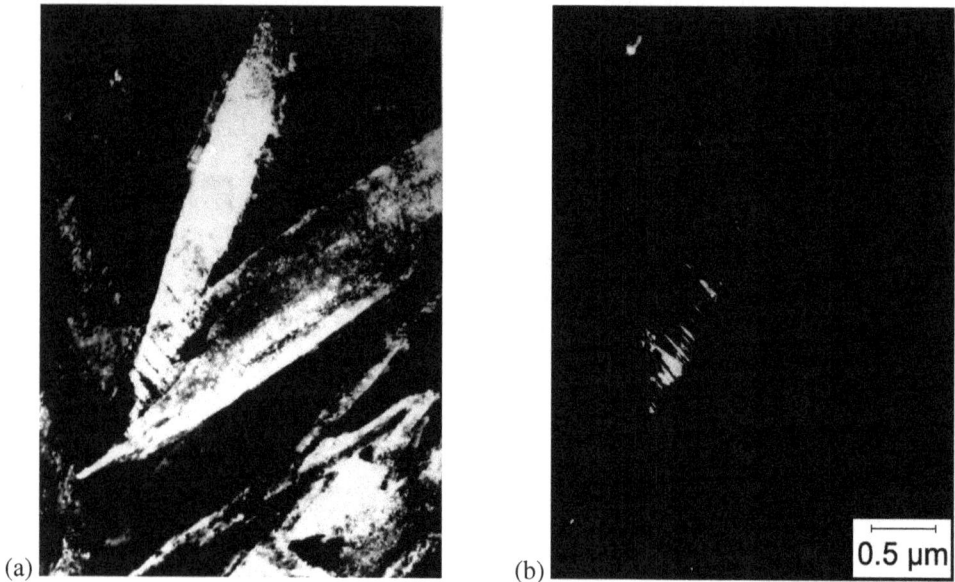

Figure 5.15 Two platelets of martensite which have collided during the course of transformation. The steel has the composition Fe-0.44C-1.74Si-0.67Mn-0.83Cr-0.39Mo-1.85Ni-0.15V wt%. (a) Bright field image showing mechanical twins in the plate that experienced the collision. (b) Corresponding dark field image highlighting the mechanical twins at the contact surface [8].

5.2.6 SUMMARY

The difficulties in reconciling the observations thus far can be summarised as follows:

(a) The lattice deformation implies the Bain orientation which is not observed experimentally. This is resolved because **B** must be combined with **R** in order to permit the existence of a line that remains invariant to the net operation. It is this rotation **R** which corrects the Bain orientation to that observed experimentally.

(b) The habit plane indices of martensite are irrational [67] and often peculiar, for example, $\approx \{3\,10\,15\}_\gamma$. Why is this?

(c) The observed shape deformation is an invariant-plane strain. However, an invariant-plane strain cannot change the austenite structure into that of α'.

(d) The microstructure of the martensite sometimes contains finely spaced transformation-twins [68] that need to be explained.

These issues were all resolved independently by Wechsler, Lieberman and Read [69] and Bowles and MacKenzie [70–72]. Their phenomenological theory of martensite crystallography is summarised next.

5.2.7 CRYSTALLOGRAPHIC THEORY

The Bain strain converts the crystal structure of the austenite into that of the martensite, but the accompanying distortions are too large to sustain so there are additional mitigating operations needed to ensure that there is a good macroscopic fit between the two phases.

Referring to Figure 5.16, **B** is combined with an appropriate rigid body rotation **R** such that **RB** is an invariant-line strain (step a to c) that passes through x and is normal to the plane of the diagram. Inconsistent with this, the observed shape deformation is an invariant-plane strain $\mathbf{P_1}$ (step a to b) but this gives the wrong crystal structure. The invariant-plane of the shape deformation is defined by xw. If, however, a second homogeneous shear $\mathbf{P_2}$ is combined with $\mathbf{P_1}$ (step b to c), then the correct structure is obtained but the wrong shape since[3]

$$\mathbf{P_1 P_2} = \mathbf{RB}.$$

The discrepancies are all resolved if the shape changing effect of $\mathbf{P_2}$ is cancelled macroscopically by an inhomogeneous lattice-invariant deformation, which may be slip or twinning. Notice that the habit plane in Figure 5.16e,f is given by a fragmentation of the original plane xw, due to the inhomogeneous lattice-invariant shear. This is why the habit plane of martensite has peculiar indices. In the absence of a lattice-invariant deformation as in the $\gamma \to \varepsilon$ transformation, the sequence stops at step b and therefore the habit plane has rational indices $\{1\,1\,1\}_\gamma$.

The theory in essence explains all the important features of the martensite crystallography. The orientation relationship is predicted by setting **R** such that **BR** is an invariant-line strain. The habit plane does not have rational indices because the amount of lattice-invariant deformation needed to recover the correct the macroscopic shape is not usually rational. A substructure is predicted, either transformation twins or slip steps, both of which have been observed experimentally. The fact that the macroscopic deformation accompanying martensitic transformation is an invariant-plane strain reduces the strain energy when compared with the Bain distortion.

If is self-evident that when the lattice-invariant deformation is slip, the translated regions retain the same correspondences and lattice deformations. When it is twinning, the twinned orientations share the same γ/α' correspondence so the twin plane in the martensite must correspond to a mirror plane in the austenite [72].

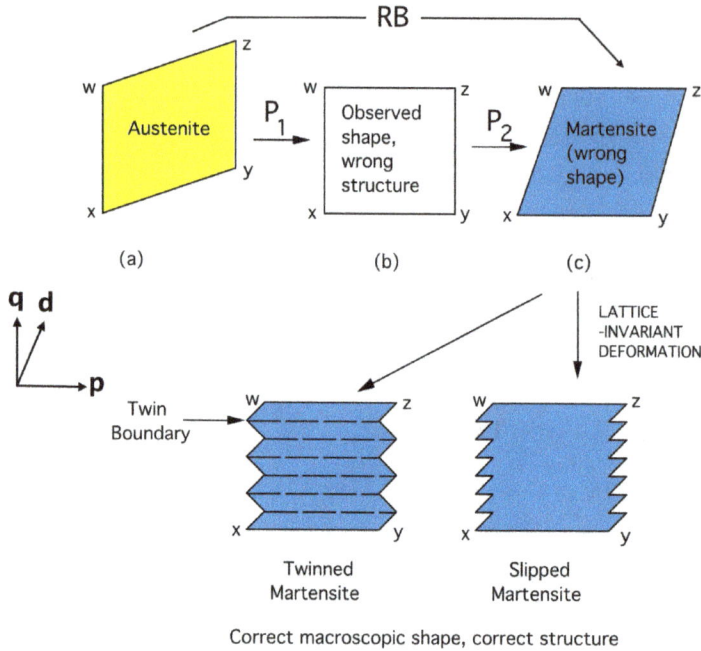

Figure 5.16 Schematic illustration of the phenomenological theory of martensite. (a) represents single crystal of austenite and (c) has a bcc structure. (b) has a structure between fcc and bcc, **p** is the habit plane unit normal and **q** is the unit normal to the plane on which the lattice-invariant shear occurs. The heavy horizontal lines in (e) are coherent twin boundaries. The vector **e** is normal to **q** but does not lie in the plane of the diagram.

5.3   QUANTITATIVE THEORY

It is important to note that the crystallographic theory places a firm mathematical connection between the habit plane, orientation relationship and shape deformation, characteristics that are not independent. The mathematical treatment requires an understanding of vector and matrix methods that are described elsewhere [73, 74].

5.3.1   FCC TO BCC MARTENSITIC TRANSFORMATION

Undistorted Vectors

The first task is to find the lines that are not distorted by the lattice deformation. The Bain strain is given by

$$(\text{F B F}) = \begin{pmatrix} \eta_1 & 0 & 0 \\ 0 & \eta_2 & 0 \\ 0 & 0 & \eta_3 \end{pmatrix}$$

where 'F' is an orthonormal basis consisting of unit basis vectors $\mathbf{f}_i$ parallel to the crystallographic axes of the conventional fcc austenite unit cell: $\mathbf{f}_1 \parallel \mathbf{a}_1$, $\mathbf{f}_2 \parallel \mathbf{a}_2$ and $\mathbf{f}_3 \parallel \mathbf{a}_3$ and the principal deformations of the Bain Strain are typically $\eta_1 = \eta_2 = 1.136071$ and $\eta_3 = 0.803324$. The strain produces a cone of undistorted lines which are generated from an original cone prior to the deformation. The concentric cones have axes parallel to $\mathbf{a}_3$. If a rigid body rotation (F J F) is added which makes these two cones touch along a line, then that line becomes invariant to the total strain (F S F). It is obvious though that there is an infinite choice of rotations that can make the cones touch along a line. The proper rotation is that which ensures that the invariant-line is at the intersection of the habit plane and the plane on which the lattice-invariant deformation occurs.[4] Similarly, the normal of the plane left invariant by (F S F) must contain the displacement directions of the lattice-invariant deformation and the shape deformation. It therefore is necessary to choose the lattice-invariant shear system. It is assumed here the lattice-invariant shear occurs on $(1\,0\,1)_F$ in the $[1\,0\,\overline{1}]_F$ direction.

Suppose the invariant-line is written as a unit vector $\mathbf{u}$. For it to lie in $(1\,0\,1)_F$ requires its components to satisfy the equation $u_1 = -u_3$. As a result of the Bain strain this becomes a new vector $\mathbf{w}$; bearing in mind that the basis F is orthonormal, the magnitude of the new vector is given by

$$
\begin{aligned}
|\mathbf{w}|^2 &= (\mathbf{w};F)[F;\mathbf{w}] \\
&= (\mathbf{u};F)(FB'F)(F\,B\,F)[F;\mathbf{u}] \\
&= (\mathbf{u};F)(F\,B\,F)^2[F;\mathbf{u}].
\end{aligned}
$$

Since the magnitude of $\mathbf{u}$ is not changed by the Bain strain, it follows that $|\mathbf{u}| = |\mathbf{w}|$, i.e.,

$$
u_1^2 + u_2^2 + u_3^2 = \eta_1^2 u_1^2 + \eta_2^2 u_2^2 + \eta_3^2 u_3^2 = 1.
$$

These equations can be solved simultaneously to give *two* solutions for undistorted lines:[5]

$$
\begin{aligned}
[F;\mathbf{u}] &= [-0.671120 \quad -0.314952 \quad 0.671120] \\
[F;\mathbf{v}] &= [-0.671120 \quad 0.314952 \quad 0.671120.]
\end{aligned}
$$

For $\mathbf{h}$ to contain the direction of the lattice-invariant shear $[1\,0\,\overline{1}]_F$, its components must satisfy

$$
h_1 = h_3 \qquad \text{and} \qquad (\mathbf{h};F^*)[F^*;\mathbf{h}] = 1.
$$

As a consequence of the Bain strain $\mathbf{h}$ becomes a new plane normal $\mathbf{l}$ and if $|\mathbf{h}| = |\mathbf{l}|$ then

$$
\begin{aligned}
|\mathbf{l}|^2 &= (\mathbf{l};F^*)[F^*;\mathbf{l}] \\
&= (\mathbf{h};F^*)(FBF)^{-1}(FB'F)^{-1}[F^*;\mathbf{h}]
\end{aligned}
$$

so that $\qquad h_1^2 + h_2^2 + h_3^2 = (l_1/\eta_1)^2 + (l_2/\eta_2)^2 + (l_3/\eta_3)^2.$

When solved simultaneously, the two solutions for the undistorted-normals are found to be

$$
\begin{aligned}
(\mathbf{h};F^*) &= (0.539127 \quad 0.647058 \quad 0.539127) \\
(\mathbf{k};F^*) &= (0.539127 \quad -0.647058 \quad 0.539127).
\end{aligned}
$$

Invariant-line strain

To convert (F B F) into an invariant-line strain (F S F) it is necessary to employ a rigid body rotation (F J F) which simultaneously brings an undistorted line (such as $\mathbf{w}$) and an undistorted normal (such as **l**) back into their original directions along **u** and **h** respectively. This is possible because the angle between **w** and **l** is the same as that between **u** and **h**:

$$
\begin{aligned}
\mathbf{l.w} &= (\mathbf{l};F^*)[F;\mathbf{w}] \\
&= (\mathbf{h};F^*)(F\,B\,F)^{-1}(F\,B\,F)[F;\mathbf{u}] \\
&= (\mathbf{h};F^*)[F;\mathbf{u}] \\
&= \mathbf{h.u.}
\end{aligned}
$$

One possibility is to rotate **l** into **h** and **w** into **u**. This choice is not unique given that there is a pair each of undistorted lines and plane normals, giving four equivalent solutions for the case where the invariant-line must lie in $(1\,0\,1)_F$ and that the invariant normal defines a plane containing $[1\,0\,\overline{1}]$. For the solution obtained using the pair **u** and **h**:

$$
\begin{aligned}
\mathbf{l} = (\mathbf{h};F^*)(F\,B\,F)^{-1} &= (\;\;0.474554\quad 0.569558\quad 0.671120) \\
\mathbf{w} = (F\,B\,F)[F;\mathbf{u}] &= [-0.762440\;-0.357809\quad 0.539127] \\
\mathbf{a} = \mathbf{u}\wedge\mathbf{h} &= (-0.604053\quad 0.723638\;-0.264454) \\
\mathbf{b} = \mathbf{w}\wedge\mathbf{l} &= (-0.547197\quad 0.767534\;-0.264454).
\end{aligned}
$$

The required rigid body rotation brings **w** back to **u**, **l** back to **h** and **b** to **a**, respectively, giving the three equations:

$$
\begin{aligned}
[F;\mathbf{u}] &= (F\,J\,F)[F;\mathbf{w}] \\
[F;\mathbf{h}] &= (F\,J\,F)[F;\mathbf{l}] \\
[F;\mathbf{a}] &= (F\,J\,F)[F;\mathbf{b}]
\end{aligned}
$$

(5.2)

which when solved gives

$$
(F\,J\,F) = \begin{pmatrix} 0.990534 & -0.035103 & 0.132700 \\ 0.021102 & 0.994197 & 0.105482 \\ -0.135633 & -0.101683 & 0.985527 \end{pmatrix}
$$

which is a rotation of $9.89°$ about $[0.602879\;-0.780887\;0.163563]_F$. The invariant-line strain $(F\,S\,F) = (F\,J\,F)(F\,B\,F)$ is therefore,

$$
(F\,S\,F) = \begin{pmatrix} 1.125317 & -0.039880 & 0.106601 \\ 0.023973 & 1.129478 & 0.084736 \\ -0.154089 & -0.115519 & 0.791698 \end{pmatrix}
$$

$$
(F\,S\,F)^{-1} = \begin{pmatrix} 0.871896 & 0.018574 & -0.119388 \\ -0.030899 & 0.875120 & -0.089504 \\ 0.165189 & 0.131307 & 1.226811 \end{pmatrix}.
$$

Orientation relationship

The coordinate transformation matrix relating the austenite and martensite (α') can be obtained using the total transformation strain $\mathbf{S}$ and the correspondence matrix $\mathbf{C}$:

$$(\alpha'\, J\, \gamma)(\gamma\, S\, \gamma) = (\alpha'\, C\, \gamma) \qquad \text{and} \qquad (\alpha'\, J\, \gamma) = (\alpha'\, C\, \gamma)(\gamma\, S\, \gamma)^{-1}$$

where $(\alpha'\, C\, \gamma)$ is the Bain correspondence.[6] Therefore,

$$(\alpha'\quad J\quad \gamma) = \begin{pmatrix} 0.902795 & -0.856546 & -0.029884 \\ 0.840997 & 0.893694 & -0.208892 \\ 0.165189 & 0.131307 & 1.226811 \end{pmatrix}$$

so that

$$(1\,1\,1)_\gamma \quad = \quad (0.010561\ 0.984639\ 0.983036)_{\alpha'}$$

$$[\overline{1}0\,1]_\gamma \quad = \quad [-0.932679 - 1.049889\ 1.061622]_{\alpha'}.$$

$(1\,1\,1)_\gamma$ is very nearly but not exactly parallel to $(0\,1\,1)_{\alpha'}$ and $[\overline{1}0\,1]_\gamma$ is about $3°$ from $[\overline{1}\,\overline{1}\,1]_{\alpha'}$. The orientation relationship is illustrated in Figure 5.17.

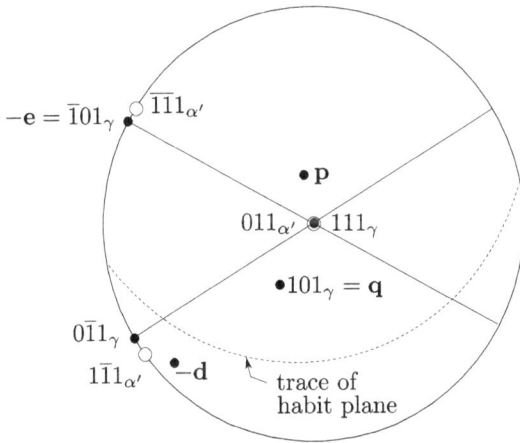

Figure 5.17 Stereographic representation of the orientation relationship between martensite and austenite. The lattice-invariant shear plane $\mathbf{q}$ and direction $-\mathbf{e}$, the habit plane $\mathbf{p}$ and unit displacement vector $\mathbf{d}$ are also shown.

Habit plane and shape deformation

It was assumed that the lattice-invariant shear is applied inhomogeneously on $(1\,0\,1)[\overline{1}0\,1]_F$, the effect of which is to cancel the shape change due to the homogeneous deformation $(\mathrm{F}\,P_2\,\mathrm{F})$ which must also occur on $(1\,0\,1)[\overline{1}0\,1]_F$. $(\mathrm{F}\,S\,\mathrm{F})$ is an invariant-line strain with the condition that the invariant-line $\mathbf{u}$ is in $(1\,0\,1)_F$ and that the plane defined by the invariant-normal $\mathbf{h}$ contains the direction $[1\,0\,\overline{1}]_F$.

The habit plane unit-normal $\mathbf{p}$ can be obtained by the factorisation $\mathbf{S} = \mathbf{P_1 P_2}$:

$$(\mathrm{F}\,S\,\mathrm{F}) \quad = \quad (\mathrm{F}\,P_1\,\mathrm{F})(\mathrm{F}\,P_2\,\mathrm{F})$$

$$= \quad \left[\mathbf{I} + m[\mathrm{F};\mathbf{d}](\mathbf{p};\mathrm{F}^*)\right]\left[\mathbf{I} + n[\mathrm{F};\mathbf{e}](\mathbf{q};\mathrm{F}^*)\right]$$

$$(F\,S\,F)^{-1} = (F\,P_2\,F)^{-1}(F\,P_1\,F)^{-1}$$

$$= \left[I - n[F;e](q;F^*)\right]\left[I - am[F;d](p;F^*)\right] \qquad (5.3)$$

where $a^{-1} = \det(F\,P_1\,F)$ and $\det(F\,P_2\,F) = 1$ since the latter is just a shear that does not cause a change in volume. Using Equation 5.3, it follows that

$$(q;F^*)(F\,S\,F)^{-1} = (q;F^*) - b(p;F^*) \qquad \text{where} \qquad b = am(q;F^*)[F;d],$$

so that

$$(p;F^*) \parallel (0.197162\ 0.796841\ 0.571115).$$

As expected, this habit plane is irrational. To completely describe the shape deformation matrix P_1 it is necessary to determine m and the unit displacement direction d. Using Equation 5.3,

$$(F\,S\,F)[F;e] = [F;e] + m[F;d](p;F^*)[F;e]$$

giving

$$[F;d] = [-0.223961\ 0.727229\ -0.648829]_F$$

and

$$m = |md| = 0.223435.$$

The magnitude m of the displacement can be factorised into a shear component s parallel to the habit plane and a dilatational component ζ normal to the habit plane. Hence, $\zeta = md.p = 0.0368161$ and $s = (m^2 - \zeta^2)^{\frac{1}{2}} = 0.220381$. These are typical values of the dilatational and shear components of the shape strain for ferrous martensites. Finally, the shape deformation matrix is given by

$$(F\,P_1\,F) = \begin{pmatrix} 0.990134 & -0.039875 & -0.028579 \\ 0.032037 & 1.129478 & 0.092800 \\ -0.028583 & -0.115519 & 0.917205 \end{pmatrix}.$$

Lattice-invariant shear

The shape deformation P_1 does not change the fcc lattice to that of bcc martensite. A further homogeneous lattice deformation P_2 is necessary to complete the structural change. Nevertheless, its effect is not visible when the shape deformation is measured experimentally. This is because the homogeneous deformation associated with P_2 is offset by an inhomogeneously applied lattice-invariant shear. Therefore, the lattice-invariant deformation must on average be the inverse of P_2, i.e., on the plane defined by the unit normal q but in the opposite direction $-e$. The magnitude of the lattice-invariant shear should on average equal that of P_2. Given that $S = P_1 P_2$, it follows that for the case at hand,

$$(F\,P_2\,F) = \begin{pmatrix} 1.132700 & 0.000000 & 0.132700 \\ 0.000000 & 1.000000 & 0.000000 \\ -0.132700 & 0.000000 & 0.867299 \end{pmatrix}$$

which is a homogeneous shear on the system $(1\ 0\ 1)[1\ 0\ \bar{1}]_F$ with a magnitude $n = 0.2654$. The lattice-invariant shear is on $(1\ 0\ 1)[\bar{1}\ 0\ 1]_F$. In practice this shear occurs as the interface moves and the interface dislocations glide with it. These dislocations might have a Burgers vector $\mathbf{b} = \frac{a}{2}[\bar{1}\,0\,1]_\gamma$ and if they are located on every K'th slip plane, then

$$n = |\mathbf{b}|/Kd = 1/K \qquad \text{so that} \qquad K = 1/0.2654 = 3.7679$$

where d is the spacing of the $(1\,0\,1)_\gamma$ planes. Because K should be an integral number of plane-spacings, the result must be taken to mean that there will on average be a dislocation located on every 3.7679 th slip plane. In reality, the dislocations will be non-uniformly placed, either 3 or 4 $(1\,0\,1)$ planes apart.

The line vector of the dislocations is the invariant-line $\mathbf{u}$ and the spacing of the intrinsic dislocations as measured on the habit plane is $Kd/(\mathbf{u} \wedge \mathbf{p}.\mathbf{q})$ where all the vectors are unit vectors. The average spacing is therefore expected to be

$$3.7679(\sqrt{2}a_\gamma)/0.8395675 = 3.1734 a_\gamma.$$

Taking $a_\gamma = 0.356$ nm gives an average spacing of 1.13 nm.

If the lattice-invariant shear is twinning rather than slip, then the martensite plate will contain finely spaced transformation twins. The mismatch between the parent and product lattices is in the slip case accommodated with the help of intrinsic dislocations, whereas for the internally twinned martensite there are no such dislocations. Each twin terminates in the interface to give a facet between the parent and product lattices, a facet that is forced into coherency. The width of the twin and the size of the facet is sufficiently small to enable this forced coherency to exist. The alternating twin-related regions therefore prevent misfit from accumulating over large distances along the habit plane.

If the (fixed) magnitude of the twinning shear is denoted s_T, then the volume fraction V_V^T of the twin orientation, necessary to cancel the effect of $\mathbf{P}_2$, is given by $V_V^T = n/s_T$, assuming that $n < s_T$. Lattice-invariant shear on $(1\,0\,1)[\bar{1}\,0\,1]_\gamma$ corresponds to $(1\,1\,2)_{\alpha'}[\bar{1}\,\bar{1}\,1]_{\alpha'}$ and twinning on this latter system involves a shear $s = 1/\sqrt{2}$, giving $V_V^T = 0.375$.

It is important to note that the twin plane in the martensite corresponds to a mirror plane in the austenite; this is a necessary condition when the lattice-invariant shear involves twinning, because the twinned and untwinned regions of the martensite must undergo Bain strain along different though crystallographically equivalent principal axes [72].

The theory predicts the fraction of twins in each martensite plate, when the lattice-invariant shear is twinning. However, the factors governing the spacing of the twins are less established. The finer the spacing of the twins, the smaller will be the strain energy associated with the matching of each twin variant with the parent lattice at the interface. On the other hand, the amount of coherent twin boundary within the martensite increases as the spacing of the twins decreases.

The lattice-invariant shear is an integral part of the transformation; it does not happen as a separate event after the lattice change has occurred. The transformation and the lattice-invariant shear occur simultaneously as the interface migrates. It is well known that in ordinary plastic deformation, twinning rather than slip tends to be the favoured deformation mode at low temperatures or when high strain rates are imposed. It often is suggested that martensite with low martensite-start temperatures will tend to be twinned rather than slipped, but this cannot be formally justified because the lattice-invariant shear is an integral part of the transformation and not a physical deformation mode on its own. Indeed, it is possible to find lattice-invariant deformation modes in martensite which do not occur in ordinary plastic deformation experiments. The reasons why some martensites are internally twinned and others slipped are not clearly understood [75]. When the spacing of the transformation twins is roughly comparable to that of the dislocations in slipped martensite, the interface energies are roughly equal. The interface energy increases with twin thickness and at the observed thicknesses is large compared with the corresponding interface in slipped martensite. The combination of the relatively large interfacial energy and the twin boundaries left in the martensite plate means that internally twinned martensite is never thermodynamically favoured relative to the slipped version. It is possible that the greater mobility of the twinned interface favours that mode when the growth rate is very large and the plate is elastically accommodated.

5.4 ε-MARTENSITE

There is interest in steels containing substantial concentrations of manganese, some of which can transform in part into ε-martensite, either during cooling or by deformation of the austenite [76]. It is worth noting, however, that because of the suppression of transformation temperatures by these large solute concentrations, the ε-hcp-phase has only ever been observed as martensite; there are no reported observations of the phase at equilibrium with the austenite, in the form of say allotri-omorphs that grow by reconstructive transformation.

The microstructure of ε-martensite appears simple because the habit plane is exactly $\{111\}_\gamma$, the octahedral plane of austenite, of which there are only four non-parallel variants, Figure 5.18. The parallel-sided plates are slender when compared with lenticular α'-martensite and there is evidence to suggest that the ε is associated with less dislocation debris and plastic accommodation in the adjacent austenite, rather like the thin-plate α'-martensite in iron-nickel alloys (p. 269). X-ray diffraction peaks from ε-martensite exhibit widths that are identical to those associated with undeformed austenite [77], an indication that the martensite is well accommodated, with a low defect density within the plates, although stacking faults have been observed parallel to the basal plane [78]. Consistent with these observations, Fe-Mn-Si alloys can exhibit a shape memory effect [79]. In mixtures of thin and thick plates, the thicker ones are less well accommodated, making it more difficult for the transformation interface to reverse direction [80]. Figure 5.18b,c shows the surface relief due to martensitic transformation, some of which is reversed on heating to induce austenite formation – the arrows indicate positions where a thin-plate has disappeared completely while the thicker ones persist.

5.4.1 CRYSTALLOGRAPHY: FCC TO HCP TRANSFORMATION

Suppose austenite transforms into ε-martensite without a change in density. The transformation strain is then a shear on the close-packed $\{1\,1\,1\}_\gamma$ plane (the invariant-plane) along $\langle 1\,1\,\overline{2}\rangle_\gamma$, Figure 5.19. The magnitude of the shear is $8^{-\frac{1}{2}}$, which is half the normal twinning shear for austenite. Using an orthonormal basis Z, consisting of unit basis vectors parallel to $[1\,0\,0]_\gamma$, $[0\,1\,0]_\gamma$ and $[0\,0\,1]_\gamma$ directions respectively, and with $(\mathbf{p};Z^*) = 3^{-\frac{1}{2}}(1\,1\,1)$, $[Z;\mathbf{d}] = 6^{-\frac{1}{2}}[1\,1\,\overline{2}]$ and $m = 8^{-\frac{1}{2}}$ gives the total

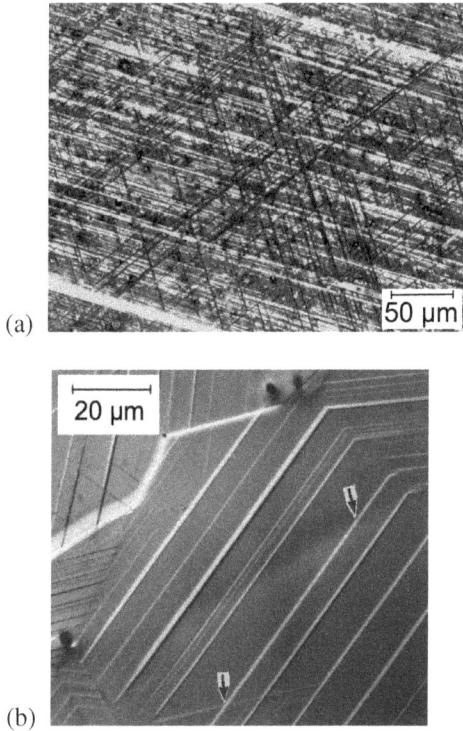

Figure 5.18 (a) ε-Martensite in a Fe-25.3Mn-2.8Si wt% alloy with $M_S^\varepsilon = 70\,^\circ$C. Micrograph courtesy of Hong-Seok Yang. (b) Fe-32.7Mn-6.2Si wt% alloy showing surface relief due to ε-martensite $M_S^\varepsilon = 23\,^\circ$C. (c) Changes in the surface relief due to reversion to austenite on heating to $600\,^\circ$C. Micrographs (b,c) courtesy of Professor Kaneaki Tsuzaki.

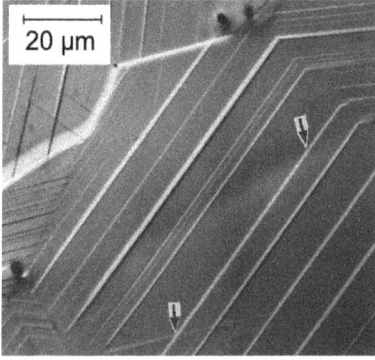

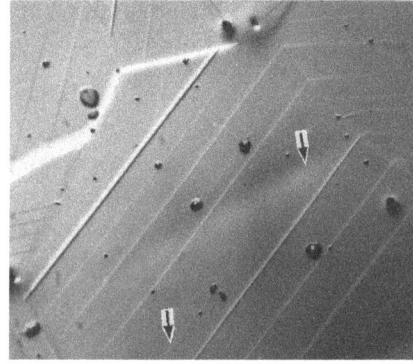

deformation, an invariant-plane strain:

$$(Z\,P\,Z) = \frac{1}{12}\begin{pmatrix} 13 & 1 & -2 \\ 1 & 13 & 1 \\ -2 & -2 & 10 \end{pmatrix}. \tag{5.4}$$

It is instructive to factorise this shear into a pure strain $(Z\,Q\,Z)$ and a rigid body rotation $(Z\,J\,Z)$. The eigenvectors of $(Z\,Q\,Z)$ represent the directions of maximum linear strain. The matrix $(Z\,T\,Z) = (Z\,P'\,Z)(Z\,P\,Z)$ is

$$(Z\,T\,Z) = \frac{1}{144}\begin{pmatrix} 174 & 30 & -6 \\ 30 & 174 & -6 \\ -6 & -6 & 102 \end{pmatrix} \tag{5.5}$$

the eigenvalues and eigenvectors of which are

$$
\begin{aligned}
\eta_1 &= 1.421535 & [Z;\mathbf{u}] &= [0.704706 \quad 0.704706 \quad -0.082341] \\
\eta_2 &= 1.000000 & [Z:\mathbf{v}] &= [0.707107 \quad -0.707107 \quad 0.000000] \\
\eta_3 &= 0.703465 & [Z;\mathbf{w}] &= [0.058224 \quad 0.058224 \quad 0.996604].
\end{aligned}
$$

The eigenvectors form an orthogonal set and since $\mathbf{v}$ lies in the invariant-plane. The vectors $\mathbf{u}$, $\mathbf{v}$ and $\mathbf{w}$ also are the eigenvectors of $(Z\,Q\,Z)$, the eigenvalues of which are the square roots of the eigenvalues of $(Z\,T\,Z)$; they are 1.192282, 1.0 and 0.838728 respectively. Therefore, the maximum extensions and contractions during the $\gamma \rightarrow \varepsilon$ martensitic transformation are less than 20%. The matrix $(Z\,Q\,Z)$ is now obtained by a similarity transformation with rotation matrices formed from

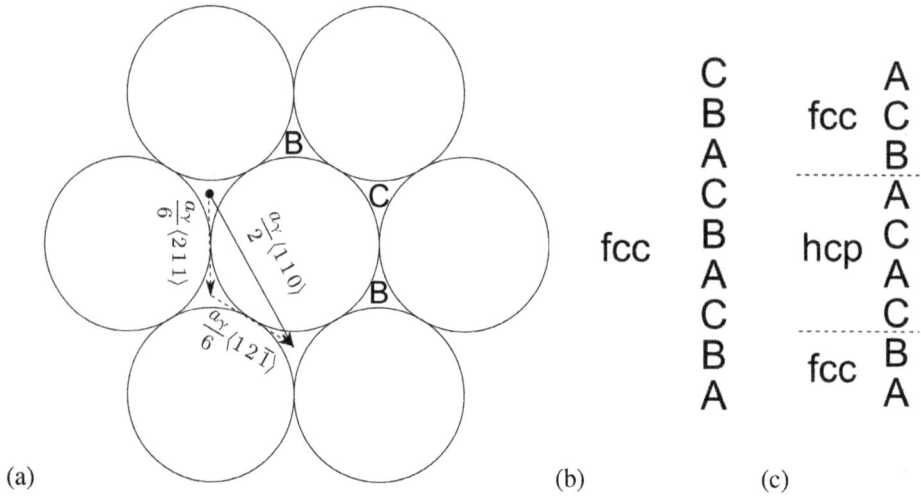

Figure 5.19 (a) $(\bar{1}11)_\gamma$ close-packed plane showing how a unit lattice dislocation can dissociate into two Shockley partials. The positions B,C mark possible positions of subsequent planes in the stacking sequence. (b) Stacking sequence of the close-packed planes in a fcc crystal. (c) The passage of a Shockley partial dislocation produces an intrinsic stacking fault which amounts to a two-layer thick region in which the stacking sequence is for the close-packed planes of the hcp crystal. Notice that for this region to truly represent hcp there has to be an additional displacement which is a contraction normal to the close-packed planes in order to obtain the correct hcp density which exceeds that of fcc-iron.

the eigenvectors:

$$
(Z\,Q\,Z) = \begin{pmatrix} 0.70471 & 0.70711 & 0.05822 \\ 0.70471 & -0.70711 & 0.05822 \\ -0.08234 & 0.00000 & 0.99660 \end{pmatrix} \times \begin{pmatrix} 1.19228 & 0 & \\ 0 & 1 & 0 \\ 0 & 0 & 0.83873 \end{pmatrix}
$$

$$
\times \begin{pmatrix} 0.70471 & 0.70471 & -0.08234 \\ 0.70711 & -0.70711 & 0.00000 \\ 0.05822 & 0.05822 & 0.99660 \end{pmatrix}
$$

$$
= \begin{pmatrix} 1.094944 & 0.094943 & -0.020515 \\ 0.094943 & 1.094944 & -0.020515 \\ -0.020515 & -0.020515 & 0.841125 \end{pmatrix}. \tag{5.6}
$$

The pure rotation can now be derived from $(Z\,J\,Z) = (Z\,P\,Z)(Z\,Q\,Z)^{-1}$

$$
(Z\quad J\quad Z) = \begin{pmatrix} 0.992365 & -0.007635 & 0.123091 \\ -0.007635 & 0.992365 & 0.123091 \\ -0.123092 & -0.123092 & 0.984732 \end{pmatrix}
$$

which is a right-handed rotation of $10.03°$ about $[1\bar{1}0]_Z$ axis.

The fcc to hcp transformation physically occurs by the movement of a single set of Shockley partial dislocations, Burgers vector $\mathbf{b} = \frac{a}{6}\langle 11\bar{2}\rangle_\gamma$ on *alternate* close-packed $\{111\}_\gamma$ planes. To produce a fair thickness of hcp martensite, a mechanism has to be sought which allows Shockley partials to be generated on every other slip plane. Motion of the partials would cause a shearing of the austenite lattice on the system $\{111\}_\gamma \langle 11\bar{2}\rangle_\gamma$. The average magnitude of the shear strain is $s = |\mathbf{b}|/2d$ where d is the spacing of the close-packed planes. This is exactly the shear system described by

(Z P Z) and its physical effect is to tilt an originally flat surface about a line given by its intersection with the habit plane, through some angle dependent on the indices of the free surface. By measuring such tilts it is possible to measure s, which is confirmed to equal half the twinning shear.

In fcc crystals, the close-packed planes have a stacking sequence $...ABCABC...$. The passage of a single Shockley partial causes the sequence to change to $...ABA...$ creating a three layer thick region of hcp phase whose close-packed planes are stacked with a periodicity of 2. It is natural therefore for the habit plane to be $\{111\}_\gamma$. If the habit plane deviates slightly from $\{111\}_\gamma$ then the interface will consist of stepped sections of close-packed plane, the steps representing the Shockley partial transformation dislocations. The spacing of the partials along $\langle 111\rangle_\gamma$ would be $2d$, the partials located on every alternate plane of the fcc crystal.

The homogeneous deformation matrix (Z P Z) is compatible with the dislocation mechanism of transformation. It predicts the correct macroscopic surface relief effect and its invariant-plane is the habit plane of the martensite. However, if (Z P Z) is considered to act homogeneously over the entire crystal, then it would carry half the atoms into the wrong positions. For instance, if the habit plane is designated A in the sequence ABC of close-packed planes, then the effect of (Z P Z) is to leave A unchanged, shift the atoms on plane C by $2sd$ and those on plane B by sd along the shear direction. This puts the atoms originally in C sites into A sites as required for hcp stacking. However, the B atoms are located at positions half way between B and C sites, through a distance $\frac{a}{12}\langle 11\overline{2}\rangle_\gamma$. Shuffles are thus necessary to bring these atoms back into the original B positions and to restore the $...ABA...$ hcp sequence. The shuffle here is a purely formal concept; consistent with the fact that the Shockley partials glide over alternate close-packed planes, the deformation (Z P Z) must in fact be considered homogeneous only on a scale of every two planes. By locking the close-packed planes together in pairs, displacement of the B atoms to the wrong positions is avoided.

The Bain strain (Figure 5.8) did not necessitate shuffles because a vector **u** defining the position of an atom in the austenite unit cell corresponds to a vector in the ferrite lattice which also terminates at an atom. The Bain correspondence thus defines the position of each and every atom in the ferrite lattice relative to the austenite lattice. It is only possible to obtain a correspondence matrix like this, when the primitive cells of each of the lattices concerned contain just one atom. If this is not the case then any lattice correspondence defines the final positions of an integral fraction of the atoms, the remainder having to shuffle into their correct positions in the product lattice. The hcp structure has two atoms in the primitive cell so only 1/2 are placed in their final positions by the homogeneous deformation, the other one having to shuffle.

5.4.2 $\alpha \rightarrow \varepsilon$ TRANSFORMATION

This transformation is of particular interest for alloys subjected to shock deformation by impact, explosive detonation, magnetic means or laser impingement [81]. The large strain rates involved ($\approx 10^6\,\mathrm{s}^{-1}$) lead to transient pressures of about 13 GPa, causing the martensitic transformation of ferrite at ambient temperature in time scales ranging from ns to μs. The release of pressure following the shock causes the $\varepsilon \rightarrow \alpha'$ reversion at a pressure of about 10 GPa, indicating hysteresis. The $\alpha' \rightarrow \varepsilon$ start pressure is in the range 8.6–15.3 GPa, and $\varepsilon' \rightarrow \alpha$ covering 7–16.2 GPa [82], depending on the monitoring technique used and the state of the sample. First principles calculations indicate an α/ε equilibrium pressure of 8.4 GPa at 0 K [82, 83]. Experimental measurements should in principle bound this value during heating and cooling because the transformations are martensitic and the rates involved are too rapid to be consistent with equilibrium.

The addition of a solute to form a binary iron alloy naturally alters the transition pressures, Figure 5.20. The general trends are consistent with those calculated using thermodynamic data [84] though that calculated for chromium is the opposite of the behaviour measured. Using data from

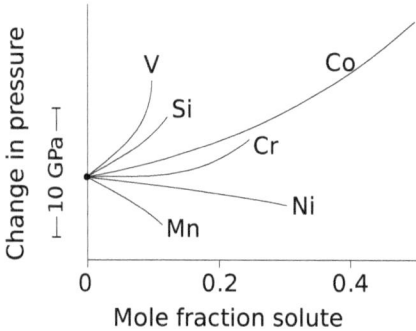

Figure 5.20 Compilation of experimental data on the effect of solutes in binary iron alloys, on the pressure for the transition $\alpha \to \varepsilon$. Adapted from Duvall and Graham [81].

shock induced $\alpha \to \varepsilon$ transformation, Forbes [85] was able to derive a relationship between the fraction transformed and the free energy of transformation:

$$1 - V_V^\varepsilon = \exp\{b_{27}(\Delta G^{\alpha \to \varepsilon} - \Delta G_{M_S^\varepsilon}^{\alpha \to \varepsilon})\}$$

which is similar in form to the Koistinen and Marburger equation (p. 284).

Given the short time scales of the $\alpha \to \varepsilon$ transformation and because the ε is not retained, there is no clear evidence for its microstructure. The orientation relationship is probably the same as that observed for zirconium [86], $(0001)_{\text{hcp}} \parallel (110)_{\text{bcc}}$, $[\bar{1}1\bar{2}0]_{\text{hcp}} \parallel [\bar{1}1\bar{1}]_{\text{bcc}}$.

5.5 MARTENSITE-START TEMPERATURE

If austenite can be supercooled sufficiently then martensite eventually forms at a start-temperature M_S. Some monitoring techniques are more sensitive than others, so M_S temperatures measured using different methods are unlikely to be identical. The uncertainty has been demonstrated to be approximately $\pm 20°$C [87]. Figure 5.21 shows some dilatometric data where the deviation of the curve from a straight line extrapolated from the fully austenitic state represents martensitic transformation. However, there are difficulties in defining the exact temperature at which the data deviate from the thermal contraction of austenite.

One way to make measurements more reproducible is the offset method, where the M_S temperature is defined by a critical strain calculated for 1 volume percent of martensitic transformation assuming that the latter occurs at room temperature, using equations for the lattice parameters of austenite and martensite. The technique preserves the notion that the early stages of martensite formation correspond to the start temperature. When reported M_S data, to state at the same time the sensitivity of the analysis technique. This concept is similar to the offset used in defining the 0.2% proof strength from a stress-strain curve for a material that shows gradual yielding.

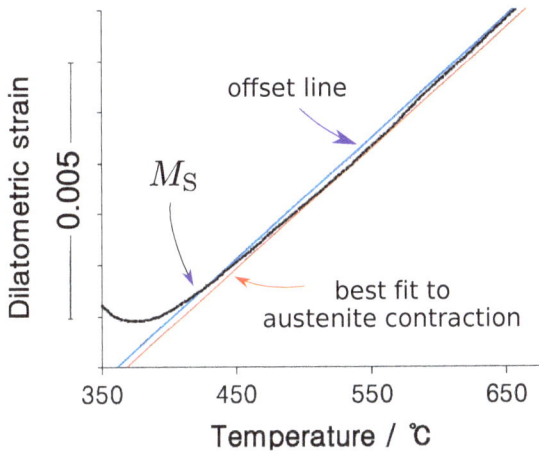

Figure 5.21 Offset method to objectively and reproducibly determine the M_S temperature from dilatometric data. The offset line is parallel to the thermal contraction line of austenite.

The M_S temperature is found not to be sensitive to the austenite grain size or the austenitisation temperature as long as the temperature is high enough to dissolve residual phases such as carbides. This is why empirical equations are able to successfully estimate M_S as a function of the chemical composition alone, for example, the classic equation due to Steven and Haynes [88]:

$$M_S^{\alpha'}/°C = 561 - 474w_C - 33w_{Mn} - 17w_{Ni} - 17w_{Cr} - 21w_{Mo} \tag{5.7}$$

	C	Mn	Ni	Cr	Mo
Minimum/wt%	0.10	0.20	0.0	0.0	0.0
Maximum/wt%	0.55	1.70	5.0	3.5	1.0

Martensitic transformation can occur only below the T_0 temperature where austenite and ferrite of the same chemical composition have identical free energies, Figure 5.22, with additional undercooling to account for nucleation and any stored energy terms due for example to the shape deformation and transformation twinning. The manner in which the martensite-start temperature can be expressed empirically as in Equation 5.7 without any dependence on time suggests that there may be a critical value of the driving force $\Delta G^{\gamma\alpha}$ that must be achieved for the onset of transformation. This critical value $\Delta G_{M_S}^{\gamma\alpha}$ is illustrated in Figure 5.22. The only effect of alloying elements would be to modify the relative thermodynamic stabilities of the austenite and martensite phases.

For dilute steels, the critical value of the driving force is found to be about $-1100\,\mathrm{J\,mol^{-1}}$, varying a little with the carbon concentration although more so with nickel [89–92]. [7]

It is surprising that $\Delta G_{M_S}^{\gamma\alpha}$ is found to be insensitive to the chemical composition. Martensitic transformation occurs by the propagation of a glissile interface. Anything that impedes the glide of the interfacial dislocations must therefore depress the transformation temperature. Solute additions generally strengthen the austenite and there is considerable evidence that the ability of austenite to resist deformation is an important factor in its decomposition to martensite [93]. Since the resistance to

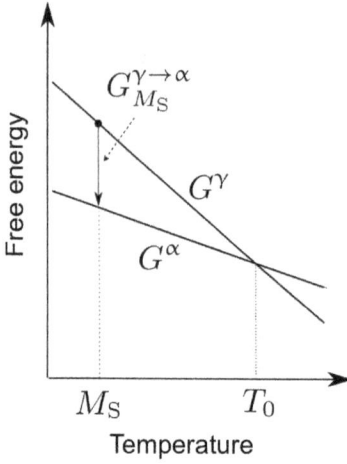

Figure 5.22 Plots of the free energies of austenite and ferrite of the same chemical composition as a function of temperature. Martensite occurs when the free energy difference $G^{\gamma \to \alpha}$ between these two phases reaches a critical value $G^{\gamma \to \alpha}_{M_S}$.

dislocation motion varies with the square root of the solute concentration, it is found that the critical driving force should vary with the concentration as follows [94]:

$$
\begin{aligned}
-\Delta G^{\gamma \to \alpha}_{M_S}/\mathrm{J\,mol^{-1}} \;=\; & 1010 + [(4009^2 x_C + 3097^2 x_N)]^{\frac{1}{2}} \\
& + \; [1879^2 x_{Si} + 1980^2 x_{Mn} + 1418^2 x_{Mo} + 1868^2 x_{Cr} + 1618^2 x_V + 1653^2 x_{Nb}]^{\frac{1}{2}} \\
& + \; [280^2 x_{Al} + 752^2 x_{Cu} + 172^2 x_{Ni} + 714^2 x_W]^{\frac{1}{2}} - 352 x_{Co}^{\frac{2}{3}}.
\end{aligned}
\tag{5.8}
$$

This gives much improved predictions of the M_S temperature when the steels is richly alloyed, in which case $|\Delta G^{\gamma \to \alpha}_{M_S}|$ can be as large in magnitude as $3000\,\mathrm{J\,mol^{-1}}$.

The development of transformation below M_S is given empirically by the Koistinen and Marburger equation [25], the derivation of which is described in Section 5.8.2:

$$
1 - V_V^{\alpha'} = \exp\{-b_2(M_S - T_q)\}.
$$

Start temperature for HCP martensite

The hexagonal close-packed form of iron is not stable under ambient conditions of temperature and pressure, but it can be stabilised by adding substantial concentrations of solutes such as manganese which reduce the stacking fault energy of the austenite. For ambient pressures, the martensite-start temperature, like that of α'-martensite, has been expressed empirically for the concentration and temperature range indicated below [95]:

$$
\begin{aligned}
M_S^{\varepsilon}/\mathrm{K} \;=\; & 576 \pm 8 - (489 \pm 31)w_C - (9.1 \pm 0.4)w_{Mn} - (17.6 \pm 2)w_{Ni} \\
& - (9.2 \pm 1)w_{Cr} + (21.3 \pm 2)w_{Al} + (4.1 \pm 1)w_{Si} - (19.4 \pm 5)w_{Mo} \\
& - (1 \pm 1)w_{Co} - (41.3 \pm 6)w_{Cu} - (50 \pm 18)w_{Nb} - (86 \pm 12)w_{Ti} \\
& - (34 + 10)w_V - (13 \pm 5)w_W
\end{aligned}
\tag{5.9}
$$

C	$0-0.35$	Co	$0-8$	Ni	$0-6.8$	Nb	$0-1.21$
Mn	$11.2-35.9$	Cu	$0-3.1$	Cr	$0-13.7$	Ti	$0-1.72$
Al	$0-5$	V	$0-2.2$	Si	$0-7.1$	W	$0-4.48$
Mo	$0-4.46$	M_S/K	$167-467$				

An analysis in terms of a critical driving force at M_S^ε has not been possible because the thermodynamic data necessary are inaccurate [95]. Figure 5.23 shows that $\Delta G_{M_S}^{\gamma\varepsilon}$ sometimes yields large positive values which oppose transformation.

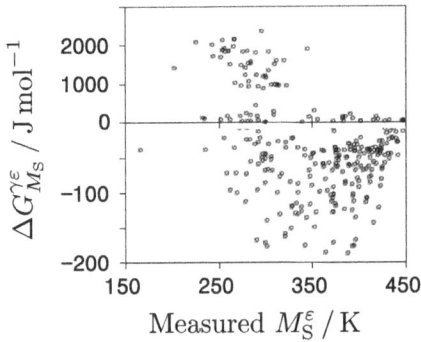

Figure 5.23 The driving force calculated at the martensite-start temperature for a variety of alloys. All values of driving force should represent a reduction in free energy, but in many cases they do not because the thermodynamic data used for the estimates are not reliable at very large solute concentrations. After Yang et al. [95].

5.5.1 EFFECT OF AUSTENITE GRAIN SIZE ON M_S

The martensite-start temperature varies with the austenite grain size [96–103]. There is, therefore, a dependence of M_S on the austenitisation temperature and time [97, 104], assuming that there are no other phases, such as undissolved carbides, present during austenitisation [105]. The stability of austenite to martensitic transformation is increased dramatically as the austenite grain size is reduced from 60 to 0.6 µm in a Fe-Ni alloy [106].

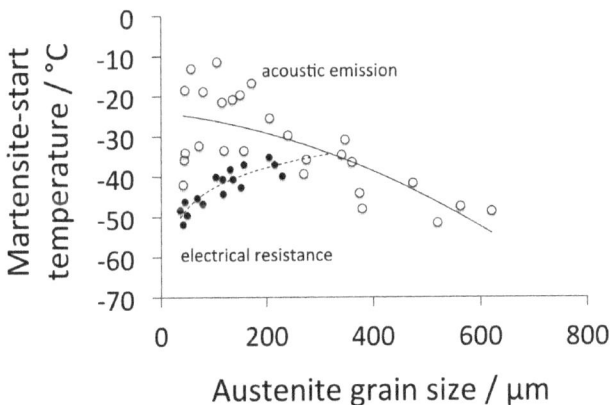

Figure 5.24 The dependence of the α'-martensite start-temperature on the γ-grain size, as measured using the acoustic emission or the less precise electrical resistance techniques [107].

All of these observations are somewhat related to the method used to measure M_S. In Figure 5.24, measurements of M_S versus $\overline{L}_\gamma$ conducted using the less sensitive electrical resistance method show a different trend than the more sensitive acoustic emission data. A technique that requires a greater amount of martensite to form before it is detected will naturally lead to a lower M_S. The austenite grain size determines the volume per plate of martensite, so more plates are needed to reach a detectable volume fraction when the grain size is reduced.

A model can be formulated on this basis, beginning with a description of how the volume per plate of martensite evolves as the austenite grain is partitioned by martensite. The change in the fraction of martensite due to a corresponding change in N_V, the number of martensite plates per unit volume, will be related to the fraction of untransformed austenite $(1 - V_V^{\alpha'})$, the volume per plate (aspect ratio z_t/z_ℓ multiplied by the average size $\overline{V}_C$ of the geometrically partitioned austenite compartment) [108]:

$$\frac{dV_V^{\alpha'}}{dN_V} = \frac{z_t}{z_\ell}(1 - V_V^{\alpha'})\overline{V}_C \equiv \frac{z_t}{z_\ell}\frac{1 - V_V^{\alpha'}}{N_V^C} \tag{5.10}$$

with $z_t/z_\ell \approx 0.05$ and N_V^C is the number of compartments per unit volume. Assuming that each plate produces one additional compartment,

$$N_V^C = \frac{1}{V^\gamma} + N_V \tag{5.11}$$

where V^γ is the average austenite grain volume. Using Equations 5.10 and 5.11,

$$N_V = \frac{1}{V^\gamma}\left[\exp\left\{-\frac{\ln(1 - V_V^{\alpha'})}{z_t/z_\ell}\right\} - 1\right]. \tag{5.12}$$

Based on experimental data, the number per unit volume can also be expressed as [109]

$$N_V \approx b_{25}[\exp\{b_{26}(M_S^\circ - T)\} - 1] \tag{5.13}$$

where M_S° is defined as a fundamental martensite-start temperature for an austenite grain size that is so large that the formation of just one martensite plate is detectable using routine methods. On combining Equations 5.12 and 5.13 gives [109]

$$M_S^\circ - T = \frac{1}{b_{25}}\ln\left[\frac{1}{b_{26}\overline{L}_\gamma^3}\left\{\exp\left(-\frac{\ln(1 - V_V^{\alpha'})}{z_t/z_\ell}\right) - 1\right\} + 1\right] \tag{5.14}$$

where $b_{25} = 0.2689$, $b_{26} = 1\,\text{mm}^3$ and the mean lineal intercept $\overline{L}_\gamma$ is in mm. M_S° is given by the point where $\Delta G^{\gamma \to \alpha} = -700\,\text{J}\,\text{mol}^{-1}$, i.e. the stored energy of martensite due to the shape deformation and twin interfaces. M_S° is therefore purely a thermodynamic quantity with no consideration given to kinetic effects. If M_S is defined to correspond to the point where the fraction $V_V^{\alpha'} = 0.01$, then by setting $T = M_S$ in Equation 5.14 gives the grain size dependence of the martensite-start temperature as illustrated in Figure 5.25.

Equation 5.14 has been shown to apply also to ε martensite, but with $b_{25} = 0.33$ and the plate aspect ratio consistent with experimental data of 0.03 [95].

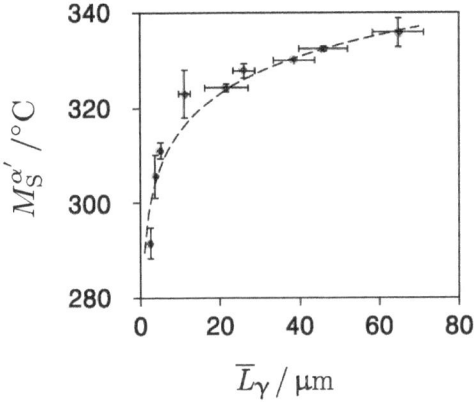

Figure 5.25 Measured variation in the martensite-start temperature as a function of austenite grain size, determined from dilatometric data (after Yang et al. [109]).

5.6 THERMODYNAMICS

There is a reduction in free energy when martensite grows from supercooled austenite; some of this energy is dissipated as heat. When the rate of transformation is rapid, the associated enthalpy change can lead to a local rise in temperature that is sufficiently large to induce a change in microstructure. Martensite plates which at first have the mode of lattice-invariant shear as twinning, change into slip as the local temperature during growth increases due recalescence associated with the rapid release of the heat of transformation, Figure 5.26 [110]. The *average* rise in the temperature due to recalescence has been measured for Fe-31Ni wt% to be about 30°C from a starting temperature of −60°C. Not all of the free energy is dissipated as heat – the latent heat of transformation is substantially less than that expected from independent thermodynamic data. A large fraction therefore remains as energy stored in the steel.

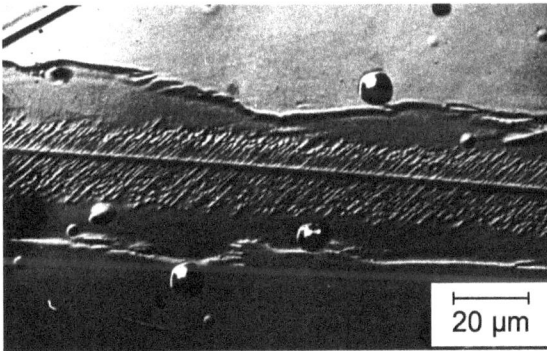

Figure 5.26 A martensite plate in Fe-32Ni wt% alloy. The central region of the martensite plate is twinned but its peripheries are not. Reproduced from Patterson and Wayman [110], with the permission of Elsevier.

A substantial fraction of the stored energy is attributed to the shape deformation, a defining feature of martensitic transformation. The strain energy per unit volume (G_V^e) when this deformation is elastically accommodated is given, for an isolated plate constrained by austenite, as [7]

$$\frac{1-v}{E_{\rm s}}G_{\rm V}^{\rm e} = \overbrace{\frac{2}{9}(1+v)\Delta^2 + \frac{\pi z_{\rm t}}{4z_\ell}\zeta^2 + \frac{\pi z_{\rm t}}{3z_\ell}(1+v)\Delta\zeta}^{\text{contribution from volume change}}$$

$$+ \quad \underbrace{\frac{\pi}{8}(2-v)\frac{z_{\rm t}}{z_\ell}s^2}_{\text{contribution due to shear}} \tag{5.15}$$

where s is the shear strain, ζ the expansion normal to the habit plane and Δ any uniform dilation.[8] $z_{\rm t}$ and z_ℓ are the thickness and diameter of the oblate spheroid shape used to represent the martensite plate, $E_{\rm s}$ is the shear modulus of the austenite and v is its Poisson's ratio. Notice that even in the absence of a shear strain, the strain energy is dependent on the plate aspect ratio $z_{\rm t}/z_\ell$ because the strain ζ is directed normal to the habit plane. In the absence of any uniform dilatation, the equation can be simplified to $G_{\rm V}^{\rm e} = E'z_{\rm t}/z_\ell$ where E' agglomerates all the other terms in Equation 5.15.

The equation was derived by Christian [7] using Eshelby's theory for a constrained transformation in which the phases maintain coherency, but there is a simple interpretation useful in understanding the shape of a martensite plate. If a tensile stress σ is applied to obtain an elastic strain ε then the strain energy per unit volume is proportional to $\sigma\varepsilon = E\varepsilon^2$. It is reasonable therefore to expect all the strain terms in Equation 5.15 to be squared. The proportionality $G_{\rm V}^{\rm e} \propto z_{\rm t}/z_\ell$ is more difficult to explain. The magnitude of the displacement vector increases with distance normal to the habit plane, Figure 5.27. When pushing against the surrounding material, it obviously is an advantage to minimise the absolute displacement. The martensite therefore adopts a thin, lenticular-plate shape which ensures a minimal displacement at the plate tip; the small $z_{\rm t}/z_\ell$ ratio in turn ensures that the displacements never become too large. These considerations apply also to mechanical twins which adopt the lenticular morphology with sharp tips.

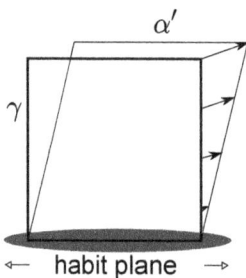

Figure 5.27 Schematic illustration of the shape deformation due to unconstrained $\gamma \rightarrow \alpha'$ transformation, and a lenticular plate when the transformation is constrained. The black arrows are displacement vectors, increasing in magnitude with vertical distance from the invariant-plane.

For representative values of the parameters in Equation 5.15, $G_{\rm V}^{\rm e}$ for an elastically accommodated plate comes to about $600\,{\rm J\,mol}^{-1}$ [111], which is less than typical values of $|\Delta G_{M_{\rm S}}^{\gamma\alpha}|$. The partitioning of the strain energy between the austenite and martensite depends on the plate aspect ratio. The shear strain is by far the dominant component of the shape deformation; considering that strain alone, the proportion of the total strain energy stored in the martensite is approximately $2z_{\rm t}/z_\ell$ [p. 467 53]. A thin plate is therefore well-accommodated and most of the strain energy is contained in the surrounding austenite. This stored energy does not contribute to the latent heat of transformation.

This becomes particularly relevant in the design of bulk nanostructured steels where the heat of transformation can otherwise frustrate attempts to generate the fine structures by suppressing the transformation to low temperatures [112, 113].

The martensite plates in many alloys contain finely spaced transformation twins resulting from the lattice-invariant deformation. The stored energy per unit volume due to this internal microstructure is approximately σ_t/d_t where $\sigma_t \approx 0.2\,\mathrm{J\,m^{-2}}$ is the coherent twin boundary energy per unit area and $d_t \approx 20\,\mathrm{nm}$ the twin spacing. This gives an energy of about $100\,\mathrm{J\,mol^{-1}}$.

Steel martensites often contain a substantial density of dislocations, which might be taken to contribute to the overall stored energy. However, the dislocations are generated when the strains due to the shape deformation are relaxed by plastic deformation. They do not therefore form a separate contribution to the stored energy because it is the shape change that drives the deformation. The stored energy due to an elastically accommodated plate should therefore form the upper limit of the stored energy even when plastic relaxation occurs [111].

The total stored energy thus amounts to about $700\,\mathrm{J\,mol^{-1}}$ which is, as it should be, less than the magnitude of the available driving force at M_S of about $1100\text{-}3000\,\mathrm{J\,mol^{-1}}$. The need to undercool the austenite to a temperature below that required to account for the stored energy comes from the nucleation stage, which is discussed later.

The $\gamma \rightarrow \varepsilon$-martensite transformation seems to occur at temperatures where the free energy change is much smaller, in the range $\Delta G_{M_S}^{\varepsilon\gamma} \approx 100 - 300\,\mathrm{J\,mol^{-1}}$ (Table 5.3). It is odd that the stored energy terms for ε-martensite must be much smaller than for α-martensite which has a smaller shear strain. This may be a consequence of the smaller aspect ratio of the ε-martensite, the thickening of which is limited by the availability of appropriately placed Shockley partials on every second close-packed plane. The absence of an easy thickening mechanism must contribute to a small aspect ratio.

Table 5.3
The driving force for the austenite to ε-martensite transformation at the martensite-start temperature. Many more data can be found in [95].

Alloy	$\Delta G_{M_S}^{\gamma\varepsilon}/\mathrm{J\,mol^{-1}}$	Ref.	Alloy	$\Delta G_{M_S}^{\gamma\varepsilon}/\mathrm{J\,mol^{-1}}$	Ref.
Fe-Ru (10-20 at.%)	-210	[114]	Fe-18Cr-12Ni wt%	-126	[115]
Fe-16Cr-13Ni wt%	-142	[115]	Fe-Mn (15-30 at.%)	-270	[116]

5.6.1 THERMOELASTIC EQUILIBRIUM

Martensitic transformation is in principle reversible by virtue of its glissile interface. A reversal of driving force should change the direction of interface motion once the intrinsic resistance of the lattice is overcome. The driving force can originate in the chemical free energy change, an applied stress or magnetic field. Perfect reversibility can be achieved when the interface can glide conservatively, i.e., the shape deformation is elastically accommodated.

In thermoelastic equilibrium, the assembly attains a minimum free energy at some finite volume fraction of martensite [111]. A change in the driving force stimulates an increment of martensite. This is illustrated in Figure 5.28 for a plate of martensite which has grown to a limiting length between hard obstacles. For any given set of conditions the plate will thicken at constant length z_ℓ until the thickness becomes $z_t = z_\ell |\Delta G| / E'$.

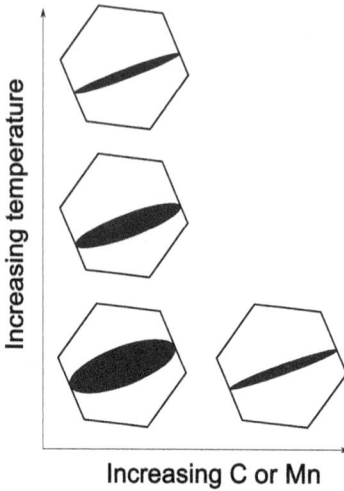

Figure 5.28 An illustration of thermoelastic equilibrium. The plate of martensite ideally shrinks when the temperature is raised or when the concentration of austenite-stabilising elements is increased. Both of these lead to a reduction in the driving force for martensitic transformation.

Whether martensitic transformation in any particular alloy will be thermoelastic or not can be predicted from an estimated stress field to see whether it is sufficiently large to induce significant plastic deformation in the adjacent austenite [117, 118]. This is illustrated vividly by iron-platinum alloys that undergo an order-disorder transformation in the austenite with an accompanying large change in the mechanical properties. The ordered austenite is better able to elastically accommodate the martensite because it has a much lower shear modulus and a high yield strength. Martensite that grows from ordered austenite is therefore thermoelastic whereas that forming in disordered austenite is not. Ling and Owen's calculations for disordered and partially ordered austenite are illustrated in Figure 5.29. It is apparent that the extent of plastic accommodation is greatest in the disordered case. Not surprisingly, martensite in the fully ordered austenite exhibits reversibility; calculations suggest a plastic field that is so small that it may not exist in practice. This criterion for thermoelasticity is unambiguous.

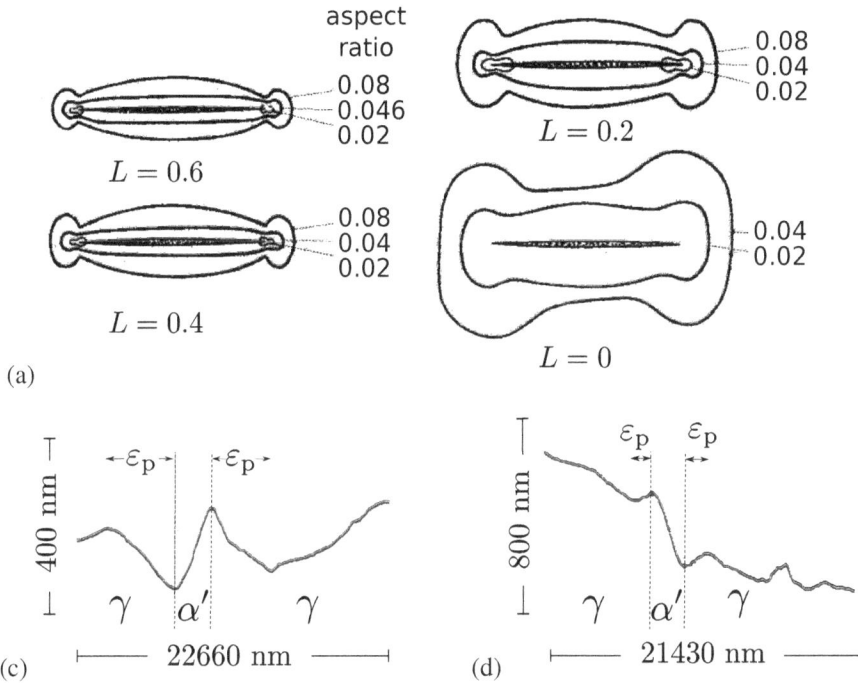

Figure 5.29 (a) Ling and Owen's calculations showing the contours where the matrix satisfies the yield criterion, as a function of the aspect ratio (z_t/z_ℓ) of the martensite plate and the degree of ordering (order parameter L). The region between each contour and the martensite plate represents plastically deformed austenite. Reproduced and adapted for clarity from [118], with the permission of Elsevier. (b,c) Atomic force microscope scans across the surface relief produced when a Fe-24Pt at% austenite is transformed partially into martensite. ε_p refers to the plastic accommodation in the austenite adjacent to the martensite plate. $L = 0$ and $L = 0.7$ for (b) and (c) respectively. After Vevecka et al. [119].

5.7 MARTENSITE: NUCLEATION MECHANISM

5.7.1 FCC TO HCP TRANSFORMATION

The onset of martensitic transformation requires a much larger driving force than can be accounted for by the stored energy terms put together. This is because of the need to nucleate martensite, a process which in the vast majority of circumstances is heterogeneous and requires activation.

Classical nucleation relies on chance fluctuations of structure and composition, induced by the natural thermal vibrations of atoms. Such events are most unlikely at the very low temperatures where martensite can be demonstrated to form readily.

The nucleation of ε-martensite from austenite is most readily visualised in terms of the dissociation of a unit lattice dislocation into a pair of Shockley partials on a close-packed plane (Figure 5.19):

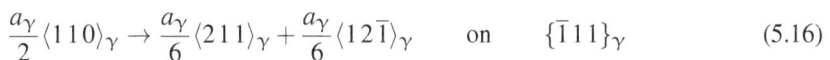

$$\frac{a_\gamma}{2}\langle 110\rangle_\gamma \rightarrow \frac{a_\gamma}{6}\langle 211\rangle_\gamma + \frac{a_\gamma}{6}\langle 12\bar{1}\rangle_\gamma \qquad \text{on} \qquad \{\bar{1}11\}_\gamma \tag{5.16}$$

The resulting fault separating the partials gives a two-layer thick region with the hcp stacking sequence of close-packed planes.

The hexagonal form has a different density to fcc iron. Just altering the stacking sequence of the close-packed planes does not achieve a change in their spacing. That the spacing does change is demonstrated by the analysis of fault contrast using transmission electron microscopy, which indicates that there is a displacement normal to the fault [120–122]. The magnitude of the displacement is proportional to the number of parallel faults and in the direction expected from the difference in the densities of the two phases. The transformation strain is therefore an invariant-plane strain rather than just a shear.[9]

The energy of the fault representing the hcp embryo has a variety of components: the chemical free energy change per mole ($\Delta G^{\gamma\varepsilon}$), the strain energy per mole (G^e) and a term $\sigma_{\gamma\varepsilon} \simeq 0.01\,\mathrm{J\,m^{-2}}$ due to the interface between the austenite and ε-martensite [123]:

$$\sigma_f = n\rho_A(\Delta G^{\gamma\varepsilon} + G^e) + 2\sigma_{\gamma\varepsilon} \qquad (5.17)$$

where σ_f is the fault energy per unit area, n is the number of planes involved in the faulting process ($n = 2$ for an intrinsic stacking fault) and ρ_A is the density of atoms in the close-packed plane in moles per unit area. The term $\sigma_{\gamma\varepsilon}$ is strictly a function of n but such data are not available in practice so a constant value has to be assumed.

The net fault energy can be zero if the magnitude of the chemical free energy change is sufficiently large. The fault would then be expected to extend spontaneously.[10] The process can begin from defects including isolated dislocations and groups of dislocations in arrays akin to symmetrical tilt boundaries that are spaced two close-packed planes apart, i.e. in a configuration suitable to generate the hcp structure from fcc. Since it is energetically more favourable for thicker faults ($n > 2$) to dissociate (Figure 5.30), it is possible for martensitic nucleation to occur at temperatures where the stacking fault energy is still positive ($n = 2$). This also means that isolated faults and dissociated dislocations can be observed in the austenite alongside larger regions of fully developed martensite. Those regions which have not yet developed into martensite nuclei, i.e. for which the fault energy σ_f is still much greater than zero, can be regarded as *embryos*, a term normally reserved in classical nucleation theory for fluctuations which have not surmounted the activation barrier.

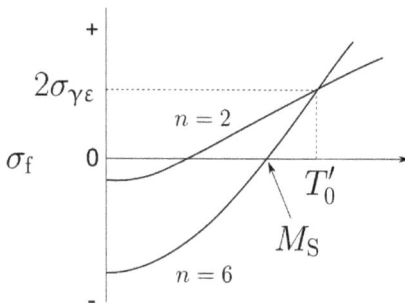

Figure 5.30 Variation in the fault energy σ_f as a function of the fault thickness n. The martensite-start temperature can therefore be higher than that at which the stacking-fault ($n = 2$) energy becomes negative [123]. The γ and ε phases have the same free energy (when strain energy is accounted for) at the T_0' temperature.

Classical structural and compositional fluctuations are not relevant in this model of martensitic nucleation based on dislocation dissociation. The concept of *barrierless nucleation* is illustrated in Figure 5.31; the embryo has an equilibrium size when $\sigma_f > 0$. Although there is a minimum in the free energy as a function of the separation of partials, there is no maximum that follows if the size increases by chance so the embryo cannot develop into a viable nucleus. On the other hand, nucleation occurs spontaneously when the fault energy becomes zero at a sufficiently low temperature. The embryo is then stimulated into growth without the need to overcome an activation barrier; this is barrierless nucleation with the property that it is athermal.

Olson and Cohen's model is consistent with experimental data when comparisons are made using reasonable values of the interfacial energies and defect number densities. The thickness of the fault at M_S for the $\gamma \rightarrow \varepsilon$ transformation is estimated to be $n = 10$, corresponding to a wall of five dislocations spaced two close-packed planes apart. There should be about $10^{12}\,\mathrm{m}^{-3}$ such defects present in annealed austenite, consistent with typical experimental data of about $2 \times 10^{12}\,\mathrm{m}^{-3}$ plates per unit volume, considered to be the quantity detectable using standard experimental techniques such as dilatometry or electrical resistance measurements [124].

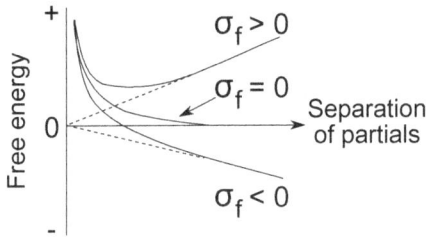

Figure 5.31 Variation in the total free energy as a function of the separation of the partial dislocations in the process leading to the nucleation of martensite [123].

That stacking faults can be regarded as potential nuclei for hcp martensite has a long history, with the destruction of the defect in the process contributing to facilitating heterogeneous nucleation [125]. In contrast, in the Olson and Cohen model, the defect evolves into a nucleus so that there is no explicit gain from the originating defect. The difference is illustrated in Figure 5.32, which compares the free energy versus size curves for homogeneous nucleation, for heterogeneous nucleation in which energy is gained by the destruction of a defect and finally, barrierless nucleation. There is no maximum illustrated in the curves in Figure 5.31.

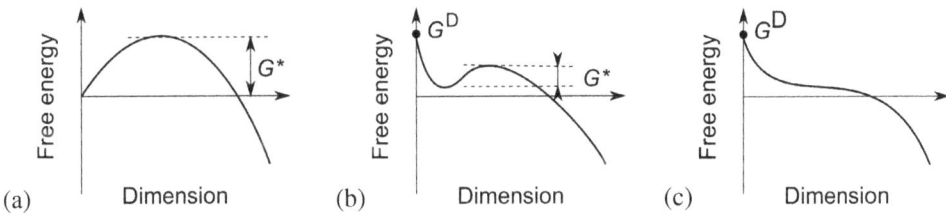

Figure 5.32 The difference between (a) homogeneous nucleation; (b) heterogeneous nucleation; (c) barrierless nucleation. After Olson and Cohen [75]. G^D is the energy of a defect which is either consumed during the nucleation process as in (b), or which evolves into the nucleus as in (c).

The fault model is verified experimentally for hcp martensite and there are many direct observations to confirm the mechanism. As might be expected, a large plate of hcp martensite has a curved interface made up of Shockley partials [34]. Hot-stage microscopy has revealed the spontaneous increase in the separation of partials as the temperature is reduced [126, 127]. As pointed out earlier, Brooks et al. have vividly demonstrated the dilatational strain associated with faulting [120–122]. And the athermal character of martensitic nucleation in appropriate circumstances is predicted naturally by the model.

5.7.2 ROLE OF THERMAL ACTIVATION

For dislocations to move, they must surmount the barriers between equilibrium positions in the perfect lattice. It follows that unless the driving force is very large, both the nucleation and growth of martensite will require some thermal activation. This is why isothermal martensite is observed at very low temperatures where even a small activation barrier becomes important.

The partials bounding a fault repel each other. Their equilibrium separation becomes infinite when $\sigma_f = 0$, assuming that the dislocations can move freely. However, even a perfect crystal resists dislocation motion since there is an activation barrier between successive equilibrium positions, making the yield strength τ_y temperature dependent (Figure 5.33a). Although the net resistance becomes smaller at high temperatures, it never vanishes but reaches a limiting value τ_μ known as the athermal resistance which scales only with the shear modulus and hence does not vary greatly with temperature. The athermal resistance arises from the long-range stress fields of obstacles. Fluctuations caused by thermal vibrations are important over distances of the order of a few atoms and hence cannot assist the dislocations to overcome any fields that extend over large distances. When the temperature is such that $\tau_y = \tau_\mu$, thermal vibrations readily overcome any short-range obstacles. This activation barrier to dislocation motion may be written G_o^* in the absence of any applied stress. However, its magnitude is reduced when an applied shear stress helps the dislocation to surmount the barrier. The applied shear stress τ promotes the dislocation up the activation barrier (Figure 5.33b) so that the activation energy is reduced [128, 129]:

$$G^* = G_o^* - (\tau - \tau_\mu)V^* \tag{5.18}$$

where V^* is an activation volume. The activation energy G^* for a unit length of dislocation is the total energy under the curve illustrated in Figure 5.33b. The effective applied stress is $\tau - \tau_\mu$. The work done by the applied stress for a unit length of dislocation is therefore $(\tau - \tau_\mu)b \times$ distance along reaction coordinate. Since the term multiplying the shear stress has units of volume for a unit length of dislocation, it is called an activation volume.

The stress τ can be provided either by applying a load or via the chemical driving force (a *transformational* stress) given by

$$\tau = -\frac{\sigma_f}{nb} \tag{5.19}$$

On substituting into Equation 5.18 for σ_f (Equation 5.17) and for τ (Equation 5.19), the activation energy is expressed in terms of the driving force:

$$G^* = G_o^* \left[\tau_\mu + \frac{\rho_A}{b} G^e + \frac{2\sigma_{\varepsilon\gamma}}{nb} \right] V^* + \frac{\rho_A V^*}{b} \Delta G^{\gamma\varepsilon} \tag{5.20}$$

This relationship is useful not only in dealing with martensite kinetics but also in the estimation of the rates of the bainite and Widmanstätten ferrite transformations. The important result is that

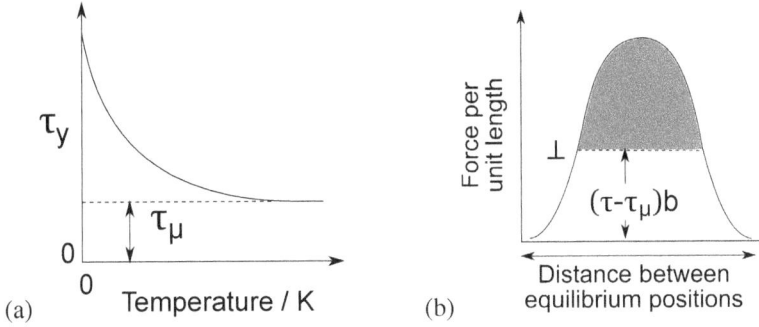

Figure 5.33 (a) Temperature dependence of the resistance to dislocation motion. (b) Schematic illustration of how an applied stress has the effect of reducing the activation barrier by the total area less the shaded area (after Conrad [128]).

the activation energy is found to be proportional directly to the driving force, which contrasts with the inverse-power relationship associated with classical nucleation theory. The link between thermally activated plastic deformation and indeed the linear dependence of the activation energy on the chemical driving force was proposed first by Magee and Paxton [124, 130].

The Olson and Cohen model is consistent with observations that martensitic transformation can sometimes occur isothermally. If τ_y is the resistance to the motion of dislocations, then nucleation is possible when $\sigma_f < -nb\tau_y$. The nucleation is athermal when the intersection between σ_f and $-nb\tau_y$ curves occurs in a regime where $\tau_y = \tau_\mu$, the resistance to dislocation motion becomes independent of temperature (Figure 5.34). For the case where the resistance τ_y is temperature dependent, the rate of motion will vary with temperature. Furthermore, because of the curvature of the σ_f function, the rate must go through a maximum; there are two intersections identified in Figure 5.34 at points x and y where the rate is zero. The model therefore predicts the experimentally observed C-curve behaviour for isothermal martensite.

In all cases, transformation will stop at a given temperature once the embryos capable of developing into nuclei are exhausted. There will therefore exist a limiting volume fraction of martensite at a given temperature for both isothermal and athermal transformations. For isothermal transformation embryos develop towards the limiting volume fraction by a process in which they are assisted over the activation barrier by thermal fluctuations so the time dependence of the reaction is well-defined in terms of a nucleation rate. For anisothermal transformation, all of the appropriate embryos would spontaneously develop into plates of martensite. However, if the rate of isothermal transformation is rapid then it becomes impossible to distinguish experimentally between isothermal and athermal transformation.

5.7.3 FCC TO BCC TRANSFORMATION

That hcp martensite nucleates by the dissociation of dislocations is now well-established both experimentally and from the viewpoint of dislocation theory. The model of the nucleus is intuitively satisfying because the hcp structure can be generated from fcc by an invariant-plane strain on the fault plane.

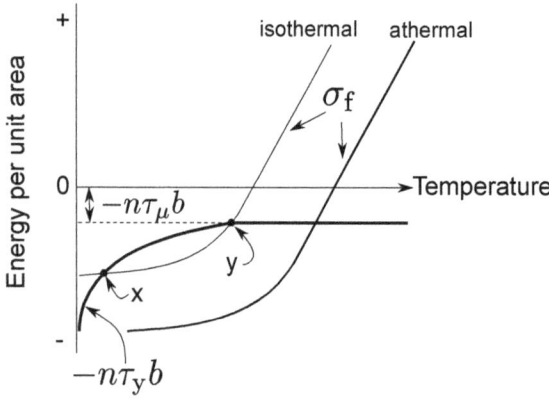

Figure 5.34 The conditions appropriate for isothermal and anisothermal martensitic nucleation [123]. The bold curve represents the total resistance to dislocation motion. This has a minimum, temperature-independent value τ_μ. Detectable nucleation becomes possible when the σ_f curve crosses below the $-n\tau_y b$ curve.

However, austenite cannot be transformed into bcc martensite by an invariant-plane strain. The nucleation process must therefore be more complex than faulting on a single plane. Since the transformation strain is an invariant-line strain at least two sets of partials with slip planes that are not coplanar are required.

Olson and Cohen emphasised faulting on the close packed plane. A possible mechanism can be explained phenomenologically as illustrated in Figure 5.35. A unit dislocation $\frac{a_\gamma}{2}[1\bar{1}0]$ dissociates into six partials via the reaction:

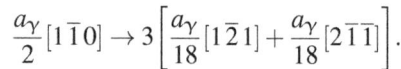

$$\frac{a_\gamma}{2}[1\bar{1}0] \rightarrow 3\left[\frac{a_\gamma}{18}[1\bar{2}1] + \frac{a_\gamma}{18}[2\bar{1}\bar{1}]\right].$$

The partials are distributed over three adjacent $(111)_\gamma$ planes (Figure 5.35b). Notice that each of the partials has a magnitude that is a third of a Shockley partial.

This faulting alone cannot lead to a bcc crystal structure. An additional invariant-plane strain on $(11\bar{1})_\gamma \parallel (011)_\alpha$ finishes the structural change (Figure 5.35c). The shear is accomplished by seven $\frac{a_\alpha}{8}[01\bar{1}]$ partial dislocations followed on the eighth plane by a $\frac{a_\alpha}{8}[0\bar{1}1]$ and a $\frac{a_\gamma}{6}[112]$ Shockley partial, all of which reside in the structure of the interface. Their dislocation lines lie along the $[1\bar{1}0]_\gamma$ direction, giving a semi-coherent embryo with the following variant of the Kurdjumov-Sachs orientation relationship:

$$(111)_\gamma \quad \parallel \quad (101)_\alpha$$

$$[1\bar{1}0]_\gamma \quad \parallel \quad [11\bar{1}]_\alpha.$$

This peculiar combination of seven partials and another pair on the eighth plane is necessary to leave the original $(111)_\gamma$ fault plane unrotated. Finally, a further set of screw dislocations is added that cancel out the remaining long-range strain field of the partial dislocations (Figure 5.35d), still leaving the original fault plane unrotated.

Obviously, the thicker embryos can develop into martensite more readily. Thicker faults can be produced by the "initial" dissociation of $\frac{a_\gamma}{2}[1\bar{1}0]$ every three planes, but this would require the existence of a suitable array of these dislocations in the untransformed austenite. Whatever the details of the proposed fault mechanism, the idea that nucleation occurs from an array of lattice dislocations seems to be justified experimentally [111] although direct evidence for the development of faults remains illusive.

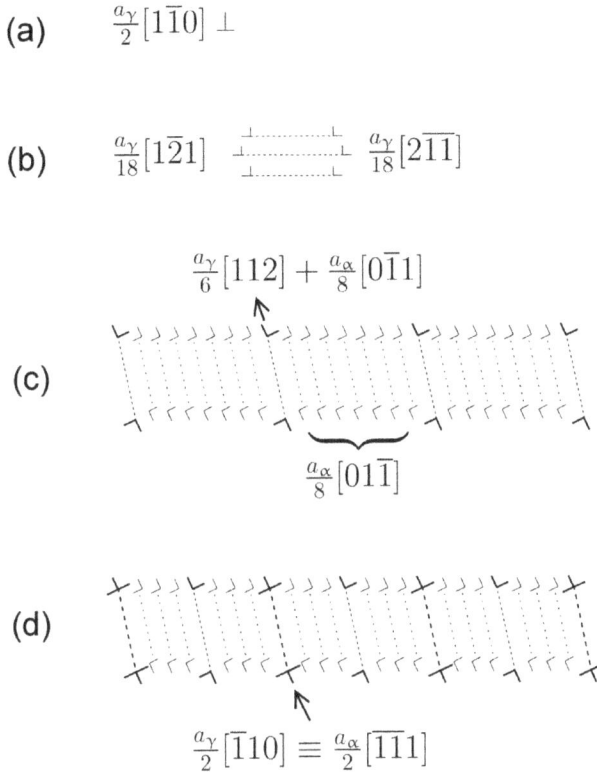

(a) $\frac{a_\gamma}{2}[1\bar{1}0]$ ⊥

(b) $\frac{a_\gamma}{18}[1\bar{2}1]$ $\frac{a_\gamma}{18}[2\bar{1}\bar{1}]$

$$\frac{a_\gamma}{6}[112] + \frac{a_\alpha}{8}[0\bar{1}1]$$

(c)

$$\frac{a_\alpha}{8}[01\bar{1}]$$

(d)

$$\frac{a_\gamma}{2}[\bar{1}10] \equiv \frac{a_\alpha}{2}[\bar{1}\bar{1}1]$$

Figure 5.35 The Olson and Cohen model for the development of a semicoherent body-centred cubic embryo from a perfect screw dislocation (a) in a face-centred cubic austenite. (b) Three-dimensional dissociation of dislocation over a set of three close-packed planes. The structure thus produced is not yet body-centred cubic. (c) Relaxation of fault to a body-centred cubic structure, involving the introduction of partial dislocations in the interface. (d) Addition of perfect screw dislocations which cancel the long-range strain field of the partial dislocations introduced in (c).

5.7.4 NUCLEATION AT LARGE DRIVING FORCES

Austenite can in principle be deformed continuously into the structure of martensite. Zener pointed out in 1948 that the elastic constant of the lattice can be used as a measure of its stability to shear deformation. In particular, for a cubic crystal, the two moduli characterising the resistance to shear deformation are C_{44} and $C' = \frac{1}{2}(C_{11} - C_{12})$. The former is a measure of the resistance to shear on the system $(010)[001]$ whereas the latter is the corresponding resistance on $(110)[1\bar{1}0]$. C_{44} is larger than C' so their ratio is a measure of the elastic anisotropy. C' sometimes softens as the transformation temperature is approached. If it actually approaches zero then the austenite becomes mechanically *unstable* and deforms spontaneously into the structure of martensite, passing through a continuous series of transition structures in the process.[11]

The onset of mechanical instability during transformation is illustrated in Figure 5.36 which shows total energy calculations for the fcc↔bcc transformation in sodium. The state of the transition is characterised by an "order parameter" which is 1 and 0 for the fcc and bcc structures respectively, with intermediate values representing transition structures. Since the calculations are for 0 K, the driving force for transformation is simulated by altering the pressure. The fcc and bcc allotropes in sodium are in stable equilibrium at $P = 87.6$ kbar. They are stable to any mechanical deformation (change in η) since they both lie in energy minima separated by an energy barrier which at equilibrium is about 150 J mol^{-1}. This intervening barrier becomes smaller for the bcc→fcc transformation as the pressure is reduced. Similarly, it becomes smaller for the reverse transformation as the pressure is increased. However, both phases remain mechanically stable when the barrier remains finite. The cases where the fcc and bcc structures become mechanically unstable are illustrated with curves a and b respectively (cf. Figure 5.32).

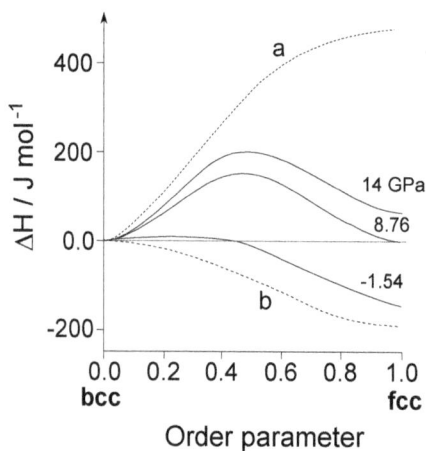

Figure 5.36 The computed free energy at 0 K for the fcc to bcc homogeneous lattice deformation in sodium for a variety of pressures (after Olson [131]). Curves a and b are schematic.

The case for iron is more complicated because magnetic properties must be accounted for. One difficulty is that many first principles calculations which use the local density approximation predict the incorrect ground state for iron. Krasko and Olson [132] avoided this by introducing a parameter fitted to experimental data, to enable ferromagnetic iron to be set as the ground state for iron at 0 K. Their results are shown in Figure 5.37 where at the equilibrium pressure (about 140 kbar), the ferrite is stable in the ferromagnetic state whereas the austenite is stable in the nonmagnetic state. The latter is used as an approximation to the paramagnetic state. There is an energy barrier of nearly 13 kJ mol^{-1} separating the ferrite and austenite which are in equilibrium, much larger than that associated with sodium. Given that the austenite and ferrite are in different magnetic states and because the minimum enthalpy curves in Figure 5.37 are calculated allowing the volume to vary, there is a discontinuity in volume at the α/γ intersection. The cusp at the intersection is quite different from the smooth maximum in the case of sodium.

This cusp-like barrier persists at all pressures where nonmagnetic austenite is more stable than ferromagnetic austenite. However, at a critical pressure of -110 kbar (equivalent to a chemical driving force of $-14.5\,\text{kJ}\,\text{mol}^{-1}$), ferromagnetic austenite has a lower free energy than nonmagnetic austenite. This abrupt change in the magnetic properties of the austenite causes elastic softening which in turn leads to mechanical instability. In fact, Krasko and Olson predict that ferromagnetic austenite in pure iron is always mechanically unstable, but at low driving forces it is the nonmagnetic form which is of lower energy and that state is not unstable. One difficulty with their calculations is that the antiferromagnetic state was not included; it is the ground state of austenite (Chapter 2).

These results demonstrate that the mechanical instability of austenite is not an important issue for all practical situations where the driving force for martensitic transformation is far less than the estimated $14.5\,\text{kJ}\,\text{mol}^{-1}$.

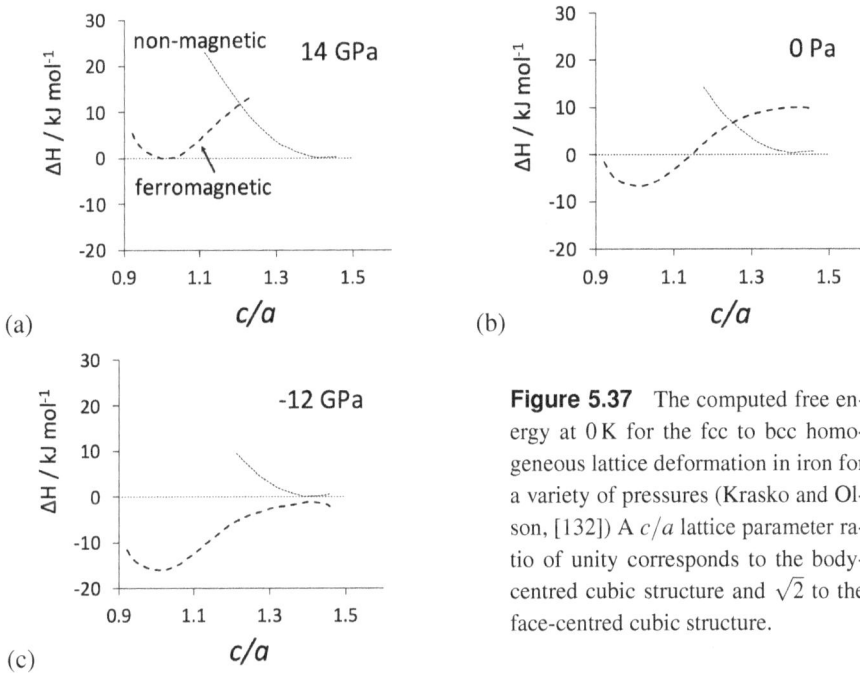

(a)

(b)

(c)

Figure 5.37 The computed free energy at $0\,\text{K}$ for the fcc to bcc homogeneous lattice deformation in iron for a variety of pressures (Krasko and Olson, [132]) A c/a lattice parameter ratio of unity corresponds to the body-centred cubic structure and $\sqrt{2}$ to the face-centred cubic structure.

As pointed out by Christian [111], it is unlikely that the martensitic transformation in steels can be described in terms of the wholesale mechanical collapse of austenite. Such a mechanism would not lead to the observed morphology which consists of well-defined plates which grow by the propagation of equally well-defined sharp interfaces that separate perfect regions of the co-existing parent and product phases.

Figure 5.38a illustrates the experimental features of a martensite plate in steel. The order parameter η changes abruptly at the interface between the fully formed martensite and the matrix austenite. Figure 5.38b shows the corresponding case where the austenite becomes mechanically unstable and evolves gradually into the structure of the martensite as the order parameter increases towards unity. The interface is now quite diffuse. It is possible that the morphology illustrated is incorrect because the transformation would tend to develop along elastically soft directions. But the final shape may not be plate-like if the austenite as a whole becomes mechanically unstable.

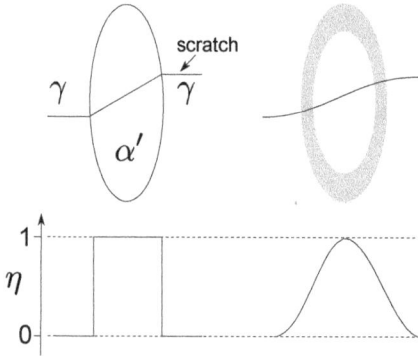

Figure 5.38 (a) Classical mechanism in which the nucleus has its final structure and a sharp boundary with the austenite. (b) Case where the structural change happens gradually with an ill-defined interface.

It is worth exploring whether mechanical instability might assist the nucleation of martensite. Figure 5.39 shows schematically the variation in the free energy as a function of temperature and order parameter for a sodium-like material. At $T = T_1$, it is possible that the deformation field of a defect can nudge a small region of the austenite to a point beyond η_1, where $\partial^2 \Delta G / \partial \eta^2$ is negative so martensite can form spontaneously in a process described as a strain spinodal by analogy with a chemical spinodal [111, 133]. There is therefore a gradient energy term to account for the cost of the soft interface due to the inhomogeneous strains [134]. Continuing the analogy with the chemical spinodal, whereas thermally activated nucleation (involving the large fluctuations needed to overcome a barrier) is necessary outside of the spinodal, spontaneous nucleation can occur within the strain spinodal because the system is then unstable to infinitesimal perturbations. The actual driving force at which spontaneous nucleation becomes possible depends on the "strength" of the nucleating defect. It is, however, reasonable to assume that in steels, mechanical instability does not play a significant role: the driving force at the M_S temperature is small when compared with that required to induce instability [135].

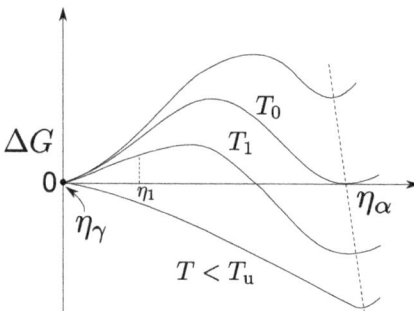

Figure 5.39 Schematic free-energy versus order parameter relations for a sodium-like material. The γ and α have the same free energy at T_0; the austenite has a higher free energy for temperatures below T_0. The austenite becomes mechanically unstable at temperatures below T_u. After Christian [111].

5.8   OVERALL TRANSFORMATION KINETICS

Three kinds of kinetic behaviour occur in iron alloys (Figure 5.40). The most common is athermal transformation in which the volume fraction is a function only of the undercooling below the martensite-start temperature.[12] Almost all commercial steels exhibit athermal martensitic transformation; these alloys tend to have an M_S temperature which is above ambient and insensitive to the cooling rate up to $\approx 50{,}000\,\mathrm{K\,s^{-1}}$. Isothermal transformation has been studied in systems with M_S temperatures well below room temperature, particularly in alloys with very small interstitial solute concentrations.

The third kind of behaviour is abrupt, with a burst of transformation that consumes a large fraction of the austenite over a narrow temperature range. Following this burst, subsequent transformation can only be promoted by further cooling. The enthalpy released during the sudden transformation can lead to recalescence and often to the emission of audible noise in the form of clicks.

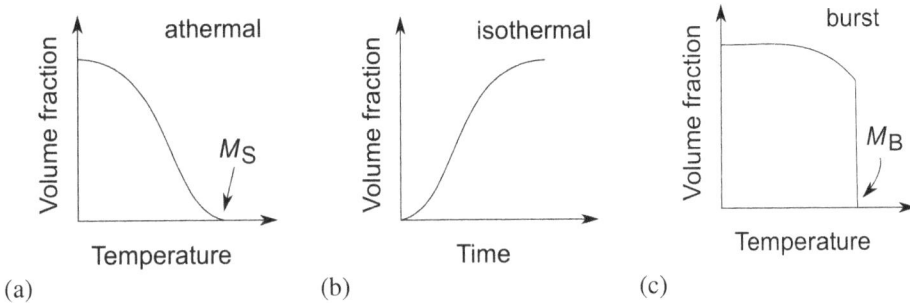

Figure 5.40 The three different kinds of kinetic behaviour observed during the formation of martensite (after Raghavan [136]), where the volume fraction is that of martensite.

5.8.1 PARTITIONING OF AUSTENITE

The essential idea of all theoretical treatments of martensite kinetics is that plates of martensite partition the austenite grains into smaller regions in which new plates form and further refine the residual austenite.[13] A typical transformation experiment might involve the measurement of the volume fraction of martensite as a function of either the temperature during continuous cooling, or the time during isothermal treatment. The growth of the first generation of martensite plates is halted by the austenite grain boundaries. Subsequent generations are limited by the size of the remaining "compartments" of austenite. The estimation of transformation curves requires a knowledge of the mean volume per plate ($\overline{V}$).

Suppose that martensite plates form at random within a given grain of austenite of volume V_γ. The first plate to form will have a volume $V_1^{\alpha'}$, resulting in a reduction in the fraction of austenite available for transformation and at the same time partitioning the austenite grain into two. Assuming that each new plate transforms a constant fraction $\psi = V_1^{\alpha'}/V_\gamma$ of the compartment of austenite, the second plate of martensite ($n = 2$) will have a volume $V_2^{\alpha'} = V_\gamma(1 - V_V^{\alpha',1})\psi/2$ where $V_V^{\alpha',1}$ is the fraction of martensite present prior to the formation of the second plate. More generally, the volume of the nth plate is given by [108, 137]

$$V_n^{\alpha'} = V_\gamma(1 - V_V^{\alpha',n-1})\frac{\psi}{n}.$$

Assuming that the formation of each plate results in one new compartment of austenite, the number of austenite compartments per unit volume is

$$N_V^c = \frac{1}{V_\gamma} + N_V^{\alpha'}$$

where the first term on the right hand side is the number of austenite grains per unit volume and $N_V^{\alpha'} = n/V_\gamma$ is the number of martensite plates per unit volume, assuming that all austenite grains behave identically. The rate of change of volume fraction with $N_V^{\alpha'}$ is

$$\begin{aligned}
\frac{\mathrm{d}V_V^{\alpha'}}{\mathrm{d}N_V^{\alpha'}} &= \psi(1 - V_V^{\alpha'})\frac{1}{N_V^c} \\
&= \psi(1 - V_V^{\alpha'}) \Big/ \left(\frac{1}{V_\gamma} + N_V^{\alpha'}\right)
\end{aligned}$$

which on integration between limit gives [137]

$$\ln\{1 - V_V^{\alpha'}\} = -\psi\ln\{1 + N_V^{\alpha'}V_\gamma\} \tag{5.21}$$

and the mean volume per plate of martensite,

$$\overline{V} = V_\gamma\left(\frac{V_V^{\alpha'}}{(1 - V_V^{\alpha'})^{-1/\psi} - 1}\right). \tag{5.22}$$

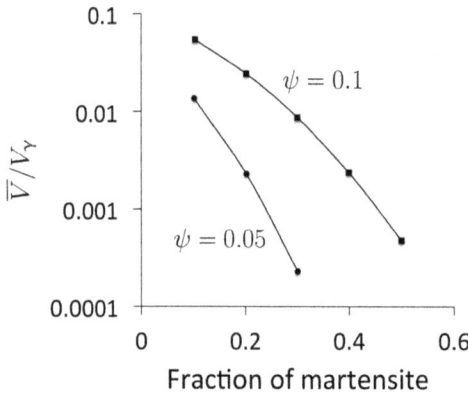

Figure 5.41 The average, normalised plate-volume as a function of the fraction of martensite (Equation 5.22). ψ is the fraction of a compartment of austenite which is transformed by the formation of a plate of martensite.

The process just described is one in which successive generations of plates "fill-in" the untransformed austenite (Figure 5.42). However, there are other circumstances where the transformation spreads from grain to grain as illustrated in Figure 5.42b, in which case the mean volume per plate does not depend on the fraction of transformation and this has been verified experimentally [137, 138]. Some quantitative data to this effect are listed in Table 5.4. It is clear that the mean volume per plate can be assumed to be approximately constant; the data also illustrate typical number densities of martensite plates and their apparent aspect ratios. The number density data are from direct metallographic measurements and compare well with values of about $10^7 \, \text{cm}^{-3}$ derived from indirect measurements, for example, by fitting to isothermal transformation data. Although the aspect ratio is approximately constant for the results presented in Table 5.4, there are other data which indicate that it increases sharply during the course of the transformation and then begins to diminish when the volume fraction is beyond about 0.6 [139, 140].

Table 5.4

Quantitative metallographic data on martensite plates in two Fe-Ni-C alloys. The aspect ratio here is the thickness of the plate divided by its length. The results are due to McMurtrie and Magee [124] whose data are more comprehensive than illustrated here.

Alloy	V_V	Aspect ratio $\frac{z_t}{z_\ell}$	$N_V^{\alpha'}/10^7 \times \text{cm}^{-3}$	$\overline{V}/10^{-9} \times \text{cm}^3$
Fe-23.8Ni-0.42C wt%	0.144	0.243	2.47	5.84
	0.177	0.207	1.8	9.85
	0.246	0.257	4.95	4.98
	0.451	0.241	6.32	7.13
	0.548	0.233	8.28	6.61
Fe-28.5Ni-0.40C wt%	0.075	0.197	1.11	6.75
	0.29	0.234	7.22	3.70
	0.51	0.163	9.03	5.65

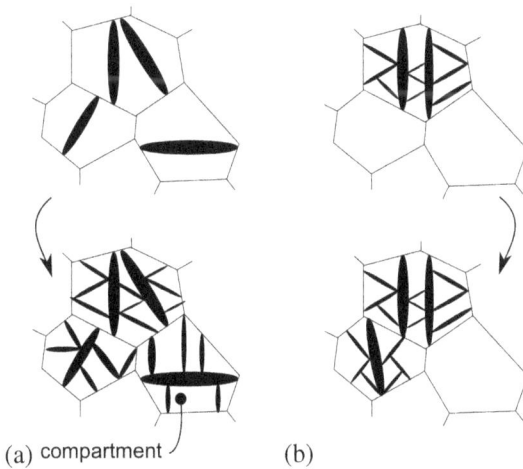

Figure 5.42 Schematic illustration of the partitioning of austenite. (a) "Fill-in" mode of transformation, where the mean volume per plate decreases as the volume fraction of martensite increases. (b) "Spreading" transformation with sequential decomposition of austenite grains, giving a constant mean volume per plate as transformation progresses.

(a) compartment   (b)

The detailed distribution in the size of martensite plates within a given austenite grain can depend on kinetic parameters such as the growth velocity and the nucleation rate. The partitioning model by Fisher and co-workers assumes that each new plate subdivides the remaining austenite. It has been shown that this leads to a size distribution which is scale-invariant [141]. The number of plates of a given size increases as the γ-grain size decreases, with an upper cut-off length $\overline{L}_\gamma$ and the lower cut-off determined by the total number of plates of martensite within the γ-grain. In the alternative extreme where all the plates are nucleated simultaneously, there is a smaller spread in the size of martensite plates and a size distribution that peaks at a value $\propto \overline{L}_\gamma/\sqrt{N}$. Assuming that nucleation occurs at random and that the rate of nucleation is constant, Rao and Sengupta argue that it should be possible in principle to use an experimental plate size distribution to derive the product of the plate growth velocity (also assumed constant) and the total time required to nucleate N plates.

5.8.2 ATHERMAL TRANSFORMATION

Athermal transformation is illustrated in Figure 5.4 which shows that the logarithm of the fraction of transformation varies linearly with the degree of undercooling $M_S - T_q$ below the martensite-start temperature. The interpretation of these experimental observations was given by Magee [124], whose analysis is presented here. If it is assumed that the change in the number of martensite plates per unit volume of austenite, dN_{V_γ}, is proportional to the increase in the driving force for transformation then:

$$dN_{V_\gamma} = -b_1 V_m d(\Delta G^{\alpha\gamma})$$

where b_1 is a positive proportionality constant and V_m is the molar volume. The corresponding change in the volume fraction of martensite is

$$dV_V^{\alpha'} = \overline{V} dN_V^{\alpha'} \tag{5.23}$$

where $\overline{V}$ is the mean volume per plate and $N_V^{\alpha'}$ is the number of plates per unit volume with $(1 - V_V^{\alpha'})dN_{V_\gamma} = dN_V^{\alpha'}$. Writing

$$d(\Delta G^{\alpha\gamma}) = \frac{d(\Delta G^{\alpha\gamma})}{dT} dT$$

and substituting into Equation 5.23 gives

$$dV_V^{\alpha'} = -\overline{V}(1 - V_V^{\alpha'})b_1 V_m \frac{d(\Delta G^{\gamma\alpha})}{dT} dT.$$

Integration from M_S to T_q assuming that $\overline{V}$, b_1 and $d(\Delta G^{\alpha\gamma})/dT$ remain constant gives

$$1 - V_V^{\alpha'} = \exp\left\{\overline{V}b_1 V_m\left(\frac{d(\Delta G^{\alpha\gamma})}{dT}\right)(M_S - T_q)\right\}.$$

There are experimental data to show that the variation in the fraction of martensite with $d(\Delta G^{\alpha\gamma})/dT)$ is correctly predicted by this equation, which compares well with the empirical relationship derived by Koistinen and Marburger (Equation 5.9), although b_2 in the latter must clearly vary with the chemical composition of the alloy. The form of the equation implies that there will *always* be some austenite retained. This is correct in practice because the build up of elastic strain energy means that plates of martensite stop growing when they are obstructed and consistent with the assumption that each plate transforms only a finite volume of austenite.

However, the assumption of a *constant* volume per plate is questionable, though Magee points out that for cases where autocatalysis[14] occurs, the transformation progresses by spreading from grain to grain rather than uniformly throughout the specimen (Figure 5.42b). Although there will be a distribution of plate sizes within any given grain, the mean volume per plate for each increment in the overall fraction of transformation will not vary when the austenite grains transform in sequence.

5.8.3 ISOTHERMAL MARTENSITE

Martensite, in the vast majority of steels, occurs by athermal transformation with the fraction of martensite depending only on the undercooling below the M_S temperature. Isothermal transformation is rare, found mostly in alloys with an M_S temperature less than about 400 K. It was discovered by Kurdjumov and Maksimova in 1948 [142] and most subsequent investigations (e.g., [143]) have supported the suggestion that because martensite plates grow rapidly, it is their nucleation that controls the rate of isothermal transformation. This does not rule out the possibility that slow growth might be the rate-limiting step; Yeo [144] reported isothermal martensite in which the growth rate was only 1×10^{-4} m s^{-1}. Figure 5.43 shows a TTT diagram for martensite where the classical C-curve characteristic of isothermal reactions is apparent. A sufficiently rapid quench can therefore completely suppress martensitic transformation, which can then be stimulated by warming to a temperature where the rate of reaction is significant [142, 145].

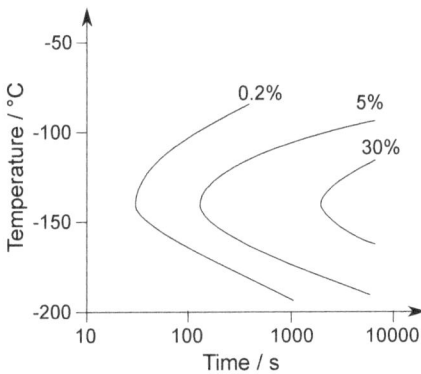

Figure 5.43 An isothermal martensitic-transformation diagram for an Fe-3.62Mn-23.2Ni wt% alloy, adapted from Shih, Averbach and Cohen [146].

Athermal martensite sometimes precedes isothermal decomposition and influences the location and form of the C-curves on the TTT diagram [146]. Figure 5.44 shows isothermal transformation curves for a Fe-15Ni wt% alloy; in each case, the fraction present at zero time is the amount of martensite that forms athermally. It is natural that when martensite forms isothermally, the temperature at which it begins to form should depend on the cooling rate to the transformation temperature. However, for alloys where athermal transformation precedes isothermal decomposition, the amount of martensite that forms above the athermal-M_S is very small indeed [147]. Solutes that increase the stability of the austenite shift the C-curves for isothermal transformation to longer times and lower temperatures. Isothermal martensite can form after bursts of martensitic transformation [145, 148]. There are contradictory reports about the shape of isothermal martensite when compared with the athermal variety (including that formed during bursts of transformation). Some studies indicate that isothermal martensite tends to be in the form of very thin plates, frequently containing a dislocation substructure rather than twinning. Others claim that there is no great difference in the two microstructures. These difficulties might be reconciled by the work of Okamoto and Oka [149], who showed that when thin-plate isothermal martensite is cooled, it thickens into lenticular plates reminiscent of athermal martensite. The final lenticular plates contain little evidence of the original thin-plate martensite. When the isothermally formed thin-plate martensite is warmed, it develops into lower bainite with the characteristic carbide microstructure. The final lenticular plate then contains the original mid-rib of the isothermal martensite surrounded by the lower bainite.

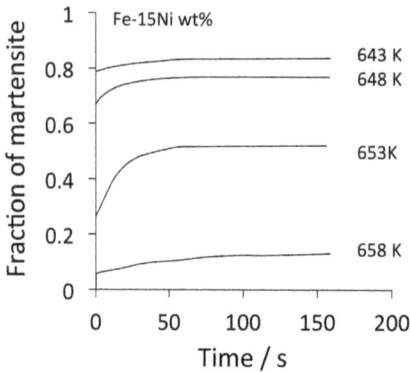

Figure 5.44 Isothermal transformation in samples that have already partially transformed athermally to martensite. The M_S temperature of the alloy is 672 K. After Tsuzaki et al. [147]

Table 5.5 lists some of the alloys where isothermal martensite has been positively identified. Although many of the alloys contain very little carbon, there are others with large concentrations of carbon. Isothermal martensite is not confined to carbon-free alloys of iron.

Table 5.5

Iron alloys in which martensite has been reported to form isothermally. In some cases, athermal transformation precedes isothermal decomposition. The chemical compositions are in wt%.

Ni	Mn	C	Other	References
	6	0.6		[142]
23	3.5			[142]
28.83		0.008		[144]
25.9	1.94	0.025		[150]
0.02	0.38	1.44	0.17Cr	[151]
23.7		0.53		[148]
20.2	5.65	0.009		[152]
22.88	3.81	< 0.003		[153, 154]
21.94	3.61	0.10		[155]
		1.1-1.8		[149]
14.99-23.46	< 0.05	< 0.005		[147]
9		1.4		[156]
18		0.7		[156]
33				[145]
23.2	2.8	0.009		[157]

Isothermal martensite is predicted by nucleation theory for regimes in which the development of the embryo requires help from thermal fluctuations (Section 5.7.2). In steels, plates of martensite plate tend to grow extremely rapidly. They are halted abruptly at hard obstacles such as grain boundaries or other plates. It is possible, therefore, to develop a kinetic theory for isothermal martensite, based on concepts of nucleation alone, with the assumption that each nucleus transforms a known quantity of austenite.

Any consideration of the nucleation rate must include the effect of autocatalysis since the number of new nuclei created per initial nucleus can be very large indeed [150, 158]. The instantaneous number density of nucleation sites N_V is therefore given by

$$N_V = \left(N_V^o + p_a \frac{V_V^{\alpha'}}{\overline{V}} - N_V' \right) (1 - V_V^{\alpha'}) \qquad (5.24)$$

where N_V^o is the initial number density of nucleation sites and p_a represents the autocatalytic additional number of sites introduced by one plate of martensite, $\overline{V}$ is the volume per martensite plate. The number of sites already used up by martensite nucleation events is N_V'. Each plate also transforms a certain volume of austenite and any of the sites in that volume are eliminated; hence the term $(1 - V_V^{\alpha'})$. The nucleation rate per unit volume at any instant is therefore given by

$$I_V = N_V \nu \exp\left\{ -\frac{G^*}{RT} \right\}$$

where ν is an empirical attempt frequency in the range 10^7–10^{13} s^{-1}. G^* is an empirical activation free energy. During a time interval $t = \tau$ to $t = \tau + d\tau$, the change in the volume fraction of martensite is, assuming a constant volume $\overline{V}$ per plate, given by

$$dV_V^{\alpha'} = I_V \overline{V} d\tau$$

This equation would need to be integrated numerically given that I_V is not constant and is a function of $V_V^{\alpha'}$. For small volume fractions of martensite:

$$V_V^{\alpha'} \propto \exp\{\Phi_6 t - 1\} \qquad \text{where} \qquad \Phi_6 = p_a \nu \exp\left\{ -\frac{G^*}{RT} \right\}$$

It appears, therefore, that the fraction transformed is dependent exponentially on the autocatalytic factor but linearly on the initial number of nucleation sites. A number of experimental observations are consistent with this conclusion [124], not surprising given the number of new sites generated by autocatalysis is many thousands of time greater than the initial number density.

A further result obtained by fitting experimental data to the kinetic theory is that the activation energy is found to be proportional to the chemical driving force (Figure 5.45). This, of course, is consistent with martensite nucleation theory, Equation 5.20.

The theory above has been useful in understanding the form of the isothermal transformations curves and indeed, of the effects of plastic strain and magnetic fields on the transformation behaviour. It has also revealed the potency of autocatalytic effects. There remain some significant assumptions. The argument that the volume per plate is constant may not in general be justified; $\overline{V}$ is expected to be a function of $V_V^{\alpha'}$. The variation is sometimes taken into account by replacing $\overline{V}$ in Equation 5.24 by the term

$$\left(\overline{V} + \frac{d\overline{V}}{d(\ln N_V)} \right)$$

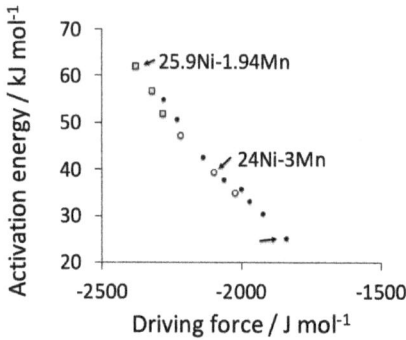

Figure 5.45 Activation energy versus the chemical free energy change for martensitic transformation [124]. The concentrations for these alloys of iron are in wt% and the data are obtained by fitting to isothermal transformation experiments using a value $p_a v = 10^{23} \, \text{cm}^{-3} \, \text{s}^{-1}$. Experimentally measured values of $\overline{V}$ were used in the analysis.

but this in turn introduces a further empirical parameter since the dependence of the volume on the number density is not generally known.

Martensite nucleation theory suggests that there is likely to be a distribution of activation energies, corresponding to that of defect potencies. Although this effect can in principle be accounted for, the potency distribution must necessarily be obtained by fitting to experimental data.

5.8.4 BURSTS OF TRANSFORMATION

The burst phenomenon (Figure 5.40) is attributed to intense autocatalysis, observed both in steels and interstitial-free iron-alloys (Table 5.6). Autocatalysis is not the complete explanation if the term simply represents the generation of additional nuclei. Additional nucleation defects are known to be generated during many martensitic reactions (e.g., isothermal) that do not transform violently. Bokros and Parker [159] studied the crystallography of the plates participating in bursts and concluded that there has to be a degree of mechanical coupling between them for a burst to operate. Only those plates that couple effectively lead to the classical zig-zag formations, consisting of four plates with nearly parallel $\{3\,10\,15\}_\gamma$ habit planes, the poles of which cluster about common $\langle 1\,1\,0 \rangle_\gamma$ directions (Figure 5.46). In subsequent work on a large series of alloys, Brook and Entwisle [160] found that pronounced bursts occurred in alloys with a martensite habit plane close to $\{2\,5\,9\}_\gamma$ but not when the habit was close to $\{2\,2\,5\}_\gamma$. This was attributed to better mechanical coupling in the former case.

Mechanical coupling relieves some of the strain energy associated with an isolated plate. This allows the martensite plates to grow with larger aspect ratios (Section 5.6.1). The plates associated with burst transformations are routinely thicker for the same length when compared with those in steels where gradual transformation occurs [140].

The hypothesis that good mechanical coupling is essential for the development of bursts of transformation seems reasonable. There is, however, the alternative possibility that the nature of autocatalytically generated nuclei might be depend on the alloy. It may be feasible that the defects created by plate formation are suitable only for the nucleation of particular variants. The nature of the defects generated might vary with the strength and stacking fault energy of the parent austenite. In such circumstances, burst formation might be better described in terms of strain-affected transformation, instead of the stress-affected transformation implicit in mechanical coupling.

Table 5.6

Examples of alloys in which martensitic transformation begins with a burst of transformation.

Alloy Composition / wt%	Reference
Fe-31.75Ni-0.01C	[159]
Fe-32Ni-0.035C	[161]
Fe-30.7Ni-0.28C	[105]
Fe-24Ni-0.5C	[160, 162]
Fe-28Ni-0.4C	[163]

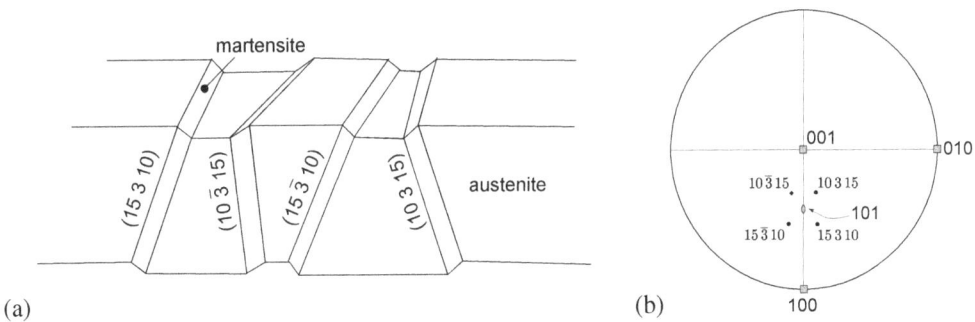

Figure 5.46 (a) An autocatalytic burst of martensitic transformation in a single crystal of austenite. There is a mechanical coupling between adjacent plates as they partly accommodate their shape deformations. Adapted from Bokros and Parker [159]. (b) Stereographic projection relative to the austenite orientation showing poles of the habit planes participating in the burst, clustered around $[101]_\gamma$.

The quantity of martensite that forms during the burst decreases as the austenite grain size is reduced. The shear stress concentration at the tip of each plate is proportional to the square of its diameter, which in turn depends on the austenite grain size; bursts are therefore easier to trigger in large-grained samples [105].[15] This also makes it difficult to spread of bursts into adjacent grains in fine-grained samples.

Bursts of transformation do not occur in all austenite grains at the same instant. However, the fraction of martensite per transforming grain is found to be larger when the γ-grains are coarse, Figure 5.47. The dependence on temperature is because the fraction of austenite transformed during the burst increases if the driving force available at M_S is large [164]. The addition of carbon to an Fe-Ni alloy, leading to a reduction in driving force at a given temperature, has been shown to lead to smaller bursts of transformation [105]. Plastic deformation of austenite prior to transformation can limit bursts due to mechanical stabilisation [159].

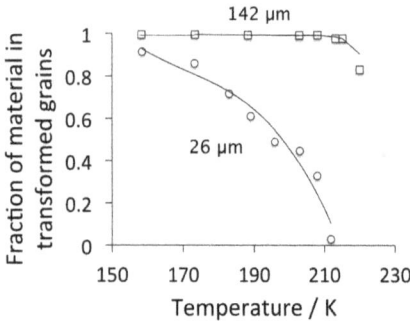

Figure 5.47 Fraction of austenite grains that show martensite as a function of the temperature to which the sample is cooled below the M_S temperature and of the austenite grain size. Selected data from [165].

5.9 STRESS AND STRAIN EFFECTS

The displacement vector and the habit plane of martensite constitute a deformation system, just as a slip plane and slip direction form a slip system. There would therefore be 24 independent deformation-systems associated with the $\gamma \to \alpha'$ and 12 for $\gamma \to \varepsilon$ transformations. During slip, the system that operates will be that which has the largest resolved shear stress on the slip plane in the slip direction [166]. An external stress imposed on austenite will similarly tend to favour those crystallographic variants that comply best with the stress [167]. This can be expressed as an interaction energy ΔG_{mech} which would add to or oppose any chemical driving force ΔG_{chem} depending on the nature of the stress. If the 3×3 tensor σ_{ij} represents the applied system of stresses then the traction $\mathbf{t}$ on the habit plane is obtained product $t_i \sigma_{ij} = p_j$ where p_j represent the direction cosines of the habit plane normal $\mathbf{p}$. The traction represents the total stress on the habit plane, consisting of a normal stress σ_N and shear stress τ with

$$\sigma_N = |\mathbf{t}| \cos\{\theta\} \qquad \tau = |\mathbf{t}| \cos\{\beta\} \cos\{\phi\}$$

where θ is the angle between the $\mathbf{p}$ and $\mathbf{t}$; β is the angle between $\mathbf{t}$ and the direction of the maximum resolved shear stress, with ϕ defining the angle between that resolved shear stress and the direction of shear of the shape deformation. The mechanical interaction energy is then

$$\Delta G_{mech} = -(\tau s + \sigma_N \zeta) \tag{5.25}$$

where the negative sign makes ΔG_{mech} a reduction if stress favours transformation. Referring to Figure 5.48, the driving force $\Delta G_{M_S}^{\gamma\alpha}$ defines the martensite-start temperature in the absence of external influence. If a stress that is below the yield strength of the austenite is applied, then the *net* sum $\Delta G_{chem} + \Delta G_{mech}$ can equal $\Delta G_{M_S}^{\gamma\alpha}$ at $M_\sigma > M_S$ because the contribution required from the chemical term is reduced.

The basic principle therefore is that the net driving force which is the sum of the chemical and mechanical terms must be equal to or greater in magnitude than $\Delta G_{M_S}^{\gamma\alpha}$. ΔG_{chem} opposes transformation at $T > T_0'$ in which case transformation depends entirely on ΔG_{mech}. Eventually, the γ yield stress is exceeded which limits the stress that can be applied and a point is reached where $\Delta G_{mech} = \Delta G_{M_S}^{\gamma\alpha}$ at a temperature M_d, beyond which martensitic transformation ceases.

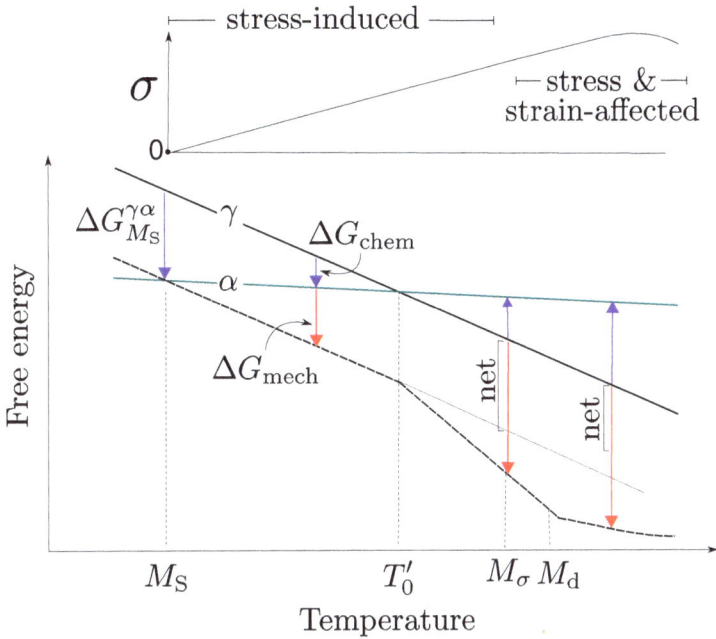

Figure 5.48 The martensite-start temperature as a function of temperature and applied stress. Note that the resistance to the plastic deformation of austenite decreases as the temperature increases.

The transformation is said to be stress-assisted in the range M_S-M_d because the stress is expressed through a purely thermodynamic term defined by Equation 5.25. Once the austenite yields prior to transformation, the thermodynamic term persists, but defects within the austenite may stimulate transformation by other mechanisms such as by increasing the number density of nucleation sites, an effect that could amplify through autocatalysis [168]. It is important to understand that the region between M_σ-M_d therefore covers the *combined* effects of stress and strain. In many cases the entire range beyond M_S can be explained by ΔG_{mech} alone because the stress during and after yielding is the highest that can be imposed on the austenite [169, 170].

To study strain-affected transformation requires the austenite to be deformed first, the stress removed and then transformed into martensite. A model for strain-induced martensite based on nucleation at deformation-band intersections due to Olson and Cohen [171] is simplified here to deal with generic defects. If the number density of nucleation sites introduced by deformation, per unit volume of austenite is N_V, the change in the number of plates of martensite per unit volume is $dN_V^{\alpha'} = p \times dN_V$, where p is the probability that the site will develop into a plate. The increment in the volume fraction of martensite is then:

$$\frac{dV_V^{\alpha'}}{1 - V_V^{\alpha'}} = \overline{V} \, dN_V^{\alpha'} = \overline{V} \, p \, dN_V$$

$$\therefore \quad V_V^{\alpha'} = 1 - \exp\{-\overline{V} \, p \, N_V\} \tag{5.26}$$

where $\overline{V}$ is the average volume per plate of martensite. The function N_V (for example its strain dependence) and p would require additional expression depending on the nature of defects involved. In the original work, the fraction of deformation bands was assumed to vary with strain as $1 - \exp\{-b_{28}\varepsilon\}$ and the number of such bands per unit volume $N_V^* = (1 - \exp\{-b_{28}\varepsilon\})/\overline{V}^*$, where $\overline{V}^*$ is the volume per shear band, and that $N_V = b_{29}(N_V^*)^{b_{30}}$. The b_i are all constants derived by fitting to experimental data due to [172], but it is noteworthy that those data included the simultaneous effects of stress and strain, so the inevitable contribution from ΔG_{mech} was neglected.

Other empirical models have been reviewed [173], one of which includes the combined effects of stress, plastic strain and the chemical driving force (and hence the composition of the steel) [174]. These have considerable value in that once the fitting parameters are established, they can be applied to a large class of steels without further measurements:

- Sugimoto [175] assumed that the change in the fraction of martensite with plastic strain is proportional to the fraction of austenite, giving

$$\ln\{V_V^{\gamma o}\} - \ln\{V_V^{\gamma}\} = b_{31}\varepsilon_p \tag{5.27}$$

  where $V_V^{\gamma o}$ is the volume fraction of austenite prior to deformation. This equation is a good representation of experimental data, but b_{31} needs to be calibrated for each new austenite composition. The equation has been applied widely in the analysis of data where both stress and strain are applied simultaneously.

- Sherif et al. [176] incorporated $\Delta G^{\alpha\gamma} = G^{\gamma} - G^{\alpha}$ into the Sugimoto framework to enable predictions as a function of the austenite composition:

$$\ln\{V_V^{\gamma o}\} - \ln\{V_V^{\gamma}\} = b_{32}\Delta G^{\alpha\gamma}\varepsilon_p \tag{5.28}$$

  where $b_{31} = 0.002017\,\text{mol J}^{-1}$

- Mukherjee et al. [174] took account of $\Delta G^{\gamma\alpha}$, $G_V^e = 1.2 \times 10^8\,\text{J m}^{-3}$ and a function $f\{\varepsilon\} = b_{33}\varepsilon^{b_{34}}$ to arrive at

$$\frac{V_V^{\gamma}}{V_V^{\gamma o}} = \frac{\Delta G^{\gamma\alpha} + G_V^e}{\Delta G^{\gamma\alpha} + G_V^e - b_{33}\varepsilon_p^{b_{34}}} \tag{5.29}$$

  with $b_{33} = 1.056 \times 10^9\,\text{J m}^{-3}$ and $b_{34} = 1.03$.

5.9.1 TRANSFORMATION TEXTURE

Figure 5.49 shows martensite induced in an initially austenitic polycrystalline sample by applying a uniaxial tension at $T > M_S$; at a small stress, this has the effect of inducing plates of martensite that form at $\approx 45°$ to the stress axis, where the shear stress would be at its maximum. Even though the individual austenite grains present different orientations, there are 24 variants possible per grain so one or more could find itself in a near-optimum orientation with respect to the stress. The amount of martensite that forms increases with stress, a reflection of the mechanical driving force (Equation 5.25). To calculate texture requires both *variant selection* and the fraction of each variant to be estimated; selection implies the favouring of some crystallographic variants over others within the set of 24 available within a single crystal of austenite.

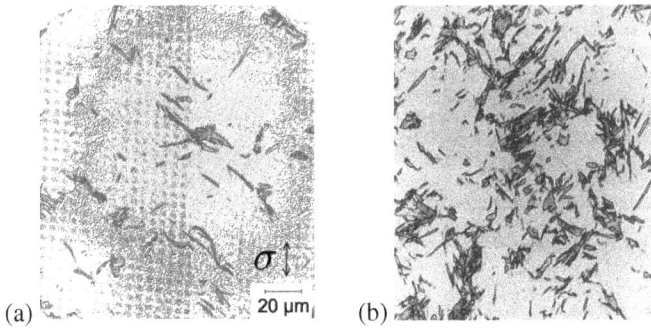

Figure 5.49 Polycrystalline austenitic Fe-28.1Ni-0.4C wt% ($M_S = -80\,^\circ$C) transformed at $-44\,^\circ$C by applying a tensile stress along the direction indicated. (a) Small stress. (b) Relatively large stress [163].

Given that each martensite plate can be defined by the mathematically connected set $\mathbf{p}$, $\mathbf{d}$ and $(\alpha\,J\,\gamma)$, it should be possible to calculate the orientations of the variants that are induced by stress given the crystallographic texture of the austenite. Each variant in a given grain of austenite will have a different M_S temperature because each variant will be associated with a different ΔG_{mech}. If each variant is identified to be different in this respect then there will be $(24 \times N_\gamma)$ martensite-start temperatures (N_γ is the number of austenite grains).

The orientations of martensite can be referred to the sample axes to present a crystallographic texture. This can be misleading because although the crystallographic distribution is determined, the volume fractions of each variant, which contribute to *intensities* in a texture plot are not calculated. This is illustrated in Figure 5.50; with this approach an austenite grain containing two variants of martensite of equal size will display the same calculated distribution of poles as that in which one variant is a hundred times larger in volume than the other. To account for the difference in volume

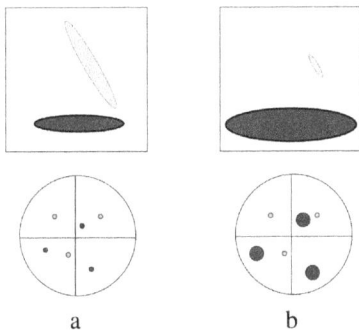

Figure 5.50 Distribution of $\langle 100 \rangle_\alpha$ poles from two martensite plates, referred to the austenite frame. (a) Pole figure for plates of identical size. (b) Pole figure showing different intensities for the large and small plate.

a   b

fractions of the distinct entities requires thermodynamic and kinetic theory to be incorporated into the texture calculation. There have been two attempts at doing this, one of which uses the Bain strain as the transformation strain [177] and the other the shape deformation [178] which is relevant to the interaction with stress and forms the basis for the discussion that follows.

The evolution of the martensite fraction as a function of undercooling below M_S is described by Equation 5.9 which can be adapted to deal with the many different transformation temperatures within the sample using the simultaneous transformations adaptation of the Avrami extended volume concept as described in Section 13.4; in a temperature interval dT [178]:

$$dV_V^{\alpha'_1} = b_2\left(1 - \sum_{i=0}^{24} V_V^{\alpha'_i}\right)dT, \quad \text{with} \quad dV_V^{\alpha'_1} = 0 \quad \text{if} \quad M_{S_1} < T$$

$$\vdots$$

$$dV_V^{\alpha'_{24}} = b_2\left(1 - \sum_{i=0}^{24} V_V^{\alpha'_i}\right)dT, \quad \text{with} \quad dV_V^{\alpha'_{24}} = 0 \quad \text{if} \quad M_{S_{24}} < T \qquad (5.30)$$

with each equation obtained by differentiating Equation 5.9. The fractions from individual austenite grains are then summed over all the N_γ austenite grains, assuming that there is no interaction between the grains. Figure 5.51a shows calculations for transformation in 500 austenite grains arranged to achieve a random γ-texture; an interesting observation is that the application of stress increases the temperature range over which the martensite forms because the dominant favoured-variants are induced at high temperatures and those that are not optimally oriented with respect to the stress form at lower temperatures than the case where there is no stress applied. The starting texture of the austenite (random, Goss, Copper and Cube) does not affect these results [178] because under stress, much of the contribution to the volume fraction is from the favoured variants.

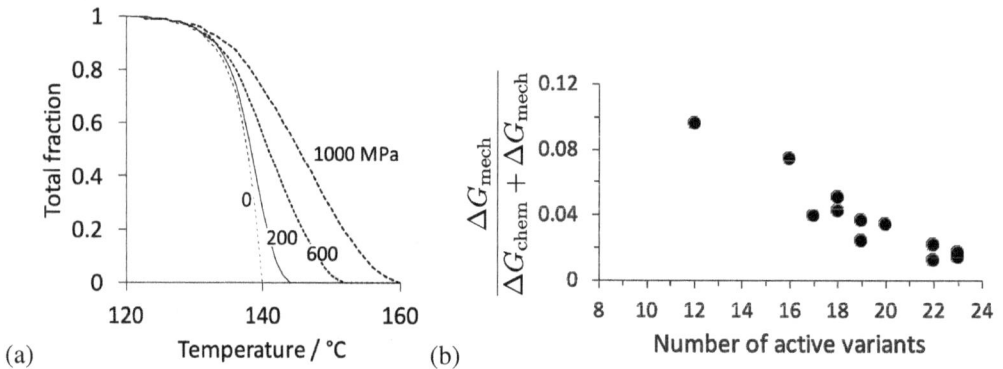

Figure 5.51 (a) The evolution of martensite content as a function of temperature and applied tensile-stress, assuming that the polycrystalline austenite has a random texture [178].

The interaction of each variant with the applied stress is well-described by ΔG_{mech} but it is useful to understand the extent of variant selection. Those "selected" can be defined as those that comply best with the applied stress and the intensity of selection can reasonably be approximated by the magnitude of the ratio $\Delta G_{mech}/(\Delta G_{chem} + \Delta G_{mech})$. It turns out that there is almost a linear variation between this ratio and the number of variants (of the 24 possible) is active in any particular grain of austenite, Figure 5.51b [179].

5.9.2 MECHANICAL STABILISATION

Whereas the defects induced by the plastic deformation of austenite can enhance nucleation rates, there comes a point where the defect density interferes with the progress of the glissile interfaces that are a necessary feature of displacive transformations, or indeed, mechanical twinning. The overall transformation rate would then decrease even though the nucleation rate might increase. A critical plastic strain is not appropriate to define the onset of *mechanical stabilisation* because the ability of the interface to absorb the defects into the product lattice and continue its motion depends also on the driving force. The balance between the force and the resistance the interface encounters forms the basis of a theory for mechanical stabilisation [180]. The stress τ_T driving the motion of the interface originates from the chemical free energy change $\Delta G^{\alpha\gamma}$ of transformation:

$$\tau_T = b_{35}\Delta G^{\alpha\gamma} \tag{5.31}$$

where b_{35} is a constant assumed to be equal to unity and $\Delta G^{\alpha\gamma} = G^\gamma - G^\alpha$, i.e., the magnitude of the driving force. The stored energy of martensite is about $600\,\mathrm{J\,mol^{-1}}$ [111], values which are subtracted from $\Delta G^{\alpha\gamma}$. When multiplied by the magnitude of the Burgers vector, b, this gives the force per unit length available for the austenite-martensite interface to move.

The plastic strain when dislocations with a density $\rho_\perp$ and Burgers vector of magnitude b translate in austenite on average through a distance $\bar{z}$ is given by [181]

$$\varepsilon_p = \rho_\perp b\bar{z}. \tag{5.32}$$

For a given dislocation density, the spacing between the dislocations is $\rho^{-\frac{1}{2}}$. The mean shear stress τ needed to force dislocations past each other is, in these circumstances:

$$\tau = \frac{E_s b\rho_\perp^{\frac{1}{2}}}{8\pi(1-\nu)} \tag{5.33}$$

where E_s is the shear modulus and ν the Poisson's ratio. On combining with Equation 5.32 and noting that the force per unit length is τb,

$$\tau b = \frac{G b^{\frac{3}{2}}}{8\pi(1-\nu)}\sqrt{\frac{\varepsilon_p}{\bar{z}}}. \tag{5.34}$$

The mean free distance $\bar{z}$ must decrease as the plastic strain increases [182]

$$\bar{z} = \frac{b_{36}\bar{L}_\gamma}{b_{36} + \bar{L}_\gamma \varepsilon_p} \tag{5.35}$$

where $\bar{L}_\gamma$ is the original grain size of austenite prior to deformation and b_{36} is a coefficient about equal to $1\,\mu\mathrm{m}$ [182]. Solid-solution hardening and other mechanisms will also contribute τ_S to the resistance to interface motion so this must be added to τ. Once this is done, the onset of mechanical stabilisation is when the stress driving the interface equals that opposing its motion:

$$\tau_T = \tau + \tau_S$$

$$b\Delta G^{\alpha\gamma} = \frac{1}{8\pi(1-\nu)}E_s b^{\frac{3}{2}}\sqrt{\frac{\varepsilon_p}{\bar{z}}} + \tau_S b \tag{5.36}$$

This equation can therefore be used to calculate the critical strain for mechanical stabilisation and examples of its application can be found elsewhere [180].

5.10   TETRAGONALITY OF MARTENSITE

Comprehensive experiments on the state of carbon within steel were reported by Abel around 1883, published formally in 1885 [183]. It was found that "the carbon in cold-rollled steel exists in the form of a definite iron carbide, approximating the formula Fe_3C or to a multiple of that formula". In the same experiments, hardened steel (presumably martensitic) "appeared to have the effect of preventing or arresting the separation of carbon, as a definite carbide". Subsequent work [184] had suggested that the carbon should be in interstitial solution in martensite; using lattice parameter determinations Honda and Nishiyama [185] demonstrated that the observed density proved that the carbon does indeed exist in interstices rather than as pairs C_2 that substitute for iron atoms. Figure 5.52a shows carbon atoms located on equivalent octahedral sites in the austenite on all three edges of the fcc unit cell. When the Bain distortion is applied as indicated, all of the carbon atoms end up in just one of three sublattices of the martensite interstices that have their shortest axes aligned to the compression axis, thereby rendering the α'-lattice tetragonal.

The tetragonality as a function of the carbon concentration of martensite is shown in Figure 5.52b; first principles calculations have shown that the trends illustrated are maintained to 3 wt% carbon [186]. Carbon causes an expansion along the shortest axis and a slight contraction along the other two orthogonal axes. Given that the Bain strain causes all of the carbon atoms in austenite to end up in just one sublattice of the martensite interstices, the lattice itself adopts a tetragonal symmetry for all concentrations of carbon greater than zero. If the temperature at which martensite forms is greater than the ordering temperature (p. 71) then there may be a subsequent redistribution of carbon atoms that makes the martensite cubic. This is one reason why low carbon martensites do not exhibit tetragonality in practice.

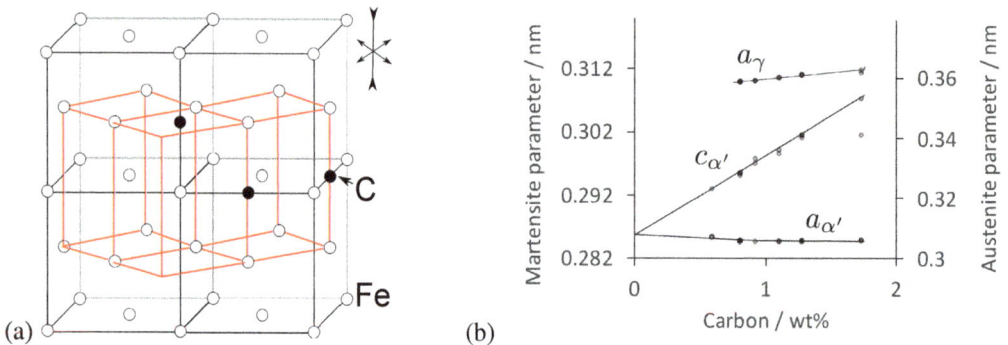

Figure 5.52 (a) Four fcc austenite unit cells (black) and the bct versions of the austenite unit cells (red). Austenite has one octahedral interstice per Fe atom whereas martensite has three. The orientation of the compression axis during the Bain distortion is illustrated. (b) Lattice parameters of martensite and austenite in Fe-C alloys, adapted using data from Honda and Nishiyama [185]. The $c_{\alpha'}/a_{\alpha'}$ ratio is $1 + 0.045w_C$.

There are other aspects of the tetragonality of ferrous martensite, reviewed by Christian [187]. The c/a ratio in Fe-Ni-C, Fe-Al-C and Fe-Al-Mn-C alloys is found to be larger for twinned thin-plate martensite than for untwinned martensite [188, 189]. The plastic accommodation of the shape change in the latter case leads to the displacement of carbon atoms from the positions inherited via the Bain deformation. Twinned martensite in Fe-7Al-2C and Fe-8Al-2C wt% alloys tend to exhibit an abnormally large c/a ratio because of the coherent precipitation of fine κ-carbide (Section 11.7) in the austenite; this coherency is inherited by the martensite. The precipitates themselves undergo the Bain deformation leaving their large density of carbon atoms along the c-axis. Other contributions to tetragonality come from specific carbon atom configurations in the austenite when the alloy contains a large aluminium concentration [190]. These explanations seem to fail for Fe-20Ni-1C wt% and similar alloys, where an anomalously large tetragonality is observed [188, 189]; recent first-principles calculations suggest instead that that the cause of the anomaly is that the nickel makes the alloy more elastically compliant than Fe-C [186].

REFERENCES

1. H. K. D. H. Bhadeshia, and R. W. K. Honeycombe: Steels: Microstructure and Properties: 4th ed., Elsevier, 2017.

2. M. Cohen, G. B. Olson, and P. C. Clapp: 'On the classification of displacive transformations (what is martensite?)', In: G. B. Olson, and M. Cohen, eds. *International Conference on Martensitic Transformations ICOMAT '79*. Massachusetts, USA: Alpine Press, 1979:1–11.

3. H. K. D. H. Bhadeshia, and C. M. Wayman: 'Phase transformations: nondiffusive': In: D. E. Laughlin, and K. Hono, eds. *Physical Metallurgy*, chap. 9. North-Holland: Elsevier B. V., 2014:1021–1072.

4. J. C. Williams, D. D. Fontaine, and N. E. Paton: 'The ω-phase as an example of an unusual shear transformation', *Metallurgical Transactions*, 1973, **4**, 2701–2708.

5. S. K. Sikka: 'Omega phase in materials', *Progress in Materials Science*, 1982, **27**, 245–310.

6. C. J. McHargue, and H. L. Yakel: 'Phase transformations in cerium', *Acta Metallurgica*, 1960, **8**, 637–646.

7. J. W. Christian: 'Accommodation strains in martensite formation, the use of the dilatation parameter', *Acta Metallurgica*, 1958, **6**, 377–379.

8. L. C. Chang, and H. K. D. H. Bhadeshia: 'Metallographic observations of bainite transformation mechanism', *Materials Science and Technology*, 1995, **11**, 106–108.

9. M. Hayakawa, K. Nishio, J. Hamakita, and T. Ondo: 'Isothermal and athermal martensitic transformations in a zirconia–yttria alloy', *Materials Science & Engineering A*, 1999, **273–275**, 213–217.

10. A. B. Greninger: 'The martensite transformation in beta copper-aluminium alloys', *Transactions of the AIME*, 1939, **133**, 204–227.

11. F. X. Kayser: 'Polymorphic transformation in ytterbium', *Physical Review Letters*, 1970, **25**, 662–664.

12. C. S. Barrett, and L. Meyer: 'Argon-nitrogen phase diagram', *The Journal of Chemical Physics*, 1965, **42**, 107–112.

13. J. T. A. Pollock, and H. W. King: 'Low temperature martensitic transformations in In/Tl alloys', *Journal of Materials Science*, 1968, **3**, 372–379.

14. C. J. Altstetter: 'Transformations in rare earth metals', *Metallurgical Transactions*, 1973, **4**, 2723–2730.

15. X. Li, J. A. Martínez-González, J. P. Hernández-Ortiz, A. Ramírez-Hernández, Y. Zhou, M. Sadati, R. Zhang, P. F. Nealey, and J. J. de Pablo: 'Mesoscale martensitic transformation in single crystals of topological defects', *Proceedings of the National Academy of Sciences*, 2017, **114**, 10011–10016.

16. R. F. Bunshah, and R. F. Mehl: 'Rate of propagation of martensite', *Transactions of the AIME*, 1953, **193**, 1251–1258.

17. S. R. Coriell, and D. Turnbull: 'Relative roles of heat transport and interface rearrangement rates in the rapid growth of crystals in undercooled melts', *Acta Metallurgica*, 1982, **30**, 2135–2139.

18. B. Wei, D. M. Herlach, and B. Feuerbacher: 'Rapid crystal growth in undercooled alloy melts', *Microgravity Quarterly*, 1993, **3**, 193–197.

19. J. Weertman: 'High velocity dislocations', In: P. G. Shewmon, and V. F. Zackay, eds. *Response of Metals to High Velocity Deformation*. New York, USA: Interscience, 1960:205–247.

20. B. Chalmers: 'The cry of tin', *Nature*, 1932, **129**, 650–651.

21. S. L. van Doren, R. B. Pond Sr, and R. E. Green Jr: 'Acoustic characteristics of twinning in indium', *Journal of Applied Physics*, 1976, **47**, 4343–4348.

22. K. Takashima, Y. Higo, and S. Nunomura: 'The propagation velocity of the martensitic transformation in 304 stainless steel', *Philosophical Magazine A*, 1984, **49**, 231–241.

23. D.-Z. Yang, and C. M. Wayman: 'Slow growth of isothermal lath martensite in an Fe-21N-4Mn alloy', *Acta Metallurgica*, 1984, **32**, 949–954.

24. R. J. Salzbrenner, and M. Cohen: 'On the thermodynamics of thermoelastic martensitic transformation', *Acta Metallurgica*, 1979, **27**, 739–748.

25. D. P. Koistinen, and R. E. Marburger: 'A general equation prescribing the extent of the austenite-martensite transformation in pure iron-carbon alloys and plain carbon steels', *Acta Metallurgica*, 1959, **7**, 59–60.

26. S. Kajiwara: 'Strengthening of austenite by reverse martensitic transformation in high nickel steels', In: *International conference on Physical Metallurgy of Thermomechanical Processing of Steels and Other Metals,*. Tokyo, Japan: Iron and Steel Institute of Japan, 1988:895–902.

27. R. P. Zerwech, and C. M. Wayman: 'On the nature of the $\alpha \rightarrow \gamma$ transformation in iron: A study of whiskers', *Acta Metallurgica*, 1965, **13**, 99–107.

28. B. P. J. Sandvik, and C. M. Wayman: 'Characteristics of lath martensite: Part II. The martensite-austenite interface', *Metallurgical transactions A*, 1983, **14**, 823–834.

29. P. M. Kelly, A. Jostsons, and R. G. Blake: 'The orientation relationship between lath martensite and austenite in low carbon, low alloy steels', *Acta Metallurgica et Materalia*, 1990, **38**, 1075–1081.

30. E. J. Efsic, and C. M. Wayman: 'Crystallography of the Fcc to Bcc martensitic transformation in an iron-platinum alloy', *Transactions of The Metallurgical Society of the American Institute of Mining, Metallurgical and Petroleum Engineers*, 1967, **239**, 873–882.

31. D. P. Dunne, and J. S. Bowles: 'Measurement of the shape strain for the (225) and (259) martensitic transformation', *Acta Metallurgica*, 1969, **17**, 201–212.

32. A. J. Morton, and C. M. Wayman: 'Theoretical and experimental aspects of the "(225)" austenite-martensite transformation in iron alloys', *Acta Metallurgica*, 1966, **14**, 1567–1581.

33. D. P. Dunne, and C. M. Wayman: 'The effect of austenite ordering on the martensite transformation in Fe-Pt alloys near the composition Fe_3Pt: II. Crystallography and general features', *Metallurgical Transactions*, 1973, **4**, 147–152.

34. J. A. Venables: 'The martensite transformation in stainless steel', *Philosophical Magazine*, 1962, **7**, 35–44.

35. G. V. Kurdjumov, and G. Sachs: 'Über den mechanismus der stahlhärtung', *Zietschrift für Physik*, 1930, **64**, 325–343.

36. Z. Nishiyama: 'X-ray investigation of the mechanism of the transformation from face centered cubic lattice to body centered cubic', *Science Reports of Tohoku Imperial University*, 1934, **23**, 637–634.

37. G. Wassermann: 'Einflusse der α-γ-umwandlung eines irreversiblen nickelstahls auf kristallorientierung und zugfestigkeit', *Archiv für das Eisenhüttenwesen*, 1933, **6**, 347–351.

38. A. B. Greninger, and A. R. Troiano: 'Crystallography of austenite decomposition', *Transactions of the AIME*, 1940, **140**, 307–336.

39. A. B. Greninger, and A. R. Troiano: 'The mechanism of martensite formation', *Transactions of the AIME*, 1949, **185**, 590–598.

40. K. M. Knowles, D. A. Smith, and W. A. T. Clark: 'On the use of geometric parameters in the theory of interphase boundaries', *Scripta Metallurgica*, 1982, **16**, 413–416.

41. G. Nolze: 'Characterisation of the fcc/bcc orientation relationship by EBSD using pole figures and variants', *Zietschrift für Metallkunde*, 2004, **95**, 744–755.

42. S. Godet, J. C. Glez, Y. He, J. J. Jonas, and P. J. Jacques: 'Grain-scale characterization of transformation texture', *Journal of Applied Crystallography*, 2004, **37**, 417–425.

43. J. W. Christian: 'Simple geometry and crystallography applied to ferrous bainites', *Metallurgical Transactions A*, 1990, **21**, 799–803.

44. J. W. Christian, and A. G. Crocker: Dislocations in Solids, ed. F. R. N. Nabarro, vol. 3, chap. 11: Amsterdam, Holland: North Holland, 1980:165–252.

45. E. C. Bain: 'The nature of martensite', *Transactions of the AIME*, 1924, **70**, 25–46.

46. T. Tadaki, and K. Shimizu: 'Electron microscope study of the martensitic transformation in ordered Fe_3Pt alloy', *Transactions of JIM*, 1970, **11**, 44–50.

47. M. Kato, R. Monzen, and T. More: 'A stress-induced martensitic transformation of spherical iron particles in a Cu-Fe alloy', *Acta Metallurgica*, 1978, **26**, 605–613.

48. G. B. Olson, and M. Cohen: 'Interphase boundary dislocations and the concept of coherency', *Acta Metallurgica*, 1979, **27**, 1907–1918.

49. J. W. Christian: 'Deformation by moving interfaces', *Metallurgical Transactions A*, 1982, **13**, 509–538.

50. B. A. Bilby: 'Types of dislocation source', In: *Bristol Conference Report on Defects in Crystalline Solids*. London, U.K.: The Physical Society, 1955:124–133.

51. J. W. Christian: Theory of Transformations in Metals and Alloys, Part II: 3 ed., Oxford, U. K.: Pergamon Press, 2003.

52. K. Ogawa, and S. Kajiwara: 'High-resolution electron microscopy study of ledge structures and transition lattices at the austenite–martensite interface in Fe-based alloys', *Philosophical Magazine*, 2004, **84**, 2919–2947.

53. J. W. Christian: Theory of Transformations in Metals and Alloys, Part I: 2 ed., Oxford, U. K.: Pergamon Press, 1975.

54. N. D. H. Ross, and A. G. Crocker: 'A generalized theory of martensite crystallography and its application to transformations in steels', *Acta Metallurgica*, 1970, **18**, 405–418.

55. D. P. Dunne, and C. M. Wayman: 'An assessment of the double shear theory as applied to ferrous martensitic transformations', *Acta Metallurgica*, 1971, **19**, 425–438.

56. F. Maresca, and W. A. Curtin: 'The austenite/lath martensite interface in steels: Structure, athermal motion, and in-situ transformation strain revealed by simulation and theory', *Acta Materialia*, 2020, **134**, 302–323.

57. M. Kehoe, and P. M. Kelly: 'The role of carbon in the strength of ferrous martensite', *Scripta Metallurgica*, 1970, **4**, 473–476.

58. L. A. Norström: 'On the yield strength of quenched low carbon lath martensite', *Scandinavian Journal of Metallurgy*, 1976, **5**, 159–165.

59. G. M. Smith: 'The microstructure and yielding behaviour of some Ti steels': Ph.D. thesis, University of Cambridge, Cambridge, U. K., 1984.

60. T. Maki, S. Shimooka, S. Fujiwara, and I. Tamura: 'Formation temperature and growth behaviour of thin plate martensite in Fe-Ni-C alloys', *Transactions of JIM*, 1975, **16**, 35–41.

61. G. Miyamoto, A. Shibata, T. Maki, and T. Furuhara: 'Precise measurement of strain accommodation in austenite matrix surrounding martensite in ferrous alloys by electron backscatter diffraction analysis', *Acta Materialia*, 2009, **57**, 1120–1131.

62. T. Maki, S. Furutani, and I. Tamura: 'Shape memory effect related to thin plate martensite with large thermal hysteresis in ausaged Fe-Ni-Co-Ti alloy', *ISIJ International*, 1989, **29**, 438–445.

63. M. Umemoto, and C. M. Wayman: 'Crystallography and morphology studies of Fe-Pt martensites lenticular to thin plate transition and thin plate morphology', *Acta Metalllurgica*, 1978, **26**, 1529–1549.

64. F. W. Schaller, and D. J. Schmatz: 'The inheritance of defects by martensite', *Acta Metalllurgica*, 1963, **11**, 1193–1194.

65. S. Kajiwara: 'Velocity of the austenite-martensite interface in reverse', *Materials Characterization*, 1995, **34**, 105–119.

66. G. Krauss, and A. R. Marder: 'The morphology of martensite in iron alloys', *Metallurgical Transactions*, 1971, **2**, 2343–2357.

67. A. B. Greninger, and A. R. Troiano: 'Orientation habit of martensite', *Nature*, 1938, **141**, 38.

68. Z. Nishiyama, and K. Shimizu: 'Direct observation of sub-structures in martensite', *Acta Metallurgica*, 1959, **7**, 432–433.

69. M. S. Wechsler, D. S. Lieberman, and T. A. Read: 'On the theory of the formation of martensite', *Transactions of the AIME Journal of Metals*, 1953, **197**, 1503–1515.

70. J. K. Mackenzie, and J. S. Bowles: 'The crystallography of martensite transformations II', *Acta Metallurgica*, 1954, **2**, 138–147.

71. J. S. Bowles, and J. K. Mackenzie: 'The crystallography of martensite transformations, part I', *Acta Metallurgica*, 1954, **2**, 129–137.

72. J. K. Mackenzie, and J. S. Bowles: 'The crystallography of martensite transformations III FCC to BCT transformations', *Acta Metallurgica*, 1954, **2**, 224–234.

73. C. M. Wayman: 'The crystallography of martensitic transformations in alloys of iron': In: H. Hermans, ed. *Advances in Materials Research*, vol. 3. John Wiley & Sons, Inc., 1968:147–304.

74. H. K. D. H. Bhadeshia: Geometry of Crystals, Polycrystals, and Phase Transformations: Florida, USA: CRC Press, ISBN 9781138070783, freely downloadable from www.phase-trans.msm.cam.ac.uk, 2017.

75. G. B. Olson, and M. Cohen: 'Theory of martensitic nucleation: a current assessment', In: M. S. H. I. Aaronson, D. E. Laughlin:, ed. *Proceedings of the International Conference on Solid→Solid Phase Transformations*. Warrendale, Pennsylvania, USA: TMS-AIME, 1981:1209–1213.

76. G. Frommeyer, U. Brüx, and P. Neumann: 'Supra-ductile and high-strength manganese-TRIP/TWIP steels for high energy absorption purposes', *ISIJ International*, 2003, **43**, 438–446.

77. M. Koyama, Y. Abe, and K. Tsuzaki: 'Split and shift of ε-martensite peak in an X-ray diffraction profile during hydrogen desorption: a geometric effect of aatomic sequence', *ISIJ International*, 2018, **58**, 1745–1747.

78. C. Hayzelden, K. Chattopadhyay, J. C. Barry, and B. Cantor: 'Transmission electron microscopy observations of the f.c.c.-to-h.c.p. martensite transformation in CoNi alloys', *Philosophical Magazine A*, 1991, **63**, 461–470.

79. Y. Tomota, and T. Maki: 'Reversibility in martensitic transformation and shape memory in high Mn ferrous alloys', *Materials Science Forum*, 2000, **327-328**, 191–198.

80. K. Tsuzaki, Y. Natsume, and T. Maki: 'Transformation reversibility in Fe-Mn-Si shape memory alloy', *Journal de Physique Colloque*, 1995, **5**, C8–409–C8–414.

81. G. E. Duvall, and R. A. Graham: 'Phase transitions under shock-wave loading', *Reviews of Modern Physics*, 1977, **49**, 523–579.

82. N. A. Zarkevich, and D. D. Johnson: 'Coexistence pressure for a martensitic transformation from theory and experiment: Revisiting the bcc-hcp transition of iron under pressure', *Physical Review B*, 2015, **91**, 174104.

83. L. Stixrude: 'Structure of iron to 1 Gbar and 40000 K', *Physical Review Letters*, 2012, **108**, 055505.

84. K. Ishida, and T. Nishizawa: 'Effect of alloying elements on stability of epsilon iron', *Transactions of the Japan Institute of Metals*, 1974, **15**, 225–231.

85. J. W. Forbes: 'Experimental investigation of the kinetics of the shock-induced alpha to epsilon phase transformation in Armco iron': Technical Report TR 77-137, Naval Surface Weapons Centre, Maryland, U.S.A., 1977.

86. W. G. Burgers: 'On the process of transition of the cubic-body-centered modification into hexagonal-close-packed modification of zirconium', *Physica*, 1934, **1**, 561–586.

87. H.-S. Yang, and H. K. D. H. Bhadeshia: 'Uncertainties in the dilatometric determination of the martensite-start temperature', *Materials Science and Technology*, 2007, **23**, 556–560.

88. W. Steven, and A. G. Haynes: 'The temperature of formation of martensite and bainite in low alloy steels', *Journal of the Iron and Steel Institute*, 1956, **183**, 349–359.

89. M. Cohen, E. S. Machlin, and V. G. Paranjpe: Thermodynamics in Physical Metallurgy: Ohio, U. S. A.: ASM, 1950.

90. L. Kaufman, and M. Cohen: 'Thermodynamics and kinetics of martensitic transformation', *Progress in Metal Physics*, 1958, **7**, 165–246.

91. H. K. D. H. Bhadeshia: 'The driving force for martensitic transformation in steels', *Metal Science*, 1981, **15**, 175–177.

92. H. K. D. H. Bhadeshia: 'Thermodynamic extrapolation and the martensite-start temperature of substitutionally alloyed steels', *Metal Science*, 1981, **15**, 178–180.

93. R. G. Davies, and C. L. Magee: 'Influence of austenite and martensite strength on martensite morphology', *Metallurgical Transactions*, 1971, **2**, 1939–1947.

94. G. Ghosh, and G. B. Olson: 'Kinetics of FCC→BCC heterogeneous martensitic nucleation', *Acta Metallurgica and Materialia*, 1994, **42**, 3361–3370.

95. H.-S. Yang, J. H. Jang, H. K. D. H. Bhadeshia, and D. W. Suh: 'Critical assessment: martensite-start temperature for the γ to ε transformation', *CALPHAD*, 2012, **36**, 16–22.

96. A. S. Sastri, and D. R. F. West: 'Effect of austenitizing conditions on the kinetics of martensite formation in certain medium alloy steels', *Journal of the Iron and Steel Institute*, 1965, **203**, 138–145.

97. O. A. Ankara, A. S. Sastri, and D. R. F. West: 'Some effects of austenitizing conditions on martensite', *Journal of the Iron and Steel Institute*, 1966, **204**, 509–511.

98. T. Maki, S. Shimooka, M. Umemoto, and I. Tamura: 'The morphology of strain-induced martensite and thermally transformed martensite in Fe-Ni-C alloys', *Transactions of JIM*, 1972, **13**, 400–427.

99. G. S. Ansell, P. J. Brofman, T. J. Nichol, and G. Judd: 'Effect of austenite strength on the transformation to martensite in Fe-Ni and Fe-Ni-C alloys', In: G. B. Olson, and M. Cohen, eds. *International Conference on Martensitic Transformations ICOMAT '79.* 1979:350–355.

100. J. Pietikainen: 'Thermal stabilization of austenite in high nickel-carbon steels', *Journal de Physique Colloque*, 1982, **43**, C4-479–C4-484.

101. P. J. Brofman, and G. S. Ansell: 'On the effect of fine grain size on the martensite-start temperaure in Fe-27Ni-0.025C alloys', *Metallurgical Transactions A*, 1983, **14A**, 1929–1931.

102. S. J. Lee, and Y. K. Lee: 'Effect of austenite grain size on martensitic transformation of a low alloy steel', *Materials Science Forum*, 2005, **475-479**, 3169–3172.

103. J. Huang, and Z. Xu: 'Effect of dynamically recrystallized austenite on the martensite start temperature of martensitic transformation', *Materials Science & Engineering A*, 2006, **A438-A440**, 254–257.

104. T. Maki, S. Shimooka, and I. Tamura: 'Martensite-start temperature and morphology of martensite in Fe-31Ni-0.23C alloy', *Metallurgical Transactions*, 1971, **2**, 2944–2955.

105. M. Umemoto, and W. S. Owen: 'Effects of austenitising temperature & austenite grain size on the formation of athermal martensite in an Fe-Ni and Fe-Ni-C alloy', *Metallurgical Transactions*, 1974, **5**, 2041–2046.

106. W. C. Leslie, and R. L. Miller: 'The stabilization of austenite by closely spaced boundaries', *ASM Transactions Quarterly*, 1964, **57**, 972–979.

107. G. B. Olson, K. Tsuzaki, and M. Cohen: 'Statistical aspects of martensitic nucleation', *Materials Research Society Symposium Proceedings*, 1987, **57**, 129–148.

108. J. C. Fisher, J. H. Hollomon, and D. Turnbull: 'Kinetics of the austenite to martensite transformation', *Metals Transactions*, 1949, **185**, 691–700.

109. H. S. Yang, and H. K. D. H. Bhadeshia: 'Austenite grain size and the martensite-start temperature', *Scripta Materialia*, 2009, **60**, 493–495.

110. R. L. Patterson, and C. M. Wayman: 'The crystallography and growth of partially-twinned martensite plates in Fe-Ni alloys', *Acta Metallurgica*, 1966, **14**, 347–369.

111. J. W. Christian: 'Thermodynamics and kinetics of martensite', In: G. B. Olson, and M. Cohen, eds. *International Conference on Martensitic Transformations ICOMAT '79*. Massachusetts, USA: Alpine Press, 1979:220–234.

112. T. Yokota, C. Garcia-Mateo, and H. K. D. H. Bhadeshia: 'Formation of nanostructured steel by phase transformation', *Scripta Materialia*, 2004, **51**, 767–770.

113. H. K. D. H. Bhadeshia: 'Nanostructured bainite', *Proceedings of the Royal Society of London A*, 2010, **466**, 3–18.

114. L. Kaufman: 'Thermodynamics of martensitic fcc⇌bcc and fcc⇌hcp transformations in the iron-ruthenium system', In: *Physical Properties of Martensite and Bainite, Special Report 93*. London, U.K.: Iron and Steel Institute, 1965:49–52.

115. G. B. Olson, and M. Cohen: 'A general mechanism of martensitic nucleation, Pt. I. general concepts and the FCC→HCP transformation', *Metallurgical Transactions A*, 1976, **7**, 1897–1904.

116. K. Ishida: 'Effect of alloying elements on the critical driving force of martensitic transformation in iron alloys', *Scripta Metallurgica*, 1977, **11**, 237–242.

117. G. B. Olson, and W. S. Owen: 'Stress field of a martensitic particle and the conditions for thermoelastic behaviour', In: *New Aspects of Martensitic Transformations*. Tokyo, Japan: Japan Institute of Metals, 1976:105–110.

118. H. C. Ling, and W. S. Owen: 'A model of the thermoelastic growth of martensite', *Acta Metallurgica*, 1981, **29**(10), 1721–1736.

119. A. Vevecka, H. Ohtsuka, and H. K. D. H. Bhadeshia: 'Plastic accommodation of martensite in disordered and ordered iron-platinum alloys', *Materials Science and Technology*, 1995, **11**, 109–112.

120. J. W. Brooks, M. H. Loretto, and R. E. Smallman: 'In situ observations of martensite formation in stainless steel', *Acta Metallurgica*, 1979, **27**, 1829–1838.

121. J. W. Brooks, M. H. Loretto, and R. E. Smallman: 'Direct observations of martensite nuclei in stainless steel', *Acta Metallurgica*, 1979, **27**, 1839–1847.

122. F. R. Chien, R. J. Clifton, and S. R. Nutt: 'Stress-induced phase transformation in single crystal titanium carbide', *Journal of the American Ceramics Society*, 1995, **78**, 1537–1545.

123. G. B. Olson, and M. Cohen: 'A general mechanism of martensitic nucleation, Parts I–III', *Metallurgical Transactions A*, 1976, **7A**, 1897–1923.

124. C. L. Magee: 'The nucleation of martensite', In: H. I. Aaronson, and V. F. Zackay, eds. *Phase Transformations*. Materials Park, Ohio, USA: ASM International, 1970:115–156.

125. J. W. Christian: 'A theory for the transformation in pure cobalt', *Proceedings of the Royal Society of London A*, 1951, **206A**, 51–64.

126. E. Votava: 'Electron microscopic investigation of the phase transformation of thin cobalt samples', *Acta Metallurgica*, 1960, **8**, 901–904.

127. E. Votava: 'The phase transformation in thin cobalt films', *Journal of the Institute of Metals*, 1961, **90**, 129–132.

128. H. Conrad: 'Thermally activated deformation of metals', *Journal of Metals*, 1964, **145**, 582–588.

129. J. E. Dorn: 'Low temperature dislocation mechanisms', In: A. R. Rosenfield, G. T. Hahn, A. L. Bement, and R. I. Jaffee, eds. *Dislocation Dynamics*. New York: McGraw Hill, 1968:27.

130. C. L. Magee, and H. W. Paxton: 'Experimental support for 'hard' martensite', *Transactions of the Metallurgical Society of AIME*, 1968, **242**, 1766–1767.

131. G. B. Olson: 'Lattice stability and the mechanism of martensitic transformations', In: *Proceeding of the International Conference on Martensitic Transformations*. Tokyo, Japan: Japan Institute of Metals, 1986:25–34.

132. G. L. Krasko, and G. B. Olson: 'Ferromagnetism and crystal lattice stability of bcc and fcc iron', *Journal of Applied Physics*, 1990, **67**, 4570–4572.

133. P. C. Clapp: 'A localized soft mode theory for martensitic transformations', *Physical Status Solidi (b)*, 1973, **57**, 561–569.

134. M. Wuttig, and T. Suzuki: 'The martensite transformation', *Materials Science & Engineering*, 1976, **25**, 135–138.

135. G. B. Olson, H. K. D. H. Bhadeshia, and M. Cohen: 'Coupled diffusional/displacive transformations, Part II: Solute trapping', *Metallurgical & Materials Transactions A*, 1990, **21A**, 805–809.

136. V. Raghavan: 'Kinetics of martensitic transformations', In: G. B. Olson, and W. S. Owen, eds. *Martensite*. Ohio, USA: ASM International, 1992:197–226.

137. D. G. McMurtrie, and C. L. Magee: 'The average volume of martensite plates during transformation', *Metallurgical Transactions*, 1970, **1**, 3185–3191.

138. N. N. Thadhani, and M. A. Myers: 'Kinetics of martensitic transformation induced by a tensile stress pulse', *Acta Metallurgica*, 1986, **34**, 1625–1641.

139. W. Y. C. Chen, and P. G. Winchell: 'Martensite plate arrangements in tool steel', *Metallurgical Transactions A*, 1976, **7**, 1177–1182.

140. J. R. C. Guimarães, and J. C. Gomes: 'Microstructural aspects of martensite transformation in coarse-grained Fe-31.1 pct Ni-0.02 pct C', *Metallurgical and Materials Transactions A*, 1979, **10**, 109–112.

141. M. Rao, and S. Sengupta: 'Kinematic theory for scale-invariant patterns in acicular martensites', *Physica A: Statistical Mechanics and Its Applications*, 1996, **224**, 403–411.

142. G. V. Kurdjumov, and O. P. Maksimova: 'Kinetics of the transformation of austenite into martensite at low temperatures', *Doklady Akademii Nauk, SSSR*, 1948, **61**, 83–86.

143. E. S. Machlin, and M. Cohen: 'Isothermal mode of the martensitic transformation', *Journal of Metals*, 1952, **4**, 489–500.

144. R. B. G. Yeo: 'Isothermal martensite transformation in iron-base alloys of low carbon content', *Transactions of the AIMME*, 1962, **224**, 1222–1227.

145. K. Ullakko, B. Sundqvist, and J. Pietikainen: 'Isothermal martensite transformation of FeNiC alloys as a function of hydrostatic pressure', *Materials Science Forum*, 1990, **56–58**, 197–200.

146. C. H. Shih, B. L. Averbach, and M. Cohen: 'Some characteristics of the isothermal martensitic transformation', *Journal of Metals*, 1955, **203**, 709–710.

147. K. Tsuzaki, T. Fukiage, T. Maki, and I. Tamura: 'Effect of Ni on the isothermal character of lath martensitic transformation in Fe-Ni alloys', *Materials Science Forum*, 1986, **56-58**, 229–234.

148. D. S. Sarma, J. A. Whiteman, and S. R. Keown: 'The structure of burst and isothermal martensites in an Fe-24Ni-0.5C wt% alloy', *Journal of Materials Science*, 1979, **14**, 693–698.

149. H. Okamoto, and M. Oka: 'Isothermal martensite transformation in a 1.80 wt% C steel', *Metallurgical & Materials Transactions A*, 1985, **16**, 2257–2262.

150. V. Raghavan, and A. R. Entwisle: 'Isothermal martensite kinetics in iron alloys', In: *Physical properties of martensite and bainite, Special Report 93*. London, U.K.: Iron and Steel Institute, 1965:29–37.

151. R. H. Edwards, and N. F. Kennon: 'The isothermal transformation of austenite at temperatures near Ms: II transformation diagram', *Journal of the Australian Institute of Metals*, 1970, **4**, 201–205.

152. K. Wakasa, and C. M. Wayman: 'Crystallography and morphology of ferrous lath martensite', *Metallography*, 1981, **14**, 49–60.

153. S. Kajiwara: 'Morphology and crystallography of the isothermal martensite in Fe Ni Mn alloys', *Philosophical Magazine*, 1981, **43**, 1483–1503.

154. S. Kajiwara: 'Continuous observation of isothermal martensite formation in Fe Ni Mn alloys', *Acta Metallurgica*, 1984, **32**, 407–413.

155. S. Kajiwara: 'Rate controlling mechanism for isothermal martensite transformation', In: *Proceeding of the International Conference on Martensitic Transformations*. Tokyo, Japan: Japan Institute of Metals, 1986:259–264.

156. K. Ullakko, M. Nieminen, and J. Pietikainen: 'Prevention of martensitic transformation during rapid cooling', *Materials Science Forum*, 1989, **56**, 225–228.

157. G. Ghosh: 'Effect of pre-strain on the kinetics of isothermal martensitic transformation', *Philosophical Magazine A*, 1995, **71**, 333–345.

158. S. R. Pati, and M. Cohen: 'Nucleation of isothermal martensitic transformation', *Acta Metallurgica*, 1969, **34**, 189–199.

159. J. C. Bokros, and E. R. Parker: 'The mechanism of the martensite burst transformation in Fe-Ni single crystals', *Acta Metallurgica*, 1963, **11**, 1291–1301.

160. R. Brook, and A. R. Entwisle: 'Kinetics of burst transformation to martensite', *Journal of the Iron and Steel Institute*, 1965, **203**, 905–912.

161. P. C. Maxwell, A. Goldberg, and J. C. Shyne: 'Stress-assisted and strain-induced martensites in Fe-Ni-C alloys', *Metallurgical Transactions*, 1974, **5**, 1305–1318.

162. D. S. Sharma, J. A. Whiteman, and S. R. Keown: 'The structure of burst and isothermal martensites in an Fe-24Ni-0.5C wt% alloy', *Journal of Materials Science*, 1979, **14**, 693.

163. H. K. D. H. Bhadeshia: 'An aspect of the nucleation of 'burst' martensite in steels', *Journal of Material Science*, 1982, **17**, 383–386.

164. A. R. Entwistle, and J. A. Feeny: 'The effect of austenitizing conditions on martensite. transformation by bursts', In: *The Mechanism of Phase Transformation in Crystalline Solids*. London, U.K.: Institute of Metals, 1969:16–161.

165. J. R. C. Guimarães, and J. C. Gomes: 'Metallographic study of the influence of austenite grain size on martensite kinetics in Fe-31.9Ni-0.02C', *Acta Metallurgica*, 1978, **26**, 1591–1596.

166. E. Schmid, and W. Boas: Plasticity of Crystals (translated from the 1935 edition of Kristalplastizitaet): London, U.K.: F. A. Hughes and Co., 1950.

167. J. R. Patel, and M. Cohen: 'Criterion for the action of applied stress in the martensitic transformation', *Acta Metallurgica*, 1953, **1**, 531–538.

168. G. B. Olson, and M. Cohen: 'Mechanism for strain-induced nucleation of martensite', *Journal of Less Common Metals*, 1972, **28**, 107–118.

169. I. Tamura: 'Deformation induced martensitic transformation and transformation-induced plasticity in steels', *Metal Science*, 1982, **16**, 245–254.

170. S. Chatterjee, and H. K. D. H. Bhadeshia: 'TRIP-assisted steels: Stress or strain-affected martensitic transformation', *Materials Science and Technology*, 2007, **23**, 1101–1104.

171. G. B. Olson, and M. Cohen: 'Kinetics of strain-induced martensitic transformation', *Metallurgical Transactions A*, 1975, **6A**, 791–795.

172. T. Angel: 'Formation of martensite in austenitic stainless steel', *Journal of the Iron and Steel Institute*, 1954, **181**, 165–174.

173. M. Mukherjee, T. Bhattacharyya, and S. B. Singh: 'Models for austenite to martensite transformation in TRIP-aided steels: A comparative study', *Material and Manufacturing Processes*, 2010, **25**, 206–210.

174. M. Mukherjee, S. B. Singh, and O. N. Mohanty: 'Deformation-induced transformation of retained austenite in transformation induced plasticity-aided steels: A thermodynamic model', *Metallurgical & Materials Transactions A*, 2008, **39**, 2319–2328.

175. K. Sugimoto, M. Kobayashi, and S. Hashimoto: 'Ductility and strain-induced transformation in a high-strength TRIP aided dual phase steel', *Metallurgical Transactions A*, 1992, **23A**, 3085–3091.

176. M. Sherif, C. Garcia-Mateo, T. Sourmail, and H. K. D. H. Bhadeshia: 'Stability of retained austenite in TRIP-assisted steels', *Materials Science and Technology*, 2004, **20**, 319–322.

177. H. N. Han, C. G. Lee, D. W. Suh, and S. J. Kim: 'A microstructure-based analysis for transformation induced plasticity and mechanically induced martensitic transformation', *Materials Science & Engineering A*, 2008, **485**, 224–233.

178. H. K. D. H. Bhadeshia: 'Displacements caused by the growth of bainite in steel', *The Banaras Metallurgist*, 2014, **19**, 1–7.

179. S. Kundu, K. Hase, and H. K. D. H. Bhadeshia: 'Crystallographic texture of stress-affected bainite', *Proceedings of the Royal Society A*, 2007, **463**, 2309–2328.

180. S. Chatterjee, H. S. Wang, J. R. Yang, and H. K. D. H. Bhadeshia: 'Mechanical stabilisation of austenite', *Materials Science and Technology*, 2006, **22**, 641–644.

181. R. W. K. Honeycombe: The Plastic Deformation of Metals: London, U.K.: Edward Arnold, 1984.

182. F. Barlat, M. V. Glazov, J. C. Brem, and D. J. Lege: 'Critical assessment of the micromechanical behaviour of dual phase and TRIP-assisted multiphase steels', *Canadian Metallurgical Quarterly*, 2002, **18**, 919–939.

183. F. Abel: 'Final report on experiments bearing upon the question of the condition in which carbon exists in steel', *Proceedings of the Institution of Mechanical Engineers*, 1885, **36**, 30–57.

184. N. Seljakov, G. V. Kurdjumov, and N. Goodtzov: 'Eine röntgenographische unterscuchung der strucktur des kohlenstoffstahls', *Zietschrift für Physik*, 1927, **45**, 384–408.

185. E. Honda, and Z. Nishiyama: 'On the nature of the tetragonal and cubic martensites', *Science Reports of Tohoku Imperial University*, 1932, **21**, 299–331.

186. R. Chentouf, S. Cazottes, F. Danoix, M. Goune, H. Zapolsky, and P. Maugis: 'Effect of interstitial carbon distribution and nickel substitution on the tetragonality of martensite: A first-principles study', *Intermetallics*, 2017, **89**, 92–99.

187. J. W. Christian: 'Tetragonal martensites in ferrous alloys – a critique', *Materials Transactions, JIM*, 1992, **33**, 208–214.

188. S. Kajiwara, T. Kikuchi, and S. Uehara: 'Origin for abnormal tetragonalities of martensite in steels', In: *Proceeding of the International Conference on Martensitic Transformations*. Tokyo, Japan: Japan Institute of Metals, 1986:301–306.

189. S. Kajiwara, and T. Kikuchi: 'On the abnormally large tetragonality of martensite in Fe-Ni-C alloys', *Acta Metallurgica et Materalia*, 1991, **39**, 1123–1131.

190. S. Uehara, S. Kajiwara, and T. Kikuchi: 'Origin of abnormally large tetragonality of martensite in high carbon iron alloys containing aluminum', *Transactions of JIM*, 1992, **33**, 220–228.

191. G. F. Bolling, and R. H. Richman: 'Continual mechanical twinning Parts III, IV', *Acta Metallurgica*, 1965, **13**, 745–757.

Notes

[1] The $\gamma \to \varepsilon$ transformation and martensitic transformation in ordered Fe_3Be (simple shearing) [191] are examples where the transformation strain is an invariant-plane strain.

[2] The assumption of the exact KS orientation means that there was no invariant line in the interface, and the martensite modelled was not a three-dimensionally enclosed plate.

[3] Notice that a combination of two non-coplanar invariant-plane strains gives an invariant-line strain, the invariant-line lying at the intersection of the two invariant-planes.

[4] This would be the invariant line in the α'/γ interface. The interfacial dislocations would have their line vectors parallel to this invariant line.

[5] This is evident from Figure 5.8b where there are two different undistorted lines in the plane of the diagram.

[6] The matrices can be converted from the basis F to γ using a similarity transformation. Since $\mathbf{f}_i \parallel \mathbf{a}_i$ we find that $(F\,S\,F) = (\gamma\,S\,\gamma)$.

[7] The actual magnitude is dependent on the exact thermodynamic model used for the calculations. This is particularly the case with martensite since the equilibrium data measured at high temperatures have to be extrapolated to much lower temperatures. Any analysis should therefore be internally consistent in the model and data used, but this has not always been the case in the published literature.

[8] We have not considered this term before. It comes from the Bowles and Mackenzie theory where a small uniform dilatation is in principle permitted. Experiments to date have not revealed such a dilation.

[9] Similar observations have been reported for martensitic transformation in TiC, Chapter 11.

[10] The fault energy would in fact have to be negative in order for the partial dislocations to overcome any lattice friction, such as that arising from the Peierls barrier to dislocation motion.

[11] A cubic structure will be mechanically stable if $C_{11} - C_{12} > 0$, $C_{11} + 2C_{12} > 0$ and $C_{44} > 0$.

[12] This statement, which is commonly made, is true only when the observations are made with an ordinary time resolution, i.e. no better than about a millisecond. More precise measurements must obviously reveal that athermal martensite takes time to form. Thus, Thadhani and Meyers [138] found that martensite normally considered to be athermal exhibits isothermal character in a microsecond regime.

[13] Not all martensite forms in this way; the morphology of lath martensite is in the form of packets of parallel plates. There do not appear to be any theoretical treatments for the kinetics of lath martensite.

[14] Autocatalysis refers to the case where the formation of one plate stimulates the nucleation of many others. This can lead to a sudden burst of transformation. The cascade of transformation usually is limited to an individual grain of austenite.

[15] This is analogous to the explanation of Hall and Petch for the decrease in yield strength at the grain size becomes coarser.

6 Bainite

6.1 MICROSTRUCTURE

Bainite is a two-phase, or sometimes a three-phase mixture, but its defining feature is the platelet of ferrite that determines the evolution of the remaining components of its structure. The platelets always are found to be thinner than the wavelength of visible light, Figure 6.1.[1]

In upper bainite (α_{ub}), the platelets may be separated in part by austenite, or if the transformation time is sufficient, by cementite. This austenite could, during cooling from the transformation temperature, transform in part into martensite.

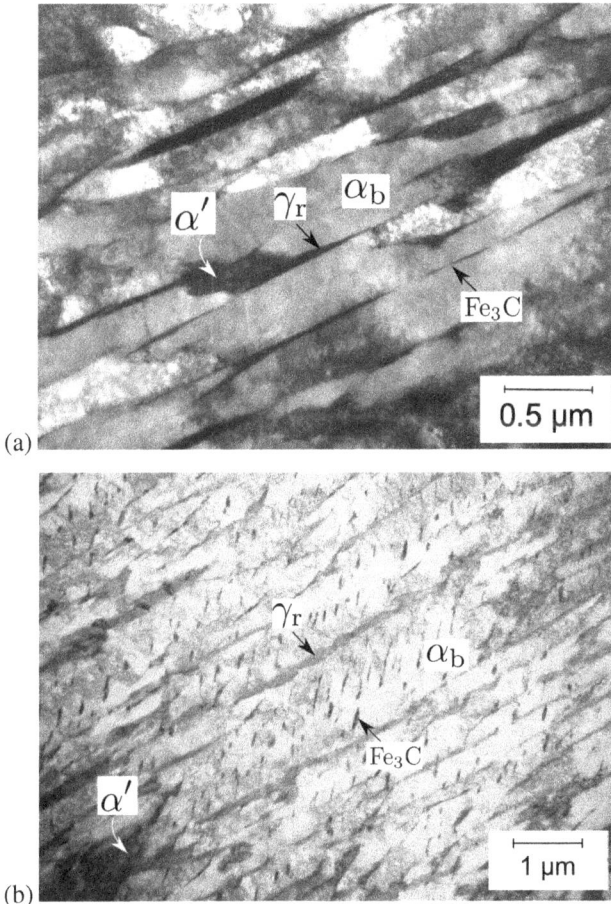

(a)

Figure 6.1 The platelets of bainite are finer than can be resolved using optical microscopy and it would not be possible to determine the phases present without diffraction data. Prolonged transformation would cause the regions between the α_b platelets to decompose into Fe_3C and ferrite depositing epitaxially on to the α_b.

(a) Structure of upper bainite in Fe-0.095C-1.63Si-1.99Mn wt% transformed at 400 °C for 800 s. Micrograph courtesy of L. C. Chang.

(b)

(b) Lower bainite in Fe-0.3C-4.08Cr wt% transformed at 435 °C for 10 min.

Lower bainite (α_{ub}) has the distinction that additional carbides are found within the platelets as well as between them. The precipitation within the platelets precedes that from austenite. This is because the driving force for precipitation from supersaturated ferrite is much greater; this also raises the possibility of metastable transition-carbides forming within α_{lb} instead of cementite, in perfect analogy to the tempering reactions that are established for martensite. Like upper bainite, the precipitation from austenite can be slow or can be suppressed entirely by appropriate alloying, in which case the austenite can be retained wholly or partially depending on its M_S temperature.

The mechanism of the bainite transformation is in essence simple, so it is appropriate to set the scene before dwelling into detail. The evidence supports a binding similarity with martensitic transformation, but moderated by the smaller driving forces and somewhat increased atomic mobility at $T > M_S$. The displacive mechanism of transformation ensures that the shape of bainitic ferrite is that of a thin plate, as established using a high-resolution two-surface analysis, Figure 6.2a [1, 2]. Experiments like these can be done using light microscopy for martensite or Widmanstätten ferrite which can form coarse plates, whereas the individual platelets of bainite are *always* fine by comparison.

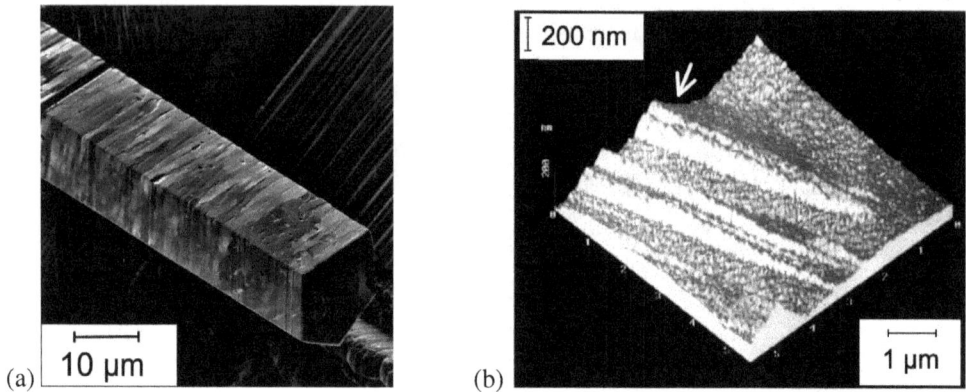

Figure 6.2 (a) A two-surface image of an ion-beam machined sample showing in three dimensions the plate-shaped bainite – reproduced with permission of Elsevier, from Costin et al. [1]. (b) Atomic force microscope image of the displacements caused by individual plates of bainite at the free surface of austenite [3]. Each plate has a lenticular shape ending at a sharp tip and is associated with a shear displacement that causes the plastic deformation in the adjacent austenite (arrowed).

Experiments to characterise the displacements accompanying the growth of individual platelets of bainite also require a high resolution technique to avoid averaging over intervening phases. Figure 6.2b shows an atomic force microscope image of a single crystal of austenite which was polished flat and then transformed into bainite. The displacements are invariant-plane strains with $s \approx 0.26$ and ζ estimated to be ≈ 0.03 [3]; given the typical aspect ratio of about 0.02 for bainite plates, the stored energy due to an elastically accommodated plate is evaluated as $\approx 400\,\mathrm{J\,mol^{-1}}$; the magnitudes of the strains are similar to those accompanying the $\gamma \rightarrow \alpha'$ martensitic transformation, Table 7.2. There is clear evidence for irregular deformation in the austenite adjacent to the α_{ub}-platelets; furthermore, Figure 6.3a shows the large dislocation density in both phases, γ and α_{ub} in contact at the interface, akin to the dislocation tangles found at α'/γ interfaces [4]. When a platelet of bainite forms, the plastic relaxation of the austenite and associated creation of defects has the effect of stifling the glissile interface so individual platelets are limited by transformation-driven mechanical stabilisation, even in the absence of impingement with hard obstacles such as grain boundaries.

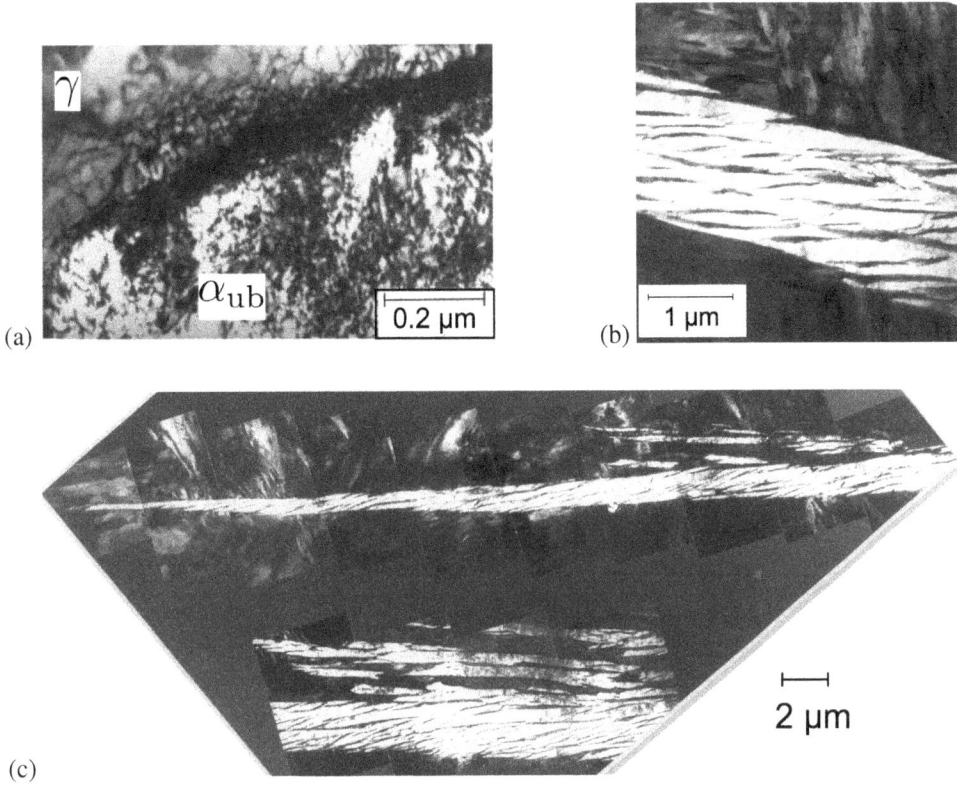

(a)

(b)

(c)

Figure 6.3 (a) Dislocation debris at the γ/α_{ub} interface due to plastic accommodation effects. (b) A sheaf of bainite consisting of microscopic platelets (often called *subunits* [5, 6]). which clearly show a lenticular shape with sharp tips consistent with the minimisation of strain energy (p. 288). The sample is transformed partially, so the dark regions are martensite and the films between the platelets are retained austenite. (c) Montage of a bainite sheaf, the upper part being a continuation of the lower image. There are myriads of platelets, all of approximately the same size and orientation [7].

The arrest of the platelets has been modelled using the mechanical stabilisation theory described in Section 5.9.2 [8]. It is predicted therefore that the subunits will tend to become more slender as the driving force is increased because the onset of mechanical stabilisation is delayed. The consequences of the growth arrests are dramatic, leading to a refinement of the effective grain size; what appears in an optical microscope to be a single plate of bainite actually is an organised cluster of platelets when examined at a greater resolution, Figure 6.3b,c [7]. The effective grain size is defined by the mean free slip-distance, which is about twice the thickness of the platelets.

The individual platelets are referred to as *subunits*, organised into clusters known as *sheafs* [5, 6]; the sheaf is not a homogeneous entity but rather a mixture with γ, θ or both located between the subunits. When the sheaf is a mixture of $\gamma + \alpha_b$, it is a bicrystal because all the α_b subunits are connected in the same crystallographic orientation and the austenite in which they grow percolates throughout the structure of the sheaf [9].

It remains to explain the organisation of the subunits within a sheaf. The process begins with heterogeneous nucleation at an austenite grain surface or an intragranular inclusion. Once the initial platelet is arrested, the local transformation event is propagated by the stimulation of further platelets, largely at the tip of the original one. This is because the strain field and better elastic accommodation near the tip of a lenticular plate favours the formation of another with the same shape deformation [10] combined with the easier dissipation of the partitioned-carbon diffusion field at the tip of the lenticular subunits may also help. That the former is probably the controlling factor is evident from observations that a subunit-sheaf structure can be induced in martensite that grows in severely deformed austenite at sub-zero temperatures, Figure 6.4 [11].

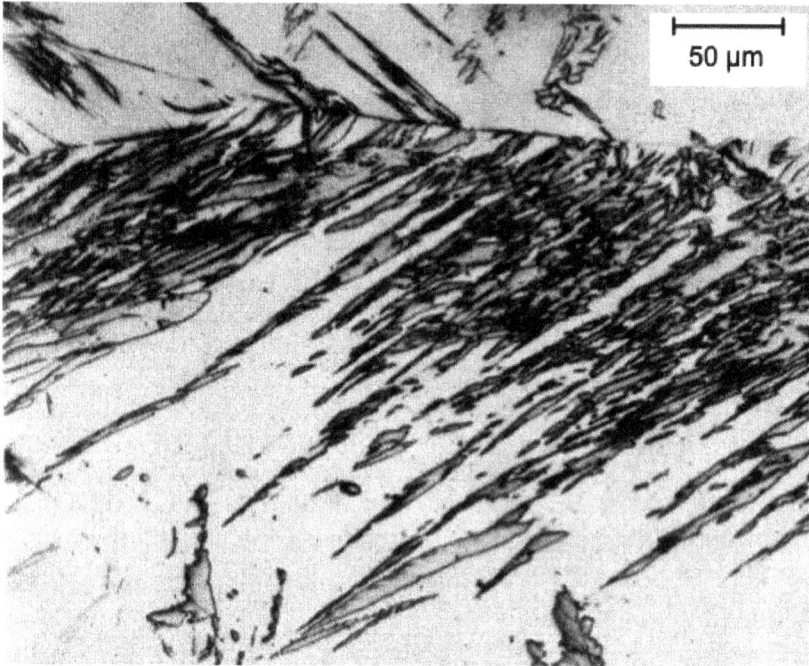

Figure 6.4 Fe-29Ni-0.26C wt% pulled at $-11\,°C$ in tension to an elongation of 71%, showing the morphology of martensite that forms in the deformed austenite. The martensite plates have broken up into subunits rather like the sheaf of bainite in Figure 6.3. The alloy transforms normally into coarse lenticular α'-plates in undeformed austenite on cooling to $-60\,°C$. Micrograph courtesy of Professor Tadashi Maki.

The thickening of a sheaf occurs by subunits stimulated at the sides of the original platelet, albeit at a smaller rate. This makes the shape highly anisotropic (Figure 6.3e), in fact plate-like in three dimensions [12], with an averaged shear strain of just 0.129 [13]. The averaging is over the composite structure of the sheaf consisting of individual α_b platelets, each of which has a much greater shear strain, and the other phases such as austenite or carbides that lie between the subunits. This constitutes a subtle mechanism of mitigating the strain energy by avoiding thick bainite platelets.

6.2 MORE ABOUT THE MECHANISM

To understand other features of the transformation, it is necessary first to consider the atomistic mechanism; the shape deformation caused by bainite growth and the transformation temperatures involved rule out the diffusion of host or substitutional solutes. This has been verified repeatedly using atom probe measurements which show a flat distribution of the Fe/X atom ratio across the α_b/γ interface [pp. 27-31, 14]. However, the same technique reveals at the point of examination that some of the carbon has partitioned into the austenite. There is no segregation of any species to the α_b/γ interfaces in the untempered state, consistent with the high level of coherency in the interface implied by the shape deformation.

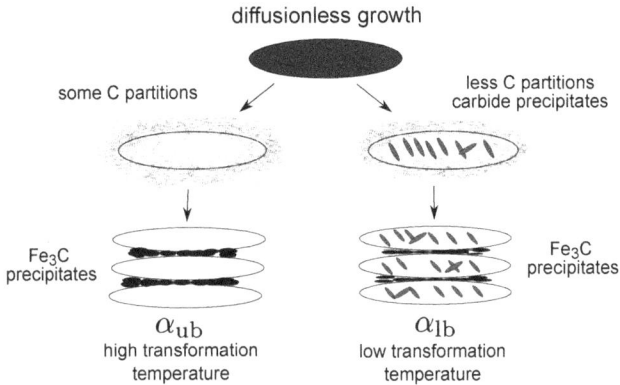

diffusionless growth

some C partitions   less C partitions
carbide precipitates

Fe₃C   Fe₃C
precipitates   precipitates

α_{ub} α_{lb}
high transformation   low transformation
temperature   temperature

Figure 6.5 Diffusionless growth is followed by the partitioning of some of the excess carbon into the residual austenite and possibly by the precipitation of carbides from supersaturated ferrite. In the latter case, the transformation results in lower bainite. When the composition of the austenite reaches an appropriate concentration, it becomes possible for cementite to precipitate between the platelets of ferrite.

Consider the case where bainite in the first instance grows without diffusion, Figure 6.5. At the temperatures where it typically forms, tempering reactions can be triggered rapidly after the initial rapid transformation event. The most rapid of these are the partitioning of carbon from the supersaturated α_b into the residual austenite and the precipitation of carbides from that ferrite. The latter reaction is common during the tempering of martensite and often involves first the precipitation of fine particles of cementite or transition carbides such as ε, η and χ (Chapter 9); the crystallography of transition carbide precipitation is identical in tempered martensite and lower bainite. Precipitation from the carbon-enriched austenite is slower,[2] the original austenite films decomposing into elongated particles of cementite and ferrite that is deposited epitaxially on to the pre-existing α_b. As shown in Figure 6.5, the upper and lower bainite are distinguished by the absence or presence respectively, of carbides within the α_b platelets, the determining factor being the competition between carbon partitioning and carbide precipitation.

The evolution of the structure of bainite shown in Figure 6.1 in not a continuous process. There are distinct stages that can be studied experimentally in isolation. Furthermore, a good deal of the technology of sophisticated steels depends on the ability to terminate the reaction at different points in the sequence:

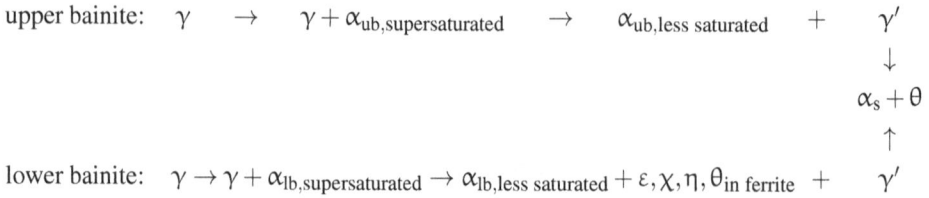

upper bainite: $\gamma \quad \rightarrow \quad \gamma + \alpha_{ub,\text{supersaturated}} \quad \rightarrow \quad \alpha_{ub,\text{less saturated}} \quad + \quad \gamma'$
$$\downarrow$$
$$\alpha_s + \theta$$
$$\uparrow$$

lower bainite: $\gamma \rightarrow \gamma + \alpha_{lb,\text{supersaturated}} \rightarrow \alpha_{lb,\text{less saturated}} + \varepsilon, \chi, \eta, \theta_{\text{in ferrite}} \quad + \quad \gamma'$

where γ' represents carbon-enriched austenite and α_s the secondary ferrite that precipitates epitaxially on to the pre-existing bainitic ferrite. The stages involved are now discussed in detail.

6.2.1 STAGE 1: VESTIGES

The first stage, the plate that forms without any change in composition, is the most difficult to isolate, but there are extraordinary vestiges of this initial event that prove the mechanism. Atom probe investigations have always revealed carbon concentrations in bainitic ferrite that are far in excess of equilibrium concentrations, thought originally to be located at dislocations [15–25]. The first observation using methods where the concentration can be measured in defect-free solid solution was by Pereloma et al. [26], followed shortly afterwards by Caballero et al. [27, 28]. The concentrations observed are far greater than expected from equilibrium. Figure 6.6a shows data from an atom probe experiment, collected from regions that are free from defects. Although the measured concentrations are less than $\bar{x}$, they are much greater than expected from equilibrium. The excess carbon persists in the ferrite even after holding at the transformation temperature of 200 °C for 250 h, so it is not the mobility of the carbon that prevents it from partitioning into the residual austenite.

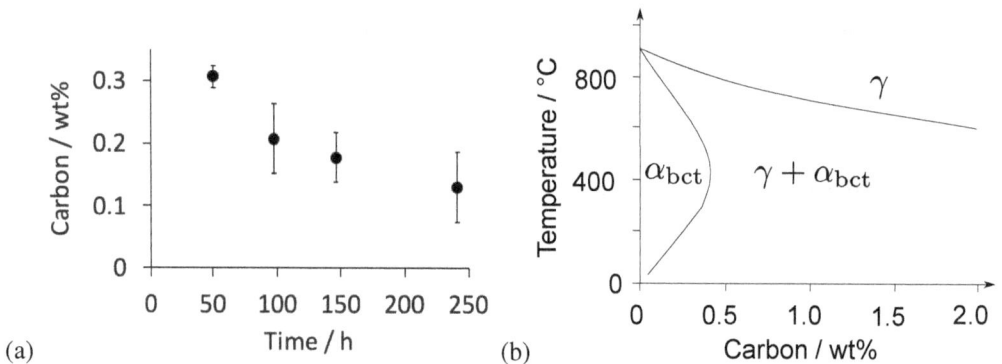

Figure 6.6 (a) Carbon concentration in solid solution within bainitic ferrite as a function of transformation time at 200 °C, Fe-1C-1.5Si-1.9Mn-1.3Cr-0.26Mo wt%. Adapted from Garcia-Mateo et al. [29]. (b) Calculated phase diagram for the equilibrium between body-centred tetragonal ferrite and austenite. After Jang et al. [30].

The explanation for the reluctance of the carbon to partition from α_b lies in the fact that the Bain strain leaves all the carbon atoms inherited by the α_b on one sublattice of the interstices, making the unit cell tetragonal in the first instance (Section 5.10); the tetragonal cell of bainite has been verified experimentally using a number of techniques [29, 31–36]. As a consequence, the equilibrium to be considered is between the body-centred tetragonal ferrite (α_{bct}) and austenite, not that between bcc-ferrite and austenite. Figure 6.6b shows a calculated Fe-C diagram for the α_{bct}/γ equilibrium [30]; it is evident that the solubility of carbon is increased significantly relative to bcc-ferrite.

An alternative explanation has been proposed that carbon atoms associate with host vacancies in the α_b and that this reduces the diffusion coefficient of carbon sufficiently to explain its retention in the plates [37]. However, it is established that the fraction of carbon atoms that pair with host vacancies is negligibly small (Section 3.5.3), so small that it cannot perceptibly influence the diffusivity of carbon or iron atoms in the ferrite [38].

6.2.2   STAGE 2A: CARBON PARTITIONING

The decarburisation of bainitic ferrite can be swift, depending on the transformation temperature and the thermodynamic state at the interface with austenite. Consider a plate of thickness z_t with a one-dimensional flux of carbon along z normal to the α/γ interface, originating at the interface and positive in the austenite [39]. Mass conservation at the point where decarburisation has been achieved gives [40]

$$\frac{1}{2}z_t(\bar{x}-x^\alpha) = \int_0^\infty \left[x^\gamma\{z,t_d\} - \bar{x}\right]\,dz \tag{6.1}$$

where x^α is the concentration that will persist in the ferrite at the termination of decarburisation, t_d the time to decarburise the ferrite plate and $x^\gamma\{z,t_d\} = x_I^\gamma$ at $z = 0$. The diffusivity of carbon in austenite is slower than in ferrite, it is assumed that the rate of decarburisation is determined by the flux in the austenite. The concentration x_I^γ in austenite at the interface is assumed to remain constant for times $0 < t < t_d$, although it must eventually decrease as homogenisation occurs. The concentration profile in the austenite is given by

$$x^\gamma = \bar{x} + (x_I^\gamma - \bar{x})\mathrm{erfc}\left\{\frac{z_t}{2(\overline{D}t_d)^{\frac{1}{2}}}\right\} \tag{6.2}$$

which on integration yields

$$t_d^{\frac{1}{2}} = \frac{z_t(\bar{x}-x_I^\alpha)\pi^{\frac{1}{2}}}{4\overline{D}^{\frac{1}{2}}(x_I^\gamma - \bar{x})}. \tag{6.3}$$

It remains to determine the terms x_I^α and x_I^γ. They could be set to the paraequilibrium concentrations $x^{\alpha\gamma}$ and $x^{\gamma\alpha}$ respectively but the former is particularly difficult to justify because the diffusivity within the ferrite is so much more rapid than in austenite; this would lead to an underestimation of t_d. A finite difference method can be used instead that allows x_I^α to vary freely down to a limit which may be the paraequilibrium solubility in ferrite or some higher concentration given that it is now known that α_b can retain a substantial excess of carbon in solution during the early stages of transformation. Figure 6.7a shows that the t_d can be very small at the typical temperatures where bainite forms; these particular data assume $x_I^\gamma = x^{\gamma\alpha}$, but this can be relaxed [41] although the outcomes are not substantially different.

The modelling of the partitioning of carbon from supersaturated plates has had a resurgence since the invention of the *quench and partitioning* process [43–45]. In this, austenite is partially trans-

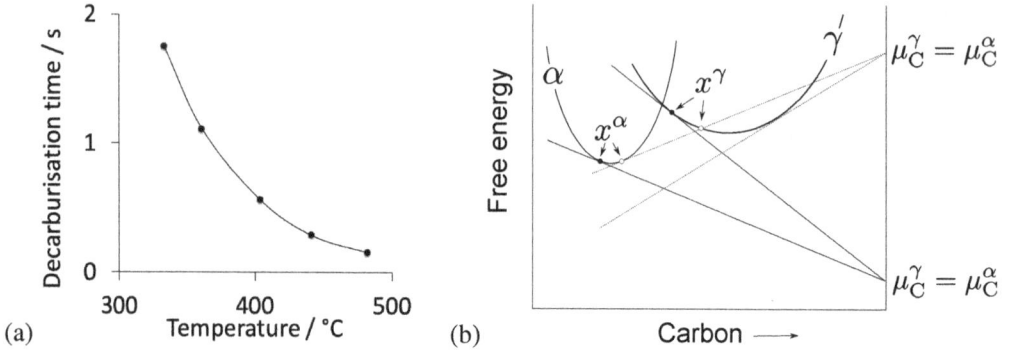

Figure 6.7 (a) The time to decarburise a plate of bainitic ferrite of thickness 0.2 μm, calculated using a finite difference method assuming that the concentration eventually diminishes to $x^{\alpha\gamma}$ [42]. (b) Multiple possibilities for selecting austenite and ferrite compositions that allow the chemical potential of carbon to be uniform.

formed into martensite and the two-phase mixture is then heated to allow carbon from the martensite to partition into the austenite,[3] the goal being to stabilise the latter. The compositions of the martensite and austenite at the interface are set to ensure that the $\mu_C^\gamma = \mu_C^\alpha$ though $\mu_{Fe}^\gamma \neq \mu_{Fe}^\alpha$. The scenario is illustrated in Figure 6.7b where it is clear that the composition-set x^α and x^γ cannot be selected uniquely. Additional conditions are therefore required to ensure that the average composition of the steel lies between those of the two phases and that mass balance is satisfied:

$$\bar{x}_C \ni x_C^\alpha \quad \text{and} \quad x_C^\gamma \tag{6.4}$$

$$\bar{N}_C = N_C^\gamma + N_C^\alpha \tag{6.5}$$

where N_C is the number of carbon atoms in the phase identified by the superscript, and $\bar{N}_C$ is the total number of carbon atoms. The condition outlined in Relation 6.4 is illustrated in Figure 6.7, and since during quenching and partitioning, N_{Fe}^α and N_{Fe}^γ are fixed, Equation 6.5 leads to

$$\frac{N_C}{N_{Fe}+N_C} = \frac{N_{Fe}^\alpha}{N_{Fe}}\left(\frac{N_C^\alpha}{N_{Fe}^\alpha+N_C^\alpha}\right) + \frac{N_{Fe}^\gamma}{N_{Fe}}\left(\frac{N_C^\gamma}{N_{Fe}^\gamma+N_C^\gamma}\right)$$

$$\text{so that} \quad \bar{x}_C = x_\alpha^\alpha x_C^\alpha + x_\gamma^\gamma x_C^\gamma \tag{6.6}$$

where x_α^α and x_γ^γ represent the mole fractions of the ferrite and austenite respectively. With these conditions, the partitioning times are not essentially different from those reported in Figure 6.7a.

6.2.3 STAGE 2B: PRECIPITATION FROM FERRITE

Matas and Hehemann [46] proposed that the transition from upper to lower bainite depends on the competition between the partitioning of excess carbon and precipitation of carbides from the supersaturated ferrite. If the former is more rapid than the precipitation reaction, an upper bainitic microstructure is inevitable (Figure 6.5). Lower bainite occurs when $t_d < t_\theta$ where t_θ is the time to precipitate a significant amount of cementite (or in the present context, transition carbides of iron). This simple criterion is able to explain quantitatively the transition between the two microstructures of bainite if t_θ can be determined [47]. Referring to Figure 6.8, suppose that in a series of experiments, the transformation temperature is reduced at a constant carbon concentration. Then three possibilities exist as the isothermal transformation-temperature is reduced:

- At low concentrations, pearlite gives way to upper bainite and then to martensite, with no intervening lower bainite. This is because the carbon concentration is small enough to let partitioning prevent precipitation, i.e. $t_\theta > t_d$ for all temperatures where bainite forms.
- In a narrow concentration range (0.34-0.39 wt%) it is possible to transform isothermally into pearlite at the highest temperature, followed by α_{ub}, then α_{lb} and finally α'.
- Upper bainite does not occur at any temperature because the carbon concentration is so large that $t_\theta < t_d$ for all concentrations greater than ≈ 0.4wt%.

These predictions are consistent with experimental observations on Fe-C alloys [48–51]. If in a low-carbon steel, the austenite is enriched in carbon during transformation, then α_{ub} forms first to be followed by α_{lb}. It is emphasised that the comparison between t_d and t_θ is not binary; the two processes occur simultaneously but in some cases one dominates over the other and hence the observed differences in microstructure.

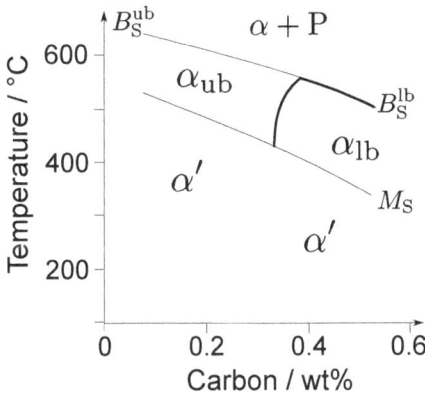

Figure 6.8 The calculated upper-bainite start, lower-bainite start and martensite-start temperatures in Fe-C as a function of the carbon concentration. After Takahashi and Bhadeshia [47].

6.2.4 STAGE 3: TERMINATION

The final stage of the reaction, $\gamma' \to \alpha_s + \theta$ is sensitive to the substitutional solute content of the steel. It is well-established that the precipitation of cementite can for all intents and purposes be suppressed completely by adding ≈ 1 wt% of silicon or aluminium to the steel. This extraordinary retardation is because the cementite is forced at low temperatures to form under paraequilibrium conditions, thereby trapping Si as it grows [52]. Given these conditions, cementite precipitation can become *impossible* if the trapping reduces the driving force towards zero [53].

The chemical composition of the cementite can change from its paraequilibrium state during subsequent heat treatment. Certain steels used in the power generation industries have very large components made from bainitic steel, which then serve at mean temperatures $\approx 560\,°C$ for periods of 25 years or more. The enrichment of cementite during service by solutes such as chromium can then be used forensically as an indicator of the remaining creep-life of the component [54–56]. The process of cementite enrichment is rather like the reverse of the partitioning of carbon from bainitic ferrite; it can therefore be treated using Equation 6.3 with appropriately modified concentration and diffusion terms [40].

The reaction $\gamma' \to \alpha_s + \theta$ can also be avoided by limiting the transformation time in alloys that do not contain significant amounts of cementite suppressing solutes. Figure 6.9 shows such a case for an alloy containing only 0.05 wt% of carbon, transformed isothermally at 600 °C for just 30 s; the bainite platelets are separated by thin films of retained austenite, which on prolonged holding at the transformation temperature decompose into cementite [57].

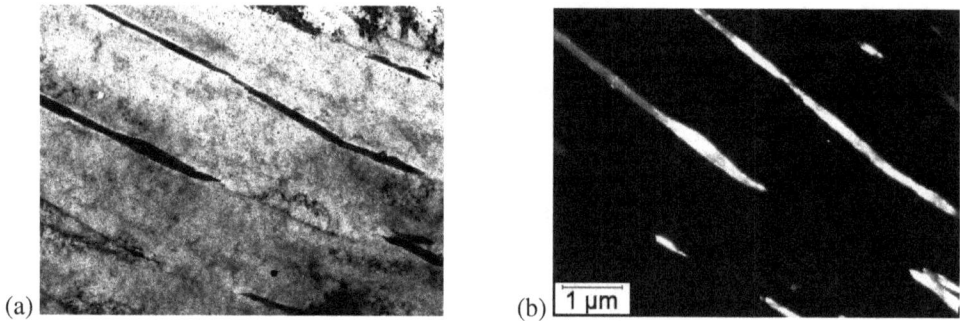

(a)   (b) 1 μm

Figure 6.9 Fe-0.05C-1.55Mn-0.12Si-0.1Nb wt% transformed isothermally at 600 °C for 30 s and quenched to ambient temperature. (a) Bright field image showing bainite plates with intervening retained austenite that is in the form of films but not continuous given the small carbon concentration. (b) Corresponding dark field image of retained austenite. After Yan and Bhadeshia [57].

From a technological point of view, it is of considerable interest to prevent the $\gamma' \to \alpha_s + \theta$ reaction. When the transformation-induced plasticity steels were first invented [58], they were rich in expensive alloying elements to ensure a fully austenitic state at ambient temperature, for example, Fe-0.3C-8Ni-4Mo-2Mn-2Si wt% [59]. With bainite in which cementite precipitation is suppressed, it becomes possible to stabilise the austenite in a low-alloy steel with the carbon that is partitioned, so that the average concentration of carbon is small even though $x_\gamma \approx 1.2$ wt%. This gave rise to commercially viable TRIP-assisted[4] steels and cast irons that have enjoyed tremendous commercial success because of their low cost and the ability to mass produce without compromising properties [14].

The theory of the bainite transformation as summarised in Figure 6.5 has been a contributing factor in the design of these affordable TRIP-assisted iron alloys. It permits the calculation of the volume fraction and chemical composition of the austenite, assuming that cementite precipitation from γ' is suppressed. In that case, the formation of bainitic ferrite must stop when diffusionless growth becomes impossible. Figure 6.10a shows that the T_0 curve is the locus of all points on the temperature versus carbon concentration plot where austenite and ferrite of the same composition have identical free energies. The T_0' curve is analogous but accounting for any stored energy, primarily due to the shape deformation. The transformation to bainite must then cease when x_γ is equal to the concentration $x_{T_0'}$ given by the T_0' curve at the transformation temperature. In practice the distribution of partitioned carbon in the austenite is not likely to be homogeneous [46, 60, 61] so there will be some scatter, but the important point is that the concentration $x_{T_0'} \ll x_{Ae_3}$ so the reaction will remain thermodynamically *incomplete*; there will be no further bainite generated by prolonged holding at the transformation temperature.[5]

The process of isothermal transformation is illustrated in Figure 6.10b, where the formation of each plate followed by partitioning means that successive plates must grow in enriched austenite, with transformation coming to a halt when $x_\gamma = x_{T_0'}$. This limit is reached at a later stage when the transformation temperature is lower, but there also is an upper limit $\bar{x} = x_{T_0'}$ beyond which no bainitic transformation can occur even though that temperature is in the two phase $\alpha + \gamma$ field.[6] Many techniques and alloys have been studied to validate this concept, the incomplete reaction phenomenon and the implications that follow on the time-temperature transformation diagram [14]. An important outcome is that it becomes possible to calculate many aspects of the microstructure in steels where the $\gamma' \to \alpha_s + \theta$ reaction is suppressed. The maximum volume fraction of bainite is then given by the simple application of the lever rule:

$$V_V^{\alpha_{ub}} \approx \frac{x_{T_0'} - \bar{x}}{x_{T_0'} - x^{\alpha_{ub}}}, \qquad V_V^{\gamma'} = 1 - V_V^{\alpha_{ub}} \tag{6.7}$$

where $V_V^{\gamma'}$ is the fraction of residual austenite at the transformation temperature, with carbon concentration $x_{T_0'}$; the amount of this retained to ambient temperature can in turn be estimated using the Koistinen and Marburger Equation 5.9, with M_S being that of the austenite with its enriched composition. Therefore, the phase fractions, phase compositions of α_{ub}, γ_r and α' can all be estimated, forming the basis of the design of TRIP-assisted steels. A similar method can be used to calculate the thickness of the austenite films between adjacent bainite platelets, by assuming that the carbon diffusion field due to an existing plate of ferrite prevents the close approach of another parallel plate because the T_0 condition is violated [62].

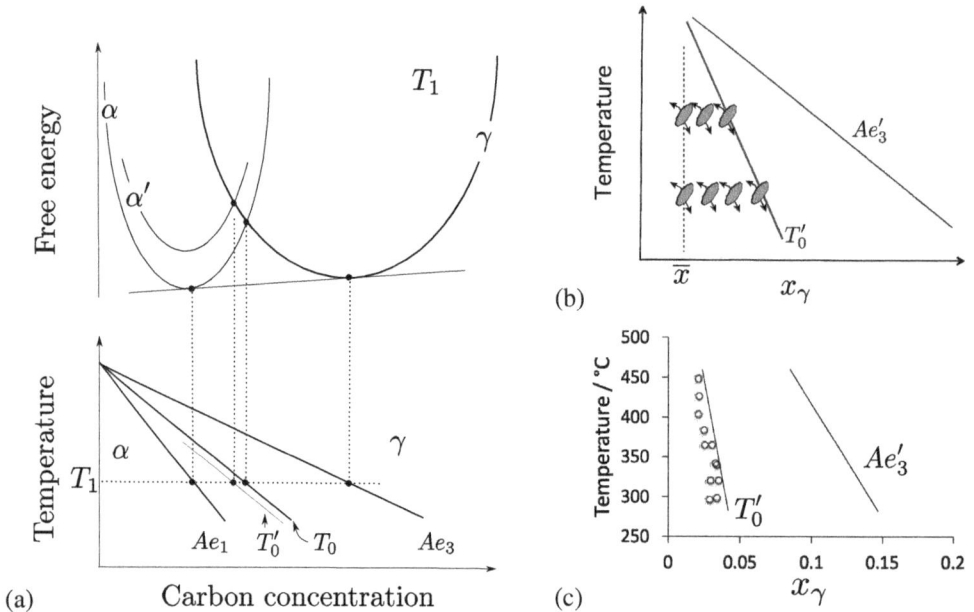

Figure 6.10   (a) Illustration of the T_0 line, which represents the locus of carbon concentrations where austenite and ferrite of the same composition have identical free energy, with the T_0' curve incorporating the effect of strain energy due to the shape deformation. Also illustrated are the Ae_1 and Ae_3 boundaries given by the compositions where the α and γ free energy curves share a common tangent. (b) A schematic plot of temperature versus the carbon concentration of austenite. $\bar{x}$ is the concentration in austenite prior to transformation. Bainite plates that grow without diffusion and then partition carbon are shown; this can only continue until $x_\gamma = x_{T_0'}$. (c) Experimental data (Fe-0.43C-3Mn-2.12Si wt%) showing the composition of the austenite at the point that reaction ceases. In (b,c), the Ae_3' curve is equivalent to Ae_3 but for the paraequilibrium state.

6.3   KINETICS

It can be more revealing to deal simultaneously with a variety of transformations rather than to consider each in isolation. For this reason, the quantitative description of the nucleation of bainite is therefore deferred to Section 7.6 where it is shown that bainite and Widmanstätten ferrite develop from the same nucleus and that it is the conditions that prevail during growth that make them distinct. They both require carbon to partition at the nucleation stage and the relationship between the activation energy for nucleation and the driving force indicates a mechanism consistent with that by which martensite nucleates, i.e., a manifestation of dislocation dissociation.

This description applies to the initial nucleation event in the development of a sheaf, which then multiplies by autocatalysis where bits of the mechanically stabilised α_b/γ interface are able to break away and develop into further lenticular subunits of transformation. The initial event can occur at the austenite grain surfaces or intragranularly on appropriate non-metallic inclusions, Figure 6.11; the sheaf morphology develops in both cases as long as the austenite grain size is large enough but both follow an identical thermodynamic framework [63].

Because bainite propagates by a subunit mechanism, the lengthening rate of a sheaf is bound to be

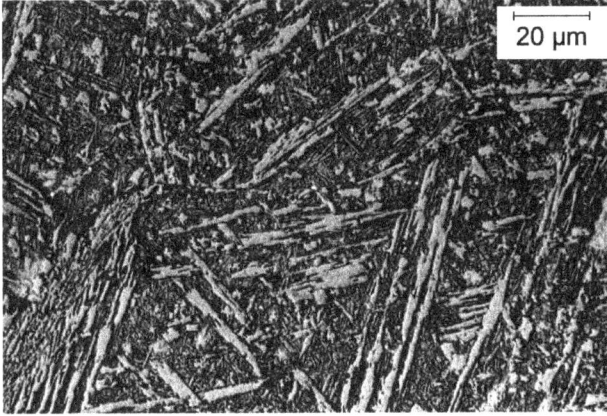

Figure 6.11 Fe-0.052C-0.44Si-1.51Mn-1.52Ni-0.41M0-0.66Cr wt% and 411 parts per million of oxygen. Partial transformation at 530 °C following austenitisation at 1350 °C. Bainite sheaves develop intragranularly following nucleation on oxide particles. Intragranularly nucleated bainite is sometimes referred to as "acicular ferrite". Micrograph courtesy of Gethin Rees.

slower than that of a subunit, with

$$v_{\text{sheaf}} = \frac{v_{\text{subunit}}}{1 + (\Delta t / t_{\text{c}})} \tag{6.8}$$

where t_{c} is the time taken for a subunit to grow to its final length and Δt is the time taken to stimulate the next subunit. v_{subunit} is the average lengthening rate of a subunit, which because of its fine scale has been measured using a technique such as photoemission electron microscopy, which has revealed the rate to be some three orders of magnitude greater than expected from carbon diffusion-controlled rate [64]. Measurements made at lower resolutions should be interpreted to represent sheaf growth; the first such data are due to Speich and Cohen [65] who used hot-stage optical microscopy to show that the lengthening rate is constant and that the width develops in proportion to the length, as might be expected from strain energy considerations. The same rationale can be applied to the observation that the aspect ratio of the sheaf increases as the transformation temperature is reduced, given the greater driving force available at lower temperatures.[7] An analysis of low-resolution observations indicates that sheaf lengthening rates also are greater than expected assuming carbon-diffusion control [66].

The growth rate of an individual subunit can be treated in the same manner as that of martensite, but is expected to be retarded by the plastic collapse that dominates the development of structure. Although martensitic transformation can be rapid, plates can grow slowly when limited by dislocation effects at small driving forces [67, 68]. Figure 6.12 compares the isothermal lengthening rates of bainite sheaves and martensite laths to show that they can be comparable. The bainite data cover a much greater range of temperatures but it has been shown in all cases that the growth rate if faster than that expected from carbon diffusion-control [66].

Although there has been no theoretical treatment of the lengthening rate as a function of the plastic accommodation, the resistance to interface motion is defined by τ_S (Equation 5.36). In diffusion-less, displacive transformations where interfacial dislocations benefit from thermal activation, the

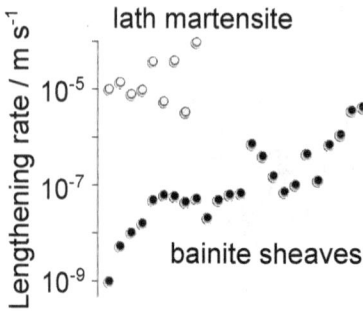

Figure 6.12 The data for the lengthening rate of bainite sheaves is from [Figure 6.16, 14] covering a range of steels. Those for lath martensite are from [67], for a Fe-21Ni-4Mn alloy transformed isothermally at $-80\,°C$.

velocity of the interface is expressed in a form similar but not identical to Equation 4.64 [69]:

$$v = v_0 \exp\left\{ -G^* \left(1 - \left[\frac{G_I}{G'_I} \right]^{\frac{1}{2}} \right) \Big/ kT \right\} \tag{6.9}$$

where v_0 is set at $30\,\mathrm{m\,s^{-1}}$ [70], $G'_I = 1.22 \times 10^{-3} E_s\ \mathrm{J\,m^{-3}}$ [69] and $G^* = 0.31 V_a$ J; all the other terms have been defined previously. If the evolution of the dislocation debris as a function of the length of the subunit is known, then the free energy driving the interface, G_I, can be adjusted accordingly to predict arrest using the additional theory outlined in Section 5.9.2.

REFERENCES

1. W. L. Costin, O. Lavigne, and A. Kotousov: 'A study on the relationship between microstructure and mechanical properties of acicular ferrite and upper bainite', *Materials Science & Engineering A*, 2016, **663**, 193–203.

2. S. Lenka: 'Bainitic steel with 30% GPa% characteristics. https://doi.org/10.17863/cam.51538': Ph.D. thesis, University of Cambridge, Cambridge, U. K., 2019.

3. E. Swallow, and H. K. D. H. Bhadeshia: 'High resolution observations of displacements caused by bainitic transformation', *Materials Science and Technology*, 1996, **12**, 121–125.

4. S. Kajiwara: 'Characteristic features of shape memory effect and related transformation behavior in Fe-based alloys', *Materials Science & Engineering A*, 1999, **273-275**, 67–88.

5. H. I. Aaronson, and C. Wells: 'Sympathetic nucleation of ferrite', *Transactions of the AIME*, 1956, **206**, 1216–1223.

6. R. F. Hehemann: 'The bainite transformation', In: H. I. Aaronson, and V. F. Zackay, eds. *Phase Transformations*. Materials Park, Ohio, USA: American Society of Materials, 1970:397–432.

7. H. K. D. H. Bhadeshia, and D. V. Edmonds: 'The mechanism of bainite formation in steels', *Acta Metallurgica*, 1980, **28**, 1265–1273.

8. S. Chatterjee, H. S. Wang, J. R. Yang, and H. K. D. H. Bhadeshia: 'Mechanical stabilisation of austenite', *Materials Science and Technology*, 2006, **22**, 641–644.

9. L. C. D. Fielding, E. J. Song, D. K. Han, H. K. D. H. Bhadeshia, and D. W. Suh: 'Hydrogen diffusion and the percolation of austenite in nanostructured bainitic steel', *Proceedings of the Royal Society of London A*, 2014, **470**, 20140108.

10. G. B. Olson, and W. S. Owen: 'Stress field of a martensitic particle and the conditions for thermoelastic behaviour', In: *New aspects of martensitic transformations*. Tokyo, Japan: Japan Institute of Metals, 1976:105–110.

11. I. Tamura, T. Maki, and H. Hato: 'Morphology of strain-induced martensite and TRIP in Fe-Ni-Cr alloy', *Trans. ISIJ*, 1970, **10**, 163–172.

12. G. R. Srinivasan, and C. M. Wayman: 'Isothermal transformation in an Fe-7.9Cr-1.1C alloy', *Trans. Met. Soc. AIME*, 1968, **242**, 78–81.

13. G. R. Srinivasan, and C. M. Wayman: 'The crystallography of the bainite transformation', *Acta Metallurgica*, 1968, **16**, 621–636.

14. H. K. D. H. Bhadeshia: Bainite in steels: theory and practice: 3rd ed., Leeds, U.K.: Maney Publishing, 2015.

15. H. K. D. H. Bhadeshia, and A. R. Waugh: 'Bainite: An atom probe study of the incomplete reaction phenomenon', *Acta Metallurgica*, 1982, **30**, 775–784.

16. H. K. D. H. Bhadeshia, and A. R. Waugh: 'An atom-probe study of bainite', In: H. I. Aaronson, D. E. Laughlin, R. F. Sekerka, and C. M. Wayman, eds. *Solid-Solid Phase Transformations*. Warrendale, Pennsylvania, USA: TMS-AIME, 1982:993–998.

17. I. Stark, G. D. W. Smith, and H. K. D. H. Bhadeshia: 'The element distribution associated with the incomplete reaction phenomenon in steels: an atom probe study', In: G. E. Lorimer, ed. *Phase Transformations '87*. London, U.K.: Institute of Metals, 1988:211–215.

18. I. Stark, G. D. W. Smith, and H. K. D. H. Bhadeshia: 'Distribution of substitutional alloying elements during the bainite transformation', *Metallurgical Transactions A*, 1990, **21**, 837–844.

19. B. Josefsson, and H.-O. Andrén: 'Microstructure of granular bainite', *Journal de Physique Colloque*, 1988, **49**, C6–293 – C6–298.

20. B. Josefsson, and H. O. Andrén: 'Microstructure and thermodynamic behaviour of a Cr-Mo submerged arc weld metal in the as-welded state', In: S. A. David, and J. M. Vitek, eds. *Recent Trends in Welding Science and Technology (TWR'89)*. Metals Park, Ohio, USA: ASM International, 1989:243–247.

21. M. Peet, S. S. Babu, M. K. Miller, and H. K. D. H. Bhadeshia: 'Three-dimensional atom probe analysis of carbon distribution in low-temperature bainite', *Scripta Materialia*, 2004, **50**, 1277–1281.

22. F. G. Caballero, M. K. Miller, S. S. Babu, and C. Garcia-Mateo: 'Atomic scale observations of bainite transformation in a high carbon high silicon steel', *Acta Materialia*, 2007, **55**, 381–390.

23. E. V. Pereloma, E. V. Timokhina, M. K. Miller, and P. D. Hodgson: 'Three-dimensional atom probe analysis of solute distribution in thermomechanically processed TRIP steels', *Acta Materialia*, 2007, **55**, 2587–2598.

24. C. Garcia-Mateo, M. Peet, F. G. Caballero, and H. K. D. H. Bhadeshia: 'Tempering of a hard mixture of bainitic ferrite and austenite', *Materials Science and Technology*, 2004, **20**, 814–818.

25. F. G. Caballero, M. K. Miller, A. J. Clarke, and C. Garcia-Mateo: 'Examination of carbon partitioning into austenite during tempering of bainite', *Scripta Materialia*, 2010, **63**, 442–445.

26. E. Pereloma, H. Beladi, L. Zhang, and I. Timokhina: 'Understanding the behavior of advanced high-strength steels using atom probe tomography', *Metallurgical & Materials Transactions A*, 2012, **43**, 3958–3971.

27. F. G. Caballero, M. K. Miller, C. Garcia-Mateo, and J. Cornide: 'New experimental evidence of the diffusionless transformation nature of bainite', *Journal of Alloys and Compounds*, 2013, **577**, S626–S630.

28. F. G. Caballero, M. K. Miller, C. Garcia-Mateo, J. Cornide, and M. J. Santofimia: 'Temperature dependence of carbon supersaturation of ferrite in bainitic steels', *Scripta Materialia*, 2012, **67**, 846–849.

29. C. Garcia-Mateo, J. A. Jimenez, H. W. Yen, L. Morales-Rivas, M. Kuntz, S. P. Ringer, J. R. Yang, and F. G. Caballero: 'Low temperature bainitic ferrite: Evidence of carbon supersaturation and tetragonality', *Acta Materialia*, 2015, **91**, 162–173.

30. J. H. Jang, H. K. D. H. Bhadeshia, and D. W. Suh: 'Solubility of carbon in tetragonal ferrite in equilibrium with austenite', *Scripta Materialia*, 2012, **68**, 195–198.

31. C. N. Hulme-Smith, I. Lonardelli, A. C. Dippel, and H. K. D. H. Bhadeshia: 'Experimental evidence for non-cubic bainitic ferrite', *Scripta Materialia*, 2013, **69**, 409–412.

32. R. K. Dutta, R. M. Huizenga, M. Amirthalingam, H. Gao, A. King, M. J. M. Hermans, and I. M. Richardson: 'Synchrotron diffraction studies on the transformation strain in a high strength quenched and tempered structural steel', *Materials Science Forum*, 2014, **777**, 231–236.

33. R. K. Dutta, R. M. Huizenga, M. Amirthalingam, A. King, H. Gao, M. J. M. Hermans, J. Sietsma, and I. M. Richardson: 'In-situ synchrotron diffraction studies on transformation strain development in a high strength quenched and tempered structural steel – Part I. Bainitic transformation', *Metallurgical & Materials Transactions A*, 2014, **45**, 218–229.

34. C. N. Hulme-Smith, M. J. Peet, I. Lonardelli, A. C. Dippel, and H. K. D. H. Bhadeshia: 'Further evidence of tetragonality in bainitic ferrite', *Materials Science and Technology*, 2015, **31**, 254–256.

35. D. A. Mirzayev, A. A. Mirzoev, I. V. Buldashev, and K. Y. Okishev: 'Thermodynamic analysis of the formation of tetragonal bainite in steels', *Fizika Metallov i Metallovedenie*, 2017, **118**, 547–553.

36. Z. Xiong, D. R. G. Mitchell, A. A. Saleh, and E. V. Pereloma: 'Tetragonality of bcc phases in a transformation-induced plasticity steel', *Metallurgical & Materials Transactions A*, 2018, **49**, 5925–5929.

37. R. Rementeria, R. Dominguez-Reyes, C. Capdevila, C. Garcia-Mateo, and F. G. Caballero: 'Positron annihilation spectroscopy study of carbon-vacancy interaction in low-temperature bainite', *Scientific Reports*, 2020, **10**, 1–6.

38. R. B. McLellan: 'The thermodynamics of interstitial-vacancy interactions in solid solutions', *Journal of Physics and Chemistry of Solids*, 1988, **49**, 1213–1217.

39. K. R. Kinsman, and H. I. Aaronson: 'Discussion on bainite', In: *Transformation and Hardenability in Steels*. Ann Arbor, Michigan, USA: Climax Molybdenum, 1967:33–37.

40. H. K. D. H. Bhadeshia: 'Theoretical analysis of changes in cementite composition during the tempering of bainite', *Materials Science and Technology*, 1989, **5**, 131–137.

41. M. Hillert, L. Hoglund, and J. Ågren: 'Escape of carbon from ferrite plates in austenite', *Acta Metallurgica and Materialia*, 1993, **41**, 1951–1957.

42. S. Mujahid, and H. K. D. H. Bhadeshia: 'Partitioning of carbon from supersaturated ferrite plates', *Acta Metallurgica and Materialia*, 1992, **40**, 389–396.

43. J. Speer, D. K. Matlock, B. C. D. Cooman, and J. G. Schroth: 'Carbon partitioning into austenite after martensite transformation', *Acta Materialia*, 2003, **51**, 2611–2622.

44. J. G. Speer, D. V. Edmonds, F. C. Rizzo, and D. K. Matlock: 'Partitioning of carbon from supersaturated plates of ferrite, with application to steel processing and fundamentals of the bainite transformation', *Current Opinion in Solid State and Materials Science*, 2004, **8**, 219–237.

45. J. G. Speer, E. de Moor, and A. J. Clarke: 'Critical assessment 7: Quenching and partitioning', *Materials Science and Technology*, 2015, **31**, 3–9.

46. S. J. Matas, and R. F. Hehemann: 'The structure of bainite in hypoeutectoid steels', *TMS-AIME*, 1961, **221**, 179–185.

47. M. Takahashi, and H. K. D. H. Bhadeshia: 'Model for transition from upper to lower bainite', *Materials Science and Technology*, 1990, **6**, 592–603.

48. M. Oka, and H. Okamoto: 'Isothermal transformations in hypereutectoid steels', In: *International Conference on Martensitic Transformations ICOMAT '86*. Tokyo, Japan: Japan Institute of Metals, 1986:271–275.

49. Y. Ohmori: 'The crystallography of the lower bainite transformation in a plain carbon steel', *Transaction of the ISIJ*, 1971, **11**, 95–101.

50. Z. Lawrynowicz: 'Transition from upper to lower bainite in Fe-C-Cr steel', *Materials Science and Technology*, 2004, **20**, 1447–1454.

51. Z. Lawrynowicz: 'Rationalisation of austenite transformation to upper or lower bainite in steels', *Advances in Materials Science*, 2014, **14**, 14–23.

52. E. Kozeschnik, and H. K. D. H. Bhadeshia: 'Influence of silicon on cementite precipitation in steels', *Materials Science and Technology*, 2008, **24**, 343–347.

53. H. K. D. H. Bhadeshia, M. Lord, and L.-E. Svensson: 'Silicon-rich bainitic steel welds', *Transactions of JWRI*, 2003, **32**, 91–96.

54. R. B. Carruthers, and M. J. Collins: 'Use of scanning transmission electron microscopy in estimation of remanent life of pressure parts', In: *Quantitative Microanalysis with High Spatial Resolution*. London, U.K.: Metals Society, 1981:108–111.

55. B. J. Cane, and R. D. Townsend: 'Prediction of remaining life in low alloy steels': Tech. Rep. PRD/L/2674/N84, Central Electricity Generating Board, Leatherhead, Surrey, U.K., 1984.

56. H. K. D. H. Bhadeshia, A. Strang, and D. J. Gooch: 'Ferritic power plant steels: Remanent life assessment and the approach to equilibrium', *International Materials Reviews*, 1998, **43**, 45–69.

57. P. Yan, and H. K. D. H. Bhadeshia: 'Mechanism and kinetics of solid-state transformation in high-temperature processed linepipe steel', *Metallurgical & Materials Transactions A*, 2013, **44**, 5468–5477.

58. E. R. Parker, and V. F. Zackay: 'Strong and ductile steels', *Scientific American*, 1968, **219**, 36–45.

59. V. F. Zackay, and E. R. Parker: 'Treatment of steel': U.S. Patent No. 3,488,231, 1970.

60. A. Schrader, and F. Wever: 'Zur frage der eignung des elektronenmikroskops fur die gefugeuntersuchung von stahlen', *Archiv für das Eisenhüttenwesen*, 1952, **23**, 489–495.

61. P. G. Self, H. K. D. H. Bhadeshia, and W. M. Stobbs: 'Lattice spacings from lattice fringes', *Ultramicroscopy*, 1981, **6**, 29–40.

62. L. C. Chang, and H. K. D. H. Bhadeshia: 'Austenite films in bainitic microstructures', *Materials Science and Technology*, 1995, **11**, 874–881.

63. G. I. Rees, and H. K. D. H. Bhadeshia: 'Thermodynamics of acicular ferrite nucleation', *Materials Science and Technology*, 1994, **10**, 353–358.

64. H. K. D. H. Bhadeshia: 'Solute-drag, kinetics and the mechanism of the bainite transformation', In: A. R. Marder, and J. I. Goldstein, eds. *Phase Transformations in Ferrous Alloys*. Ohio, USA: TMS-AIME, 1984:335–340.

65. G. R. Speich, and M. Cohen: 'The growth rate of bainite', *TMS-AIME*, 1960, **218**, 1050–1059.

66. A. Ali, and H. K. D. H. Bhadeshia: 'Growth rate data on bainite in alloy steels', *Materials Science and Technology*, 1989, **5**, 398–402.

67. D.-Z. Yang, and C. M. Wayman: 'Slow growth of isothermal lath martensite in an Fe-21N-4Mn alloy', *Acta Metallurgica*, 1984, **32**, 949–954.

68. M. Villa, M. F. Hansen, K. Pantleon, and M. A. J. Somers: 'Thermally activated growth of lath martensite in Fe–Cr–Ni–Al stainless steel', *Materials Science and Technology*, 2015, **31**, 115–122.

69. G. B. Olson, H. K. D. H. Bhadeshia, and M. Cohen: 'Coupled diffusional/displacive transformations, Part II: Solute trapping', *Metallurgical & Materials Transactions A*, 1990, **21A**, 805–809.

70. M. Grujicic, G. B. Olson, and W. S. Owen: 'Mobility of martensitic interfaces', *Metallurgical & Materials Transactions A*, 1985, **16**, 1713–1722.

71. E. S. Davenport, and E. C. Bain: 'Transformation of austenite at constant subcritical temperatures', *Transactions of the American Institute of Mining and Metallurgical Engineers*, 1930, **90**, 117–154.

72. S. Matas, and R. F. Hehemann: 'Retained austenite and the tempering of martensite', *Nature*, 1960, **187**, 685–686.

Notes

[1]Historical and evolutionary aspects of bainite have been described elsewhere [14, 71] in considerable depth so will not be reproduced here.

[2]A simple calculation of free energy changes explains this slow rate. For a Fe-0.4C wt% alloy transformed at 400 °C, $\Delta G^{\alpha' \rightarrow \alpha + \theta} = -688\,\mathrm{J\,mol^{-1}}$. Assuming that the carbon-enriched residual-austenite has a much greater carbon concentration of 1.2 wt%, $\Delta G^{\gamma \rightarrow \gamma' + \theta} = -482\,\mathrm{J\,mol^{-1}}$, which is a significantly smaller magnitude. The difference between the two calculations becomes much greater as the transformation temperature is increased. At 500 °C there is no driving force for the precipitation of cementite from the austenite.

[3]Matas and Hehemann first attempted to partition carbon from martensite into residual austenite by tempering at a relatively low temperature [46, 72].

[4]The term *TRIP-assisted* is used to indicate that the austenite retained in such alloys is a minority phase, typically below a volume fraction of 0.2.

[5]The T_0 concept explains incomplete reaction, but none of the theories that x^α increases steadily towards $\bar{x}$ as the driving force increases can explain premature reaction termination. If the ferrite does form with a partial supersaturation ($x^{\alpha\gamma} \leq x^\alpha \leq \bar{x}$) and the driving force to sustain that supersaturation diminishes to zero at some point, then there is nothing to stop the composition of the ferrite adjusting towards paraequilibrium so that the transformation can continue. Growth with partial supersaturation is therefore inconsistent with incomplete reaction. Solute drag arguments fail on the same basis. The same argument applies to the use of the NP-LE curve as a limit because as transformation progresses, tie-line shifting leads to equilibrium (Chapter 4).

[6]In fact, the B_S temperature may fall below the condition satisfying $\bar{x} = x_{T_0'}$ if the nucleation requirement is not met, Section 7.6.

[7]Notice that in the diffusion-controlled lengthening of plates, the thickness increases parabolically with time because of the sideways accumulation of solute.

7 Widmanstätten ferrite

7.1 INTRODUCTION

Widmanstätten ferrite (α_W) gets its name from the plate-shaped crystals that are arranged in patterns reminiscent of the macrostructures found in some meteorite specimens. The patterns were noticed in 1808 by Alois de Widmanstätten of the Imperial Porcelain Works in Vienna, on the Hradschina meteorite. In 1900, Osmond reported similar patterns on the head of a steel ingot [1]. This may be where the analogy between meteoritic α_W and that in steels ends, because the chemical compositions of the two varieties are so dramatically different. For obvious reasons, the evolutionary state of the Widmanstätten ferrite in meteorites is ill-defined; is it the case that the substitutional solute content was altered during cooling after the growth of the giant plates of meteoric α_W?

In steels, the plates that nucleate at the austenite grain boundaries are designated *primary* whereas others that develop from ferrite allotriomorphs are called *secondary*. Widmanstätten ferrite plates have a lenticular edge with the overall shape replicating a thin wedge. They are typically a few micrometres in width and lack internal structure; this is why in etched metallographic samples they usually appear the same white colour as allotriomorphic ferrite (Figure 7.1). In contrast pearlitic and bainitic microstructures have, on the same length scale, considerable substructure not resolved in an optical microscope, causing them to etch dark.

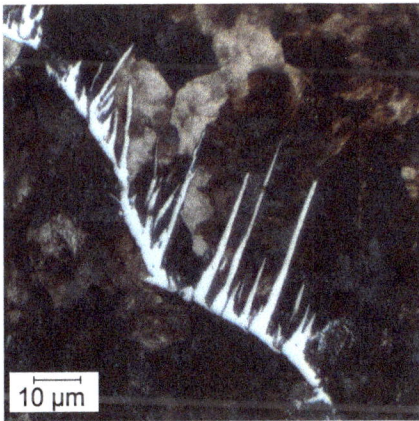

Figure 7.1 Widmanstätten ferrite plates emanating from one side of a prior-austenite grain boundary with the remaining austenite transforming into pearlite during cooling. Micrograph courtesy of Rolando M. Núñez Monrroy of the Pontifical Catholic University of Peru.

Widmanstätten ferrite is structurally homogeneous. The mechanism of transformation is displacive so the plates exhibit a reproducible combination of irrational habit-plane, orientation relationship and shape deformation with the parent austenite. Secondary α_W plates therefore can initiate only from allotriomorphic ferrite (α) that happens to be appropriately orientated with the austenite. The habit plane, orientation relationship and shape deformation are mathematically connected (Chapter 5) so the habit plane is defined uniquely once the variant of the orientation relationship is fixed by the α. Therefore, a particular allotriomorph of ferrite can develop into just one, or at most two degenerate variants of Widmanstätten ferrite. This is why colonies of parallel plates form at allotriomorphs. Observations using crystallographic orientation imaging show that there frequently are low-misorientation boundaries between plates of Widmanstätten ferrite and the allotriomorphs from which they grow [2], because unlike the α_W/γ orientation, the α/γ relationship is not constrained by the need to ensure a glissile interface during transformation. For the same reason, misorientations within a colony of parallel plates tend to be negligibly small.

The α/γ orientation relationship fixes the habit plane of the α_W that can develop from the allotriomorph, because $\mathbf{p}$, (α_W J γ) and $\mathbf{P}_1$ are not independent (p. 272). To mitigate interference with the diffusion field of the allotriomorph, $\mathbf{p}$ must lie at a large angle to the α/γ interface, Figure 7.2. If it does not, then that part of the α/γ fails to stimulate secondary α_W. This also explains why the secondary Widmanstätten ferrite forms in sets of parallel plates because the habit plane $\mathbf{p}$ is fixed by the orientation of the allotriomorph with the austenite. This condition is relaxed as the transformation temperature is reduced, because the greater driving force available can permit α_W to penetrate the diffusion field of the allotriomorph even when $\mathbf{p}$ has a shallow inclination to the α/γ interface.

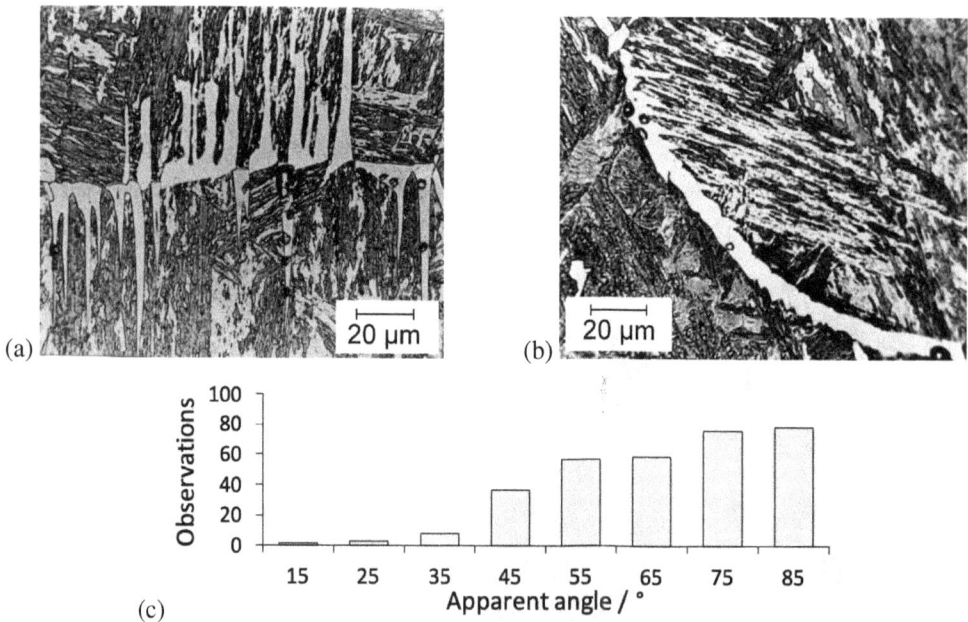

Figure 7.2 Fe-0.26C-0.23Mn wt% steel transformed isothermally. (a) Widmanstätten plates generally lengthen at a large angle to austenite grain boundaries. (b) Plates suppressed when the habit plane makes a shallow angle with the allotriomorphic ferrite. (c) Number of observations of the apparent angle between the plates and the α/γ grain boundary. Selected data from Fong and Glover, [3]. Angles plotted represent means within a range of 10°. Micrographs courtesy of S. G. Glover.

It will be evident in what follows that Widmanstätten ferrite in *steels* grows under paraequilibrium conditions, i.e., the iron to substitutional solute ratio remains constant across the transformation interface, but subject to that constraint, carbon partitions to maintain a uniform chemical potential in both phases in contact at the interface. It is emphasised, however, that Widmanstätten ferrite is common in *interstitial-free* alloys of iron, Table 7.1, during transformation at temperatures below

T_0. Plates of α_W remain supersaturated with substitutional solute in Fe-Ni-Cu when other transformation products show clear arrays of copper precipitation [4]. An important conclusion, therefore, is that the displacive transformation mechanism is independent of the presence of carbon. Figure 7.3 shows primary Widmanstätten ferrite plates in a high-purity Fe-Cr-Ni alloy in a partially transformed sample.

Table 7.1
Widmanstätten ferrite in interstitial-free iron alloys.

Alloy (wt%)	Description	Ref.
Fe-9.6Cr	Secondary α_W generated by transformation in the range 670 °C to 700 °C. Particularly clear characterisation showing Widmanstätten plates in a partially transformed specimen.	[5]
Fe-7.4Cr-1.8Ni	Primary and secondary α_W generated by transformation in the range 600 °C to 650 °C, later verified to be below T_0 [6]. Particularly revealing characterisation over range of scales, with clear α_W plates shown in a partially transformed specimen.	[7]
Fe-1.8Cu-4.8Ni	α_W generated by transformation at 600 °C. Interesting observation that α_W retained Cu in solution.	[4]
Fe-9.14Ni	α_W generated in the temperature range 495 °C to 565 °C with $T_0 = 623 \pm 5$ °C.	[8, 9]
Fe-15Ni	α_W generated in the temperature range 372 °C to 352 °C with $T_0 = 503 \pm 5$ °C.	[10]

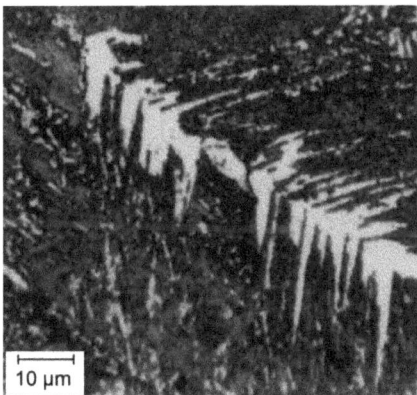

Figure 7.3 Widmanstätten ferrite plates in a partially transformed sample of interstitial-free Fe-7.4Cr-1.8Ni wt% alloy. Reprinted by permission from Springer Nature, Metallurgical Transactions A, copyright 1981, Ricks et al. [7].

7.2 CRYSTALLOGRAPHY

Widmanstätten ferrite shows all the characteristics of a displacive transformation in which only the carbon is partitioned during growth. As Christian pointed out [11], the movement of interstitial atoms such as carbon need not affect the crystallography of the deformation that transforms the host lattice; it is obvious, for example, that the presence of hydrogen in austenite does not influence the crystallography of the martensite that forms subsequently. The invariant-plane strain shape deformation (Figure 7.4) occurs on a habit plane that has irrational indices, thereby constraining the morphology to that of lenticular plates.[1] The crystallographic measurements by Watson and McDougall [12] are summarised below; as already emphasised, a principal feature of the crystallographic theory of martensite (Chapter 5) is that the habit plane, orientation relationship, shape deformation and lattice-invariant deformation are not independent variables. The following results are particularly important in that the variables form a related set, i.e., not measured in independent experiments as is often the case:

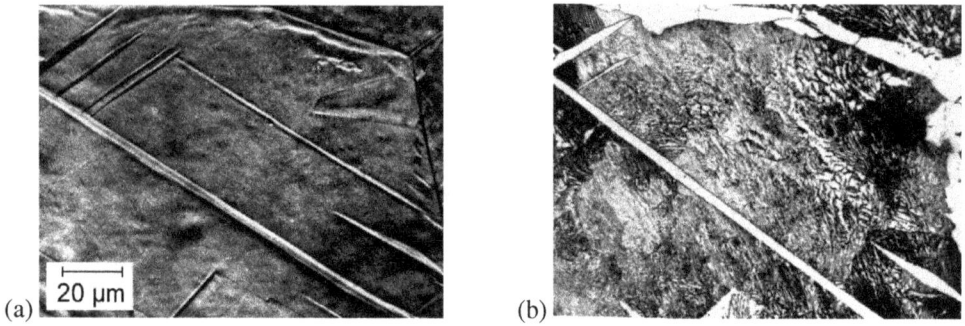

(a) 20 μm   (b)

Figure 7.4   (a) The shape deformation due to Widmanstätten ferrite, as viewed on a polished and transformed sample; the deformation reveals the lenticular shape of the unhindered plates. (b) Corresponding optical micrograph after light polishing and etching. Notice also the lenticular shape of the Widmanstätten ferrite plates, tapering to a sharp edge, and the lack of the shape deformation in regions containing allotriomorphic ferrite or pearlite. After Watson and McDougall [12], reproduced with permission of Elsevier.

For the specific variant of the lattice correspondence

$$(\alpha_W \quad C \quad \gamma) = \begin{pmatrix} 1 & 0 & 1 \\ 1 & 0 & \bar{1} \\ 0 & 1 & 0 \end{pmatrix}$$

the habit plane normal is

$$(\mathbf{p};\gamma^*) = (0.5057 \quad 0.4523 \quad 0.7346),$$

with the orientation relationship being irrational but close to Kurdjumov-Sachs (Figure 7.5):

$$(1\,0\,1)_{\alpha_W} \quad \| \quad (0.5916\, 0.5772\, 0.5628)_\gamma$$

$$[1\,1\,\bar{1}]_{\alpha_W} \quad \| \quad [0.6984\, \overline{0.7157}\, 0.0001]_\gamma$$

and the average magnitude of the shape deformation and direction:

$$m \;=\; 0.36$$

$$[\gamma;\mathbf{d}] \;-\; [\overline{0.8670}\, 0.1143\, 0.2770].$$

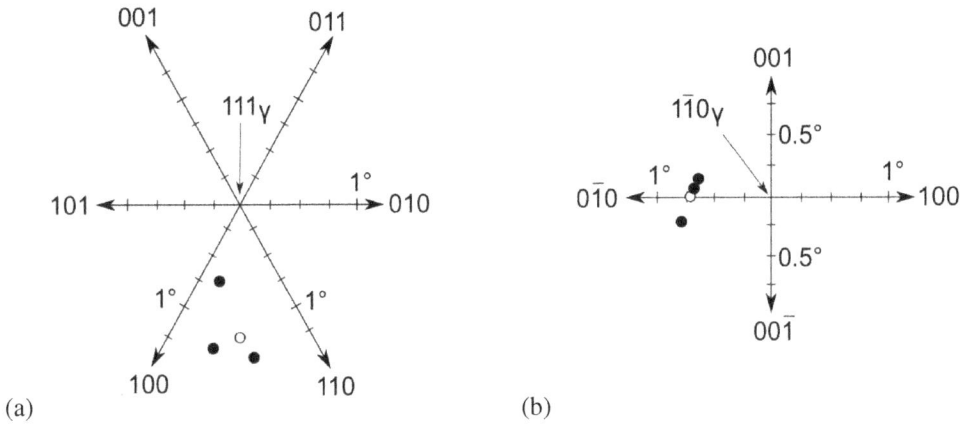

Figure 7.5 Stereographic projection, referred to austenite, with an angular span less than $2°$, showing the irrational orientation relationship between austenite and Widmanstätten ferrite, with respect to the standard correspondence. (a) Centred on $(111)_\gamma$ with the filled points representing the orientations of $(110)_{\alpha_W}$ from three plates. The mean orientation of $(110)_{\alpha_W}$ is plotted as an open circle. (b) Centred on $(1\bar{1}0)_\gamma$, with the circles representing the $(11\bar{1})_{\alpha_W}$ with the interpretation of filled and open circles as in (a). Reproduced with permission of Elsevier, from [12].

The displacement vector **d** does not lie precisely within the habit plane **p** because it describes both the shear and dilatational strains, the latter being directed normal to the habit plane.

The measurements reported by Watson and McDougall have the precision to prove that consistent with the crystallographic theory of martensite, neither the habit plane, nor the orientation relationship, is rational. The orientation relationship is irrational to ensure an invariant-line can exist in the interface; the habit plane is irrational because a lattice-invariant deformation has to be incorporated in order to render the macroscopically observed shape deformation an invariant-plane strain. A further result is that the lattice-invariant deformation probably is not a simple shear, but consists of a combination of shears on $(1\,1\,1)_\gamma$ and $(\bar{1}\,\bar{1}\,1)_\gamma$ in the common direction $[1\,\bar{1}\,0]_\gamma$ with the shear on $(1\,1\,1)_\gamma$ being dominant. An alternative common shear direction $[\bar{1}\,1\,0]_\gamma$ was also found to be compatible with the results, in which case the shears would be shared nearly equally between the two planes (cf. two lattice-invariant shears theory, p. 267).

In all their experiments, there was a specific association between a given variant of the habit plane and a given variant of the orientation relationship. This follows necessarily from the crystallographic theory of martensite.

In summary, the growth of Widmanstätten ferrite is accompanied by a change in the shape of the transformed region, characterised as an invariant-plane strain that has a shear component that is large in comparison with other transformations in steels (Table 7.2). This means that there is a co-ordinated motion of iron atoms during the transformation, but given that α_W can occur above the T_0 temperature, carbon must partition, making this a truly paraequilibrium transformation. The displacive mechanism is compatible with the partitioning of carbon during growth because the carbon is in interstitial solution and therefore does not affect the observable consequences of the coordinated transfer of Fe and X atoms across the transformation interface [11]. The transformation interface must be glissile but its velocity will be limited by the diffusion of carbon in the austenite ahead of the plate tip. When Widmanstätten ferrite occurs in interstitial-free iron alloys, it grows below the T_0 temperature.

Table 7.2

Shape change due to transformation. An invariant-plane strain here implies a large shear component as well as a dilatational strain normal to the habit plane. The strains s and ζ refer to the shear and dilatational components respectively. The values stated are approximate and will vary slightly as a function of lattice parameters and the details of crystallography. Only mechanical twins occur in ferrite. *Thin plates* **refers to lenticular plates with sharp edges. The question marks indicate data that are not experimentally verified.**

	Shape Change	s	ζ	Morphology	
α	Volume change	0.00	0.02	Irregular	[13]
P	Volume change	0.00	0.03	Spheroidal colonies	[14]
α_W	IPS	0.36	0.02	Thin plates	[12]
α_b	IPS	0.26	0.03	Thin plates	[15]
α'	IPS	0.20	0.02	Thin plates	[16]
θ	IPS ?	0.21 ?	0.16 ?	Thin plates	
α-twin	Shear	$1/\sqrt{2}$	0.00	Thin plates	
γ annealing-twin	–	0.00	0.00	Facetted	
γ mechanical-twin	Shear	$1/\sqrt{2}$	0.00	Thin plates	
ε	IPS	$1/\sqrt{8}$	0.02	Thin plates	[17]
TiC	IPS	$1/\sqrt{8}$	−0.08	Thin plates	[18]
η-carbide	IPS	0.21	0.157	Thin plates	[19]

7.3   ACCOMMODATION OF SHAPE DEFORMATION

Plates of Widmanstätten ferrite taper to a fine edge (Figure 7.4) in order to accommodate the elastic strain associated with the shape deformation accompanying the change in crystal structure. The experiments by Watson and McDougall [12] have demonstrated that in the transformed regions of pre-polished samples of austenite, the surface remains plane but is tilted about the lines of intersection of the Widmanstätten ferrite plates with the free surface. This tilt is an invariant-plane strain on the Widmanstätten ferrite habit plane, in which the shear strain is by far the largest component.

The yield strength of austenite is typically less than 100 MPa at the elevated temperatures where Widmanstätten ferrite grows. Not surprisingly, there is considerable evidence that the shape deformation induces plastic strain in the adjacent austenite, usually localised on one side of the plate. Whereas the surface corresponding to the transformed region remains planar during tilting, the plastic strain in the austenite is not uniform, curving away from the plate. However, when the consequences of the transformation strain are mitigated by the adjacent formation of accommodating

(a)

(b)

(c)

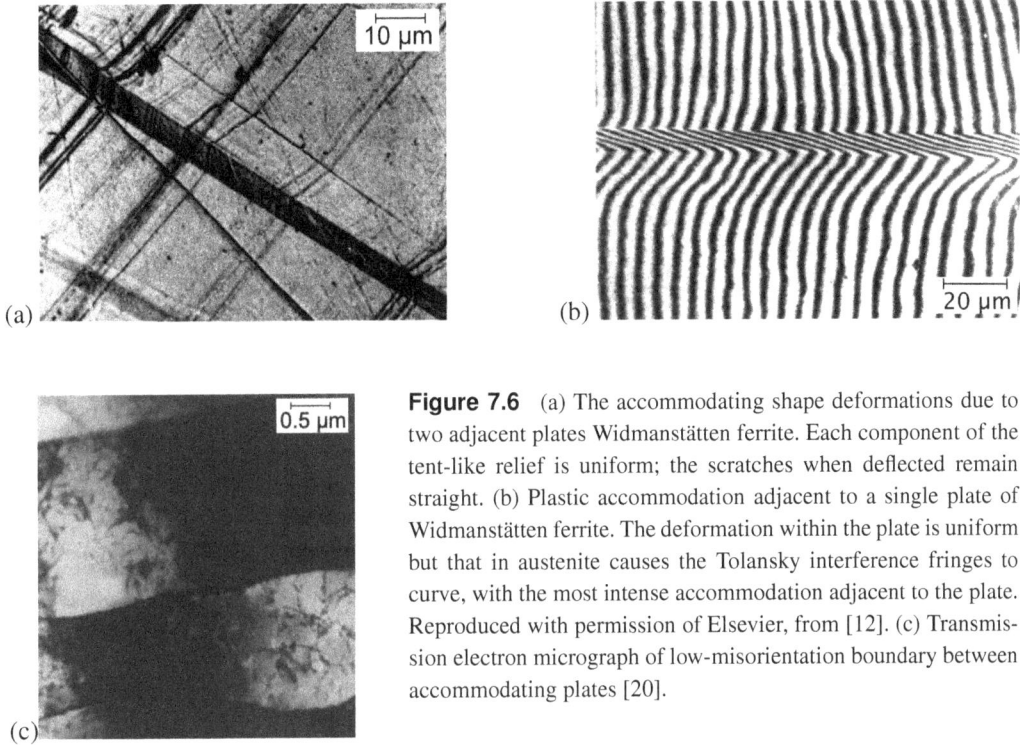

Figure 7.6 (a) The accommodating shape deformations due to two adjacent plates Widmanstätten ferrite. Each component of the tent-like relief is uniform; the scratches when deflected remain straight. (b) Plastic accommodation adjacent to a single plate of Widmanstätten ferrite. The deformation within the plate is uniform but that in austenite causes the Tolansky interference fringes to curve, with the most intense accommodation adjacent to the plate. Reproduced with permission of Elsevier, from [12]. (c) Transmission electron micrograph of low-misorientation boundary between accommodating plates [20].

plates, each component of the doubly tilted surface remains planar (Figure 7.6). These plates compensate for each other's shape deformations, thereby obviating the need for plastic relaxation in the surrounding austenite. These self-accommodating plates are in similar though not identical orientation in space so there is a low-misorientation interface between them [20]. The probability of simultaneously nucleating an appropriate pair of plates is smaller than that of an isolated plate so the α_W microstructure tends to be coarse. What appears on an optical microscopy scale to be a single plates in fact consists of a pair of plates with different habit planes and shape deformations, leading to the typical wedge shape of Widmanstätten ferrite.

Strains due to the displacements can also be relieved by the reconstructive formation of ferrite in the influenced region. Watson and McDougall elegantly demonstrated such diffusional transformation localised at the tips of Widmanstätten ferrite plates and confirmed that this additional transformation does not exhibit any surface relief.

Because of the variety of ways in which the elastic strains due to the shape deformation is accommodated, the stored energy of Widmanstätten ferrite is much smaller than might otherwise be expected, at about $50\,\mathrm{J\,mol^{-1}}$ [20]. This is why the plates are able to form at such small undercoolings below the equilibrium Ae_3 temperature.

7.3.1 INTERFACIAL STRUCTURE

Transmission electron microscopy of the interface between Widmanstätten ferrite and retained austenite by Rigsbee and Aaronson [21] has revealed two sets of linear defects, steps that are mono- or tri-atomic in height and intrinsic dislocations. These observations were suggested originally to rule out a glissile interface but are correctly interpreted in terms of the framework defined by Christian and Crocker as follows [22–25].

A semi-coherent interface containing a single array of intrinsic dislocations is considered to be glissile when the dislocations are able to move conservatively as the interface migrates. The intrinsic dislocations can therefore be pure screws or have Burgers vectors that do not lie in the interface plane.[2] The α_W/γ orientation relationship of a reference variant selected for discussion is approximately $[\bar{1}10]_\gamma \parallel [\bar{1}1\bar{1}]_{\alpha_W}$, $(111)_\gamma \parallel (110)_{\alpha_W}$. For this reference variant, the interface plane characterised has the indices $(15\ 21\ 9)_\gamma$ with the intrinsic dislocations having a Burgers vector $\mathbf{b} \parallel [\bar{1}10]_\gamma$, consistent with the first set of data in [21]. Since the interface is $18°$ from $(111)_\gamma$, the Burgers vector of the intrinsic dislocations does not lie in the interface plane. A glissile interface also requires that the glide planes of these dislocations associated with the α_W meet those in the γ edge to edge at the interface along the dislocation lines. This condition is also satisfied because $(111)_\gamma \parallel (110)\alpha_W$. Such glide would inhomogeneously shear the volume of material swept by the interface without altering the parent or product structures; the intrinsic dislocations represent the lattice-invariant deformation.

The change in structure as the interface is translated is achieved by the motion of the small steps that have a mono- or tri-atomic height. These steps can be generated by a series of virtual operations that associate them with a dislocation character and a strain field. They are not pure steps, nor can their Burgers vector content be identified with a lattice vector. Their translation is conservative and are best described as coherency dislocations (Section 5.2.3) which accomplish transformation and can glide without creating or destroying lattice sites.

The Widmanstätten ferrite interface therefore has all the features necessary to achieve displacive transformation without requiring diffusion in order to move. The next section describes how the motion of the glissile interface can be stifled by obstacles. This feature, commonly referred to as mechanical stabilisation, is unique to displacive transformations.

It is noted in passing that in zirconium, the hydride grows by a displacive paraequilibrium mechanism which is identical to that of Widmanstätten ferrite in steels. The hydride has a crystallography consistent with the theory of martensite, grows as pairs of accommodating plates, with hydrogen diffusion towards the hydride during the growth process [26, 27].

7.3.2 MECHANICAL STABILISATION

The observed shape deformation and other characteristics of Widmanstätten ferrite indicate a glissile interface between Widmanstätten ferrite and austenite. The interfacial dislocations are able to glide, leaving a transformed region in their wake and causing the deformation. It is natural to expect an interface like that to be hindered by obstacles in the form of precipitates or other dislocations present in the austenite. It follows that the growth process during displacive transformation can be disrupted or suppressed when it occurs in plastically deformed austenite. This is the mechanical stabilisation referred to in Section 5.9.2.

The Widmanstätten ferrite transformation can be mechanically stabilised [28, 29]. Not only is the plate-like microstructure of the Widmanstätten ferrite refined and disrupted when it grows in deformed austenite, but the total quantity obtained during isothermal transformation is also reduced, because the extent of transformation per nucleus is also reduced. Allotriomorphic ferrite growth is accelerated due to the free energy reduced by the elimination of excess defects in the process, whereas the same defects hinder the growth of Widmanstätten ferrite by obstructing its glissile interface, Figure 7.7.

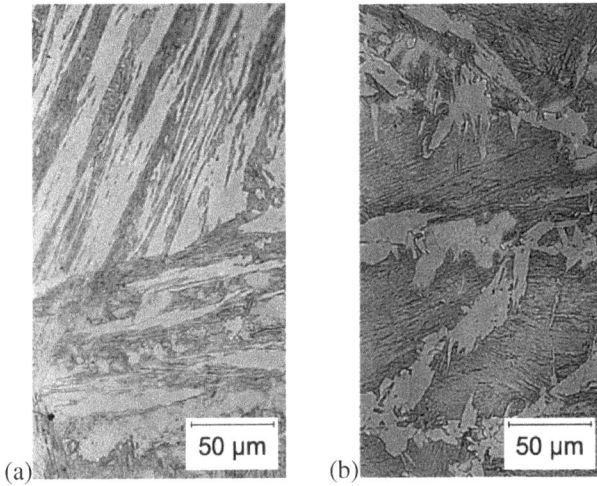

Figure 7.7 Widmanstätten ferrite in Fe-0.059C-1.96Si-2.88Mn wt% transformed isothermally at 560 °C. (a) Structure from undeformed austenite. (b) Structure from austenite subjected to 40% strain, showing how α_W is suppressed. Reproduced from Larn and Yang [29], with permission from Elsevier.

7.4 TRANSFORMATION-START TEMPERATURE

Time-temperature-transformation diagrams consist essentially of two C-curves (Figure 7.8), one of which describes reconstructive transformations where all of the atoms move in an uncoordinated manner. At temperatures below those at which individual atoms are mobile, the change in crystal structure is achieved by a homogeneous deformation of the austenite. This displacive transformation to Widmanstätten ferrite and bainite is represented by the lower temperature C-curve.

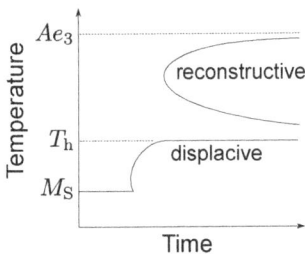

Figure 7.8 Schematic time-temperature-transformation diagram illustrating the two C-curves and the T_h temperature.

The lower C-curve has a characteristic flat top at the temperature T_h corresponding to the highest temperature at which displacive transformations occur at a detectable rate. This temperature must be controlled by nucleation which always is more difficult than growth because of the disproportionate amount of free energy consumed in creating new interfaces when compared with that released by the phase change. It therefore is instructive to examine the magnitude of the free energy change available at the temperature T_h for a variety of steels, assuming implicitly that the nucleation rate at that temperature is identical for all alloys.

Figure 7.9 shows two sets of data, with each point representing a steel with a unique chemical composition [20]. The first plot is a calculation of the maximum free energy change ΔG_m (p. 177) accompanying the paraequilibrium nucleation of ferrite at the temperature T_h. Carbon is therefore allowed to partition at will between the ferrite and austenite. The second set is for the case where nucleation occurs without any composition change. Since the driving force for transformation must always be negative, it can be concluded that the nucleation of Widmanstätten ferrite or bainite cannot occur without the partitioning of carbon.

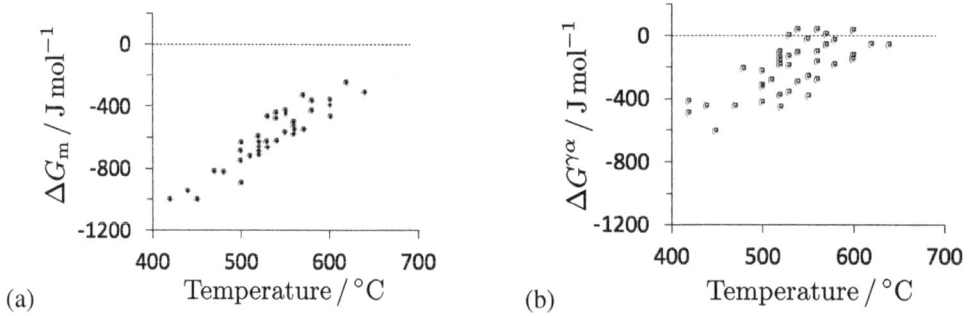

Figure 7.9 The free energy change at the temperature T_h with each point representing a different steel chemical composition. (a) The free energy change ΔG_m calculated assuming paraequilibrium nucleation. (b) The free energy change $\Delta G^{\gamma\alpha}$ calculated assuming that nucleation occurs without any change in chemical composition.

A useful result which follows from the analysis is a method for the estimation of T_h for any steel. The best fit linear relationship describing the data in Figure 7.9a is as follows:

$$G_N = 3.637\,T_h - 2540\,\mathrm{J\,mol^{-1}} \tag{7.1}$$

where T_h is expressed in °C. G_N defines the free energy change necessary to nucleate Widmanstätten ferrite or bainite in any steel. For a particular alloy, the free energy change driving nucleation, ΔG_m, will be a function of temperature T, the slope of which will in general be greater than that of G_N. The intersection of ΔG_m with G_N on a plot of free energy versus temperature then yields T_h for that particular steel. The accuracy of this calculation can be assessed from the scatter illustrated in Figure 7.9a.

G_N is a *universal nucleation function* which is independent of the steel composition [20], whereas the dependence of ΔG_m on temperature is determined by the steel composition. Although it is empirical and limited to low-alloy steels, it nevertheless is possible to interpret the function in terms mechanism, as shown in the next section.

An important outcome of this analysis is that the dependence of W_S or B_S on alloy composition cannot be explained on thermodynamic grounds alone. There is a much greater deviation of the transformation-start temperature from the equilibrium Ae_3 temperature as the transformation is suppressed by alloying to low temperatures, where nucleation becomes more difficult, Figure 7.10.

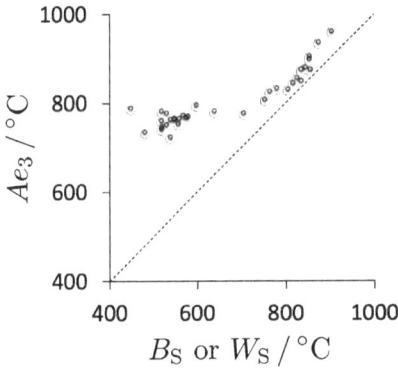

Figure 7.10 Variation of the temperature $T_h \equiv W_S$ or $T_h \equiv B_S$ with the Ae_3 temperature. T_h is the highest temperature at which displacive transformation occurs.

7.5 MECHANISM OF NUCLEATION

A phase that has a tendency to transform does not necessarily achieve that change in a reasonable time because of the cost of creating the interface surrounding the product particle. This cost scales with the surface to volume ratio and therefore is largest at the point of genesis when the particle is small. This presents a barrier to the growth of the embryo, a barrier which in classical nucleation theory must be surmounted by random fluctuations of structure and composition. These chance events are called *heterophase fluctuations*. A few of these fluctuations might be large enough to lead to a reduction in free energy as the particle size increases, in which case they may progress along that path. The barrier to the sustained growth of the fluctuation is the activation energy (Section 4.4). However, the detailed characteristics of the fluctuations responsible for the nucleation of Widmanstätten ferrite remain to be determined although there are clues in the linearity of the plot of G_N versus T_h shown in Figure 7.9a. The nucleation rate per unit volume will depend on an activation energy G^*:

$$I_V \propto v \exp\left\{ -\frac{G^*}{kT} \right\} \tag{7.2a}$$

where v is an attempt frequency and all the other terms have their usual meanings. It follows that

$$-G^* \propto \beta T \qquad \text{where} \qquad \beta = k \ln\{I_V/v\}. \tag{7.2b}$$

Bearing in mind that the temperature T_h corresponds to the highest temperature on a time-temperature-transformation (TTT) diagram where displacive transformation occurs, it may be assumed that the nucleation rate has a fixed value for any steel at its T_h temperature. The term β therefore becomes a constant if T is replaced by T_h giving

$$-G^* \propto T_h. \tag{7.2c}$$

The activation energy is, as expected, a function of the available driving force which at T_h has the value G_N. This leads to the important result that the relationship between G_N and T_h can only be linear if $G^* \propto G_N$.

Figure 4.8 in Section 4.4 shows that for classical theory, an activation barrier G^* to nucleation by heterophase fluctuations. It was deduced there that

$$G^* \propto \frac{\sigma_{\alpha\gamma}^3}{(\Delta G_{chem} + \Delta G_{strain})^2}.$$

Classical nucleation theory based on heterophase fluctuations is therefore unable to explain the linear dependence illustrated in Figure 7.9a. A consideration of alternative shapes or the inclusion of strain energy in the calculation of G^* does not change its inverse dependence on ΔG_m although the exponent can then reach (-4).

An alternative possibility is that nucleation occurs by the dissociation of existing arrays of glissile dislocations rather than by heterophase fluctuations (e.g., Olson and Cohen [30]). The activation energy for nucleation is then that for the motion of the dislocations. This activation energy and its dependence on driving force can be investigated by resorting to standard deformation theory.[3]

Dislocations in any crystal occupy equilibrium positions. The successive equilibrium positions are separated by energy barriers (G_0^*), the effect of which is to introduce a friction stress the magnitude of which diminishes with increasing temperature as thermal fluctuations assist the dislocation to overcome the barriers. However, the stress required for deformation does not decrease to zero because of an athermal friction stress due to very long range interactions (Figure 7.11). The application of an external shear stress τ_A decreases the magnitude of the activation energy to a new value

$$G^* \simeq G_0^* - (\tau_A - \tau_\mu)V^* \tag{7.3a}$$

where V^* is the activation volume.

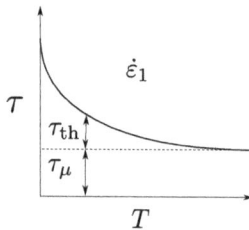

Figure 7.11 Temperature dependence of the stress needed to move a dislocation at a specified strain rate $\dot{\varepsilon}_1$. The stress at any temperature has two components, thermal (τ_{th}) and athermal (τ_μ).

The faults created by the dissociation of the array of dislocations have an energy per unit area, which by analogy with Equation 5.17, is given by:

$$\sigma_f = n\rho_A(\Delta G_m + G^e) + 2\sigma_{\gamma\alpha} \tag{7.3b}$$

where ρ_A is the moles per unit area of fault plane, G^e is the strain energy per mole and ΔG_m, the chemical component of the molar free energy change. It is possible for the fault energy to be negative as ΔG_m becomes negative for metastable austenite. The free energy change then effectively becomes the equivalent of the applied shear stress τ_A on the dislocation array, favouring their dissociation with

$$\tau_A = -\frac{\sigma_f}{nb}. \tag{7.3c}$$

Substituting Equations 7.3b and 7.3c into 7.3a gives

$$\underbrace{G^*}_{\text{activation energy}} = G_O^* + V^* \left[\tau_\mu + \underbrace{\frac{\rho_A}{b} \Delta G_m + \frac{\rho_A}{b} G^e + \frac{2\sigma_{\gamma\alpha}}{nb}}_{\text{driving force term}} \right]. \tag{7.3d}$$

This demonstrates that unlike classical nucleation theory, the activation energy is here proportional to the driving force. Since ΔG_m is negative below the equilibrium transformation temperature, G^* decreases as the driving force increases. This direct proportionality was recognised by Magee [31] in his analysis of the Koistinen and Marburger's equation for the martensite-start temperature.

It appears therefore that Widmanstätten ferrite nucleates by a mechanism consistent with the dissociation of glissile arrays of dislocations (the so-called pre-existing embryos). This mechanism is displacive in character.

7.6   RATIONALISATION OF SHEAR TRANSFORMATIONS

The analysis of the driving force available at T_h, Figure 7.9a, does not distinguish between bainite and Widmanstätten ferrite, and yet the data all fall on the same trend line. This means that the two transformations have the same nucleation mechanism involving the partitioning of carbon, i.e., paraequilibrium nucleation. It is appropriate then to identify T_h with either the Widmanstätten ferrite-start temperature (W_S) or the bainite-start (B_S) temperature, depending on the free energy available to sustain the growth of each structure.

Widmanstätten ferrite and bainite have the same nucleus; if the available driving force can sustain the diffusionless growth of this nucleus, then $T_h = B_S$; otherwise $T_h = W_S$. Nucleation will occur at a detectable rate when the temperature is T_h is reached because the driving force available for nucleation then becomes less than G_N:

$$\Delta G_m \leq G_N. \tag{7.4}$$

The nucleated phase can develop into Widmanstätten ferrite if a further condition is satisfied, that the driving force for paraequilibrium growth exceeds the stored energy of Widmanstätten ferrite, which amounts to about $50\,\text{J}\,\text{mol}^{-1}$ [20, 32]. If, on the other hand, at the temperature T_h, the driving force for partitionless growth exceeds the stored energy for bainite, then Widmanstätten ferrite is not favoured and $T_h = B_S$ (Figure 7.12). These conditions can be expressed in terms of the nucleation function G_N and the stored energy terms for Widmanstätten ferrite ($G_{SW} \approx 50\,\text{J}\,\text{mol}^{-1}$) and bainite ($G_{SB} \approx 400\,\text{J}\,\text{mol}^{-1}$) as follows:

$$\left. \begin{array}{l} \Delta G_m \qquad\quad < G_N \\ \Delta G^{\gamma \rightarrow \alpha + \gamma'} \;\; < G_{SW} \end{array} \right\} \qquad \text{defines onset of } \alpha_W \tag{7.5}$$

$$
\left.\begin{array}{ll}
\Delta G_{\mathrm{m}} & < G_{\mathrm{N}} \\
\Delta G^{\gamma\alpha} & < G_{\mathrm{SB}}
\end{array}\right\} \qquad \text{defines onset of } \alpha_{\mathrm{b}} \qquad (7.6)
$$

$$
\Delta G^{\gamma\alpha} \qquad < G_{\mathrm{N}}^{\alpha'} \qquad \text{defines onset of } \alpha' \qquad (7.7)
$$

$\Delta G^{\gamma\to\alpha+\gamma'}$ is the free energy change accompanying the paraequilibrium growth of α_{W}.[4] The stored energies have their origins primarily in the strain energies due to the shape deformation. The requirement for martensite, $G_{\mathrm{N}}^{\alpha'}$, has been described in Section 5.5 and could additionally include the contributions due to internal twinning. It is assumed here that for common steels, $G_{\mathrm{N}}^{\alpha'} \approx 1100 \, \mathrm{J\,mol^{-1}}$, but as described by Equation 5.8, it depends also on the strength of the austenite, a factor that becomes important when the solute concentrations are large.

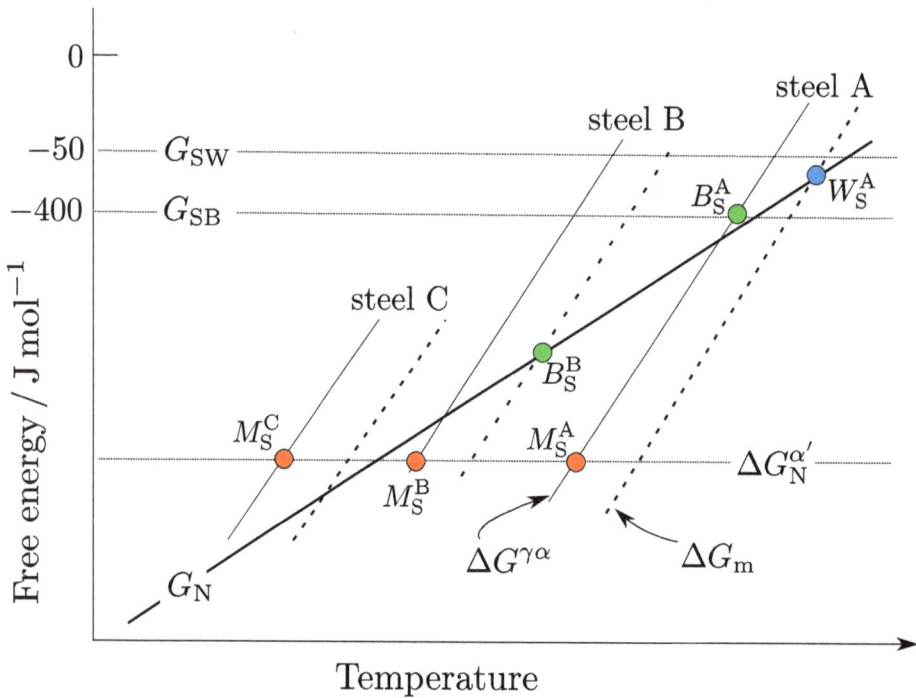

Figure 7.12 A pair of free energy curves (ΔG_{m} and $\Delta G^{\gamma\alpha}$) for each of a low (A), medium (B) and high (C) alloy steel. The intersections of these curves with the functions defining nucleation and growth yield the transformation-start temperatures Widmanstätten ferrite ●, bainite ● and martensite ●. They reveal that not all transformations are possible in all steels.

Figure 7.12 is particularly interesting in that it predicts that not all of the transformations considered can occur in all steels. In alloy A which at any temperature has a large driving force because of its low solute content, α_{W}, α_{b} and α' can all form at successively lower transformation temperatures. This is because Widmanstätten ferrite is able to nucleate at a temperature where its growth can also be sustained since the driving force at that temperature can account for the stored energy G_{SW}. Given that Widmanstätten ferrite is able to nucleate, is follows that so can bainite, making it possible to stimulate bainite when the driving force for diffusionless transformation exceeds the stored energy term G_{SB}. Martensite is triggered eventually at a sufficient undercooling to account for its diffusionless nucleation and growth at M_S^A.

In contrast, when nucleation becomes possible for alloy B, there is sufficient driving force for its diffusionless growth into bainite. It is predicted therefore that Widmanstätten ferrite cannot form in that alloy, but further undercooling triggers the martensitic transformation.

In alloy C, both Widmanstätten ferrite and bainite are suppressed because at the nucleation condition for these phases is only satisfied in the domain where martensitic transformation becomes possible. This steel can therefore only be martensitic.

These concepts have been applied to study whether it is possible to obtain bainite at about $200\,^\circ C$ in low-carbon steel by alloying with substitutional solutes. Both calculations and experiments prove that the B_S and M_S temperatures merge as the substitutional solute content is increased, making it impossible to obtain bainite in the system studied at temperatures less than $300\,^\circ C$, Figure 7.13.

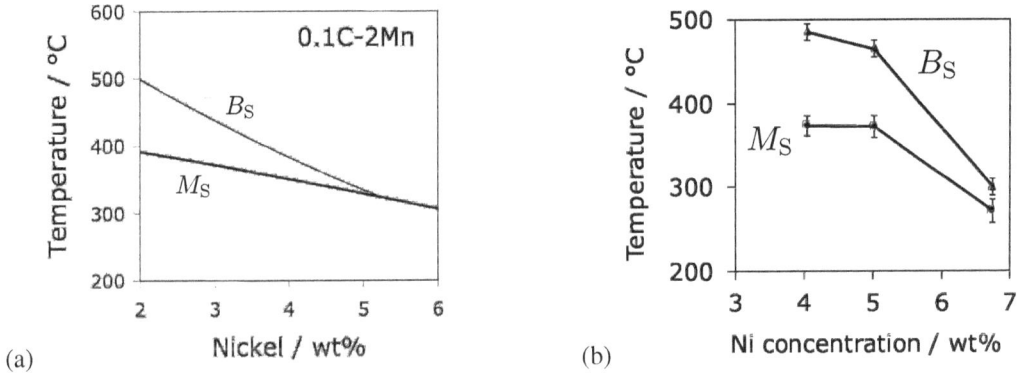

Figure 7.13   (a) Calculations using the method illustrated in Figure 7.12 illustrating how the bainite-start and martensite-start temperatures merge as the substitutional solute content is increased [33]. (b) Experimental data illustrating the same effect [34]. Note that because these are experimental alloys, the concentrations of carbon and manganese are not entirely constant across the range of nickel concentrations. The three alloys have the compositions Fe-0.13C-2.28Mn-4.03Ni, Fe-0.13C-2.27Mn-5.02Ni and Fe-0.20C-2.5Mn-6.74Ni wt%.

7.7   NUCLEATION RATE

The nucleation rate can be expressed as a function of the activation energy using the Arrhenius type equation which is characteristic of a thermally activated process:

$$I = b_8 \exp\left\{-\frac{G^*}{RT}\right\} = b_8 \exp\left\{-\frac{b_9 + b_{10}\Delta G_m}{RT}\right\} \tag{7.8a}$$

where the b_i are constants and the equation recognises the linear dependence of the activation energy on the driving force according to Equation 7.3d. Noting the earlier assumption that the nucleation rate at T_h is identical for all steels, the equation can be rewritten in terms of the undercooling below

T_h:

$$I = I_{T_h} \exp\left\{ -\frac{b_9 \Delta T}{RTT_h} - \frac{b_{10}}{R}\left(\frac{\Delta G_m}{T} - \frac{G_N}{T_h}\right) \right\} \tag{7.8b}$$

where I_{T_h}, b_9 and b_{10} are constants to be determined by fitting to experimental data. $\Delta T = T_h - T$ is the undercooling below the T_h temperature.

Consider now two steels designated A and B; using Equation 7.8a the ratio of their Widmanstätten ferrite nucleation rates at their respective start temperatures is given by

$$\frac{I_A}{I_B} = \exp\left\{ -\frac{(b_9 - b_{10}b_7)(W_{SB} - W_{SA})}{RW_{SB}W_{SA}} \right\}. \tag{7.9}$$

Since both steels are at their W_S temperatures, the ratio must be unity in which case $b_9 = b_7 \times b_{10}$. There are therefore only two unknowns in Equation 7.9 which can be reduced to

$$I = b_{11} \exp\left\{ -\frac{b_9}{RT} - \frac{b_9 \Delta G_m}{b_7 RT} \right\}. \tag{7.10}$$

When I represents the nucleation rate per unit area of austenite grain surface, the constants obtained by fitting to experimental data are $b_{11} \approx 7.38 \times 10^8 \, \text{s}^{-1} \, \text{m}^{-2}$, $b_9 \approx 2.065 \times 10^4 \, \text{J} \, \text{mol}^{-1}$ and $b_7 \approx 2540 \, \text{J} \, \text{mol}^{-1}$ [35].

7.8 CAPILLARITY

The growing tip of a Widmanstätten ferrite plate is curved. The addition of new atoms will therefore lead to an increase in interfacial area that has to be provided for from the available free energy. This is equivalent to a relative change in the positions of the free energy curves as illustrated in Figure 7.14, where $\sigma \, dO/dn$ is the additional energy due to the new α/γ surface created as an atom is added to the α particle. The equilibrium between the ferrite and austenite therefore changes with the new phase compositions identified by the subscript r for curved interfaces. This is known as the Gibbs-Thomson capillarity effect [23].

From the approximately similar triangles (ABC and DEF),

$$\frac{\mu_{C,r}^\gamma - \mu_C^\gamma}{\sigma(dO/dn)} = \frac{1 - x^{\alpha\gamma}}{x^{\gamma\alpha} - x^{\alpha\gamma}} \tag{7.11a}$$

where x is the mole fraction of carbon. The chemical potential of carbon in austenite is $\mu_C = \mu^\circ + RT \ln\{\Gamma_C^\gamma x\}$ where $\Gamma_C^\gamma\{x\}$ is the activity coefficient of carbon in austenite containing a concentration x of carbon. It follows that

$$\mu_{C,r}^\gamma - \mu_C^\gamma = RT \ln\left\{ \frac{\Gamma_C^\gamma\{x_r^{\gamma\alpha}\}x_r^{\gamma\alpha}}{\Gamma_C^\gamma\{x^{\gamma\alpha}\}x^{\gamma\alpha}} \right\} \tag{7.11b}$$

and

$$\frac{\Gamma_C^\gamma\{x_r^{\gamma\alpha}\}}{\Gamma_C^\gamma\{x^{\gamma\alpha}\}} = \left[\Gamma_C^\gamma\{x^{\gamma\alpha}\} + (x_r^{\gamma\alpha} - x^{\gamma\alpha})\frac{d\Gamma_C^\gamma}{dx} \right] \Big/ \Gamma_C^\gamma\{x^{\gamma\alpha}\}$$

$$= 1 + (x_r^{\gamma\alpha} - x^{\gamma\alpha})\frac{d\ln\{\Gamma_C\{x^{\gamma\alpha}\}\}}{dx}$$

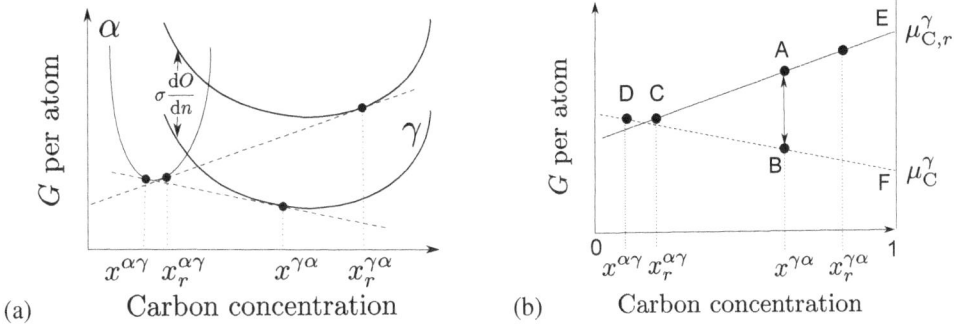

Figure 7.14 (a) An illustration of the Gibbs-Thomson effect. $x^{\alpha\gamma}$ and $x^{\gamma\alpha}$ are the equilibrium compositions of the ferrite and austenite respectively when the two phases are connected by a planar interface. The subscript r identifies the corresponding equilibrium compositions when the interface is curved. (b) A magnified view. μ_C^{γ} is the chemical potential of carbon in the austenite which is in equilibrium with ferrite at a flat interface. $\mu_{C,r}^{\gamma}$ is the corresponding chemical potential when the interface has a radius of curvature r. The distance $AB \simeq \sigma(dO/dn)$.

For a finite plate tip radius,

$$x_r^{\gamma\alpha} = x^{\gamma\alpha}[1 + (\Psi/r)] \tag{7.11c}$$

where Ψ is called the capillarity constant [23]. It follows from these relations that

$$\Psi = \frac{\sigma V_a^{\alpha}}{kT} \frac{1 - x^{\gamma\alpha}}{x^{\alpha\gamma} - x^{\gamma\alpha}} \left[1 + \frac{d(\ln \Gamma_C\{x^{\gamma\alpha}\})}{d(\ln x^{\gamma\alpha})} \right]^{-1}. \tag{7.12}$$

This assumes that the α composition is unaffected by capillarity, a reasonable approximation given that $x^{\alpha\gamma}$ always is small.

7.9 GROWTH OF WIDMANSTÄTTEN FERRITE

The partitioning of carbon is a thermodynamic necessity when Widmanstätten ferrite forms in steels, even though the ratio of the iron to substitutional solute atoms remains constant throughout transformation. The displacive mechanism implies a glissile interface so the rate controlling factor during growth is expected and found to be the diffusion of the partitioned carbon in the austenite ahead of the interface [32, 36–39]. Particle dimensions during diffusion-controlled growth vary parabolically with time when the extent of the diffusion field in the matrix increases with particle size. The growth rate then decreases with time. The diffusion-controlled lengthening of plates or needles can, however, occur at a constant rate when solute can be partitioned to the sides of the plates or needles. Thickening on the other hand, does follow parabolic kinetics due to the sideways accumulation of solute.

A simple model for the growth of plates can be derived by assuming that the carbon concentration gradient in the austenite ahead of the plate tip is constant and that the extent of the diffusion field (z_d, Figure 7.15) is equal approximately to the radius at the plate tip. Clearly, more carbon is partitioned into the austenite as the interface advances; if the carbon concentration $x^{\gamma\alpha}$ is to remain unchanged, then the partitioned carbon has to be carried away by diffusion at a rate consistent with that of partitioning. This condition at the moving interface can be expressed as follows:

$$(x^{\gamma\alpha} - x^{\gamma\alpha})v_\ell = \overline{D}_C^\gamma(\bar{x} - x^{\alpha\gamma})/z_d \tag{7.13}$$

where v_ℓ is the rate at which the plate lengthens. The diffusion coefficient $\overline{D}_C^\gamma$ is the integrated average diffusion coefficient of carbon in austenite, the integration being over the range of carbon concentration present in the austenite, $\bar{x}$ to $x_r^{\gamma\alpha}$. On substituting the plate tip radius r for z_d, it follows that $v_\ell = f_1\overline{D}_C^\gamma/r$, where f_1 incorporates the concentration terms in Equation 7.13; the calculated velocity then increases without limit as the tip radius decreases. However, this neglects the capillarity effect due to the curvature in the growing interface.

Zener's model assumes a hemispherical tip to the plate; for a spherical particle $dO/dn = 2V_a^\alpha/r$ where V_a^α is the volume per atom of ferrite. The driving force consumed by capillarity will be proportional to r^{-1}. If the tip radius becomes a critical value r_c, then all of the available driving force is dissipated in creating new interface so the growth velocity becomes zero. It follows that the ratio of the driving force consumed in the capillarity correction to the total available is simply r_c/r:

$$\therefore v_\ell = \frac{f_1\overline{D}_C^\gamma}{r}\left(1 - \frac{r_c}{r}\right). \tag{7.14}$$

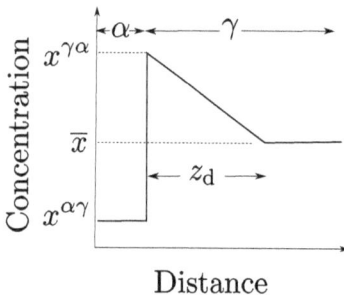

Figure 7.15 The gradient of concentration ahead of the interface is constant in this Zener approximation, so it becomes zero beyond the diffusion distance z_d.

Equation 7.14 gives the velocity as a function of the plate tip radius; the function is illustrated in Figure 7.16. In the absence of capillarity, the velocity increases indefinitely as the plate tip radius becomes finer since the diffusion distance becomes correspondingly smaller. If capillarity is taken into account then the velocity goes through a maximum at $r = 2r_c$. This simple analysis of plate growth is due to Zener and is based on a plate of constant thickness with a hemicylindrical edge of radius r.

One difficulty with the Zener analysis is that the velocity is found to be proportional to the supersaturation f_1 whereas for diffusion-controlled growth it should tend to infinity as $f_1 \to 1$ because solute partitioning becomes unnecessary in the limit that $x^{\alpha\gamma} = \bar{x}$. Hillert [40] gave an improved solution by assuming that the concentration in the plane tangent to the plate tip decays exponentially with distance normal to the habit plane of the plate. The resulting "Zener-Hillert" equation [41] has the characteristics expected physically, that at low supersaturations the velocity is proportional to the concentration term but tends to infinity as the supersaturation tends to unity:

$$f_1\left(1 - \frac{r_c}{r}\right) = 4p(1+4p)^{-1} \tag{7.15}$$

with　　$f_1(1 - \frac{r_c}{r}) = p$　as　$f_1(1 - \frac{r_c}{r}) \rightarrow 0$

and　　$f_1(1 - \frac{r_c}{r}) = 1 - \frac{1}{4p}$　as　$f_1(1 - \frac{r_c}{r}) \rightarrow 1$

where the Péclet number p, which is a dimensionless velocity, is given by $p = v_\ell r / 2\overline{D}_{11}$. With this equation the Péclet number tends to infinity as the effective supersaturation f_1 tends to unity.

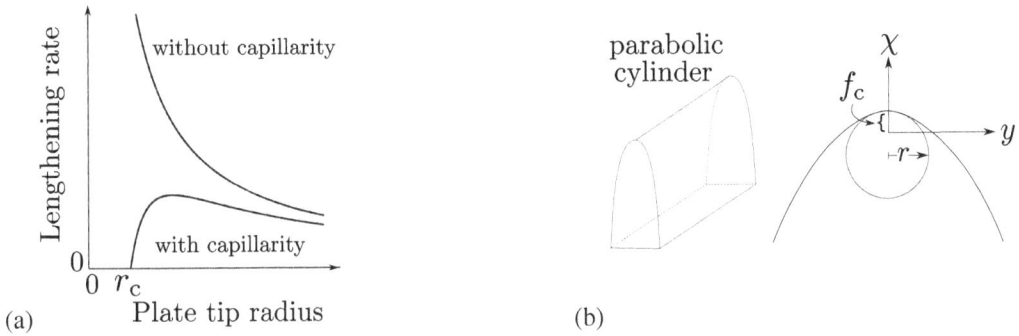

Figure 7.16 (a) The plate lengthening rate as a function of the plate-tip radius. (b) An illustration of the shape of a parabolic cylinder, definitions of the tip radius r, the focal distance f_c and the coordinates.

In fact, Ivantsov [42] gave an exact solution for diffusion-controlled growth of an interface which advances at a uniform rate in the shape of a parabolic cylinder (Figure 7.16), i.e. for the lengthening rate of a plate:

$$f_1 = (\pi p)^{0.5} \exp\{p\} \text{erfc}\{p^{0.5}\} \tag{7.16}$$

but this neglects capillarity effects.

Trivedi proposed a solution which does allow for capillarity and is able to deal with a variation in the composition along the interface of the plate [43]. The assumed shape is again that of a parabolic cylinder. As with Ivantsov's method, the solution is shape preserving so the plate thickens and lengthens at the same time. The plate lengthening rate at a temperature T for steady-state growth is obtained by solving the equation:

$$f_1 = (\pi p)^{0.5} \exp\{p\} \text{erfc}\{p^{0.5}\} \left[1 + \frac{v_\ell}{v_\text{I}} f_1 S_1\{p\} + \frac{r_c}{r} f_1 S_2\{p\}\right]. \tag{7.17}$$

The function $S_2\{p\}$ depends on the Péclet number (Figure 7.17); it corrects for variation in composition due to changing curvature along the interface and has been evaluated numerically by Trivedi. The term containing S_1 is prominent when growth is not diffusion-controlled; v_I is the interface-controlled growth velocity of a flat interface. For diffusion-controlled growth, which is discussed first, v_I is large when compared with v_ℓ and the term containing it can be neglected. Trivedi's solution for diffusion-controlled growth assumes a constant shape, in this case a parabolic cylinder,

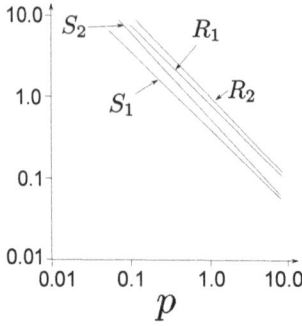

Figure 7.17 Dependence of Trivedi's functions S_1, S_2, R_1, and R_2 on the Péclet number p.

but because $x_r^{\gamma\alpha}$ varies over the surface of the parabolic cylinder the solution is not quite shape-preserving. Trivedi claims that the variation in $x_r^{\gamma\alpha}$ has a negligible effect provided the tip radius is greater than $3r_c$.

Equation 7.17 on its own relates velocity and tip radius but does not allow these quantities to be fixed; additional theory is needed to enable the choice of a particular r, and hence fix v_ℓ. Small tip radii favour fast growth due to the point effect of diffusion, but this is opposed by the capillarity effect since the driving force for growth tends to zero as $r \to r_c$. Zener proposed that the plate should tend to adopt a tip radius that allows v_ℓ to be maximised but there is no fundamental justification nor experimental evidence to support this hypothesis. Work on the dendritic growth of solid from liquid, formally an almost identical problem, has demonstrated conclusively that the dendrites do not attain a radius consistent with the maximum velocity hypothesis, the actual tip radius being determined by a shape stability criterion [44, 45]. If these results can be extrapolated to solid-state transformations, then any calculated velocities would be less than those given by the Zener hypothesis.

In the absence of reliable data on plate tip radii, the Zener hypothesis provides an upper limit for v_ℓ. A comparison with experimental data shows that the actual α_w lengthening rates in Fe-C alloys always *exceed* the Zener maximum values [32]. This discrepancy can only increase if the maximum velocity hypothesis is not valid, or if any appreciable part of the free energy is dissipated in interface processes. It is possible that the shape of Widmanstätten ferrite may deviate from that of a parabolic cylinder. It therefore is useful to compare experimental data with needle lengthening rates, even though a needle shape is a poor representation of Widmanstätten ferrite with its definite habit plane. Unlike the Zener-Hillert models, the ratio r/r_c does not in the Trivedi model have a fixed value but is a function of the supersaturation (Figure 7.18). However, there is an interesting consequence if the maximum velocity criterion is applied. Figure 7.18 shows that the ratio r/r_c is a function g of f_1 so that $r/r_c = g\{f_1\}$. From Equation 7.12, it is seen that

$$r_c = \Psi x^{\gamma\alpha}/(\bar{x}-x^{\gamma\alpha}) \qquad \text{so that} \qquad r = \Psi x^{\gamma\alpha} g\{f_1\}(\bar{x}-x^{\gamma\alpha}). \tag{7.18}$$

Substitution of r in Equation 7.11c then shows that $x_r^{\gamma\alpha}$ does not depend on Ψ, and hence is not a function of the interfacial energy σ when the maximum velocity criterion is used. Nevertheless, σ affects the velocity v_ℓ because

$$v_\ell = 2\overline{D}_C^\gamma p/r = 2\overline{D}_C^\gamma p(\bar{x}-x^{\gamma\alpha})/\Psi\bar{x}^{\gamma\alpha}g\{f_1\}. \tag{7.19}$$

Since p depends only on f_1 and r/r_c, it follows (within the maximum velocity hypothesis) that for a constant supersaturation, the lengthening rate is inversely proportional to σ. Doubling σ halves v_ℓ, and so the effect of interface energy on any lengthening rate calculated on the basis of the Zener hypothesis is easily deduced.

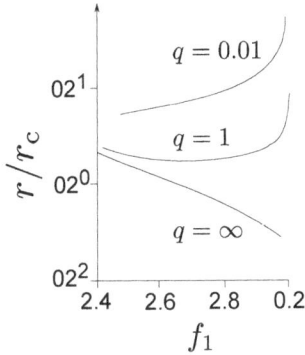

Figure 7.18 Dependence of the ratio r/r_c with the normalised supersaturation f_1, as a function of the parameter q, for the case where the plates lengthen at the maximum rate. Diffusion-controlled growth is the case where $q = \infty$ meaning that the interface is very mobile (Equation 7.21).

Trivedi also has obtained a steady-state solution for the diffusion-controlled growth of paraboloids of revolution (i.e. needles):

$$f_1 = p\exp\{p\}Ei\{p\}[1 + \frac{v_\ell}{v_I}f_1R_1\{p\} + \frac{r_c}{r}f_1R_2\{p\}] \tag{7.20}$$

where the function $R_2\{p\}$ corrects for the variation in composition due to the changing curvature of the interface and has been evaluated numerically by Trivedi (Figure 7.17). The critical radius r_c is found to be twice as large as that for plates. A comparison of experimental data with maximum calculated needle lengthening rates shows that the plate model is a somewhat better representation of Widmanstätten ferrite lengthening (Figure 7.19).

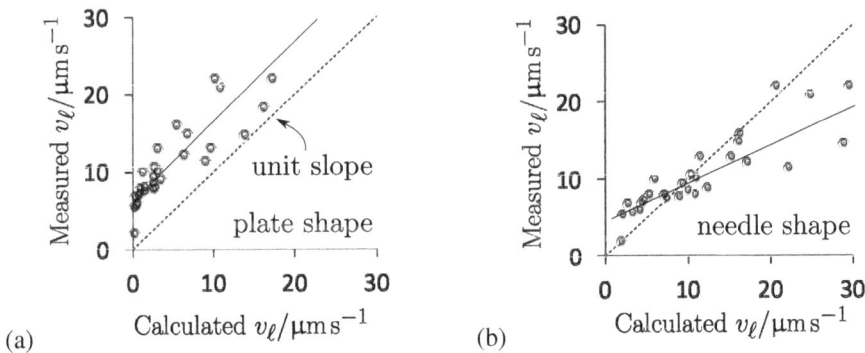

(a)   (b)

Figure 7.19 Plots of the experimental versus calculated lengthening rates for Widmanstätten ferrite, assuming (a) a plate shape and (b) needle shape. The calculations are based on the maximum velocity hypothesis and the α_w/γ interface energy is taken to be 0.2 J m^{-2}. The details of the calculations are given by Bhadeshia [32] but the experimental data are due to [36–38].

7.9.1 INTERFACE-CONTROLLED GROWTH

Trivedi [43] has also given a more elaborate theory to take account of any free energy dissipated in interface processes. In this, a second set of functions ($S_1\{p\}$ for plates and $R_1\{p\}$ for needles, Figure 7.17) is introduced to allow for the variation in dissipation due to the changing orientation of the interface. The velocity v_I describing the case where growth is interface controlled is, for a flat interface, written

$$v_I = b_{12}(x^{\gamma\alpha} - \bar{x})$$

where b_{12} is described as an interface kinetics coefficient. When growth is not diffusion-controlled, the concentration of carbon at the interface will be depressed when compared with the case where it is diffusion-controlled. The ratio of this depression to that due to capillarity at the interface is given by p/q, where

$$q = \frac{b_{12}(x^{\gamma\alpha} - \bar{x})}{2\bar{D}/r_c}.$$

It is likely, because of the glissile nature of the transformation interface, that the growth of Widmanstätten ferrite is in the vast majority of cases controlled by the diffusion of carbon in the austenite ahead of the interface. The exception to this would be when α_W grows in interstitial-free iron alloys or in steels containing very little carbon. The diffusion-controlled lengthening rate becomes indefinitely large as $\bar{x} \to x^{\alpha\gamma}$ so interfacial processes may then become rate-controlling. In typical welding alloys used for steels, v_ℓ can exceed $500\ \mu m\,s^{-1}$ during continuous cooling transformation [46].

7.9.2 GROUPS OF PLATES

Several parallel plates of Widmanstätten ferrite often grow together, forming *packets* that originate at an austenite grain surface. It is possible that in such cases the diffusion fields of adjacent plates interfere and affect the growth kinetics described above. Trivedi and Pound [41] used the Ivantsov solution to obtain an estimate of the effective diffusion distance z_{dn} at the tip of a plate in a direction normal to the lengthening direction. Soft impingement is expected if the separation z_s of adjacent plates is less than $2z_{dn}$. The concentration field in the matrix is given by

$$c^\gamma - \bar{c} = (c^{\gamma\alpha} - \bar{c})[\text{erfc}\{p^{0.5}\chi_n\}/\text{erfc}\{p^{0.5}\}]$$

$$\text{where} \quad \chi_n = [(\chi^2 + y^2)^{1/2} + \chi]^{1/2}$$

is a normalised parabolic coordinate and χ and y are dimensionless moving Cartesian coordinates illustrated in Figure 7.16,[5] with origin at the focus of the parabolic interface, a distance f_c below the plate tip. The moving coordinates are related to the fixed cartesian coordinates X and Y by

$$\chi = (X - v_\ell t)/r \quad \text{and} \quad y = Y/r.$$

Assuming that soft impingement becomes important when $c^\gamma - \bar{c} = 0.1(c^{\gamma\alpha} - \bar{c})$, the effective diffusion distance is obtained by solving

$$\mathrm{erfc}\left\{p^{\frac{1}{2}}\left[\left(\frac{1}{4}+\left(\frac{z_{dn}}{r}\right)^2\right)^{\frac{1}{2}}+\frac{1}{2}\right]^{\frac{1}{2}}\right\}=0.1\,\mathrm{erfc}\{p^{\frac{1}{2}}\}$$

bearing in mind that the tip of the plate is located a distance $f_c = \frac{1}{2}r$ from the focus of the parabola. Figure 7.20 shows the variation in the effective diffusion distance as a function of the Péclet number. The diffusion distance at small and typical values of p is large compared with the observed plate half-spacings, meaning that the effects of soft-impingement cannot be neglected. The observed growth rate should therefore be smaller than predicted for the growth of an isolated plate in an infinite medium, but this contradicts experimental data (Figure 7.19) which show the measured rates to be somewhat greater than predicted. One possibility is that the shape of Widmanstätten ferrite is not strictly that of a parabolic cylinder and that its shape may not be preserved as the plate lengthens. Metallographic evidence shows that plates in colonies tend to advance as cusps which are quite unlike parabolic cylinders. Thus, their major radii of curvature lie in the adjacent austenite (e.g., Figure 27, Townsend and Kirkaldy, [37]). There is a tendency also to adapt to soft-impingement by changing shape. The effect of soft-impingement could be mitigated by favourable strain interactions between adjacent plates.

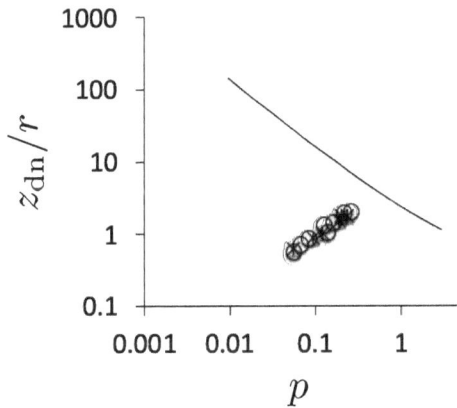

Figure 7.20　The curve is a plot of the diffusion distance z_{dn} between adjacent plates, normalised with the plate tip radius r, against the Péclet number. Soft-impingement effects can be neglected when adjacent parallel plates are spaced a distance greater than $2z_{dn}$. The data are from Townsend and Kirkaldy [37] who measured the Widmanstätten ferrite plate spacings for Fe-0.385C wt% and Fe-0.405C wt% steels. Their measurements have been divided by two in order to compare against z_{dn}/r.

It is surprising at first sight that the calculation of effective diffusion distance is independent of the plate length. This is because only the region at the plate tip is considered. Since the solution for the growth of a parabolic cylinder is shape-preserving, the plate is expected to thicken at its origin; indeed, plates are observed experimentally to thicken. It is likely therefore that its maximum thickness will eventually exceed the estimated value of z_{dn} so it is steric hindrance that prevents the formation of an adjacent plate rather than the extent of the diffusion field.

7.10 PLATE THICKENING

It has been emphasised that the Ivantsov and Trivedi solutions are shape-preserving; this is illus-trated in Figure 7.21, with a series of identical parabolae which show the sequence of growth. The parabolae are equidistant along X to illustrate a constant rate of lengthening. An observer located at the origin would also notice that the plate thickens along Y but at an ever decreasing rate. It follows necessarily from the parabolic shape that the thickness must increase with $t^{\frac{1}{2}}$. The position of the interface with respect to the moving coordinates χ, y (Figure 7.16) is given by setting $\chi_n = 1$:

$$\chi_n = [(\chi^2 + y^2)^{\frac{1}{2}} + \chi]^{\frac{1}{2}} = 1.$$

Making the substitutions $X = 0$ and $Y = Y_t$ gives

$$\left\{ \left[\frac{v_\ell^2 t^2}{r^2} + \frac{Y_t^2}{r^2} \right]^{\frac{1}{2}} - \frac{v_\ell t}{r} \right\}^{\frac{1}{2}} = 1, \qquad \therefore \quad Y_t = (r^2 + 2v_\ell t r)^{\frac{1}{2}}$$

where Y_t is the plate half-thickness along Y at $X = 0$. The plate thickness will therefore vary parabol-ically with time at a rate that is smaller than v_ℓ.

Figure 7.21 The translation of the same parabola along X in the sequence 1-6 illustrates the nature of the shape-preserving solution. The distance be-tween adjacent parabolae is identi-cal along X to represent the constant growth rate in that direction.

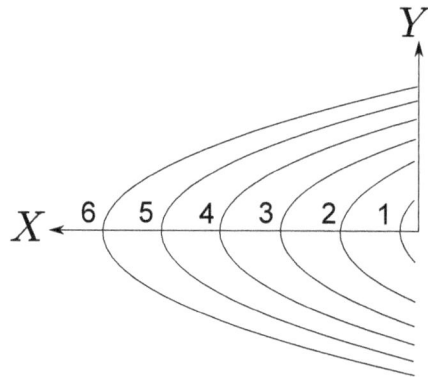

Figure 7.22a shows how the plate aspect ratio $a = Y_t/v_\ell t$ decreases as the plate lengthens. The for-mulation of the theory requires that the plate exists at time zero, with length $X = 0.5r$ and $Y_t = r$ because the focus is located at $X = 0$; the initial aspect ratio is therefore 2, which is an artefact of the model, though the decrease in aspect ratio as the plate lengthens is observed experimentally. The Widmanstätten ferrite plates start off rather stubby and become slender as they grow (Figure 7.1). Consistent with many observations, the theory also predicts smaller aspect ratios at lower transfor-mation temperatures.

The rate at which the plates thicken should compare against that at which allotriomorphs of fer-rite thicken during diffusion-controlled growth. The thickness Z of the allotriomorph is given by (Chapter 4)

$$Z = \alpha_{1d} t^{0.5} \tag{7.21}$$

where α_{1d} is the parabolic-thickening rate constant for one-dimensional growth; using this relation,

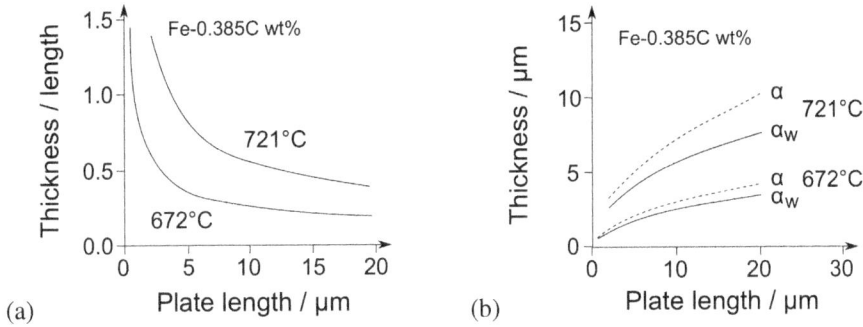

Figure 7.22 (a) The calculated aspect ratios of Widmanstätten ferrite plates as a function of the plate length and transformation temperature. (b) A comparison of the parabolic thickening of Widmanstätten ferrite versus that of allotriomorphic ferrite under identical conditions. The growth of Widmanstätten ferrite is modelled as in [32], but for comparison with allotriomorphic ferrite, $50\,\text{J}\,\text{mol}^{-1}$ of stored energy associated with Widmanstätten ferrite is neglected; the thickness of the allotriomorphic ferrite is calculated at the time corresponding to a particular length of the α_W. The neglect of stored energy does not affect the relative positions of any of the curves. The plate lengthening rates are 0.76×10^{-6} and $0.098 \times 10^{-6}\,\text{m}\,\text{s}^{-1}$ for 672 and 721°C respectively.

the ratio of plate to allotriomorph thickening rates is given by

$$\frac{\mathrm{d}Y_t/\mathrm{d}t}{\mathrm{d}Z/\mathrm{d}t} \simeq \frac{\sqrt{2v_\ell r}}{\alpha_{1\mathrm{d}}} \qquad \text{when} \qquad v_\ell t \gg r.$$

This ratio is found to be less than unity (Figure 7.22b) because of the dependence of the plate thickening rate on the tip radius. The parabolic shape is geometrically constrained by the radius at the tip; a finer radius leads to a smaller thickening rate.

The model described above can be compared against measurements [47] that confirm that the thickening rate always decreases with time Figure 7.23. In some cases the plates thickened smoothly (curve B) whereas in others it increased abruptly between sequences of smooth thickening (curves D and E). The discontinuous thickening was attributed to irregularities in the shape of the interface. Consistent with theory, the plate thickness was in many cases smaller than expected for an allotriomorph (curves B–C), but the opposite result was found in other instances (curve E), with very large variations even within the same austenite grain. In some cases (curve C) the plates simply stopped thickening. These detailed variations cannot be explained but might have something to do with the fact that the growth theory neglects the effect of elastic and plastic strains caused by the displacive mechanism of transformation, on the shape of the transformation product.

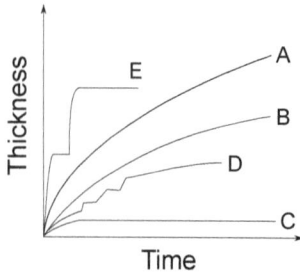

Figure 7.23 Schematic illustration of the different Widmanstätten ferrite plate thickening behaviours reported by Kinsman, Eichen and Aaronson [47]. Curve A represents the one-dimensional carbon diffusion-controlled thickening of allotriomorphic ferrite.

7.11 GROWTH FROM ALLOTRIOMORPHIC FERRITE

Widmanstätten ferrite frequently grows from layers of allotriomorphic ferrite in circumstances where the latter has an orientation with the austenite that is consistent with the displacive mechanism by which α_W plates grow. The allotriomorph orientation must be in the "Bain region", which implies that no plane or direction can be rotated by more than about 11° from the Bain orientation [48]. The actual condition may be more stringent since for Widmanstätten ferrite the close-packed $\{1\,1\,1\}_\gamma$ and $\{0\,1\,1\}_\alpha$ planes are unlikely to deviate by more than 1° from parallelism, and the corresponding close-packed directions $\langle \bar{1}\,1\,0 \rangle_\gamma$ and $\langle 1\,\bar{1}\,1 \rangle_\alpha$ are unlikely to be more than about 6° from parallelism.

In some classic work, C. S. Smith [49] considered that allotriomorphic ferrite is likely to have a good-fit orientation relationship with one of the austenite grains with which it is in contact; after all, a small interfacial energy is conducive to easier nucleation. It is that side of the allotriomorph which might develop into Widmanstätten ferrite (Figure 7.1).

It has been argued that the evolution of Widmanstätten ferrite from allotriomorphic ferrite can be described in terms of an instability of the transformation interface during diffusion-controlled growth [37]. If by chance a small region of the growing α advances ahead of the main interface then the concentration gradient ahead of that perturbation increases, assuming that the ferrite grows with a constant composition. The perturbation will therefore grow faster, making the interface unstable; the instability is opposed by the capillarity because of the curvatures associated with the perturbation. There is, therefore, a critical size of perturbations below which the interface will remain stable. The general theory for the reaction of an interface to perturbations was given first by Mullins and Sekerka [50, 51]. It is introduced here in the form applied by Townsend and Kirkaldy neglecting elastic strain, anisotropy of interfacial energy and assuming local equilibrium all along the interface. Instability is investigated by invoking a sinusoidal perturbation of wavelength λ at the planar interface and determining whether it grows or decays. The perturbation that grows fastest is assumed eventually to become the dominant one, with a wavelength λ given by

$$\lambda = 2\pi\sqrt{3}\left[\Psi D_C^\gamma \frac{c^{\gamma\alpha}}{v(c^{\alpha\gamma} - c_r)} \right]^{1/2} \tag{7.22}$$

where v is the velocity of the mean allotriomorphic ferrite/austenite interface. The dependence on this velocity arises because the size of the critical fluctuation (as opposed to the one that grows fastest) decreases as the concentration gradient at the interface increases. The velocity is of course time dependent:

$$v = \frac{\alpha_{1d}}{2t^{\frac{1}{2}}},$$

and is large at first with λ correspondingly small, so that the interface is unstable as soon as it moves. This is inconsistent with experimental evidence. Townsend and Kirkaldy used half the measured "incubation time" for Widmanstätten ferrite formation to set the velocity. Although they then obtained good agreement between λ and the measured spacing between adjacent plates of Widmanstätten ferrite, the procedure for setting v remains unjustified and makes the concept dependent of prior measurement.

There is an important difference in the mechanism by which Widmanstätten ferrite and allotriomorphic ferrite form. Unlike the α/γ interface, the α_w/γ interface must be glissile. Consequently, even when the allotriomorphic ferrite is suitably oriented, there must be a change in interface structure for Widmanstätten ferrite to form. This "nucleation" barrier may not differ from the dislocation-dissociation mechanism discussed in Section 7.5 in which case the same thermodynamic criteria must apply. In these circumstances, the presence of a solute diffusion profile ahead of the allotriomorphic ferrite will tend to oppose the formation of Widmanstätten ferrite because the partitioned solute would tend to stabilise the austenite. There are observations of Widmanstätten ferrite-depleted zones at allotriomorph boundaries, in circumstances where the former can nucleate both at the allotriomorphic ferrite and intragranularly at non-metallic inclusions [52].

But the basic concept of Widmanstätten ferrite simply being a perturbation of the allotriomorphic form is deeply flawed. The fundamental difference in mechanism and the neglect of the fact that the plate shape is determined by the minimisation of strain energy due to the shape deformation is ignored in the application of perturbation theory. All models based on interface instability rely on carbon diffusion fields, whereas Widmanstätten ferrite also occurs in interstitial-free iron alloys where there are no diffusion fields. These are some of the reasons why the phase field models to date simply do not capture the physics of the problem. The basic theory of the phase field method has been described in Section 4.6.4.

The first treatment of the problem using phase field modelling, generated a two-dimensional plate representation by imposing very strong anisotropy of α_W/γ interfacial energy in a Fe-C alloy, with plates initiated at a thin layer of allotriomorphic ferrite [53].[6] That anisotropy of interfacial energy cannot explain the shape of α_W is obvious given that similarly oriented allotriomorphs or idiomorphs can grow to a substantial size without developing plate shapes. Furthermore, both primary and secondary Widmanstätten ferrite plates grow in interstitial-free iron alloys, Table 7.1. Primary α_W plates are not associated with any diffusion fields in interstitial-free alloys so the question of morphological stability does not arise.[7]

There is a second type of application of phase field models where a plate exists, for example as a half-ellipsoid, from time zero and is then allowed develop into one of much smaller aspect ratio. The shape is controlled by introducing assumed anisotropy, in the form of interfacial energy or strain energy. These models essentially treat the growth of individual plates and can reveal phenomena that are difficult to access experimentally. In a generic treatment of Widmanstätten-like structures, Cottura et al. considered that the levels of interfacial-energy anisotropy used to generate shape in phase field models is unrealistic, so elastic anisotropy may be the root cause of the Widmanstätten morphology [54]. As an example, they considered an invariant-plane strain in which there is only one non-zero eigenstrain (0.03) so the elastic energy is zero when the strain in normal to the habit plane and maximum when it lies in the habit plane. The transformation strain matrix for more complex problem of the $\gamma \to \alpha$ transformation was generated by assuming a Bain strain combined with a rotation of 5.26° about the $[100]_\gamma$ or $[010]_\gamma$ directions to comply with the Kurdjumov-Sachs orientation. However, this is not the correct set of operations that leads to the orientation relationship.[8] An inconsistency is that there are just four minima in the energy density plot; the number of habit plane variants is in fact 24. It is emphasised also that the shape of α_W is determined by the shape deformation, of which there are also 24 variants. Nevertheless, the conclusion that the elastic energy plays the determining role in generating the shape is an advance in the modelling.[9] In some later work, Cottura et al. [55] incorporated viscoelasticity to model the relaxation of strains, which still led to a constant though reduced lengthening rate associated with a larger tip radius; however, this is not the same as plastic relaxation but rather, involves diffusion of all atoms and therefore does not represent plastic relaxation due to the dislocation-driven deformation of austenite. Diffusional relaxation is not expected within the time scale required to generate Widmanstätten ferrite.

REFERENCES

1. N. T. Belaiew: 'On the genesis of the Widmanstätten structure in meteorites and in iron-nickel and iron-carbon alloys', *Mineralogical Magazine and Journal of the Mineralogical Society*, 1924, **20**, 173–185.
2. D. Phelan, and R. Dippenaar: 'Widmanstätten ferrite plate formation in low-carbon steels', *Metallurgical & Materials Transactions A*, 2004, **35**, 3701–3706.
3. H. S. Fong, and S. G. Glover: 'Angle between grain-boundary Widmanstätten ferrite and grain boundaries in carbon steel', *Metal Science*, 1982, **16**, 239–244.
4. R. A. Ricks, P. R. Howell, and R. W. K. Honeycombe: 'Formation of supersaturated ferrite during decomposition of austenite in iron-copper and iron-copper-nickel alloys', *Metal Science*, 1980, **14**, 562–568.
5. J. V. Bee, and R. W. K. Honeycombe: 'Isothermal decomposition of austenite in a high purity iron chromium binary alloy', *Metallurgical Transactions A*, 1978, **9**, 587–593.
6. H. K. D. H. Bhadeshia: 'Fe-7.4Cr-1.8Ni.txt': 2020: URL https://www.phase-trans.msm.cam.ac.uk/2020/Fe-7.4Cr-1.8Ni.txt.
7. R. A. Ricks, J. V. Bee, and P. R. Howell: 'The decomposition of austenite in a high purity iron-chromium-nickel alloy', *Metallurgical Transactions A*, 1981, **12**, 1587–1594.
8. S. H. Chong, A. Sayles, R. Keyse, J. D. Atkinson, and E. A. Wilson: 'Examination of microstructures and microanalysis of an Fe-9% Ni alloy', *Materials Transactions, JIM*, 1998, **39**, 179–188.
9. E. A. Wilson, and S. H. Chong: 'Isothermal transformations in an Fe-9Ni wt% alloy', *Metallurgical & Materials Transactions A*, 2002, **33**, 2425–2431.

10. E. A. Wilson, D. V. Shatansky, and Y. Ohmori: 'A study of transformations in continuously cooled Fe-15Ni alloys', *ISIJ International*, 2001, **41**, 866–875.

11. J. W. Christian: 'The origin of surface relief effects in phase transformations', In: V. F. Zackay, and H. I. Aaronson, eds. *Decomposition of Austenite by Diffusional Processes*. New York, USA: Interscience, 1962:371–386.

12. J. D. Watson, and P. G. McDougall: 'The crystallography of Widmanstätten ferrite', *Acta Metallurgica*, 1973, **21**, 961–973.

13. T. Koseki: 'Allotriomorphic ferrite: no surface relief': http://youtu.be/OQ5lVjYssko, 2013.

14. S. W. Seo, H. K. D. H. Bhadeshia, and D. W. Suh: 'Pearlite growth rate in Fe-C and Fe-Mn-C steels', *Materials Science and Technology*, 2015, **31**, 487–493.

15. E. Swallow, and H. K. D. H. Bhadeshia: 'High resolution observations of displacements caused by bainitic transformation', *Materials Science and Technology*, 1996, **12**, 121–125.

16. D. P. Dunne, and C. M. Wayman: 'The crystallography of ferrous martensite', *Metallurgical Transactions*, 1971, **2**, 2327–2341.

17. J. W. Brooks, M. H. Loretto, and R. E. Smallman: 'Direct observations of martensite nuclei in stainless steel', *Acta Metallurgica*, 1979, **27**, 1839–1847.

18. F. R. Chien, R. J. Clifton, and S. R. Nutt: 'Stress-induced phase transformation in single crystal titanium carbide', *Journal of the Americal Ceramics Society*, 1995, **78**, 1537–1545.

19. K. A. Taylor, G. B. Olson, M. Cohen, and J. Vander Sande: 'Carbide precipitation during stage 1 tempering of FeNiC martensite', *Metallurgical Transactions A*, 1989, **20**, 2749–2765.

20. H. K. D. H. Bhadeshia: 'Rationalisation of shear transformations in steels', *Acta Metallurgica*, 1981, **29**, 1117–1130.

21. J. M. Rigsbee, and H. I. Aaronson: 'Interfacial structure of the broad faces of ferrite plates', *Acta Metallurgica*, 1979, **27**, 365–376.

22. J. W. Christian: 'Deformation by moving interfaces', *Metallurgical Transactions A*, 1982, **13**, 509–538.

23. J. W. Christian: Theory of Transformations in Metals and Alloys, Part I: 2 ed., Oxford, U. K.: Pergamon Press, 1975.

24. J. W. Christian, and A. G. Crocker: Dislocations in Solids, ed. F. R. N. Nabarro, vol. 3, chap. 11: Amsterdam, Holland: North Holland, 1980:165–252.

25. H. K. D. H. Bhadeshia: 'The structure of the broad faces of ferrite plates', *Scripta Metallurgica*, 1983, **17**, 1475–1479.

26. M. P. Cassidy, and C. M. Wayman: 'Crystallography of hydride formation in Zr: 1. $\delta \rightarrow \gamma$ transformatoin', *Metallurgical Transactions A*, 1980, **11**, 47–56.

27. M. P. Cassidy, and C. M. Wayman: 'Crystallography of hydride formation in Zr: 2. $\delta \rightarrow \epsilon$ transformatoin', *Metallurgical Transactions A*, 1980, **11**, 57–67.

28. P. H. Shipway, and H. K. D. H. Bhadeshia: 'Mechanical stabilisation of bainite', *Materials Science and Technology*, 1995, **11**, 1116–1128.

29. R. H. Larn, and J. R. Yang: 'Effect of compressive deformation of austenite on the Widmanstätten ferrite transformation in Fe-Mn-Si-C steel', *Materials Science & Engineering A*, 1999, **264**, 139–150.

30. G. B. Olson, and M. Cohen: 'A general mechanism of martensitic nucleation, Parts I–III', *Metallurgical Transactions A*, 1976, **7A**, 1897–1923.

31. C. L. Magee: 'The nucleation of martensite', In: H. I. Aaronson, and V. F. Zackay, eds. *Phase Transformations*. Materials Park, Ohio, USA: ASM International, 1970:115–156.

32. H. K. D. H. Bhadeshia: 'Critical assessment: Diffusion-controlled growth of ferrite plates in plain carbon steels', *Materials Science and Technology*, 1985, **1**, 497–504.

33. H. K. D. H. Bhadeshia: 'Hard bainite', In: J. M. Howe, D. E. Laughlin, J. K. Lee, U. Dahmen, and W. A. Soffa, eds. *Solid-Solid Phase Transformations, TME-AIME, Warrendale, USA*, vol. 1. Warrendale, Pennsylvania, USA: TMS-AIME, 2005:469–484.

34. H.-S. Yang, and H. K. D. H. Bhadeshia: 'Designing low-carbon, low-temperature bainite', *Materials Science and Technology*, 2008, **24**, 335–342.

35. G. I. Rees, and H. K. D. H. Bhadeshia: 'Bainite transformation kinetics, Part I, modified model', *Materials Science and Technology*, 1992, **8**, 985–993.

36. M. Hillert: 'The growth of ferrite, bainite and martensite': Technical Report, Swedish Institute for Metals Research, Stockholm, Sweden, 1960.

37. R. D. Townsend, and J. S. Kirkaldy: 'Widmanstätten ferrite formation in Fe-C alloys', *Trans. A. S. M.*, 1968, **61**, 605–619.

38. E. P. Simonen, H. I. Aaronson, and R. Trivedi: 'Lengthening kinetics of ferrite and bainite sideplates', *Metallurgical Transactions*, 1974, **4**, 1239–1245.

39. N. Oku, K. Asakura, J. Inoue, and T. Koseki: 'In-situ observation of ferrite plate formation in low carbon steel during continuous cooling process', *Tetsu-to-Hagane*, 2008, **94**, B33–B38.

40. M. Hillert: 'Role of interfacial energy during solid-state phase transformations', *Jernkontorets Annaler*, 1957, **141**, 757–789.

41. R. Trivedi, and G. M. Pound: 'Growth kinetics of plate like precipitates', *Journal of Applied Physics*, 1969, **40**, 4293–4300.

42. G. P. Ivantsov: 'Temperature field around spherical, cylindrical and acircular crystal growing in a supercooled melt', *Doklady Akademii Nauk, SSSR*, 1947, **58**, 567–569.

43. R. Trivedi: 'Role of interfacial free energy and interface kinetics during the growth of precipitate plates and needles', *Metallurgical Transactions*, 1970, **1**, 921–927.

44. M. E. Glicksman, R. J. Schaefer, and J. D. Ayres: 'Dendritic growth – a test of theory', *Metallurgical Transactions A*, 1976, **7**, 1747–1759.

45. J. S. Langer, and H. Muller-Krumbhaar: 'Theory of dendritic growth. I. Elements of a stability analysis', *Acta Metallurgica*, 1978, **26**, 1681–1687.

46. H. K. D. H. Bhadeshia, L.-E. Svensson, and B. Gretoft: 'Model for the development of microstructure in low alloy steel (Fe-Mn-Si-C) weld deposits', *Acta Metallurgica*, 1985, **33**, 1271–1283.

47. K. R. Kinsman, E. Eichen, and H. I. Aaronson: 'Thickening kinetics of proeutectoid ferrite plates in Fe-C alloys', *Metallurgical Transactions A*, 1975, **6**, 303–317.

48. A. Crosky, P. G. McDougall, and J. S. Bowles: 'The crystallography of the precipitation of alpha rods from beta Cu Zn alloys', *Acta Metallurgica*, 1980, **28**, 1495–1504.

49. C. S. Smith: 'Microstructure', *Transaction of the ASM*, 1953, **45**, 533–575.

50. W. W. Mullins, and R. F. Sekerka: 'Morphological stability of a particle growing by diffusion or heat flow', *Journal of Applied Physics*, 1963, **34**, 323–329.

51. W. W. Mullins, and R. F. Sekerka: 'Stability of a planar interface during solidification of a dilute binary alloy', *Journal of Applied Physics*, 1964, **35**, 444–451.

52. A. Ali, and H. K. D. H. Bhadeshia: 'Microstructure of high strength steel refined with intragranularly nucleated Widmanstätten ferrite', *Materials Science and Technology*, 1991, **7**, 895–903.

53. I. Loginova, J. Ågren, and G. Amberg: 'Widmanstätten ferrite in binary Fe-C phase-field approach', *Acta Materialia*, 2004, **52**, 4055–4063.

54. M. Cottura, B. Appolaire, A. Finel, and Y. Le Bouar: 'Phase field study of acicular growth: Role of elasticity in Widmanstätten structure', *Acta Materialia*, 2014, **72**, 200–210.

55. M. Cottura, B. Appolaire, A. Finel, and Y. Le Bouar: 'Plastic relaxation during diffusion-controlled growth of Widmanstätten plates', *Scripta Materialia*, 2015, **108**, 117–121.

56. A. Yamanaka, T. Takaki, and Y. Tomita: 'Phase-field simulation of austenite to ferrite transformation and Widmanstätten ferrite formation in Fe-C alloy', *Materials Transactions*, 2006, **47**, 2725–2731.

57. A. Bhattacharya, K. Ankit, and B. Nestler: 'Phase-field simulations of curvature-induced cascading of Widmanstätten-ferrite plates', *Acta Materialia*, 2016, **123**, 317–328.

58. J. W. Christian: 'Simple geometry and crystallography applied to ferrous bainites', *Metallurgical Transactions A*, 1990, **21**, 799–803.

59. H. K. D. H. Bhadeshia: Geometry of Crystals, Polycrystals, and Phase Transformations: Florida, USA: CRC Press, ISBN 9781138070783, freely downloadable from www.phasetrans.msm.cam.ac.uk, 2017.

60. C. Lin, J. Wan, and H. Ruan: 'Phase field modeling of Widmanstätten ferrite formation in steel', *Journal of Alloys and Compounds*, 2018, **769**, 620–630.

Notes

[1] It is common though inaccurate to state that α_W grows on $\{111\}_\gamma$, originating from the early descriptions of the structure. To quote Belaiew [1], "… a certain amount of the excess element will be forced to crystallise out 'on the spot', that is, not on the boundary, but in the middle of the grain. Such separation will necessarily follow the crystallographic planes in a face-centred cubic lattice, these planes will be the four octahedral planes. The separation of the excess element will follow in space these planes. On a secant plane such octahedral sections will appear as *Widmanstätten figures*".

[2] The terms *interface* and *interface plane* means the average plane determined on a macroscopic scale rather than a particular component of its structure.

[3] Some of the basics are described already in Section 5.7.2, but are repeated here for clarity and context.

[4] $\Delta G^{\gamma \to \alpha + \gamma'}$ is smaller in magnitude than ΔG_m as illustrated in Figure 4.10, because in the former case, the change in the composition of the austenite as the ferrite grows cannot be neglected.

[5] The equation for χ_n is stated incorrectly by Trivedi and Pound.

[6] There have been many similar simulations based essentially on the same principle of stability and induced anisotropy to generate the shape, for example, [56, 57].

[7] A *preferred* habit plane can arise only if there is sufficient anisotropy in the interface energy or elastic properties [58]. If an enclosed particle is generated by a displacive mechanism then it is not possible for components of the interface to be incoherent while others have continuity of planes and vectors. It is wrong, as is often quoted in phase field based models of

α_W, to state that the growing tip of the plate will be 'disordered' or incoherent. In fact, the major weakness of phase field models is the lack of structure in interfaces which are set to unrealistic thicknesses.

[8] The crystallographic theory of martensite deals with this [p. 196, 59].

[9] A similar conclusion was reached by Lin et al. in their phase field modelling of iron containing carbon and nitrogen [60]. The transformation strain used was identical to [54]. It has the same limitations.

8 Cementite

In the binary Fe-C system, cementite has a composition that approximates Fe_3C. It was regarded originally as a phase that *cements* the structure of steel [1, 2]. In mineralogy, the carbide is known as *cohenite* $(Fe,Ni,Co)_3C$, after Emil Cohen, who was investigating material of meteoric origin [3]. Together with ferrite, it is a most prolific phase to be found in steels and of course, is seminal to the structure of pearlite. The morphology of cementite when in steel is illustrated in the context of the other transformations, but Figure 8.1a shows that there is nothing particularly unusual about the microstructure of polycrystalline cementite as a single phase. There are the grain boundary junctions and the boundaries themselves are not strongly facetted even though the crystal structure of cementite is anisotropic. Figure 8.1b shows dislocation arrays within deformed cementite; it is not the lack of dislocations that makes cementite hard, but as seen later, the unit cell is large and primitive which means that the Burgers vector of lattice dislocations must also be large, making it difficult for dislocations to move. For the most common slip system in cementite, $(010)[001]$, the energy (neglecting elastic constant differences) of a dislocation will be roughly three times greater than in ferrite. This comparison is valid because cementite is in fact metallic.

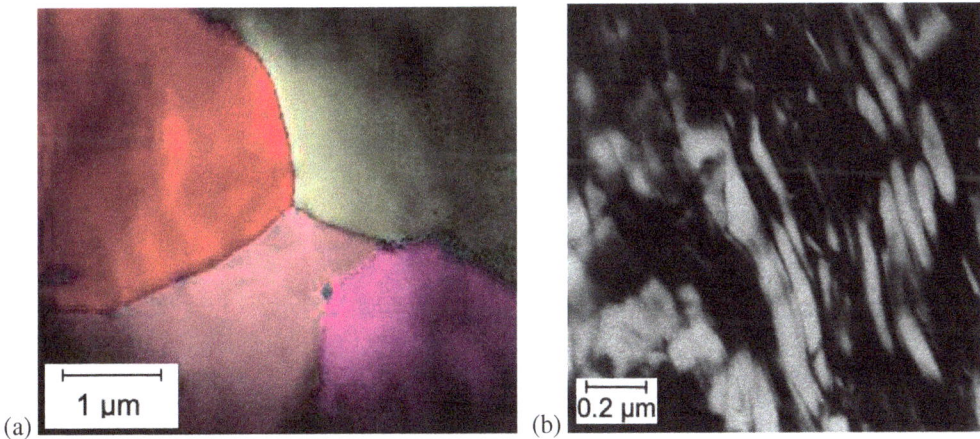

(a) 1 µm   (b) 0.2 µm

Figure 8.1 (a) Orientation image of synthesised cementite showing grain boundaries and their junctions. Reprinted from Mussi et al. [4] by permission of Taylor & Francis Ltd, http://www.tandfonline.com. (b) Dislocation arrangement in cementite when a 1.25C wt% steel is deformed 92%. Reprinted from Inoue et al. [5] with the permission of Elsevier.

Cementite often is said to be metastable with respect to graphite but this represents the case where it is in contact with ferrite; when isolated, it is more stable than graphite, Figure 8.2a,b. This is why during the carburisation of iron, cementite in contact with ferrite decomposes into a mixture of ferrite and graphite, whereas it does not do so when lodged within coke [6]. Tiny particles of cementite that are surrounded by a thin shell of carbon remain stable as cementite during heat-treatment at 700°C [7]. Graphite cannot dissolve iron; when it encapsulates cementite it would be necessary to nucleate ferrite for the cementite to decompose but it is hard to imagine that ferrite could nucleate in the carbon-rich environment. So although the reaction $Fe_3C \rightarrow 3Fe_\alpha + graphite$ leads to a reduction in free energy, iron must be able to nucleate for the reaction to be possible at all.

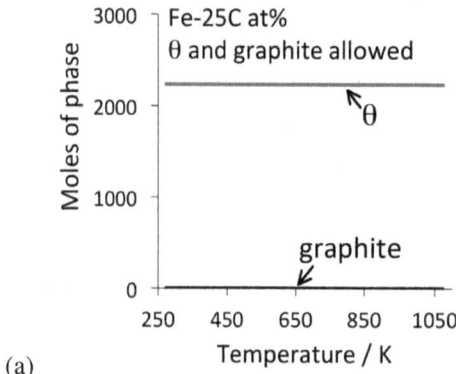

(a)

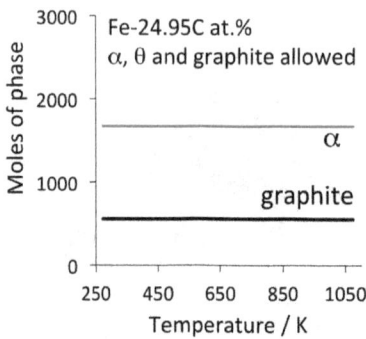

(b)

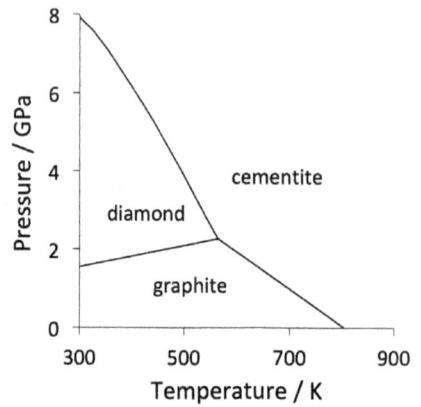

(c)

Figure 8.2 Phase diagram calculations for 100 kg total weight, using MTDATA [8] and the SGTE thermodynamic database. (a) Fe-25C at.%, permitting only cementite and graphite to co-exist. (b) The average carbon concentration is reduced slightly to allow ferrite to appear, in which case the most stable mixture becomes that of ferrite and graphite [9]. (c) Stability fields for diamond, graphite and cementite, when present in iron. Adapted from Lipschutz and Anders [10].

Figure 8.2c is an interesting phase diagram in that it is valid only in the presence of ferritic iron as solvent. As expected, cementite is not stable relative to graphite under ambient conditions and at very high pressures, but cementite dominates at high temperatures. At very high pressures, it is diamond that replaces graphite, with the same tendency for the broader cementite phase field. The diagram was constructed in the context of studies on meteorites, to explain the occurrence of diamonds in some meteorites [10]. There are indications that the diamonds formed from cementite under the shock associated with the impact of the meteorite with the Earth. Diamond has been observed to grow from graphite during the detonation of cast iron [11].

It is difficult to explain on the basis of Figure 8.2c why cementite is present in meteorites, that after all cool very slowly. It is speculated that the impurities in cementite give it a greater stability. Meteorites, in addition to iron and carbon, contain nickel at concentrations in the range 5.5-8 wt%. It seems that to retain cementite it must form at a temperature between about 640-610 °C. That which forms above the upper limit decomposes on cooling and cementite does not precipitate below the lower limit. The temperature at which it actually forms is controlled by the nickel concentration in meteoric iron. It will, for example, be retained when the average nickel content is 7Ni wt% but not if it is 5 or 9 wt%, although the concentration of nickel in the cementite will be smaller than average [12].

8.1 CRYSTAL STRUCTURE OF CEMENTITE

Cementite has an orthorhombic unit cell with lattice parameters as $a = 0.50837$ nm, $b = 0.67475$ nm and $c = 0.45165$ nm, corresponding to the space group *Pnma*. There are twelve atoms of iron in the unit cell and four of carbon, as illustrated in Figure 8.3. Four of the iron atoms are located on mirror planes whereas the other eight are at general positions (point symmetry 1).

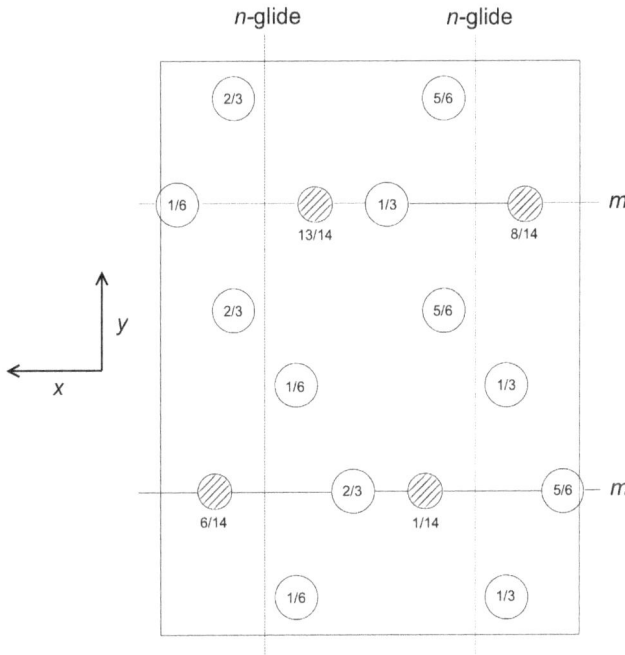

Figure 8.3   The crystal structure of cementite, consisting of twelve iron atoms (large) and four carbon atoms (small, hatched pattern). The fractional z coordinates of the atoms are marked. Notice that four of the iron atoms are located on mirror planes, whereas the others are at general locations where the only point symmetry is a monad. The pleated layers parallel to (100) are in $...ABABAB...$ stacking with carbon atoms occupying interstitial positions at the folds within the pleats, with all carbon atoms located on the mirror planes. There are four Fe_3C formula units within a given cell.

The lattice type is primitive (P). There are *n*-glide planes normal to the *x*-axis, at $\frac{1}{4}x$ and $\frac{3}{4}x$ involving translations of $\frac{b}{2} + \frac{c}{2}$. The mirror planes are normal to the *y*-axis, and *a*-glide planes normal to the *z*-axis, at heights $\frac{1}{4}z$ and $\frac{3}{4}z$ with fractional translations of $\frac{a}{2}$ parallel to the *x*-axis. The space group symbol is therefore *Pnma* [13].

Table 8.1

Wyckoff positions for space group $Pnma$ **[14]. Each Wyckoff position in this space group is labelled with a letter (Table 8.1); thus, the eight iron atoms in general positions are labelled with the letter "d", and the remaining four on mirror planes with the letter "c"; the number preceding the letter, for example, the "8" in 8d, denotes the number of equivalent positions in the cell.**

Multiplicity	Wyckoff letter	Site symmetry	Coordinates
8	d	1	(x,y,z) $(-x+\frac{1}{2},-y,z+\frac{1}{2})$ $(-x,y+\frac{1}{2},-z)$ $(x+\frac{1}{2},-y+\frac{1}{2},-z+\frac{1}{2})$ $(-x,-y,-z)$ $(x+\frac{1}{2},y,-z+\frac{1}{2})$ $(x,-y+\frac{1}{2},z)$ $(-x+\frac{1}{2},y+\frac{1}{2},z+\frac{1}{2})$
4	c	.m.	$(x,\frac{1}{4},z)$ $(-x+\frac{1}{2},\frac{3}{4},z+\frac{1}{2})$ $(-x,\frac{3}{4},-z)$ $(x+\frac{1}{2},\frac{1}{4},-z+\frac{1}{2})$
4	b	-1	$(0,0,\frac{1}{2})$ $(\frac{1}{2},0,0)$ $(0,\frac{1}{2},\frac{1}{2})$ $(\frac{1}{2},\frac{1}{2},0)$
4	a	-1	$(0,0,0)$ $(\frac{1}{2},0,\frac{1}{2})$ $(0,\frac{1}{2},0)$ $(\frac{1}{2},\frac{1}{2},\frac{1}{2})$

8.1.1 TYPES OF INTERSTITIAL SITES

There are prismatic, octahedral and three kinds of tetrahedral interstices between the iron atoms in the cementite unit cell, Figure 8.4; the space available within each is defined from the centre of the interstice to the boundary of the nearest iron atom; the sizes are therefore 0.71, 0.53, 0.34, 0.26 and 0.28 Å [15]. The centres of the prismatic interstices lie on mirror planes so there are four per cell (4c, Table 8.1) and they all are filled with carbon atoms in the stoichiometric form of cementite [16]. The smaller octahedral interstices, of which there are four per cell (4a, Table 8.1), are empty in pure cementite unless the carbon concentration exceeds 25 at.%; the tetrahedral interstices are too small to be occupied by carbon. When hydrogen enters the cementite lattice, it locates in the octahedral [17] interstices because the prismatic ones are occupied by carbon.

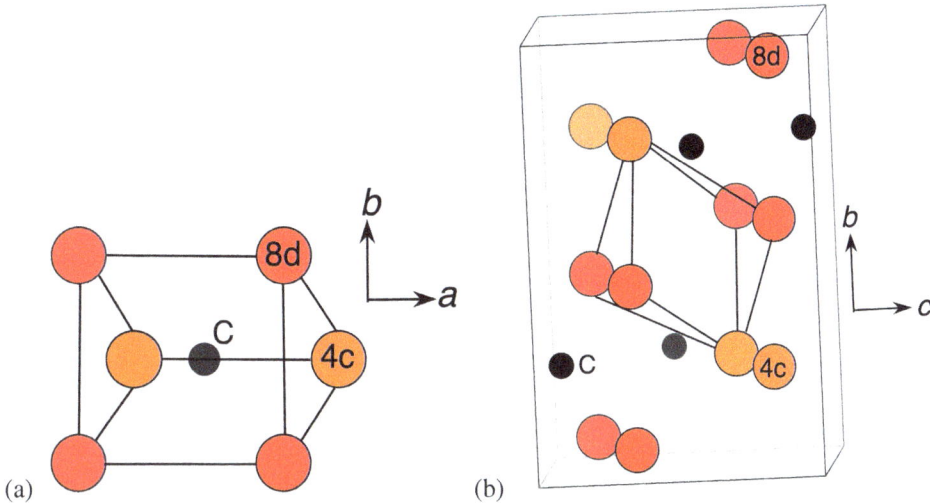

Figure 8.4 Two kinds of interstices in the cementite unit cell. (a) Prismatic. The apex of the prism of iron atoms is above the square base. (b) Octahedral. The red and orange coloured atoms are iron, with the latter located on mirror planes.

8.2 CARBON CONTENT OF CEMENTITE

It has been known since 1855 that the carbide common in steels approximates to the formula Fe_3C or to a multiple of that formula [18]. But the structure of cementite could only be determined much later to reveal that the carbon atoms in cementite are located in interstitial sites [16, 19]. Because of the size of the carbon atom relative to the interstices available, deviations from the 3:1 Fe:C atom ratio lead to lattice parameter changes [20], which therefore can be used to assess the concentration of carbon. Thus, the specific volume of cementite in equilibrium with ferrite at ambient temperature is greater than expected assuming Fe_3C, indicating a deficit of carbon [21, 22]. Calculations similarly show how the density changes with the carbon concentration [23]. Furthermore, gradients of carbon concentration spanning tens of micrometres have been measured within cementite during carburisation experiments [24]. All of these observations indicate that there is some flexibility in the 3Fe:1C ratio that normally is associated with cementite.

These and other results form the basis of the thermodynamically assessed phase boundaries for the equilibrium between cementite (θ) and α or γ, Figure 8.5a. Cementite has sometimes been depicted as a line compound, but Figure 8.5b shows that the variation in free energy as a function of composition has a broad minimum which at high temperatures is not located at 25 at.% of carbon [25]. The fact that ferrite can precipitate from cementite equilibrated at elevated temperatures establishes the increase in its carbon concentration on cooling, Figure 8.6 [26]. When pure cementite is decarburised, its density and Curie temperature change due to the deviation from the stoichiometric composition [27].

Any deviations from stoichiometry are nevertheless limited because the bond energy between a carbon atom and iron is greater than that between two iron atoms, so a deficit of carbon would lead to a reduction in cohesion [28]. A carbon concentration beyond 25 at.% is less likely because the excess would need to be accommodated in the less-favoured interstices within the cementite lattice.[1]

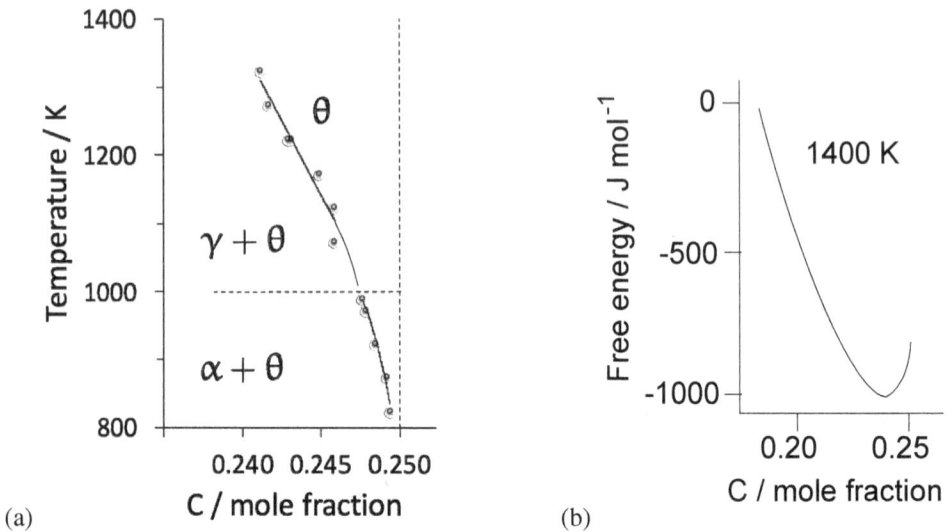

(a)

(b)

Figure 8.5 (a) The composition of cementite that is in equilibrium with austenite or with ferrite in an Fe-C alloy. The data are due to Leineweber et al. [23], determined by measuring the lattice parameters of cementite following quenching from the appropriate temperature. (b) Free energy curve of cementite as a function of chemical composition (referred to γ-Fe and graphite). After Gohring et al. [25].

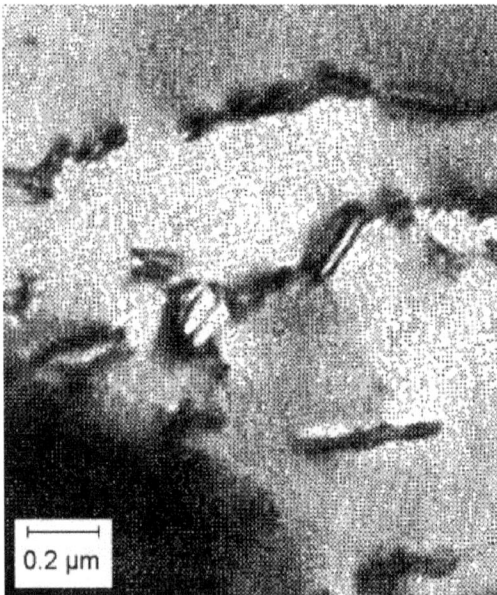

Figure 8.6 Precipitation of fine ferrite-platelets from cementite. Reproduced with permission of Taylor and Francis from [26]. The crystallography of α precipitation in θ is reported separately on a different alloy, with the major growth direction of the ferrite being $\langle 111 \rangle_\alpha \parallel [100]_\theta$ and with the habit plane close to $(010)_\theta$ in the *Pnma* setting [29].

8.3 MAGNETIC PROPERTIES

Cementite under ambient conditions is a metallic ferromagnet that becomes paramagnetic beyond the Curie temperature T_C of $\approx 186\,°C$, Figure 8.7 [30]. The average magnetisation at 0 K is about 1.86 μ_B. Calculations of the local magnetic moments on the four iron atoms located on mirror planes (4c) and at the eight at general positions (8d) give estimates within the ranges 1.92-2.01 and 1.74-1.957 μ_B respectively [30–33] at 0 K. The uncertainty has it origin on the size of the region ("muffin tin") over which the moment is calculated, together with numerical inaccuracies in the methods used; the moment summed over the unit cell is nevertheless about the same in all the studies.

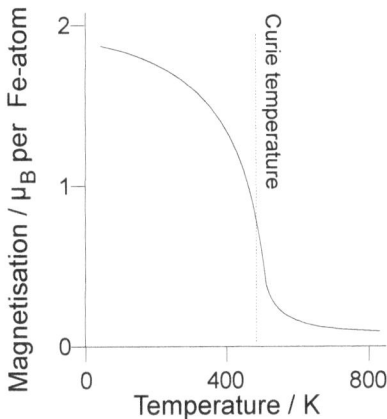

Figure 8.7 Calculated magnetisation of cementite as a function of temperature, adapted from Dick et al. [30].

There is a transition from ferromagnetic to nonmagnetic states at 25 GPa pressure and 300 K [34]. The term *nonmagnetic* is used here because it is not clear whether the magnetic collapse corresponds to a loss of spin correlation or to a transition from a high-spin to a low-spin state. There is a volume contraction of 2-3 % following the transition from the ferromagnetic state. The cementite structure, with its orthorhombic symmetry, is magnetically anisotropic, with the ease of magnetisation increasing in the order $[100]_\theta$, $[010]_\theta$ and $[001]_\theta$ [35–37]. The magnetocrystalline anisotropy energy is $334\pm20\,kJ\,m^{-3}$ [37]. The dominant domain walls lie in the $(001)_\theta$ plane, Figure 8.8 [38]; Hillert and Lange first observed magnetic domains in cementite [39].

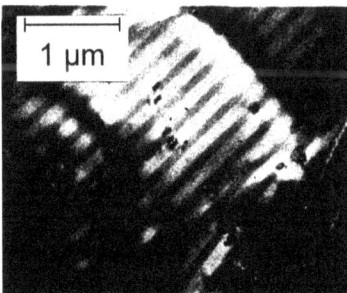

Figure 8.8 Ferromagnetic domain structure of cementite, observed using transmission electron microscopy. Reprinted from [38], with the permission of AIP Publishing.

Substitutional solutes affect the magnetic properties of cementite. Nickel reduces the saturation magnetisation because it replaces the iron atoms that have a greater magnetic-moment per atom.

The same explanation applies in part to Mn and Cr, consistent with the average alloy magnetic-moment per atom to be expected from the Slater-Pauling curve. The alloying has a minor effect on the intrinsic magnetic moment of the iron atoms [40]. Manganese in cementite makes it magnetically softer, i.e., reduces its coercivity [41]. The influence of substitutional solutes on the magnetic moment of iron is, naturally, site-specific (Table 8.2).

Table 8.2
Magnetic moments (in units of μ_B per iron atom) as a function of a silicon atom substituted into an 8d or 4c iron site. Data from Jang et al. [33]. Similar site-specific data for chromium in cementite are available in Medvedeva et al. [42].

	Fe_3C	$(Fe_{11}Si^{4c})C_4$	$(Fe_{11}Si^{8d})C_4$
Fe(4c)	2.059	2.021	1.881
Fe(8d)	1.957	1.793	1.852

The dilution of the magnetic moment is not the only consequence of manganese additions to cementite [43]. In $(Fe_{1-x}Mn_x)_3C$ at 0 K, the spins on manganese atoms that locate on 8d positions adopt an antiferromagnetic alignment, whereas the Fe and Mn at 4d positions have identical spins (Figure 8.9). The total magnetisation per unit cell is then reduced as the Mn concentration is increased. If the cell contains eight or more Mn atoms, the 8d layer assumes perfect antiferromagnetic arrangement with the remaining atoms in the 4c positions in a ferromagnetic alignment [43]. Pure Mn_3C has an antiferromagnetic ground-state structure with a [001] magnetisation direction [44].

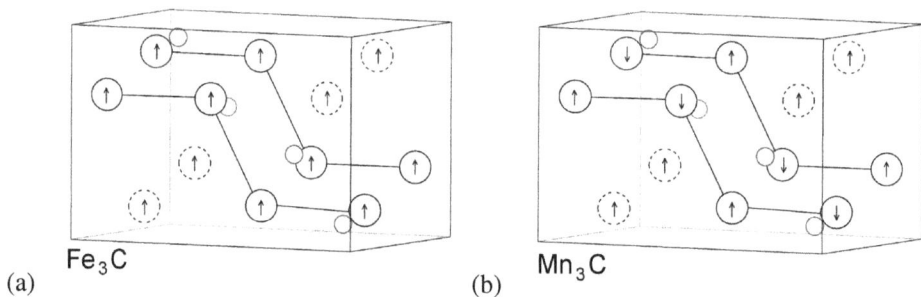

Figure 8.9 Orthorhombic unit cell with eight metal atoms in the 8d positions (circles), four in the 4c locations (dashed circles) and four carbon atoms (small circles). The magnetic structures are from calculations representative of 0 K. (a) Ferromagnetic Fe_3C. (b) Mn_3C. The 8d layers are perfectly antiferromagnetic, whereas the four atoms at 4c locations have aligned spins making Mn_3C a ferrimagnet. Adapted from Appen et al. [43].

Cementite exhibits a magnetocaloric effect [45]. The alignment of magnetic spins is reduced during adiabatic demagnetisation. Since the total entropy remains constant during the adiabatic conditions, the increase in magnetic entropy on the removal of the applied field is compensated for by a decrease in temperature. If demagnetisation occurs isothermally, then the change in magnetic entropy leads to a corresponding change in total entropy. Measurements indicate an adiabatic change in temperature of 1.76 ± 0.01 K during a field change of 2 T. When the magnetic field is changed from 0 T to 20 T, there is an entropy change under isothermal conditions of 3 J K^{-1} kg^{-1}.

The is some uncertainty about the Curie temperature, T_C, which depends on the carbon concentration of cementite, Table 8.3, where the data represent the average compositions of samples synthesised using mixtures of iron and graphite powders by heating to 1110 °C under a pressure of 1 GPa. It is noteworthy that the Curie measurements are due to cementite alone. There is a pressure dependence, with T_C reduced to below ambient temperature in stoichiometric cementite for pressures in excess of 6 GPa, whereas carbon-rich cementite remains ferromagnetic to higher pressures ($\approx$7 GPa).

Table 8.3
Ambient pressure measurements of the Curie temperature of cementite as a function of its carbon concentration. Data from Walker et al. [46], determined by making cementite as a part of a transformer. Choe [35] reports a somewhat lower T_c of 167.6 °C determined using a superconducting quantum interference magnetometer for ambient pressure.

Phases present	Nominal at% C	T_C / °C
cementite, graphite	26	174
cementite	25	186
cementite	25	187
cementite	23	173
cementite, Fe	22	173

An interpretation [47] of the change in magnetic properties with pressure attributes the phenomenon to the volume dependent two-state theory for the high magnetic-moment to small-volume low moment transition. Using an X-ray technique and diamond anvil equipment, it has been determined experimentally that the loss of ferromagnetism occurs at about 10 GPa. The change in volume required to induce the magnetic transition is about 5% [47, 48].

8.4 THERMAL PROPERTIES

The average thermal expansion coefficient of polycrystalline cementite changes from 6.8×10^{-6} K^{-1} to 16.2×10^{-6} K^{-1} as the sample is heated to beyond the Curie temperature, Figure 8.10 [49].

Figure 8.11 shows diffraction data [50–52] for each of the lattice parameters of cementite as a function of temperature, of which a is most sensitive to the change from the ferromagnetic to paramagnetic state, characterised by a contraction during heating within the ferromagnetic range. It is not clear why the a parameter is most sensitive to the magnetic transition.

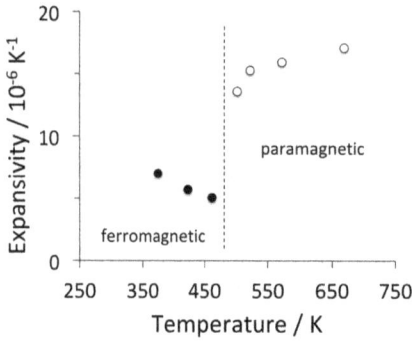

Figure 8.10 The linear thermal expansion coefficient of polycrystalline cementite as a function of temperature and magnetic state. Adapted using data from Umemoto et al. [49].

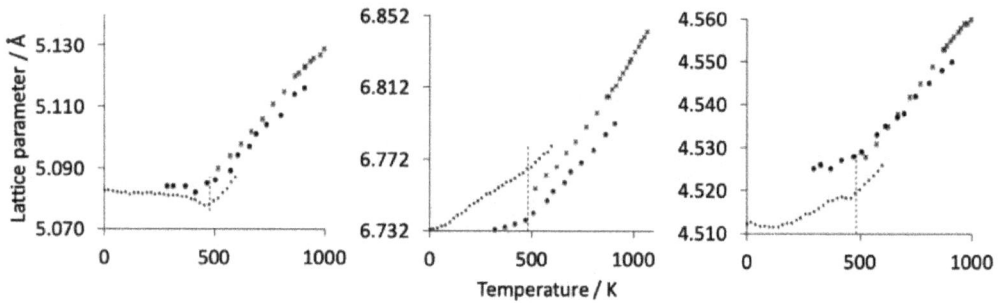

Figure 8.11 Neutron and X-ray diffraction data on the three lattice parameters a, b and c of cementite as a function of temperature. Data from [51] (small circles with error bars), [50] (filled circles) and [52] (crosses). The dashed line in each case identifies the Curie temperature. The calculated pressure dependencies of the lattice parameters are as follows [53]: $\Delta a = 0.0041 \times P$, $\Delta b = 0.00578 \times P$ and $\Delta c = 0.00374 \times P$ Å, where the pressure P is in GPa.

8.5 SURFACE ENERGY

Cementite is found experimentally to cleave on the $\{101\}$, (001) and $\{102\}$ planes [54]. While this might seem inconsistent with the ranking of energies in Table 8.4, fracture in fact depends on the *difference* between the energies of exposed and internal surfaces [55].

Table 8.4
Calculated surface energies of cementite in a vacuum.

Crystallographic indices	Surface energy / J m^{-2}	Method	Reference
(001)	2.34	Interatomic potentials	[56]
(001)	2.47	First principles	[57]
(010)	2.00	Interatomic potentials	[56]
(010)	2.26	First principles	[57]
(100)	1.96	Interatomic potentials	[56]
(100)	2.05	First principles	[57]

8.6 ELASTIC PROPERTIES OF SINGLE CRYSTALLINE CEMENTITE

Calculated elastic moduli are summarised in Table 8.5. The anisotropy is compared against Mn_3C which is isomorphous with Fe_3C, and ferrite, in Figure 8.12. The orthorhombic carbides are as expected, more anisotropic than the cubic-α, but it is particularly noticeable that the modulus C_{44} is small in cementite. A crystal subjected to an elastic strain is mechanically stable only if there is a resulting increase in its internal energy [58]. For an orthorhombic crystal, this stability criterion manifests as follows [59]:

$$C_{22} + C_{33} - 2C_{23} > 0$$
$$C_{11} + C_{22} + C_{33} + 2C_{12} + 2C_{13} + 2C_{23} > 0$$
$$C_{11} > 0, \ C_{22} > 0, \ C_{33} > 0, \ C_{44} > 0, \ C_{55} > 0, \ C_{66} > 0$$

Hydrostatic compression leads to an increase in stiffness due to the increase in density accompanying pressurisation, Figure 8.13.

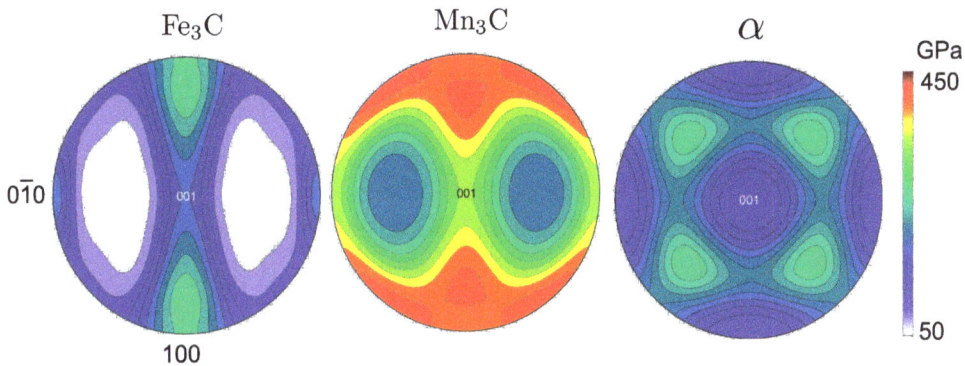

Figure 8.12 Stereographic projections showing the variation of calculated single-crystal elastic moduli as a function of orientation, for the setting *Pnma*. Fe_3C and Mn_3C are based on data from [60] and lattice parameters from [61] and α-iron. Plots courtesy of Shaumik Lenka.

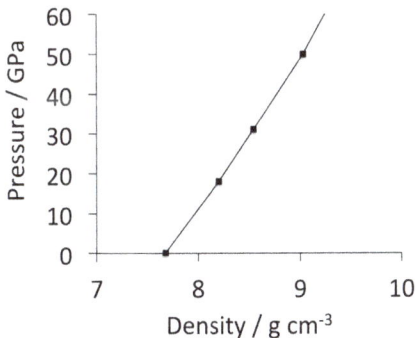

Figure 8.13 Experimentally determined plot of pressure versus density for polycrystalline cementite. Data from Fiquet et al. [67].

There is experimental evidence [68] that first principles calculations relevant for 0 K overestimate the moduli of cementite, bearing in mind that finite temperatures are expected to stiffen the ce-

Table 8.5

Modulus data in GPa for cementite at 0 K and zero pressure unless otherwise indicated. $\Delta C_{ij} = C_{ij}\{400\,\text{K}\} - C_{ij}\{0\,\text{K}\}$ **is the change in heating from 0 K to 400 K. A more comprehensive listing is available in [9].**

	C_{11}	C_{12}	C_{13}	C_{22}	C_{23}	C_{33}	C_{44}	C_{55}	C_{66}	K	
					Fe_3C						
	383	162	156	344	162	300	28	134	135		[62]
ΔC_{ij}	39	10	6	19	−4	10	7	−11	−7		[62]

Cementite containing substitutional solutes

	C_{11}	C_{12}	C_{13}	C_{22}	C_{23}	C_{33}	C_{44}	C_{55}	C_{66}	K	
$(\text{Fe}_2\text{Mn})\text{C}$	402	165	155	418	168	398	68	154	99		[63]
$(\text{Fe}_2\text{Mn})\text{C}$	266	105	58	286	115	263	44	135	144	150	[64]
$(\text{Mn}_2\text{Fe})\text{C}$	480	219	210	407	176	486	16	170	174	284	[64]
Mn_3C	544	241	167	504	187	432	62	200	179		[60]
$(\text{CoFe}_2)\text{C}$	375	164	128	295	136	334	−3	133	137		[65]
$(\text{Co}_2\text{Fe})\text{C}$	374	138	131	299	118	327	−75	129	93		[65]
$(\text{NiFe}_2)\text{C}$	398	16	111	238	104	261	−9	113	80		[65]
$(\text{Ni}_2\text{Fe})\text{C}$	354	127	134	256	115	278	−69	103	46		[65]
$(\text{Fe}_2\text{Cr})\text{C}$	472	111	130	315	117	352	13	176	166	200	[64]
$(\text{Cr}_2\text{Fe})\text{C}$	452	179	220	443	162	450	123	128	186	273	[64]

Fe_3C moduli at non-zero pressures

	C_{11}	C_{12}	C_{13}	C_{22}	C_{23}	C_{33}	C_{44}	C_{55}	C_{66}	K	
$P = 15\,\text{GPa}$	459	216	177	413	238	353	25	148	158	276	[66]
$P = -6\,\text{GPa}$	358	131	145	299	137	285	−3	122	118	196	[66]
$P = -11\,\text{GPa}$	305	104	108	260	110	234	−13	112	104	160	[66]

Single-crystal Young's modulus E / GPa

Orientation	Measured	Calculated
[100]	262 ± 32	287
[001]	213 ± 45	221

mentite [62]. Experimental measurements on single crystals of cementite prove that the Young's modulus is smaller than that calculated theoretically, even though the sample contained some manganese [68]; the discrepancy between experiment and theory could therefore be larger given that according to theory, manganese increases the modulus of cementite (Table 8.5).

8.7   ELASTIC PROPERTIES OF POLYCRYSTALLINE CEMENTITE

The pressure dependence of the bulk modulus of cementite is of importance in understanding the iron-rich phases (including cementite) that may exist within a radius of 1200 km from the centre of the earth. Measurements made using diamond anvil cells subjected to synchrotron X-rays to determine the pressure-volume relationship with the data fitted to modified Birch-Murnaghan equation of state as follows [69]:

$$
P = \left\{ P_r - \frac{1}{2}(3K_r - 5P_r) \left[1 - \left(\frac{V}{V_r} \right)^{-2/3} \right] \right.
$$
$$
\left. + \frac{9}{8} K_r \left(K_r' - 4 + \frac{35 P_r}{9 K_r} \right) \left[1 - \left(\frac{V}{V_r} \right)^{-2/3} \right]^2 \right\} \left(\frac{V}{V_r} \right)^{-5/3} \tag{8.1}
$$

where V_r is the selected reference volume, and P_r, K_r and K_r' are the pressure, isothermal bulk modulus and pressure dependence of that bulk modulus respectively, all at the reference volume, respectively [70, 71]:

Magnetic state	K_r/GPa	V_r/Å^3 atom^{-1}	P_r/GPa	K_r'
Nonmagnetic, 300 K, $25 \geq P \leq 187$ GPa	290 ± 13	9.341	0.0 ± 1.6	3.76 ± 0.18

The unmodified form of the Birch-Murnaghan equation is [70, 72]:

$$
P = \frac{3}{2} K_o \left[\left(\frac{V}{V_0} \right)^{-\frac{7}{3}} - \left(\frac{V}{V_0} \right)^{-\frac{5}{3}} \right] \left\{ 1 - \frac{3}{4}(4 - K_o') \left[\left(\frac{V}{V_0} \right)^{-\frac{2}{3}} - 1 \right] \right\} \tag{8.2}
$$

where $V_0 = 155.28$ Å^3 [70] and K_o are the volume and isothermal bulk modulus at 1 bar and 300 K respectively, and K_o' is the first pressure derivative of K_o at 300 K [70]. The measured data using this equation are in Table 8.6.

The calculated data from single-crystal elasticity can be used to estimate the elastic properties of polycrystalline cementite by assuming uniform stress (Reuss) or uniform strain (Voigt) throughout

Table 8.6

Measured equation of state data [34]. There are three sets of values stated for the paramagnetic state studies by Litasov et al. [73] corresponding to different equations of state used to analyse the experimental data.

Magnetic state	K_o / GPa	K_o'	Reference
Ferromagnetic	179.4 ± 7.8	4.8 ± 1.6	[34]
Ferromagnetic	175.4 ± 3.5	5.1 ± 0.3	[74]
"Nonmagnetic"	288 ± 42	4	[34]
Ferromagnetic	175	5	[73]
Paramagnetic	190	4.8	[73]
Paramagnetic	191	4.68	[73]
Paramagnetic	194	4.6	[73]

the cementite [75]:

$$K_{\text{Reuss}} = [S_{11} + S_{22} + S_{33} + 2(S_{12} + S_{23} + S_{13})]^{-1}$$

$$K_{\text{Voigt}} = [C_{11} + C_{22} + C_{33} + 2(C_{12} + C_{23} + C_{13})/9$$

$$E_{\text{s,Reuss}} = \frac{15}{4(S_{11} + S_{22} + S_{33} - S_{12} - S_{23} - S_{13}) + 3(S_{44} + S_{55} + S_{66})}$$

$$E_{\text{s,Voigt}} = \frac{C_{11} + C_{22} + C_{33} - C_{12} - C_{23} - C_{13}}{15} + \frac{C_{44} + C_{55} + C_{66}}{5}$$

$$E = 9KE_s/(3K + E_s) \quad \text{and} \quad v = (3K/[2 - E_s])/(3K + E_s)$$

(8.3)

where S represents a compliance, E, K and E_s are the Young's, bulk and shear moduli, v is the Poisson's ratio; the absence of a subscript indicates an average of the Reuss and Voigt values. Using Jiang et al.'s single crystal data (Table 8.5) gives $K = 227$ GPa, $E_s = 75$ GPa, $E = 203$ GPa and $v = 0.35$ for zero Kelvin. The Young's modulus of pure polycrystalline cementite has been measured to be 196 GPa, but can be as high as 245 GPa when alloyed with solutes such as chromium and manganese [49]. Measurements on thin (210 nm), polycrystalline films of cementite indicate a Young's modulus of 177 GPa, which gives a shear modulus of 70 GPa assuming that the Poisson's ratio is 0.26 within isotropic elasticity [76, 77]. The Poisson's ratio measured on samples of cementite containing 28% porosity has been reported to decrease almost linearly from 0.254 to 0.246 as the temperature is increased from 95 to 290 K [78].

8.8 SUBSTITUTIONAL SOLUTES

Alloying Fe_3C with manganese increases its stability with respect to graphite [79]; it has long been known that cementite becomes more stable when it "unites with manganese", sometimes resulting in the growth of robust single-crystals known as *Speigeleisenkristall* [80]. Figure 8.14a shows that the addition of manganese permits cementite to co-exist with graphite and ferrite, whereas in the same circumstances, a Fe-25C at.% steel would, at equilibrium, consist only of a mixture of ferrite and graphite. The cementite in the Fe-C-Mn alloy contains manganese, the equilibrium composition of which at low temperatures is more akin to Mn_3C than Fe_3C (Figure 8.14b).

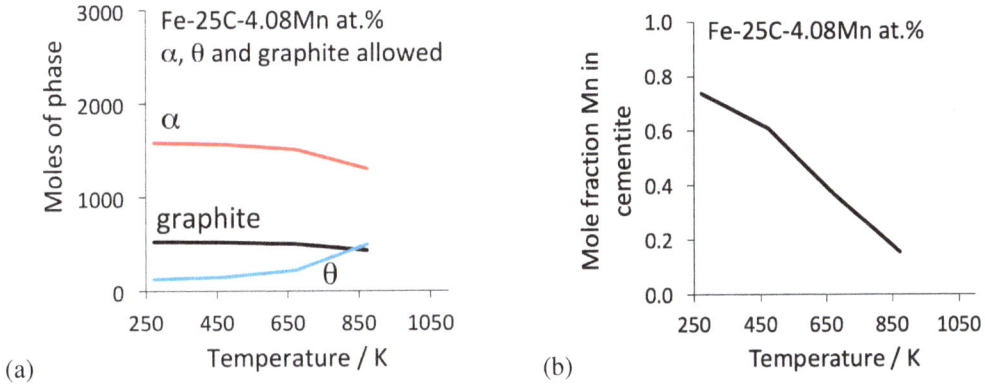

Figure 8.14   Phase diagram calculations for 100 kg total weight, using MTDATA [8] and the SGTE thermodynamic database. Fe-25C-4.08Mn at.%, permitting cementite, graphite and ferrite to co-exist. (a) Equilibrium phase mixture as a function of temperature. (b) The equilibrium manganese concentration in cementite for the calculations presented in (a).

Substituting an atom with a lower magnetic moment than iron reduces the saturation magnetisation [40, 81]. Iron atoms in cementite have local magnetic moments of $1.97\,\mu_B$ or $1.88\,\mu_B$ per atom, depending on whether they are located on the mirror or general positions [82]; the corresponding values for manganese and nickel are about 0.8 and $0.6\,\mu_B$ respectively when substituted onto the mirror sites. Nickel therefore reduces the saturation magnetisation of the alloyed cementite, but the Curie temperature, which depends on the coupling between the magnetic ions, increases [83]. Experimental data on Ni in cementite are limited because the θ tends to be unstable when nickel is forced into its lattice by mechanical alloying, which leads first to the formation of an amorphous phase, followed by the crystallisation of Ni-rich cementite its decomposition [84]. First principles calculations show that the substitution of nickel (or cobalt) makes the cementite less stable with respect to a mixture of α-iron and graphite [85]. This is not the case for all concentrations because the free energy of formation over the range 10-50 at.% Ni is favourable, Figure 8.15.

Chromium has an affinity for carbon consistent with the free energy of formation decreasing systematically with concentration [42, 87]. Manganese too is a carbide former and once some complex magnetic effects (Section 8.3) are accounted for, stabilises cementite. A compilation of data on a variety of solutes affecting the formation energy of cementite at 0 K is presented in Figure 8.16. It would be reasonable to assume that the uncertainty in the calculations is indicated by the scatter in ΔF for pure cementite. The tendencies are that scandium, titanium, vanadium, zirconium, and niobium substitutions into cementite make it more stable relative to its pure form [88], but their efficacy in this context may be compromised by the limits of solubility or the tendency to form other

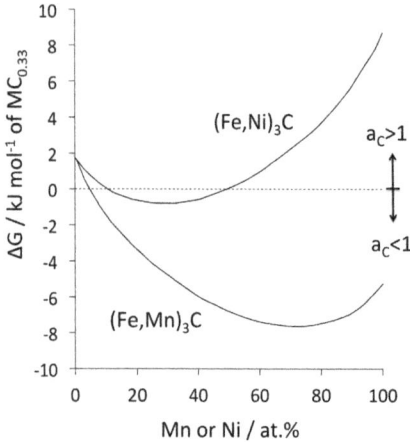

Figure 8.15 The free energy of formation associated with the reaction $M + \frac{1}{3}C \rightarrow M_3C$ occurring at $650\,^\circ C$, as a function of the manganese or nickel concentrations. Adapted from Grabke et al. [86]. a_C stands for the activity of carbon in cementite.

compounds.

Figure 8.16 shows that silicon reduces the stability of cementite, as is well-known in the design of steels and cast irons. It is added to steel to delay cementite precipitation but at concentrations ($\approx 1\,\text{wt}\%$) small enough that graphite is avoided; therefore, carbon partitioned during ferrite formation enriches the residual austenite, permitting it to be retained untransformed at ambient temperature. The influence of silicon on the precipitation of cementite is substantially greater when the matrix phase is supersaturated austenite, because the driving force for precipitation from supersaturated ferrite is greater [89] – many applications that exploit the role of silicon in modern steels are reviewed elsewhere [90–92].

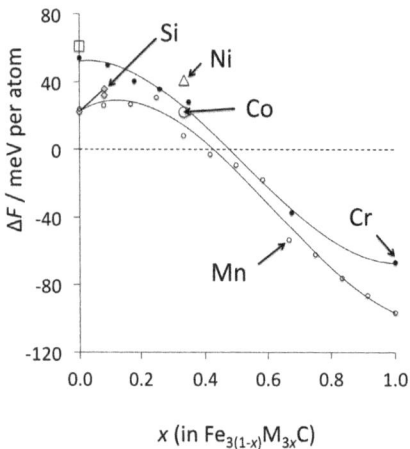

Figure 8.16 The calculated formation energy ΔF of cementite for the reaction $\{F_{\text{Fe}_{3(1-x)}M_{3x}C} - [3(1-x)F_{\text{Fe}} + 3xM + F_{\text{graphite}}]\}/4$, where 'M' stands for a metal atom other than iron. Compilation of data from [33, 43, 85, 87, 93].

Boron lodges within the prismatic interstices when it substitutes for carbon, but is a larger atom so there is a net increase in volume but the lattice parameters change in a nonuniform manner [94]. Figure 8.17 shows that large concentrations of boron can be introduced into cementite without changing its orthorhombic symmetry; the saturation magnetisation and the Curie temperature increase. Helium atoms are relatively rigid so they tend to substitute for iron in cementite; the energy needed to substitute C, Fe_{8d} and Fe_{4c} are 5.07, 3.34 and 3.52 eV respectively [95]. In contrast, the most stable location for a hydrogen atom is an octahedral interstice, surrounded by six iron atoms, with a lower energy than the corresponding interstice in ferrite [96]. Changes in the lattice parameters of cementite due to the substitution of Mo, Mn, Cr and Ni are listed in Table 8.7 [97].

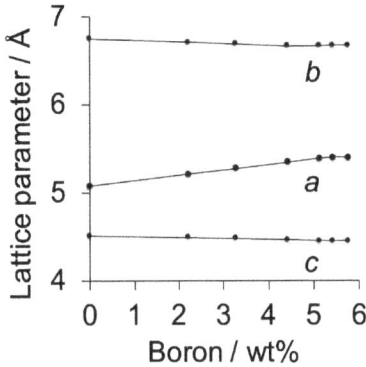

Figure 8.17 The lattice parameters of Fe_3C and $Fe_3(B,C)$ as boron substitutes for carbon in the cementite unit cell. Adapted from Nicholson [94].

Table 8.7

Change in the lattice parameters of cementite (*Pnma*) as a function of the concentration. The coefficients are derived from the work of Kagawa and Okamoto [97].

Solute	Δa / Å wt%$^{-1}$	Δb / Å wt%$^{-1}$	Δc / Å wt%$^{-1}$	Concentration limit / wt%
Mo	0.001276732	0.007352941	0.002538071	3.2
Mn	−0.002061431	0	−0.001239669	4.85
Cr	−0.002328289	−0.001445087	−0.000874126	3.49
Ni	−0.001637331	−0.000814332	−0.000404858	2.08

8.8.1 PRECIPITATES INSIDE CEMENTITE

Ferrite has been shown to precipitate in cementite when its carbon concentration in equilibrium with α enriches as the temperature is reduced, Section 8.2. This probably happens only when the cementite size is large because otherwise the ferrite can precipitate at the interface with pre-existing α. Other precipitates possible include copper or copper-rich particles that form within cementite during tempering [98]. Figure 8.18 shows copper precipitates in identical orientation in the cementite that forms a part of pearlite; it is likely that precipitate at the transformation front [99]. When both austenite and copper particles are found within Widmanstätten cementite, they have a common orientation relationship with the cementite so it might be safe to assume that the θ/Cu crystallography is identical to that of θ/γ [100].

Whereas copper precipitates in both the ferrite and cementite of pearlite by an interphase mechanism, presumably because it has a low solubility in both phases, vanadium carbides similarly precipitate but only in the ferritic component of pearlite [101, 102]. This is because vanadium has a greater solubility in cementite.

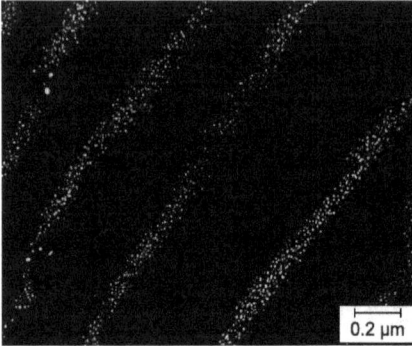

Figure 8.18 Dark-field electron microscope image of fine copper precipitates within the cementite component of pearlite. Fe-1.45C-1.96Cu wt% transformed at 670 °C. After Chairuangsri and Edmonds [99], reproduced with the permission of Elsevier.

8.9 THERMODYNAMIC PROPERTIES

8.9.1 HEAT CAPACITY

There are estimates using a combination of density functional theory and quantum Monte Carlo methods of the heat capacity of cementite [30]. These allow the individual contributions of phonon, electronic and magnetic components to be studied and are consistent with conventional thermodynamic assessments, Figure 8.19 [93]. The behaviour illustrated is typical with the electronic component making only a minor contribution because only those electrons near the Fermi energy can be promoted to unoccupied states. The vibrational component increases smoothly with temperature and forms the majority contribution to the total heat capacity. The cusp in the total heat capacity curve is therefore entirely due to magnetic changes in the vicinity of the Curie temperature of cementite.

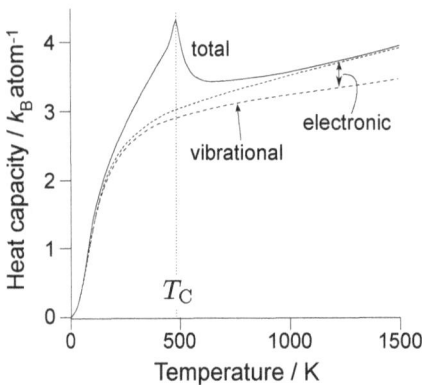

Figure 8.19 The calculated components of the heat capacity of cementite as a function of temperature at zero pressure; adapted from Dick et al. [30].

8.9.2 EQUILIBRIUM BETWEEN CEMENTITE AND MATRIX DISLOCATIONS

The theory of the yield point effect in ferritic steel relied on the reduction in energy when a misfitting carbon-atom that is in solid solution reduces its energy by interacting with the strain field of a dislocation [103]. This reduction, ≈ 0.5 eV, is so large that the carbon atom become more stable at a dislocation core than in carbides such as ε and θ [104]. Very severe deformation can therefore lead to the dissolution of cementite with the carbon relocating to dislocations. The reduction in energy as the carbon segregates to dislocations can be incorporated into phase diagram calculations; Figure 8.20 shows how the solubility of carbon in a mixture of ferrite and dislocations, that is in equilibrium with cementite, is sensitive to the density of dislocations [105].

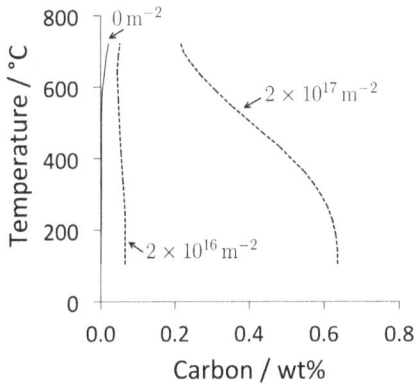

Figure 8.20 Phase diagram calculations showing how the $\alpha + \perp/\theta$ phase boundary is affected by dislocations present in the ferrite, where $\perp$ represents dislocations the density of which is indicated numerically near each curve.

There is another reason why cementite can dissolve when the steel is severely deformed. The plastic deformation thins the cementite by elongation, so increasing the θ/α interfacial area per unit volume; the cementite may also be refined by the deformation cutting the particles as they pass through [106]. Both of these factors reduce its stability until it is more energetically favourable for the cementite to dissolve.

8.9.3 GRAPHITISATION AND SYNTHESIS

Cementite can be metastable relative to graphite in Fe-C binary alloys in the presence of austenite or ferrite, Figure 1.12 [107–109].[2] Cementite presumably is easier to nucleate in the solid-state than graphite, hence its ubiquitous presence in most steels. Graphite does not have a good fit with the ferrite lattice so the structural component of interfacial energy is expected to be large [110]. Whereas there are reproducible orientation relationships between cementite, α and γ, there are none when it comes to graphite precipitation inside steel. Indeed, iron that forms inside a carbon nanotube seems to solidify in random orientations [111]. The fact that the formation of graphite from cementite leads to a large expansion in volume must add to the difficulties of solid-state nucleation:

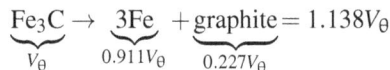

$$\underbrace{Fe_3C}_{V_\theta} \rightarrow \underbrace{3Fe}_{0.911V_\theta} + \underbrace{graphite}_{0.227V_\theta} = 1.138V_\theta$$

where V_θ is the volume of cementite. It would be necessary for iron to diffuse in order to accommodate the growing graphite [112].

One consequence of the metastability of cementite and its relative ease in nucleation, is the phenomenon of *metal dusting*, in which cementite is generated during the desorption of gases such as CO on the steel surface when the activity of carbon in the gas is sufficiently large [86]. The oxygen partial pressure should be small enough to minimise oxidation. The cementite then decomposes into fine particles of iron and graphite, i.e., the dust.

Cementite can be synthesised by gas ($CH_4 + H_2 + Ar$) carburising of iron oxides at about 750 °C. Figure 8.21 shows the thermal stability of such cementite in the form of time-temperature-transformation curves, when the carbide is reheated to a variety of temperatures. The rate at which it decomposes is a lot slower when made using titanomagnetite, attributed to titanium dissolving in cementite and stabilising it [113]. However, phase equilibrium calculations show that there is a negligible solubility of titanium in cementite Longbottom et al. [113]. On the other hand, when pure cementite is in equilibrium with iron containing dissolved titanium, the cementite for some reason becomes stable to the formation of graphite.

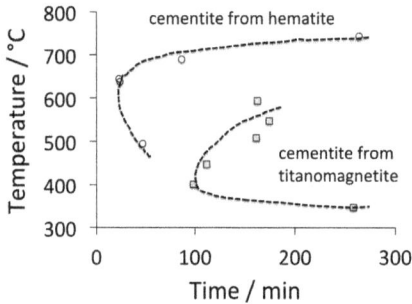

Figure 8.21 Time, temperature and 50% transformation diagrams for the decomposition of cementite into elemental iron and carbon. In one case, the cementite is made by carburising hematite ore (Fe_2O_3), and in the other by similarly carburising titanomagnetite ($Fe_{(1-x)}Ti_xO_4$). Selected data from Longbottom et al. [113].

8.10 CEMENTITE PRECIPITATION IN METALLIC GLASS

Amorphous alloys of iron precipitate cementite when their carbon concentration is sufficiently large; it is difficult to be specific because there is no phase diagram relating to the equilibrium between cementite and the glassy alloy, or even whether an equilibrium mixture of glass and cementite is possible. Figure 8.22a shows cementite and ferrite obtained by the devitrification of a binary glassy-steel 500 nm thick film during heat treatment at just 300°C. It is not clear why the cementite is heavily faulted but its shape indicates that the growth process is reconstructive in nature. This would require the diffusion of iron atoms, and indeed, the rate of transformation is found to be slower than expected from the diffusion of carbon alone [114].

Metallic glasses are configurationally frozen at the glass transition temperature and hence have a greater free volume than supercooled liquid. It is expected therefore that diffusion should be easier than in the crystalline version of the material. The measurement of diffusion coefficients is complicated because the glass may simultaneously undergo structural changes such as relaxation and ultimately, devitrification. Experiments on the diffusion of iron in a variety of metallic glasses in their relaxed condition show that there is indeed an enhanced diffusivity in the glassy state when compare with that in ferrite, although this does not account for the possible effects of chemical composition (Figure 8.22b [115]). It is entirely conceivable that solutes like boron (frequently a component of metallic glasses) will lead to a general expansion of the mean inter-atomic spacing (p. 396).

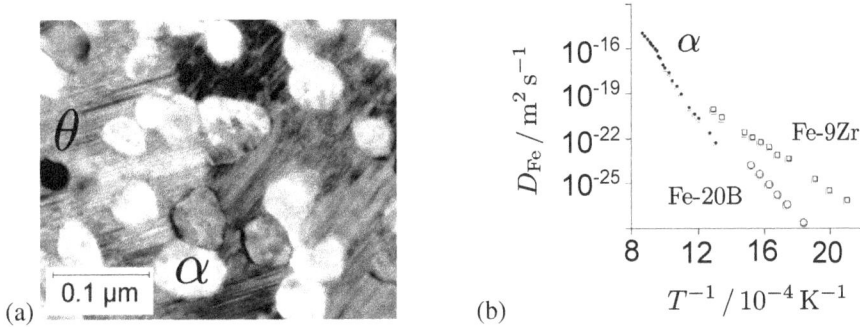

Figure 8.22   (a) Cementite (majority phase, containing planar faults) and equiaxed ferrite, crystallised from metallic glass films of composition Fe-13.6C at% by heat treatment at 300 °C for 3 h. Reproduced from Fillon et al. [114] with the permission of Elsevier. (b) ^{59}Fe tracer diffusion coefficients in Fe-Zr and Fe-B glassy metals in the relaxed state. Self-diffusion data for iron in ferrite are included from [116, 117], for comparison purposes. Selected data on the amorphous alloys from Horvath et al. [115].

8.11   CARBON NANOTUBES – ROLE OF CEMENTITE

The growth of carbon nanotubes from gaseous hydrocarbons is catalysed by fine particles of transition metals, particularly iron. The size of catalyst particles correlates with the diameters of the nanotubes generated [e.g., 118]. Bulky iron or thin films of iron are not as effective as dispersed particles [119]. Flat surfaces do not form good templates for the growth of *tubes*.

There has been discussion [120] about whether it is the iron particles or the cementite particles that form subsequently, catalyse the carbon nanotubes. Environmental transmission electron microscopy has provided clear evidence for "graphitic networks" forming first on cementite particles, followed by the genesis of carbon nanotube growth [121, 122]. The process of carbon depositing on the cementite particle is not uniform, so carbon diffuses through the cementite from the graphite-rich region to uncoated areas, leading to the expulsion of carbon filaments [119].

Not everyone accepts these conclusions about the role of cementite; Tessonnier et al. [123] comment on electron beam induced artefacts and the possible role of surface diffusion. Nevertheless, X-ray diffraction experiments involving nanotube formation in a fluidised bed where a mixture of ethylene, hydrogen and nitrogen is catalysed to decompose by iron supported on alumina, indicate that the iron is converted into metastable cementite which then decomposes into a more stable mixture of iron and carbon-nanotubes, rather like the ordinary process of graphitisation [124]. Mössbauer spectroscopy and transmission electron microscopy of nanotubes synthesised by the catalytic decomposition of acetylene on iron particles have shown that while α-iron is the active centre for the breakdown of acetylene, it is cementite formation that induces the growth of the carbon nanotubes [125].

The presence of α-iron or cementite particles within carbon nanotubes can add a magnetic function that has the potential for exploitation in devices. Tubes synthesised by the pyrolysis of liquid hydrocarbon in a mixture containing ferrocene [Fe(C$_2$H$_5$)$_2$] end up with some 90% of the enclosed particles in cementite acting as single-domain ferromagnets [126].

8.12   DISPLACIVE MECHANISM OF CEMENTITE PRECIPITATION

Cementite can precipitate from supersaturated ferrite at temperatures below 200 °C, in time periods long enough for carbon to partition but for iron to be immobile [127]. The resulting fine platelets of cementite should then grow by a paraequilibrium mechanism with the rate controlled by the diffusion of carbon in the supersaturated-ferrite, towards the growing cementite. This mechanism would lead to the observed reproducible orientation-relationship between the cementite and the matrix from which it precipitates.

The commonly stated α/θ orientation relationship associated with bainite and tempered martensite is due to Bagaryatskii [128, 129]:

$$[1\,0\,0]_\theta \quad \| \quad [1\,\overline{1}\,\overline{1}]_\alpha \, \| \, \mathbf{z}_1$$

$$[0\,1\,0]_\theta \quad \| \quad [2\,1\,1]_\alpha \, \| \, \mathbf{z}_2$$

$$[0\,0\,1]_\theta \quad \| \quad [0\,\overline{1}\,1]_\alpha \, \| \, \mathbf{z}_3. \tag{8.4}$$

Here, the orthonormal basis 'Z' is formed by the unit vectors $\mathbf{z}_1$, $\mathbf{z}_2$ and $\mathbf{z}_3$. Figure 8.23 illustrates the relationship between rows of atoms parallel to $[111]_\alpha$ which transform into zig-zags with the average direction $[\overline{1}\,0\,0]_\theta$ in the cementite. The displacements that can in principle produce the zig-zags are all parallel to $[1\,\overline{1}\,0]_\alpha$ in the plane $(1\,1\,\overline{2})_\alpha$. Any homogeneous deformation that transforms the crystal structure of ferrite to that of cementite would therefore need to be accompanied by the shuffle of atoms to their ultimate positions (page 255).

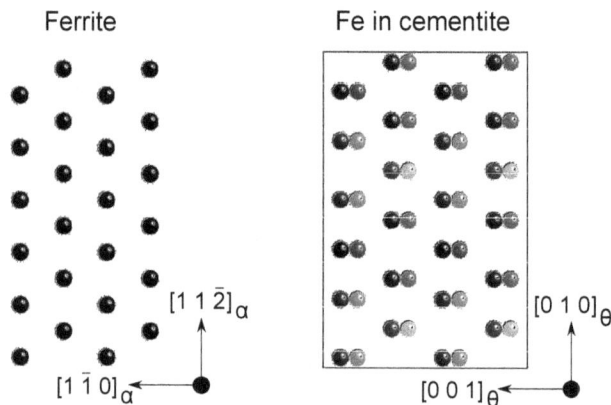

Figure 8.23 The image on the left represents the disposition of iron atoms in ferrite, with rows of atoms going into the plane of the diagram along $[111]_\alpha$. The figure on the right is the cementite cell in the Bagaryatskii relationship with respect to the ferrite. The rows along $[1\,\overline{1}\,\overline{1}]_\alpha$ turn into zig-zags with the average direction $[\overline{1}\,0\,0]_\theta$ into the plane of the diagram.

The first attempt at a displacive transformation model for cementite precipitating in supersaturated ferrite was due to Andrews [127], which assumes the Bagaryatskii orientation relationship with the pure deformation (Z S Z) referred to the orthonormal basis Z defined according to the identities in Equations 8.4:

$$(Z\,S\,Z) = \begin{pmatrix} k = 1.024957 & 0 & 0 \\ 0 & g = 0.960242 & 0 \\ 0 & 0 & m = 1.116120 \end{pmatrix} \qquad (8.5)$$

where k, g and m are the principal deformations. This homogeneous deformation on its own does not generate the cementite lattice, shuffles are necessary to bring the iron atoms into positions consistent with the space group *Pnma*. Since two of these principal deformations are expansions and the third a contraction, it is not possible to find an invariant line between the two lattices without adding a rigid body rotation as an additional deformation. However, any such rigid body rotation would alter the orientation relationship from the observed Bagaryatskii relation.[3] It follows that the Andrews deformation cannot lead to a glissile interface between the ferrite and cementite, violating a fundamental requirement for displacive transformation.

The θ/α orientation relationship found by Isaichev [130], using rational indices, can be expressed as follows:

$$\langle 1\,0\,0\rangle_\theta \quad \| \quad \langle 1\,\bar{1}\,\bar{1}\rangle_\alpha$$
$$\{0\,3\,1\}_\theta \quad \approx\| \quad \{1\,0\,1\}_\alpha. \qquad (8.6)$$

It is important to note the approximation sign, because these planes are not exactly parallel, as pointed out in the original paper by Isaichev [130]. There is an angle of some $1.5\text{-}2°$ between $\{0\,3\,1\}_\theta$ and $\{1\,0\,1\}_\alpha$.

In fact, the orientation relationship is likely to be irrational if the cementite grows without the diffusion of the larger atoms. The Isaichev orientation relationship is close to that of Bagaryatskii making them difficult to distinguish using conventional electron diffraction. It deviates from Bagaryatskii by a rotation of $3.8°$ about the a-axis of the cementite [131]. Accurate measurements on tempered martensite have repeatedly identified the Isaichev orientation relationship and this has led to the suggestion that the Bagaryatskii orientation does not exist [132, 133]. It turns out the deformation described in Equation 8.5, when combined with a rigid body rotation that converts the Bagaryatskii orientation into that of Isaichev, renders the combination an invariant-line strain [134]:

$$(Z\,S_I\,Z) = \underbrace{\begin{pmatrix} 1.024957 & 0 & 0 \\ 0 & 0.960242 & 0 \\ 0 & 0 & 1.116120 \end{pmatrix}}_{\text{`Bain strain'}} \times \underbrace{\begin{pmatrix} 1 & 0 & 0 \\ 0 & 0.9978 & -0.0663 \\ 0 & 0.0663 & 0.9978 \end{pmatrix}}_{\text{`rigid body rotation'}}$$

$$= \underbrace{\begin{pmatrix} 1.02496 & 0 & 0 \\ 0 & 0.958129 & -0.063664 \\ 0 & 0.073999 & 1.113660 \end{pmatrix}}_{\text{`Isaichev strain'}}. \qquad (8.7)$$

The eigenvectors and eigenvalues (λ_i) for (Z S$_I$ Z) are

$$[0 \; -0.486527 \; 0.873665]_Z \qquad \lambda_1 = 1.07245$$

$$[1 \; 0 \; 0]_Z \qquad \lambda_2 = 1.02496$$

$$[0 \; 0.839485 \; -0.543383]_Z \qquad \lambda_3 = 0.999338.$$

The third eigenvector is invariant because its magnitude is essentially unchanged; the maximum elongation has been reduced to 7.2% compared with the 11.6% associated with the (Z S Z) Bagaryatskii orientation.

The matrix (Z S$_I$ Z) can be converted to the cementite basis using a similarity transformation:

$$
(\theta \; S_I \; \theta) = \begin{pmatrix} a & 0 & 0 \\ 0 & b & 0 \\ 0 & 0 & c \end{pmatrix} \begin{pmatrix} 1.02496 & 0 & 0 \\ 0 & 0.958129 & -0.063664 \\ 0 & 0.073999 & 1.113660 \end{pmatrix} \begin{pmatrix} 1/a & 0 & 0 \\ 0 & 1/b & 0 \\ 0 & 0 & 1/c \end{pmatrix}
$$

$$
= \begin{pmatrix} 1.02496 & 0 & 0 \\ 0 & 0.958129 & -0.063664\frac{b}{c} \\ 0 & 0.073999\frac{c}{b} & 1.113660 \end{pmatrix}
$$

$$
(\theta \; S_I \; \theta)^{-1} = \begin{pmatrix} 0.975648 & 0 & 0 \\ 0 & 1.039110 & 0.0594024\frac{b}{c} \\ 0 & -0.0690456\frac{c}{b} & 0.893993 \end{pmatrix}.
$$

The coordinate transformation matrix for the Isaichev relationship becomes

$$
\underbrace{(\alpha \; J \; \theta)}_{\text{coordinate}} = \underbrace{(\alpha \; C \; \theta)}_{\text{correspondence}} (\theta \; S_I \; \theta)^{-1}
$$

$$
= \begin{pmatrix} 1 & 2 & 0 \\ -1 & 1 & -1 \\ -1 & 1 & 1 \end{pmatrix} \begin{pmatrix} 0.975648 & 0 & 0 \\ 0 & 0.999569 & 0.0571418\frac{b}{c} \\ 0 & -0.664178\frac{c}{b} & 0.859971 \end{pmatrix}
$$

$$
= \begin{pmatrix} 0.975648 & 2.078220 & 0.177038 \\ -0.975648 & 1.085445 & -0.805474 \\ -0.975648 & 0.992774 & 0.982512 \end{pmatrix}
$$

$$
(\theta \; J \; \alpha) = \begin{pmatrix} 0.341653 & -0.341653 & -0.341653 \\ 0.319377 & 0.207124 & 0.112254 \\ 0.0165554 & -0.548553 & 0.565107 \end{pmatrix}.
$$

The process described above for cementite, is analogous to that for the martensitic transformation of austenite, where the Bain strain [135] changes the lattice but does not leave any line invariant, and the orientation relationship implied by the Bain strain is not that observed. The correct irrational orientation relationship that is observed, is obtained by adding a precise rigid body rotation that in combination with the Bain strain becomes an invariant-line strain.

The Bagaryatskii deformation as described by Andrews does not leave any vector invariant, but when combined with a rigid body rotation that generates the Isaichev orientation, the resulting total deformation is an invariant-line strain. Furthermore, the principal deformations associated with this invariant-line strain are substantially smaller than those of the Bagaryatskii deformation. This explains the occurrence of the Isaichev orientation relationship.

The analogy with the martensitic transformation of austenite (γ) is clear; the η_i are the principal distortions: The calculations will depend on the lattice parameters of cementite and ferrite, but as long

Transformation	Pure deformation	Pure deformation + rigid body rotation	Final orientation
$\gamma \to \alpha$	Bain strain $\eta_i = 1.136,\ 1.136,\ 0.803$	Invariant-line strain $\eta_i = 1.124,\ 1,\ 0.922$	Kurdjumov-Sachs type
$\alpha \to \theta$	Bagaryatskii $\eta_i = 1.116,\ 1.024,\ 0.960$	Invariant-line strain $\eta_i = 1.073,\ 1.025,\ 1$	Isaichev

as the parameters are known as a function of temperature and composition, they are straightforward to repeat.

The following observations now become compatible with the paraequilibrium, displacive precipitation of cementite from supersaturated ferrite at temperatures where the mobility of iron and substitutional solutes is limited:

- It is possible to define a homogeneous deformation which is an invariant-line strain for the $\alpha \to \theta$ transformation. This is a minimum condition for the existence of a glissile interface between the parent and product lattices.
- Cementite variant selection occurs when elastically loaded martensite is tempered [136]. Such selection is characteristic of a strong interaction of the shape deformation accompanying transformation, with the applied stress.
- The displacive precipitation of cementite would require the diffusion of carbon. However, the iron to substitutional solute ratio must remain constant. This has been verified using the atom-probe technique [137].

First principles calculations of displacive transformation

It is possible to calculate using first principles methods, the free energy of a phase as a function of the configuration, usually for 0 K. These methods consider unconstrained transformation where each state is estimated in isolation, i.e., strain energy due to the obvious displacements that accompany transformations is ignored and interfacial structures are treated over very short ranges using rational planes, so the defect structure and mobility of the interface cannot be assessed with confidence.

One recent calculation due to Zhang et al. [82] describes an intermediate phase that connects austenite, ferrite and cementite during transformations between pairwise combinations of these phases. The intermediate structure is meant to exist only at the transformation front. This interpretation is essentially another way of representing the strain fields of coherency dislocations (p. 267) which accomplish the lattice change.

8.13 α/θ ORIENTATION RELATIONS IN PEARLITE

The level of reproducibility that characterises displacive transformations is not essential when the mechanism is reconstructive. Focussing on data collected with sufficient precision to permit distinctions, many orientation relationships have been observed between cementite and ferrite that form pearlite [138]:

<table>
<tr><td colspan="4" align="center">"New-2"</td><td colspan="4" align="center">"New-3"</td></tr>
<tr><td>$(0\,3\,\bar{1})_\theta$</td><td>$\parallel$</td><td></td><td>$(\bar{1}\,0\,1)_\alpha$</td><td>$(0\,3\,\bar{1})_\theta$</td><td>$\parallel$</td><td></td><td>$(\bar{1}\,0\,1)_\alpha$</td></tr>
<tr><td>$[1\,0\,0]_\theta$</td><td>8.5° from</td><td></td><td>$[1\,3\,1]_\alpha$</td><td>$[\bar{2}\,2\,0]_\theta$</td><td>2.4° from</td><td></td><td>$[1\,\bar{3}\,1]_\alpha$</td></tr>
<tr><td>$[\bar{1}\,1\,3]_\theta$</td><td>$\parallel$</td><td></td><td>$[1\,\bar{1}\,1]_\alpha$</td><td>$[1\,1\,3]_\theta$</td><td>$\parallel$</td><td></td><td>$[1\,\bar{1}\,1]_\alpha$</td></tr>
</table>

<table>
<tr><td colspan="4" align="center">"New-4"</td><td colspan="4" align="center">"New-5"</td></tr>
<tr><td>$(1\,0\,2)_\theta$</td><td>$\parallel$</td><td></td><td>$(\bar{1}\,0\,1)_\alpha$</td><td>$(0\,3\,\bar{1})_\theta$</td><td>$\parallel$</td><td></td><td>$(\bar{1}\,0\,1)_\alpha$</td></tr>
<tr><td>$[1\,1\,3]_\theta$</td><td>5.95° from</td><td></td><td>$[1\,0\,1]_\alpha$</td><td>$[1\,1\,3]_\theta$</td><td>8.5° from</td><td></td><td>$[1\,\bar{1}\,1]_\alpha$</td></tr>
<tr><td>$[2\,1\,\bar{1}]_\theta$</td><td>$\parallel$</td><td></td><td>$[1\,3\,1]_\alpha$</td><td>$[1\,0\,0]_\theta$</td><td>$\parallel$</td><td></td><td>$[1\,3\,1]_\alpha$</td></tr>
</table>

The New-2, New-3 and New-5 relations are close to but significantly different from the classical Pitsch-Petch [139, 140] orientation:

$$(0\,1\,0)_\theta \quad \parallel \quad (5\,\bar{2}\,\bar{1})_\alpha$$

$$[0\,0\,1]_\theta \quad 2.6°\text{ from} \quad [1\,3\,\bar{1}]_\alpha$$

$$[1\,0\,0]_\theta \quad 2.6°\text{ from} \quad [1\,1\,3]_\alpha$$

but as Zhang and Kelly [138] pointed out, neither the Pitsch-Petch, nor the Bagaryatskii orientations were found, and suggested that they may not actually exist if accurate measurements are made; only the Isaichev relation was observed in addition to the new orientations relationships listed above.

The mechanisms required to explain these observations depend on the carbon concentration of the steel [138]. In hypoeutectoid steel, the ferrite precedes the cementite and therefore influences its nucleation, leading to the Isaichev orientation relationship. In hypereutectoid steels, the cementite nucleates first while in contact with austenite, leading to the New-5 orientation relationship with the ferrite; in this case the orientation of the ferrite within the pearlite is determined by its relationship with the austenite. The mechanisms determining the New-2, New-3 and New-4 relationships are not as clear, but apparently occur in eutectoid steel and are dominated by the relationship of the closest packed planes with the cementite with the closest-packed planes in austenite and ferrite.

8.14 PITSCH γ/θ ORIENTATION RELATIONSHIP

The Pitsch orientation relationship [140, 141] between austenite and cementite in the *Pnma* setting is given by

$$[100]_\theta \quad \| \quad [110]_\gamma$$

$$[010]_\theta \quad \| \quad [\bar{2}25]_\gamma$$

$$[001]_\theta \quad \| \quad [5\bar{5}4]_\gamma.$$

The $[100]_\theta$ vector is not only parallel to $[110]_\gamma$, but its length is almost identical, making it an invariant line between the two lattices. For $a_\gamma = 0.3568\,\text{nm}$, $a_\theta = 0.50837\,\text{nm}$, $b_\theta = 0.67475\,\text{nm}$, $c_\theta = 0.45165\,\text{nm}$, the coordinate transformations become

$$(\gamma\, J\, \theta) \;=\; \begin{pmatrix} 1.007488 & -0.658402 & 0.791147 \\ 1.007488 & 0.658402 & -0.791147 \\ 0 & 1.646005 & 0.632918 \end{pmatrix}$$

$$(\theta\, J\, \gamma) \;=\; \begin{pmatrix} 0.496284 & 0.496284 & 0 \\ -0.184101 & 0.184101 & 0.460251 \\ 0.478783 & -0.478783 & 0.383027 \end{pmatrix}.$$

Using the correspondence matrix due to Sleeswyk [142], configured to the *Pnma* space-group setting, the deformation that converts the austenite lattice to that of cementite is given by

$$(\gamma\, S\, \gamma) \;=\; (\gamma\, J\, \theta)(\theta\, C\, \gamma)$$

$$= \begin{pmatrix} 1.007488 & -0.658402 & 0.791147 \\ 1.007488 & 0.658402 & -0.791147 \\ 0 & 1.646005 & 0.632918 \end{pmatrix} \times \frac{1}{6} \begin{pmatrix} 4 & 2 & 0 \\ 1 & -1 & -3 \\ -3 & 3 & -3 \end{pmatrix}$$

$$= \begin{pmatrix} 0.166352 & 0.841136 & -0.066373 \\ 1.176970 & -0.169478 & 0.066373 \\ -0.042124 & 0.042124 & -1.139460 \end{pmatrix}.$$

The eigenvectors and eigenvalues (λ_i) for $(\gamma\, S\, \gamma)$ are given approximately by

$$[0.344239\ 0.46953\ 1]_\gamma \qquad \lambda_1 = -1.17374$$

$$[1\ 1\ 0]_\gamma \qquad \lambda_2 = 1.00749$$

$$[-1.60845\ 2.26399\ 1]_\gamma \qquad \lambda_3 = -0.976336.$$

The alternative correspondence defined by Sleeswyk [142]:

$$(\theta\, C\, \gamma) = \frac{1}{12} \begin{pmatrix} 7 & 5 & 3 \\ 1 & -1 & -6 \\ -6 & 6 & -6 \end{pmatrix}$$

leads to the eigenvectors and eigenvalues for the deformation matrix as follows:

$$[0.26009 \ -0.553681 \ 1]_\gamma \qquad \lambda_1 = -1.17374$$

$$[1 \ 1 \ 0]_\gamma \qquad \lambda_2 = 1.00749$$

$$[-1.8993 \ 1.97315 \ 1]_\gamma \qquad \lambda_3 = -0.976336.$$

The deformations λ_i therefore are not different and it remains the case that the $[100]_\theta \parallel [110]_\gamma$ is unrotated and almost unchanged in length, an invariant line.

REFERENCES

1. F. Osmond, and J. Werth: 'Theorie cellulaire des propriétés de l'acier', *Annales des Mines*, 1885, **8**, 5–84.

2. F. Osmond, and J. Werth: 'Structure cellulaire de l'acier foundu', *Comptes Rendus*, 1885, **100**, 450–452.

3. S. B. Hendricks: 'XXXVI. The crystal structure of cementite', *Zietschrift für Kristallographie*, 1930, **74**, 534–545.

4. A. Mussi, P. Cordier, S. Ghosh, N. Garvik, B. C. Nzogang, P. Carrez, and S. Garruchet: 'Transmission electron microscopy of dislocations in cementite deformed at high pressure and high temperature', *Philosophical Magazine*, 2016, **96**, 1773–1789.

5. A. Inoue, T. Ogura, and T. Muramatsu: 'Burgers vectors of dislocations in cementite crystal', *Scripta Metallurgica*, 1977, **11**, 1–5.

6. J. Zhang, A. Schneider, and G. Inden: 'Cementite decomposition and coke gasification in He and H_2-He gas mixtures', *Corrosion Science*, 2004, **46**, 667–669.

7. X. Lei, W. Wang, Z. Ye, N. Zhao, and H. Yang: 'High saturation magnetization of Fe_3C nanoparticles synthesized by a simple route', *Dyes and Pigments*, 2017, **139**, 448–452.

8. NPL: 'MTDATA': Software, National Physical Laboratory, Teddington, U.K., 2006.

9. H. K. D. H. Bhadeshia: 'Cementite', *International Materials Reviews*, 2020, **65**, 1–27.

10. M. E. Lipschutz, and E. Anders: 'The record in the meteorites – IV: Origin of diamonds in iron meteorites', *Geochimica et Cosmochimica Acta*, 1961, **24**, 83–105.

11. V. V. Sobolev, Y. N. Taran, and S. I. Gubenko: 'Synthesis of diamond in cast iron', *Metal Science and Heat Treatment*, 1993, **35**, 3–9.

12. R. Brett: 'Cohenite: its occurrence and a proposed origin', *Geochimica et Cosmochimica Acta*, 1967, **31**, 143–159.

13. E. J. Fasiska, and G. A. Jeffrey: 'On the cementite structure', *Acta Crystallographica*, 1965, **19**, 463–471.

14. 'Table of space group symbols, http://www.cryst.ehu.es/cgi-bin/cryst/programs/nph-wp-list': 2018: URL http://www.cryst.ehu.es/cgi-bin/cryst/programs/nph-wp-list.

15. C. Jiang, S. A. Maloy, and S. G. Srinivasan: 'A computational method to identify interstitial sites in complex materials', *Scripta Materialia*, 2008, **58**, 739–742.

16. B. G. Lyashchenko, and L. M. Sorokin: 'Determination of the position of carbon in cementite by the neutron diffraction method', *Soviet Physics - Crystallography*, 1963, **8**, 300–304.

17. K. Kawakami, and T. Matsumiya: '*Ab-initio* investigation of hydrogen trap state by cementite in bcc-Fe', *ISIJ international*, 2013, **53**, 709–713.

18. F. Abel: 'Final report on experiments bearing upon the question of the condition in which carbon exists in steel', *Proceedings of the Institution of Mechanical Engineers*, 1885, **36**, 30–57.

19. H. Lipson, and N. J. Petch: 'The crystal structure of cementite Fe_3C', *Journal of the Iron and Steel Institute*, 1940, **142**, 95P–103P.

20. N. J. Petch: 'Interpretation of the crystal structure of cementite', *Journal of the Iron and Steel Institute*, 1944, **149**, 143–150.

21. F. X. Kayser, and Y. Sumitomo: 'On the composition of cementite in equilibrium with ferrite at room temperature', *Journal of Phase Equilibria*, 1997, **18**, 458–464.

22. L. Battezzati, M. Baricco, and S. Curiotto: 'Non-stoichiometric cementite by rapid solidification of cast iron', *Acta Materialia*, 2005, **53**, 1849–1856.

23. A. Leineweber, S. L. Shang, and Z. K. Liu: 'C-vacancy concentration in cementite, Fe_3C_{1-z}, in equilibrium with α-Fe[C] and γ-Fe[C]', *Acta Materialia*, 2015, **86**, 374–384.

24. A. Schneider, and M. Palm: 'Characterisation of non-stoichiometric cementite (Fe_3C)', *CALPHAD*, 2020, **68**, 101689.

25. H. Göhring, A. Leineweber, and E. J. Mittemeijer: 'A thermodynamic model for non-stoichiometric cementite; the Fe–C phase diagram', *CALPHAD*, 2016, **52**, 38–46.

26. T. Okamoto, and H. Matsumoto: 'Precipitation of ferrite from cementite', *Metal Science*, 1975, **9**, 8–12.

27. W. Stuckens, and A. Michel: 'Variations in the stoichiometry of pure cementite', *Comptes rendus de l'Académie des Sciences, Paris*, 1961, **253**, 2358–2360.

28. A. H. Cottrell: 'A theory of cementite', *Materials Science and Technology*, 1993, **9**, 277–280.

29. D. V. Shtansky, K. Nakai, and Y. Ohmori: 'Mechanism and crystallography of ferrite precipitation from cementite in an Fe-Cr-C alloy during austenitizatio', *Philosophical Magazine A*, 1999, **79**, 1655–1669.

30. A. Dick, F. Körmann, T. Hickel, and J. Neugebauer: '*Ab initio* based determination of thermodynamic properties of cementite including vibronic, magnetic, and electronic excitations', *Physical Review B*, 2011, **84**, 125101.

31. J. Häglund, G. Grimvall, and T. Jarlberg: 'Electronic structure, X-ray photoemission spectra, and transport properties of Fe_3C (cementite)', *Physical Review B*, 1991, **44**, 2914–2919.

32. A. K. Arzhnikov, L. V. Dobysheva, and C. Demangeat: 'Structural peculiarities of cementite and their influence on magnetic characteristics', *Journal of Physics – Condensed Matter*, 2007, **19**, 196214.

33. J. H. Jang, I. G. Kim, and H. K. D. H. Bhadeshia: 'Substitutional solution of silicon in cementite: a first-principles study', *Computational Materials Science*, 2009, **44**, 1319–1326.

34. J. F. Lin, V. V. Struzhkin, H. K. Mao, and R. J. Hemley: 'Magnetic transition in compressed Fe_3C from X-ray emission spectroscopy', *Physical Review B*, 2004, **70**, 212405.

35. H. J. Choe, T. Terai, T. Fukuda, T. Kakeshita, S. Yamamoto, and M. Yonemura: 'Easy axis of magnetization of Fe_3C prepared by an electrolytic extraction method', *Journal of Magnetism and Magnetic Materials*, 2016, **417**, 1–5.

36. B. Reznik, A. Kontny, M. Uehara, J. Gattacceca, P. Solheid, and M. Jackson: 'Magnetic domains and magnetic stability of cohenite from the Morasko iron meteorite', *Journal of Magnetism and Magnetic Materials*, 2017, **426**, 594–603.

37. S. Yamamoto, T. Terai, T. Fukuda, K. Sato, T. Kakeshita, S. Horii, M. Ito, and M. Yonemura: 'Magnetocrystalline anisotropy of cementite pseudo single crystal fabricated under a rotating magnetic field', *Journal of Magnetism and Magnetic Materials*, 2018, **451**, 1–4.

38. A. S. Keh, and C. A. Johnson: 'Ferromagnetic domain structures in cementite', *Journal of Applied Physics*, 1963, **34**, 2670–2676.

39. M. Hillert, and N. Lange: 'Direct observation of magnetic domains in cementite', *Journal of Applied Physics*, 1959, **10**, 945.

40. L. V. Dobysheva: 'First-principles calculations for alloyed cementite $(Fe–Ni)_3C$', *Bulletin of the Russian Academy of Sciences: Physics*, 2017, **81**, 798–802.

41. A. I. Ul'yanov, A. A. Chulkina, V. A. Volkov, E. P. Elsukov, A. V. Zagainov, A. V. Protasov, and I. A. Zykina: 'Structural state and magnetic properties of cementite alloyed with manganese', *Physics of Metals and Metallography*, 2012, **113**, 1134–1145.

42. N. I. Medvedeva, I. R. Shein, M. A. Konyaeva, and A. L. Ivanovskii: 'Effect of chromium on the electronic structure and magnetic properties of cementite', *Physics of Metals and Metallography*, 2008, **105**, 568–573.

43. J. von Appen, B. Eck, and R. Dronskowski: 'A density-functional study of the phase diagram of cementite-type $(Fe, Mn)_3C$ at absolute zero temperature', *Journal of Computational Chemistry*, 2010, **31**, 2620–2627.

44. V. I. Razumovskiy, and G. Ghosh: 'A first-principles study of cementite (Fe_3C) and its alloyed counterparts: Structural properties, stability, and electronic structure', *Computational Materials Science*, 2015, **110**, 169–181.

45. B. Kaeswurm, K. Friemert, M. Gürsoy, K. P. Skokov, and O. Gutfleisch: 'Direct measurement of the magnetocaloric effect in cementite', *Journal of Magnetism and Magnetic Materials*, 2016, **410**, 105–108.

46. D. Walker, J. Li, B. Kalkan, and S. M. Clark: 'Thermal, compositional and compressional demagnetization of cementite', *American Mineralogist*, 2015, **100**, 2610–2624.

47. E. Duman, M. Acet, and E. F. Wassermann: 'Magnetic instabilities in Fe_3C cementite particles observed with Fe K-Edge X-ray circular dichroism under pressure', *Physical Review Letters*, 2005, **94**, 075502.

48. S. Khmelevskyi, A. V. Ruban, and P. Mohn: 'Electronic structure analysis of the pressure induced metamagnetic transition and magnetovolume anomaly in Fe_3C – cementite', *Journal of Physics – Condensed Matter*, 2005, **17**, 7345–7352.

49. M. Umemoto, Z. G. Liu, K. Masuyama, and K. Tsuchiya: 'Influence of alloy additions on production and properties of bulk cementite', *Scripta Materlulla*, 2001, **45**, 391–397.

50. R. C. Reed, and J. H. Root: 'Determination of the temperature dependence of the lattice parameters of cementite by neutron diffraction', *Scripta Materialia*, 1997, **38**, 95–99.

51. I. G. Wood, L. Vočadlo, K. S. Knight, D. P. Dobson, W. G. Marshall, G. D. Price, and J. Brodholt: 'Thermal expansion and crystal structure of cementite, Fe_3C, between 4 and 600 K determined by time-of-flight neutron powder diffraction', *Journal of Applied Crystallography*, 2004, **37**, 82–90.

52. K. D. Litasov, S. V. Rashchenko, A. N. Shmakov, Y. N. Palyanov, and A. G. Sokol: 'Thermal expansion of iron carbides, Fe_7C_3 and Fe_3C, at 297–911 K determined by in situ X-ray diffraction', *Journal of Alloys and Compounds*, 2015, **628**, 102–106.

53. S. Gorai, P. S. Ghosh, C. Bhattacharya, and A. Arya: 'Ab-initio study of pressure evolution of structural, mechanical and magnetic properties of cementite (Fe_3C) phase', In: *AIP Conference Proceedings*, vol. 1942. New York, USA: AIP Publishing, 2018:030015.

54. A. Inoue, T. Ogura, and T. Masumoto: 'Deformation and fracture behaviours of cementite', *Transactions of JIM*, 1976, **17**, 663–672.

55. P. M. Anderson, J.-S. Wang, and J. R. Rice: 'Thermodynamic and mechanical models of interfacial embrittlement', In: G. B. Olson, M. Azrin, and E. S. Wright, eds. *Innovations in Ultrahigh-Strength Steel Technology*. Massachusetts, USA: US Army Materials Technology Laboratory, 1987:619–650.

56. M. Ruda, D. Farkas, and G. Garcia: 'Atomistic simulations in the Fe-C system', *Computational Materials Science*, 2009, **45**, 550–560.

57. W. C. Chiou Jr., and E. A. Carter: 'Structure and stability of Fe_3C cementite surfaces from first principles', *Surface Science*, 2003, **530**, 87–100.

58. M. Born: 'On the stability of crystal lattices. I', *Mathematical Proceedings of the Cambridge Philosophical Society*, 1940, **36**, 160–172.

59. S. K. R. Patil, S. V. Khare, B. R. Tuttle, J. K. Bording, and S. Kodambaka: 'Mechanical stability of possible structures of PtN investigated using first-principles calculations', *Physical Review B*, 2006, **73**, 104118.

60. G. Ghosh: 'A first-principles study of cementite (fe3c) and its alloyed counterparts: Elastic constants, elastic anisotropies, and isotropic elastic moduli', *AIP Advances*, 2015, **5**, 087102.

61. H. Dierkes, and R. Dronskowski: 'High-resolution powder neutron diffraction on Mn_3C', *Zeitschrift für anorganische und allgemeine Chemie*, 2014, **640**, 3148–3152.

62. L. Mauger, J. E. Herriman, O. Hellman, S. J. Tracy, M. S. Lucas, J. A. M. noz, Y. Xiao, J. Li, and B. Fultz: 'Phonons and elasticity of cementite through the Curie temperature', *Physical Review B*, 2017, **95**, 024308.

63. L. Huang, Y. Tu, X. Wang, X. Zhou, F. Fang, and J. Jiang: 'Site preference of manganese in mn-alloyed cementite', *Physica Status Solidi B*, 2016, **253**, 1623–1628.

64. Z. Q. Lv, W. T. Fu, S. H. Sun, X. H. Bai, Y. Gao, Z. H. Wang, and P. Jiang: 'First-principles study on the electronic structure, magnetic properties and phase stability of alloyed cementite with Cr or Mn', *Journal of Magnetism and Magnetic Materials*, 2011, **323**, 915–919.

65. C. X. Wang, Z. Q. Lv, W. T. Fu, Y. Li, S. H. Sun, and B. Wang: 'Electronic properties, magnetic properties and phase stability of alloyed cementite (Fe,M)$_3$C (M=Co,Ni) from density-functional theory calculations', *Solid State Sciences*, 2011, **13**, 1658–1663.

66. M. Nikolussi, S. L. Shang, T. Gressmann, A. Leineweber, E. J. Mittemeijer, Y. Wang, and Z. K. Liu: 'Extreme elastic anisotropy of cementite, Fe$_3$C: First-principles calculations and experimental evidence', *Scripta Materialia*, 2008, **59**, 814–817.

67. G. Fiquet, J. Badro, E. Gregoryanz, Y. Fei, and F. Occelli: 'Sound velocity in iron carbide (Fe$_3$C) at high pressure: Implications for the carbon content of the earth's inner core', *Physics of the Earth and Planetary Interiors*, 2009, **172**, 125–129.

68. B. W. Koo, Y. J. Chang, S. P. Hong, C. S. Kang, S. W. Jeong, W. J. Nam, I. J. Park, Y. K. Lee, K. H. Oh, and Y. W. Kim: 'Experimental measurement of Young's modulus from a single crystalline cementite', *Scripta Materialia*, 2014, **82**, 25–28.

69. N. Sata, G. Shen, M. L. Rivers, and S. R. Sutton: 'Pressure-volume equation of state of the high-pressure B2 phase of NaCl', *Physical Review B*, 2002, **65**, 104114.

70. J. Li, H. K. Mao, Y. Fei, E. Gregoryanz, M. Eremets, and C. S. Zha: 'Compression of Fe$_3$C to 30 GPa at room temperature', *Physics and Chemistry of Minerals*, 2002, **29**, 166–169.

71. N. Sata, K. Hirose, G. Shen, Y. Nakajima, Y. Ohishi, and N. Hirao: 'Compression of FeSi, Fe$_3$C, Fe$_{0.95}$O, and FeS under core pressures and implication for light element in the earth's core', *Geophysical Research: Solid Earth*, 2010, **115**, B09204.

72. F. Birch: 'Equation of state and thermodynamic parameters of NaCl to 300 kbar in the high-temperature domain', *Journal of Geophysical Research*, 1986, **91**, 4949–4954.

73. K. D. Litasov, I. S. Sharygin, P. I. Dorogokupets, A. Shatskiy, P. N. Gavryushkin, T. S. Sokolova, E. Ohtani, J. Li, and K. Funakoshi: 'Thermal equation of state and thermodynamic properties of iron carbide Fe$_3$C to 31 GPa and 1473 K', *Journal of Geophysical Research: Solid Earth*, 2013, **118**, 1–11.

74. H. P. Scott, Q. Williams, and E. Knittle: 'Stability and equation of state of cementite to 73 GPa: implications for carbon in the earth's core', *Geophysical Research Letters*, 2001, **28**, 1875–1878.

75. K. B. Panda, and K. S. Ravi Chandran: 'First principles determination of elastic constants and chemical bonding of titanium boride (TiB) on the basis of density functional theory', *Acta Materialia*, 2006, **54**, 1641–1657.

76. S. J. Li, M. Ishihara, H. Yumoto, T. Aizawa, and M. Shimotomai: 'Characterisation of cementite films prepared by electron-shower-assisted PVD method', *Thin Solid Films*, 1998, **316**, 100–104.

77. H. Mizubayashi, S. J. Li, H. Yumoto, and M. Shimotomai: 'Young's modulus of single phase cementite', *Scripta Materialia*, 1999, **40**, 773–777.

78. S. P. Dodd, G. A. Saunders, M. Cankurtaran, B. James, and M. Acet: 'Ultrasonic study of the temperature and hydrostatic pressure dependences of the elastic properties of polycrystalline cementite (Fe$_3$C)', *Physica Status Solidi A*, 2003, **198**, 272–281.

79. M. Umemoto, Y. Todaka, T. Takahashi, P. Li, R. Tokumiya, and K. Tsuchiya: 'High temperature deformation behavior of bulk cementite produced by mechanical alloying and spark plasma sintering', *Materials Science & Engineering A*, 2004, **375-377**, 894–898.

80. S. Shimura: 'A study on the structure of cementite', *Proceedings of the Imperial Academy*, 1930, **6**, 269–271.

81. X. Wang, X. Chen, H. Ding, and H. Yang: 'Synthesis and magnetic properties of Fe_3C doped with Mn or Ni for applications as adsorbents', *Dyes and Pigments*, 2017, **144**, 76–79.

82. X. Zhang, T. Hickel, J. Rogal, S. Fähler, R. Drautz, and J. Neugebauer: 'Structural transformations among austenite, ferrite and cementite in Fe-C alloys: A unified theory based on ab initio simulations', *Acta Materialia*, 2015, **99**, 281–289.

83. T. Shigematsu: 'Invar properties of cementite $(Fe_{1-x}Me_x)_3C$, Me= Cr, Mn, Ni', *Journal of the Physical Society of Japan*, 1975, **39**, 915–920.

84. A. I. Ulyanov, A. A. Chulkina, V. A. Volkov, A. L. Ulyanov, and A. V. Zagainov: 'Structure and magnetic properties of mechanically synthesized $(Fe_{1-x}Ni_x)_{75}C_{25}$ nanocomposites', *Physics of Metals and Metallography*, 2017, **118**, 691–699.

85. X. Wang, and M. Yan: 'Effect of cobalt and nickel on the structural stability for Fe_3C: first-principles calculations', *International Journal of Modern Physics B*, 2009, **23**, 1135–1140.

86. H. J. Grabke, R. Krajak, and J. C. Nava Paz: 'On the mechanism of catastrophic carburization: 'metal dusting'', *Corrosion Science*, 1993, **35**, 1141–1150.

87. M. A. Konyaeva, and N. I. Medvedeva: 'Electronic structure, magnetic properties and stability of binary and ternary $(Fe,Cr)_3C$ and $(Fe,Cr)_7C_3$', *Physics of the Solid State*, 2009, **51**, 2084–2089.

88. I. R. Shein, N. I. Medvedeva, and A. L. Ivanovskii: 'Electronic structure and magnetic properties of Fe_3C with 3d and 4d impurities', *Physica Status Solidi B*, 2007, **244**, 1971–1981.

89. E. Kozeschnik, B. Sonderegger, I. Holzer, J. Rajek, and H. Cerjak: 'Computer simulation of the precipitate evolution during industrial heat treatment of complex alloys', *Materials Science Forum*, 2007, **539**, 2431–2436.

90. K. B. Rundman, D. J. Moore, K. L. Hayrynen, W. J. Dubensky, and T. N. Rouns: 'Microstructure and mechanical properties of austempered ductile iron', *Journal of Heat Treating*, 1988, **5**, 79–95.

91. P. Jacques, Q. Furnémont, T. Pardoen, and F. Delannay: 'The role and significance of martensite on the mechanical properties of TRIP-assisted multiphase steels', In: D. Miannay, P. Costa, D. Francois, and A. Pineau, eds. *Advances in Mechanical Behaviour, Plasticity and Damage*. Netherlands: Elsevier, 2000:823–828.

92. H. K. D. H. Bhadeshia: Bainite in steels: theory and practice: 3rd ed., Leeds, U.K.: Maney Publishing, 2015.

93. B. Hallstedt, D. Djurovic, J. von Appen, R. Dronskowski, A. Dick, F. Körmann, T. Hickel, and J. Neugebauer: 'Thermodynamic properties of cementite (Fe_3C)', *Computer Coupling of Phase Diagrams and Thermochemistry*, 2010, **34**, 129–133.

94. M. E. Nicholson: 'Solubility of boron in Fe_3C and variation of saturation magetization, Curie temperature, and lattice parameter of $Fe_3(C,B)$ with composition cementite', *Journal of Metals*, 1957, **9**, 1–6.

95. B. L. He, D. H. Ping, and W. T. Geng: 'First-principles study of helium trapping in cementite Fe_3C', *Journal of Nuclear Materials*, 2014, **444**, 368–372.

96. K. Kawakami, and T. Matsumiya: '*Ab initio* investigation of hydrogen trap state by cementite in bcc-Fe', *ISIJ International*, 2013, **53**, 709–713.

97. A. Kagawa, and T. Okamoto: 'Lattice parameters of cementite in Fe–C–X (X= Cr, Mn, Mo, and Ni) alloys', *Transactions of JIM*, 1979, **20**, 659–666.

98. J. Zelenty, G. D. W. Smith, K. Wilford, J. M. Hyde, and M. P. Moody: 'Secondary precipitation within the cementite phase of reactor pressure vessel steels', *Scripta Materialia*, 2016, **115**, 118–122.

99. T. Chairuangsri, and D. V. Edmonds: 'The precipitation of copper in abnormal ferrite and pearlite in hyper-eutectoid steels', *Acta Materialia*, 2000, **48**, 3931–3949.

100. G. Fourlaris, A. J. Baker, and G. D. Papadimitriou: 'Microscopic characterisation of ε-Cu interphase precipitaiton in hypereutectoid Fe-C-Cu alloys', *Acta Metallurgica et Materalia*, 1995, **43**, 2589–2604.

101. G. L. Dunlop, C.-J. Carlsson, and G. Frimodig: 'Precipitation of VC in ferrite and pearlite during direct transformation of a medium carbon microalloyed steel', *Metallurgical & Materials Transactions A*, 1978, **9**, 261–266.

102. B. I. Izotov: 'Precipitation of disperse vanadium carbides at the interphase boundary upon the pearlitic transformation of a steel', *Physics of Metals and Metallography*, 20111, **11**, 592–597.

103. A. H. Cottrell, and B. A. Bilby: 'Dislocation theory of yielding and strain ageing of iron', *Proceedings of the Physics Society A*, 1949, **62**, 49–62.

104. D. Kalish, M. Cohen, and S. A. Kulin: 'Strain tempering of bainite in 9Ni-4Co-0.45C steel', *Journal of Materials*, 1970, **5**, 169–183.

105. M. Maalekian, and E. Kozeschnik: 'A thermodynamic model for carbon trapping in lattice defects', *CALPHAD*, 2008, **32**, 650–654.

106. J. Languillaume, G. Kapelski, and B. Baudelet: 'Cementite dissolution in heavily cold drawn pearlitic steel wires', *Acta Materialia*, 1997, **45**, 1201–1212.

107. C. P. Yap, and C. L. Liu: 'The free energy, entropy and heat of formation of iron carbide (Fe_3C)', *Transactions of the Faraday Society*, 1934, **28**, 788–797.

108. L. S. Darken, and R. W. Gurry: 'Free energy of formation of cementite and the solubility of cementite in austenite', *Transactions of the AIME (Journal of Metals)*, 1951, **3**, 1015–1018.

109. J. Chipman: 'Thermodynamics and phase diagram of the Fe-C system', *Metallurgical Transactions*, 1972, **3**, 55–64.

110. Y. Inokuti: 'Formation of graphite on the surface of cold rolled low carbon steel sheet during annealing', *Transactions of the Iron and Steel Institute of Japan*, 1974, **15**, 314–323.

111. D. Goldberg, M. Mitome, C. Müller, C. Tang, A. Leonhardt, and Y. Bando: 'Atomic structures of iron-based single-crystalline nanowires crystallized inside multi-walled carbon nanotubes as revealed by analytical electron microscopy', *Acta Materialia*, 2006, **54**, 2567–2576.

112. A. Okamoto: 'Graphite formation in high-purity cold-rolled carbon steels', *Metallurgical Transactions A*, 1989, **20**, 1917–1927.

113. R. J. Longbottom, O. Ostrovski, J. Zhang, and D. Young: 'Stability of cementite formed from hematite and titanomagnetite ore', *Metallurgical & Materials Transactions B*, 2007, **38**, 175–184.

114. A. Fillon, X. Sauvage, B. Lawrence, C. Sinclair, M. Perez, A. Weck, E. Cantergiani, T. Epicier, and C. P. Scott: 'On the direct nucleation and growth of ferrite and cementite without austenite', *Scripta Materialia*, 2015, **95**, 35–38.

115. J. Horvath, J. Ott, K. Pfahler, and W. Ulfert: 'Tracer diffusion in amorphous alloys', *Materials Science & Engineering*, 1988, **97**, 409–413.

116. F. S. Buffington, K. Hirano, and M. Cohen: 'Self diffusion in iron', *Acta Metallurgica*, 1961, **9**, 434–439.

117. R. J. Borg, and C. E. Birchenall: 'Self diffusion in alpha-iron', *Transactions of the AIME*, 1960, **218**, 980–984.

118. V. Jourdain, and C. Bichara: 'Current understanding of the growth of carbon nanotubes in catalytic chemical vapour deposition', *Carbon*, 2013, **58**, 2–39.

119. M. A. Ermakova, D. Y. Ermakov, A. L. Chuvilin, and G. G. Kuvshinov: 'Decomposition of methane over iron catalysts at the range of moderate temperatures: The influence of structure of the catalytic systems and the reaction conditions on the yield of carbon and morphology of carbon filaments', *Journal of Catalysis*, 2001, **201**, 183–197.

120. A. C. Dupuis: 'The catalyst in the CCVD of carbon nanotubes – a review', *Progress in Materials Science*, 2005, **50**, 929–961.

121. A. K. Schaper, H. Hou, A. Greiner, and F. Phillipp: 'The role of iron carbide in multiwalled carbon nanotube growth', *Journal of Catalysis*, 2004, **222**, 250–254.

122. H. Yoshida, S. Takeda, T. Uchiyama, K. Kohno, and Y. Homma: 'Atomic-scale in-situ observation of carbon nanotube growth from solid state iron carbide nanoparticles', *Nano Letters*, 2008, **8**, 2082–2086.

123. J. P. Tessonnier, and D. S. Su: 'Recent progress on the growth mechanism of carbon nanotubes: a review', *ChemSusChem*, 2011, **4**, 824–847.

124. L. Pellegrino, M. Daghetta, R. Pelosato, A. Citterio, and C. V. Mazzocchia: 'Searching for rate determining step in CNT formation:the role of cementite', *Chemical Engineering Transactions*, 2013, **32**, 739–744.

125. M. Pérez-Cabero, J. B. Taboada, A. Guerrero-Ruiz, A. R. Overweg, and I. Rodríguez-Ramos: 'The role of alpha-iron and cementite phases in the growing mechanism of carbon nanotubes: A ^{57}Fe Mössbauer spectroscopy study', *Physical Chemistry and Chemical Physics*, 2006, **8**, 1230–1235.

126. A. S. Basaev, B. B. Bokhonov, O. F. Demidenko, V. A. Lubanov, G. I. Makovetskii, E. L. Prudnikova, A. A. Reznev, A. N. Saurov, V. M. Fedosyuk, Y. A. Fedotova, B. G. Shuliskii, and K. I. Yanushkevich: 'Synthesis and properties of magnetically functionalized carbon nanotubes', *Nanotechnologies in Russia*, 2008, **3**, 184–190.

127. K. W. Andrews: 'The structure of cementite and its relation to ferrite', *Acta Metallurgica*, 1963, **11**, 939–946.

128. Y. A. Bagaryatskii: 'Possible mechanism of martensite decomposition', *Doklady Akademii Nauk, SSSR*, 1950, **73**, 1161–1164.

129. D. N. Shackleton, and P. M. Kelly: 'Morphology of bainite', In: *Physical Properties of Martensite and Bainite, Special Report 93*. London, U.K.: Iron and Steel Institute, 1965:126–134.

130. I. V. Isaichev: 'Orientation of cementite in tempered carbon steel (orientatsiya tsementita v otpushchennoi uglerodistoi stali)', *Zhurnal Tekhncheskoi Fiziki*, 1947, **17**, 835–838.

131. F. G. Wei, and K. Tsuzaki: 'Crystallography of $[011]/54.7°$ lath boundary and cementite in tempered 0.2C steel', *Acta Materialia*, 2005, **53**, 2419–2429.

132. M.-X. Zhang, and P. M. Kelly: 'Determination of carbon content in bainitic ferrite and carbon distribution in austenite by using CBKLDP', *Materials Characterization*, 1998, **40**, 159–168.

133. M.-X. Zhang, and P. M. Kelly: 'Crystallography of spheroidite and tempered martensite', *Acta Materialia*, 1998, **46**, 4081–4091.

134. H. K. D. H. Bhadeshia: 'Solution to the Bagaryatskii and Isaichev ferrite-cementite orientation relationship problem', *Materials Science and Technology*, 2018, **34**, 1666–1668.

135. E. C. Bain: 'The nature of martensite', *Transactions of the AIME*, 1924, **70**, 25–46.

136. J. W. Stewart, R. C. Thomson, and H. K. D. H. Bhadeshia: 'Cementite precipitation during tempering of martensite under the influence of an externally applied stress', *Journal of Materials Science*, 1994, **29**, 6079–6084.

137. S. S. Babu, K. Hono, and T. Sakuri: 'APFIM studies on martensite tempering of Fe-C-Si-Mn steel', *Applied Surface Science*, 1993, **67**, 321–327.

138. M.-X. Zhang, and P. M. Kelly: 'Accurate orientation relationships between ferrite and cementite in pearlite', *Scripta Materialia*, 1997, **37**, 2009–2015.

139. N. J. Petch: 'The orientation relationships between cementite and α-iron', *Acta Crystallographica*, 1953, **6**, 96–96.

140. W. Pitsch: 'Der orientierungszusammenhang zwischen zementit und austenit', *Acta Metallurgica*, 1962, **10**, 897–900.

141. K. H. Yang, and W. K. Choo: 'Variants of the orientation relationship between austenite and cementite in Fe-30Mn-1C wt % alloy', *Acta Metallurgica et Materalia*, 1994, **42**, 263–269.

142. A. W. Sleeswyk: 'Crystallography of the austenite-cementite transition', *Philosophical Magazine*, 1966, **13**, 1223–1237.

143. H. S. Kitaguchi, S. Lozano-Perez, and M. P. Moody: 'Quantitative analysis of carbon in cementite using pulsed laser atom probe', *Ultramicroscopy*, 2014, **147**, 51–60.

144. M. H. Hong, W. T. R. Jr., T. Tarui, and K. Hono: 'Atom probe and transmission electron microscopy investigations of heavily drawn pearlitic steel wire', *Metallurgical & Materials Transactions A*, 1999, **30**, 717–727.

Notes

[1]The atom probe permits the composition of cementite to be measured directly using time-of-flight mass spectroscopy. There are, nevertheless, difficulties in measuring the carbon concentration of cementite [143]. It has not yet been possi-

ble to demonstrate small deviations from stoichiometry using such these high-resolution methods. Extremely small (4 nm) cementite particles in severely deformed mixtures of ferrite and cementite, apparently contain only 16 at% of carbon, a concentration that recovers to the 25 at% when the mixture is annealed to reduce the defect density and coarsen the cementite [144]; the deformation is said to introduce defects into the cementite, with this somehow leading to the reduction in carbon concentration. It is important to note, however, that the particles containing such a large deviation from stoichiometry have not been proven to retain the orthorhombic crystal structure.

[2]However, if cementite and α-iron can somehow coexist at temperatures above the Fe-C eutectoid, then free energy of formation data indicate that the mixture would be stable relative to α-iron+graphite, Figure 8.24.

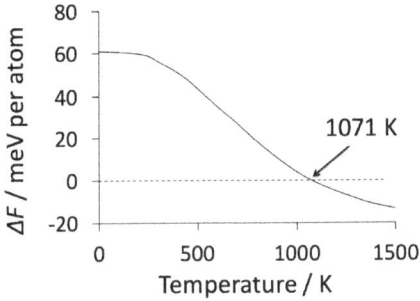

Figure 8.24 The formation energy ΔF of cementite for the reaction $F_{Fe_3C} - [3F_{Fe} + F_{graphite}]/4$. Data from CALPHAD assessment by Hallstedt et al. [93]. A negative value implies that cementite becomes stable relative to the mixture of α and graphite.

[3]Notice that in the case of the martensitic transformation of austenite, the pure Bain strain generates the correct lattice but does not lead to the necessary invariant line. However, the Bain orientation is not that observed experimentally. When a rigid body rotation is added that makes the combined deformation an invariant line strain, it generates the observed orientation relationship.

9 Other Fe-C carbides

9.1 GENERIC CONSIDERATIONS

There are several carbides in the binary Fe-C system that are richer in carbon than cementite. This includes the ε-carbide [1], χ-carbide (Hägg) [2] and η-carbide [3]. In martensitic or bainitic steels, these carbides are generally considered to be metastable relative to cementite, because the following sequence of precipitation, or a part of the sequence, is observed experimentally [4]:

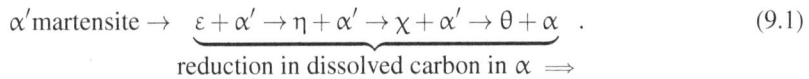

$$\alpha' \text{martensite} \rightarrow \underbrace{\varepsilon + \alpha' \rightarrow \eta + \alpha' \rightarrow \chi + \alpha' \rightarrow \theta + \alpha}_{\text{reduction in dissolved carbon in } \alpha \implies} . \tag{9.1}$$

During carburisation, when carbon is introduced, from an external source, into the steel via its surface from an external source, the transition χ-carbide can overlay cementite when the activity of carbon becomes sufficiently large [5]. Similarly, when carbon is ion-implanted into iron, only η-carbide forms at the highest ion-fluence, with χ-carbide at intermediate fluences and cementite at the lowest fluence [6]. It is particularly interesting that in these ion-implantation experiments, all of the carbides (χ, η, θ) could form at $-70\,°C$, i.e., without the long-range diffusion of iron [6]. Heating cementite in a hydrogen-CO mixture can also convert it into χ-carbide [7].[1]

Observations like these suggested domains on the iron-carbon *equilibrium* phase diagram where mixtures of $\chi + \alpha$ (or $\eta + \alpha$) may have greater thermodynamic stability than $\alpha + \theta$ [8, 9]. The modified Fe-C phase diagram shown in Figure 9.1 [10, 11] implies that if a mixture of $\alpha + \theta$ is cooled and held at 400 °C, then the cementite should transform into χ-carbide, and similar considerations apply to the $\eta + \alpha$ phase field. Although this has never been observed in practice, it could be argued that the driving force for that transformation and the carbon concentration of the ferrite are both very small for η to be observed in a reasonable time period [12]. At large driving forces, such as during precipitation from high-carbon martensite, both χ and η carbides do form, but it cannot be claimed that they may not ultimately be replaced by cementite.

There is evidence that the modified phase diagram is not correct. Table 9.1 lists cases where cementite precipitation is observed in domains where Figure 9.1 would imply that other carbides are thermodynamically more stable. Pearlite that has been aged for two thousand years under ambient conditions remains as a mixture of ferrite and cementite [13]; mixtures of ferrite and cementite are found in meteorites that have cooled at extraordinarily slow rates, some $10\,K$ per million years [14]. χ-carbide has been shown to begin to transform into cementite at 300 °C in iron implanted with carbon ions [6]. Mechanically synthesised mixtures of ε, χ and cementite decompose to just cementite when heated to 400 °C [15].

First principles calculations of individual crystals can sometimes indicate an order of thermodynamic stability although they do not consider coexistence in equilibrium with another phase, nor do they include constraint when the precipitate is enclosed by a matrix. Table 9.2 represents the change in internal energy when each of the transition carbides (and cementite) is formed from elemental

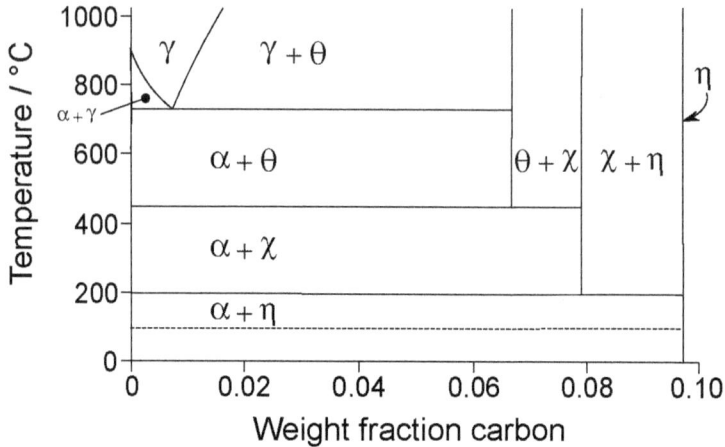

Figure 9.1 Calculated Fe-C equilibrium phase diagram [10, 11]. The dashed line represents a temperature below which a Fe-C solid solution would tend to undergo the clustering of carbon atoms which in turn may lead to a conditional spinodal. ε-carbide is missing from this diagram because of the lack of appropriate thermodynamic data.

iron and carbon. Unfortunately, the discrepancies are sufficiently large to make the relative stabilities of the carbides difficult to assess with confidence. These calculations usually are representative of 0 K and zero pressure whereas the formation energies in practice are sensitive to temperature [26]. Temperature-dependent calculations that account for magnetic and other heat capacity terms show that η-carbide is less stable than cementite beyond 57 °C [27].

9.1.1 OBSERVATIONS OF χ, η, ε AND θ CARBIDES IN TEMPERED Fe-C

Figure 9.2 shows specifically data from binary Fe-C steels using a Larson-Miller parameter [36] to represent the kinetic strength of the heat treatment. Although empirical, the parameter is acknowledged widely to rationalise the combined effects of time and temperature during tempering [e.g., 37–39].

Two conclusions can be drawn from Figure 9.2. First, that cementite is the most stable phase over a wide temperature range when the strength of the heat treatment is greatest within the scope of the dataset. Secondly, that cementite is the only precipitate when the excess carbon concentration is small. This would be expected for a phase of high stability because a small carbon concentration in solution corresponds to a small driving force for precipitation. In such circumstances, transition phases (which by definition lead to a smaller reduction in free energy) would not be able to precipitate.

Table 9.1

Steels in which cementite is observed experimentally to precipitate at low temperatures. Only the maximum tempering temperatures are listed from the data by Langer [16], although cementite could be observed mixed with ε-carbide to tempering temperatures as low as 166 °C [16]. The bainitic microstructures were all generated isothermally, whereas both the ferrite and martensite were tempered isothermally.

Steel composition / wt%	Heat treatment	Structure	Reference
0.05C-0.0019N-1.5Al	260 °C, 76 min; 1400 min	ferritic	[16]
0.02C-0.0034N-0.003Al	250 °C, 15-250 min	ferritic	[16]
0.02C-0.0186N-0.37Mn	260 °C, 187 min	ferritic	[16]
	260 °C, 60 min; 250 °C, 362-5706 min;	ferritic	[16]
0.02C-0.0010N-0.015Al-1.0Ni	220 °C, 435-5650 min; 200 °C, 2807 min	ferritic	[16]
	190 °C, 7980 min	ferritic	[16]
0.014C	150 °C, 170 h; 260 °C, 3-10 min	ferritic	[17]
0.49C-1.18Si-0.15Mn-1.22Cr-3.57Ni-0.26Mo	290 °C 5 h	bainitic	[18]
1.8C	150 °C, 5 days	bainitic	[19]
1.8C	200 °C, 55 h	bainitic	[20]
0.43C-3Mn-2.12Si	300 °C, 30 min; 247 °C, 30 min	bainitic	[21]
Fe-0.8C	316 °C, 60 min	martensitic	[22]
Fe-0.8C	427 °C, 60 min	martensitic	[22]
Fe-0.43C-3Mn-2.12Si	350 °C, 30 min	martensitic	[23]
Fe-1.30C	350 °C, 21 days	martensitic	[24]
Fe-0.11C-0.21Si-28Ni	150 °C, 48 h	martensitic	[25]

Table 9.2

First principles calculations of the change in internal energy ΔU at 0 K and zero pressure for the reaction $[\Delta U = U_{\mathrm{Fe}_n\mathrm{C}_m} - nU_{\mathrm{Fe}} - mU_{\mathrm{C}}]/(n+m)$. These are calculations that consider the formation of the carbide as an isolated phase from the constituent atoms.

Carbide	ΔU / kJ mol^{-1}	Reference
Cementite Fe$_3$C	5.38	[28–30]
Cementite Fe$_3$C	5.89	[26]
Cementite Fe$_3$C	5.60	[31]
Cementite Fe$_3$C	5.21	[31]
Cementite Fe$_3$C	2.51	[32]
ε-carbide Fe$_{2.4}$C	6.23	[33]
ε-carbide Fe$_{2.2}$C	4.26	[34]
η-carbide Fe$_2$C	126.1	[35]
η-carbide Fe$_2$C	4.00	[34]
η-carbide Fe$_2$C	1.68	[32]
Häag χ-carbide Fe$_5$C$_2$	2.45	[32]

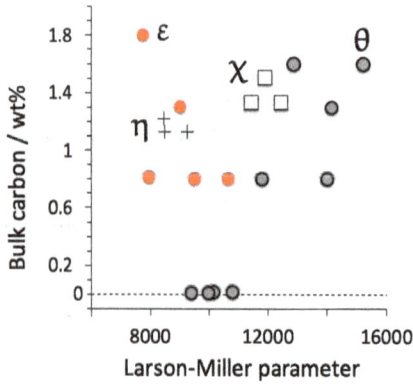

Figure 9.2 The Larson-Miller parameter $T(\log t + 20)$, where T is the absolute tempering-temperature and t the tempering time in hours. The data represent the evolution of carbides in Fe-C binary steels as a function of the strength of the tempering treatment.

9.2 χ-CARBIDE

Hägg-carbide, also known as χ-carbide, has the approximate chemical composition $Fe_{2.2}C$, often written as Fe_5C_2. The unit cell is monoclinic (space group C2/c, four formula units in the cell) with $a = 1.1588$ nm, $b = 0.4579$ nm, $c = 0.5059$ nm and $\beta = 97.75°$ [40–43]. The structure is illustrated in Figure 9.3, with the b axis unique so the lattice points in the unit cell are at $0,0,0$ and $\frac{1}{2},\frac{1}{2},0$. The coordinates of the atoms are given in Table 9.3.

Table 9.3

Coordinates of atoms in the χ-carbide unit cell. Adapted data from [44]. There are twenty iron atoms and eight carbon atoms in the monoclinic cell, $4 \times Fe_5C_2$.

Atom	Multiplicity	Wyckoff letter	Site symmetry	Coordinates		
				x	y	z
Fe	8	f	1	0.097125	0.084171	0.417689
Fe	8	f	1	0.214975	0.58659	0.30585
Fe	4	c	2	0.214975	0.58659	0.30585
C	8	f	1	0.11732	0.30626	0.08025

The calculated single-crystal elastic properties (GPa) are as expected, anisotropic [44]:

C_{11}	C_{22}	C_{33}	C_{44}	C_{55}	C_{66}	C_{12}	C_{13}	C_{23}	C_{46}	C_{15}	C_{25}	C_{35}
349	341	410	139	132	35	188	151	164	−11.5	−10.7	22.3	−1.4

Some of the steels in which χ-carbide has been observed experimentally are listed in Table 9.4. Häag carbide is apparently more stable than either ε or η-carbide when in equilibrium with ferrite; both ε and η are replaced by Hägg during the tempering of martensite. The ε-carbide that forms during the implantation of high-carbon bearing steels with carbon transforms into the more stable χ-carbide on annealing [45]. At carbon concentrations greater than that of cementite, χ-carbide can exist in stable equilibrium with cementite. Indeed, during carburisation reactions where

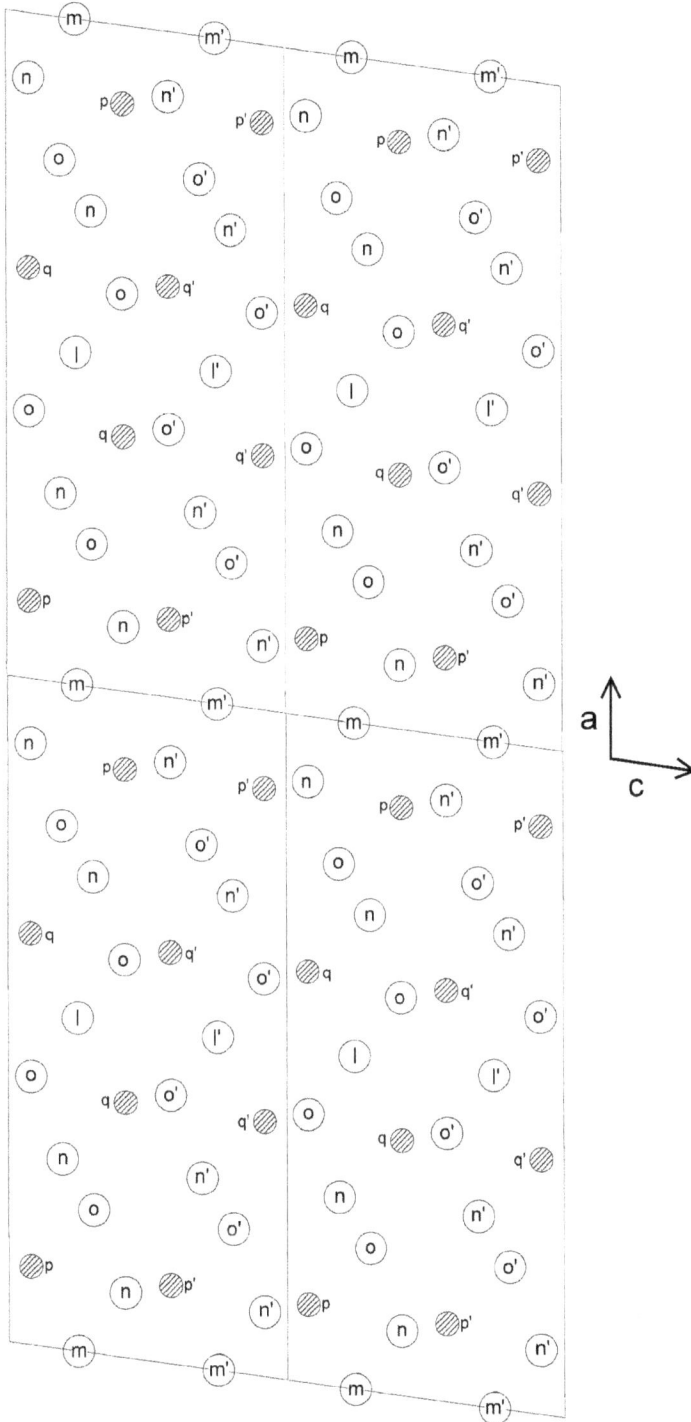

Figure 9.3 Four unit cells of the monoclinic cell of χ-carbide projected on the (010) plane; this makes it easier to see the symmetry elements. The fractional heights of the atoms along the b-axis are l= 0.07732, $l' = 1-l$, m= 0.57732, $m' = 1-m$, n= 0.08417, $n' = 1-n$, o= 0.59659, $o' = 1-o$, p=0.30626, $p' = 1-p$, q= 0.80626, $q' = 1-q$. The c-glide plane is parallel to the plane of the diagram at half height and the diad is located at the iron atoms (heights m,m') on the cell edges.

the activity of carbon is sufficiently large, Hägg carbide precipitates in preference to cementite [5]. High-resolution electron microscopy of tempered martensite has demonstrated that what at first sight appears to be faulted cementite, in fact consists of interpenetrating layers of cementite and χ, described as *microsyntactic intergrowth* [46–49]. The $\{2\,0\,0\}_\chi$ planes are found to be parallel to the $\{0\,1\,0\}_\theta$ planes (in the setting *Pnma*) of different spacing (0.57 and 0.67 nm respectively). Thus, the faults in the cementite are regions of χ, each a few interplanar spacings thick and this intimate mixture of cementite and χ consequently has a non-stoichiometric overall composition expressed by $Fe_{2n+1}C_n$, where $n \geq 3$. Similar observations have been reported in [50], but for cementite in both tempered martensite and lower bainite, in a Fe-0.7C wt% steel (Figure 9.4). In both cases, the cementite particles contained regions of χ-carbide, lending support to the idea that the transformation from Hägg to cementite occurs by intergrowth [51]. When isolated χ-carbide is heat treated to induce cementite, magnetic measurements indicate that the transition involves mixtures of the two phases in changing proportions [52], consistent with the intergrowth mechanism.

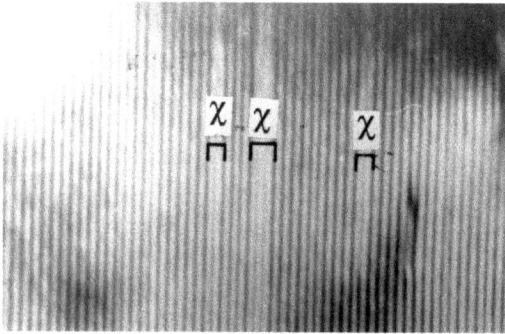

Figure 9.4 Lattice image showing the intergrowth of layers of (010) cementite (*Pnma* setting, spacing 0.67 nm) and (200) χ-carbide (spacing 0.57 nm) formed during the tempering of cementite. Image courtesy of Professor Y. Ohmori, further details in [50].

Table 9.4

Steels in which χ-carbide has been observed experimentally. The Ohmori experiments [51] used virgin martensite samples heated at 5 °C min^{-1} to a particular temperature and then quenched to ambient temperature.

Steel	Heat treatment	Reference
Fe-1.22C wt%	300 °C, 350 °C, 1 h	[53]
Fe-1.22C-0.26Si-0.6Mn wt%	300-450 °C, anisothermal	[51, 54]
Fe-1.0-1.7C-0.28Si-0.18Mn wt%	247-397 °C, 1-2 h	[46]
Fe-1.5C wt%	277 °C, 48 h	[47]
Fe-0.25C-16.08Co-4.97-Ni-0.71Cr-2.82Mo wt%	510 °C, 5-15 min	[48, 49]

The Curie temperature of χ-carbide is 529 K [55] but also has been quoted as 520 K [56, 57] with the average magnetic moment per iron atom calculated to be 1.69 μ_B [44], which compares reasonably well with measurements in the range 1.72-1.75 μ_B for χ, ϵ and θ carbides [58].

In the context of steels, χ-carbide is known primarily for its precipitation during the tempering of high-carbon martensite, tending to locate at transformation twin boundaries, plate boundaries and sometimes within the matrix [53, 59]. The orientation that develops is interesting [51]:

$$(100)_\chi \parallel \{11\bar{2}\}_\alpha \qquad (010)_\chi \parallel \{110\}_\alpha \qquad [001]_\chi \parallel \{1\bar{1}1\}_\alpha$$

because the twin plane of the martensite on which χ forms, is not in fact the $\{11\bar{2}\}_\alpha \parallel (100)_\chi$, the twin boundaries act simply as heterogeneous nucleation sites [51].

χ-carbide is of particular importance amongst the iron carbides as a catalyst in the Fischer-Tropsch process in which hydrocarbon products are synthesised from mixtures of carbon-monoxide and hydrogen. Surface energies and the structure of relaxed surfaces have been calculated [60] but the variations as a function of the crystallographic orientation of the surface are not large enough to reach conclusions about the surface best suited for catalytic activity. The addition of platinum atoms to the χ-carbide surface can promote catalytic activity; the $(100)_\chi$ surface has a low activation barrier for surface diffusion, making it possible for the Pt atoms to aggregate. In contrast, $(111)_\chi$ leaves the platinum atoms in a dispersed state because they are trapped by large barriers associated with the structure of that surface [61].

9.3 η-CARBIDE

The tempering of martensite at low temperatures is associated with the precipitation of η-carbide (Fe_2C) [62], the crystal structure of which is orthorhombic [3], space group *Pnnm* with lattice parameters $a = 0.4704$ nm, $b = 0.4318$ nm and $c = 0.2830$ nm [4], Figure 9.5a. Table 9.5 lists some of the steels where η-carbide has been observed experimentally, in general verified using electron or X-ray diffraction or Mössbauer spectroscopy. The crystallographic orientation between the carbide and martensite is as follows [63]:

$$(010)_\eta \parallel \{011\}_\alpha \qquad (001)_\eta \parallel \{100\}_\alpha \qquad [100]_\eta \parallel \{0\bar{1}1\}_\alpha$$

Table 9.5

Steels in which η-carbide has been observed experimentally. The cryogenic treatment refers to cooling the sample to $-180\,°C$ prior to the tempering heat-treatment.

Steel	Heat treatment	Reference
Fe-1.13C wt%	120 °C, 1-100 days	[3]
Fe-1.22C wt%	125 °C, 16 h	[62]
Fe-0.39C-24.9Ni wt%	100, 150 °C, 1 h	[64]
Fe-0.88C-14.8Ni wt%	100, 150 °C, 1 h	[64]
Fe-1.44C-0.3Si-0.4Mn-12.2Cr-0.84Mo-0.43V wt%	cryogenic + 180 °C, 30 min	[65]
Fe-0.28C-3Mo wt%	200 °C, 10 min	[66]
Fe-14.7Ni-0.92C wt%	25 °C, 2-3 years	[67]

The calculated elastic properties are as follows (GPa) [68]:

$$\begin{array}{ccccccccc} C_{11} & C_{22} & C_{33} & C_{44} & C_{55} & C_{66} & C_{12} & C_{13} & C_{23} \\ 323 & 340 & 378 & 110 & 97 & 136 & 189 & 136 & 158 \end{array}$$

$$K = 223 \qquad E_s = 147 \qquad E_s = 362 \qquad v = 0.23$$

These have been used to estimate the fraction of η-carbide in a gear steel assuming that the Young's modulus of a composite mixture of martensite, austenite and the carbide scales with the respective volume fractions [68]. In that study, η-carbide was induced to precipitate in a carburised steel by cooling to 77 K, which is too low to permit any diffusion of substitutional atoms including iron. The change in crystal structure to η is therefore achieved by a deformation according to the crystallographic theory [69–72] with the following parameters [73], referred to the martensite lattice:

Habit plane of η	$\mathbf{p} = (0.4049\ 0.0627\ -0.9122)$
Displacement direction	$\mathbf{d} = [-0.4919\ 0.0597\ -0.8686]$
Magnitude of shape deformation	$m = 0.2626$
Magnitude of lattice-invariant deformation	$n = 0.1855$
Shear strain on habit plane	$s = 0.211$
Dilatational strain normal to habit plane	$\zeta = 0.157$

Figure 9.5b shows the morphology of thin η-carbide precipitates; it is telling that they form as lenticular plates with sharp tips, because this is the shape that minimises the elastic strain energy due to the shape deformation [74]. High resolution imaging also shows evidence within the η-plates of fine structure corresponding to the lattice-invariant deformation [73].

The Curie temperature of η-carbide is 267 °C [52].

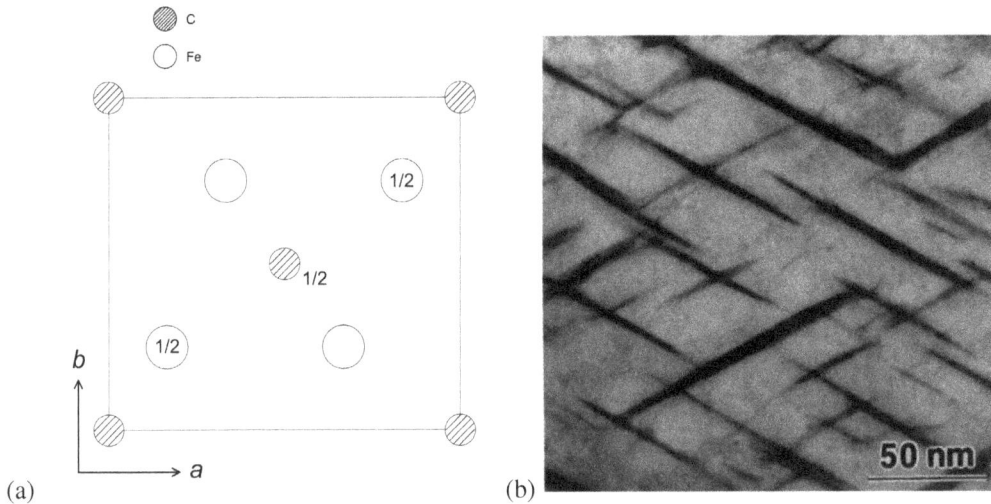

Figure 9.5 (a) The orthorhombic unit cell of η-carbide projected on to the $\{001\}$ plane. The fractional coordinates along the c axis are marked for those atoms not located at heights 0,1. The four iron atoms are located at coordinates $\frac{1}{6}, \frac{1}{4}, \frac{1}{2}$ and at all other symmetry related positions, whereas the two carbon atoms are located at 0,0,0 and equivalent positions. (b) η-carbide plates in an Fe-25Ni-1C wt% transformed into martensite and then tempered at 150°C for 1 h. Micrograph reprinted by permission of Springer Nature from [73], https://link.springer.com/article/10.1007/BF02670168.

9.4 ε-CARBIDE

The crystal structure of ε-carbide is best described by the space group $P6_322$ with lattice parameters $a = 0.4767$ nm and $c = 0.4354$ nm [75]. The six iron atoms are in a close-packed hexagonal stacking at the 6g locations whereas the carbon atoms occupy octahedral interstices (Table 9.6). There are three sets of octahedral sites, labelled 2b, 2c and 2d in the Wyckoff notation; the carbon atoms order on the 2d sites because that leads to the greatest C-C distance. It has been shown using first principles calculations that a disordering of the carbon atoms, i.e. a distribution over the 2b, 2c and 2d sites is not favoured [76]. Even the 2d sites will not be fully occupied because the chemical composition of ε-carbide is approximately $Fe_{2.4}C$ as identified using the atom probe, where peak concentrations of carbon at just under 30 at.% have been measured in precipitates assumed to be ε [76, 77]. It follows that there will be some carbon atoms in addition to those in the occupied sublattice of octahedral sites illustrated in Figure 9.6, to account for the Fe:C ratio of 2.4.

If the carbon atoms are distributed at random on all of the octahedral sites, ε-carbide is indexed as hexagonal close-packed with the space group $P6_3/mmc$ and unit cell parameters $a_h = 0.2752$ nm (so that $a = \sqrt{3}a_h$) and $c_h = c = 0.4354$ nm. The difference is illustrated in Figure 9.7. With the point group $P6_322$, only one sublattice of octahedral interstices is occupied, thus reducing the symmetry. In Figure 9.7b, it is assumed that the carbon atoms are dispersed at random on all three sublattices of octahedral interstices, making the overall symmetry is greater, and indeed, consistent with that of hcp-iron. As pointed out by Nagakura [75], not all of the observed diffracted intensity is explained by indexing the structure as $P6_3/mmc$, and furthermore, first principles calculations indicate that it

Table 9.6

Wyckoff positions for space group $P6_322$ of ε-carbide. Adapted from [75]. The octahedral interstices consist of three independent Wyckoff sites, 2b, 2c, 2d, each of which has a multiplicity of two. Here, only the 2d sites are occupied because the carbon atoms are then the most widely separated.

Multiplicity	Wyckoff letter	Site symmetry	Coordinates
6	g	.2.	$\frac{1}{3}, 0, 0; \quad 0, \frac{1}{3}, 0; \quad \bar{\frac{1}{3}}, \bar{\frac{1}{3}}, 0$ $\bar{\frac{1}{3}}, 0, \frac{1}{2}; \quad 0, \bar{\frac{1}{3}}, \frac{1}{2}; \quad \frac{1}{3}, \frac{1}{3}, \frac{1}{2}$
2	d	3.2	$\frac{1}{3}, \frac{2}{3}, \frac{3}{4}; \quad \frac{2}{3}, \frac{1}{3}, \frac{1}{4}$
2	c	3.2	$\frac{1}{3}, \frac{2}{3}, \frac{1}{4}; \quad \frac{2}{3}, \frac{1}{3}, \frac{3}{4}$
2	b	3.2	$0, 0, \frac{1}{4}; \quad 0, 0, \frac{3}{4}$

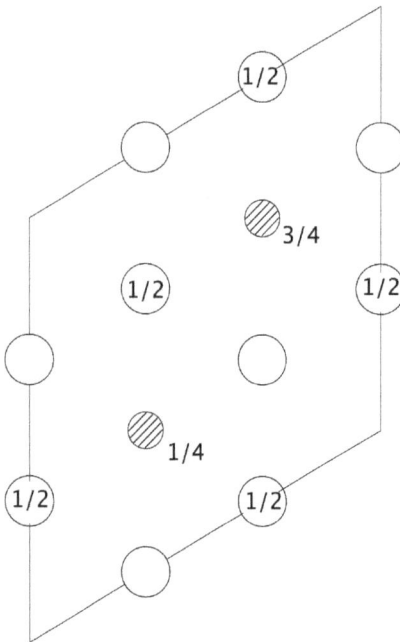

Figure 9.6 Projection on to the basal plane of the unit cell of ε-carbide, consistent with the space group $P6_322$. The carbon atoms are shaded and occupy octahedral interstices. In practice, there will be an additional small quantity of carbon in other octahedral interstices so that the chemical composition complies with $Fe_{2.4}C$.

is energetically favourable for the carbon atoms to reside in a single sublattice of octahedral sites, although it is emphasised that the calculations refer to 0 K [78].

ε-carbide often precipitates during the tempering of martensite in steels [24, 79–81]; it also is common within lower bainitic ferrite [22, 76, 82–93]. The orientation relationship between the carbide

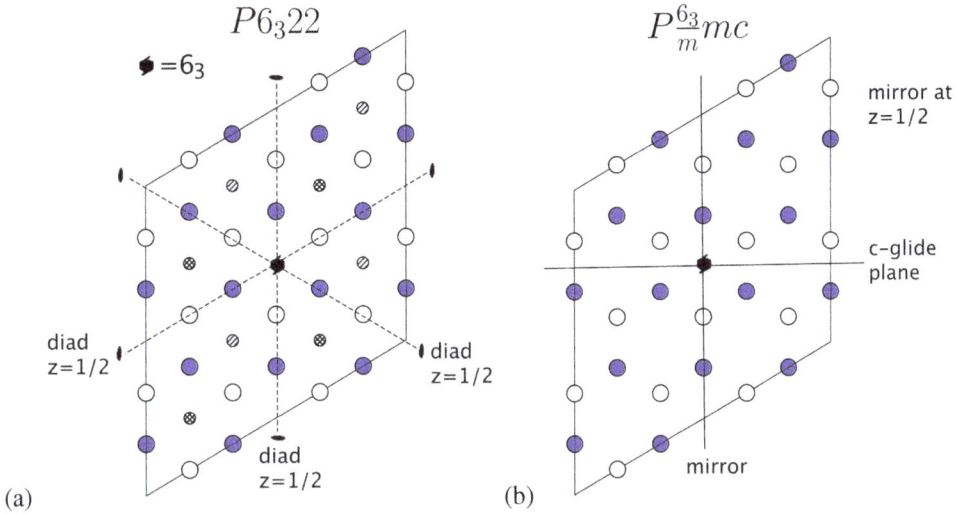

Figure 9.7 2×2 cell structure-projection on to the basal plane. The iron atoms coloured blue are at height $\frac{1}{2}$ whereas the smaller carbon atoms that are hatched and cross-hatched are at heights $\frac{3}{4}$ and $\frac{1}{4}$ respectively. (a) Space group $P6_322$ with one set of octahedral interstices (2d) occupied by carbon and some further carbon (not illustrated) in the 2b sites which would be 25% occupied to make the overall structure consistent with the formula $F_{2.4}C$. (b) The carbon atoms (not illustrated) are now randomly distributed in any of the three kinds of octahedral interstices, so that the overall symmetry becomes $P6_3/mmc$.

and martensite is reported as [94]

$$(2\bar{1}1)_{\alpha'} \quad \| \quad (10\bar{1}0)_{\varepsilon} \equiv (100)_{\varepsilon}, \qquad (100)_{\varepsilon} \| [210]_{\varepsilon} \equiv [10\bar{1}0]_{\varepsilon}$$

$$(11\bar{1})_{\alpha'} \quad \| \quad (\bar{1}2\bar{1}0)_{\varepsilon} \equiv (\bar{1}20)_{\varepsilon}, \qquad (\bar{1}20)_{\varepsilon} \| [010]_{\varepsilon} \equiv \frac{1}{3}[\bar{1}2\bar{1}0]_{\varepsilon}$$

$$\{011\}_{\alpha'} \quad \| \quad (0001)_{\varepsilon} \equiv (001)_{\varepsilon}, \qquad (001)_{\varepsilon} \| [001]_{\varepsilon} \equiv [0001]_{\varepsilon} \qquad (9.2)$$

stated to an accuracy of about $\pm 2°$. The coordinate transformation matrices for converting the indices of directions $\mathbf{u}$ or plane normals $\mathbf{h}$ between the two lattices are as follows:

$$(\varepsilon\, \mathrm{J}\, \alpha') = \begin{pmatrix} \frac{2k}{3} & -\frac{k}{3} & \frac{k}{3} \\ \frac{g}{3} + \frac{k}{3} & \frac{g}{3} - \frac{k}{6} & \frac{k}{6} - \frac{g}{3} \\ 0 & \frac{m}{2} & \frac{m}{2} \end{pmatrix} \qquad \begin{aligned} [\varepsilon; \mathbf{u}] &= (\varepsilon\, \mathrm{J}\, \alpha')\,[\alpha'; \mathbf{u}] \\ (\mathbf{h}; \alpha') &= (\mathbf{h}; \varepsilon)\,(\varepsilon\, \mathrm{J}\, \alpha') \end{aligned}$$

$$(\alpha'\, \mathrm{J}\, \varepsilon) = \begin{pmatrix} \frac{2g-k}{2gk} & \frac{1}{g} & 0 \\ -\frac{g+k}{2gk} & \frac{1}{g} & \frac{1}{m} \\ \frac{g+k}{2gk} & -\frac{1}{g} & \frac{1}{m} \end{pmatrix} \qquad \begin{aligned} [\alpha'; \mathbf{u}] &= (\alpha'\, \mathrm{J}\, \varepsilon)\,[\varepsilon; \mathbf{u}] \\ (\mathbf{h}; \varepsilon) &= (\mathbf{h}; \alpha')\,(\alpha'\, \mathrm{J}\, \varepsilon) \end{aligned}$$

where the constants are ratios of lattice parameters, $k = \sqrt{6}a_{\alpha'}/\sqrt{3}a_{\varepsilon}$, $g = \sqrt{3}a_{\alpha'}/a_{\varepsilon}$ and $m = \sqrt{2}a_{\alpha'}/c_{\varepsilon}$. The orientation relationship is close to that proposed by Jack [24].

The pure deformation that converts the bcc structure of ferrite to that of ε-carbide can be derived assuming the correspondence proposed for a similar transformation in titanium [95]:

$$\mathbf{z}_1 \parallel [100]_{\alpha'} \quad \rightarrow \quad [110]_{\varepsilon} \equiv [11\bar{2}0]_{\varepsilon}$$

$$\mathbf{z}_2 \parallel [01\bar{1}]_{\alpha'} \quad \rightarrow \quad [1\bar{1}0]_{\varepsilon} \equiv [1\bar{1}00]_{\varepsilon}$$

$$\mathbf{z}_3 \parallel [011]_{\alpha'} \quad \rightarrow \quad [001]_{\varepsilon} \equiv [0001]_{\varepsilon} \tag{9.3}$$

where $\mathbf{z}_1$, $\mathbf{z}_2$, and $\mathbf{z}_3$ define an orthonormal basis Z whose basis vectors are parallel to the principal axes of the deformation (Figure 9.8a). Using the lattice parameters $a_{\alpha'} = 0.2867$ nm, $a_{\varepsilon} = 0.4767$ nm and $c_{\varepsilon} = 0.4354$ nm, the pure deformation is given by

$$(\mathrm{Z\,S\,Z}) = \begin{pmatrix} \eta_1 = 0.959888 & 0 & 0 \\ 0 & \eta_2 = 1.175716 & 0 \\ 0 & 0 & \eta_3 = 1.073855 \end{pmatrix}$$

where the η_i are the distortions, i.e. the ratio of the length of the vector in α' following the deformation to its original magnitude.

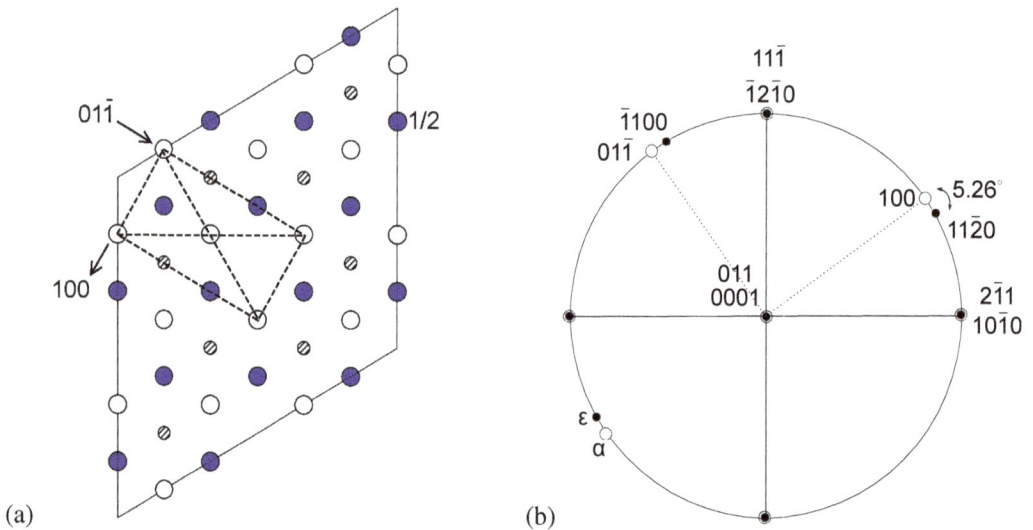

Figure 9.8 (a) Projection of the ε-carbide structure on the basal plane, with superimposed $\{011\}_{\alpha'}$ plane (dashed lines) to illustrate the lattice correspondence. The iron atoms in blue are located height half. Adjacent carbon atoms are at heights $1/4$ and $3/4$ respectively. (b) Stereographic projection to illustrate the rotation required to supplement the pure deformation so that the total is consistent with the observed orientation relationship.

In the crystallographic basis, this becomes:

$$(\alpha' \, S \, \alpha') \;=\; (\alpha' \, J \, Z)(Z \, S \, Z)(Z \, J \, \alpha')$$

$$= \frac{1}{a_{\alpha'}} \begin{pmatrix} 1 & 0 & 0 \\ 0 & -\frac{1}{\sqrt{2}} & \frac{1}{\sqrt{2}} \\ 0 & \frac{1}{\sqrt{2}} & \frac{1}{\sqrt{2}} \end{pmatrix} \begin{pmatrix} \eta_1 & 0 & 0 \\ 0 & \eta_2 & 0 \\ 0 & 0 & \eta_3 \end{pmatrix} a_{\alpha'} \begin{pmatrix} 1 & 0 & 0 \\ 0 & -\frac{1}{\sqrt{2}} & \frac{1}{\sqrt{2}} \\ 0 & \frac{1}{\sqrt{2}} & \frac{1}{\sqrt{2}} \end{pmatrix}$$

$$= \begin{pmatrix} 0.959888 & 0 & 0 \\ 0 & 1.124790 & -0.050931 \\ 0 & -0.050931 & 1.124790 \end{pmatrix}.$$

This pure deformation corresponds to the relations identified in Equation 9.3 but does not represent the observed orientation relationship in Equation 9.2. An additional right-handed, rigid body rotation of $5.26°$ about $[0\overline{1}\overline{1}]_{\alpha'}$ is required to restore the observed orientation relationship (Figure 9.8b), so the total deformation becomes

$$(\alpha' \, S_2 \, \alpha') \;=\; \underbrace{\begin{pmatrix} 0.959888 & 0 & 0 \\ 0 & 1.124790 & -0.050931 \\ 0 & -0.050931 & 1.124790 \end{pmatrix}}_{\text{pure deformation}} \underbrace{\begin{pmatrix} 0.995789 & 0.064824 & -0.064824 \\ -0.064824 & 0.997894 & 0.002106 \\ 0.064243 & 0.002106 & 0.997895 \end{pmatrix}}_{\text{rigid rotation}}$$

$$= \begin{pmatrix} 0.955846 & 0.062224 & -0.062224 \\ -0.076185 & 1.122310 & -0.048454 \\ 0.075561 & -0.048454 & 1.122310 \end{pmatrix}. \tag{9.4}$$

The eigenvectors and eigenvalues (λ_i) for $(\alpha' \, S_2 \, \alpha')$ are

$$[-0.805796 \;\; -0.985771 \;\; 1]_{\alpha'} \qquad \lambda_1 = 1.10919$$

$$[0 \;\; 1 \;\; 1]_{\alpha'} \qquad \lambda_2 = 1.07386$$

$$[-2.04386 \;\; -1.0226 \;\; 1]_{\alpha'} \qquad \lambda_3 = 1.01742. \tag{9.5}$$

These eigenvalues are consistent with the relatively low theoretical density of ε-carbide ($6.96\,\mathrm{g\,cm^{-3}}$) but the deformations listed are large when compared with cementite precipitation in ferrite (page 477). Furthermore, the principal deformations are all positive, making it impossible to find any lines that could be invariant between ε and ferrite. The large principal distortions probably can only be accommodated when the particle size is small. This may explain why ε-carbide is always observed as fine particles visible only using transmission electron microscopy. The carbide precipitates as platelets [87]. The habit plane when precipitated in martensite has been reported to be close to $\{001\}_{\alpha'}$ [79].

ε-Carbide in steel, given time, definitely transitions into cementite; the evidence supports this over a large temperature range. One example given in Figure 9.9 represents a time-temperature-transformation diagram for this transition [16]. Given that ε-carbide is less stable than cementite, and that the principal deformations needed to generate ε are larger than for cementite, it is not clear why ε-carbide is a precursor to cementite. Possible explanations are as follows:

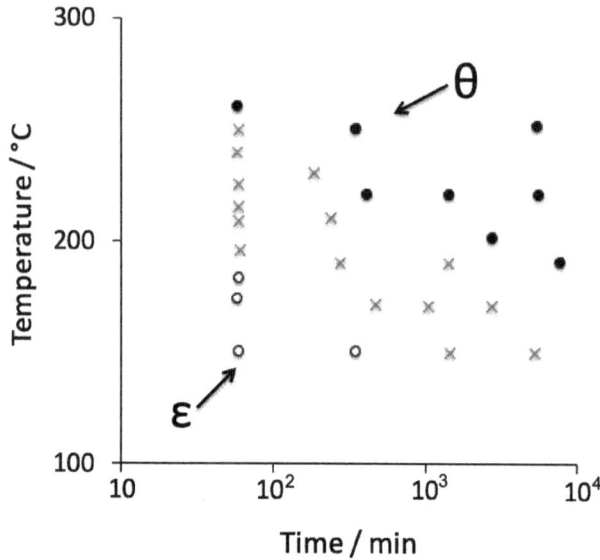

Figure 9.9 Carbide precipitation during the tempering of a 0.02C-1Ni wt% steel. Precipitation begins with ε which is then replaced gradually by cementite. The crosses refer to the case where both carbides can be observed simultaneously. Adapted using data from Langer [16].

- It is energetically more favourable for carbon atoms to segregate to dislocations rather than to precipitate [96]; the binding energy of a carbon atom to the core of a dislocation in iron is about 0.5 eV, whereas that with ε-carbide is much smaller [97] and with cementite slightly less than 0.5 eV [98]. Figure 9.10 shows calculations for both cementite and ε-carbide, where carbon concentrations below each line would be accommodated entirely at dislocations. For example, ε-carbide will not precipitate if the concentration in the matrix is less than 0.2 wt% at a dislocation density of 2×10^{16} m^{-2}, and the threshold of concentration would be smaller with respect to cementite precipitation. This would explain why ε-carbide mostly in steels containing large carbon concentrations (Table 9.7 and 9.8).

- Processes that precede the precipitation of ε-carbide. Carbon in solution within martensite will tend to form clusters to reduce the overall strain energy. The clusters are anisotropic, forming along specific crystallographic orientations. Multiplets, i.e., clusters with many aligned carbon atoms, are described as having a base-centred monoclinic structure of chemical composition Fe_3C, which after the further migration of carbon into the monoclinic multiplet, undergoes an *in situ* transformation into ε-carbide of composition Fe_9C_4 [99, 100]. Such a transformation leads to the correct orientation relationship of the ε-carbide with the martensite matrix.

- Most calculations of the relative stabilities of ε-carbide and cementite deal with the binary Fe-C system. It is feasible that other solutes affect this. Silicon inhibits cementite precipitation from ferrite or austenite but thermodynamic calculations show that ε-carbide is even more inhibited than cementite; it has been argued that the effect of silicon is to alter the lattice parameters in a manner that enhances coherency [33]. Both manganese and aluminium additions to steels enhance the stability of ε-carbide [33].

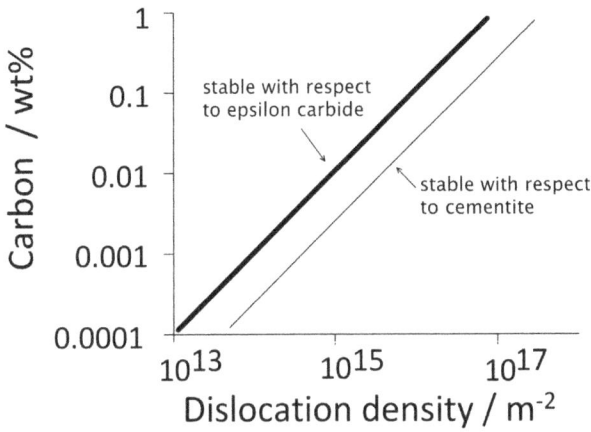

Figure 9.10 The dislocation density required to prevent the precipitation of either ε-carbide or cementite; this would happen if the carbon concentration in the matrix falls below the line. After Kalish and Cohen [96].

ε-carbide in austenite

There is a single study showing the homogeneous precipitation of ε-carbide in carbon-rich austenite [107], Figure 9.11, with the orientation relationship:

$$\{111\}_\gamma \quad \| \quad (0001)_\varepsilon$$
$$\langle 1\bar{1}0 \rangle_\gamma \quad \| \quad (1\bar{2}10)_\varepsilon.$$

The precipitates are in the form of fine, coherent particles homogeneously distributed throughout the austenite and form in at least three variants of the orientation relationship (Figure 9.11). When the austenite transforms into martensite, only two of these variants adopt the Jack orientation relationship with the martensite.

9.4.1 ELASTIC MODULI OF ε-CARBIDE

Experimental single-crystal modulus data are not available so it is necessary to resort to first-principles calculations limited to 0 K and 0 pressure. For a hexagonal crystal, mechanical stability requires that [108–110]

$$C_{11} > 0, \; C_{11} > |C_{12}|, \; C_{44} > 0$$
$$(C_{11} + C_{12})C_{33} > 2C_{13}^2$$

One difficulty in doing the calculations is that the carbon atoms are not distributed at random and the overall composition is not precisely fixed. The ε-carbide structure then corresponds to the $P6_322$

Table 9.7

Compositions of steels (wt%) in which ε-carbide has been found in lower bainite. The carbon concentration quoted for the alloy studied by [90] represents an estimate of the concentration in the austenitic matrix of an austempered ductile cast iron. That for [76] also represents the estimated carbon in the austenite prior to transformation [101], neglecting that locked in proeutectoid cementite. The temperature refers to that at which bainite was generated. There are also four alloys with carbon concentrations less than 0.05 wt%, in which ε-carbide is obtained on tempering below 250 °C; these are listed in Table 9.1.

C	Si	Mn	Ni	Cr	Mo	V	$T/°C$	Reference
0.87	-	-	-	-	-	-	260	[22, 102]
0.95	0.22	0.60	3.27	1.23	0.13	-	α_{lb}	[82]
0.60	2.00	0.86	-	0.31	-	-	α_{lb}	[82]
1.00	0.36	-	0.20	1.41	-	-	α_{lb}	[82]
0.58	0.35	0.78	-	3.90	0.45	0.90	α_{lb}	[82]
1.00	2.15	0.36	-	-	-	-	400	[83]
0.60	2.00	0.86	-	0.31	-	-	350	[84]
0.60	2.00	-	-	-	-	-	350	[85]
0.41	1.59	0.79	1.85	0.75	0.43	0.08	300	[86]
0.54	1.87	0.79	-	0.30	-	-	275	[87]
0.74	2.40	0.51	-	0.52	-	-	380	[89]
1.30	3.09	0.17	-	-	-	-	400	[90]
0.40	2.01	-	4.15	-	-	-	215	[91]
0.71	1.83	0.52	0.02	0.5	0.19	-	400	[92]
0.57	1.39	0.74			0.44		320	[93]
0.77	0.30	0.23	0.07	1.38	0.02		260	[76]
1.8C							150	[19]

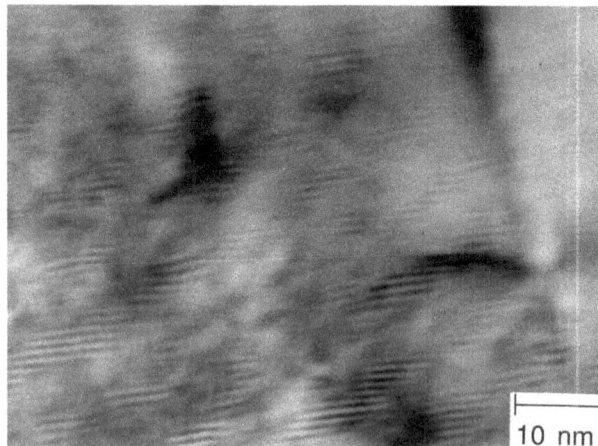

Figure 9.11   The homogeneous precipitation of ε-carbide in austenite. Micrograph courtesy of Dr Isabel Gutierrez.

Table 9.8

Compositions of steels (wt%) in which ε-carbide has been found in martensite tempered at the temperature indicated. The ≤ sign indicates tempering by continuous heating to the temperature indicated. The temperatures marked by an asterisk represent special heat treatments including autotempering or isothermal martensitic transformation. Q& P stands for a quench and partitioning heat treatment [103] with the temperature of the partitioning component included in brackets.

C	Si	Mn	Ni	Cr	Other	$T/°C, t/h$	Reference
1.30	-	-	-	-	-	120, 960	[24]
≈0.8	-	-	-	-	-	204, 1; 260, 1	[22]
0.11	0.21	-	28.00	< 0.10	-	100, 48	[25]
0.52	0.04	0.14	19.1	-	-	135, 1	[94]
0.49	3.88	0.10	-	-	-	520, 0.17	[94]
0.81	< 0.02	< 0.02	< 0.02	-	-	125, 1	[94]
1.00	< 0.35	0.4	-	1.4	-	100, 200	[104]
1.80	0.05	0.01	-	-	-	73*, 240	[19]
0.69	-	-	-	-	-	≤ 150	[81]
1.03	0.36	0.68	-	-	-	≤ 150	[81]
0.49	1.19	1.20	0.98	-	0.21Nb	Q&P (400)	[105]
0.41	1.27	1.30	1.01	0.56	-	Q&P (180)	[106]
0.28	-	-	-	-	3.0Mo	200	[66]

space group, covering the compositions Fe_6C_2, $Fe_{12}C_5$, and Fe_6C_3. When C is located at the partially occupied octahedral interstices, additional Fe-C bonds are formed with six nearest Fe atoms. Therefore, the Fe atoms are grouped into atoms without (Fe1) and with (Fe2) nearest bonding to C located at partially occupied positions. Table 9.9 shows the results [111]. An increase in the carbon concentration causes a stiffening of the ε-carbide. The values obtained indicate that all three compositions of the ε-carbide are mechanically stable at zero Kelvin and zero pressure. Unlike cementite [112], the Voigt averaged Young's modulus of ε-carbide exceeds that of iron. In a mixture of ferrite and carbide, a greater carbide-modulus would make it more difficult for the carbide to participate in plastic deformation [113] whereas the obvious plasticity of cementite in steel is the basis of the design of strong pearlitic-steel ropes.

The ductility of relatively brittle materials is often modelled in terms of the ratio of the shear to bulk modulus (E_s/B) on the grounds that the resistance to plastic deformation is related to E_s whereas the brittle-fracture strength is proportional to B [114–116]. A low value of E_s/B is then an indication of malleability whereas a high value corresponds to brittleness. For all compositional varieties of ε-carbide, E_s/B is found to be much greater at 0.48 than is the case for cementite at 0.33. In the case of α-iron and nickel the ratios are 0.33 and 0.34 respectively [115, 116]. Therefore, ε-carbide is expected to be brittle, more so than cementite which has demonstrated capabilities of plastic deformation either in its pure state [117] or when incorporated in steel [118]. There is no direct experimental evidence on the brittleness of ε-carbide, presumably because of the difficulties in synthesising large enough samples. The only evidence is anecdotal, that ε-carbide causes a deterioration in the toughness of steel when the maximum carbide-size becomes about 200 nm [119].

Table 9.9

Elastic constants (C_{ij} / GPa) and bulk modulus (B / GPa) , shear modulus (E_s / GPa), Young's modulus (E / GPa) and Poisson's ratio (v) of ε-carbide [111]. Key: ^{a} **[122],** ^{b} **[123],** ^{c} **[124]**

	Fe_6C_2	$Fe_{12}C_5$	Fe_6C_3
C_{11}	324.3 (325^a, 322^b)	356.3 (337^c)	378.7
C_{33}	332.6 (339^a, 332^b)	375.9 (348^c)	379.2
C_{44}	131.1 (121^a, 129^b)	137.5 (121^c)	147.2
C_{12}	154.9 (144^a, 138^b)	137.6 (157^c)	144.8
C_{13}	156.9 (134^a, 140^b)	189.2 (173^c)	208.4
B	213.2 (202^a, 201^b)	235.6 (226^c)	251.1
E_s	103.6	115.1 (99^c)	120.6
E	267.4	296.9 (260^c)	311.9
v	0.29	0.29 (0.3^c)	0.29
$c_{11} - c_{12}$	169.4	218.7	233.9
E_s/B	0.486	0.489	0.480
$c_{12} - c_{44}$	23.8	0.1	-2.4

The Poisson's ratio of ε-carbide (0.35) is greater than that of cementite (0.30 [120]), the effect of which would be to partition more of the strain energy associated with the precipitation into the carbide with a smaller portion residing in the surrounding matrix [pp. 469-470, ref. 121]. Therefore, the coherency strain fields associated with ε-carbide should have a smaller extent than is the case for cementite.

REFERENCES

1. L. J. E. Hofer, E. M. Cohn, and W. C. Peebles: 'The modifications of the carbide, Fe_2C: Their properties and identification', *Journal of the American Chemical Society*, 1949, **71**, 189–185.

2. G. Hägg: 'Pulverphotogramme eines neuen eisencarbides', *Zeitschrift für Kristallographie*, 1934, **89**, 92–94.

3. Y. Hirotsu, and S. Nagakura: 'Crystal structure and morphology of the carbide precipitated from martensitic high-C steel during the 1st stage of tempering', *Acta Materialia*, 1972, **20**, 645–655.

4. M. Dirand, and L. Afqir: 'Identification structural precise des carbures precipites dans lest aciers faiblement allies aux divers stades due revenu. Mechanismes de precipitation', *Acta Metallurgica*, 1983, **31**, 1089–1107.

5. A. Schneider, and G. Inden: 'Carbon diffusion in cementite (Fe_3C) and Hägg carbide (Fe_5C_2)', *Computer Coupling of Phase Diagrams and Thermochemistry*, 2007, **31**, 141–147.

6. A. Königer, C. Hammerl, M. Zeitler, and B. Rauschenbach: 'Formation of metastable iron carbide phases after high-fluence carbon ion implantation into iron at low temperatures', *Physical Review B*, 1997, **55**, 8143–8147.

7. H. E. du Plessis: 'The crystal structures of the iron carbides': Ph.D. thesis, University of Johannesburg, Johannesburg, South Africa, 2008.

8. J. Chipman: 'Thermodynamics and phase diagram of the Fe-C system', *Metallurgical Transactions*, 1972, **3**, 55–64.

9. A. Schneider, and G. Inden: 'Thermodynamics of Hägg carbide (Fe_5C_2) formation', *Steel Research*, 2001, **72**, 503–507.

10. R. Naraghi, M. Selleby, and J. Ågren: 'Thermodynamics of stable and metastable structures in the Fe-C system', *Computer Coupling of Phase Diagrams and Thermochemistry*, 2014, **46**, 148–158.

11. H. K. D. H. Bhadeshia, A. R. Chintha, and S. Lenka: 'Critical assessment 34: are Hägg, eta and epsilon carbides transition-phases relative to cementite in steels?', *Materials Science and Technology*, 2019, **35**, 1301–1305.

12. J. Ågren: 'Private communication to H. K. D. H. Bhadeshia', 2019: Transformation from cementite to χ-carbide.

13. E. Bravo Munõz, J. Chamón Fernández, J. Guzmán Arasanz, R. Arévalo Peces, A. Javier Criado, J. Antonio Martínez, and A. J. Criado Portal: 'Archaeologic analogues: Microstructural changes by natural ageing in carbon steels', *Journal of Nuclear Materials*, 2006, **349**, 1–5.

14. A. E. Ringwood: 'Cohenite as a pressure indicator in iron meteorites', *Geochimica et Cosmochimica Acta*, 1960, **20**, 155–158.

15. F. Miani, P. Matteazzi, and D. Basset: 'Thermal stability of nanocrystalline iron-iron carbide composites. A M′ossbauer study', *Hyperfine Interactions*, 1994, **94**, 2219–2222.

16. F. Miani, P. Matteazzi, and D. Basset: 'An investigation of carbide precipitation in iron', *Metal Science Journal*, 1968, **2**, 59–66.

17. W. C. Leslie, R. M. Fisher, and N. Sen: 'Morphology and crystal structure of carbides precipitated from solid solution in alpha iron', *Acta Metallurgica*, 1959, **7**, 632–644.

18. J. O. V. Dias: 'The first high-strength bainitic steel designed for hydrogen embrittlement resistance': Ph.D. thesis, University of Cambridge, Cambridge, U. K., 2018.

19. H. Okamoto, and M. Oka: 'Isothermal martensite transformation in a 1.80 wt% C steel', *Metallurgical & Materials Transactions A*, 1985, **16**, 2257–2262.

20. M. Oka, H. Okamoto, and K. Ishida: 'Transformation of lower bainite in hypoeutectoid steels', *Metallurgical Transactions A*, 1990, **21**, 845–851.

21. H. K. D. H. Bhadeshia, and D. V. Edmonds: 'The bainite transformation in a silicon steel', *Metallurgical Transactions A*, 1979, **10A**, 895–907.

22. A. E. Austin, and C. M. Schwartz: 'Electron diffraction study of iron carbides in bainite and tempered martensite', *Proc. ASTM*, 1952, **52**, 592–596.

23. S. S. Babu, K. Hono, and T. Sakuri: 'APFIM study of partitioning of substitutional elements during tempering of low alloy steel martensite', *Metallurgical & Materials Transactions A*, 1994, **25**, 499–508.

24. K. H. Jack: 'Structural transformations in the tempering of high-carbon martensitic steels', *Journal of the Iron and Steel Institute*, 1951, **169**, 26–36.

25. C. J. Barton: 'The tempering of low-carbon internally twinned martensite', *Acta Metallurgica*, 1969, **17**, 1085–1093.

26. B. Hallstedt, D. Djurovic, J. von Appen, R. Dronskowski, A. Dick, F. Körmann, T. Hickel, and J. Neugebauer: 'Thermodynamic properties of cementite (Fe_3C)', *Computer Coupling of Phase Diagrams and Thermochemistry*, 2010, **34**, 129–133.

27. C. M. Fang, M. H. F. Sluiter, M. A. van Huis, C. K. Ande, and H. W. Zandbergen: 'Origin of predominance of cementite among iron carbides in steel at elevated temperatures', *Physical Review Letters*, 2010, **105**, 055503.

28. J. H. Jang, I. G. Kim, and H. K. D. H. Bhadeshia: 'Substitutional solution of silicon in cementite: a first-principles study', *Computational Materials Science*, 2009, **44**, 1319–1326.

29. J. H. Jang, I. G. Kim, and H. K. D. H. Bhadeshia: 'First-principles calculations and the thermodynamics of cementite', *Materials Science Forum*, 2010, **638-642**, 3319–3324.

30. I. G. Kim, G. Rahman, J. H. Jang, Y. Y. Song, S. W. Seo, H. K. D. H. Bhadeshia, A. J. Freeman, and G. B. Olson: 'A systematic study on iron carbides from first principles', *Materials Science Forum*, 2010, **654-656**, 47–50.

31. M. A. Konyaeva, and N. I. Medvedeva: 'Electronic structure, magnetic properties and stability of binary and ternary $(Fe,Cr)_3C$ and $(Fe,Cr)_7C_3$', *Physics of the Solid State*, 2009, **51**, 2084–2089.

32. C. M. Fang, M. A. van Huis, and H. W. Zandbergen: 'Stability and structures of the ε-phases of iron nitrides and iron carbides from first principles', *Scripta Materialia*, 2011, **64**, 296–299.

33. J. H. Jang, I. G. Kim, and H. K. D. H. Bhadeshia: 'ε-carbide in alloy steels: first-principles assessment', *Scripta Materialia*, 2010, **63**, 121–123.

34. X. W. Liu, Z. Cao, S. Zhao, R. Gao, Y. Meng, J. X. Zhu, R. Cameron, C. F. Huo, Y. Yang, Y. W. Li, and X. D. Wen: 'Iron carbides in Fischer-Tropsch synthesis: theoretical and experimental understanding in epsilon-iron carbide phase assignment', *The Journal of Physical Chemistry C*, 2017, **121**, 21390–21396.

35. H. I. Faraoun, Y. D. Zhang, C. Esling, and H. Aourag: 'Crystalline, electronic, and magnetic structures of θ-Fe_3C, χ-Fe_5C_2, and η-Fe_2C from first principle calculation', *Journal of Applied Physics*, 2006, **99**, 093508.

36. F. R. Larson, and J. Miller: 'Time-temperature relationships for rupture and creep stresses', *Transactions of ASME*, 1952, **74**, 765–781.

37. G. J. Heimerl: 'Time-temperature parameters and an application to rupture and creep of aluminum alloys': Technical Report 3195, National Advisory Committee for Aeronautics, Washington, USA, 1954.

38. L. C. Canale, X. Yao, J. Gu, and G. E. Totten: 'A historical overview of steel tempering parameters', *International Journal of Microstructure and Materials Properties*, 2008, **3**, 474–525.

39. E. Virtanen, C. J. V. Tyne, B. S. Levy, and G. Brada: 'The tempering parameter for evaluating softening of hot and warm forging die steels', *Journal of Materials Processing Technology*, 2013, **213**, 1364–1369.

40. J. P. Sénateur, R. Fruchart, and A. Michel: 'Minerale-formule et structure cristalline due carbure de Hägg', *Comptes Rendus Hebromadaires des Seances de L'Académie des Sciences*, 1962, **255**, 1615–1616.

41. M. J. Duggin, and L. J. E. Hofer: 'Nature of χ-iron carbide', *Nature*, 1966, **212**, 248–250.

42. K. H. Jack, and S. Wild: 'Nature of χ-carbide and its possible occurrence in steels', *Nature*, 1966, **212**, 248–250.

43. J. Retief: 'Powder diffraction data and Rietveld refinement of Hägg-carbide, χ-Fe_5C_2', *Powder Diffraction*, 1999, **14**, 130–132.

44. A. Leineweber, S. Shang, Z. K. Liu, M. Widenmeyer, and R. Niewa: 'Crystal structure determination of Hägg carbide, χ-Fe_5C_2 by first-principles calculations and Rietveld refinement', *Zietschrift für Kristallographie*, 2012, **227**, 207–220.

45. K. Kobs, H. Dimigen, C. J. M. Gerritsen, J. Politiek, L. J. van Ijzendoorn, R. Oechsner, A. Lluge, and H. Ryssel: 'Friction reduction and zero wear for 52100 bearing steel by high-dose implantation of carbon', *Applied Physics Letters*, 1990, **57**, 1622–1624.

46. S. Nagakura, S. Suzuki, and M. Kusunoki: 'Structure of the precipitated particles at the third stage of tempering of martensitic iron-carbon steel studied by high resolution electron microscopy', *Transactions of the Japan Institute of Metals*, 1981, **22**, 699–709.

47. Y. Nakamura, T. Mikami, and S. Nagakura: '*In situ* high temperature electron microscopic study of the formation and growth of cementite particles in the third stage of tempering', *Transactions of the Japan Institute of Metals*, 1985, **26**, 876–885.

48. G. Ghosh, C. E. Campbell, and G. B. Olson: 'Analytical electron microscopy study of paraequilibrium cementite precipitation in ultra-high strength steel', *Metallurgical & Materials Transactions A*, 1999, **30**, 501–512.

49. G. Ghosh, and G. B. Olson: 'Precipitation of paraequilibrium cementite: experiments, thermodynamic and kinetic modelling', *Acta Materialia*, 2002, **50**, 2099–2119.

50. Y. Ohmori: 'Precipitation of iron carbides in lower bainite and tempered martensite in Fe-C alloys', In: *International Conference on Martensitic Transformations ICOMAT '86*. Tokyo, Japan: Japan Institute of Metals, 1986:5878–594.

51. Y. Ohmori: 'χ-carbide and its transformation into cementite during the tempering of martensite', *Transactions of the Japan Institute of Metals*, 1972, **13**, 119–127.

52. E. M. Cohn, and L. J. E. Hofer: 'Mode of transition from Hägg iron carbide to cementite', *Journal of the American Chemical Society*, 1950, **72**, 4662–4664.

53. C. B. Ma, T. Ando, D. L. Williamson, and G. Krauss: 'Chi-carbide in tempered high carbon martensite', *Metallurgical & Materials Transactions A*, 1983, **14**, 1033–1045.

54. Y. Ohmori, and S. Sugisawa: 'The precipitation of carbides during tempering of high carbon martensite', *Transactions of JIM*, 1971, **12**, 170–178.

55. P. H. Dünner, and S. Müller: 'Thermomagnetic investigations on a pure iron-carbon alloy in the 3rd tempering stage', *Acta Metallurgica*, 1965, **13**, 25–36.

56. E. M. Cohn, and L. J. E. Hofer: 'Some thermal reactions of higher iron carbides', *Journal of Chemical Physics*, 1953, **21**, 354–359.

57. V. A. Barinov, A. V. Protasov, and V. T. Surikov: 'Studying mechanosynthesized Hägg carbide (χ-Fe$_5$C$_2$)', *Physics of Metals and Metallography*, 2015, **116**, 835–845.

58. L. J. E. Hofer, and E. M. Cohn: 'Saturation magnetizations of iron carbides', *Journal of the American Chemical Society*, 1959, **81**, 1576–1582.

59. Y. Ohmori, A. T. Davenport, and R. W. K. Honeycombe: 'Tempering of a high-C martensite', *Trans. ISIJ*, 1972, **12**, 112–117.

60. P. J. Steynberg, J. A. van den Berg, and W. Janse van Rensburg: 'Bulk and surface analysis of Hägg Fe carbide (Fe$_5$C$_2$): a density functional theory study', *Journal of Physics – Condensed Matter*, 2008, **20**, 064238.

61. Y. He, P. Zhao, W. Guo, Y. Yang, C. F. Huo, Y. W. Li, and X. D. Wen: 'Hägg carbide surfaces induced Pt morphological changes: a theoretical insight', *Catalysis Science & Technology*, 2016, **6**, 6276–6738.

62. D. A. Williamson, S. Nagazawa, and G. Krauss: 'A study of the early stages of tempering in an Fe-1.2C pct alloy', *Metallurgical transactions A*, 1979, **10**, 1351–1363.

63. Y. Imai: 'Phases in quenched and tempered steels', *Transactions of JIM*, 1975, **16**, 721–734.

64. K. A. Taylor, G. B. Olson, M. Cohen, and J. Vander Sande: 'Spinodal decomposition during aging of FeNiC martensites', *Metallurgical Transactions A*, 1989, **20**, 2717–2736.

65. F. Meng, K. Tagashira, R. Azuma, and H. Sohma: 'Role of eta-carbide precipitations in the wear resistance improvements of Fe-12Cr-Mo-V-1.4C tool steel by cryogenic treatment', *ISIJ International*, 1994, **34**, 205–210.

66. T. P. Hou, Y. Li, J. J. Zhang, and K. M. Wu: 'Effect of magnetic field on the carbide precipitation during tempering of a molybdenum-containing steel', *Journal of Magnetism and Magnetic Materials*, 2012, **324**, 857–861.

67. W. Lu, M. Herbig, C. H. Liebscher, L. Morsdorf, R. K. W. Marceau, G. Dehm, and D. Raabe: 'Formation of eta carbide in ferrous martensite by room temperature ageing', *Acta Materialia*, 2018, **158**, 297–312.

68. A. Oila, C. Lung, and S. Bull: 'Elastic properties of eta carbide (η-Fe$_2$C) from ab initio calculations: application to cryogenically treated gear steel', *Journal of Materials Science*, 2014, **49**, 2383–2390.

69. J. S. Bowles, and J. K. Mackenzie: 'The crystallography of martensite transformations, part I', *Acta Metallurgica*, 1954, **2**, 129–137.

70. J. K. Mackenzie, and J. S. Bowles: 'The crystallography of martensite transformations II', *Acta Metallurgica*, 1954, **2**, 138–147.

71. J. K. Mackenzie, and J. S. Bowles: 'The crystallography of martensite transformations III FCC to BCT transformations', *Acta Metallurgica*, 1954, **2**, 224–234.

72. M. S. Wechsler, D. S. Lieberman, and T. A. Read: 'On the theory of the formation of martensite', *Transactions of the AIME Journal of Metals*, 1953, **197**, 1503–1515.

73. K. A. Taylor, G. B. Olson, M. Cohen, and J. Vander Sande: 'Carbide precipitation during stage 1 tempering of FeNiC martensite', *Metallurgical Transactions A*, 1989, **20**, 2749–2765.

74. J. W. Christian: 'Accommodation strains in martensite formation, the use of the dilatation parameter', *Acta Metallurgica*, 1958, **6**, 377–379.

75. S. Nagakura: 'Study of metallic carbides by electron diffraction part III. Iron carbides', *Journal of the Physical Society of Japan*, 1959, **14**, 186–195.

76. W. Song, J. von Appen, P. Coi, R. Dronskowski, D. Raabe, and W. Bleck: 'Atomic-scale investigation of ε and θ precipitates in bainite in 100Cr6 bearing steel by atom probe tomography and ab initio calculations', *Acta Materialia*, 2013, **61**, 7582–7590.

77. F. G. Caballero, M. K. Miller, and C. Garcia-Mateo: 'Atom probe tomography analysis of precipitation during tempering of a nanostructured bainitic steel', *Metallurgical & Materials Transactions A*, 2011, **42**, 3660–3668.

78. W. Song, P.-P. Choi, G. Inden, U. Prahl, D. Raabe, and W. Belck: 'On the spheroidized carbide dissolution and elemental partitioning in high carbon bearing steel 100Cr6', *Metallurgical & Materials Transactions A*, 2014, **45**, 595–606.

79. M. G. H. Wells: 'An electron microscope study of tempering in an Fe-Ni-C alloy', *Acta Metallurgica*, 1964, **12**, 389–399.

80. G. R. Speich, and W. C. Leslie: 'Tempering of steel', *Metallurgical Transactions*, 1972, **3**, 1043–1054.

81. Y. Ohmori, and I. Tamura: 'ε-carbide precipitation during tempering of plain carbon martensite', *Metallurgical Transactions A*, 1992, **23**, 2737–2751.

82. S. J. Matas, and R. F. Hehemann: 'The structure of bainite in hypoeutectoid steels', *TMS-AIME*, 1961, **221**, 179–185.

83. J. Deliry: 'Noveau carbure de fer transformation bainitique dans les aciers au carbone silicium', *Les Mémoires Scientifiques de la Revue de Métallurgie*, 1965, **62**, 527–550.

84. J. M. Oblak, and R. F. Hehemann: 'Structure and growth of Widmanstätten ferrite and bainite', In: *Transformation and Hardenability in Steels*. Michigan, USA: Climax Molybdenum, 1967:15–38.

85. R. F. Hehemann: 'The bainite transformation', In: H. I. Aaronson, and V. F. Zackay, eds. *Phase Transformations*. Materials Park, Ohio, USA: American Society of Materials, 1970:397–432.

86. G. Y. Lai: 'On the precipitation of epsilon-carbide in lower bainite', *Metallurgical Transactions A*, 1975, **6**, 1469–1471.

87. D. H. Huang, and G. Thomas: 'Metallography of bainitic transformation in silicon containing steels', *Metallurgical Transactions A*, 1977, **8**, 1661–1674.

88. E. Dorazil, and J. Svejcar: 'Study of the upper bainite in silicon steel', *Archiv für das Eisenhüttenwesen*, 1979, **50**, 293–298.

89. B. J. P. Sandvik: 'The bainite reaction in Fe-Si-C alloys: the primary stage', *Metallurgical Transactions*, 1982, **13A**, 777–787.

90. W. J. Dubensky, and K. B. Rundman: 'An electron microscopy study of carbide formation inaustempered ductile iron', *AFS Transactions*, 1985, **93**, 389–394.

91. V. T. T. Miihkinen, and D. V. Edmonds: 'Microstructural examination of two experimental high strength bainitic low alloy steels containing silicon', *Materials Science and Technology*, 1987, **3**, 422–431.

92. Y. C. Jung, Y. Ohmori, K. Nakai, and H. Ohtsubo: 'Bainite transformation in a silicon steel', *ISIJ International*, 1997, **37**, 789–796.

93. J. Liu, and C. P. Luo: 'Precipitation behavior of the lower bainitic carbide in a medium-carbon steel containing Si, Mn and Mo', *Materials Science & Engineering A*, 2006, **438-440**, 153–157.

94. S. Murphy, and J. A. Whiteman: 'The precipitation of epsilon-carbide in twinned martensite', *Metallurgical Transactions*, 1970, **1**, 843–848.

95. A. Kelly, and G. W. Groves: Crystallography and Crystal Defects: London, U.K.: Longmans, 1970.

96. D. Kalish, M. Cohen, and S. A. Kulin: 'Strain tempering of bainite in 9Ni-4Co-0.45C steel', *Journal of Materials*, 1970, **5**, 169–183.

97. R. A. Arndt, and A. C. Damask: 'Kinetics of carbon precipitation in irraditated iron – III. calorimetry', *Acta Metallurgica*, 1964, **12**, 341–353.

98. L. S. Darken, and R. W. Gurry: Physical Chemistry of Metals: New York, USA: McGraw-Hill, 1953.

99. S. Ito, N. Tsushima, and H. Muro: 'Accelerated rolling contact fatigue test by a cylinder-to-ball rig', In: J. J. C. Hoo, ed. *Rolling Contact Fatigue Testing of Bearing Steels*. Warrendale, Pennsylvania, USA: ASTM, 1982:125–135.

100. J. M. R. Génin: 'The clustering and coarsening of carbon multiplets during the aging of martensite from Mössbauer spectroscopy: The preprecipitation stage of epsilon carbide', *Metallurgical & Materials Transactions A*, 1987, **18**, 1371–1388.

101. H. K. D. H. Bhadeshia: 'Steels for bearings', *Progress in Materials Science*, 2012, **57**, 268–435.

102. A. E. Austin, and C. M. Schwartz: 'Decomposition of austenite and martensite', *Proceeding of the American Society for Testing and Materials*, 1955, **55**, 623–625.

103. D. V. Edmonds, K. He, F. C. Rizzo, B. C. D. Cooman, D. K. Matlock, and J. G. Speer: 'Quenching and partitioning martensite – a novel steel heat treatment', *Materials Science & Engineering A*, 2006, **438-440**, 25–34.

104. V. H. Borchers, and K. Doffin: 'Kinetik der bildung des karbids Fe_2C in stahl 100Cr6', *Archiv für das Eisenhüttenwesen*, 1969, **40**, 493–498.

105. X. D. Wang, W. Z. Xu, Z. H. Guo, L. Wang, and Y. H. Rong: 'Carbide characterization in a Nb-microalloyed advanced ultrahigh strength steel after quenching-partitioning-tempering process', *Materials Science & Engineering A*, 2010, **527**, 3373–3378.

106. H. Y. Li, X. W. Lu, W. J. Li, and X. J. Jin: 'Microstructure and mechanical properties of an ultrahigh-strength 40SiMnNiCr steel during the one-step quenching and partitioning process', *Metallurgical & Materials Transactions A*, 2010, **41**, 1284–1300.

107. I. Gutierrez, J. Aranzabal, F. Castro, and J. J. Urcola: 'Homogeneous formation of epsilon carbides within austenite during isothermal transfm. of a ductile iron at 410°C', *Metallurgical & Materials Transactions A*, 1995, **26**, 1045–1060.

108. S. K. R. Patil, S. V. Khare, B. R. Tuttle, J. K. Bording, and S. Kodambaka: 'Mechanical stability of possible structures of PtN investigated using first-principles calculations', *Physical Review B*, 2006, **73**, 104118.

109. O. Beckstein, J. E. Klepeis, G. L. W. Hart, and O. Pankratov: 'First-principles elastic constants and electronic structure of α-Pt_2Si and PtSi', *Physical Review B*, 2001, **63**, 134112.

110. Z. Q. Chen, Y. S. Peng, M. Hu, C. M. Li, and Y. T. Luo: 'Elasticity, hardness, and thermal properties of ZrB_n ($n = 1, 2, 12$)', *Ceramics International*, 2016, **42**, 6624–6631.

111. J. H. Jang, S. J. Park, T. H. Lee, and H. K. D. H. Bhadeshia: 'First-principles calculations of elastic constants for epsilon-carbide and the consequences', *Materials Science and Technology*, 2020, **36**, 615–622.

112. H. K. D. H. Bhadeshia: 'Cementite', *International Materials Reviews*, 2020, **65**, 1–27.

113. Y. Tomota, K. Kuroki, T. Mori, and I. Tamura: 'Tensile deformation of two ductile phase alloys: flow curves of α/γ Fe-Cr-Ni alloys', *Materials Science & Engineering*, 1976, **24**, 85–94.

114. S. F. Pugh: 'Relations between the elastic moduli and the plastic properties of polycrystalline pure metals', *The London, Edinburgh, and Dublin Philosophical Magazine and Journal of Science*, 1954, **45**, 823–843.

115. S. S. Hecker, D. L. Rohr, and D. F. Stein: 'Brittle fracture in iridium', *Metallurgical & Materials Transactions A*, 1978, **9**, 481–488.

116. A. H. Cottrell: 'Strengths of grain boundaries in pure metals', In: J. A. Charles, and G. C. Smith, eds. *Advances in Physical Metallurgy*. London, U.K.: The Institute of Metals, 1990:181–187.

117. T. Terashima, Y. Tomota, M. Isaka, T. Suzuki, M. Umemoto, and Y. Todaka: 'Strength and deformation behavior of bulky cementite synthesized by mechanical milling and plasma-sintering', *Scripta Materialia*, 2006, **54**, 1925–1929.

118. J. Gil-Sevillano: 'Room temperature plastic deformation of pearlitic cementite', *Materials Science & Engineering*, 1975, **21**, 221–225.

119. X. T. Deng, T. L. Fu, Z. D. Wang, R. D. K. Misra, and G. D. Wang: 'Epsilon carbide precipitation and wear behaviour of low alloy wear resistant steels', *Materials Science and Technology*, 2016, **32**, 320–327.

120. M. Umemoto, S. E. Kruger, and H. Ohtsuka: 'Ultrasonic study on the change in elastic properties of cementite with t temperature and mn content using nearly full density polycrystalline bulk samples', *Materials Science & Engineering A*, 2019, **742**, 162–168.

121. J. W. Christian: Theory of Transformations in Metals and Alloys, Part I: 3 ed., Oxford, U. K.: Pergamon Press, 2003.

122. L. Hui, Z.-Q. Chen, Z. Xie, and C. Li: 'Stability, magnetism and hardness of iron carbides from first-principles calculations', *Journal of Superconductivity and Novel Magnetism*, 2018, **31**, 353–364.

123. Z. Lv, S. Sun, P. Jiang, B. Wang, and W. Fu: 'First-principles study on the structural stability, electronic and magnetic properties of Fe_2C', *Computational Materials Science*, 2008, **42**(4), 692–697.

124. R. Salloom, and S. Srinivasan: 'Elastic constants and structural stability of non-stoichiometric epsilon ε-Fe2.4C carbide', *Materials Chemistry and Physics*, 2019, **228**, 210–214.

Notes

[1]Therefore, the concept of stability as implied by Equation 9.1 does depend on conditions, in this case the activity of carbon in the gas.

10 Nitrides

The concentration of nitrogen in typical steels tends to be small, typically 10-30 parts per million, so much of the information about nitrides in steels comes from studies of nitrogen-enriched surfaces. Nitrogen can be introduced into the surface of iron within practical timescales at temperatures in excess of about $500\,°C$, with the aim of coating it with hard nitrides ($\approx 1000\,HV$) that resist wear [1, 2]. The layer that forms normally consists of the ε-$Fe_{3-z}N$ ($0 \leq z \leq 1$) on top with cubic γ'-Fe_4N underneath, followed by nitrogen in solid solution within the ferrite. There are significant structural differences between these two nitrides, neither of their compositions are exactly stoichiometric. The nitrogen atoms are not necessarily located in the ideal positions described below, there is a degree of disorder, that is greater for the ε than γ' nitrides. This is part of the reason why the self-diffusion coefficient of nitrogen is much greater in ε-nitride than in γ' [1]. There is a further nitride, the orthorhombic ζ, which has a composition close to Fe_2N [3]. All of the nitrides, with the exception of ζ, are ferromagnetic at ambient temperature.

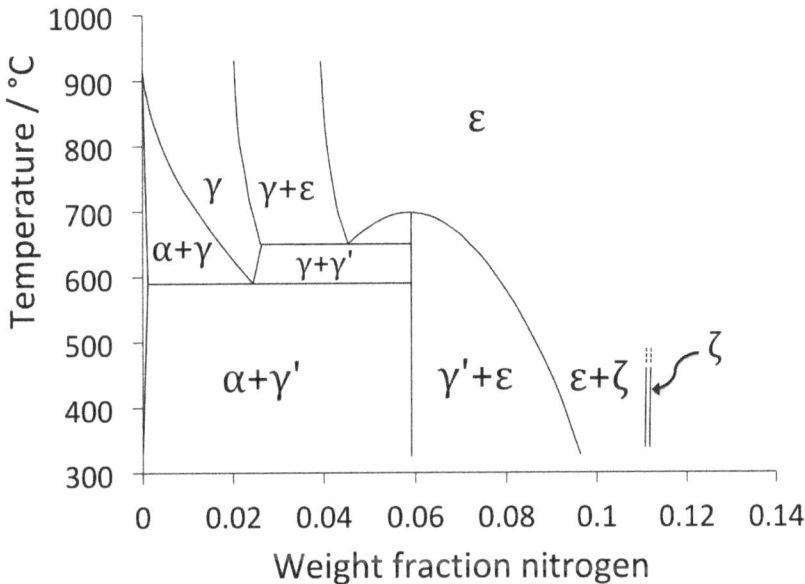

Figure 10.1 Part of the iron-nitrogen phase diagram calculated using the TCFE8 thermodynamic database. This database does not contain information about ζ-nitride, which is incorporated instead from Wriedt et al. [4]. The calculations treat γ' as a stoichiometric compound, but its composition can accommodate nitrogen over the range 5.6-5.9 wt% (19.3-20 at.%) [4]. The illustrated extent of the ζ phase field needs verification. Calculations courtesy of Shaumik Lenka.

The iron-nitrogen phase diagram in Figure 10.1 shows that ε-nitride exists over a large composition range, essentially from Fe_3N to Fe_2N, whereas γ' and ζ nitrides have more narrowly defined compositions. Nevertheless, γ' is not exactly stoichiometric with its concentration of nitrogen increasing at low temperatures. As a result, plates of ferrite precipitate from γ' during the cooling of nitrided specimens [5].

10.1 γ'-NITRIDE

The composition of γ'-nitride assuming stoichiometry is Fe_4N, consistent with the primitive cubic crystal structure, space group $Pm\bar{3}m$. The four iron atoms occupy the corners and face-centres of the unit cell, with the single nitrogen atom located in the octahedral interstice at the centre of the cell (Figure 10.2a). In this ordered state, the octahedral interstices at the cell edges ($\frac{1}{2},0,0$; $0,\frac{1}{2},0$; and $0,0,\frac{1}{2}$), are unoccupied. If the actual composition of the γ' deviates from 20 at.% nitrogen, the lattice parameter becomes a function of the nitrogen concentration [6, 7]:

$$a_{\gamma'}/\text{nm} = 0.37988 + 0.095315\left(y_N - \frac{1}{4}\right)$$

where y_N is the occupied fraction of the sublattice of octahedral sites. The linear thermal expansion coefficient is $(7.62 \pm 0.75) \times 10^{-6}\,\text{K}^{-1}$.

The lattice parameter of γ' is similar to that of austenite; it is found therefore to exhibit a cube-cube orientation relationship when it precipitates in austenite in a Fe-9.8N at% alloy during ageing at 210 °C [8]. The resulting local depletion of nitrogen stimulates the austenite to transform into ferrite accompanied by the further precipitation of γ' (Figure 10.2b). An approximate orientation relationship between ferrite and the nitride has the closest packed planes parallel [9]:

$$(1\bar{1}1)_{\gamma'} \parallel (110)_\alpha \qquad [011]_{\gamma'} \parallel \underbrace{[0.49\ \overline{0.49}\ \overline{0.72}]_\alpha}_{10° \text{ from } [1\bar{1}\bar{1}]_\alpha}.$$

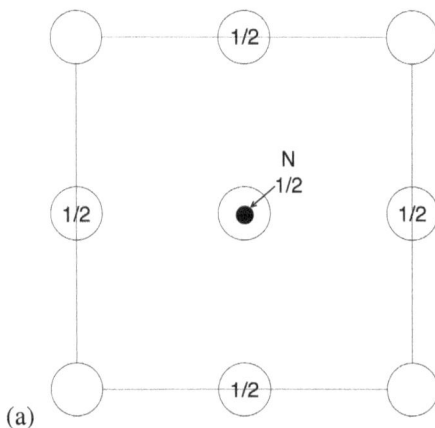

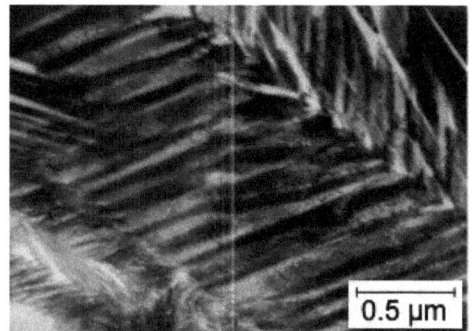

(a)   (b)

Figure 10.2 (a) Projection of the crystal structure of cubic Fe_4N on to $\{100\}$, with fractional heights indicated for all atoms other than those at heights 0,1. (b) Fe-N alloy transformed at 225 °C, consisting of α and γ'. After Jiao et al. [9], reproduced with permission from Elsevier.

The nitride is ferromagnetic below the Curie temperature of 767 ± 10 K with an average magnetic moment per iron atom of 2.2 μ_B [10–12], similar to that of α-iron even though its structure could be considered to be similar to that of austenite expanded by nitrogen. However, this is consistent with the fact that the two-state model of austenite where the high-volume form is indeed ferromagnetic (p. 31).

The development of a nitrided surface on steel begins with the nucleation and growth of γ' at the surface, followed by that of the more nitrogen-rich ε-nitride on top of the γ'. The thickening of the γ' layer is treated assuming nitrogen diffusion control, a flat interface with the steel and the absence of a defect structure [13, 14]. The α/γ' equilibrium-solubility of nitrogen in ferrite is much less than in the nitride, so the diffusion flux through the nitride determines the thickening kinetics. If the concentration of nitrogen in equilibrium with the surface (S) in contact with the active atmosphere is $c^{\gamma'S}$, and that in equilibrium with ferrite is $c^{\gamma'\alpha}$, then assuming a constant concentration gradient through the nitride,

$$(c^{\gamma'\alpha} - c^{\alpha\gamma'})\frac{\partial Z}{\partial t} = D_N^{\gamma'}\frac{\partial c}{\partial z}\bigg|_{z=Z} \approx D_N^{\gamma'}\frac{c^{\gamma'S} - c^{\gamma'\alpha}}{Z} \tag{10.1}$$

which gives and $Z = \alpha_{1d}\sqrt{t}$ (Equation 4.12), i.e. parabolic thickening.[1] When nitriding pure iron, the region near the surface will consist of ε whereas that underneath in contact with the ferrite will be the γ' nitride with its smaller nitrogen concentration; most nitriding conditions make the ε the thermodynamically more stable phase so the γ' is essentially a transition phase, Figure 10.3. The ε will therefore grow into the γ' as nitriding progresses. It is necessary then to consider the two nitride-layers and ferrite to be connected in series with local equilibria maintained at contact surfaces; this can be implemented numerically but the overall outcome still results in a parabolic thickening of each of the components of the compound layer [13, 15]. This is to be expected with the assumption of diffusion-controlled growth because the diffusion distance increases with the layer thickness making the gradients shallower as thickening proceeds. Figure 10.3 also illustrates porosity in the ε, which develops when the nitrogen potential of the active gas is greater than can be absorbed by the growth of the nitride itself [16].

The kinetic theory described above is incomplete because there is no treatment of the nucleation of the nitrides, nor of the time taken for the dissociation reactions that must occur at the surface of the iron. The quality of the iron surface has a role in promoting such reactions [17]. On clean surfaces, nucleation of the nitride not surprisingly begins at grain boundaries, with growth leading to allotriomorphs that decorate the boundaries. A layer of wuestite on the surface of the iron can promote the nucleation of particular nitrides but the reasons for this are not clear [18]. The incubation time towards the establishment of a layer depends also on the surface reactions. Most nitriding steels will contain solutes such as Cr or Al, both of which are strong nitride formers; the analysis does not account for substitutional solute diffusion.

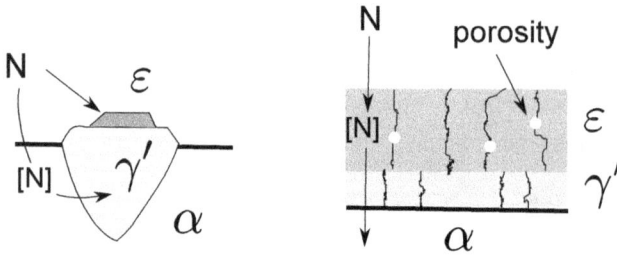

Figure 10.3 Schematic illustration of the development of a polycrystalline compound layer during nitriding. [N] indicates dissolved nitrogen. Adapted from Somers and Mittemeijer [13].

10.2 ε-NITRIDES

ε-Iron nitride (Fe_3N) has a primitive hexagonal crystal structure, space group $P6_322$, with the nitrogen atoms occupying octahedral interstices as illustrated in Figure 10.4a [19]. The nitrogen-rich phase Fe_2N has the structure shown in Figure 10.4b, with a reduced symmetry given by the space group $P\bar{3}1m$. The lattice parameters as a function of concentration are given by [20, 21]

$$a/\text{nm} = 0.44652 + 0.06851y_N \quad \text{slowly cooled nitride}$$
$$c/\text{nm} = 0.42433 + 0.03903y_N$$

$$a/\text{nm} = 0.44542 + 0.07111y_N \quad \text{quenched from 573 K}$$
$$c/\text{nm} = 0.42535 + 0.03662y_N$$

where y_N is the occupied fraction of the sublattice of octahedral sites, with $y_N = \frac{1}{2}$ for Fe_2N and $\frac{1}{3}$ for Fe_3N. These equations are derived from data covering the range $0.33 < y_N < 0.47$ ($0.0 < x < 0.4$). The quenched samples are said to represent the state of equilibrium at 573 K with a greater disorder in the distribution of nitrogen amongst the variety of octahedral interstices. The equations can be used to estimate the nitrogen content of the ε-nitrides, as long as the cooling conditions involved in the preparation of the nitride are accounted for [21]. An intermediate nitride where the composition is not stoichiometric, Fe_2N_{1-x} has the space group $P312$. All of these structures are based on the iron atoms forming a hexagonal close-packed cell with nitrogen in the octahedral interstices. The ε-nitride Fe_2N may, over a narrow composition range close to the stoichiometric ratio, transform into an orthorhombic structure (ζ) [19], which represents a distortion of the ε due to a rearrangement of the nitrogen atoms.

Fe_2N is ferromagnetic with a Curie temperature and magnetic moment per iron atom that is sensitive to the exact nitrogen concentration (Figure 10.5) and can contain carbon.

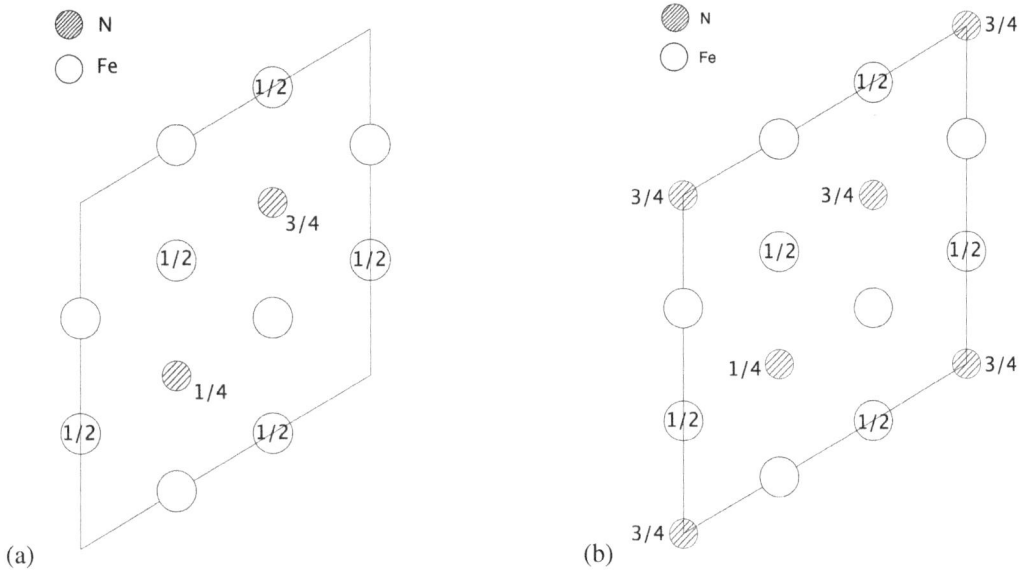

Figure 10.4 (a) The structure of Fe_3N (space group $P6_322$), (b) Fe_2N (space group $P\bar{3}1m$); for a nitride with a composition between Fe_2N and Fe_3N; the sites at $0.0.\frac{3}{4}$ would be occupied partially so the space group would reduce to $P312$. In both the cases, the structure is projected on to the basal plane, with the fractional coordinates along the c-axis indicated. Structures according to Jack [19].

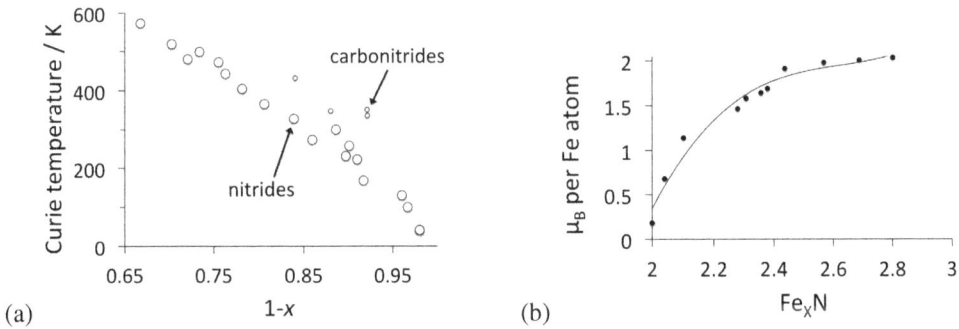

Figure 10.5 (a) The measured Curie temperatures for ε-nitrides $Fe_2N_{(1-x)}$ and carbonitrides. Data on pure nitrides from [22–24] and for the carbonitrides from [23, 25]. (b) The magnetic moment per iron atom, as a function of the nitrogen concentration for the hexagonal iron nitrides. Data adapted from Wold et al. [26].

Table 10.1

Fractional coordinates of iron atoms and octahedral interstices in the hexagonal representation of the unit cells of ε-nitrides, presented here with a consistent origin. There are six iron atoms in the cell which has lattice parameters that depend on the nitrogen concentration, but are approximately $a = 0.48$ nm and $c = 0.44$ nm. Two of the structures are illustrated in Figure 10.4.

Fractional coordinates				Occupied interstices		
Iron atoms			Octahedral interstices	$P6_322$	$P\bar{3}1m$	$P312$
$\frac{1}{3},0,0$	$\frac{1}{3},0,1$	$\frac{2}{3},0,\frac{1}{2}$	$\frac{1}{3},\frac{2}{3},\frac{1}{4}$	Fully	Fully	Fully
$0,\frac{1}{3},0$	$0,\frac{1}{3},1$	$\frac{1}{3},\frac{1}{3},\frac{1}{2}$	$\frac{1}{3},\frac{2}{3},\frac{3}{4}$	–	–	–
$1,\frac{1}{3},0$	$1,\frac{1}{3},1$	$0,\frac{2}{3},\frac{1}{2}$	$\frac{2}{3},\frac{1}{3},\frac{1}{4}$	–	–	–
$\frac{2}{3},\frac{2}{3},0$	$\frac{2}{3},\frac{2}{3},1$	$1,\frac{2}{3},\frac{1}{2}$	$\frac{2}{3},\frac{1}{3},\frac{3}{4}$	Fully	Fully	Fully
$\frac{1}{3},1,0$	$\frac{1}{3},1,1$	$\frac{2}{3},1,\frac{1}{2}$	$0,0,\frac{1}{4}$	–	–	–
			$0,0,\frac{3}{4}$	–	Fully	Partly

10.3 ζ-Fe$_2$N

ζ-Nitride crystallises with the space group $Pbcn$, the unit cell containing four Fe$_2$N formula units, and lattice parameters $a = 0.44373$ nm, $b = 0.55413$ nm and $c = 0.48429$ nm [27]. The structure is illustrated in Figure 10.6. Unlike γ' and ε nitrides, ζ is not ferromagnetic and has a small magnetic moment per iron atom of just $\approx 0.19\,\mu_B$ [26]. Although orthorhombic, the structure can still be conceived as consisting of a close-packed hexagonal stacking of iron atoms with nitrogen residing in octahedral interstices, but the arrangement of nitrogen atoms is different from ε which can have a similar chemical formula. ζ and ε nitrides are neighbours on the phase diagram, Figure 10.1; the former can be induced to transform into the latter at 1600 K and 15 GPa pressure, even though ε-Fe$_2$N is less dense than ζ-nitride [28]. The high pressure in those experiments was necessary to stop the compounds from releasing nitrogen; the experiments revealed the pressure dependence (Equation 8.2) of the bulk moduli to be $K_0 = 162$ GPa, $K_0' = 5.24$ and $V_0 = 118.09$ Å^3, for η, and $K_0 = 172$ GPa and $K_0' = 5.7$ for ε-Fe$_3$N$_{1.08}$ [28].

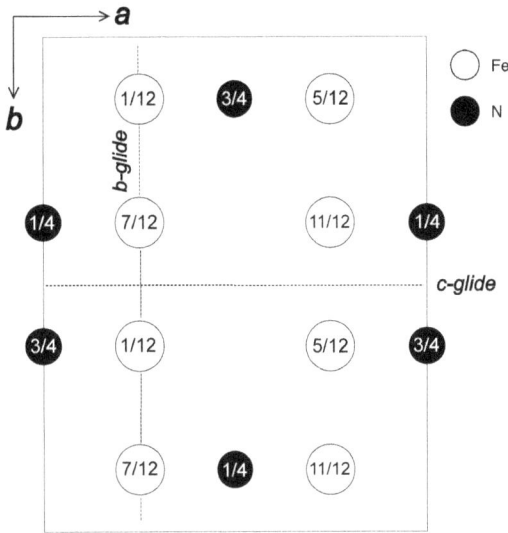

Figure 10.6 A projection of the crystal structure of ζ-nitride, with fractional heights along the c-axis marked. The space group is *Pbcn*; the n-glide planes are parallel to the plane of the diagram and at height $\frac{1}{4}$ and $\frac{3}{4}$.

10.4 α''-Fe$_{16}$N$_2$

Precipitation of α'' occurs when nitrogen-rich austenite is quenched to generate body-centred tetragonal martensite, followed by tempering [29]. There is a giant magnetic moment of $3\,\mu_B$ associated with its iron atoms [30]; although the extent of the exceptional magnetic properties remains a subject for debate, it is clear that its saturation magnetisation at $222\,\mathrm{emu\,g^{-1}}$ is greater than that of α-iron ($192\,\mathrm{emu\,g^{-1}}$) [31]. The Curie temperature for nanoparticles of α'' is about 634 K [32] but has been reported to be 813 K when the nitride is in the form of thin films [33]. Isolated α'' tends to decompose into a mixture of γ'-Fe$_4$N and α-Fe at temperatures as low as 460 K [32, 34, 35].

The crystal structure has the tetragonal space group $I4/mmm$, with the lattice parameters $a = 0.57176\,\mathrm{nm}$ and $c = 0.6288\,\mathrm{nm}$ [19, 35]. The fractional coordinates of the atoms in the cell are listed in Table 10.2. The structure, Figure 10.7, can be regarded as a cell consisting of a distorted set of eight body-centred cubic cells of α-iron [19]. A careful scrutiny of Figure 10.7b shows why the observed orientation relationship between α'' and α is as follows [34]:

$$\{100\}_\alpha \parallel \{001\}_{\alpha''} \quad \text{and} \quad \langle 100 \rangle_\alpha \parallel \langle 001 \rangle_{\alpha''}.$$

The good fit between the $\{100\}_\alpha$ and $\{002\}_{\alpha''}$ planes is not matched in other orientations so the nitride forms as thin plates on the cube planes of the ferrite, as illustrated in Figure 10.7c.

Table 10.2
Fractional coordinates of atoms in the unit cell of α''-$Fe_{16}N_2$ [35].

Atom	Multiplicity	Wyckoff letter	Fractional coordinates
Fe	4	e	$0, 0, 0.208$
Fe	4	d	$\frac{1}{2}, 0, \frac{1}{4}$
Fe	8	h	$\frac{1}{4}, \frac{1}{4}, 0$
N	2	b	$0, 0, \frac{1}{2}$

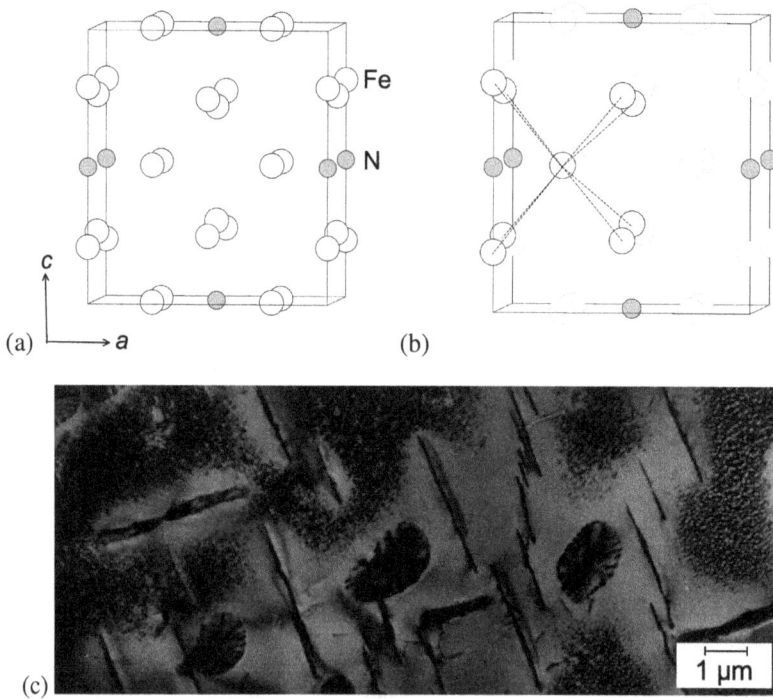

(a)   (b)

(c)

Figure 10.7 (a) The tetragonal unit cell of α''-$Fe_{16}C_2$. The nitrogen atoms are in irregular octahedral interstices. (b) Same as (a) but highlighting a distorted body-centred cubic cell. (c) α'' plates in ferrite, after Hayes et al. [34], reproduced with permission of Elsevier.

10.5 Z-PHASE

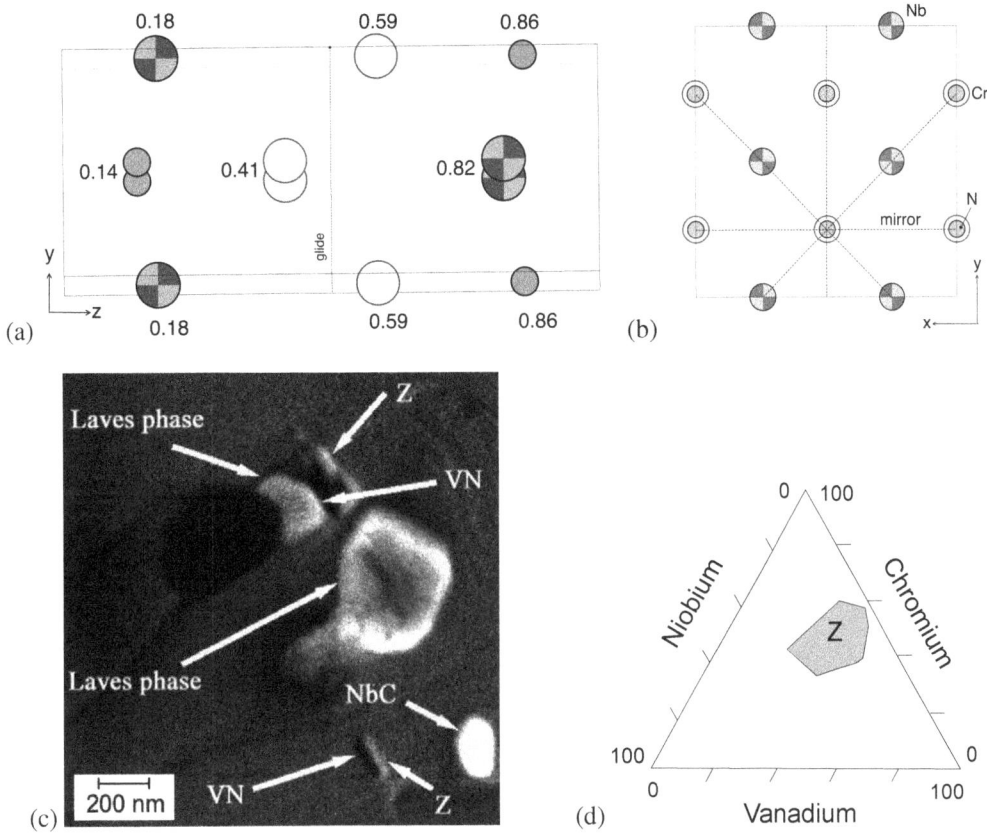

Figure 10.8 The structure of Z-phase, space group $P\frac{4}{n}mm$ [36]. (a) The fractional distance along z is indicated, together with the glide plane at $0,0,\frac{1}{2}$, identified by the dashed line. Note that the y-axis is not quite in the plane of the diagram. (b) Projection of $2 \times 2 \times 1$ unit cells on to (001); traces of mirrors are indicated by dashed lines. (c) Fe-0.09C-12.2Cr-0.35Mo-1.97W-0.25V-0.06Nb-0.06N wt% aged at 660 °C, showing Z-phase forming on vanadium nitride. Reproduced with permission from Elsevier, from Gopayegani et al. [37]. (d) Ternary diagram showing the composition range over which Z-phase forms at 600 °C, in creep-resistant steels, adapted from [38, 39].

Z-phase is a nitride of chromium and niobium (CrNbN), known originally to precipitate in austenitic stainless steels [40]. It became particularly prominent after it was discovered that its precipitation as Cr(Nb,V)N during the long-term testing of 12CrMoVNb martensitic steels leads to microstructural instability with consequent reduction in creep rupture strength [41]. This is because it is a more stable phase than common strengthening precipitates such as nitrogen-rich MX and M_2X, which tend to dissolve when Z-phase forms. Its precipitation at the austenite grain boundaries causes the decomposition of the strengthening precipitates which focuses the deformation at the weakened grain boundary zone, leading to accelerated creep rupture [38]. Steels which perform well in creep for many tens of thousands of hours, weaken dramatically with the onset of Z-phase precipitation. The crystal structure of CrNbN has the space group $P\frac{4}{n}mm$ with lattice parameters $a = 0.304$ nm and

$c = 0.740$ nm, Figure 10.8. It often nucleates on VN when present because there is a good fit between the (001) planes of Z-phase and VN [37]. Chromium is known to increase the thermodynamic stability of Z-phase relative to NbN and VN; it substitutes into the Z-phase as (Cr,V,Nb)N [42], but the relative stabilities of these nitrides can depend on the overall chemical composition; for example, Z-phase loses its thermodynamic advantage relative to NbN if the niobium to nitrogen ratio of the steel is reduced [43].

10.6 Nb(CN), CrN, TiN, Cr$_2$N, AlN

Niobium is relevant particularly as a microalloying addition to steels, because its precipitation during hot deformation helps control the austenite grain size. Nb(C,N) can be thought of as a solid solution between NbC and NbN, both of which have a cubic-F lattice with the space group $Fm\bar{3}m$ with interstitial atoms located in the octahedral spaces between the Nb atom framework, Figure 10.9a. Indeed, there is a linear relationship between the lattice parameter of Nb(CN), NbC (0.4469 nm) and NbN (0.4389 nm) as a function of the ratio of carbon and nitrogen concentrations. The orientation relationship when precipitated in ferrite is like that of Bain (often designated Baker-Nutting) with $[001]_{Nb(C,N)} \parallel [001]_\alpha$, $[100]_{Nb(C,N)} \parallel [110]_\alpha$, and a cube-cube orientation when the carbonitride precipitates within austenite.

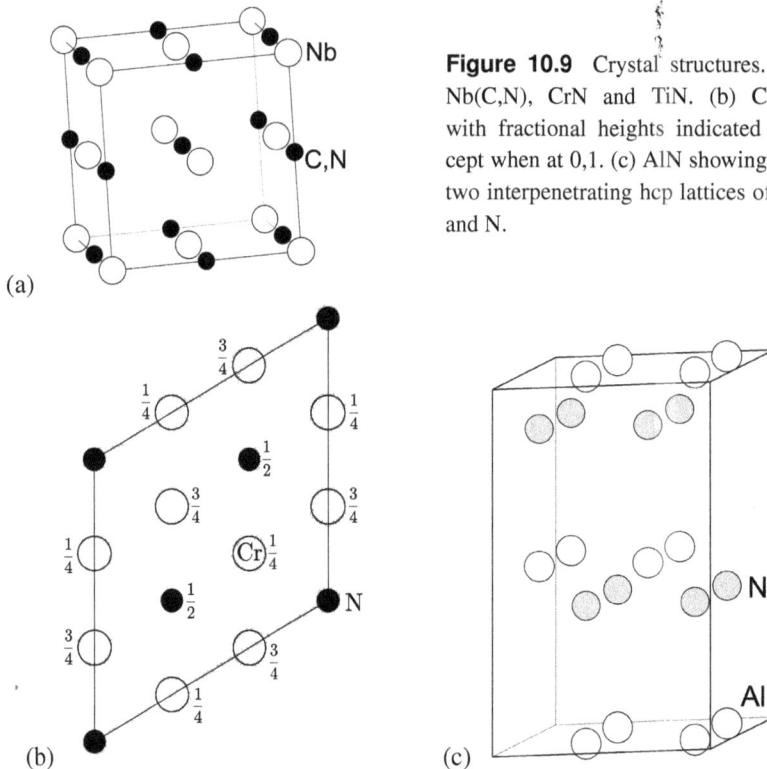

(a)

Figure 10.9 Crystal structures. (a) Nb(C,N), CrN and TiN. (b) Cr$_2$N with fractional heights indicated except when at 0,1. (c) AlN showing the two interpenetrating hcp lattices of Al and N.

(b)

(c)

Titanium often is used in boron-containing steels to getter nitrogen in order to allow boron to remain in solid solution so that it can segregate to austenite grain boundaries and increase the overall hardenability. It similarly is added to low-carbon steels in order to remove carbon from solid solution, for example in the so-called interstitial-free steels or in titanium-stabilised stainless steels. At typical concentrations, it has a much stronger affinity for interstitial elements than does niobium. The crystal structure of TiN is as shown in Figure 10.9a, with a lattice parameter of 0.427 nm. The nitride has a golden-yellow colour which becomes more copper-like when the interstitial sites contain a mixture of nitrogen and carbon [44]. The colour comes from the electronic structure of TiN. Certain wavelengths of white light are absorbed by inducing electronic transitions in the TiN, and hence are absorbed so the reflected components appears coloured. The band gap between electronic transitions depends on the composition and lattice parameter of TiN, so the colour can be controlled by manipulating these parameters. Titanium carbide also is coloured, Figure 11.2a.

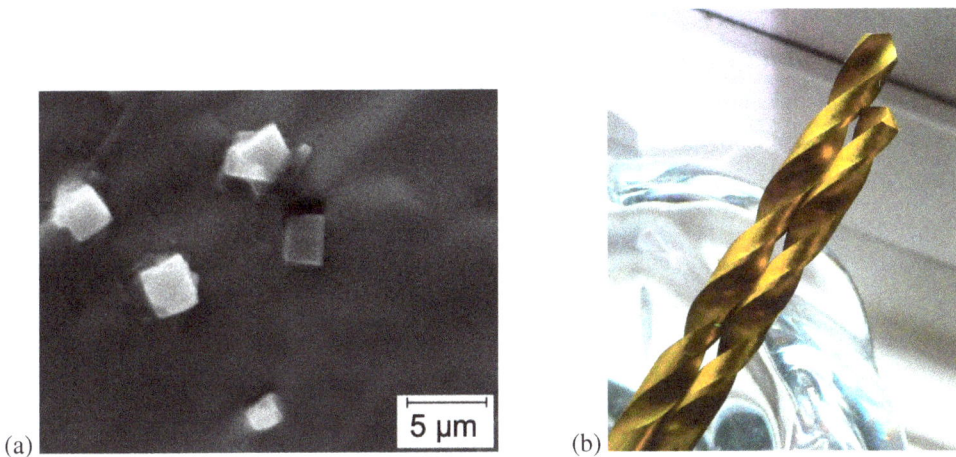

Figure 10.10 (a) Morphology of titanium nitride precipitates in steel. Reproduced from Fu et al. [45] with permission of Elsevier. (b) Golden colour of drill bits coated with titanium nitride.

TiN precipitates tend to be cuboidal in shape (Figure 10.10. First principles calculations of the surface energy of the (001) surface of TiN at 0 K, when in equilibrium with N_2, indicate a surface energy of $1.30 \, J \, m^{-2}$; the nitrogen terminated and titanium terminated (111) surface energies are 1.36 and $5.54 \, J \, m^{-2}$, respectively [46]. While these calculations are not representative of precipitation from steel, the order of interfacial energies is consistent with the observed cuboidal shape.

High nitrogen stainless steels, although difficult to produce, have significant applications including impellers and high-performance bearings [47, 48]. They contain up to about 0.6 wt% of nitrogen so are prone to the precipitation of chromium nitrides following heat treatment. The less common CrN has a cubic crystal structure with space group $Fm\bar{3}m$ and lattice parameter 0.294 nm, Figure 10.9a; it does not ordinarily occur in steels but may have uses as coatings for its hardness, friction characteristics and oxidation resistance. Nitrided layers have been shown to precipitate thin platelets of CrN on $\{100\}_\alpha$ planes in the Bain orientation with the ferrite [49], Figure 10.11. When the alloy to be nitrided contains both Cr and Al, a mixed $Cr_{1-x}Al_xN$ with the same crystal structure as CrN forms with a Bain orientation relationship with the ferrite [50].

The nitride Cr_2N with its lower nitrogen concentration can precipitate directly from ferrite or austen-

Figure 10.11 Thin platelets of CrN in nitrided Fe-3Cr wt% alloy. There are three orthogonal variants, the one parallel to the thin foil is just visible as discs. After Skiba et al. [49], reproduced with the permission of Elsevier.

ite; its crystal structure is trigonal ($P\bar{3}1m$, $a = 0.4796$ nm, $c = 0.4470$ nm) when the nitrogen atoms are ordered into three of the six sublattices of octahedral interstices within the hexagonal close-packed arrangement of Cr atoms [51], Figure 10.9b. If the nitrogen atoms are randomly dispersed on the interstices, the space group becomes $P6_3/mmc$ with lattice parameters $a = 0.2748$ nm and $c = 0.4438$ nm. When Cr_2N precipitates in duplex stainless steels, it acts as a nucleant for austenite. Depending on the nitrogen concentration and heat treatment, the precipitation can occur by a cellular mechanism $\gamma \rightarrow \gamma_{depleted} + Cr_2N$ or as discrete particles with both mechanisms occurring simultaneously, Figure 10.12. The cellular reaction, also known as discontinuous precipitation, occurs when two phases grow cooperatively at a common transformation front, but one of them is the same as the parent, albeit with a different chemical composition and possibly, crystallographic orientation. The orientation relationship between the depleted austenite and Cr_2N is [52, 53][2]

$$(0001)_{Cr_2N} \parallel (111)_{\gamma_{depleted}} \qquad [\bar{1}100]_{Cr_2N} \parallel [1\bar{1}0]_{\gamma_{depleted}}.$$

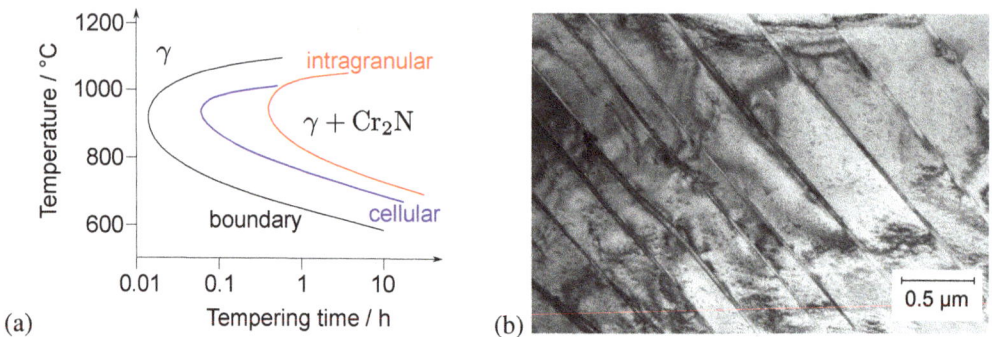

Figure 10.12 (a) Time-temperature-precipitation diagram for chromium nitride $(Cr_{0.73}Fe_{0.17}Mn_{0.10})_2N$ precipitation in Fe-19Cr-5Mn-5Ni-3Mo-0.024C-0.69N wt%, adapted from [54]. (b) Cellular precipitation of Cr2N. There are alternating lamellae of nitride and austenite. Micrograph courtesy of Patricia Almeida Carvalho [55].

Small quantities of aluminium and nitrogen in steels, often present as residuals in parts-per-million, from the steelmaking process where aluminium is used to deoxidise the melt, can influence the development of austenite grain structure. AlN precipitates are stable at high temperatures and therefore effective in pinning grain boundaries, Figure 10.13. The unit cell has the space group $P6_3mc$

with lattice parameters $a = 0.3111$ nm and 0.4978 nm [56], Figure 10.9c. The structure consists of two interpenetrating hexagonal close-packed lattices consisting of aluminium and nitrogen atoms respectively. Although aluminium carbonitrides can be made [56], it is likely that the precipitates found in steels are pure nitrides because their formation depends critically on the nitrogen concentration in the steel. Diffraction data are also inconsistent with the carbonitrides that have different crystallographic space groups.

Cubic AlN (space group $Fm\bar{3}m$) is sometimes observed to precipitate in a Bain orientation relationship with ferrite on a habit plane with the approximate indices $\{100\}_\alpha$ [57, 58]. There is a tendency to transform into the hexagonal form, although the kinetics of the transformation to the more stable nitride may be constrained by difficulty of nucleating the hexagonal form.

Aluminium nitride has a large misfit with the ferrite lattice so its precipitation is relatively slow when compared with chromium nitride even though aluminium has a greater affinity for nitrogen than chromium [59].

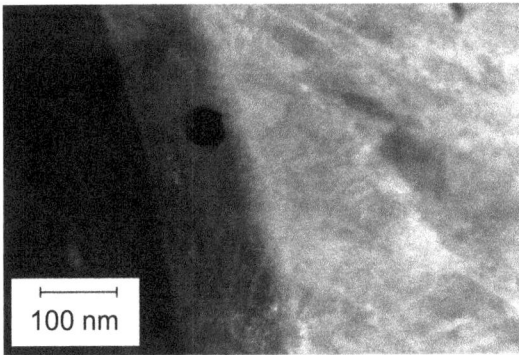

Figure 10.13 A thin, hexagonal plate of aluminium nitride that has precipitate at a prior austenite grain boundary. Micrograph courtesy of Hector Pous.

REFERENCES

1. M. A. Somers, and E. J. Mittemeijer: 'Layer-growth kinetics on gaseous nitriding of pure iron: Evaluation of diffusion coefficients for nitrogen in iron nitrides', *Metallurgical & Materials Transactions A*, 1995, **26**, 57–74.

2. S. Ooi, and H. K. D. H. Bhadeshia: 'Duplex hardening of aerospace bearings', *ISIJ International*, 2012, **52**, 1927–1934.

3. K. H. Jack: 'Binary and ternary interstitial alloys I. the iron-nitrogen system: the structures of Fe_4N and Fe_2N', *Proceedings of the Royal Society A*, 1948, **195**, 34–40.

4. H. A. Wriedt, N. A. Gokcen, and R. H. Natziger: 'The Fe-N (iron-nitrogen) system', *Bulletin of Alloy Phase Diagrams*, 1987, **8**, 355–377.

5. X. C. Xiong, A. Redja'imia, and M. Gouné: 'Transmission electron microscopy investigation of acicular ferrite precipitation in γ'-Fe_4N nitride', *Materials Characterization*, 2010, **61**, 1245–1251.

6. M. A. J. Somers, N. M. van der Pers, D. Schalkoord, and E. J. Mittemeijer: 'Dependence of the lattice parameter of γ' iron nitride, Fe_4N_{1-x}, on nitrogen content: accuracy of the nitrogen absorption data', *Metallurgical Transactions A*, 1989, **20**, 1533–1539.

7. T. Woehrle, A. Leineweber, and E. J. Mittemeijer: 'The shape of nitrogen concentration-depth profiles in γ'-Fe_4N_{1-z} layers growing on α-Fe substrates: the thermodynamics of γ'-Fe_4N_{1-z}', *Metallurgical and Materials Transactions A*, 2012, **43**, 610–618.

8. J. Foct, P. Rochegude, and A. Hendry: 'Low temperature ageing of Fe-N austenite', *Acta Metallurgica*, 1988, **36**, 501–505.

9. D. Jiao, C. P. Luo, and J. Liu: 'Isothermal transformation of high-nitrogen austenite', *Scripta Materialia*, 2007, **56**, 613–616.

10. C. Guillaud, and H. Creveaux: 'Magnetisme-preparation et proprietes magnetiques du compose defini Fe_4N', *Comptes Rendus Hebromadaires des Seances de L'Académie des Sciences*, 1946, **222**, 1170–1172.

11. G. Shirane, W. J. Takei, and S. L. Ruby: 'Mössbauer study of hyperfine fields and isomer shifts in Fe_4N and $(F\ e, Ni)_4N$', *Physical Review*, 1962, **126**, 49–52.

12. B. C. Frazer: 'Magnetic structure of Fe_4N', *Physical Review*, 1958, **112**, 751–754.

13. M. A. J. Somers, and E. J. Mittemeijer: 'Modelling the kinetics of nitriding and nitrocarburizing of iron', In: *17th ASM Heat Treatment Society Conference*. Materials Park, Ohio, USA: ASM International, 1998:321–330.

14. M. Keddam, M. E. Djeghlal, and L. Barrallier: 'A simple diffusion model for the growth kinetics of γ' iron nitride on the pure iron substrate', *Applied Surface Science*, 2005, **242**, 369–374.

15. M. Keddam, M. E. Djeghlal, and L. Barrallier: 'A diffusion model for simulation of bilayer growth $\varepsilon/\gamma')$ of nitrided pure iron', *Materials Science & Engineering A*, 2004, **378**, 475–478.

16. C. Middendorf, and W. Mader: 'Growth and microstructure of iron nitride layers and pore formation in ε-Fe_3N', *Zeitschrift für Metallkunde*, 2003, **94**, 333–340.

17. M. A. J. Somers: 'Thermodynamics, kinetics and microstructural evolution of the compound layer; a comparison of the states of knowledge of nitriding and nitrocarburising', *Heat Treatment of Metals*, 2000, **4**, 92–101.

18. P. B. Friehling, F. W. Poulsen, and M. A. J. Somers: 'Nucleation of iron nitrides during gaseous nitriding of iron; effect of a preoxidation treatment', *Zietschrift für Metallkunde*, 2001, **92**, 589–595.

19. K. H. Jack: 'The iron-nitrogen system: the crystal structures of the ε-phase iron nitrides', *Acta Crystallographica*, 1952, **5**, 404–411.

20. M. A. J. Somers, B. J. Kooi, L. Maldzinski, E. J. Mittemeijer, A. A. Van der Horst, A. M. Van der Kraan, and N. M. Van der Pers: 'Thermodynamics and long-range order of interstitials in an hcp lattice: Nitrogen in ε-Fe_2N_{1-z}', *Acta Materialia*, 1997, **45**, 2013–2025.

21. T. Liapina, A. Leineweber, E. J. Mittemeijer, and W. Kockelmann: 'The lattice parameters of ε-iron nitrides: lattice strains due to a varying degree of nitrogen ordering', *Acta Materialia*, 2004, **52**, 173–180.

22. R. J. Bouchard, C. G. Frederick, and V. Johnson: 'Preparation and properties of submicron hexagonal Fe_xN, $2 < x < 3$', *Journal of Applied Physics*, 1974, **45**, 4067–4070.

23. A. Leineweber, H. Jacobs, F. Hüning, H. Lueken, and W. Kockelmann: 'Nitrogen ordering and ferromagnetic properties of ε-Fe_3N_{1+x} $(0.10 \leq x \leq 0.39)$ and ε-$Fe_3(N_{0.8}C_{0.2})_{1.38}$', *Journal of Alloys and Compounds*, 2001, **316**, 21–38.

24. M. Kano, T. Nakagawa, T. A. Yamamoto, and M. Katsura: 'Magnetism, crystal structure and nitrogen content near the ε-ζ phase boundary of iron nitrides', *Journal of Alloys and Compounds*, 2001, **327**, 43–46.

25. B. K. Brink, K. Stáhl, T. L. Christiansen, C. Fransden, M. F. Hansen, P. Beran, and M. A. J. Somers: 'Effect of carbon on interstitial ordering and magnetic properties of ε-$Fe_2(N,C)_{1-z}$', *Journal of Alloys and Compounds*, 694, **2017**, 282–291.

26. A. Wold, R. J. Arnott, and N. Menyuk: 'Hexagonal iron nitrides', *The Journal of Physical Chemistry*, 1961, **65**, 1068–1069.

27. D. Rechenbach, and H. Jacobs: 'Structure determination of ζ-Fe_2N by neutron and synchrotron powder diffraction', *Journal of Alloys and Compounds*, 1996, **235**, 15–22.

28. U. Schwarz, A. Wosylus, M. Wessel, R. Dronskowski, M. Hanfland, D. Rau, and R. Niewa: 'High-pressure-high-temperature behavior of ζ-Fe_22N and phase transition to ε-$Fe_3N_{1.5}$', *European Journal of Inorganic Chemistry*, 2009, **12**, 1634–1639.

29. K. H. Jack: 'α''-$Fe_{16}C_2$: A giant magnetic moment material?', *Materials Science Forum*, 2000, **325-326**, 91–98.

30. T. K. Kim, and M. Takahashi: 'New magnetic material having ultrahigh magnetic moment', *Applied Physics Letters*, 1972, **20**, 492–494.

31. P. Palade, C. Plapcianu, I. Mercioniu, C. Comanescu, G. Schinteie, A. Leca, and R. Vidu: 'Structural, magnetic, and mössbauer investigation of ordered iron nitride with martensitic structure obtained from amorphous hematite synthesized via the microwave route', *Industrial and Engineering Chemistry Research*, 2017, **56**, 2958–2966.

32. I. Dirba, C. A. Schwöbel, L. V. B. Diop, M. Duerrschnabel, L. M.-L. , K. Hofmann, P. Komissinskiy, H. J. Kleebe, and O. Gutfleisch: 'Synthesis, morphology, thermal stability and magnetic properties of α''-$Fe_{16}C_2$ nanoparticles obtained by hydrogen reduction of γ-Fe_2O_3 and subsequent nitrogenation', *Acta Materialia*, 2017, **123**, 214–222.

33. Y. Sugita, K. Mitsuoka, M. Komuro, H. Hoshiya, Y. Kozono, and M. Hanzono: 'Giant magnetic moment and other magnetic properties of epitaxially grown $Fe_{16}N_2$ single-crystal films (invited)', *Journal of Applied Physics*, 1991, **70**, 5978–5982.

34. P. Hayes, W. Roberts, and P. Grieveson: 'A HVEM study of the precipitation and dissolution of iron nitrides in ferrite', *Acta Metalllurgica*, 1975, **23**, 849–854.

35. M. Widenmeyer, T. C. Hansen, and R. Niewa: 'Decomposition of metastable α''-$Fe_{16}C_2$ from in situ powder neutron diffraction and thermal analysis', *Zeitschrift für Anorganische und Allgemeine Chemie*, 2013, **639**, 2851–2859.

36. A. Jain, S. Ong, G. Hautier, W. Chen, W. Richards, S. Dacek, S. Cholia, D. Gunter, D. Skinner, G. Ceder, and K. A. Persson: 'The materials project: A materials genome approach to accelerating materials innovation', *APL Materials*, 2013, **1**, 011002.

37. A. Golpayegani, H.-O. Andrén, H. Danielsen, and J. Hald: 'A study on Z-phase nucleation in martensitic chromium steels', *Materials Science & Engineering A*, 2008, **489**, 310–318.

38. K. Sawada, H. Kushima, K. Kimura, and M. Tabuchi: 'TTP diagrams of Z phase in 9–12% Cr heat-resistant steels', *ISIJ International*, 2017, **47**, 733–739.

39. K. Sawada, H. Kushima, M. Tabuchi, and K. Kimura: 'Microstructural degradation of Gr.91 steel during creep under low stress', *Materials Science & Engineering A*, 2011, **528**, 5511–5518.

40. H. Gerlach, and E. Schmidtmann: 'Einfluß von kohlenstoff, stickstoff und bor auf das ausscheidungsverhalten eines austenitischen stahles mit rd. 16% Cr, 2% Mo, 16% Ni und niob', *Archiv für das Eisenhüttenwesen*, 1968, **39**, 139–149.

41. A. Strang, and V. Vodarek: 'Z-phase formation in a martensitic 12CrMoVNb steel', *Materials Science and Technology*, 1996, **12**, 552–556.

42. H. K. Danielsen: 'Review of Z phase precipitation in 9–12wt-%Cr steels', *Materials Science and Technology*, 2016, **32**, 126–137.

43. J. Svoboda, and H. Riedel: 'Modeling of spontaneous transformation of nitrides to Z phase accounting for stress relaxation by diffusion and interface activity', *Computational Materials Science*, 2019, **161**, 24–34.

44. F. A. Bannister: 'Osbornite, meteoritic titanium nitride', *Mineralogical Magazine and Journal of the Mineralogical Society*, 1941, **26**, 36–44.

45. H. Fu, Y. Miao, X. Chan, and B. Qiao: 'Effect of austempering on the structures and performances of cast high carbon Si-Mn steel', *Steel Research International*, 2007, **78**, 358–363.

46. D. Gall, S. Kodambaka, M. A. Wall, I. Petrov, and J. E. Greene: 'Pathways of atomistic processes on TiN(001) and (111) surfaces during film growth: an *ab initio* study', *Journal of Applied Physics*, 2003, **93**, 9086–9094.

47. J. W. Simmons: 'Overview: high-nitrogen alloying of stainless steels', *Materials Science & Engineering A*, 1996, **207**, 159–169.

48. H. K. D. H. Bhadeshia: 'Steels for bearings', *Progress in Materials Science*, 2012, **57**, 268–435.

49. O. Skiba, A. Redjaïmia, J. Dulcy, J. Ghanbaja, G. Marcos, N. Cladeira-Meulnotte, and T. Czerwiec: 'A proper assessment of TEM diffraction patterns originating from CrN nitrides in a ferritic matrix', *Materials Characterization*, 2018, **144**, 671–677.

50. A. R. Clauss, E. Bischoff, S. S. Hozmani, R. E. SchacherL, and E. J. Mittemeijer: 'Crystal structure and morphology of mixed $Cr_{1-x}A_l xN$ nitride precipitates: Gaseous nitriding of a Fe-1.5 wt pct Cr-1.5 wt pct Al alloy', *Metallurgical & Materials Transactions A*, 2009, **40**, 1923–1934.

51. T. H. Lee, S. J. Kim, and S. Takaki: 'On the crystal structure of Cr_2N precipitates in high-nitrogen austenitic stainless steel. II. Order–disorder transition of Cr_2N during electron irradiation', *Acta Crystallographica B*, 2006, **62**, 190–196.

52. D. B. Rayaprolu, and A. Hendry: 'Cellular precipitation in a nitrogen alloyed stainless steel', *Materials Science and Technology*, 1989, **5**, 328–332.

53. P. A. Carvalho, I. F. Machado, G. Solórzano, and A. F. Padilha: 'On Cr_2N precipitation mechanisms in high-nitrogen austenite', *Philosophical Magazine*, 2008, **88**, 229–242.

54. J. W. Simmons: 'Mechanical properties of isothermally aged high-nitrogen stainless steel', *Metallurgical & Materials Transactions A*, 1995, **26**, 2579–2995.

55. I. F. Machado, P. A. Carvalho, and A. F. Padilha: 'Austenite instability and precipitation behavior of high nitrogen stainless steels', In: A. Pramanik, and B. Kumar, eds. *Stainless Steel: Microstructure, Mechanical Properties and Methods of Application*. New York, USA: Nova Science Publishers, 2016:1–36.

56. G. A. Jeffrey, and V. Y. Wu: 'The structures of the aluminium carbonitrides', *Acta Crystallographica*, 1963, **16**, 559–566.

57. M. H. Biglari, C. M. Brakman, and E. J. Mittemeijer: 'Crystal structure and morphology of AlN precipitating on nitriding of an Fe-2at.% Al alloy', *Philosophical Magazine A*, 1995, **72**, 1281–1299.

58. M. Sennour, and C. Esnouf: 'Contribution of advanced microscopy techniques to nano- precipitates characterization: case of AlN precipitation in low-carbon steel', *Acta Materialia*, 2003, **51**, 943–957.

59. K. S. Jung, R. E. Schacherl, E. Bischoff, and E. J. Mittemeijer: 'Nitriding of ferritic Fe–Cr–Al alloys', *Surface and Coatings Technology*, 2010, **204**, 1942–1946.

60. A. Marciniak: 'Equilibrium and non-equilibrium models of layer formation during plasma and gas nitriding', *Surface Engineering*, 1985, **1**, 283–288.

61. F. Vanderschaeve, R. Taillard, and J. Foct: 'Discontinuous precipitation of Cr_2N in a high nitrogen, chromium-manganese austenitic stainless steel', *Journal of Material Science*, 1995, **30**, 6035–6046.

Notes

[1] $D_N^{\gamma'} = 1.675 \times 10^{-9} \exp\{-64000\,J\,mol^{-1}/RT\}\,m^2\,s^{-1}$, assumed to be concentration independent [60]. $D_N^{\varepsilon} = 2.1 \times 10^{-8} \exp\{-93517\,J\,mol^{-1}/RT\}\,m^2\,s^{-1}$ [15].

[2] The orientation relationship between the depleted austenite and Cr_2N in [61] does not seem to be correct in terms of the angles involved.

11 Substitutionally alloyed precipitates

11.1 TiC, NbC

There are many kinds of metallic carbides that exist in pure forms or as precipitates in a variety of alloy systems, but the focus here is on alloys of iron. Ti, Zr, Hf, V, Nb and Ta when dissolved in austenite or ferrite, have a tendency to precipitate as carbon-rich MC carbides with the rock salt crystal structure (space group $Fm\bar{3}m$). In this structure, there is one octahedral interstitial site per metal atom, which when fully occupied, yields the composition MC; when half the sites are vacant, the carbide becomes M_2C. And there can be a significant deviation from these stoichiometric compositions.[1] Although MC-type plutonium and uranium carbides have not been shown to *precipitate* in steels, when they function as fuel for fast breeder nuclear reactors, they are clad in stainless steel. They can then release carbon into the stainless steel, thereby compromising its ductility.

The binary carbides all have a metallic character, are paramagnetic and melt at very high melting temperatures. It is quite common therefore for titanium or niobium carbides to precipitate in liquid steel prior to its solidification. TiC and NbC melt at 3250 °C and 3500 °C respectively. In hardfacing alloys, the deliberately large volume fractions of carbides form first in liquid. The shape of the carbides is close to their equilibrium morphology when they form unconstrained by the liquid. Figure 11.1 shows the three-dimensional shape of a NbC dendrite that is strongly facetted, reflecting its cubic symmetry; when observed in cross-sections, the shapes still reveal some features of cubic symmetry and strong faceting. In contrast, TiC when it forms from liquid maintains a cuboidal shape [1]; however, dendritic TiC is observed when the cooling rate is large [2], indicating perhaps a role of constitutional supercooling.

When precipitating in steels, the metal sublattice may contain a limited mixture of metal atoms, for example, (Ti,Mo)C, (Ti,Nb)C, (Ti,V)C [3] and (Ti,Cr)C [4]. The chemical composition is not usually stoichiometric because some of the octahedral interstices are vacant. The vacant interstitial sites may be randomly distributed or in an ordered array; in the latter case, they are known as constitutional vacancies because of their influence on crystal symmetry. The range of compositions possible in TiC_{1-x} with $0.52 < x < 1$ as a single phase rather than as a precipitate in steel, leads to order-disorder transformations in $TiC_{0.59-0.62}$ [5]:

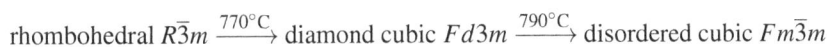

$$\text{rhombohedral } R\bar{3}m \xrightarrow{770°C} \text{diamond cubic } Fd3m \xrightarrow{790°C} \text{disordered cubic } Fm\bar{3}m$$

There is an interesting consequence of a flexible titanium carbide composition in the manufacture of ceramic matrix composite with steel [6]. When $TiC_{0.57}$ is heat treated in contact with a eutectoid steel, the pearlite fraction in the vicinity of the contact surface decreases dramatically. The carbide absorbs carbon as it approaches a state of equilibrium with the steel. When a similar treatment is

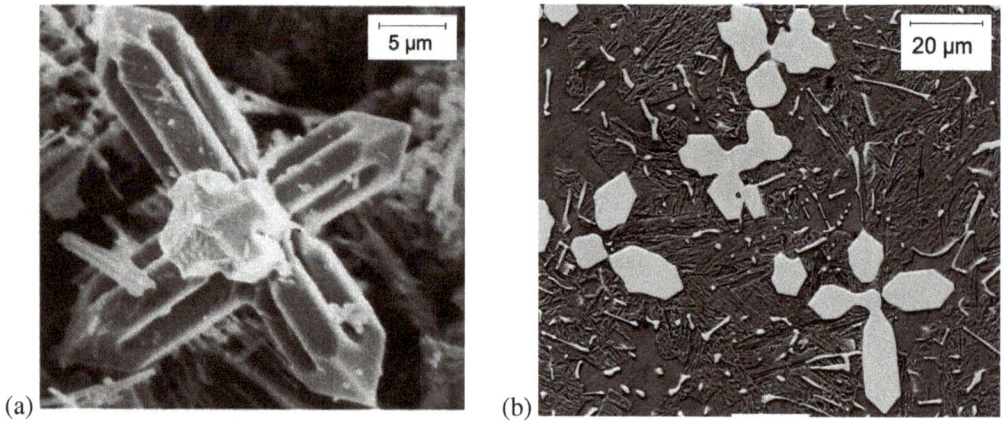

(a)   (b)

Figure 11.1 (a) Three-dimensional form of a dendrite of NbC in Fe-34Cr-5Nb-4.5C wt%. Micrograph courtesy of Berit Gretoft. (b) Cross sections of NbC dendrites in Fe-1.4C-6Cr-8Nb-1Si wt%. Micrograph courtesy of Mario C. Cordero-Cabrera. In both cases, the alloys were deposited using arc welding.

carried out with low carbon steel, TiC dissipates titanium into the adjacent steel in order to achieve a composition closer to its stoichiometric concentration. TiC is therefore defined to be chemically active because when bonded to steel, it stimulates ferrite formation from austenite, by locally decarburising the austenite, Figure 11.2a [7, 8]. Given its wide range of composition, the carbon in TiC can be displaced by neutron irradiation of a stainless steel, thereby inducing the precipitation of chromium-rich $M_{23}C_6$, which then grows to engulf the titanium carbide, Figure 11.2b [9].

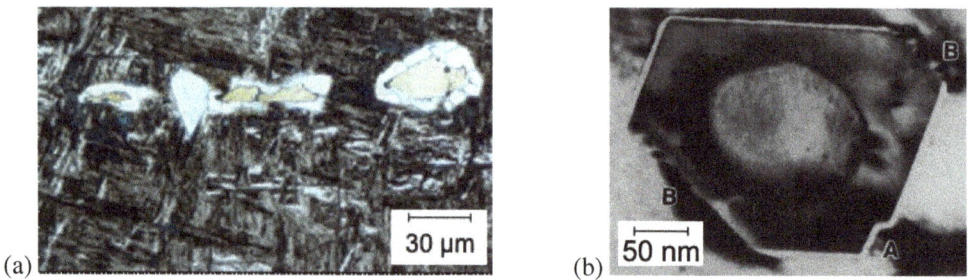

(a)   (b)

Figure 11.2 (a) TiC particles bonded between steel samples stimulate the formation of ferrite between the TiC/γ interfaces. The ferrite engulfs the carbide particles, although it is notable that the layers of ferrite are polycrystalline. Micrograph courtesy of Toshiko Koseki. (b) Spheroidal TiC particle surrounded by octahedral $M_{23}C_6$ precipitate in an irradiated stainless steel. Reproduced from Kestermich and Nandedkar [9], with permission of Elsevier.

Non-stoichiometric niobium carbide (NbC_y) can have structural modifications due to the ordering of carbon atoms. There is a disorder-order transformation from cubic $NbC_{0.83}$ ($Fm\bar{3}m$, $a = 0.4458$ nm) to a monoclinic Nb_6C_5 ($C2/m$, $a = c = 0.546$ nm, $b = 0.9458$ nm, $\beta = 109.47°$) on cooling below 1300 K [10]. Other ordered forms of niobium carbide have been proposed but the space group $C2/m$ is the only one that has been verified experimentally [11, 12].

Nb_2C has an orthorhombic structure (orthorhombic $Pnma$ $a = 0.1091$ nm, $b = 0.30954$ nm, $c = $

0.49746 nm) that transforms into the hexagonal form ($P\bar{3}1m$, $a = b = 0.54169$ nm, $c = 0.49719$ nm, $\alpha = 120°$) at about 1470 °C [13]. The carbide is rare in steels [14] but can be found in association with NbC in coatings containing gradients in the concentration of carbon [15, 16].

Titanium carbide can be induced into martensitic transformation during deformation, Figure 11.3 [17]. Plates of martensite develop on the $\{111\}_F$ planes, where the subscript refers to the form with space group $Fm\bar{3}m$ ($a = 0.4327$ nm), growing along the $\langle 11\bar{2}\rangle_F$. The transformed structure is hexagonal (designated H), with $a = 0.306$ nm, $c = 1.838$ nm, $P6_3/mmc$, and $(1\bar{1}1)_F \parallel (0001)_H$, $[121]_F \parallel [01\bar{1}0]_H$. There is a large volume contraction of 8% normal to the habit plane and a shear along $[121]_F$ of 0.7677. The habit plane of the martensite is perfectly coherent with its parent phase. The shape deformation is [17]

$$(\text{F P F}) = \begin{pmatrix} 1.14 & 0.14 & 0.14 \\ 0.14 & 1.14 & 0.14 \\ -0.36 & -0.36 & 0.64 \end{pmatrix} \quad \text{for} \quad \mathbf{p} = \frac{1}{\sqrt{3}}(111), \quad m\mathbf{d} = [0.242\ 0.242\ \overline{0.623}],$$

where $\mathbf{p}$ and $\mathbf{d}$ are the unit normal to the habit plane and unit displacement vector respectively, and m is the magnitude of the shape deformation.

The martensite forms with a thickness of four layers, each layer consisting of titanium atoms parallel to $\{111\}_F$; such "quadlayers" have a rhombohedral structure with $\alpha = \beta = \gamma = 63.73°$ and $a = b = c = 0.29$ nm, $R\bar{3}m$. It is when coarsening occurs by further martensitic transformation with adjacent quadlayers in alternating twin orientations (all parallel to the same habit plane) that the structure is interpreted as the hexagonal form with the large c-parameter.

The ability for different solutes to substitute on to the metal sublattice in TiC is of considerable technological significant because the lattice parameter of the carbide can be manipulated. In one variety of automotive steel, following hot deformation the steel transforms into allotriomorphic ferrite with the interphase precipitation of fine arrays of (Ti,Mo)C [18, 19], Figure 11.4a. Coarsening can occur after coiling the hot steel but this can be reduced by adjusting the Ti/Mo ratio so that the interfacial energy is reduced. First principles calculations relevant for 0 K indicate that the solution of molybdenum in TiC is not favoured from a thermodynamic point of view, the introduction of molybdenum reduces the lattice parameter of (Ti,Mo)C relative to TiC, permitting a better match [20]. Substitution of tungsten can achieve the same goal [21], as can the mixing on the same lattice of the carbide with certain nitrides [22], Figure 11.4b.

Both TiC and NbC adopt a cube-cube orientation relationship when precipitating in austenite. If the austenite subsequently transforms into ferrite then the carbide inherits the γ/α crystallography. However, when nanometre-sized precipitation occurs at an γ/α interface, TiC adopts the Bain orientation with the ferrite, but on coarsening, the particles rotate to the Nishiyama-Wasserman relationship [23]; this observation is particularly interesting because it demonstrates that the Bain orientation, which may be forced during the nucleation of a small particle, must rotate as the particle grows in order to minimise strain energy. The Bain orientation also is observed when fine niobium carbide precipitates in ferrite, but it would be interesting to see whether that is maintained when particles coarsen. Coarse NbC particles in coatings have been identified to have a Kurdjumov-Sachs orientation relationship with the ferrite [16].

(a)

(b)   (c)   (d)

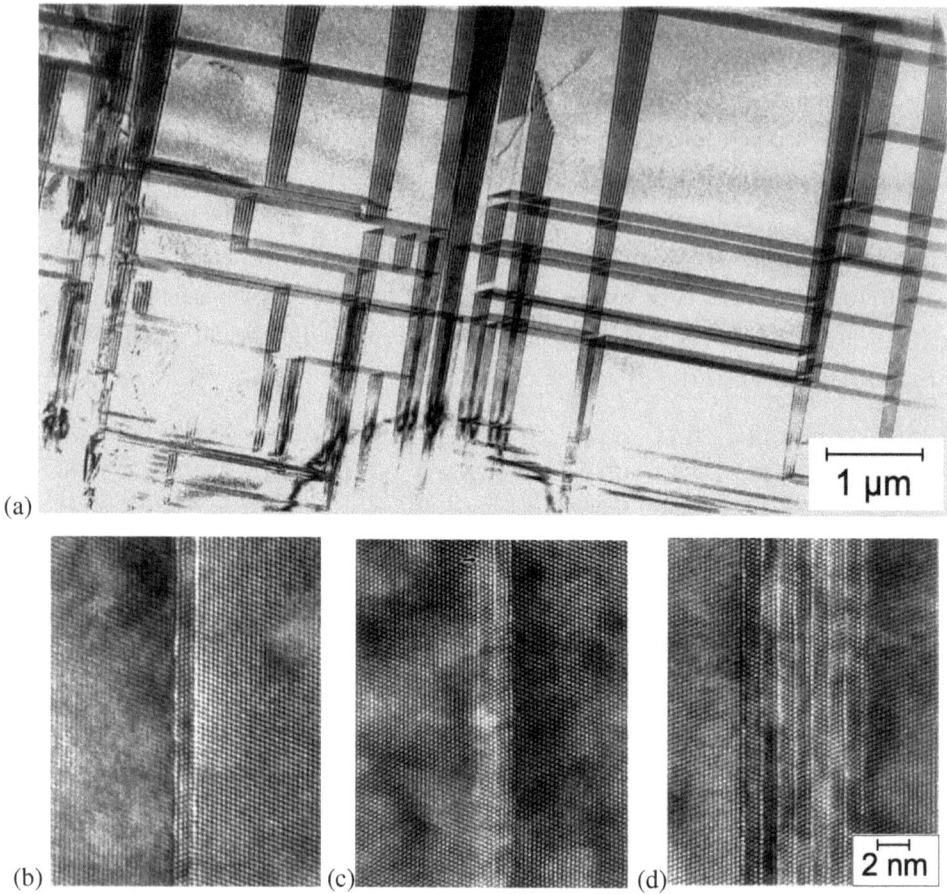

Figure 11.3   (a) Transmission electron micrograph showing martensite plates in a single crystal of TiC, the transformation established using detailed diffraction analysis. (b–d) High-resolution images of transformed plates in edge-on orientation, with one, two and seven quadlayers, respectively. Reproduced from Chien, Clifton and Nutt, [17], with the permission of the American Ceramic Society and Blackwell Publishing.

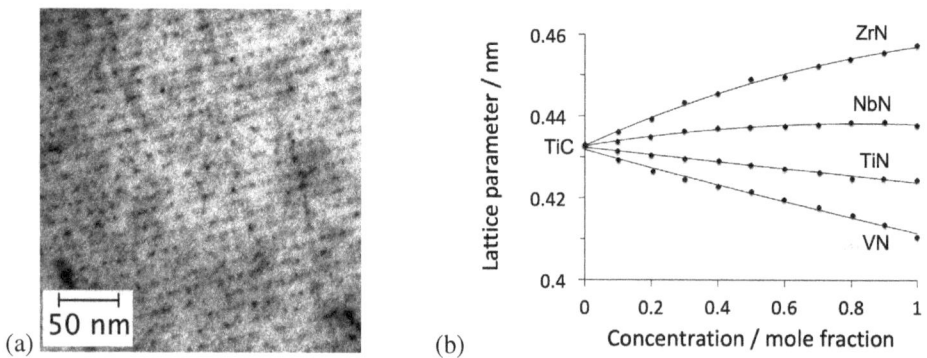

(a)   (b)

Figure 11.4   (a) Interphase precipitation of (Ti,Mo)C in ferrite in Fe-0.06C-1.5Mn-0.1Si-0.1Ti-0.2Mo wt% (after Yen et al. [19], reproduced with the permission of Elsevier). (b) Lattice parameter of (Ti,M)(C,N), from pureTiC to the various pure nitrides. The concentration axis therefore represents the proportions of the end-phases. Adapted using data from Duwez and Odell [22].

11.2 VANADIUM CARBIDE

Vanadium carbides form as fine, square-shaped platelets that precipitate on the $\{100\}_\alpha$ planes, Figure 11.5. The shape at the early stages of precipitation in austenite is octahedral with facets parallel to $\{111\}_\gamma$ [24].

The fine particles that form in ferrite or tempered martensite are associated with coherency strain fields that not only harden the steel but can trap hydrogen, thereby mitigating embrittlement due to the ingress of diffusible hydrogen into the metal. It is diffusible hydrogen that is responsible for the embrittlement of steel [25, 26].

100 nm

Figure 11.5 Three crystallographic variants of vanadium carbide forming on the cube planes of the ferrite. The plates have square shapes. After Yamasaki and Bhadeshia [27].

The vanadium carbide that precipitates in steel was thought originally to be V_4C_3 [28] with space group $Fm\bar{3}m$ and lattice parameter 0.4157 nm. However, a detailed examination of electron diffraction patterns, both new and those published in the past, has demonstrated that it is V_6C_5 with a monoclinic structure containing ordered vacancies in the carbon sublattice (Figure 11.6) [29, 30]. The space group is $C2/m$, with lattice parameters $a = 0.509$ nm, $b = 0.882$ nm, $c = 0.1018$ nm, $\beta = 109.47°$.

The lattice parameters of the monoclinic form are related to those of the V_4C_3 as follows:

$$a_{V_6C_5} = \underbrace{\sqrt{\frac{3}{2}} a_{V_4C_3}}_{=|\frac{1}{2}[11\bar{2}]a_{V_4C_3}|} \qquad b_{V_6C_5} = \underbrace{\frac{3}{\sqrt{2}} a_{V_4C_3}}_{=|\frac{3}{2}[1\bar{1}0]a_{V_4C_3}|} \qquad c_{V_6C_5} = \underbrace{\sqrt{6} a_{V_4C_3}}_{=|[112]a_{V_4C_3}|} .$$

Given that the literature is full of interpretations based on V_4C_3, the following conversions apply, noting that the basis symbols 'M' and 'C' refer to the monoclinic and cubic forms respectively, and that $[M; \mathbf{u}] = (M \, J \, C)[C; \mathbf{u}]$ and $(\mathbf{h}; M^*) = (h; C^*)(C \, J \, M)$:

$$(M \, J \, C) \; = \; \begin{pmatrix} 0.500365 & 0.500365 & -0.500365 \\ -0.333430 & 0.333430 & 0.000000 \\ 0.250182 & 0.250182 & 0.250182 \end{pmatrix} \tag{11.1a}$$

$$(C \, J \, M) \; = \; \begin{pmatrix} 0.499635 & -1.499563 & 0.999271 \\ 0.499635 & 1.499563 & 0.999271 \\ -0.999271 & 0.000000 & 1.998542 \end{pmatrix} \tag{11.1b}$$

$(001)_{V_6C_5} \parallel (111)_{V_4C_3}$ $[\overline{2}01]_{V_6C_5} \parallel [001]_{V_4C_3}$

$(010)_{V_6C_5} \parallel (\overline{1}10)_{V_4C_3}$ $[0.500365 \; 0.333430 \; 0.250182]_{V_6C_5} \parallel [010]_{V_4C_3}$

$(100)_{V_6C_5} \parallel (11\overline{1})_{V_4C_3}$ $[0.500365 \; \overline{0.333430} \; 0.250182]_{V_6C_5} \parallel [100]_{V_4C_3}$.

Therefore, the classical Baker-Nutting orientation relationship between vanadium carbide and ferrite,

$$\langle 001 \rangle_{V_4C_3} \quad \parallel \quad \langle 001 \rangle_\alpha \qquad \{110\}_{V_4C_3} \parallel \{010\}_\alpha$$

becomes $\quad \langle \overline{2}01 \rangle_{V_6C_5} \quad \parallel \quad \langle 001 \rangle_\alpha \qquad \{132\}_{V_6C_5} \parallel \{010\}_\alpha$.

Vanadium carbide at the early stages of precipitation in austenite has the following orientation relationship:

$$\langle 001 \rangle_{V_4C_3} \quad \parallel \quad \langle 001 \rangle_\gamma \qquad \{010\}_{V_4C_3} \parallel \{010\}_\gamma$$

becomes $\quad \langle \overline{2}01 \rangle_{V_6C_5} \quad \parallel \quad \langle 001 \rangle_\gamma \qquad \{132\}_{V_6C_5} \parallel \{010\}_\gamma$.

Another vanadium carbide V_8C_7 has a cubic crystal structure ($P4_332$, $a = 0.832\,\text{nm}$) when the carbon vacancies are ordered. The disordered form also is cubic but with about half the lattice parameter; the order-disorder temperature is about $1125\,^\circ\text{C}$ [31]. V_8C_7 occurs only in vanadium-rich steels such as those used in the manufacture of dies [32].

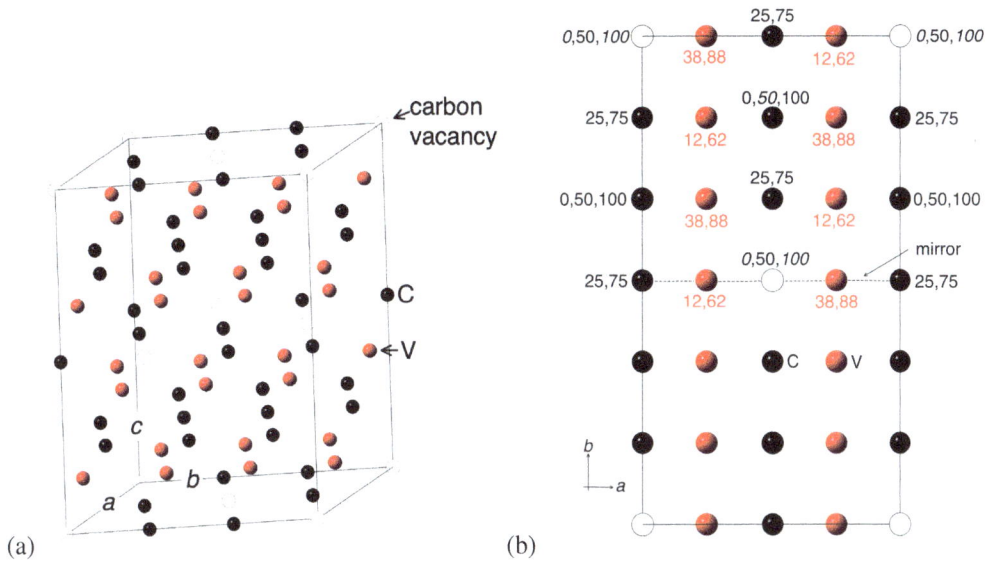

Figure 11.6 (a) Monoclinic unit cell of V_6C_5. The unique axis b is normal to both a and c. The open circles represent ordered vacancies in the carbon sites. (b) Projection of the cell on the $\{001\}$ plane, with percentage heights of the atoms indicated. The values in italics refer to the vacancies. There are alternating planes of carbon atoms with and without vacancies. The carbon atoms are octahedrally coordinated by iron atoms.

11.3 $M_{23}C_6$

The chromium carbide $Cr_{23}C_6$ is rarely observed in its pure form within steels because many other metal atoms ('M') such as Fe, Co, Mo, Mn, are able to substitute for Cr into its structure. Indeed, it is possible to obtain $M_{23}C_6$ in chromium-free steels, for example, in ternary Fe-Mo-C and Fe-W-C alloys [33]. The carbide nucleates most easily at grain boundaries (Figure 11.7), followed by less-coherent twin boundaries, coherent twin boundaries and other defects such as dislocations [34]. In many steels that are welded, it is important to control or avoid the precipitation of this carbide in the heat-affected zone of the weld, because it results in chromium-depleted regions [35] that then become susceptible to intense, localised corrosion, often referred to as *weld decay*. In alloys that at ambient temperature consist of ferrite and $M_{23}C_6$, heating to sufficiently high temperatures where austenite formation is possible, causes at first the formation of particles of M_6C in the ferrite, by the diffusion of carbon from $M_{23}C_6$. At a sufficiently high temperature, $M_{23}C_6$ can decompose into a eutectoid mixture of M_6C and austenite [33].

The lattice parameter of $Cr_{23}C_6$ is very nearly three times that of austenite, both of which have a cubic-F lattice. So when the carbide precipitates in austenite, it adopts a cube-cube orientation relationship, so the cube edges of the two unit cells are parallel [36]. Its orientation relationship with ferrite is the same as that between austenite and ferrite [37]. It follows that the three-phase crystallography, when the carbide precipitates in contact with both ferrite and austenite is given

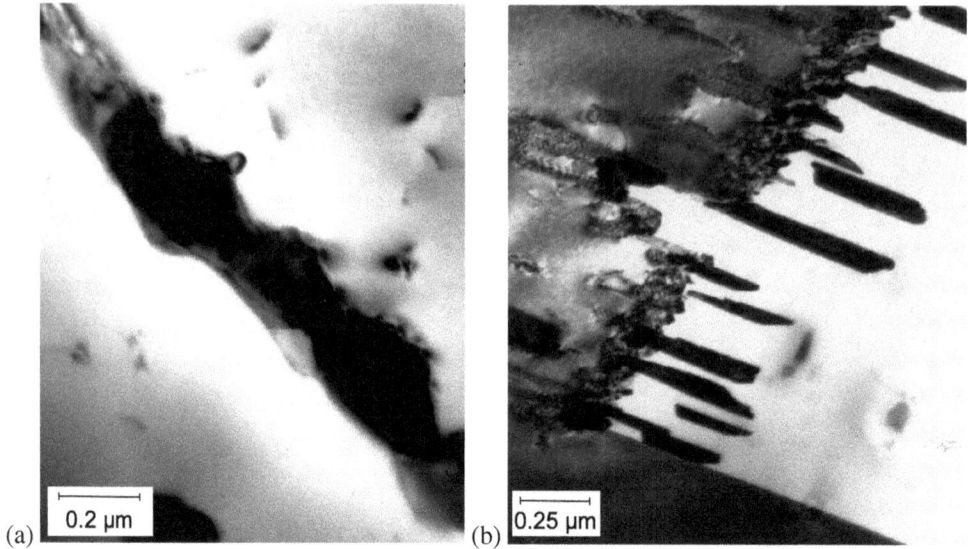

(a) 0.2 μm   (b) 0.25 μm

Figure 11.7 Creep-resistant austenitic stainless steel ($\approx$Fe-20Cr-25Ni-1Mn-1.5Mo-0.26Nb-0.06C wt%), aged 100 h at 1023 K. (a) Precipitation of $M_{23}C_6$ at a grain boundary in an austenitic stainless steel. The boundary is almost completely decorated by $M_{23}C_6$. (b) Platelets of $M_{23}C_6$ at a less-coherent twin boundary. Micrographs courtesy of Thomas Sourmail.

approximately by [38]

$$\{111\}_\gamma \quad \| \{1\,1\,0\}_\alpha \quad \| \{1\,1\,1\}_{M_{23}C_6}$$

$$\langle 1\,\overline{1}\,0 \rangle_\gamma \quad \| \langle 1\,\overline{1}\,1 \rangle_\alpha \quad \| \langle 1\,\overline{1}\,0 \rangle_{M_{23}C_6}.$$

and a similar orientation is observed when the carbide precipitates directly from ferrite [37]. A greater variety of orientations has been observed between $M_{23}C_6$ particles precipitated at boundaries and sub-boundaries in tempered martensite [39], presumably because the local crystallography of the boundary influences that of the carbide.

Table 11.1

Locations of atoms in the unit cell of $Cr_{23}C_6$**, space group** $Fm\overline{3}m$**, lattice parameter 1.065 nm [40]. For chromium atoms,** $x = 0.38199$**,** $y = 0.16991$ **and for carbon atoms,** $x = 0.2751$ **[41].**

Atom	Multiplicity	Symmetry	Fractional coordinates	
Cr	48	m.m2	$(0, y, y)$	...
Cr	32	.3m	(x, x, x)	...
Cr	8	$\overline{4}3m$	$(\frac{1}{4}, \frac{1}{4}, \frac{1}{4})$	...
Cr	4	$m\overline{3}m$	$(0, 0, 0)$	
C	24	4m.m	$(x, 0, 0)$	...

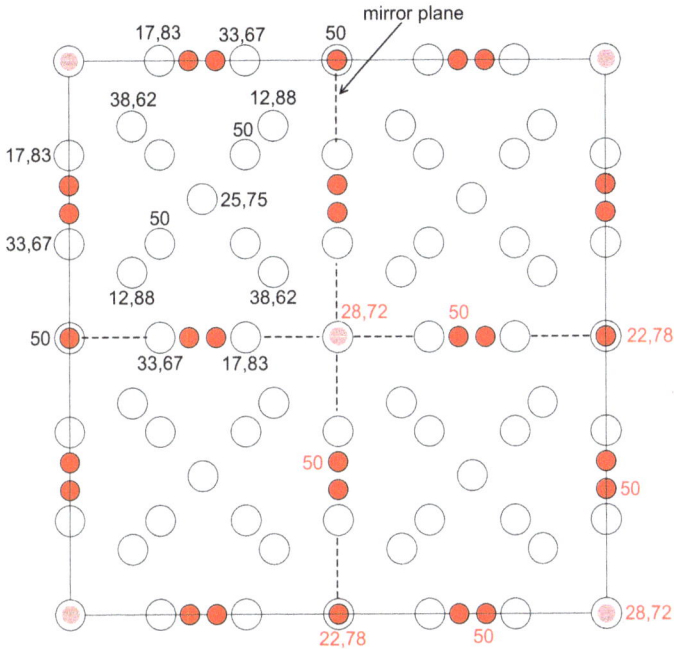

Figure 11.8 Projection on {001} of the unit cell of $Cr_{23}C_6$, divided into four quadrants related by mirror planes. The numbers indicate the percentage height along the z-coordinate normal to the diagram. The z-coordinate is identical for all quadrants (top left for Cr atoms, bottom right for C atoms) defined by the mirror planes. Within the labelled quadrants, the atoms at 0 or 100% height do not have their z-coordinates listed. Some of the carbon atoms are lightly coloured because they are hidden by Cr atoms.

11.4 M_6C

M_6C often nucleates on $M_{23}C_6$; the two structures have similar lattice parameters although their crystal symmetries are $Fd\bar{3}m$ (Figure 11.9, Table 11.2) and $Fm\bar{3}m$ respectively. In low-alloy bainitic or martensitic steels such as the creep-resistant varieties used in the power plant industries, M_6C is not an equilibrium phase [43] but forms as a transition carbide before being replaced by $M_{23}C_6$ [44]. This sequence is reversed in austenitic stainless steels that contain large concentrations of chromium and very little carbon so $M_{23}C_6$ precipitation precedes that of M_6C, with both carbides enclosed by austenite [45–47]:

$$(1\,1\,1)_\gamma \quad \| \, (1\,1\,1)_{M_{23}C_6} \quad \| \, (1\,1\,1)_{M_6C}$$

$$[0\,1\,0]_\gamma \quad \| \, [0\,1\,0]_{M_{23}C_6} \quad \| \, [\bar{1}\,2\,2]_{M_6C}.$$

However, on direct precipitation from austenite or from ferrite the M_6C it adopts a cube-cube orientation relationship [46],[2] although there are many variants reported [48] indicating perhaps that the interfacial energy during nucleation is insensitive to orientation. Consistent with this, Figure 11.10a illustrates the shapes of the precipitates, which do not seem to be strongly facetted. Some of the

Table 11.2

Locations of atoms in the unit cell of Cr_6C, space group $Fd\bar{3}m$, lattice parameter approximately 1.1 nm. For chromium atoms, $x = 0.1978$ and $y = -0.1703$ [42].

Atom	Multiplicity	Symmetry	Fractional coordinates
Cr	48	2.mm	$(x, 0, 0)$ ...
Cr	32	.3m	(y, y, y) ...
Cr	16	$\bar{3}m$	$(\frac{5}{8}, \frac{5}{8}, \frac{5}{8})$ ...
C	16	$\bar{3}m$	$(\frac{1}{8}, \frac{1}{8}, \frac{1}{8})$ ...

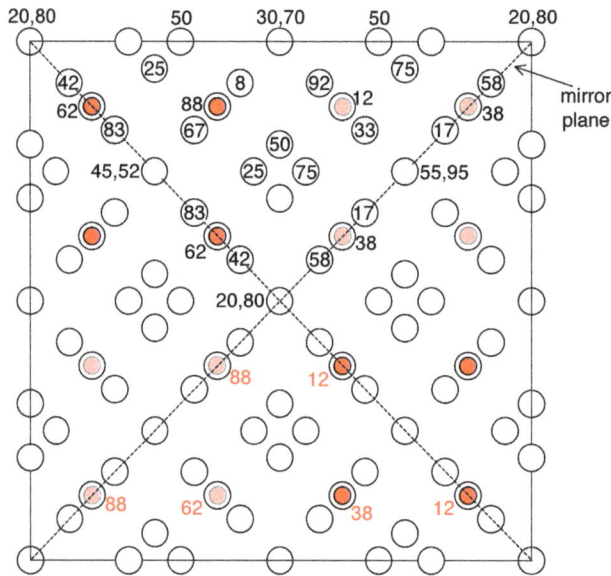

Figure 11.9 Projection on $\{001\}$ of the unit cell of Cr_6C, divided into four quadrants related by mirror planes. The numbers indicate the percentage height along the z-coordinate normal to the diagram. The cell is divided into four segments related by mirror planes. Within the labelled segments, the atoms at 0 or 100% height do not have their z-coordinates listed. Some of the carbon atoms are lightly coloured because they are hidden by Cr atoms.

other orientation relationships include

$$\{\bar{1}\bar{1}1\}_\gamma \quad \| \{\bar{2}11\}_{M_6C}$$

$$\langle \bar{1}10 \rangle_\gamma \quad \| \langle 01\bar{1} \rangle_{M_6C}$$

$$\{\bar{1}\bar{1}1\}_\gamma \quad \| \{111\}_{M_6C} \| \{111\}_{M_6C \text{ twin}}$$

$$\langle \bar{1}10 \rangle_\gamma \quad \| \langle 3\bar{2}\bar{1} \rangle_{M_6C} \| \langle \bar{3}21 \rangle_{M_6C \text{ twin}}.$$

The carbide sometimes contains twins (Figure 11.10b) which have a shape that indicates they are growth twins rather than the mechanical variety. For reasons that are not clear, when the carbide is internally twinned, neither of its component orientations has a cube-cube or twin relationship with the austenite [49].

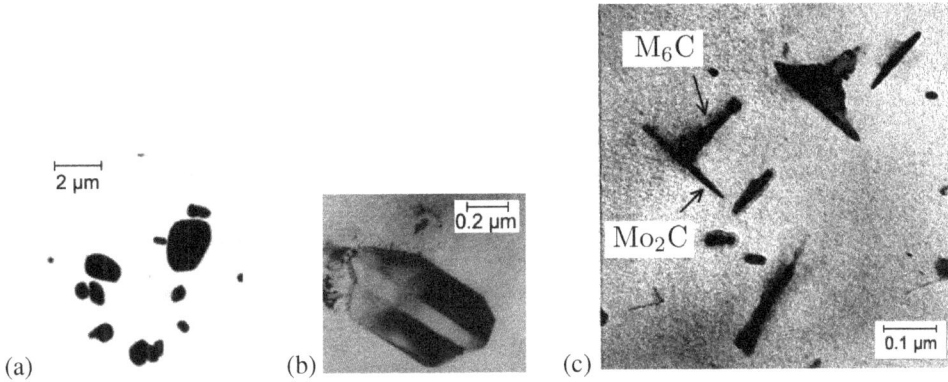

Figure 11.10 (a) M_6C in Fe-16Cr-0.8Nb wt% ferritic alloy following heat treatment at $1000\,°C$ for $100\,h$. Micrograph courtesy of Nobuhiro Fujita. (b) Twin inside a M_6C particle. After Peng and Chou [49], reproduced with the permission of Elsevier. (c) M_6C growing from metastable Mo_2C. After Kurzydłowski and Zieliński [50] with the permission of Taylor and Francis.

In ferritic steels or composites, when Mo_2C is metastable relative to M_6C but forms at an earlier stage, the M_6C nucleates on the needles of molybdenum carbide/ferrite interface [50, 51], Figure 11.10c.

11.5 M_7C_3

M_7C_3 has an orthorhombic crystal structure, space group *Pmcn* with lattice parameters $a = 0.701\,nm$, $b = 1.2142\,nm$ and $c = 0.4526\,nm$ [52], Figure 11.11. Because the ratio b/c is almost equal to $\sqrt{3}$, the structure can also be represented as a hexagonal cell with space group $P6_3mc$, $a = b = 1.402\,nm$ and $c = 0.4526\,nm$. The transformation of coordinates between the orthorhombic (basis O) and hexagonal (basis H) using the three index system) is given by

$$(O\,J\,H) = \begin{pmatrix} 1 & 1 & 0 \\ -1 & 1 & 0 \\ 0 & 0 & 1 \end{pmatrix} \qquad (H\,J\,O) = \begin{pmatrix} \frac{1}{2} & -\frac{1}{2} & 0 \\ \frac{1}{2} & \frac{1}{2} & 0 \\ 0 & 0 & 1 \end{pmatrix} \qquad (11.2)$$

with $[H;\mathbf{u}] = (H\,J\,O)[O;\mathbf{u}]$ and $(\mathbf{h};H^*) = (h;O^*)(O\,J\,H).^3$ When M_7C_3 forms at the interface between cementite and ferrite, the orientation relationship is [53]

$$(001)_{M_7C_3} \equiv (0001)_{M_7C_3} \parallel (0\bar{1}1)_{M_3C} \parallel (011)_\alpha$$

$$(010)_{M_7C_3} \equiv (11\bar{2}0)_{M_7C_3} \parallel (100)_{M_3C} \parallel (01\bar{1})_\alpha.$$

When austenite transforms into pearlite consisting of $\alpha + M_7C_3$, a number of other orientation relationships are observed between the ferrite and the carbide [54], but it is not clear whether these are coincidental, governed by the first phase to nucleate.

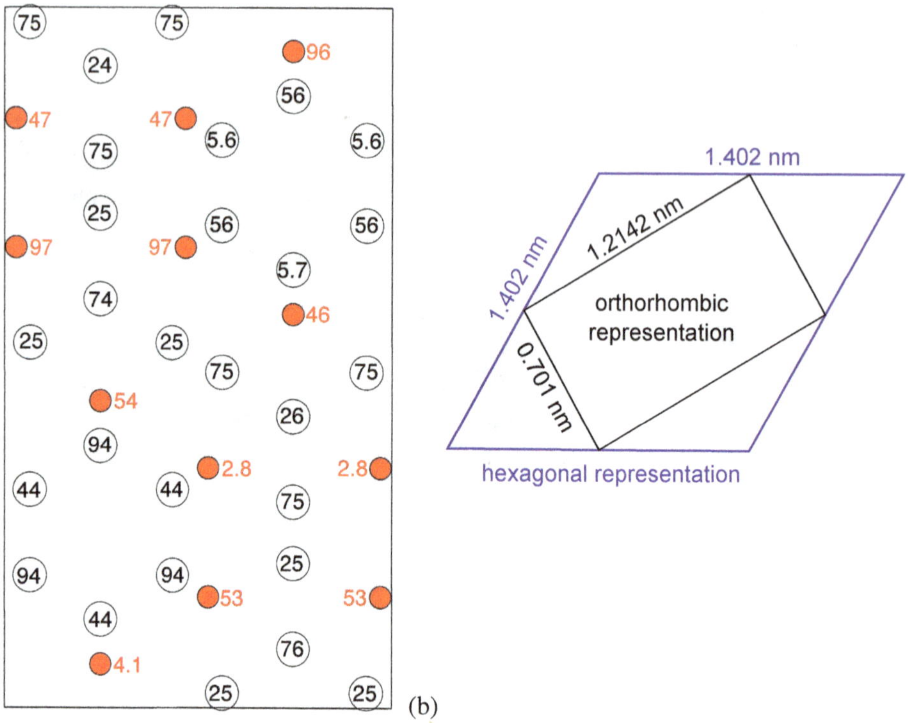

Figure 11.11 (a) Projection of the structure of M_7C_3 on to the (001) plane with a and b along the horizontal and vertical axes respectively. The percentage height along z is indicated for each atom. There are two mirror planes parallel to the bc face, intersecting a at $\frac{1}{4}a$ and at $\frac{3}{4}a$. The red atoms represent carbon and the others are metal atoms. (b) Relationship between hexagonal and orthorhombic representations of crystal structure. (b) The structure can be represented either as a hexagonal cell or one that is orthorhombic as long as the ratio in the latter case of $b/c = \sqrt{3}$.

When $b/a \neq \sqrt{3}$, the M_7C_3 subdivides into domains none of which have a six-fold axis of symmetry with domain boundaries on $\{110\}$ planes [55]. There often are faults (Figure 11.12) in the structure on $\{110\}_{M_7C_3}$ and if these faults occur in an ordered sequence then the carbide is said to develop *polytypes*; the symmetries consistent with the periodic faulting have been identified experimentally as the space groups $P2_1/b$ and $Pbca$ [56].

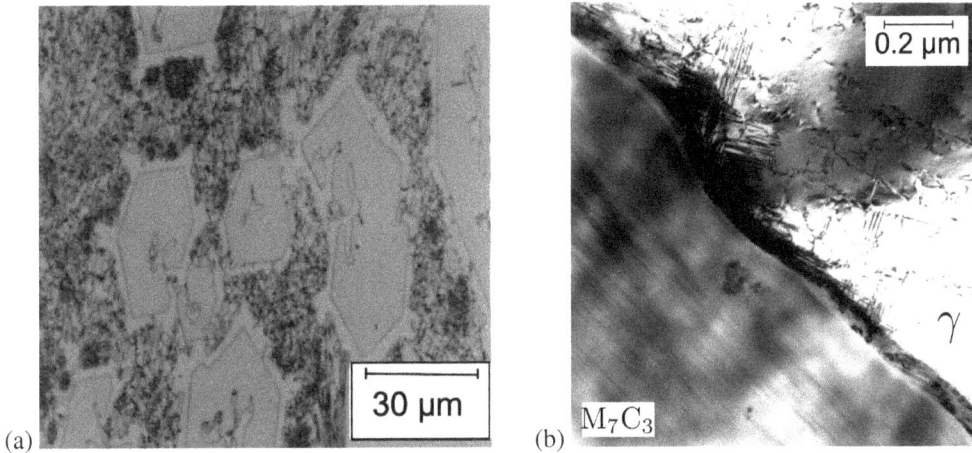

Figure 11.12 (a) Large, facetted primary M_7C_3 carbides in a hardfacing alloy. (b) Transmission electron micrograph showing streaks which are faults within the M_7C_3. More detail in [57].

11.6 Mo_2C

The Mo_2C that precipitates in steels has been described as a hexagonal close-packed arrangement of molybdenum atoms with half the octahedral interstitial sites between the close-packed planes occupied by carbon. The lattice parameters are $a_h = 0.2861$ nm and $c_h = 0.4726$ nm [58, 59]. This structure was deduced originally by Westgren and Phragmen [60] using X-ray diffraction. Assuming that the space group is $P6_3/mmc$, the Mo atoms would be located at $\frac{1}{3}, \frac{2}{3}, \frac{1}{4}$ and $\frac{2}{3}, \frac{1}{3}, \frac{3}{4}$; the carbon atoms would be at $0,0,0$ and $0,0,\frac{1}{2}$. Carbon atoms have a weak X-ray scattering power so their positions were not identified experimentally in the original work. Neutron diffraction does not have this difficulty because the scattering powers of carbon and molybdenum atoms are similar. Using neutron diffraction, it has been demonstrated that the structure is in fact orthorhombic with space group $Pbcn$, in which the molybdenum atoms are located at 8d positions and the carbon at 4c (Table 11.3); the lattice parameters are $a = 0.4724$ nm, $b = 0.6004$ nm and $c = 0.5199$ nm [61]. In both the hexagonal and orthorhombic representations, the relative positions of the metal atoms are identical.

The structures are illustrated in Figure 11.13. The relationship between the hexagonal and orthorhombic basis is given by $a \approx c_h$, $b \approx 2a_h$ and $c \approx \sqrt{3}a_h$. Noting that the basis symbols 'H' and 'O' refer to the hexagonal and orthorhombic forms respectively, and that $[\mathrm{H};\mathbf{u}] = (\mathrm{H\,J\,O})[\mathrm{O};\mathbf{u}]$ and $(\mathbf{h};\mathrm{H}^*) = (\mathrm{h};\mathrm{O}^*)(\mathrm{O\,J\,H})$,

$$(\mathrm{H\,J\,O}) = \begin{pmatrix} 0.000000 & 2.098567 & 1.049159 \\ 0.000000 & 0.000000 & 2.098319 \\ 0.999577 & 0.000000 & 0.000000 \end{pmatrix}$$

$$(\mathrm{O\,J\,H}) = \begin{pmatrix} 0.000000 & 0.000000 & 1.000423 \\ 0.476516 & -0.238258 & 0.000000 \\ 0.000000 & 0.476572 & 0.000000 \end{pmatrix}.$$

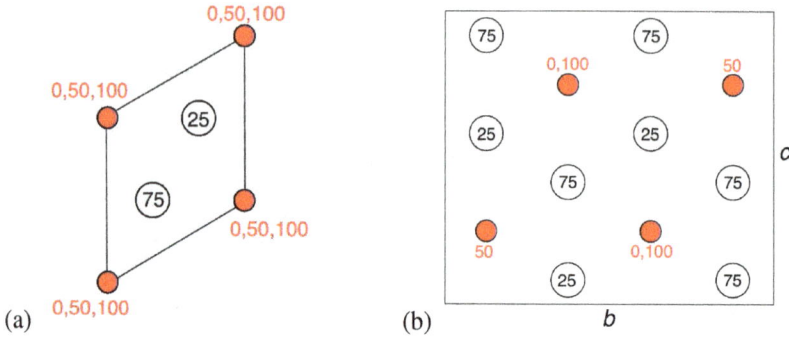

Figure 11.13 Crystal structure of Mo_2C that precipitates in steels with percentage heights indicated. (a) The hexagonal cell with space group $P6_3mmc$ but with the carbon atom (red) positions unverified. Projection on $\{0001\}$. (b) The correct representation, an orthorhombic cell with space group $Pbcn$. Projected on (100).

Table 11.3

Wyckoff positions for atoms in orthorhombic Mo_2C with space group $Pbcn$.

Multiplicity	Wyckoff letter	Site symmetry	Coordinates
8	d	1	$(\frac{1}{4},\frac{1}{8},\frac{1}{12})$ $(-\frac{1}{4}+\frac{1}{2},-\frac{1}{8}+\frac{1}{2},\frac{1}{12}+\frac{1}{2})$
			$(-\frac{1}{4},\frac{1}{8},-\frac{1}{12}+\frac{1}{2})$ $(\frac{1}{4}+\frac{1}{2},-\frac{1}{8}+\frac{1}{2},-\frac{1}{12})$
			$(-\frac{1}{4},-\frac{1}{8},-\frac{1}{12})$ $(\frac{1}{4}+\frac{1}{2},\frac{1}{8}+\frac{1}{2},-\frac{1}{12}+\frac{1}{2})$
			$(\frac{1}{4},-\frac{1}{8},\frac{1}{12}+\frac{1}{2})$ $(-\frac{1}{4}+\frac{1}{2},\frac{1}{8}+\frac{1}{2},\frac{1}{12})$
4	c	.2.	$(0,\frac{3}{8},\frac{1}{4})$ $(\frac{1}{2},-\frac{3}{8}+\frac{1}{2},\frac{3}{4})$
			$(0,-\frac{3}{8},\frac{3}{4})$ $(\frac{1}{2},\frac{3}{8}+\frac{1}{2},\frac{1}{4})$

The orientation relationship between ferrite and Mo_2C is close to [58, 59]

$$\langle 100\rangle_\alpha \parallel [2\overline{1}\overline{1}0]_H \equiv [010]_O \qquad \{011\}_\alpha \parallel (0001)_H \equiv (100)_O.$$

The coordinate transformation matrices for this orientation relationship are therefore

$$(\alpha\,J\,O) = \begin{pmatrix} 0.000000 & 2.094175 & 0.000000 \\ 1.165111 & 0.000000 & 1.282263 \\ 1.165111 & 0 & -1.282263 \end{pmatrix}$$

$$(O\,J\,\alpha) = \begin{pmatrix} -0.000000 & 0.429144 & 0.429144 \\ 0.477515 & -0.000000 & 0.000000 \\ 0.000000 & 0.389936 & -0.389936 \end{pmatrix}.$$

The correspondence matrix $(O\,C\,\alpha)$ [p.164 62] can be written by inspection of $(O\,J\,\alpha)$ as follows:

$$\begin{pmatrix} 0 & \frac{1}{2} & \frac{1}{2} \\ \frac{1}{2} & 0 & 0 \\ 0 & \frac{1}{2} & -\frac{1}{2} \end{pmatrix} \quad \text{and since} \quad (\alpha\,S\,\alpha) = (\alpha\,J\,O)(O\,C\,\alpha)$$

where (α S α) is the deformation that carries the carbide into the ferrite structure for the observed orientation relationship. It follows that

$$(\alpha \, S \, \alpha) \; = \; \begin{pmatrix} 0.000000 & 2.094175 & 0.000000 \\ 1.165111 & 0.000000 & 1.282263 \\ 1.165111 & 0 & -1.282263 \end{pmatrix} \begin{pmatrix} 0 & \frac{1}{2} & \frac{1}{2} \\ \frac{1}{2} & 0 & 0 \\ 0 & \frac{1}{2} & -\frac{1}{2} \end{pmatrix}$$

$$= \begin{pmatrix} 1.04709 & 0.00000 & 0.00000 \\ 0.00000 & 1.22369 & -0.05858 \\ 0.00000 & -0.05858 & 1.22369 \end{pmatrix}.$$

The eigenvectors and the eigenvalues for this deformation are

$$[0\,\overline{1}\,1]_\alpha \qquad \lambda_1 = 1.28227$$

$$[0\,1\,1]_\alpha \qquad \lambda_2 = 1.16511$$

$$[1\,0\,0]_\alpha \qquad \lambda_3 = 1.04709$$

which means that there cannot exist an invariant line between the two lattices. However, the distortion along $[1\,0\,0]_\alpha \parallel [010]_O$ is small in comparison with those along the orthogonal directions. It is not surprising therefore that Mo_2C grows as needles in order to minimise the strain energy by limiting dimensions along the directions where the greatest distortions arise. The needles grow along the $\langle 100 \rangle_\alpha$ directions so there are three crystallographic variants in a crystal of ferrite (Figure 11.14a). This growth direction with respect to ferrite, corresponds specifically to the $[010]_O \parallel [2\overline{1}\overline{1}0]_H$ direction according to the analysis presented above. It is notable that when Mo_2C is synthesised for the purpose of catalysis, the particles do not have a needle shape, rather, they approximate spheroids [63]; Figure 11.14b shows the particles are spheroidal, or in the form of facetted-plates, when unconstrained by precipitation from matrix-α, proving that the needle shape is a consequence of lattice fit with the ferrite. Similarly, Mo_2C that grows during the chemical vapour deposition of graphene on liquid copper contained in a molybdenum crucible, does so as well-defined facetted-platelets [64].

The orientation relationship with austenite has never been measured because as Figure 11.15 shows, at typical steel compositions, the carbide is not stable in a mixture with austenite alone; it has been assumed that $\{111\}_\gamma \parallel (0001)_H$ and $\langle 1\overline{1}0 \rangle_\gamma \parallel \langle 2\overline{1}\overline{1}0 \rangle_H$ [65]. Molybdenum carbide sometimes precipitates as long fibres, ordinarily interpreted as growing during the advance of the austenite/ferrite interface. However, the orientation relation observed is not that between α and Mo_2C, nor can it be explained assuming a KS orientation between α and γ. It was concluded therefore that the carbide forms in the austenite; the growth directions of the fibres was not parallel to $\langle 100 \rangle_\alpha$. Direct observations of precipitation in austenite are not available.

Although not conclusive, atom-probe data indicate that there is no detectable clustering of molybdenum and carbon, or of molybdenum atoms by themselves, in austenite [66].

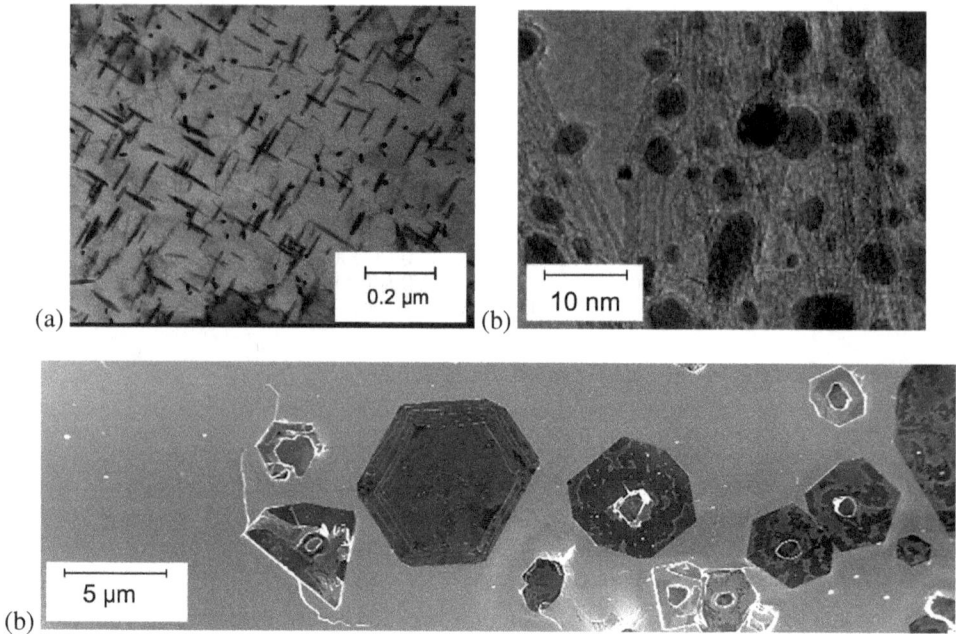

Figure 11.14 (a) Needles of molybdenum carbide. They grow along the $\langle 100 \rangle_\alpha$ with the thin foil parallel to $\{100\}_\alpha$. There are, therefore, two orthogonal needles, and the third variant is normal to the plane of the foil so has circular cross-sections. Micrograph courtesy of Shingo Yamasaki. (b) Synthetic crystals of Mo_2C on carbon substrate. Reprinted (adapted) with permission from Fei et al. [63], copyright (2016) American Chemical Society. (c) Molybdenum carbide platelets that form during the vapour deposition of graphene on liquid copper contained in a molybdenum crucible. Micrograph courtesy of Maryam Adnan Saeed.

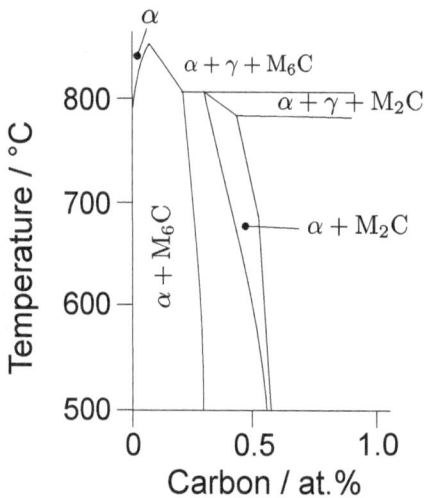

Figure 11.15 Vertical section of a part of the Fe-2Mo-C at%, adapted from Andersson [67].

11.7 κ-CARBIDE

The simplest way to reduce the density of steel is to add solutes that either increase the lattice parameter of pure iron or reduce the average atomic mass of the alloy, or both. The earliest work fo-

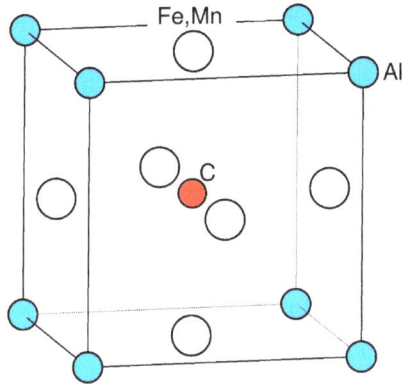

Figure 11.16 The primitive cubic unit cell of κ-carbide. There are in principle three different kinds of octahedral interstices available for carbon. Each octahedron has three dumbbells of paired atoms. The three types are therefore, Mn-Mn and two Fe-Fe pairs (thermodynamically most stable) [71]; Al-Al and two Fe-Fe pairs; Mn-Mn, Al-Al and Fe-Fe pairs.

cused on reduced-density steels by alloying with aluminium, comes from Russia [68, 69], although aluminium has long been considered as a solute that enhances the oxidation resistance of steel [70]. Aluminium reduces the stability of austenite and can eliminate the austenite phase field entirely at large enough concentrations, thus negating the possibility of controlling the microstructure using allotropic transformations. Low-density alloys therefore contain significant concentrations of manganese and carbon to balance the effect of aluminium by increasing the thermodynamic stability of austenite. The Russian low-density steels contained typically Fe-25Mn-10Al-1C wt%; it is notable that Mn and C, being lighter elements than iron, both lead to small but significant reductions in density, so the Fe-Mn-Al-C alloy ends up with a density of about 6.8 g cm^{-3} [68].

The use of relatively large concentrations of manganese and aluminium can lead to the formation of κ-carbide, which usually adversely affects the mechanical properties of the alloys. The carbide has the chemical composition Fe_3AlC, $(Fe,Mn)_3AlC$ or Mn_3AlC, with a primitive cubic crystal structure,[4] space group $Pm\overline{3}m$ (Figure 11.16). It is possible that exact stoichiometry is not maintained with vacancies on the octahedral interstices; indeed, it has been proposed on the basis of experiments that in Fe-Al-C alloys the phase should be regarded as a solid solution rather than a carbide, with a composition $Fe_{4-y}Al_yC_x$ with $x = 0.42$-0.71, $y = 0.8$-1.2 corresponding to $Fe_{3.2}Al_{0.8}C_{0.71}$ and $Fe_{2.8}Al_{1.2}C_{0.42}$ respectively [72]. In other words, aluminium can substitute on to the Fe sites, and when it does so, it makes the octahedral interstice at the centre less favourable for carbon, thus reducing its concentration. The lattice parameter at ambient temperature is a function of chemical composition [71, 72]:

$$a_\kappa / nm = 0.3625 + 0.00014 a_{Al} + 0.00072 a_C + 0.000145 a_{Mn}$$

where a_{Al} etc. represent the atomic percentages of the solute concerned. The effect of manganese is deduced from calculations for 0 K and 0 pressure. The same calculations show that the addition of manganese reduces the formation energy of κ-carbide, and that the octahedral interstice at the centre of the unit cell is the most favoured for carbon [71].

Not surprisingly, given the lattice parameter and structure of κ-carbide, it orientation relationship with the austenite in which it precipitates is cube-cube (unit cell edges parallel). When it precipitates from ferrite, the orientation relationship has been measured accurately to be Kurdjumov-Sachs [73].

11.8 LAVES PHASE

Laves phase in its idealised form has the composition Fe_2X, where X may be Cr, Nb, Si, Mo or W, but usually has multiple solutes depending on the steel concerned, for example, $(Fe,Ni,Cr)_2(Nb,Si)$ [74] or $(Fe,Cr)_2(W,Mo)$ [75]. The phase is more difficult to nucleate than many of the carbide precipitation reactions that precede it so it tends to be coarse. This can lead to embrittlement and a loss in properties that rely on solid-solution strengthening; there are circumstances, such as interphase precipitation, where the Laves phase can be evenly precipitated as a fine dispersion, in which case it improves creep resistance [76]. The crystal structure when Laves phase precipitates in common steels is primitive hexagonal (space group $P6_3/mmc$), Figure 11.17, with the basal plane usually parallel to the close-packed plane of austenite, and the x and y cell-edges parallel to $\langle 11\bar{2}\rangle_\gamma$ directions in the austenite [77]:

$$(0001)_{\text{Laves}} \parallel (111)_\gamma \qquad [2\bar{1}\bar{1}0]_{\text{Laves}} \parallel [\bar{1}2\bar{1}]_\gamma.$$

Laves phase also precipitates in tempered martensite of the type common in creep-resistant power plant steels, where a myriad of other precipitates usually precede the formation of Laves particles. Figure 11.17c illustrates an association between $M_{23}C_6$ which forms first during tempering and Laves phase. Laves precipitates are stimulated to nucleate in the chromium-depleted regions near the chromium-rich carbides or nitrides, regions which at the same time are enriched in molybdenum [78].

When nucleating at grain boundaries in tempered martensite, Laves precipitates show similar tendencies to allotriomorphic ferrite (Figure 4.7), that they adopt a more coherent orientation with respect to the grain in which they grow the least. The orientation relationship is found to be [79]

$$(\bar{1}013)_{\text{Laves}} \parallel (110)_\alpha \qquad [2\bar{1}\bar{1}0]_{\text{Laves}} \parallel [\bar{1}13]_\alpha.$$

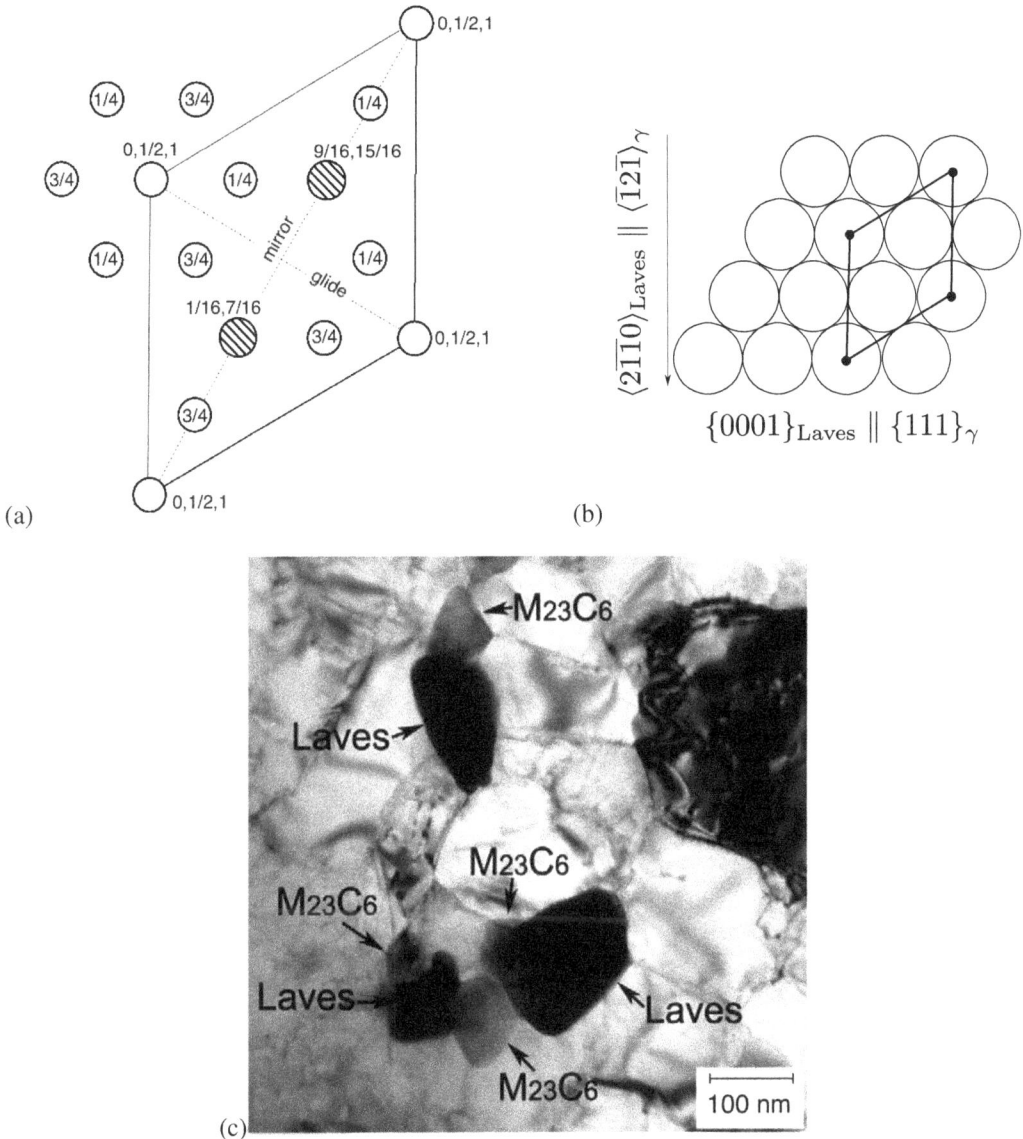

Figure 11.17 (a) The crystal structure of Laves phase, in this case Fe_2Nb, projected on to the basal plane of the hexagonal unit cell with lattice parameters $a = 0.4835$ nm and $c = 0.7881$ nm; the fractional heights of atoms along the c-axis are labelled. The open circles represent iron atoms, and those shaded represent niobium atoms. The space group of the hexagonal unit cell is $P6_3/mmc$. The screw axis is located at the corner of the cell and there are mirror planes normal to that axis at heights $\frac{1}{4}$ and $\frac{3}{4}$. (b) Positions of atoms on the basal plane of Laves phase relative to those on the close-packed plane of austenite. (c) The association of Laves phase precipitates with chromium rich $M_{23}C_6$ particles (reproduced from [75] with permission of Elsevier).

11.9 OTHER INTERMETALLIC COMPOUNDS

Table 11.4 lists some of the intermetallic precipitates that form in alloys of iron. $Fe_3(Al,Zr)$ has a lattice parameter that is approximately four times that of ferrite; it precipitates in a cube-cube orientation with the ferrite to form a dispersion that resists coarsening because of the low solubility of Zr in the matrix, and possibly the low interfacial energy between the precipitate and ferrite [80].

11.9.1 NiAl

NiAl has a primitive cubic structure with a motif of a Ni atom at 0,0,0 and aluminium atom at $\frac{1}{2},\frac{1}{2},\frac{1}{2}$ associated with each lattice point. Plates of the precipitate are reported to adopt a Kurdjumov-Sachs orientation relationship with austenite [81] but this would result in 24 possible crystallographic variants in any given austenite grain; it is surprising therefore that only two variants seem to form in each austenite grain. In fact, the reported orientation, e.g. $(1\bar{2}1)_{\text{NiAl}} \parallel (\bar{1}12)_{\gamma}$ and $[0\bar{1}1]_{\text{NiAl}} \parallel [\bar{1}1\bar{1}]_{\gamma}$ is incorrect since the direction $[0\bar{1}1]_{\text{NiAl}}$ does not lie in $(1\bar{2}1)_{\text{NiAl}}$. Other work has identified a cube-cube orientation relationship between NiAl and ferrite [82, 83].

Both Laves phase and NiAl can precipitate simultaneously, and NiAl need not be plate-shaped. Figure 11.18 shows an atom-probe tomograph of a precipitation-hardened maraging alloy of iron, that has been heat treated to form fine dispersions of both of these phases. Fe, Cr and Mo have been found to dissolve in NiAl that forms in complex ferritic alloys [84].

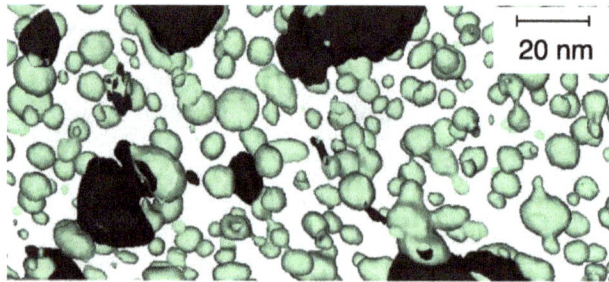

Figure 11.18 Atom-probe tomograph showing extremely fine Laves phase (molybdenum-rich, dark) and NiAl (green) precipitates in a tempered martensitic microstructure – the image is a projection of a three-dimensional sample so the volume fractions are much smaller than appear at first sight. The alloy is Fe-3.58Al-7.91Co-9.72Cr-1.17Mo-8.71Ni-0.6W at.% aged at 540 °C for 7.5 h [85].

11.9.2 σ-PHASE

σ-phase in binary alloys is an approximately equiatomic mixture of iron and a substitutional solute, such as, FeCr, FeMo, FeRe and FeW, but many other solutes can substitute in higher order alloys; the phase can occur in stainless steels designed to be exceptionally corrosion resistant, with substantial additions of chromium and molybdenum. It generally is regarded as being detrimental to the properties of steels, especially when present in a coarse form (Figure 11.19a). The crystal structure has the tetragonal space group $P4_2/mnm$; the unit cell contains thirty atoms and has the lattice parameters $a = 0.88$ nm, $c = 0.4544$ nm (which will vary with chemical composition) [86]. Figure 11.19b shows a projection of the atom positions on to the (001) plane, but it does not distinguish between the different species of atoms. The corresponding Figure 11.19c has the Wyckoff letters along each atom to help identify those that share common point symmetries. In FeCr, the 4f sites are predominantly occupied by Cr atoms whereas the 2a and 8id sites are mostly iron atoms

[87]. Glissile dislocations have never been observed in σ [88], which may explain why it is hard and brittle [89].

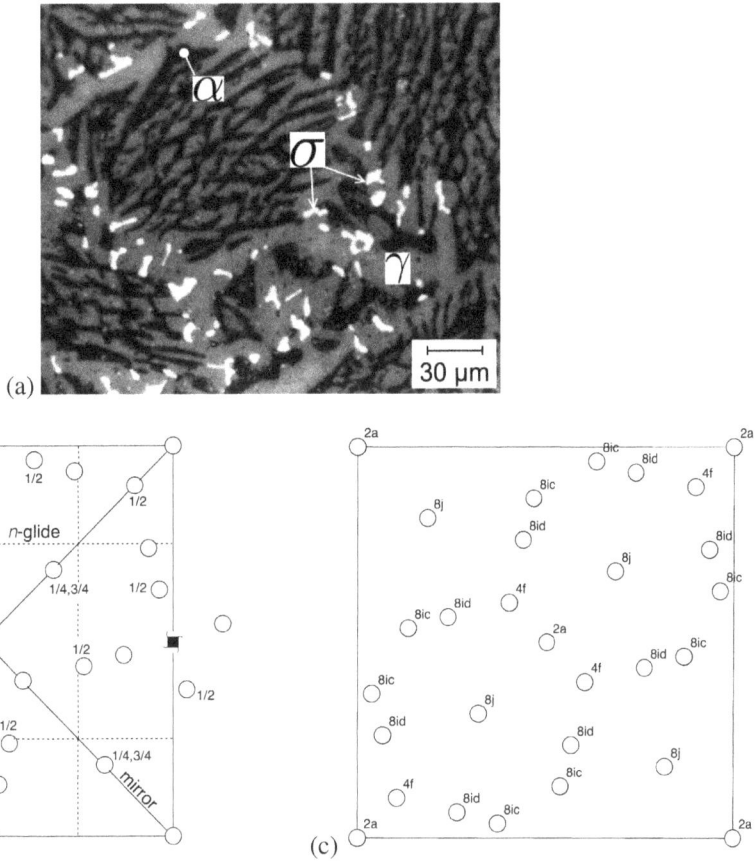

Figure 11.19   (a) σ-Phase in a super-duplex stainless steel. Micrograph courtesy of S. Sharafi [90]. (b) Projection of the tetragonal unit cell of σ-phase on (001). The fractional heights of atoms along [001] are indicated except when they are located at heights 0,1. There is a mirror plane at height $\frac{1}{2}$, normal to the 4_2 screw axis. (c) Atoms with the same Wyckoff letter are located at positions with identical point symmetry.

The $\alpha \rightarrow \sigma$ transformation in chromium-rich alloys is sluggish even when the composition of ferrite is close to that of the precipitate. It is possible that the partial ordering required controls the growth process. It has been known for some time that deformation accelerates the transformation; this is why the solid-state friction stir welding of stainless steel, which introduces ferrite in an ordinarily austenitic steel, in combination with severe localised deformation, greatly accelerates σ phase formation [91]. The orientation relationship $(110)_\sigma \parallel (110)_\alpha$ and $[001]_\sigma \parallel [002]_\alpha$ [82]. The orientation relationship when precipitating from austenite does not appear to be unique:

$$(111)_\gamma \parallel (140)_\sigma \qquad [1\bar{1}0]_\gamma \parallel [001]_\sigma \qquad [92]$$
$$(110)_\gamma \parallel (110)_\sigma \qquad [1\bar{1}2]_\gamma \parallel [1\bar{1}3]_\sigma \qquad [93]$$
$$(111)_\gamma \parallel (001)_\sigma \qquad [1\bar{1}0]_\gamma \parallel [140]_\sigma \qquad [94\text{--}96]$$
$$(111)_\gamma \parallel (001)_\sigma \qquad [1\bar{1}0]_\gamma \parallel [110]_\sigma \qquad [94, 97]$$

Figure 11.20 shows that the last two of these orientation relationships are related by a rotation of approximately $1°$ about $[001]_\sigma$ [93, 95, 98]. The preponderance of orientation relationships suggests that there is little coherence between the phases. It is likely therefore that the nucleation of σ is difficult due to a large interfacial energy, contributing to the observed slow rate of transformation and the fact that it occurs at the late stages of heat treatment after other transition precipitates that presumably are easier to nucleate. In duplex steels, it tends to nucleate heterogeneously at the most incoherent α/γ interfaces [99, 100].

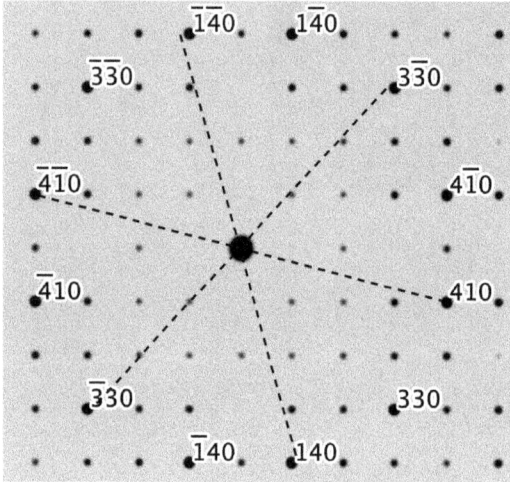

Figure 11.20 Indexed electron diffraction pattern of σ-phase with the beam direction along $[001]_\sigma \parallel \{111\}_\gamma$. The dashed lines represent the directions along which $\langle 110 \rangle_\gamma$ reciprocal lattice vectors lie.

11.9.3 μ-PHASE

μ-phase has the formula Fe_7M_6 where the metal atom 'M' is usually tungsten or molybdenum. The space group is $R\bar{3}m$ and in its hexagonal unit cell notation contains 39 atoms, with $a, b = 0.47402$ nm and $c = 2.6003$ nm (Figure 11.21). The atom positions can be summarised as follows assuming $Fe_{21}Mo_{18}$ [101]:

Atom	Fractional coordinates	Wyckoff symbol	Point symmetry
Fe	$0, 0, 0$	3a	$\bar{3}m$
Fe	$0.332237, 0.166123, 0.256750$	18h	$.m$
Mo	$0, 0, 0.34837$	6c	$3m$
Mo	$0, 0, 0.45128$	6c	$3m$

μ-Phase usually precipitates at grain boundaries in creep-resistant steels subjected either to prolonged heat treatment or extensive service at elevated temperatures. In carbon-free alloys μ-phase reinforces the grain boundaries leading to an increase in creep ductility [102]. Plasma-facing nuclear-fusion reactor devices are subjected to high temperatures and displacements per atom; a candidate composite for this purpose consists of a creep-resistant steel and tungsten, but experiments suggest that Fe_7W_6 (enriched in Cr) can form during manufacture, leading to a reduction in thermal conductivity which would compromise the application [103]. In general, when considering a macroscopic

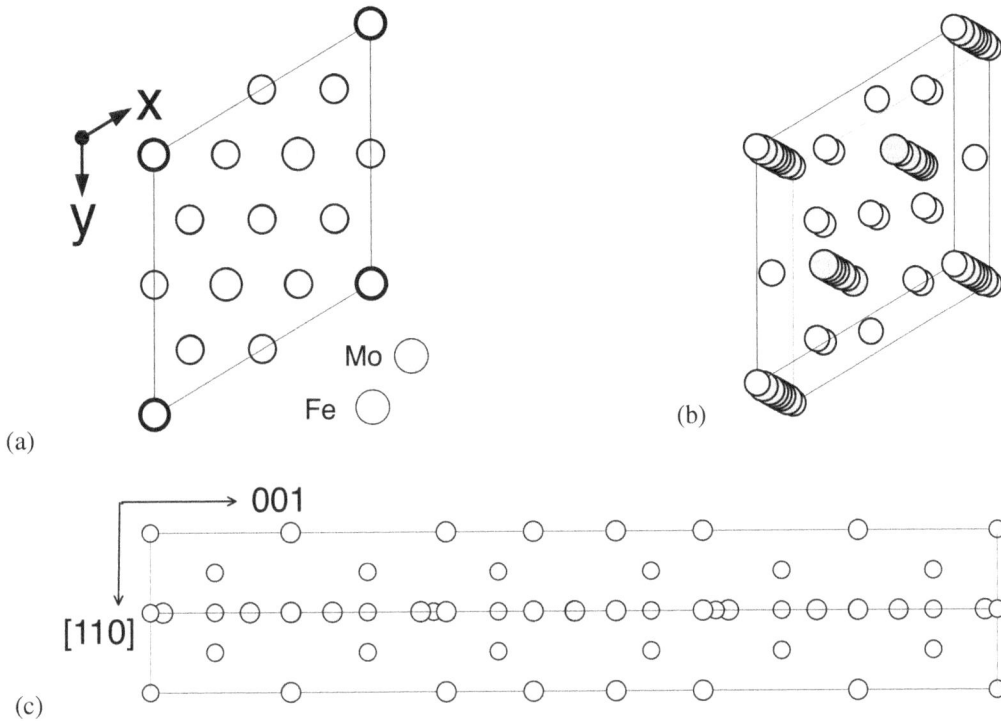

Figure 11.21   Crystal structure of μ-phase. (a) Projection on (001). The inversion triad is through the origin, parallel to the c-axis. (b) Slightly inclined to show the complexity of the underlying structure. (c) Projection on $(1\bar{1}0)$ plane – the mirror plane is (110) which is normal to [110]. There is a centre of symmetry at $\frac{1}{2}, \frac{1}{2}, \frac{1}{2}$.

scale, the compound is regarded as brittle, a property attributed to its complex crystal structure. Nevertheless, indentation experiments indicate significant ductility at ambient temperature in spite of its strength (> 3 GPa) [103]; this would imply some dislocation mobility. Indentation and micropillar experiments indicate that slip in Fe_7Mo_6 occurs predominantly on the basal planes, but prismatic slip can be triggered when the resolve shear stress on the basal planes is inadequate [104].

11.9.4   IRON-ZINC COMPOUNDS

There are a number of intermetallic compounds in the iron-zinc system that are commercially important because they form by the reaction of steel dipped into a Zn-rich liquid at about 460 °C during the galvanising process. The resulting coating is not uniform, but consists of multiple phases that are richer in zinc on progressing further from the interface with steel. The galvanising bath usually is not pure zinc but may contain aluminium, silicon and phosphorus as deliberate additions. The hot-dip galvanised coatings may be annealed to encourage the reactions that generate the intermetallic compounds, because they can offer improved corrosion resistance and other properties associated with painting and spot-welding, when compared against pure zinc coatings. The reactions with iron also help wet the steel and bind the coating to the substrate.

Figure 11.22 shows a part of the iron-zinc phase diagram with a number of intermetallic "com-

Table 11.4

Intermetallic precipitates found in alloys of iron. The lattice parameters are approximate and the compounds may contain other solutes. Exact alloy compositions can be obtained from the original sources.

Compound	Space group	Lattice parameters / nm	Alloy	Ref.
$(Fe,Ni,Cr)_2(Nb,Si)$ (Laves phase)	$P6_3/mmc$	$a = 0.4835$ $c = 0.7881$	Fe-20Cr-30Ni-2Nb-5Al wt%	[74]
$Fe_3(Al,Zr)$	$Fd3m$	0.1169	Fe-2Al-5Cr-0.5Zr at.%	[80]
NiAl	$Pm\bar{3}m$	0.2880	Fe-21Ni-14Cr-4.5Al-0.12Zr-0.02Y-0.025C wt%	[81]
σ-CrFe	$P4_2/mnm$	$a,b = 0.8799$ $c = 0.4556$	Fe-13Cr-15Ni wt%	[82]
μ-phase	$R\bar{3}m$	$a,b = 0.47402$ $c = 2.6003$	Fe-17Cr-14Ni-7W wt%	[102]
$Ni_3(Al,Ti,Si)$	$Pm3m$	0.3565	Fe-0.5Si-13Ni-17Cr-2.5Mo wt%	[105]
Ni_3Nb	$I\frac{4}{m}mm$	$a = 0.362$ $c = 0.741$	Fe-20Cr-30Ni-2Nb at.%	[106]
Fe_3Zn_{10} Γ-phase	$I\bar{4}3m$	0.9018		[107]
$Fe_{11}Zn_{39}$ Γ_1	$F\bar{4}3m$	1.7963		[108]
$FeZn_{10}$ δ_{1p}	$P6_3/mmc$	$a = 1.2787$ $c = 5.7222$ nm		[109]
$FeZn_7$ δ_{1k}	$P6_3/mcm$	$a = \sqrt{3} \times 1.2787$ $c = 5.7222$ nm		[110]
$FeZn_7$ δ_{1k}	$R3c$	$a = \sqrt{3} \times 1.2787$ $c = 3 \times 5.7222$ nm		[110]
$FeZn_{13}$ ζ	$C2/m$	$a = 1.08721$ $b = 0.76050$ $c = 0.50742$ $\beta = 100.82°$		[111]

pounds" identified; each of the phases actually has a range of compositions which is not particularly narrow. Therefore, although the phases are identified in Table 11.4 with particular ratios of Fe to Zn atoms, these may in fact vary. Recent work claims that $FeZn_{10}$ may be better represented as $Fe_{13}Zn_{126}$ [112], but this is based on the study of isolated crystals. When the compounds exist in equilibrium or local equilibrium with other Fe-Zn phases, Figure 11.22 shows that they do not have a fixed chemical composition, which may also vary with temperature.

The δ in Figure 11.22 is in a single phase field whereas Table 11.4 indicates that there is more than one crystalline form of this phase. The phase δ_{1k} is an ordered form of δ_{1p} with the two space groups listed corresponding to different forms of order [110, 113].

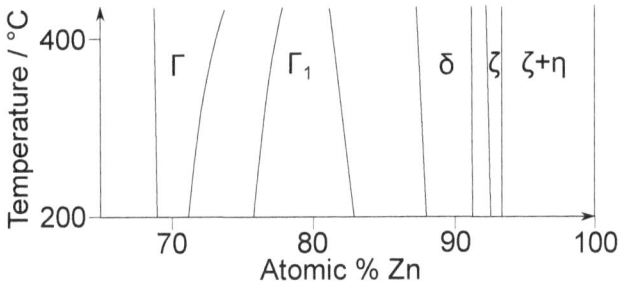

Figure 11.22   The Zn-rich region of the Fe-Zn equilibrium phase diagram, adapted from [114, 115]. The phase field identified here as δ does not distinguish between δ_{1p} and δ_{1k}, but the former (disordered version) is believed to be stable at temperatures above those illustrated here.

The crystal structures of the Fe-Zn compounds are important in the sense that they determine some of the mechanical properties of the coating. The $FeZn_{13}$ ζ-phase has a monoclinic structure with space group $C2/m$. Micropillar deformation experiments suggest that it slips on $\{1\bar{1}0\}_{\zeta}\langle 112\rangle_{\zeta}$ but since there are only two variants of the slip system, the phase shows limited plasticity in its polycrystalline state ($\approx 0.5\%$ plastic strain) whereas 20% plastic strain has been recorded in its single-crystal form [116]. The Γ-phase with its cubic symmetry ($I\bar{4}3m$), is the richest in iron content and in direct contact with the steel. It exhibits the greatest ductility, which must help retain the coating on the steel when the latter is deformed during forming operations. The other compounds, δ_{1p} and Γ_1 are found to be brittle; both of these phases have extremely large lattice parameters (Table 11.4) which may contribute to the poor plasticity since any Burgers vectors of dislocations would scale with the parameters. Laves phase, which exhibits a limited amount of ductility, has the same space group as δ_{1p} but much smaller lattice parameters.

11.10   COMPETITIVE PRECIPITATION AND DISSOLUTION

Most steels destined for long-term mechanical stability at elevated temperatures contain a variety of solutes designed to precipitate substitutionally alloyed phases. Because their formation requires atomic mobility for iron and substitutional solutes, they are introduced by tempering the steel at around 600 °C, making the microstructure resistant to change when the steel is applied at somewhat lower temperatures. Such secondary-hardened steels are of vital importance in power plant and steam turbines. Nevertheless, the service lives of such engineering structure extends over decades so changes are expected, and can be estimated as long as two parameters, the interfacial energy and number density of nucleation sites, are available for each of the precipitate phases that can in principle form. Table 11.5 lists, for creep-resistant steels, the parameters obtained empirically by fitting to experimental data.

This emphasises the limitations of kinetic models since the parameters are not universal; the number density of nucleation sites depends much on the state of the overall microstructure. The use of a single parameter for interfacial energy is obviously not sound given that a precipitate will have an enclosing set of interfaces with different energies, and furthermore, the energy may depend on site, size and chemical composition. The method still is useful because the parameters must be chosen in such a way that the correct sequence of precipitation is predicted. A combination of a large interfacial energy and low number density of nucleation sites will greatly retard precipitation, which may reflect reality if equilibrium phases such as Laves are to form at the late stages of service. Figure 11.23 shows some example calculations, where obtaining the fitting parameters for one tempering temperature enables the estimation of behaviour for a different temperature. The method can therefore be used to calculate TTT diagrams for precipitation.

Table 11.5
Carbide-ferrite interfacial energy and site density data for some precipitate phases, derived by fitting to experimental data for creep-resistant steels [43].

Phase	$\sigma_{\text{carbide}\,\alpha}$ / $\text{J}\,\text{m}^{-2}$	Nucleation site density / m^{-3}
M_2X	0.248	2.7×10^{16}
$M_{23}C_6$	0.269	2.7×10^{15}
Laves	0.331	2.7×10^{8}

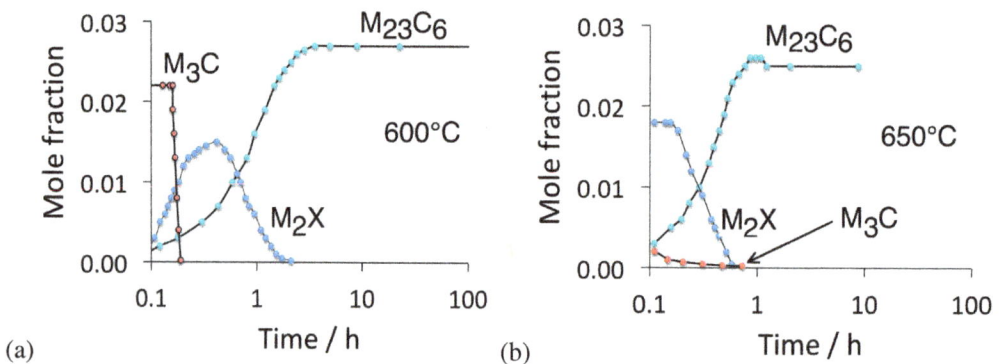

Figure 11.23   Changes in carbide precipitates during the tempering of a 10CrMoW steel at the temperatures indicated. After Brun et al. [117].

REFERENCES

1. K. I. Parashivamurthy, P. Sampathkumaran, and S. Seetharamu: 'In-situ tic precipitation in molten fe-c and their characterisation', *Crystal Research and Technology*, 2008, **43**, 674–678.

2. W. H. Chen Y: 'Growth morphology and mechanism of primary TiC carbide in laser clad TiC/FeAl composite coating', *Materials Letters*, 2003, **57**, 1233–1238.

3. G. L. Dunlop, and P. J. Turner: 'Atom-probe field-ion microscopy of mixed vanadium–titanium carbides in a low-alloy steel', *Metal Science*, 1975, **9**, 370–374.

4. H.-O. Andrén, A. Henjered, and H. Nordén: 'Composition of mc precipitates in titanium stabilised austenitic stainless steel', *Journal of Material Science*, 1980, **15**, 2365–2368.

5. M. Y. Tashmetov, V. T. Em, C. H. Lee, H. S. Shim, Y. N. Choi, and J. S. Lee: 'Neutron diffraction study of the ordered structures of nonstoichiometric titanium carbide', *Physica B*, 2002, **311**, 318–325.

6. N. Frage, L. Levin, E. Manor, R. Shneck, and J. Zabicky: 'Iron-titanium-carbon system. II. Microstructure of titanium carbide (TiCx) of various stoichiometries infiltrated with iron-carbon alloy', *Scripta Materialia*, 1996, **35**, 799–803.

7. C. H. Lee, S. Nambu, J. Inoue, and T. Koseki: 'Ferrite formation behaviors from B1 compounds in steels', *ISIJ International*, 2011, **51**, 2036–2041.

8. C. H. Lee, S. Nambu, J. Inoue, and T. Koseki: 'Ferrite formation behavior from non-metallic compounds in steels', In: S. Babu, H. K. D. H. Bhadeshia, C. E. Cross, S. A. David, T. DebRoy, J. N. DuPont, T. Koseki, and S. Liu, eds. *Trends in Welding Research*. Ohio, U. S. A.: ASM International, 2013:43–47.

9. W. Kestermich, and R. V. Nandedkar: 'Coprecipitation of $M_{23}C_6$ and MC type carbide under the influence of irradiation', *Journal of Nuclear Materials*, 1991, **179**, 1015–1018.

10. A. I. Guzev, and A. A. Rempel: 'Order-disorder phase transition channel in niobium carbide', *Physica Status Solidi A*, 1986, **93**, 71–80.

11. M. G. Kostenko, A. V. Lukoyanov, and A. A. Valeeva: 'Vacancy ordered structures in a nonstoichiometric niobium carbide NbC', *Mendeleev Communications*, 2019, **29**, 707–709.

12. M. G. Kostenko, A. V. Lukoyanov, and A. Gusev: 'Ordering sequence in strongly nonstoichiometric niobium carbide with the formation of Nb_6C_5-type superstructures', *Journal of Experimental and Theoretical Physics*, 2019, **129**, 863–876.

13. B. Lönnberg, and T. Lundström: 'Thermal expansion and phase analytical studies of Nb_2C', *Journal of Less Common Metals*, 1985, **113**, 261–268.

14. V. Kuzucu, M. Aksoy, M. H. Korkut, and M. M. Yildirim: 'The effect of niobium on the microstructure of ferritic stainless steel', *Materials Science & Engineering A*, 1997, **230**, 75–80.

15. U. Sen: 'Kinetics of niobium carbide coating produced on AISI 1040 steel by thermo-reactive deposition technique', *Materials Chemistry and Physics*, 2004, **86**, 189–194.

16. Z. Zhao, P. Hui, F. Liu, Y. Xu, L. Zhong, and M. Zhao: 'Fabrication of niobium carbide coating on niobium by interstitial carburization', *International Journal of Refractory Metals & Hard Materials*, 2020, **88**, 105187.

17. F. R. Chien, R. J. Clifton, and S. R. Nutt: 'Stress-induced phase transformation in single crystal titanium carbide', *Journal of the Americal Ceramics Society*, 1995, **78**, 1537–1545.

18. Y. Funakawa, T. Shiozaki, K. Tomita, Y. Yamamoto, and E. Maeda: 'Development of high strength hot-rolled sheet steel consisting of ferrite and nanometer-sized carbides', *ISIJ International*, 2004, **44**, 1945–1951.

19. H.-W. Yen, P.-Y. Chen, C.-Y. Huang, and J.-R. Yang: 'Interphase precipitation of nanometer-sized carbides in a titanium–molybdenum-bearing low-carbon steel', *Acta Materialia*, 2011, **59**, 6264–6274.

20. J. H. Jang, C. H. Lee, Y. U. Heo, and D. W. Suh: 'Stability of (Ti, M) C (M= Nb, V, Mo and W) carbide in steels using first-principles calculations', *Acta Materialia*, 2012, **60**, 208–217.

21. S. Park, J. Jung, S. Kang, B. W. Jeong, C. K. Lee, and J. Ihm: 'The carbon nonstoichiometry and the lattice parameter of $(Ti_{1-x}W_x)C_{1-y}$', *Journal of the European Ceramic Society*, 2010, **30**, 1519–1526.

22. P. Duwez, and F. Odell: 'Phase relationships in the binary systems of nitrides and carbides of zirconium, columbium, titanium and vanadium', *Journal of the Electrochemical Society*, 1950, **97**, 299–304.

23. H. W. Yen, C. Y. Chen, T. Y. Wang, C. Y. Huang, and J. R. Yang: 'Orientation relationship transition of interphase-precipitated nanometer-sized TiC carbides in a Ti-bearing steel', *Materials Science and Technology*, 2010, **26**, 421–430.

24. Y. Yazawa, T. Furuhara, and T. Maki: 'Effect of matrix recrystallization on morphology, crystallography and coarsening behavior of vanadium carbide in austenite', *Acta Materialia*, 2004, **52**, 3727–3736.

25. W. H. Johnson: 'On some remarkable changes produced in iron and steel by the action of hydrogen and acids', *Proceedings of the Royal Society of London*, 1875, **23**, 168–179.

26. H. K. D. H. Bhadeshia: 'Prevention of hydrogen embrittlement in steels', *ISIJ International*, 2016, **56**, 24–36.

27. S. Yamasaki, and H. K. D. H. Bhadeshia: 'Precipitation during tempering of Fe-C-Mo-V and relationship to hydrogen trapping', *Proceedings of the Royal Society of London A*, 2006, **462**, 2315–2330.

28. R. G. Baker, and J. Nutting: 'Precipitation processes in steels': Special Report 64, Iron and Steel Institute, London, U.K., 1959.

29. J. Billingham, P. S. Bell, and M. H. Lewis: 'A superlattice with monoclinic symmetry based on the compound V_6C_5', *Philosophical Magazine*, 1972, **25**, 661–671.

30. T. Epicier, D. Acevedo, and M. Perez: 'Crystallographic structure of vanadium carbide precipitates in a model Fe-C-V steel', *Philosophical Magazine*, 2008, **88**, 31–45.

31. V. N. Lipatnikov, W. Lengauer, P. Ettmayer, E. Keil, G. Froboth, and E. Kny: 'Effects of vacancy ordering on structure and properties of vanadium carbide', *Journal of Alloys and Compounds*, 1997, **261**, 192–197.

32. Y. Qi, J. Li, and C. Shi: 'Characterization on microstructure and carbides in an austenitic hot-work die steel during ESR solidification process', *ISIJ International*, 2018, **58**, 2079–2087.

33. D. V. Shtansky, and G. Inden: 'Eutectoid reaction of $M_{23}C_6$ carbide decomposition during austenitization', *Acta Materialia*, 1997, **45**, 2879–2895.

34. T. Sourmail: 'Precipitation in creep resistant austenitic stainless steel', *Materials Science and Technology*, 2001, **17**, 1–14.

35. E. C. Bain, R. H. Aborn, and J. J. B. Rutherford: 'The nature and prevention of intergranular corrosion in austenitic stainless steels', *Transactions of the American Society for Steel Treating*, 1933, **21**, 481–509.

36. M. H. Lewis, and B. Hattersley: 'Precipitation of $M_{23}C_6$ in austenitic steels', *Acta Metallurgica*, 1965, **13**, 1159–1168.

37. K. H. Kuo, and C. L. Jia: 'Crystallography of $M_{23}C_6$ and M_6C precipitated in a low alloy steel', *Acta Metallurgica*, 1985, **33**, 991–996.

38. P. R. Howell, J. V. Bee, and R. W. K. Honeycombe: 'The crystallography of the austenite-ferrite/carbide transformation in Fe-Cr-C alloys', *Metallurgical Transactions A*, 1979, **10**, 1213–1222.

39. A. Kipelova, A. Belyakov, and R. Kaibyshev: 'The crystallography of $M_{23}C_6$ carbides in a martensitic 9% Cr steel after tempering, aging and creep', *Philosophical Magazine*, 2013, **93**, 2259–2268.

40. A. L. Bowman, G. P. Arnold, E. K. Storms, and N. G. Nereson: 'The crystal structure of $M_{23}C_6$', *Acta Crystallographica B*, 1972, **28**, 3102–3103.

41. 'Co-Cr-Mo alloy; M23C6 (Cr7MoCo15C6) crystal structure: Datasheet from "PAULING FILE Multinaries edition – 2012" in SpringerMaterials (https://materials.springer.com/isp/crystallographic/docs/sd_1928254)': 2016: Accessed 2019-08-28.

42. Q.-B. Yang, and S. Andersson: 'Application of coincidence site lattices for crystal structure description. Part I: $\sigma = 3$', *Acta Crystallographica B*, 1987, **43**, 1–14.

43. J. D. Robson, and H. K. D. H. Bhadeshia: 'Kinetics of precipitation in power plant steels', *CALPHAD*, 1996, **20**, 447–460.

44. M. C. Tsai, and J. R. Yang: 'Microstructural degeneration of simulated heat-affected zone in modified 2.25Cr1Mo steel during high temperature exposure', *Journal of Materials Science*, 2003, **38**, 2373–2391.

45. T. F. Liu, S. W. Peng, Y. L. Lin, and C. C. Wu: 'Orientation relationships among $M_{23}C_6$, M_6C, and austenite in an Fe-Mn-Al-Mo-C alloy', *Metallurgical Transactions A*, 1990, **21**, 567–574.

46. S.-W. Peng, and C.-P. Chou: 'Orientation relationships between M_6C carbide and its matrices in an Fe-24.6Mn-6.6Al-3.1Mo-1C alloy', *Scripta Metallurgica et Materialia*, 1992, **26**, 243–248.

47. T. H. Lee, S. J. Kim, and Y. C. Jung: 'Crystallographic details of precipitates in Fe-22Cr-21Ni-6Mo-(N) superaustenitic stainless steels aged at 900°C', *Metallurgical & Materials Transactions A*, 2000, **31**, 1713–1723.

48. S. W. Peng, T. Y. Tsai, and C. P. Chou: 'A new orientation relationship between M_6C carbide and the austenite matrix in an Fe-Mn-Al-Mo-C alloy', *Metallurgical Transactions A*, 1993, **24**, 1671–1673.

49. S.-W. Peng, and C.-P. Chou: 'Twinned structure of M_6C in an Fe-Mn-Al-Mo-C alloy', *Scripta Metallurgica*, 1992, **27**, 1173–1178.

50. K. J. K. owski, and W. Zieliński: '$Mo_2C \rightarrow M_6C$ carbide transformation in low alloy Cr-Mo ferritic steels', *Metal Science*, 1984, **18**, 223–224.

51. R. K. Sadangi, B. H. Kear, and L. E. McCandish: 'Synthesis and processing of nanograined Fe - (Fe, Mo)$_6$C composite powders', *MRS Proceedings*, 1994, **351**, 219–225.

52. F. Z. Kayser: 'A re-examination of Westbrooks's X-ray diffraction pattern for C_7C_3', *Materials Research Bulletin*, 1996, **31**, 635–638.

53. A. Inoue, S. Arakawa, and T. Masumoto: 'In situ transformation of cementite to M_7C_3 and internal defects of M_7C_3 in high carbon-chromium steel by tempering', *Transactions of JIM*, 1978, **19**, 11–17.

54. D. V. Shtansky, K. Nakai, and Y. Ohmori: 'Crystallography and interface boundary structure of pearlite with M_7C_3 carbide lamellae', *Acta Materialia*, 1999, **47**, 1105–1115.

55. A. Q. He, H. Q. Ye, G. V. Tendeloo, and K. H. Kuo: 'Direct observation of domain structures and interface dislocations in M_7C_3 by high-resolution electron microscopy', *Philosophical Magazine A*, 1991, **63**, 1327–1333.

56. M. Kowalski: 'Polytypic structures of $(Cr,Fe)_7C_3$ carbides', *Journal of Applied Crystallography*, 1985, **18**, 430–435.

57. L.-E. Svensson, B. Gretoft, B. Ulander, and H. Bhadeshia: 'Fe-Cr-C hardfacing alloys for high-temperature applications', *Journal of Materials Science*, 1985, **21**, 1015–1019.

58. D. J. Dyson, S. R. Keown, D. Raynor, and J. A. Whiteman: 'The orientation relationship and growth direction of Mo_2C in ferrite', *Acta Metallurgica*, 1966, **14**, 867–875.

59. Y.-N. Shi, and P. M. Kelly: 'The crystallography and morphology of Mo_2C in ferrite', *Journal of Materials Science*, 2002, **37**, 2077–2085.

60. A. Westgren, and V. Phragmén: 'Rñtgenanalyse der systems wolfram-kohlenstoff und molybeän-kohlenstoff', *Zietschrift für anorganische und allgemeine Chemie*, 1926, **156**, 27–36.

61. E. Parthé: 'The structure of dimolybdenum carbide by neutron diffraction technique', *Acta Crystallographica*, 1963, **16**, 202–205.

62. H. K. D. H. Bhadeshia: Geometry of Crystals, Polycrystals, and Phase Transformations: Florida, USA: CRC Press, ISBN 9781138070783, freely downloadable from www.phase-trans.msm.cam.ac.uk, 2017.

63. L. Fei, S. M. Ng, W. Lu, M. Xu, L. Shu, W. B. Zhang, T. Sun, C. H. Lam, C. W. Leung, and C. L. Mak: 'Atomic-scale mechanism on nucleation and growth of Mo_2C nanoparticles revealed by in situ transmission electron microscopy', *Nano Letters*, 2016, **14**, 7875–7881.

64. M. Saeed, J. D. Robson, I. A. Kinloch, B. Derby, C. D. Liao, S. Al-Awadhi, and E. Al-Nasrallah: 'The formation mechanism of hexagonal Mo_2C defects in CVD graphene grown on liquid copper', *Physical Chemistry and Chemical Physics*, 2020, **22**, 2176–2180.

65. F. G. Berry, A. T. Davenport, and R. W. K. Honeycombe: 'The isothermal decomposition of alloy austenite', In: *Mechanism of Phase Transformations in Crystalline Solids*. London, U.K.: Institute of Materials, 1969:288–292.

66. I. Stark, and G. D. W. Smith: 'A FIM/atom probe study of phase transformations in Mo steels', *J. de Physique IV*, 1987, **C48**, 447–452.

67. J.-O. Andersson: 'A thermodynamic evaluation of the Fe-Mo-C system', *CALPHAD*, 1988, **12**, 9–23.

68. G. L. Kayak: 'Fe-Mn-Al precipitation hardening austenitic alloys', *Metallovedenie i Termicheskaya Obrabotka Metallov*, 1969, **No. 2**, 13–16.

69. M. F. Alekseenko, G. S. Krivonogov, L. G. Kozyreva, I. M. Kachanova, and L. V. Arapova: 'Phase composition, structure and properties of low-density steel 9G28Yu9MVB', *Metallovedenie i Termicheskaya Obrabotka Metallov*, 1972, **No. 3**, 2–4.

70. D. J. Schmatz: 'Formation of beta manganese-type structure in iron-aluminium-manganese alloys', *Trans. Metall. Soc. AIME*, 1959, **215**, 112–114.

71. S.-W. Seo: 'First principles calculations on thermodynamic properties of κ-carbide and Monte-Carlo cell gas model': Master's thesis, Pohang University of Science and Technology, Pohang, Republic of Korea, 2010.

72. M. Palm, and G. Inden: 'Experimental determination of phase equilibria in the Fe-Al-C system', *Intermetallics*, 1995, **3**, 443–454.

73. G. Y. Hoon, J. S. Lee, and Y. U. Heo: 'Accurate determination of orientation relationship between ferrite and $(Fe, Mn)_3AlC$ in an Fe-Mn-Al-C eutectoid alloy', *Materials Letters*, 2020, **272**, 127849.

74. G. Trotter, and I. Baker: 'Orientation relationships of Laves phase particles in AFA stainless steel', *Philosophical Magazine*, 2015, **95**, 4078–4094.

75. H. Cui, F. Sun, K. Chen, L. Zhang, R. Wan, A. Shan, and J. Wu: 'Precipitation behavior of Laves phase in 10%Cr steel X12CrMoWVNbN10-1-1 during short-term creep exposure', *Materials Science & Engineering A*, 2010, **527**, 7505–7509.

76. S. Kobayashi: 'Formation of the Fe_2Hf Laves phase through eutectoid type reaction of $\delta \rightarrow \gamma + Fe_2Hf$ in ferritic heat resistant steels', *MRS Proceedings*, 2015, **1760**, 2–8.

77. A. Denham, and J. Silcock: 'Precipitation of Fe_2Nb in a 16 wt% cr steel and the effect of Mn and Si additions', *Journal of the Iron and Steel Institute*, 1969, **207**, 582–592.

78. B. A. Senior: 'The precipitation of laves phase in 9Cr1Mo steels', *Materials Science & Engineering A*, 1989, **119**, L5–L9.

79. Q. Li: 'Precipitation of Fe_2W Laves phase and modelling of its direct influence on the strength of a 12Cr-2W steel', *Metallurgical and Materials Transactions A*, 2006, **37**, 89–97.

80. D. G. Morris, M. A. Muñoz-Morris, and L. M. Requejo: 'New iron–aluminium alloy with thermally stable coherent intermetallic nanoprecipitates for enhanced high-temperature creep strength', *Acta Materialia*, 2006, **54**, 2335–2341.

81. D. V. Satyanarayana, G. Malakondaiah, and D. S. Sarma: 'Characterization of the age-hardening behavior of a precipitation-hardenable austenitic steel', *Materials Characterization*, 2001, **47**, 61–65.

82. W. Xi, S. Yin, and H. Lai: 'Microstructure and mechanical properties of stainless steel produced by centrifugal-SHS process', *Journal of Materials Processing Technology*, 2003, **30**, 1–4.

83. T. H. Simm, L. Su, D. R. Galvin, P. Hill, M. Rawson, S. Birosca, E. P. Gilbert, H. K. D. H. Bhadeshia, and K. Perkins: 'Effect of a two-stage heat-treatment on the microstructural and mechanical properties of a maraging steel', *Materials*, 2017, **10**, 1346.

84. Z. Sun, C. H. Liebscher, S. Huang, Z. Teng, G. Song, G. Wang, M. Asta, M. Rawlings, M. E. Fine, and P. K. Liaw: 'New design aspects of creep-resistant NiAl-strengthened ferritic alloys', *Scripta Materialia*, 2013, **68**, 384–388.

85. T. H. Simm, L. Sun, D. R. Galvin, E. P. Gilbert, D. A. Venero, Y. Li, T. L. Martin, P. A. G. Gabot, M. P. Moody, P. Hill, H. K. D. H. Bhadeshia, S. Birosca, M. J. Rawson, and K. Perkins: 'SANS and APT study of precipitate evolution and strengthening in a maraging steel', *Materials Science & Engineering A*, 2017, **702**, 414–424.

86. G. Bergman, and D. P. Shoemaker: 'The determination of the crystal structure of the σ phase in the iron-chromium and iron-molybdenum systems', *Philosophical Magazine*, 1954, **7**, 1025–1059.

87. H. L. Yakel: 'Atom distributions in sigma phases. I. Fe and Cr atom distributions in a binary sigma phase equilibrated at 1063, 1013 and 923 K', *Acta Crystallographica B*, 1983, **39**, 20–28.

88. M. J. Marcinkowski, and D. S. Miller: 'A study of defect sub-structures in the Fe-Cr sigma phase by means of transmission electron microscopy', *Philosophical Magazine*, 1962, **78**, 1025–1059.

89. E. O. Hall, and S. H. Algie: 'The sigma phase', *Metallurgical Reviews*, 2013, **11**, 61–88.

90. S. Sharafi: 'Microstructure of super-duplex stainless steels': Ph.D. thesis, University of Cambridge, Cambridge, U. K., 1993.

91. S. H. C. Park, Y. S. Sato, H. Kokawa, K. Okamoto, S. Hirano, and M. Inagaki: 'Rapid formation of sigma phase in 304 stainless steel during friction stir welding', *Scripta Materialia*, 2003, **49**, 1175–1180.

92. K. Shinohara, T. Seo, and K. Kumada: 'Recrystallization and sigma phase formation as concurrent and interacting phenomena in 25%Cr-20%Ni steel', *Transactions of the Japan Institute of Metals*, 1979, **20**, 713–723.

93. T. H. Lee, S. J. Kim, and Y. C. Jung: 'Crystallographic details of precipitates in Fe-22Cr-21Ni-6Mo-(N) superaustenitic stainless steels aged at 900°C', *Metallurgical & Materials Transactions A*, 2000, **31**, 1713–1723.

94. S. Nenno, M. Tagaya, and Z. Nishiyama: 'Orientation relationships between gamma (f.c.c.) and sigma phases in iron-chromium-nickel alloys', *Transactions of JIM*, 1962, **3**, 82–94.

95. M. H. Lewis: 'Precipitation of (Fe, Cr) sigma phase from austenite', *Acta Metallurgica*, 1966, **14**, 1421–1418.

96. B. F. Weiss, and R. Stickler: 'Phase instabilities during high temperature exposure of 316 austenitic stainless steel', *Metallurgical and Materials Transactions B*, 1972, **3**, 851–866.

97. F. R. Beckitt: 'The formation of sigma-phase from delta-ferrite in a stainless steel', *Journal of the Iron and Steel Institute*, 1969, **207**, 632–638.

98. A. A. Popov, A. S. Bannikova, and S. V. Belikov: 'Precipitation of sigma phase in high-alloy austenitic chromium-nickel-molybdenum alloys', *The Physics of Metals and Metallography*, 2009, **108**, 586–592.

99. Y. S. Sato, and H. Kokawa: 'Preferential precipitation site of sigma phase in duplex stainless steel weld metal', *Scripta Materialia*, 1999, **40**, 659–663.

100. N. Haghdadi, D. Abou-Ras, P. Cizek, P. D. Hodgson, A. D. Rollett, and H. Beladi: 'Austenite-ferrite interface crystallography dependence of sigma phase precipitation using the five-parameter characterization approach', *Materials Letters*, 2017, **196**, 264–268.

101. K. Lejaeghere, S. Cottenier, S. Claessens, M. Waroquier, and V. V. Speybroeck: 'Assessment of a low-cost protocol for an ab initio based prediction of the mixing enthalpy at elevated temperatures: The Fe-Mo system', *Physical Review B*, 2011, **83**, 184201.

102. T. Matsuo, G. Gao, Y. Kondo, and R. Tanaka: 'Effects of tungsten on high temperature creep properties of 17Cr-14Ni steel', *Tetsu-to-Hagané*, 1985, **71**, 869–876.

103. J. Matějíček, B. Nevriá, J. Čech, M. Vilémová, V. Klevarov, and P. Haušild: 'Mechanical and thermal properties of individual phases formed in sintered tungsten-steel composites', *Acta Physica Polonica A*, 2015, **28**, 718–721.

104. S. Schröders, S. Sandlöbes, C. Birke, M. Loeck, L. Peters, C. Tromas, and S. Korte-Kerzel: 'Room temperature deformation in the Fe_7Mo_6 μ-phase', *International Journal of Plasticity*, 2018, **108**, 125–143.

105. H. R. Brager, and F. A. Garner: 'Swelling as a consequence of gamma prim (γ') and $m_{23}(c,si)_6$ formation in neutron irradiated 316 stainless steel', *Journal of Nuclear Materials*, 1978, **73**, 9–19.

106. Y. Yamamoto, M. Takeyama, Z. P. Lu, C. T. Liu, N. D. Evans, P. J. Mziasz, and M. P. Brady: 'Alloying effects on creep and oxidation resistance of austenitic stainless steel alloys employing intermetallic precipitates', *Intermetallics*, 2008, **16**, 453–462.

107. J. K. Brandon, R. Y. Brizard, P. C. Chieh, R. K. McMillan, and W. B. Pearson: 'New refinements of the gamma brass type structures Cu_5Zn_8, Cu_5Cd_8 and Fe_3Zn_{10}', *Acta Crystallographica B*, 1974, **30**, 1412–1417.

108. A. S. Koster, and J. C. Schoone: 'Structure of cubic iron-zinc phase $Fe_{22}Zn_{78}$', *Acta Crystallographica B*, 1981, **37**, 1905–1907.

109. C. H. E. Belin, and R. C. H. Belin: 'Synthesis and crystal structure determinations in the γ and δ phase domains of the iron–zinc system: Electronic and bonding analysis of $Fe_{13}Zn_{39}$ and $FeZn_{10}$, a subtle deviation from the hume–rothery standard?', *Journal of Solid State Chemistry*, 2000, **151**, 85–95.

110. N. L. Okamoto, A. Yasuhara, and H. Inui: 'Order–disorder structure of the $δ_{1k}$ phase in the Fe-Zn system determined by scanning transmission electron microscopy', *Acta Materialia*, 2014, **81**, 345–357.

111. Y. Liu, X. P. Su, F. C. Yin, Z. Li, and Y. H. Liu: 'Experimental determination and atomistic simulation on the structure of $FeZn_{13}$', *Journal of Phase Equilibria and Diffusion*, 2008, **29**, 488–492.

112. N. L. Okamoto, K. Tanaka, A. Yasuhara, and H. Inui: 'Structure refinement of the δ_{1p} phase in the Fe-Zn system by single-crystal X-ray diffraction combined with scanning transmission electron microscopy', *Acta Crystallographica B*, 2013, **70**, 275–282.

113. M. H. Hong, and H. Saka: 'Transmission electron microscopy of the iron-zinc δ_1 intermetallic phase', *Scripta Materialia*, 1997, **36**, 1423–1429.

114. X. Su, N.-Y. Tang, and J. M. Toguri: 'Thermodynamic evaluation of the Fe-Zn system', *Journal of Alloys and Compounds*, 2001, **325**, 129–136.

115. K. han, I. Ohnuma, K. Okuda, and R. Kainuma: 'Experimental determination of phase diagram in the Zn-Fe binary system', *Journal of Alloys and Compounds*, 2018, **737**, 490–504.

116. N. L. Okamoto, D. Kashioka, M. Inomoto, H. Inui, H. Takebayashi, and S. Yamaguchi: 'Compression deformability of γ and ζ Fe–Zn intermetallics to mitigate detachment of brittle intermetallic coating of galvannealed steels', *Scripta Materialia*, 2013, **69**, 307–310.

117. F. Brun, T. Yoshida, J. D. Robson, V. Narayan, H. K. D. H. Bhadeshia, and D. J. C. MacKay: 'Theoretical design of ferritic creep resistant steels using neural network, kinetic and thermodynamic models', *Materials Science and Technology*, 1999, **15**, 547–554.

118. N. A. Dubrovinskaia, L. S. Dubrovinsky, S. K. Saxena, R. Ahuja, and B. Johansson: 'High-pressure study of titanium carbide', *Journal of Alloys and Compounds*, 1999, **289**, 24–27.

119. B. Winkler, E. A. Juarez-Arellano, A. Friedrich, L. Bayarjargal, J. Yan, and S. M. Clark: 'Reaction of titanium with carbon in a laser heated diamond anvil cell and reevaluation of a proposed pressure-induced structural phase transition of TiC', *Journal of Alloys and Compounds*, 2009, **478**, 392–397.

120. C. J. Tillman, and D. V. Edmonds: 'Alloy carbide precipitation and aging during high temperature isothermal decomposition of an Fe-4Mo-0.2C alloy steel', *Metals Technology*, 1974, **1**, 456–461.

Notes

[1] Indications that cubic-TiC tends to undergo a transformation into a rhombohedral lattice when subjected to hydrostatic pressures beyond 18 GPa at 300 K [118] do not seem to be correct [119].

[2] The Kurdjumov-Sachs and Nishiyama-Wasserman orientation relationships have also been found between ferrite and M_6C during interphase precipitation [120], but this could be a consequence of the precipitate adopting a cube-cube orientation with the austenite.

[3] In the four index *hkil* notation for planes $i = -(h+k)$. The conversion for directions is a little more involved:

$$
\begin{array}{lll}
U - J = u & u = 2U + V & U = \frac{1}{3}(2u - v) \\
V - J = v \quad \rightarrow & v = U + 2V \quad \text{and} & V = \frac{1}{3}(2v - u) \\
W = w & w = W & J = -\frac{1}{3}(u + v)
\end{array}
\tag{11.3}
$$

[4] Not face-centred cubic as is often stated in the literature.

12 Pearlite

12.1 SHAPE

The classic metallographic observations of pearlite are based on polished and etched planar sections which reveal a lamellar mixture of ferrite and cementite, Figure 12.1. This mixture acts like a diffraction grating so interference colours are observed when viewed in white light. The colours vary as a function of the apparent spacing of the lamellae at the surface, giving rise to iridescence.[1] The name *pearlite* is an adaptation of the fact that natural pearls that have layered structures also exhibit iridescence.

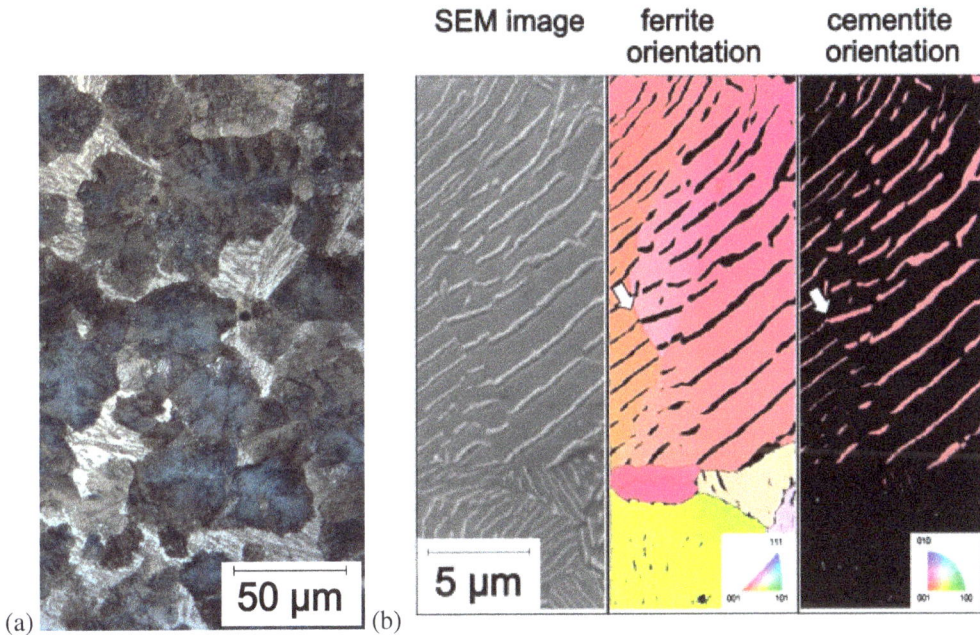

Figure 12.1 (a) Nodules of pearlite separated by lighter regions containing a mixture of bainite and martensite (micrograph courtesy of Hala Salman Hasan). The lamellar structure is not clearly resolved. The iridescent colour of the pearlite nodules is evident. (b) The structure of a colony of pearlite, showing that the crystallographic orientations of the ferrite and cementite are essentially identical at all locations. The arrows show a low-misorientation boundary between adjacent colonies, with a change in the direction in which the cementite grows. Reproduced with the permission of Elsevier from Koga et al. [1].

The two phases within pearlite are each connected in three dimensions as was demonstrated by Hillert [2] using serial sectioning, and by Dippenaar and Honeycombe [3] who showed that an individual pearlite colony is best visualised as an interpenetrating bicrystal of cementite and ferrite, Figure 12.1b. The term 'colony' is sometimes used specifically to refer to a region of a pearlite 'nodule' within which the ferrite and cementite grow in the same direction. A nodule, which is roughly spherical in shape, may therefore contain several colonies with different growth directions.

The ferrite orientations may differ slightly between the adjacent colonies within a nodule, and within a colony itself [4]. Micrographs, such as Figure 12.1, of pearlite show a great deal of detail that is not included in theoretical treatments of pearlite, for example:

- Nodules of pearlite have irregular shapes that deviate significantly from idealised spheres. The shape necessitates the branching of cementite as the nodule size increases. Growth occurs in all directions so in the absence of impingement, it would be necessary for the cementite lamellae to branch in order to maintain a spacing that is on average, uniform.
- Individual cementite lamellae are not uniform in thickness and exhibit significant perturbations in shape.
- In two-dimensional sections there are discontinuities in the cementite lamellae, even though there is no change in spacing ahead of the break. It is likely that some of these represent holes in the sheets of cementite. Such holes are believed to be imperfections in the lamellar growth process [5].

Notwithstanding the complexity of the structure of pearlite, the defining characteristic of pearlite is that the product phases grow cooperatively, i.e., the carbon partitioned as the ferrite grows is absorbed by the adjacent cementite so the diffusion distance within the austenite is limited to some fraction of the interlamellar spacing. This implies that in a binary Fe-C system, there is no net change in the chemical composition of the austenite so the pearlite can continue forming until all of the austenite is consumed. This is not the case with substitutionally alloyed steels when transformation occurs in a three-phase field where austenite, ferrite and cementite can coexist in equilibrium.

The shape of pearlite is not always nodular. Figure 12.2a–c shows the appearance of pearlite following partial transformation so that the shapes can be examined without the consequences of impingement, in a eutectoid steel [6]. Transformation at 684 °C leads to the classical nodules, but the pearlite adopts a spiky form at lower temperatures, that can be mistaken for bainite. This unusual shape represents the breakdown of cooperative growth, the ferrite advances more rapidly into the austenite than the cementite which becomes discontinuous so the structure no longer is lamellar, Figure 12.2d. It is important to understand that the spiked pearlite remains a reconstructive transformation. Figure 12.2e,f shows that there are no displacements recorded in the nodular or spiked pearlitic regions, when a polished sample of austenite is transformed at 550 °C; in contrast, bainite plates are seen clearly to lead to a surface relief that is an invariant-plane strain with a large shear component.

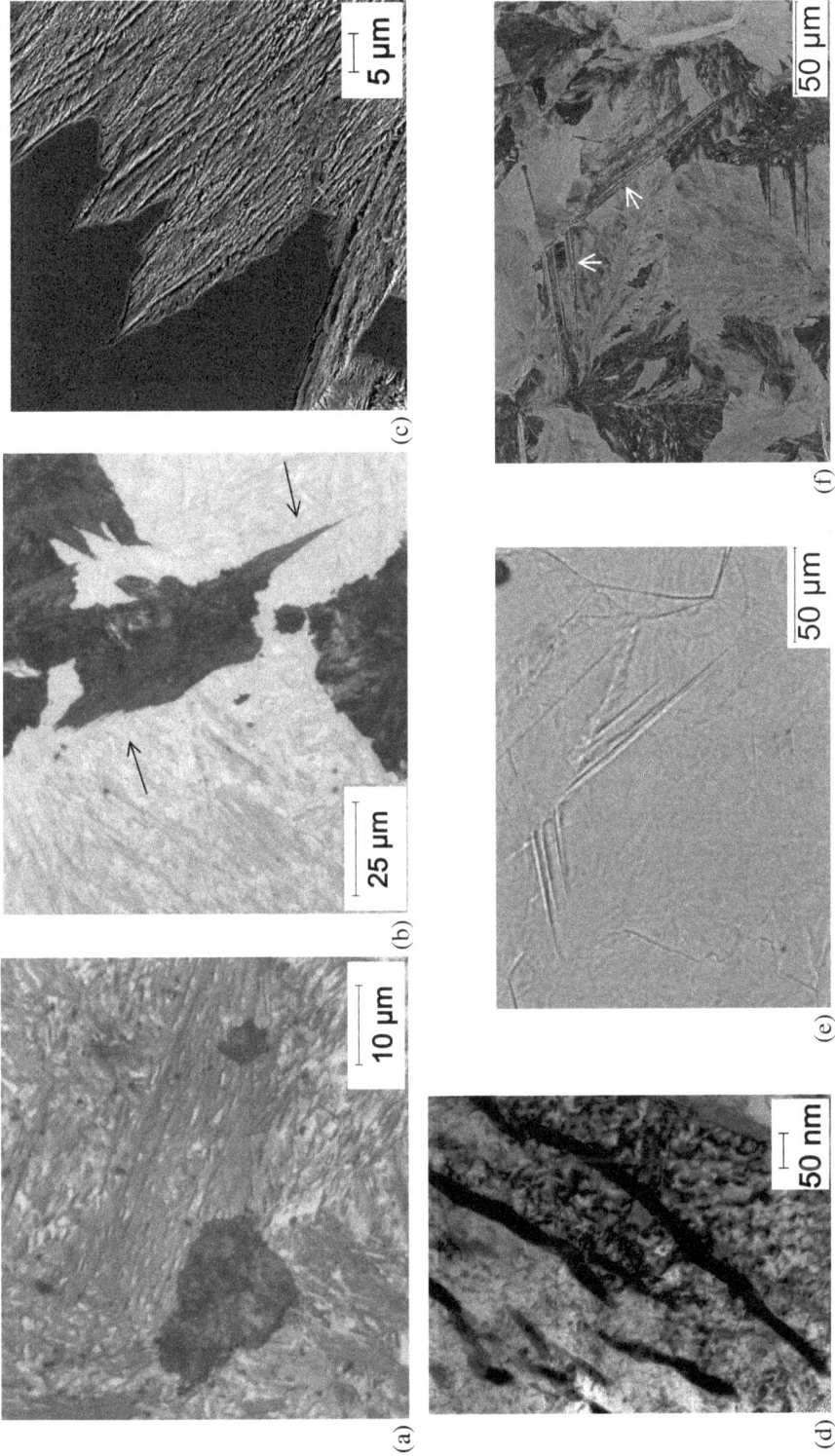

Figure 12.2 Partial isothermal-transformation to pearlite, followed by quenching so the residual austenite transforms into martensite [6]. The steel has the composition Fe-0.79C-0.98Mn wt%. (a) Transformed at 684°C. (b–d) Transformed at 550°C. (e,f) Corresponding images from the surface relief experiments. Micrograph (e) was obtained by polishing, austenitising and then transforming, and the sample is unetched. The dark-etching spiky form of pearlite where the ferrite and cementite do not grow at a common transformation front does not exhibit any surface upheavals. Nor does any of the pearlite. It is only the few plates of lighter-etching bainite (arrowed) that show the surface relief. (a) Unetched sample, (b) after light etching.

Spiky pearlite tends to form at low transformation temperatures [7, 8]. The Gibbs free energy changes due to the decomposition of γ into (a) a mixture of $\alpha + \theta$; (b) an equilibrium mixture of α and carbon-enriched γ, and (c) an equilibrium mixture of θ and carbon-depleted γ are illustrated in Figure 12.3. The reduction in free energy is greatest for reaction (a), which clearly is advantageous when ferrite and cementite to grow together. However, at greater undercoolings, the advantage in cooperative growth is diminished because the reaction (b) leads to a smaller though comparable free energy reduction. So the motivation for cooperative growth diminishes, explaining why spiky pearlite is generated at large undercoolings.

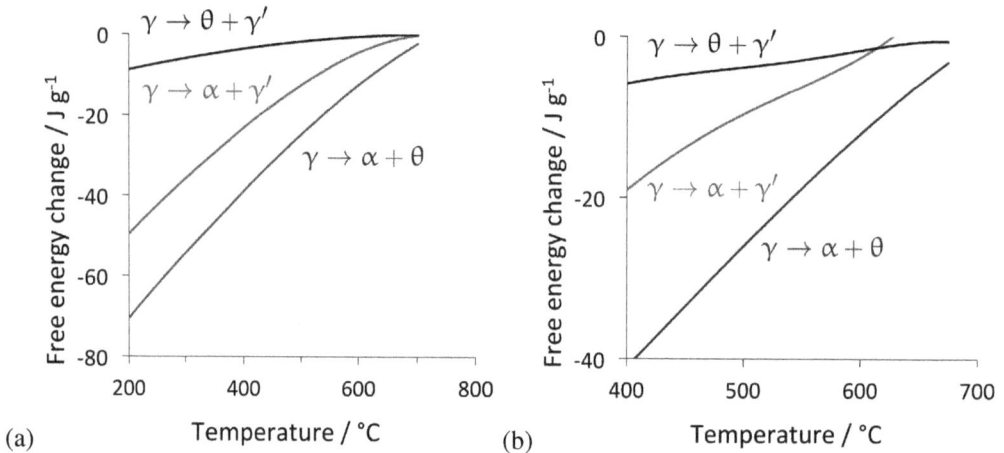

Figure 12.3 Gibbs free energy change associated with the decomposition of austenite, as a function of temperature, in a binary Fe-C alloy. γ' refers to austenite with composition altered due to the formation of α or θ. (a) Eutectoid composition, 0.76 wt% carbon. (b) Hypereutectoid steel containing 1 wt%.

12.2 NUCLEATION

Pearlite is of course, a mixture of two phases, both of which are required to establish the cooperative growth that leads to the lamellar structure. There are several scenarios to consider with respect to the initiation of pearlite:

- the predominant sites for the heterogeneous nucleation of pearlite in most steels are the austenite grain boundaries. It would be reasonable to assume that the process begins with the nucleation of ferrite in hypoeutectoid alloys and cementite in hypereutectoid steels. Figure 12.3a shows that up to the eutectoid composition, the driving force for ferrite precipitation from austenite is far greater than that for cementite. Since most steels are hypoeutectoid, it is assumed during the mathematical modelling of microstructural evolution that the nucleation rate for pearlite can be taken to be identical to that for allotriomorphic ferrite [9].
- In hypereutectoid steels, there is no driving force for the precipitation of ferrite alone from austenite at small undercoolings, but with increasing undercooling, $|\Delta G^{\gamma \rightarrow \gamma' + \alpha}| > |\Delta G^{\gamma \rightarrow \gamma' + \theta}|$, Figure 12.3b.

- In steels where transformation is not from the fully austenitic state, the presence of prior phases must influence the evolution of the pearlite. Figure 12.4a shows cementite, as a component of what eventually becomes pearlite, evolving from an allotriomorph of cementite. The bulbous cementite particle has the same crystallographic orientation as the allotriomorph of cementite, which is related crystallographically to the austenite on the other side (γ_1) [3]. It is presumed that the advance of cementite by a reconstructive mechanism is easier into γ_2, with which it has a less-coherent and hence less-mobile interface. Figure 12.4b,c illustrates two further scenarios developing in the same sample during isothermal transformation, with pearlite nucleating from the α/γ interface, from an isolated cementite particle that is in contact with the austenite.

- Mixtures of bainitic ferrite and carbon-enriched austenite generally contain two morphologies of the latter phase, film and blocks of retained austenite. When tempered at an elevated temperature to induce the austenite to decompose, it is only the larger blocky regions that transform into pearlite, whereas discrete precipitation of cementite (accompanied by epitaxial growth of ferrite) is observed when the thin films of austenite decompose. This is because there is no opportunity to establish cooperative growth if the austenite is limited in size [p.95, 10].

Although there are now numerous determinations of the $\alpha/\theta_{\text{pearlite}}$ orientation relationship [3, 12, 13], there has been no serious attempt to discover any consequences on the practical applications of pearlite. This is in contrast to models for estimating the transformation texture due to the $\gamma \to \alpha$ transformation [14]. Some of the studies using conventional electron diffraction or electron backscattered diffraction are not of sufficient accuracy to establish crystallography. The popular α/θ Bagaryatskii orientation relationship (page 403) seems to be an imprecise representation of the closely-related Isaichev orientation [15–17]. When pearlite begins with the nucleation of cementite, with the ferritic component forming on this cementite, the orientation relationship is that due to Isaichev [18]. In hypoeutectoid steels where ferrite nucleates first, other orientation relationships occur (Section 8.13).

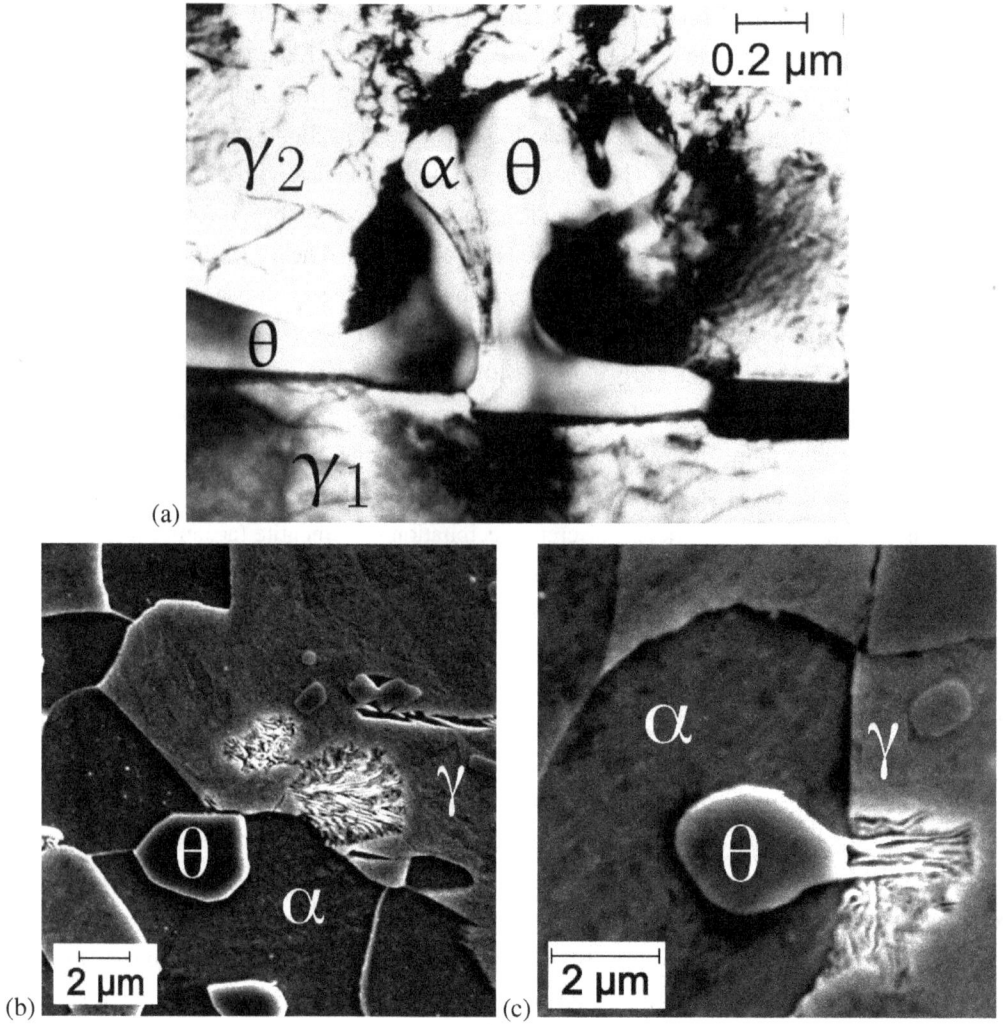

Figure 12.4 Evolution of a pearlite colony. (a) From proeutectoid cementite layer present at the austenite grain boundary, prior to the onset of pearlite (micrograph courtesy of R. Dippenaar) [11]. (b) From pre-existing ferrite/austenite interface. (c) From pre-existing cementite/austenite interface. Micrographs courtesy of Hala Salman Hasan.

12.3 GROWTH

The growth models to be described here are necessarily simplistic. The morphology of pearlite is complex [2, 3, 5, 19]. Growth occurs in three-dimensions whereas the models treat the problem essentially as the one-dimensional advance of the transformation front. It therefore is necessary for the cementite to branch when the interlamellar spacing deviates sufficiently from the 'ideal', meaning that there will be perturbations in the growth rate. The cementite and ferrite within pearlite are known to be feature-rich, with curvature, striations, discontinuities and holes [5, 19]. Some of the complexity is illustrated in Figure 12.5; it is noteworthy that the ferrite bulges out at the transformation front, presumably because $|\Delta G^{\gamma \to \gamma' + \alpha}|$ is greater than the corresponding driving force for cementite precipitation (Figure 12.3).

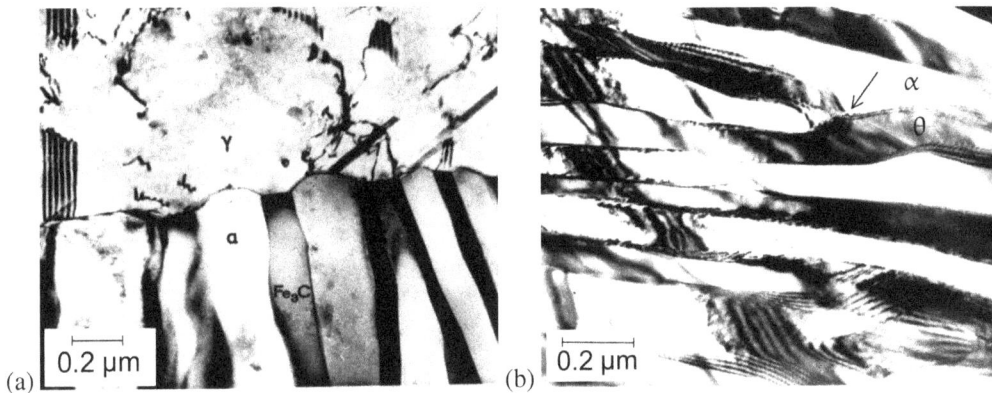

(a)   (b)

Figure 12.5 Transmission electron micrographs of pearlite in a Fe-0.79C-11.9Mn wt% steel transformed partially at 630 °C. (a) The shape of the transformation front with austenite, and an illustration of the fact that the local spacing at the interface with austenite is not uniform. (b) The arrow indicates a cementite branching event. The α/θ interface-plane can be identified by both depth fringes and the interfacial dislocation structure. The interface clearly is not flat along the length of a single θ-lamella. Micrographs courtesy of Professor R. Dippenaar [11].

The pearlite transformation in steels is unique in having both a source and a sink for solute that is partitioned into the austenite. Cementite (θ) is rich in carbon whereas ferrite (α) accommodates very little when it is in equilibrium with either cementite or austenite (γ). It therefore is necessary for carbon to be redistributed at the common transformation front that the α, θ have with the parent austenite. In a binary Fe-C system, the cementite would absorb all the carbon that is partitioned into the austenite by the growth of ferrite. This can happen by diffusion through the adjacent austenite, in a direction parallel to the transformation front, and also by a flux of carbon through the transformation interface. There is therefore no net accumulation of carbon in the austenite, leading to a constant growth rate [20, 21].

The process is illustrated in a simplified manner in Figure 12.6, where all the fluxes are essentially parallel to the transformation front, J_V being via the austenite ahead of the front, and J_B through the averaged interface between the austenite and pearlite. The proportion of ferrite and cementite within the colony is determined by the average carbon concentration of the austenite $\bar{c}^\gamma$ together with the equilibrium concentrations $c^{\alpha\theta}$ and $c^{\theta\alpha}$. Using the lever rule, it follows that

$$\frac{S_I^\alpha}{S_I} = \frac{c^{\theta\alpha} - \bar{c}^\gamma}{c^{\theta\alpha} - c^{\alpha\theta}}, \qquad \frac{S_I^\theta}{S_I} = \frac{\bar{c}^\gamma - c^{\alpha\theta}}{c^{\theta\alpha} - c^{\alpha\theta}}, \qquad V_V^\theta = \frac{S_I^\theta}{S_I} \qquad (12.1)$$

where V_V^θ is the fraction of cementite within the pearlite. Given that ferrite and cementite both

growth direction ⟶

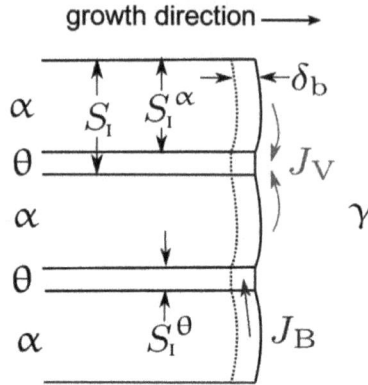

Figure 12.6 Schematic representation of a growing pearlite colony. The dashed arrows indicate the diffusion paths for the carbon, either through the austenite ahead of and parallel to the interface, or through the γ/θ and γ/α interfaces. The interfaces are assumed to be of uniform thickness δ_b. S_I is the interlamellar spacing. This diagram represents a simplified view of the three-dimensional structure of pearlite.

are necessary to form pearlite, the austenite must be metastable with respect to both of the product phases; this means that the following condition would need to be satisfied: $c^{\gamma\theta} \leq \bar{c}^\gamma \leq c^{\gamma\alpha}$ (Figure 12.7a) – this condition defines what is sometimes called the Hultgren extrapolation region. As long as this is true, it becomes possible for the austenite to transform completely into pearlite, even though $\bar{c}^\gamma$ does not correspond to the eutectoid composition. In a time-temperature-transformation diagram for a hypoeutectoid steel, ferrite would not have to precede pearlite once an undercooling is achieved whereby $c^{\gamma\theta} \leq \bar{c}^\gamma \leq c^{\gamma\alpha}$ (Figure 12.7b).

The diffusion distance parallel to the interface can be approximated as $b_{19}S$ where b_{19} is a fraction usually set to $\frac{1}{2}$, on the basis that diffusion occurs in both directions parallel to the transformation front; the average diffusion distance must in practice be smaller than S given that the ferrite and cementite are in a lamellar arrangement. Assuming at first that the diffusion flux is entirely through the volume of the austenite (i.e., J_V), the rate at which solute is partitioned per unit area of the transformation front (taking account of the fractional areas presented by each phase at the transformation front) from α and absorbed into θ must be equal,

$$\underbrace{v\frac{S_I^\alpha(\bar{c}^\gamma - c^{\alpha\gamma})}{S_I}}_{\text{partitioning of C into austenite}} = \underbrace{v\frac{S_I^\theta(c^{\theta\gamma} - \bar{c}^\gamma)}{S_I}}_{\text{absorption of C from austenite}}$$

where v is the speed of the growth front. It can be shown using Equations 12.1 that each of these

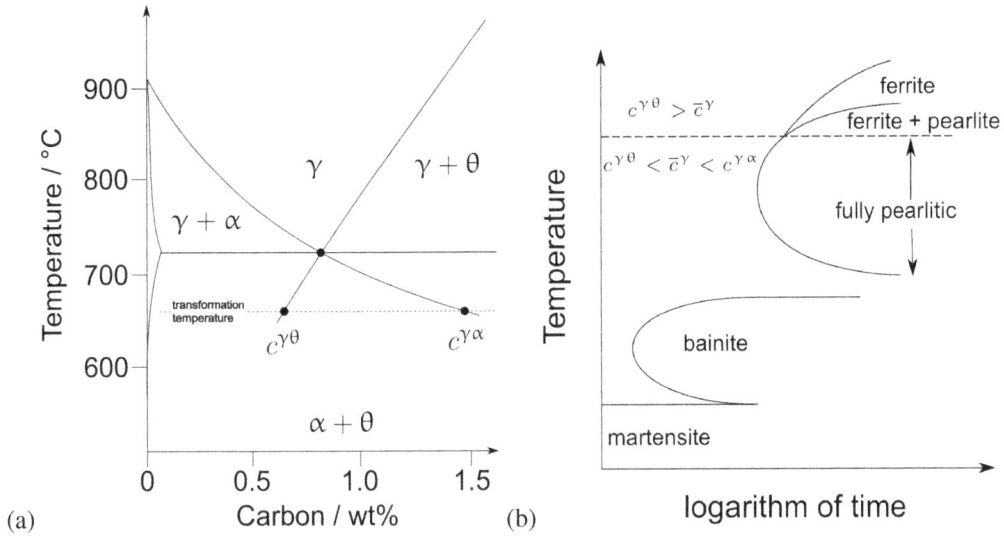

Figure 12.7 (a) Phase diagram with extrapolated phase boundaries to identify the concentrations in the austenite which is in equilibrium with cementite or ferrite. The region below the eutectoid temperature where the thermodynamic condition for the simultaneous formation of both cementite and ferrite is satisfied, is often referred to as the Hultgren extrapolation. (b) Schematic time-temperature-transformation diagram for a hypoeutectoid steel, showing the undercooling required if it is to transform into a fully pearlitic microstructure even though the carbon concentration of austenite may differ from the eutectoid composition.

terms is equivalent to

$$\frac{v S_I^\theta S_I^\alpha}{S_I^2}(c^{\theta\alpha} - c^{\alpha\theta}).$$

Since the rate of solute partitioning must equal the flux of solute at the interface, it follows that

$$\frac{v S_I^\theta S_I^\alpha}{S_I^2}(c^{\theta\alpha} - c^{\alpha\theta}) = D_C^\gamma \frac{(c^{\gamma\alpha} - c^{\gamma\theta})}{b_{19} S_I}$$

$$v = D_C^\gamma \frac{S}{b_{19} S_I^\theta S_I^\alpha} \frac{c^{\gamma\alpha} - c^{\gamma\theta}}{c^{\theta\alpha} - c^{\alpha\theta}} \tag{12.2}$$

where D_C^γ is the diffusivity of carbon in austenite.[2] Notice that this equation gives only the velocity as a function of the interlamellar spacing and furthermore, the velocity is predicted to increase indefinitely as the spacing decreases. This is because the diffusion distance decreases with S_I.

However, S_I cannot decrease without constraint, because the creation of α/θ interfaces consumes energy. The minimum value of interlamellar spacing possible is a critical spacing $S_I^C = 2\sigma^{\alpha\theta}/\Delta G$ where $\sigma^{\alpha\theta}$ is the interfacial energy per unit area and ΔG is the magnitude of the driving force for transformation in Joules per unit volume. The effective driving force is therefore given by

$$\Delta G_{net} = \Delta G - \frac{2\sigma^{\alpha\theta}}{S_I} \tag{12.3}$$

and since $\Delta G_{net} = 0$ when $S_I = S_I^C$, it follows that

$$\Delta G_{net} = \frac{2\sigma^{\alpha\theta}}{S_I^C} - \frac{2\sigma^{\alpha\theta}}{S_I} \qquad \text{so that} \qquad \frac{\Delta G_{net}}{\Delta G} = 1 - \frac{S_I^C}{S_I}. \tag{12.4}$$

Therefore, to allow for the energy consumed in the process of interface creation, Equation 12.2 is modified by a term $1 - [S_I^C/S_I]$ as follows:

$$v = D_C^\gamma \frac{S_I}{b_{19} S_I^\theta S_I^\alpha} \frac{c^{\gamma\alpha} - c^{\gamma\theta}}{c^{\theta\alpha} - c^{\alpha\theta}} \left(1 - \frac{S_I^C}{S_I}\right). \tag{12.5}$$

Figure 12.8 plots the growth rates using Equations 12.2 and 12.5 for two values of α/θ interfacial energy and reasonable values of all the other terms. The curve corresponding to Equation 12.2 has $\sigma^{\alpha\theta} = 0\,J\,m^{-2}$, whereas the one with a maximum in the growth rate has $\sigma^{\alpha\theta} = 0.3\,J\,m^{-2}$. The physical reason for the maximum in the growth rate expressed as a function of interlamellar spacing, is that a large spacing S_I increases the diffusion distance, whereas a small S_I means that a greater amount of the driving force is consumed in creating α/θ interfaces, making growth impossible when $S_I = S_I^C$.

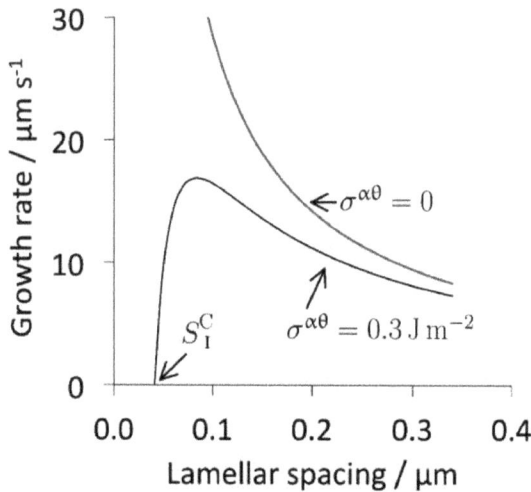

Figure 12.8 The calculated growth rate of pearlite, controlled by the diffusion of carbon in the austenite ahead of the transformation front. The growth rate increases indefinitely as the interlamellar distance is reduced, if the α/θ interfacial energy is zero. For the case where $\sigma^{\alpha\theta} = 0.3\,J\,m^{-2}$, the critical interlamellar spacing when the growth rate becomes zero is 0.042 µm.

The growth equations clearly do not have a unique solution for the growth rate, but rather, as a function of the interlamellar spacing. One assumption is that the spacing will correspond to that consistent with the maximum growth rate, i.e., when $S_I = 2S_I^C$ [20]. Alternatively, during isothermal transformation, the pearlite may adopt a spacing consistent with the maximum entropy production rate [22]; for isothermal transformation, this is of course equivalent to the maximum free energy dissipation rate. The rate of entropy production is [22, 23]

$$\dot{S} = \frac{v \Delta G_{net}}{T}. \tag{12.6}$$

At small undercoolings, the free energy change $\Delta G \approx \Delta H \Delta T / T_E$, where ΔH is the enthalpy change, T_E the eutectoid temperature and $\Delta T = (T_E - T)/T_E$. Using Equation 12.4, the rate of entropy production is

$$
\begin{aligned}
\dot{S} &= v \frac{\Delta H \Delta T}{T_E}\left(1 - \frac{S_I^C}{S_I}\right) \\
&= D_C^\gamma \frac{S_I}{b_{19} S_I^\theta S_I^\alpha} \frac{c^{\gamma\alpha} - c^{\gamma\theta}}{c^{\theta\alpha} - c^{\alpha\theta}} \frac{\Delta H \Delta T}{T_E}\left(1 - \frac{S_I^C}{S_I}\right)^2.
\end{aligned} \tag{12.7}
$$

The maximum in entropy production $\dot{S}$ occurs at $S_I = 3S_I^C$. The two criteria, maximum growth rate and maximum entropy production rate, yield different interlamellar spacings so the relevant criterion must be selected by experiment.

Supposing now that in addition to the flux through the austenite, there is additional solute transport through the averaged interface of thickness δ^b between the pearlite and austenite, defined by an overall boundary diffusion coefficient D_C^b. An additional flux must therefore be considered in Equation 12.2, leading to [24]

$$
v = \left(\frac{D_C^\gamma}{b_{19}} + \frac{\delta^b D_C^b}{b_{20} S}\right) \frac{S_I}{S_I^\theta S_I^\alpha} \frac{c^{\gamma\alpha} - c^{\gamma\theta}}{c^{\theta\alpha} - c^{\alpha\theta}}\left(1 - \frac{S_I^C}{S_I}\right). \tag{12.8}
$$

The concentration terms for the boundary are not in general expected to be identical to those determined by local equilibrium between the product phases and austenite. One assumption is that the concentration in the boundary is still represented by the phase diagram but scaled by an empirical *distribution coefficient*; since this coefficient is assumed to apply to all of the concentration terms, it can simply be incorporated into b_{20}. Figure 12.9 shows how the flux through the boundary becomes the dominant mechanism of carbon transport during pearlite growth at temperatures not far below T_E. For the calculations illustrated, because there now are two diffusion paths, the interlamellar spacing based on a maximum $\dot{S}$ is between $2.01\text{-}2.17 S_I^C$, and the corresponding values for the maximum growth rate criterion are $1.36\text{-}1.53 S_I^C$ [24].

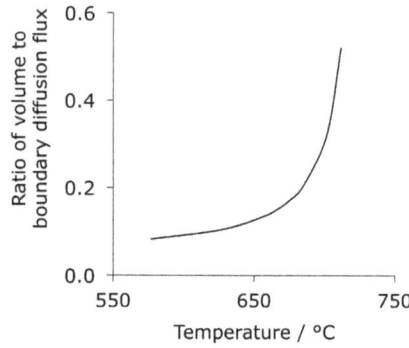

Figure 12.9 Relative contributions of volume diffusion through austenite, and boundary diffusion fluxes, during the growth of pearlite in a Fe-0.8C wt% steel. The details of the parameters used in the calculations can be found in [24].

12.4 Fe-C-X: GROWTH WITH LOCAL EQUILIBRIUM

The well-known complication is that the diffusivities of the substitutional and interstitial solutes are vastly different, so unlike the case for binary steels, it becomes necessary to discover conditions where the two or more solute fluxes can keep pace whilst maintaining equilibrium locally at the interface [22, 25–28]. And there is no doubt that substitutional solutes are partitioned between the phases at all temperatures where pearlite is observed [29–31].

Fe-C-X steels contain a substitutional solute (X) in addition to interstitial carbon. Local equilibrium requires the compositions at the interface to be maintained at levels that are consistent with a tie-line of the Fe-C-X phase diagram. At a constant temperature, this is in general not possible to achieve for the tie line passing through $\overline{c}_{Mn}, \overline{c}_c$ because the rate at which each solute is partitioned must then equal that at which it is carried away from the interface by diffusion. It is necessary therefore that

$$\text{at } \alpha/\gamma \text{ interface:} \quad \begin{cases} v(c_C^{\gamma\alpha} - c_C^{\alpha\gamma}) = -D_C^\gamma \dfrac{\partial c_C}{\partial z} \\[2mm] v(c_{Mn}^{\gamma\alpha} - c_{Mn}^{\alpha\gamma}) = -D_{Mn}^\gamma \dfrac{\partial c_{Mn}}{\partial z} \end{cases} \tag{12.9}$$

$$\text{at } \theta/\gamma \text{ interface:} \quad \begin{cases} v(c_C^{\gamma\theta} - c_C^{\theta\gamma}) = -D_C^\gamma \dfrac{\partial c_C}{\partial z} \\[2mm] v(c_{Mn}^{\gamma\theta} - c_{Mn}^{\theta\gamma}) = -D_{Mn}^\gamma \dfrac{\partial c_{Mn}}{\partial z} \end{cases} \tag{12.10}$$

where the subscripts identify the solute. Given that $D_{Mn}^\gamma \ll D_C^\gamma$, it becomes impossible to simultaneously satisfy either Equation sets 12.9 or 12.10 if the tie-line passing through $\overline{c}_{Mn}, \overline{c}_c$ is selected.

The solution of the conditions appropriate for the interface requires the following two equations [6]:

$$v_C = \left(2\frac{D_C^{\gamma}}{b_{19}} + 12\frac{D_C^b \delta^b}{b_{20} S}\right) \frac{S_I}{S^{\alpha} S^{\theta}} \left(\frac{c_C^{\gamma\alpha} - c_C^{\gamma\theta}}{c_C^{\theta\gamma} - c_C^{\alpha\gamma}}\right) \left(1 - \frac{S_I^C}{S_I}\right)$$

$$v_{Mn} = \left(2\frac{D_{Mn}^{\gamma}}{b_{19}} + 12\frac{D_{Mn}^b \delta^b}{b_{20} S_I}\right) \frac{S_I}{S^{\alpha} S^{\theta}} \left(\frac{c_{Mn}^{\gamma\alpha} - c_{Mn}^{\gamma\theta}}{c_{Mn}^{\theta\gamma} - c_{Mn}^{\alpha\gamma}}\right) \left(1 - \frac{S_I^C}{S_I}\right) \qquad (12.11)$$

where the velocities v_C and v_{Mn} are calculated on the basis of the diffusion of only carbon or only manganese, respectively. Clearly, since there is only one transformation front, the equations must be solved such that $v_C = v_{Mn}$. Bearing in mind that the interlamellar spacing is also identical in these equations, a further condition arises that

$$\frac{D_C}{D_{Mn}} = \frac{R_{Mn}}{R_C} \qquad \text{with} \qquad \begin{cases} D_i \equiv \dfrac{D_i^{\gamma}}{b_{19}} + \dfrac{6 D_i^b \delta^b}{b_{20} S_I} \\[2mm] R_i \equiv \dfrac{c_i^{\gamma\alpha} - c_i^{\gamma\theta}}{c_i^{\theta\gamma} - c_i^{\alpha\gamma}} \end{cases} \qquad (12.12)$$

The R_i condition ensures that the weighted average of the ferrite and cementite yields the mean composition of the steel, assuming that there is no long-range diffusion of manganese into the bulk of the parent phase. With these two constraints and in addition the local equilibrium condition, it becomes possible to find unique interface compositions at the growth front by coupling the conditions and the velocity equations to thermodynamic data using the following procedures:

- a trial θ/γ interface composition is set, selected from possible such tie-lines for the given transformation temperature.
- The α/γ interface composition tie-line is selected such that $\overline{c_{C,Mn}^{\alpha\gamma}, c_{C,Mn}^{\theta\gamma}} \ni \bar{c}_{C,Mn}$, where $\overline{c_{C,Mn}^{\alpha\gamma}, c_{C,Mn}^{\theta\gamma}}$ means the line connecting the compositions of ferrite and cementite. $\bar{c}_{C,Mn}$ is average composition in the system.
- If Equation 12.12 is not satisfied by these choices then the process is repeated until a solution is found.
- This solution provides the interface compositions to substitute into Equation 12.11 to calculate the single velocity $v = v_C = v_{Mn}$ of the transformation interface.

One of the difficulties in applying the type of theory described here is data on boundary-diffusion coefficients. For substitutional solutes, a good source is an assessment by Fridberg et al. [32], and the corresponding diffusion coefficient for carbon has been derived to be [6]

$$D_b^C = 1.84 \times 10^{-3} \exp\left(-\frac{124995\,\mathrm{J\,mol^{-1}}}{RT}\right) \qquad \mathrm{m^2\,s^{-1}}$$

Figure 12.10 shows that an order of magnitude agreement can be achieved between the calculated [6] and measured [33] pearlite growth rates in Fe-Mn-C steels, assuming that the flux of each of the solutes occurs both through the volume ahead of the transformation front and via the boundaries associated with the front itself. The chemical compositions of the cementite and ferrite are also well predicted [6]. Just an order of magnitude closure between theory and experiment might be disconcerting, but there are so many approximations involved: usually unspecified errors in the experimental data, the oversimplification of shape, a lack of information on interfacial energies, uncertain criteria regarding interlamellar spacing, and the assumption of local equilibrium at all interfaces. It is difficult in these circumstances to use the theory in alloy development, so most development programmes rely simply on the well-established qualitative trends when solutes are added to steels. Indeed, it is not possible to identify a single case where the theory has been applied in the design of new steels.

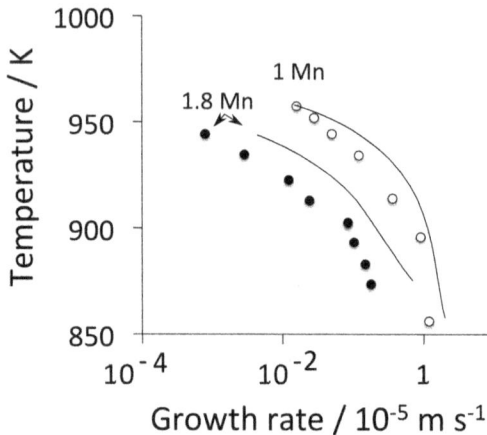

Figure 12.10 Curves showing the calculated diffusion-controlled growth rate of pearlite, in 1.0 and 1.8 wt% manganese eutectoid steels as a function of the transformation temperature [6]. The solute flux is both through the austenite ahead of the interface and through the interface itself. The corresponding data are from [33].

It is possible in a ternary system for the austenite to be supercooled into the three phase $\alpha + \gamma + \theta$ phase field where they can exist together in equilibrium. The pearlite that grows does not then have the average composition of the austenite, but carbon and the substitutional solute partition between the parent and product. The resulting change in the composition of the austenite reduces the driving force, eventually to zero when equilibrium is reached and transformation stops before all of the austenite is consumed. The reduction in driving force as equilibrium is reached also leads to a progressive change in the interlamellar spacing to ever larger values, resulting in what is known as *divergent pearlite* [34], Figure 12.11.

12.4.1   LOCAL EQUILIBRIUM?

Whereas it is well established that both the cementite and ferrite associated with the bainite reaction inherits the substitutional content of the austenite, such that the substitutional solute to iron ratio remains constant during transformation, pearlite shows different characteristics even when the transformation temperature is identical to that for bainite.

Tsivinsky et al. [35] first reported that chromium and tungsten partitioned from austenite into cementite during the growth of pearlite, but not during that of bainite. Chance and Ridley [30] found that for upper bainite in a Fe-0.81C-1.41Cr wt% alloy, the partition coefficient k_{Cr}, defined as (wt% Cr in θ)/(wt% Cr in α), could not be distinguished from unity (Figure 12.12). Chance and Ridley suggested that partitioning occurs during the pearlite reaction but at the same temperature does not

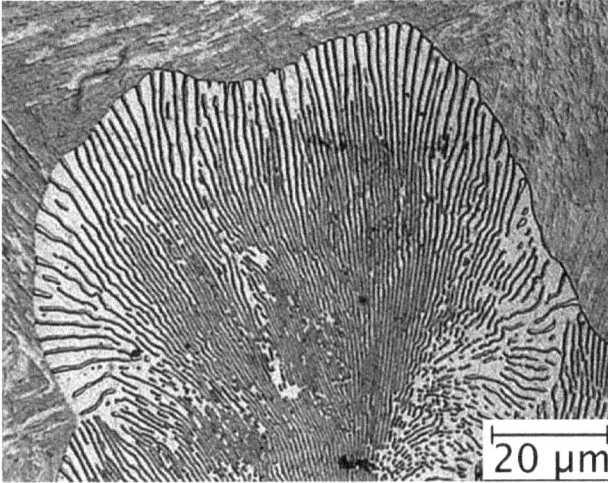

Figure 12.11 Fe-2.5C-5.4Mn at.% steel transformed from austenite supercooled into the $\alpha + \gamma + \theta$ phase field. The interlamellar spacing increases as the colony enlarges. After Hutchinson et al. [31], reproduced with the permission of Elsevier.

occur with bainite because there is a fast diffusion path along the incoherent interface for pearlite. This could be paraphrased as the difference between the reconstructive transformation mechanism for pearlite and the displacive mechanism for bainite. Figure 12.12 emphasises that the partition coefficient for pearlite is vastly different from that indicated by equilibrium; this means that when pearlite forms at low temperatures, the condition for local equilibrium at the transformation interfaces does not in fact hold. The solution to this must lie in including interface response functions that deal with solute trapping in the analysis (p. 217).

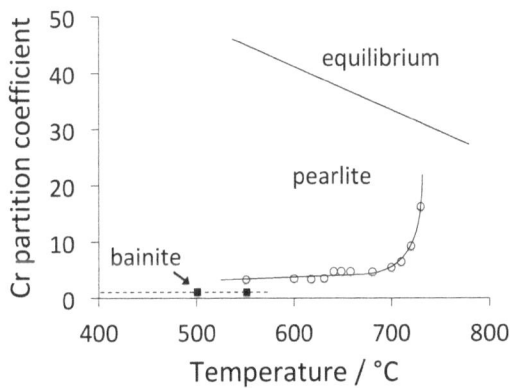

Figure 12.12 Measured ratio of chromium in cementite to that in ferrite (i.e., partition coefficient based on wt%), when the cementite is a part of bainite or pearlite [30].

12.5 FORCED VELOCITY PEARLITE

When a cylinder of austenite traverses a temperature gradient, on reaching an undercooled domain, pearlite grows along the direction of heat flow until the traverse-speed reaches some critical value. Beyond that value, the transformation front is unable to match the traverse speed so the pearlite forms the usual colonies with multi-directional growth [36]. These experiments have been revealing in that the interlamellar spacing during forced-velocity growth matches that in regular pearlite when the comparison is made at the same speed. Later work established the temperature at which the pearlite forms, thus providing comprehensive velocity, temperature and interlamellar spacing information for Fe-C system [37]. Including data from isothermal experiments, the relationship between the undercooling below the eutectoid temperature and the interlamellar spacing is found to be $S_I T_E = 8.02 \, \mu m \, °C$. This relationship of course applies to binary steels, whereas all practical alloys contain other solutes. Takahashi [38] derived the following empirical equations from published experimental data to cover a range of substitutionally alloyed steels as a function of temperature (°C):

$$\log\{S_I/\mu m\} = -2.21358 + 0.09863 \underbrace{w_{Mn}}_{0\text{-}1.8} - 0.05427 \underbrace{w_{Cr}}_{0\text{-}9}$$

$$+ 0.03367 \underbrace{w_{Ni}}_{0\text{-}3} - \log\left\{\frac{T_E - T}{T_E}\right\} \tag{12.13}$$

with temperature in the range 600-790 °C. There clearly will be other limits to applying this equation which assumes linear combination and unjustified logarithmic terms.

12.6 PEARLITE NOT CONTAINING CEMENTITE

In some steels containing relatively large concentrations of strong carbide-forming substitutional solutes such as chromium, it is possible to generate lamellar pearlite consisting of a mixture of an alloy carbide and ferrite that grow cooperatively [39, 40]. The transformation temperature must be sufficiently high to permit the diffusion of substitutional solutes. A Fe-11.8Cr-0.23 wt% alloy transformed from austenite in the range 775-700 °C leads to the formation of pearlite in which the carbide is $M_{23}C_6$, with the lamellar spacing decreasing as the transformation temperature is reduced, Figure 12.13. The alloy pearlite shows many of the characteristics of normal pearlite; the crystallographic orientation of $M_{23}C_6$ within a colony is mostly uniform, although several orientations have been reported for the fibrous form of the precipitate [41]; there is evidence of branching, and a common transformation front. The overall shape of a colony can be nodular but tends to be spiky at lower temperatures just as in Fe-C pearlite; transformation at low temperatures also requires a much longer time to accomplish given that the diffusion of substitutional atoms is then sluggish. Figure 12.13 illustrates these features using two different steels. Given that both chromium and carbon partition during the process, the transformation cannot reach completion if the temperature is in the three-phase $M_{23}C_6 + \alpha + \gamma$ field.

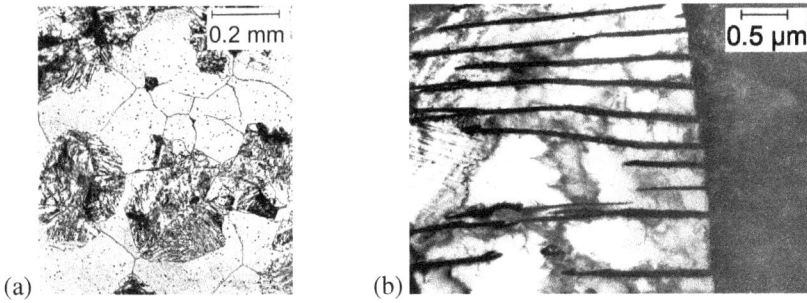

Figure 12.13 Pearlite containing alloy carbides. (a) Fe-0.23C-11.8Cr wt%. Transformed isothermally at 775 °C for 30 min; pearlite with $M_{23}C_6$ as the carbide component. (b) Fe-0.23C-11.8Cr wt%. Transformed isothermally at 750 °C for 15 min illustrating the cooperative growth of $M_{23}C_6$ and ferrite from austenite. Reproduced from Campbell and Honeycombe [40] with permission from Taylor and Francis.

Alloy-pearlite which is a mixture of ferrite and M_7C_3 has been observed during the transformation of austenite in Fe-8.2Cr-0.96C wt% steel [42, 43], Figure 12.14.

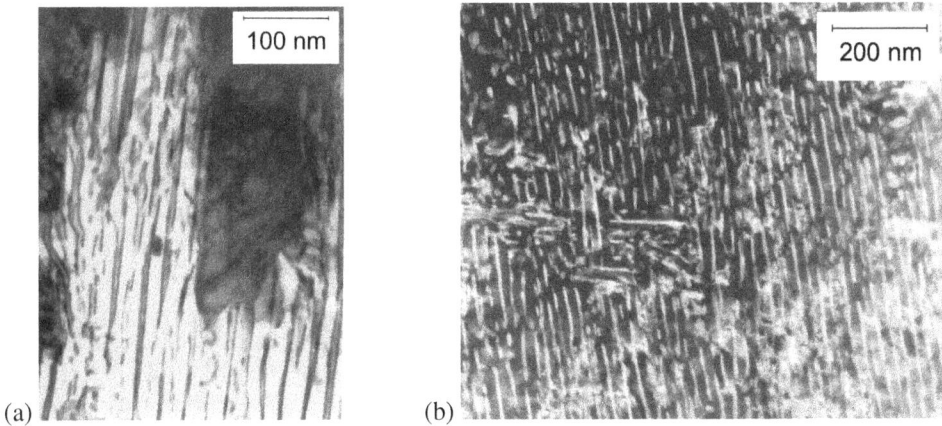

Figure 12.14 Fe-0.3C-4.08Cr wt% transformed isothermally at 478 °C for 43 days. The carbide component is now M_7C_3. (a) Illustrates the transformation front showing cooperative growth. (d) The overall microstructure where carbide continuity is sometimes lost, and occasional branching is observed.

12.7 DIVORCED EUTECTOID TRANSFORMATION

When particles of proeutectoid cementite exist in austenite, cooling to a temperature where eutectoid transformation become possible can lead to *divorced pearlite*. In this, the pre-existing cementite particles simply grow to absorb the excess carbon partitioned at the γ/α transformation front. There is, therefore, no cooperative growth of the ferrite and cementite as occurs in conventional pearlite. This kind of transformation is important as a rapid mechanism of obtaining a spheroidised instead of lamellar cementite [44].

The mechanism [45] is illustrated schematically in Figure 12.15. The mixture of austenite and proeutectoid cementite when cooled below Ae_1 leads to the flux of carbon towards cementite particles in both the austenite and ferrite, Figure 12.15. To maintain equilibrium at the α/γ interface as it advances, the carbon partitioned must equal that absorbed in the cementite:

$$(c^{\gamma\alpha} - c^{\alpha\gamma})v = D_\gamma \frac{c^{\gamma\alpha} - c^{\gamma\theta}}{\lambda_\gamma} + D_\alpha \frac{c^{\alpha\gamma} - c^{\alpha\theta}}{\lambda_\alpha} \tag{12.14}$$

where the distances $\lambda_{\gamma,\alpha}$ are defined in Figure 12.15. If ΔT is the undercooling below the temperature at which ferrite may first form, an approximate equation for the velocity of the α/γ interface is given by[3]

$$v \approx \frac{2D_\alpha}{\lambda_\gamma + \lambda_\alpha} \frac{\frac{\Delta T}{27} \left[\frac{0.28}{D_\alpha/D_\gamma} + 0.009 \right]}{0.75 + \frac{\Delta T}{27} \times 0.225} \tag{12.15}$$

where v is the velocity of the α/γ interface. By comparing this velocity against that for the growth of lamellar pearlite, it is possible to identify the domains in which divorced pearlite can be favoured over the lamellar form, Figure 12.15.

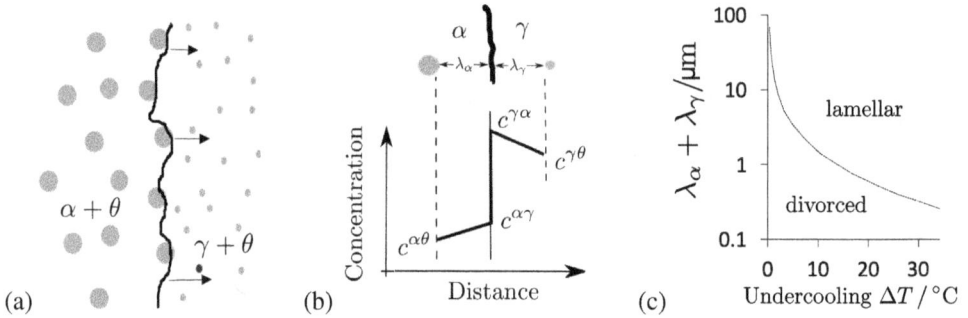

Figure 12.15 (a) Mechanism of the divorced eutectoid transformation of a mixture of austenite and fine cementite [46]. (b) Carbon fluxes in this model. (c) Low undercoolings below the eutectoid temperature and fine spacings between cementite particles favour the formation of divorced pearlite. The calculations [46] are for a plain-carbon eutectoid steel.

The phase field simulation illustrated in Figure 12.16 [47] confirms the basic circumstances by which divorced pearlite can be induced, but revealed in addition that it is possible for cementite particles both at the γ/α interface and those ahead of it to absorb partitioned carbon.

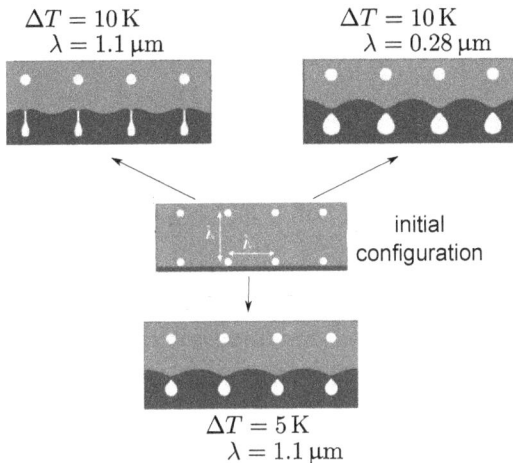

Figure 12.16 Phase field simulation showing the transition from lamellar to divorced eutectoid as a function of spacing and undercooling below the eutectoid temperature. Reproduced with permission of Elsevier, from Kumar et al. [47].

REFERENCES

1. M. Koga, E. C. Santos, T. Honda, K. Kida, and T. Shibukawa: 'Observation of wear in induction-heat-treated bearing steel bars under reciprocating motion', *Advanced Materials Research*, 2012, **457-458**, 504–510.

2. M. Hillert: 'The formation of pearlite', In: V. F. Zackay, and H. I. Aaronson, eds. *Decomposition of Austenite by Diffusional Processes*. New York, USA: Interscience, 1962:197–237.

3. R. J. Dippenaar, and R. W. K. Honeycombe: 'The crystallography and nucleation of pearlite', *Proceedings of the Royal Society A*, 1973, **333**, 455–467.

4. A. Walentek, M. Seefeldt, B. Verlinden, E. Aernoudt, and P. V. Houtte: 'Electron backscatter diffraction on pearlite structures in steel', *Journal of Microscopy*, 2006, **224**, 256–263.

5. E. Werner: 'The growth of holes in plates of cementite', *Materials Science & Engineering A*, 1991, **132**, 213–223.

6. S. W. Seo, H. K. D. H. Bhadeshia, and D. W. Suh: 'Pearlite growth rate in Fe-C and Fe-Mn-C steels', *Materials Science and Technology*, 2015, **31**, 487–493.

7. H. Jolivet: 'Transformation of austenite on cooling: morphology and genesis of the aggregates formed', *Journal of the Iron and Steel institute*, 1939, **140**, 95–114.

8. N. Ridley: 'Partitioning of alloying additions during pearlite growth in eutectoid steels', In: M. S. H. I. Aaronson, D. E. Laughlin:, and C. M. Wayman, eds. *Solid-Solid Phase Transformations*. Materials Park, Ohio, USA: TMS-AIME, 1981:807–811.

9. S. Jones, and H. K. D. H. Bhadeshia: 'Kinetics of the simultaneous decomposition of austenite into several transformation products', *Acta Materialia*, 1997, **45**, 2911–2920.

10. H. K. D. H. Bhadeshia: Bainite in Steels: Theory and Practice: 3rd ed., Leeds, U.K.: Maney Publishing, 2015.

11. R. Dippenaar: 'Decomposition of austenite in alloy steels': Ph.D. thesis, University of Cambridge, Cambridge, U. K., 1970.

12. S. A. Hackney, and G. J. Shiflet: 'The pearlite-austenite growth interface in an Fe-0.8C-12Mn alloy', *Acta Materialia*, 1987, **35**, 1007–1017.

13. A. Durgaprasad, S. Giri, S. Lenka, S. Kundu, S. Mishra, S. Chandra, R. D. Doherty, and I. Samajdar: 'Defining a relationship between pearlite morphology and ferrite crystallographic orientation', *Acta Materialia*, 2017, **129**, 278–289.

14. H. K. D. H. Bhadeshia: 'Problems in the calculation of transformation texture in steels', *ISIJ International*, 2010, **50**, 1517–1522.

15. M.-X. Zhang, and P. M. Kelly: 'Determination of carbon content in bainitic ferrite and carbon distribution in austenite by using CBKLDP', *Materials Characterization*, 1998, **40**, 159–168.

16. M.-X. Zhang, and P. M. Kelly: 'Crystallography of spheroidite and tempered martensite', *Acta Materialia*, 1998, **46**, 4081–4091.

17. H. K. D. H. Bhadeshia: 'Solution to the Bagaryatskii and Isaichev ferrite-cementite orientation relationship problem', *Materials Science and Technology*, 2018, **34**, 1666–1668.

18. S. Kante, and A. Leineweber: 'Two-phase and three-phase crystallographic relationships in white-solidified and nitrided Fe-C-Si cast iron', *Acta Materialia*, 2019, **170**, 240–252.

19. F. C. Frank, and K. E. Puttick: 'Cementite morphology in pearlite', *Acta Metallurgica*, 1956, **4**, 206–210.

20. C. Zener: 'Kinetics of the decomposition of austenite', *Transactions of the American Institute of Mining and Metallurgical Engineers*, 1946, **167**, 550–595.

21. M. Hillert: 'Role of interfacial energy during solid-state phase transformations', *Jernkontorets Annaler*, 1957, **141**, 757–789.

22. J. S. Kirkaldy, and R. C. Sharma: 'Stability principles for lamellar eutectoid(ic) reactions', *Acta Metallurgica*, 1980, **28**, 1009–1021.

23. J. W. Cahn, and W. C. Hagel: 'Theory of pearlite reaction', In: V. F. Zackay, and H. I. Aaronson, eds. *Decomposition of Austenite by Diffusional Processes*. New York: Intersciences, 1962:131–192.

24. A. S. Pandit, and H. K. D. H. Bhadeshia: 'Mixed diffusion-controlled growth of pearlite in binary steel', *Proceedings of the Royal Society A*, 2011, **467**, 508–521.

25. M. Hillert: 'Paraequilibrium': Technical Report, Swedish Institute for Metals Research, Stockholm, Sweden, 1953.

26. G. R. Purdy, D. H. Weichert, and J. S. Kirkaldy: 'The growth of proeutectoid ferrite in ternary Fe-C-Mn austenites', *Trans. TMS-AIME*, 1964, **230**, 1025–1034.

27. D. E. Coates: 'Diffusional growth limitation and hardenability', *Metallurgical Transactions*, 1973, **4**, 2313–2325.

28. H. K. D. H. Bhadeshia: 'Diffusional formation of ferrite in iron and its alloys', *Progress in Materials Science*, 1985, **29**, 321–386.

29. S. A. Al-Salman, G. W. Lorimer, and N. Ridley: 'Pearlite growth kinetics and partitioning in a Cr-Mn eutectoid steel', *Metallurgical Transactions A*, 1979, **10A**, 1703–1709.

30. J. Chance, and N. Ridley: 'Chromium partitioning during isothermal transformation of a eutectoid steel', *Metallurgical Transactions A*, 1981, **12A**, 1205–1213.

31. C. R. Hutchinson, R. E. Hackenberg, and G. J. Shiflet: 'The growth of partitioned pearlite in Fe-C-Mn steels', *Acta Materialia*, 2004, **52**, 3565–3585.

32. J. Fridberg, L.-E. Torndähl, and M. Hillert: 'Diffusion in iron', *Jernkontorets Annaler*, 1969, **153**, 263–276.

33. N. A. Razik, G. W. Lorimer, and N. Ridley: 'An investigation of manganese partitioning during the austenite-pearlite transformation using analytical electron microscopy', *Acta Metallurgica*, 1974, **22**, 1249–1258.

34. J. W. Cahn, and W. C. Hagel: 'Divergent pearlite in a manganese eutectoid steel', *Acta Metallurgica*, 1963, **11**, 561–574.

35. S. V. Tsivinsky, L. I. Kogan, and R. I. Entin: 'Investigation of the distribution of chromium and tungsten during the decomposition of austenite using the radioactive tracer method', In: *Problems of Metallography and the Physics of Metals*. Moscow, Russia: State Scientific Press, ed. B. Ya Lybubov, 1955:185–199.

36. G. F. Bolling, and R. H. Richman: 'Forced velocity pearlite', *Metallurgical Transactions*, 1970, **1**, 2095–2104.

37. D. D. Pearson, and J. D. Verhoeven: 'Forced velocity pearlite in high purity Fe-C alloys: Part 1. experimental', *Metallurgical Transactions A*, 1984, **15**, 1037–1045.

38. M. Takahashi: 'Reaustenitisation from bainite in steels': Ph.D. thesis, University of Cambridge, http://www.phase-trans.msm.cam.ac.uk/2000/phd.html#Takahashi, 1992.

39. M. Mannerkoski: 'The mechanism of formation of a periodic eutectoid structure at low temperatures in plain chromium steel', *Metal Science*, 1969, **3**, 54–55.

40. K. Campbell, and R. W. K. Honeycombe: 'The isothermal decomposition of austenite in simple chromium steels', *Metal Science*, 1974, **8**, 197–203.

41. P. R. Howell, J. V. Bee, and R. W. K. Honeycombe: 'The crystallography of the austenite-ferrite/carbide transformation in Fe-Cr-C alloys', *Metallurgical Transactions A*, 1979, **10**, 1213–1222.

42. J. V. Bee, and D. V. Edmonds: 'Metallographic study of the high-temperature decomposition of austenite in alloy steels containing Mo and Cr', *Metallography*, 1979, **12**, 3–21.

43. D. V. Shtansky, K. Nakai, and Y. Ohmori: 'Crystallography and interface boundary structure of pearlite with M_7C_3 carbide lamellae', *Acta Materialia*, 1999, **47**, 1105–1115.

44. H. K. D. H. Bhadeshia: 'Steels for bearings', *Progress in Materials Science*, 2012, **57**, 268–435.

45. J. D. Verhoeven: 'The role of the divorced eutectoid transformation in the spheroidization of 52100 steel', *Metallurgical & Materials Transactions A*, 2000, **31**, 2431–2438.

46. J. D. Verhoeven, and E. D. Gibson: 'The divorced eutectoid transformation in steel', *Metallurgical & Materials Transactions A*, 1998, **29**, 1181–1189.

47. A. Kumar, R. Mukherjee, T. Mittnacht, and B. Nestler: 'Deviations from cooperative growth mode during eutectoid transformation: Insights from a phase-field approach', *Acta Materialia*, 2014, **81**, 204–210.

48. L. E. Samuels: Lightmicroscopy of Carbon Steels: Metals Park, Ohio, USA: ASM International, 1999.

Notes

[1] Sorby first observed the mother-of-pearl appearance of what Howe later named as *pearlyte, perlite* and finally *pearlite* [48].

[2] The term $S_I/S_I^\theta S_I^\alpha$ is proportional to $1/S_I$ because both S_I^θ and S_I^α are in turn proportional to S_I when the volume fraction of cementite is constant.

[3] The terms deduced from the phase diagram, for 700°C, are $c^{\gamma\alpha} - c^{\gamma\theta} \approx \Delta T(0.28/0.27)$, $c^{\alpha\gamma} - c^{\alpha\theta} \approx \Delta T(0.009/27)$, $c^{\gamma\alpha} - c^{\alpha\gamma} \approx 0.75 + \Delta T(0.225/27)$.

13 Aspects of kinetic theory

13.1 GRAIN GROWTH

A real, three-dimensional grain structure cannot ever be in equilibrium because the space-filling grain-shape prevents the balancing of interfacial tensions. A uniform, hexagonal grain structure in two dimensions can, on the other hand, be metastable because the boundary triple-points exhibit three-fold symmetry so the tensions are balanced, assuming that the interfacial energy is identical for all boundaries. In a single-phase material, there will exist a distribution of grains sizes with some grains having a greater number of sides than others; the distribution of sizes is in general unimodal with a maximum that is twice as large as the mean value [1]. In such a structure, annealing enables the larger grains to grow at the expense of those that are smaller, but the unimodal distribution of sizes is maintained even though the mean size increases. The process is driven entirely by the excess energy present in the structure due to interfaces. On a local scale, grain boundaries tend to migrate towards their centre of curvature as atoms located at a curved boundary move into positions where they have more correctly positioned near-neighbours. The attempt at balancing tensions at grain boundary junctions leads to curvature that in turn promotes the growth of the larger grains, Figure 13.1.

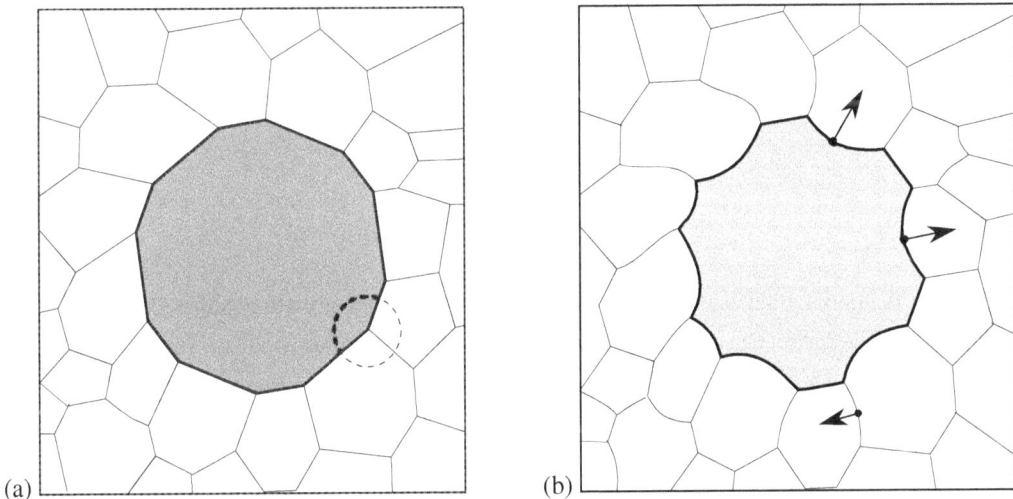

Figure 13.1 (a) An exaggerated illustration in which the grains all have flat faces so interfacial tensions are not balanced at grain boundary junctions. The dashed circle highlights one such junction where if the boundary energies are all identical, then the forces at the junction are not balanced. (b) A relaxed grain configuration where boundary segments are curved to help maintain a semblance of balance at grain boundary junctions. They therefore migrate towards their centres of curvature.

When the grain size $\overline{L}$ is measured as a mean lineal intercept, the boundary surface per unit volume is given by $2/\overline{L}$, so the excess energy locked up in the form of grain boundaries is $2\sigma/\overline{L}$. This excess energy drives the coarsening of the three-dimensional grain structure; σ is the grain boundary energy per unit area, assumed to be single-valued.

The microstructure might contain obstacles to grain boundary motion, for example, precipitates. Suppose that there is a random array of particles, volume fraction V_V with N_V uniformly-sized, spherical particles per unit volume, each of radius r then $N_V = 3V_V/4\pi r^3$ and the number of particles intersected by a unit area of boundary is $2rN_V$ so the force opposing boundary motion is $3\sigma V_V/2r$. This force is the Zener drag on the boundary [reviewed in 2]. A limiting grain size $\overline{L}_{\lim}$ can be defined when the particle pinning force equals the driving force for grain growth, i.e., when $2\sigma/\overline{L} = 3V_V/2r$, so that $\overline{L}_{\lim} = 4r/3V_V$.

For small driving forces, the average boundary migration velocity is equal to a boundary mobility M_b multiplied by the net driving force (Equation 4.64), an approximation justified at small driving forces:

$$v = M_b 2\sigma \left(\frac{1}{\overline{L}} - \frac{1}{\overline{L}_{\lim}} \right).$$

If the boundary velocity is written as $d\overline{L}/dt$ and the temperature dependence of the mobility can be separated as $M_b = M_{bo} \exp(-Q/RT)$, then it follows that [3]

$$\frac{d\overline{L}}{dt} = 2\sigma M_{bo} \exp\left\{ -\frac{Q}{RT} \right\} \left[\frac{1}{\overline{L}} - \frac{1}{\overline{L}_{\lim}} \right] \tag{13.1}$$

where t is the time during isothermal growth and Q is an activation energy for grain growth. If $\overline{L}_{\lim} = \infty$ then integration would lead to a relationship where $\overline{L}$ varies with $\sqrt{t}$ assuming that the initial grain size before coarsening is neglected.[1] But in general, integration of Equation 13.1 gives [3]

$$-\overline{L}\overline{L}_{\lim} - \overline{L}_{\lim}^2 \ln\left\{ 1 - \frac{\overline{L}}{\overline{L}_{\lim}} \right\} + \overline{L}_o \overline{L}_{\lim} + \overline{L}_{\lim}^2 \ln\left\{ 1 - \frac{\overline{L}_o}{\overline{L}_{\lim}} \right\}$$
$$= 2\sigma M_{bo} \exp\left\{ -\frac{Q}{RT} \right\} t \tag{13.2}$$

where $\overline{L}_o$ is the initial grain size. When considering austenite grain growth in large components, the time taken to reach the heat-treatment temperature can be taken into account. The steel becomes fully austenitic at a temperature Ac_3 so $\overline{L}_o$ is set to that austenite grain size that exists at that temperature. During continuous heating followed by isothermal holding at a temperature T' for time t, the term on the right hand side of Equation 13.2 becomes [4]

$$\ldots = 2\sigma M_{bo} \left[\exp\left\{ -\frac{Q}{RT'} \right\} t + \frac{1}{\dot{T}} \int_{Ac_3}^{T'} \exp\left\{ -\frac{Q}{RT} \right\} dT \right]$$

where $\dot{T}$ is the heating rate between Ac_3 and T'. The activation energy Q is about $200\,\mathrm{kJ\,mol^{-1}}$, somewhat less than that for the self-diffusion of iron; much larger values have been reported but are a consequence of overfitting to limited experimental data [4].

There are circumstances in which some grains grow much more rapidly than the general coarsening described above. Figure 13.2 illustrates a steel containing boundary-pinning AlN particles that are stable at the lower austenitisation temperature of 840 °C, but begin to dissolve in the austenite at 940 °C. In this latter regime of borderline particle stability, some regions will become depleted of particles before others, where relatively rapid grain growth can occur locally, leading to a bimodal distribution of grain sizes. Other heterogeneities that promote *abnormal* grain growth include anisotropy in crystal orientations, in boundary energy or boundary mobility. The conditions lead-

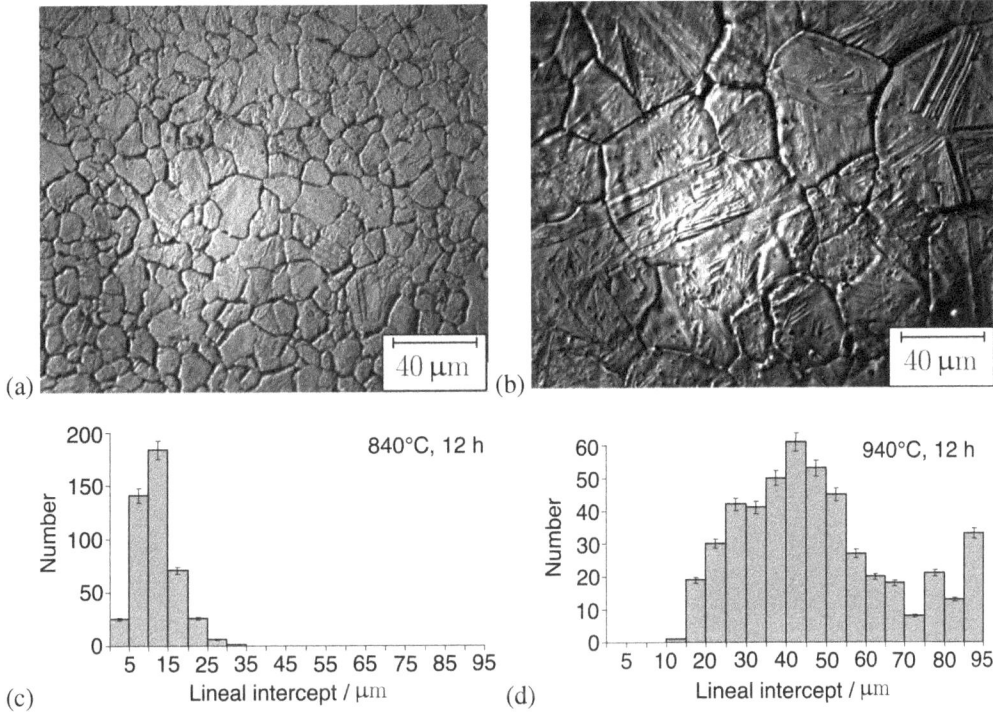

Figure 13.2 Austenite grain growth in a nuclear pressure vessel steel containing a small fraction of aluminium nitride precipitates. The austenite grains are revealed using thermal grooving that occurs during austenitisation as grain boundary and surface tensions are balanced. (a) Following austenitisation at 840 °C, (b) following austenitisation at 940 °C. (c, d) Respective distributions of lineal intercepts [4], showing that there is a bimodal distribution for the sample austenitised at 940 °C.

ing to abnormal grain growth have been modelled [1, 5], suggesting that grains with a size about 1.4 times the average will tend to grow abnormally. The practical application of this requires some mechanism by which such a size *anomaly* can appear in the microstructure.

13.2 RECRYSTALLISATION

In normal circumstances, any reduction in the density of dislocations, introduced during plastic deformation, is small during the process of recovery. The deformed grain structure is also largely unaffected by recovery. It takes the growth of new grains to initiate a much larger change, i.e., recrystallisation. New grains are stimulated in regions where the dislocation density is large (Figure 13.3). This can happen when an existing grain boundary bows into the grain containing a somewhat greater density of dislocations, in which case the crystallographic orientation of the grain with the lower dislocation density is maintained in the recrystallised region, with consequences on the development of a recrystallisation texture.

If the steel contains relatively hard particles that resist changes in shape, then deformation can lead to plastic strain gradients around the particles. When these gradients are large, they can lead to particle-stimulated recrystallisation where the new grain may have a large misorientation relative to its surroundings. Large particles are more effective since there are greater deformation gradients around them and hence are expected to be more effective in inducing recrystallisation. The deformation gradients extend approximately to the size of the coarse particles [6].

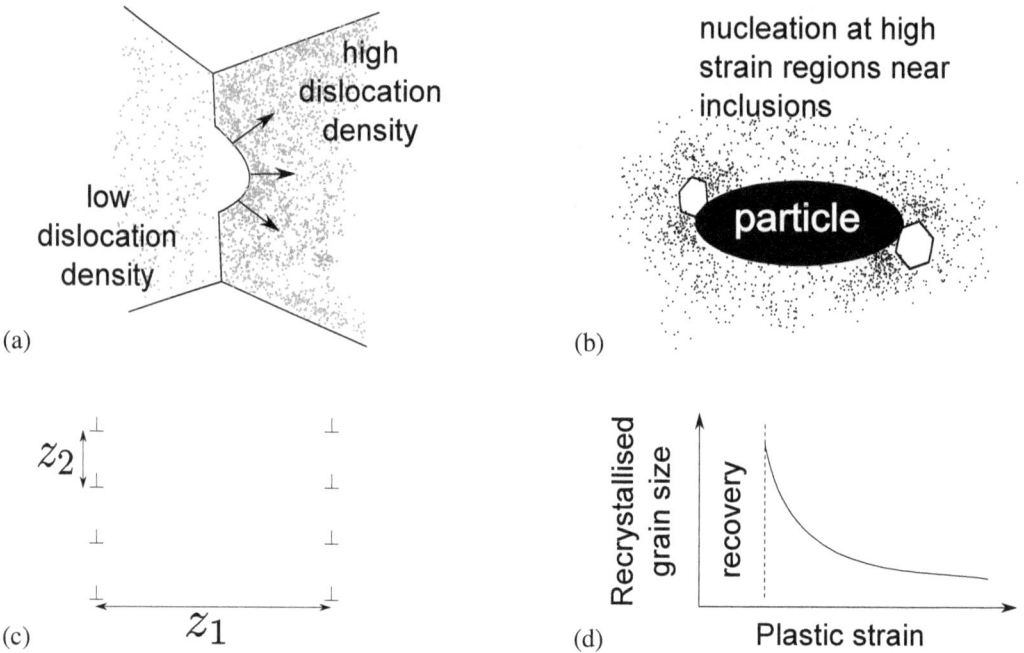

Figure 13.3 Nucleation of recrystallisation, (a) by grain boundary bowing that propagates the orientation of the grain on the left, (b) particle stimulated nucleation, which may lead to a highly misoriented grain. (c) Development of misorientation by polygonisation. (d) Variation in recrystallised grain size as a function of the amount of deformation prior to isothermal annealing.

Both of the mechanisms described require a certain misorientation to develop within the deformed matrix in order to define the genesis of a recrystallisation nucleus. Suppose that nucleation begins in a jumble of dislocations. The local rearrangement of free dislocations can lead to the creation of a region that is essentially free from dislocations, Figure 13.3c. If it is assumed that a new grain forms when a region accumulates a misorientation $\theta = 10° \simeq 0.2\,\mathrm{rad}$ with its neighbours and a size z_1, both arrived at by polygonisation, then:

$$\theta = \frac{b}{z_2} = \simeq \frac{3 \times 10^{-10}\,\mathrm{m}}{z_2} \simeq 0.2\,\mathrm{rad} \tag{13.3}$$

so that $z_1 = 15 \times 10^{-10}\,\mathrm{m}$. Given that z_1 is typically 0.1-1 μm, the critical dislocation density required to generate the misorientation, $\rho_\perp^* = 1/z_1 z_2 \simeq 10^{15} \to 10^{16}\,\mathrm{m}^{-2}$. The actual dislocation density required has to be somewhat larger if some of the defects are annihilated during recovery. Large dislocation density differences are usually only to be found in localised regions.

A greater nucleation rate leads ultimately to a finer recrystallised grain size (Figure 13.3d) during isothermal annealing. There is a level of deformation below which recrystallisation does not occur because recovery processes reduce the defect density. While the extent of the plastic strain prior to annealing is a key parameter influencing recrystallisation behaviour, other factors affect the nature of the defects introduced during deformation:

- changes to the shapes of the grain following deformation. For example, during plane-strain deformation (rolling), austenite grains are deformed in the rolling direction more severely than in other directions. This *pancaking* that increases the amount of boundary per unit volume from S_{V_0} in the undeformed state to S_V following plastic deformation, which adds to the stored energy of the material. For grains that initially are equiaxed, in the form of space-filling tetrakaidecahedra [7, 8]:

$$\frac{S_V}{S_{V_0}} = \frac{\eta_{11} + 3(\eta_{11}\sqrt{1+2\eta_{33}^2} + \sqrt{\eta_{11}^2 + 2\eta_{33}^2} + \eta_{33}\sqrt{2(1+\eta_{11}^2)}}{3(2\sqrt{3}+1)}$$

$$\frac{L_V}{L_{V_0}} = \frac{1 + \eta_{11} + 2\sqrt{1+\eta_{11}^2 + 2\eta_{33}^2}}{6}$$

where η_{11} and η_{33} are the principal distortions with the true strains along the principal directions given by $\varepsilon_{ii} = \ln \eta_{ii}$; L_V is the grain edge length per unit volume, with initial value L_{V_0}.

Because plastic deformation is not homogeneous, boundaries are roughened by slip [9] and heterogeneous slip introduces deformation bands. There may also exist annealing twins within the austenite grains. These effects are not predictable, but certainly affect the stored energy. When considering austenite grains, the effects are represented empirically; the additional S_V due to twins and bands is taken to be proportional to ε_p^2 [10]. If the flow stress σ_P is known, then the dislocation density is often taken to be proportional to $(\sigma_P/E_s b)^2$, with the proportionality constant containing the Taylor factor and an empirical constant [11].

- If a hole in a matrix is sheared, then the distortion accompanying the hole can be restored to its original shape by an appropriate deformation of the matrix, or by local matrix rotations [12, 13]; this process in effect models what happens when a material containing a rigid inclusion is deformed. The lattice rotations may be expressed in the form of subgrains which deviate sufficiently from the original orientation [14].

The deformation field around the inclusion is heterogeneous and results in plastic strain gradients that play a role in recovery and recrystallisation. However, it is possible for such deformation to relax, for example by the dissipation of defects into the surroundings, especially when the inclusions are small in size. For large inclusions, features such as dislocation loops cannot be removed from the neighbourhood of the particles so relaxation processes become less effective. As a consequence, particles that are generally several micrometres in size act to stimulate recrystallisation. Therefore, fine dispersions, such as those associated with microalloyed steels, retard recrystallisation by pinning boundaries or dislocations, rather than accelerating it.

- Given the mechanism by which recrystallisation nucleates, i.e., grain boundary bowing or polygonisation, it is natural that the crystallographic character of boundaries in the deformed state will influence the development of crystallographic texture in the recrystallised form. It has, for example, been found that in recrystallisation experiments on ferritic bicrystals, an initial "γ-fibre"[2] texture results in recrystallised grains that are significantly rotated but still on the γ-fibre of the orientation distribution function [14].

13.2.1 PHENOMENOLOGICAL TREATMENT OF RECRYSTALLISATION

Thermomechanical processing is routine in the production of structural steels, with the aim of refining the austenite grain size and hence the ferrite grain size following transformation. But the rate of production is impressively large; the time spent within the austenite phase field during rolling may be less than two seconds during each rolling reduction, although the delay between rolling-passes may be of the order of two minutes. The structural changes during these short time scales are special and yet relatively simple to model empirically [15].

During hot deformation, the stress required to deform the steel is a function of the plastic strain (ε_p), the plastic-strain rate ($\dot{\varepsilon}_p$) and temperature. That stress is a function $f\{\varepsilon_p\}$ of plastic strain, as is well-understood from any tensile test of a steel. Zener and Holloman proposed [16, 17] that the effects of strain rate and temperature can be combined by writing $\sigma = f\{\varepsilon, Z_k\}$ with Z_k, now known as the Zener-Hollomon parameter, defined as

$$Z_k = \dot{\varepsilon}_p \exp\left(\frac{Q}{RT}\right) \quad \text{s}^{-1} \tag{13.4}$$

where Q is an unspecified heat of activation since most rates are associated with an activated event. The material work hardens during hot-rolling but softens beyond a critical strain ε_p^* corresponding to the recrystallisation of austenite, with

$$\varepsilon_p^* = b_{22}(\overline{L}_o)^{b_{23}} Z_k^{b_{24}} \tag{13.5}$$

where $\overline{L}_o$ is the austenite grain size prior to deformation and b_i are empirical constants. Typical values of the empirical constants are $Q = 312 \, \text{kJ} \, \text{mol}^{-1}$; $b_{22} = 6.97 \times 10^{-4}$; $b_{23} = 0.3$ when the grain size has units of micrometres, and $b_{24} = 0.17$ when the strain rate has units of reciprocal seconds [18].

If the strain ε_p^* is reached while the steel is still being rolled, then the process of change is known as *dynamic recrystallisation*. On the other hand, *metadynamic recrystallisation* is said to occur when recrystallisation follows immediately after a rolling pass when the strain retained in the austenite exceeds that needed to induce recrystallisation. The recrystallised austenite grain size will in general be smaller than the initial size, and grain growth during the interval between passes is inevitable, as illustrated in Figure 13.4 [19]. The full theory for recrystallisation and grain growth is not presented here because it tends to be alloy specific, but can be accessed from extensive literature on the subject [20, 21].

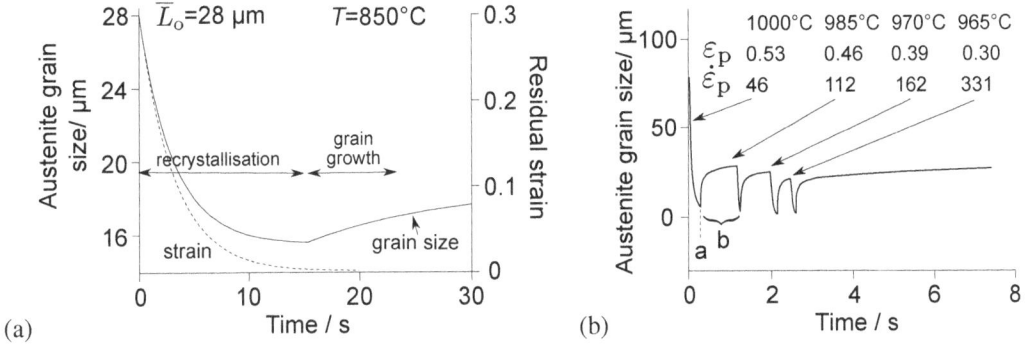

Figure 13.4 (a) Influence of a single pass of rolling deformation to a strain $\varepsilon_p = 0.3$, on the austenite grain size and residual strain (0.3 multiplied by the fraction of unrecrystallised austenite) (b) Influence of four rolling-passes on the austenite grain size and residual strain. In each pass, recrystallisation is complete at points such as 'a', followed by grain growth in regions such as 'b'. The units of $\dot{\varepsilon}_p$ are in s^{-1}. Selected data from Bombac et al. [19].

The general features of controlled rolling are summarised in Figure 13.4. Really quite sophisticated process models now exist to treat the entire sequence of rolling [22, 23], microstructural development [24] and properties [25, 26], so much so that some of these are now used in the on-line control [27] of rolling mills using machine learning methods [28] to ensure product uniformity.

13.2.2 THERMOMECHANICAL PROCESSING: LIMITS TO GRAIN REFINEMENT

Grain size refinement using thermomechanical processing is an important method for improving both the strength and toughness of steels. It is useful therefore to consider the smallest ferrite grain size that can be achieved using this manufacturing method, by balancing the driving force for transformation from austenite to ferrite against the stored energy due to grain boundaries [29]:

$$|\Delta G_V^{\gamma\alpha}| \geq \sigma_\alpha S_V^\alpha - \sigma_\gamma S_V^\gamma \tag{13.6}$$

which for equiaxed grains becomes

$$|\Delta G_V^{\gamma\alpha}| \geq \frac{2\sigma_\alpha}{\bar{L}_\alpha} - \frac{2\sigma_\gamma}{\bar{L}_\gamma}. \tag{13.7}$$

It follows that the smallest ferrite grain size that can be achieved is when all of $G_V^{\gamma\alpha}$ is used up in creating α/α grain boundaries:

$$\bar{L}_\alpha^{min} = \frac{2\sigma_\alpha}{|\Delta G_V^{\gamma\alpha}| + 2\sigma_\gamma/\bar{L}_\gamma}. \tag{13.8}$$

The term $2\sigma_\gamma/\overline{L}_\gamma$ supplements the driving force the ferrite when it forms, eliminates the austenite grain boundaries. Obviously, a reduction in the austenite grain size should always lead to finer ferrite grains but the magnitude of the change depends also on $|\Delta G_V^{\gamma\alpha}|$, i.e., on the undercooling at which the $\gamma \rightarrow \alpha$ transformation occurs. The austenite grain size becomes less important at large undercoolings.[3]

Figure 13.5 shows the ferrite grain size ($\overline{L}_\alpha^{min}$) as a function of the driving force using Equation 13.7 with $\sigma_\alpha = 0.6\,\mathrm{J\,m^{-2}}$. Also illustrated are data from industrial processing. At large undercoolings, the size achieved is much bigger than expected theoretically. This is because of recalescence caused by the larger enthalpy of transformation at greater undercoolings, which heats the steel to higher than intended temperatures, thereby reducing $\Delta G_V^{\gamma\alpha}$. Once this is accounted for, the analysis indicates that it probably is not possible to obtain allotriomorphic ferrite grain-sizes much smaller than 1 μm using thermomechanical processing of the type used in mass production.

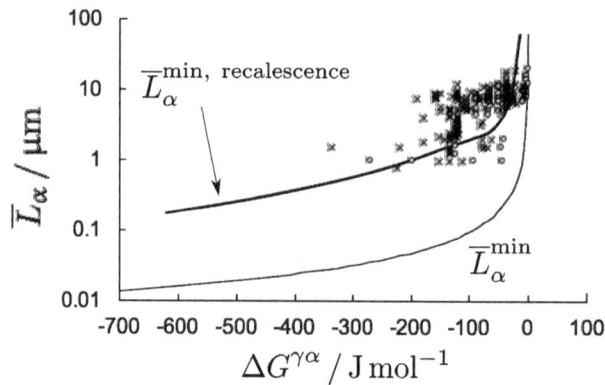

Figure 13.5 Plot of the logarithm of ferrite grain size versus the free energy change at Ar$_3$. The ideal curve represents the values of $\overline{L}_\alpha^{min}$. The points are experimental data; in some cases it is assumed that the grain size quoted in the literature corresponds to the mean lineal intercept. The sources of the data are quoted in [29].

13.3 OVERALL TRANSFORMATION KINETICS

13.3.1 ISOTHERMAL TRANSFORMATION

The evolution of phase fraction as a function of parameters such as temperature, free energy, chemical composition is important in the design of steels. To estimate the necessary transformation kinetics requires an understanding of the nucleation and growth mechanisms and rates, but impingement between precipitates that originate from different locations must be taken into account in order to calculate the phase fraction. This can be accomplished using the extended volume concept, where impingement is ignored so transformation can occur in all regions of the original matrix, irrespective of whether a particular region already contains a precipitate [30–34]. Figure 13.6 shows two precipitates existing at time t; a small interval Δt later, new particles a, b form, and c identifies increments in the sizes of the original pair. The net increase in the extended volume dV_e^α therefore comes $a + b + c$, where α refers to the precipitate phase. However, only those newly transformed regions that lie in previously untransformed matrix can contribute to a change in the real volume dV^α of the product phase:

$$dV^\alpha = \left(1 - \frac{V^\alpha}{V}\right) dV_e^\alpha \qquad (13.9)$$

where it is assumed that the microstructure develops randomly, and V is the total volume. Multiplying the change in extended volume by the probability of finding untransformed regions has the effect of excluding regions such as b, which clearly cannot contribute to the real change in volume of the product. For a random distribution of precipitated particles, this equation can easily be integrated to obtain the real volume fraction:

$$\frac{V^\alpha}{V} = 1 - \exp\left\{-\frac{V_e^\alpha}{V}\right\}. \qquad (13.10)$$

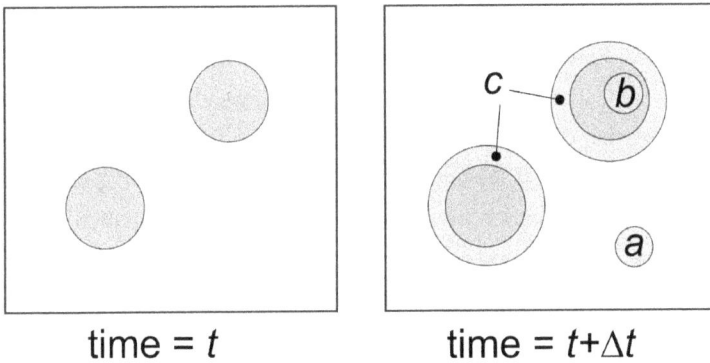

Figure 13.6 The concept of extended volume. Two precipitates have nucleated together and grown to a finite size in the time t. During a further time increment Δt, particles a and b form, of which b lies in a region that already is transformed. The original particles have an increment of growth during Δt, indicated by the regions c.

The extended volume V_e^α is straightforward to calculate using nucleation and growth models while neglecting any impingement effects. Suppose the α grows isotropically at a constant rate v and where the constant nucleation-rate per unit volume is I_V; the volume of a particle nucleated at time $t = \tau$ is given by

$$v_\tau = \frac{4}{3}\pi v^3 (t - \tau)^3. \qquad (13.11)$$

The change in extended volume over the interval τ and $\tau + d\tau$ is

$$dV_e^\alpha = \frac{4}{3}\pi v^3 (t - \tau)^3 \times I_V \times V \times d\tau. \qquad (13.12)$$

Substituting into Equation 13.10 and writing $\xi = V^\alpha/V$ gives

$$dV^\alpha = \left(1 - \frac{V^\alpha}{V}\right)\frac{4}{3}\pi v^3 (t - \tau)^3 I_V V \ d\tau$$

so that

$$-\ln\{1 - \xi\} = \frac{4}{3}\pi v^3 I_V \int_0^t (t - \tau)^3 \ d\tau$$

and

$$\xi = 1 - \exp\{-\pi v^3 I_V t^4/3\}. \qquad (13.13)$$

This equation has been derived for the specific assumptions of random nucleation, a constant nucleation rate and a constant growth rate. There are different possibilities but they often reduce to the general form:

$$\xi = 1 - \exp\{-k_A t^n\} \qquad (13.14)$$

where k_A and n characterise the reaction as a function of time, temperature and other variables. This equation often is used empirically, but only as an economic way of representing experimental data rather than having a significant predictive capability. The temptation to deduce mechanistic information from an empirical application of the Avrami equation should be avoided even if it accurately fits the data, since the fitting parameters can be ambiguous. The fact that Equation 13.14 is applied empirically is sometimes interpreted to be a weakness of the theory, but this is misleading because the application can be rigorous as shown for specific circumstances in the derivation of Equation 13.13.

The example considered here is particularly simple because the nucleation and growth rates are constant. A constant nucleation rate is possible only if the number density of nucleation sites is unlimited and assuming that the interfacial energy and driving force for transformation remain unchanged throughout the process. If the composition of the product phase is identical to that of the parent, then the latter does not enrich or deplete in solute as transformation proceeds. Examples include martensitic transformation and pearlite formation in binary Fe-C alloys.

Many scenarios cannot in reality be handled using analytical equations because the boundary conditions change during the course of transformation. For reactions involving the partitioning of solute, transformation will stop before all of the austenite is consumed, i.e., when an equilibrium fraction of ferrite has been achieved. Given that the whole of the sample volume (V) cannot then transform, $\xi \neq V^\alpha / V$, but rather, $\xi = V^\alpha / V^{\alpha\gamma}$, where $V^{\alpha\gamma}$ is the volume of ferrite that is in equilibrium with austenite within the total volume V. Partitioning of solute leads to changes in the boundary conditions for both growth and nucleation as transformation proceeds, leading to soft-impingement, which often is accounted for by a mean-field approximation whereby the composition of the transforming matrix is averaged incrementally as transformation proceeds.

13.4 SIMULTANEOUS TRANSFORMATIONS

A simple modification is necessary when applying the extended-space concept to the simultaneous formation of two precipitates, say α and β. Equation 13.10 becomes a coupled set of two equations,

$$\mathrm{d}V^\alpha = \left(1 - \frac{V^\alpha + V^\beta}{V}\right)\mathrm{d}V_e^\alpha \qquad \text{and} \qquad \mathrm{d}V^\beta = \left(1 - \frac{V^\alpha + V^\beta}{V}\right)\mathrm{d}V_e^\beta. \qquad (13.15)$$

This can be done for any number of reactions happening together [35, 36]. The resulting set of equations must in general be solved numerically, although a few analytical solutions are possible for special cases [36, 37], with the severe limitation that the different transformation products are related linearly.

A suitably complex scenario closer to reality involves the formation of allotriomorphic ferrite, Widmanstätten ferrite and pearlite forming simultaneously during continuous cooling [36]. Allotriomor-

phic ferrite nucleates at austenite grain boundaries; it is necessary therefore to consider the impingement of α-grains along the austenite grain surfaces, together with the gradual elimination of the
nucleation sites themselves. Cahn [38] treated this problem by a double application of the Avrami
extended space concept. If O_B is the total γ-γ grain boundary area within the sample, in a system
of n precipitating phases, then the total area intersected by the ith phase on a plane parallel to the
boundary (also of area O_B) at a distance y normal to that boundary is, for the jth phase:

$$\Delta O_{j,y} = \underbrace{\left(1 - \frac{\Sigma_{i=1}^{n} O_{i,y}}{O_B}\right)}_{\text{unused fraction of plane at } y} \Delta O_{j,y}^{e} \qquad (13.16)$$

where $\Delta O_{j,y}$ and $\Delta O_{j,y}^{e}$ represent the change in real area intersected by the plane at y by the phase
j, and the corresponding change in extended area, respectively. These changes are during the time
interval mt to $mt + \Delta t$ where m is an integer and $m\Delta t$ the current time. So the net contribution to the
extended area on the plane located at distance y, due to phase j at the time $m\Delta t$ is

$$\Delta O_{j,y}^{e} = O_B \Sigma_{k=0}^{m} (I_{j,k} \Delta \tau A_{j,k,y} \Delta t)$$

where $I_{j,k}$ is the nucleation rate per unit time per unit area, of phase j during the interval $k\Delta\tau$ to
$(k+1)\Delta\tau$, where so that the number of particles nucleated during $\Delta\tau$ is $O_B I_{j,k}\Delta\tau$ assuming extended
space. Using Equation 13.16 to convert extended area to real area, the updated total real-area of
intersection of phase j with the plane at y at time $t + \Delta t = (m+1)\Delta t$ is

$$O_{j,y,t+\Delta t} = O_{j,y,t} + \Delta O_{j,y}.$$

The change in extended volume of phase j, on one side of the austenite grain boundary, over the
time interval t to $t + \Delta t$, is given by integrating numerically:

$$\Delta V_j^{e} = \Delta y \Sigma_{y=0}^{y'} \Delta O_{j,y} \qquad \text{with} \qquad \Delta V_j = \left(1 - \frac{\Sigma_{i=1}^{n} V_i}{V}\right) \Delta V_j^{e} \qquad (13.17)$$

where y' is the maximum in y where phase j intersects the test plane. In this way, the real volume of
each phase can be calculated as a function of time increments as long as the nucleation and growth
functions of each phase are known.

13.4.1   BASIS OF MICROSTRUCTURAL MODELS

It is seems obvious that the application of kinetic theory requires an understanding of the mechanisms of transformations that occur in steels. This is particularly so when several transformations
occur together. Table 13.1 summarises the key features of each of the major transformation products,
with more elaborate explanations in the appropriate chapters that precede this one. Any mathematical model should ideally be consistent with these features in order to embed known phenomena.

13.4.2   ALLOTRIOMORPHIC FERRITE

Classical theory is often used to express the heterogeneous nucleation rate of allotriomorphic ferrite. Whereas the theory can be quite sophisticated in its detail [40, 41], there will be for practical

Table 13.1

Key transformation characteristics in steels. The designations are martensite α', lower bainite α_{lb}, upper bainite α_{ub}, acicular ferrite α_a, Widmanstätten ferrite α_W, allotriomorphic ferrite α, idiomorphic ferrite α_i, pearlite P, substitutional solutes X. Acicular ferrite is the term used for intragranularly nucleated bainite. Consistency of a comment with the transformation concerned is indicated by =, inconsistency by $\neq$; a bullet • identifies the case where the comment is only sometimes consistent with the transformation. The term *parent* γ implies the γ grain from which the product phase grows. Adapted from [39].

Comment	α'	α_{lb}	α_{ub}	α_a	α_W	α	α_i	P
Nucleation and growth reaction	=	=	=	=	=	=	=	=
Plate shape	=	=	=	=	=	$\neq$	$\neq$	$\neq$
IPS shape change with large shear	=	=	=	=	=	$\neq$	$\neq$	$\neq$
Lattice correspondence during growth	=	=	=	=	$\neq$	$\neq$	$\neq$	$\neq$
Co-operative growth of ferrite and cementite	$\neq$	$\neq$	$\neq$	$\neq$	$\neq$	$\neq$	$\neq$	=
High dislocation density	=	=	=	=	•	$\neq$	$\neq$	$\neq$
Necessarily has a glissile interface	=	=	=	=	=	$\neq$	$\neq$	$\neq$
Always has an orientation within the Bain region	=	=	=	=	=	$\neq$	$\neq$	$\neq$
Grows across austenite grain boundaries	$\neq$	$\neq$	$\neq$	$\neq$	$\neq$	=	=	=
High interface mobility at low temperatures	=	=	=	=	=	$\neq$	$\neq$	$\neq$
Acoustic emissions during transformation	=	=	=	$\neq$	$\neq$	$\neq$	$\neq$	$\neq$
Reconstructive diffusion during growth	$\neq$	$\neq$	$\neq$	$\neq$	$\neq$	=	=	=
Bulk redistribution of X atoms during growth	$\neq$	$\neq$	$\neq$	$\neq$	$\neq$	•	•	•
Displacive transformation mechanism	=	=	=	=	=	$\neq$	$\neq$	$\neq$
Reconstructive transformation mechanism	$\neq$	$\neq$	$\neq$	$\neq$	$\neq$	=	=	=
Diffusionless nucleation	=	$\neq$	$\neq$	$\neq$	$\neq$	$\neq$	$\neq$	$\neq$
Only carbon diffuses during nucleation	$\neq$	=	=	=	=	$\neq$	$\neq$	$\neq$
Reconstructive diffusion during nucleation	$\neq$	$\neq$	$\neq$	$\neq$	$\neq$	=	=	=
Often nucleates intragranularly on defects	=	$\neq$	$\neq$	=	$\neq$	$\neq$	=	$\neq$
Diffusionless growth	=	=	=	=	$\neq$	$\neq$	$\neq$	$\neq$
Local equilibrium at interface during growth	$\neq$	$\neq$	$\neq$	$\neq$	$\neq$	•	•	•
Local paraequilibrium at interface during growth	$\neq$	$\neq$	$\neq$	$\neq$	=	$\neq$	$\neq$	$\neq$
Diffusion of carbon during transformation	$\neq$	$\neq$	$\neq$	$\neq$	=	=	=	=
Carbon diffusion-controlled growth	$\neq$	$\neq$	$\neq$	$\neq$	=	•	•	•
Incomplete reaction phenomenon	$\neq$	=	=	=	$\neq$	$\neq$	$\neq$	$\neq$

purposes some fitting parameters due to the inability to predict for real materials, features such as the number density of nucleation sites, the interfacial energy and its anisotropy with any confidence. The fitting constants (b_i) hopefully have generic value over a wide range of steels. If transients are neglected, the steady state γ-grain boundary nucleation rate per unit area is given by

$$I_A^\alpha = b_{20} \frac{2}{\overline{L}_\gamma} \frac{kT}{h} \exp\left\{ -\frac{G^* + Q}{RT} \right\} \tag{13.18}$$

where b_{20} is an empirical constant, $2/\overline{L}_\gamma$ gives the austenite grain boundary area per unit volume; $\overline{L}_\gamma$ is the mean lineal intercept defining the austenite grain size. The activation energy for the barrier to the transfer of atoms across the interface is $Q \approx 200\,\mathrm{J\,mol^{-1}}$, that for nucleation is $G^* = b_{21}\sigma^3/\Delta G^2$, where $\sigma \approx 0.022\,\mathrm{J\,m^{-2}}$ is an effective interfacial energy per unit area.

The emphasis on the interfacial energy term is because the actual scenario is far more complicated. Given the anisotropy of the crystal structures of ferrite and austenite, the interfacial energy will depend on the usual five degrees of freedom and hence cannot be assumed to be constant. For heterogeneous nucleation at the austenite grain boundary, the balancing of interfacial tensions may need to be accounted for and it may not be reasonable to assume that the energy is independent of the size and shape of the embryo. None of this information is available. In real materials, there may be segregation of solutes to the interfaces, which would be expected to change the interfacial energies. The shape of the nucleus will determine how much of the austenite grain boundary is consumed (an energy gain) when the ferrite nucleates so the rate equation should strictly include several interfacial energies including that of the austenite grain boundary, which itself will be a function of the five degrees of freedom and hence of the crystallographic texture. The activation energies, and number densities of nucleation sites, for grain face, edge and corner nucleation are expected to be different, with all three contributing to the overall nucleation rate [38]. There is no way in practice of accounting for these complexities so the fitted σ must be regarded as a global approximated value that hopefully represents a large class of steels if the analysis is to be of value as a predictive tool, albeit with some uncertainty. Similarly, the strain energy associated with the nucleation event is an unknown dependent on similar variables (shape, mechanism, role of heterogeneous nucleation site, diffusional relaxation etc.). A thorough treatment of nucleation theory is given in [40, 41].

The allotriomorphic ferrite often is approximated as growing under paraequilibrium conditions, in the form of discs that grow on both sides of the austenite grain boundary. The ferrite thickness Z normal to the boundary increases parabolically with time as $Z = \alpha_{1d}(t - \tau)^{1/2}$ and the dimension parallel to the boundary is taken to be ηZ, where $\eta \approx 3$ [42]. For anisothermal growth, the change during a time interval dt for a particle nucleated at time $t = \tau = k\Delta\tau$ is therefore

$$\mathrm{d}Z = \frac{1}{2}\alpha_{1d}(t - \tau)^{\frac{1}{2}},$$

which in a form for numerical solution becomes

$$Z_{(m+1)\Delta t} = Z_{m\Delta t} + \frac{1}{2}\alpha_{1d}(m\Delta t - k\Delta\tau)^{-\frac{1}{2}}\Delta t. \tag{13.19}$$

The rate of change of the area of intersection on a plane y, of a disc of allotriomorphic ferrite nucleated at time $k\Delta\tau$, at time $m\Delta T$, is given by

$$\begin{aligned}
\dot{A}_{k,y}^{\alpha} &= \pi\alpha_{1d}^2\eta^2 & Z_{(m+1)\Delta t} > y \\
\dot{A}_{k,y}^{\alpha} &= (\pi\eta^2 Z_{(m+1)\Delta t}^2)/\Delta t & Z_{(m+1)\Delta t} = y \\
\dot{A}_{k,y}^{\alpha} &= 0 & Z_{(m+1)\Delta t} < y
\end{aligned}$$

Equation 13.17 can then be used to calculate the change in extended volume, but this only represents half the change since the ferrite grows into both of the adjacent austenite grains.

13.4.3 WIDMANSTÄTTEN FERRITE AND PEARLITE

During unhindered growth, the length Z of a plate of α_W, nucleated at time τ, is given by $v_\ell(t - \tau)$ (Chapter 7). For anisothermal transformation,

$$Z_{(m+1)\Delta t} = Z_{m,\Delta t} + v_\ell \Delta t.$$

If the shape of α_W is modelled as a tetragonal prism of thickness to length ratio $\eta \approx 0.02$, then

$$\begin{aligned}
\dot{A}_{k,y}^{\alpha_W} &= 2\eta v_\ell^2(m\Delta t - k\Delta\tau) & Z_{(m+1)\Delta t} > y \\
\dot{A}_{k,y}^{\alpha_W} &= \eta Z_{(m+1)\Delta t}^2/\Delta t & Z_{(m+1)\Delta t} = y \\
\dot{A}_{k,y}^{\alpha_W} &= 0 & Z_{(m+1)\Delta t} < y.
\end{aligned}$$

The transformation is displacive so the plates are confined to the γ-grains in which they grow, which means that Equation 13.17 can be applied directly to calculate the change in extended volume as a function of the nucleation and growth rates.

When the shape of pearlite is approximated as a disc which has the same length as diameter ($\eta = 1$), its half thickness when nucleated at time τ assuming a constant growth rate is given by

$$Z = v_P(t - \tau) \qquad \text{with} \qquad Z_{(m+1)\Delta t} = Z_{m\Delta t} + v_P\Delta t,$$

so that

$$\begin{aligned}
\dot{A}_{k,y}^{P} &= 2\pi\eta^2 v(m\Delta t - k\Delta\tau) & Z_{(m+1)\Delta t} > y \\
\dot{A}_{k,y}^{P} &= \pi\eta^2 Z_{(m+1)\Delta t}^2/\Delta t & Z_{(m+1)\Delta t} = y \\
\dot{A}_{k,y}^{P} &= 0 & Z_{(m+1)\Delta t} < y.
\end{aligned}$$

Any error in the calculation of the first phase to form propagates into subsequent transformation products; an exaggerated value of allotriomorphic ferrite fraction naturally leads to an underestimation of the Widmanstätten ferrite content. It is noticeable that pearlite forms over a narrow temperature range because once the austenite composition falls within the Hultgren extrapolation region (page 504), the cooperative growth rate of ferrite and cementite is relatively large.

Following validation against experimental data, Figure 13.7 [36], typical calculations are illustrated in Figure 13.8a–d to show the influence of austenite grain size and cooling rate on the development of microstructure. At a given austenite grain size, a greater cooling rate suppresses transformation

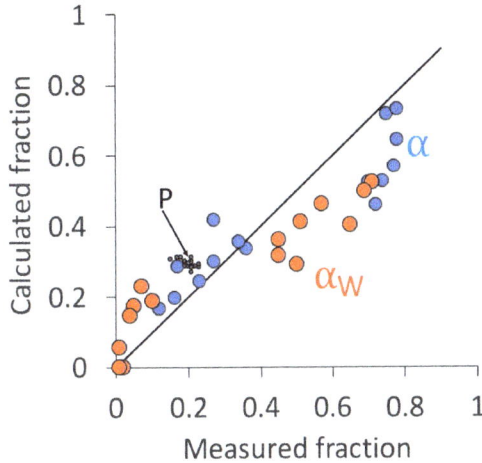

Figure 13.7 Calculated microstructures for a variety of steels, austenite grain sizes and cooling rates, compared against experimental data due to Bodnar and Hansen [43].

to lower temperatures, thus increasing the proportion of Widmanstätten ferrite. A smaller austenite grain size, on the other hand, results in a significantly smaller amount of Widmanstätten ferrite. These results are consistent with experimentally observed trends.

Idiomorphic ferrite can be induced on heterogeneous nucleation sites that are introduced deliberately into the austenite [44–46]. The resulting refinement of microstructure improves toughness [47]. There is then a competition between grain boundary nucleated allotriomorphic ferrite, and intragranularly nucleated idiomorphic ferrite. This kind of a competition is ideally suited for the simultaneous transformations method; some calculations for isothermal transformation are illustrated in Figure 13.9, for a case where idiomorphic ferrite nucleates on boron nitride particles. The relative fractions of allotriomorphic and idiomorphic ferrite are related to the corresponding ratio of grain boundary and intragranular nucleation sites. A coarse austenite grain structure therefore favours idiomorphic ferrite. Nucleation at the austenite grain boundaries is usually associated with a lower activation energy than from intragranular sites. When $\overline{L}_\gamma$ is relatively small, the initial rapid formation of allotriomorphic ferrite partitions carbon into the austenite, which makes the formation of idiomorphic ferrite even more difficult. As a result, the fraction of idiomorphic ferrite is considerably reduced.

In some steels, nucleation at the austenite grain boundaries can be rendered ineffective, either by poisoning the sites due to the segregation of solutes such as boron, or by introducing a thin layer of slow-growing allotriomorphic ferrite. Figure 13.10 shows such a case where the continuous layer of α leaves only the intragranular sites where α_W can form [48]. The competition between grain boundary and intragranularly nucleation apply also to the bainite transformation [49].

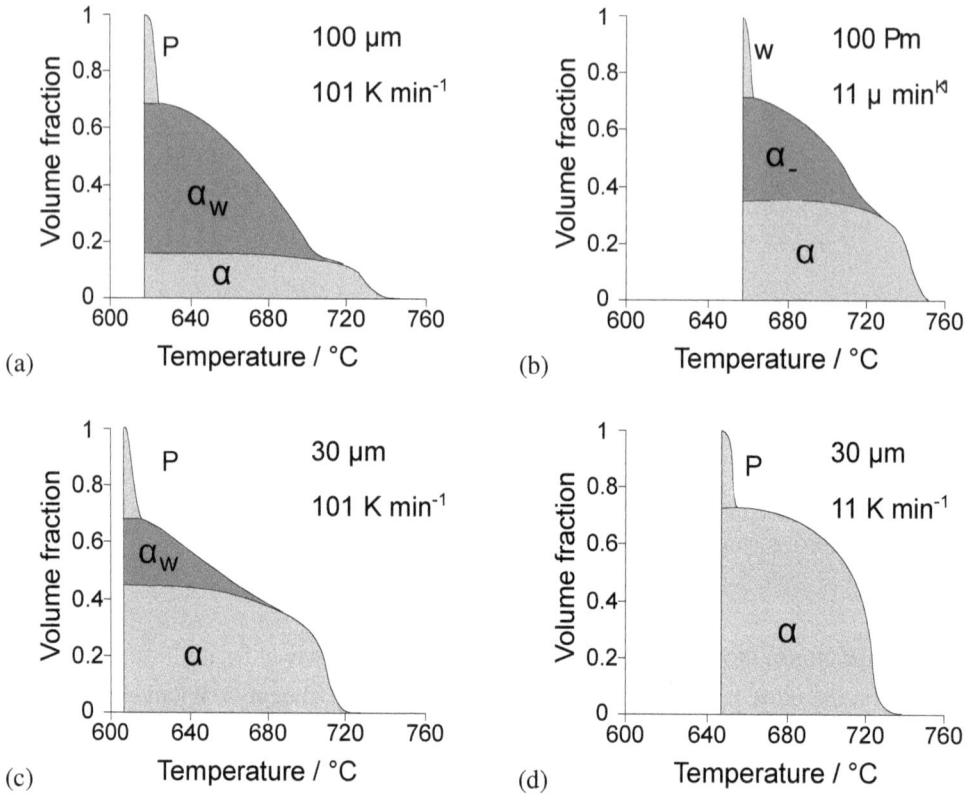

Figure 13.8 Calculations showing the evolution of microstructure; (a-d) from austenite in a Fe-0.18C-1.15Mn wt% steel as a function of the austenite grain size and constant cooling rate [36].

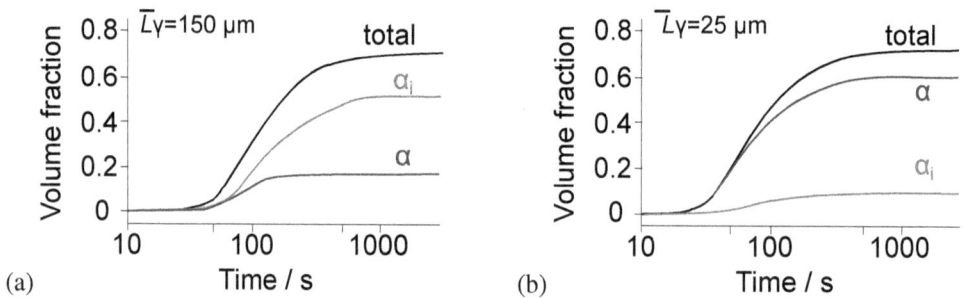

Figure 13.9 Microstructural evolution during isothermal transformation at 720°C in Fe-0.13C-0.23Si-1.44Mn-0.003B-0.0074N wt%, for two different austenite grain sizes [47].

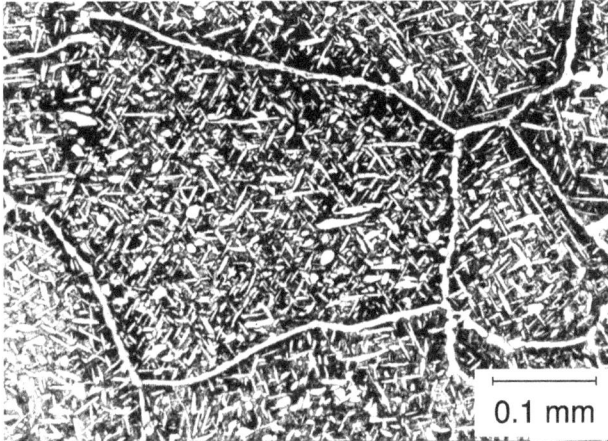

Figure 13.10 Intragranular Widmanstätten ferrite stimulated to form intragranularly by the complete decoration of austenite grain boundaries by slow-thickening, thin layers of allotriomorphic ferrite. The original austenite grain boundaries are eliminated as potential nucleation sites by the formation of allotriomorphic ferrite in the manner illustrated [48].

13.5   TIME-TEMPERATURE-TRANSFORMATION DIAGRAMS

Isothermal transformation diagrams contain curves, each of which represents a specific degree of transformation for a specific product, as a function of temperature, time and the austenite grain structure (usually represented as a grain size). The curves have a characteristic C-shape because driving force is limited at higher temperatures whereas mobility is limited at low temperatures. The maximum rate is therefore obtained at some intermediate temperature. The corresponding equilibrium phase diagram sets the thermodynamic limits for the decomposition of austenite so TTT diagrams and equilibrium phase diagrams are somewhat connected (Figure 13.11); it might be expected that all C-curves for α and α_I begin at the Ae_3 temperature. The upper limit of the temperature at which pearlite can form is set by the requirement that cementite and ferrite must both be able to precipitate from austenite (Figure 12.7). The comparison with the equilibrium phase diagram fails for the displacive transformations which require even greater undercoolings; Widmanstätten ferrite, bainite and martensite C-curves have their maximum limits defined by the W_S, B_S and M_S temperatures respectively (Figure 13.11).

Figure 13.11 also shows the isothermal transformation curve for austenite formation. The transformation rate increases indefinitely with superheating because both the diffusion coefficients and driving force for austenite formation increase when $T > Ae_3$.

An essential feature of accurately determined TTT diagrams is that they can be divided into two regimes, the higher temperature one corresponding to the domain where iron and substitutional solutes are able to diffuse over the length scales of the microstructure, and a lower temperature regime where displacive transformations have a kinetic advantage. The diagram therefore consists of two main C-curves, one set for reconstructive and the other for displacive transformations (Figure 13.12). Solutes that decrease the driving force for the decomposition of austenite retard the rate

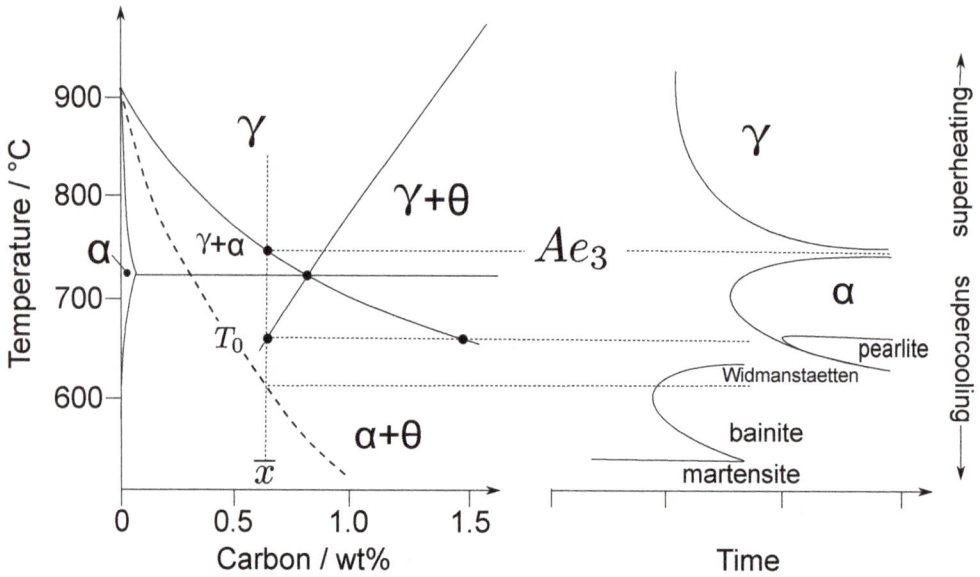

Figure 13.11 The relationship between a TTT diagram for a hypoeutectoid Fe-C alloy with a concentration $\bar{x}$ of carbon, and the corresponding binary phase diagram; equilibrium phase boundaries are continuous lines, whereas the T_0 curve is dashed to indicate that it is not an equilibrium boundary except in the phase diagram for pure iron. The TTT diagram illustrates the transformation of supercooled austenite, or during superheating, the formation of austenite.

of transformation and cause both of the C-curves to be displaced to longer times. At the same time they depress the martensite-start temperature. The retardation is always more pronounced for reconstructive reactions where all atoms have to diffuse over distances comparable to the size of the transformation product. This diffusional lag exaggerates the effect of solutes on the upper C-curve relative to the lower C-curve.

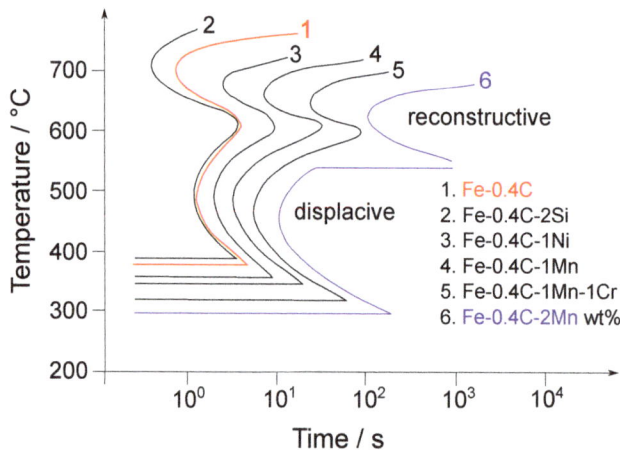

Figure 13.12 Calculated TTT diagrams showing the C-curves for the initiation of transformation, for a variety of steels.

As predicted by Zener [50], when the two curves can be distinguished clearly, the lower C-curve has a flat top. This can be identified with the Widmanstätten-start or bainite-start temperature, whichever is the larger in magnitude [51].

13.5.1 CONTINUOUS COOLING TRANSFORMATION DIAGRAMS

Figure 13.8 represents the evolution of transformation that occurs during continuous cooling. The ability to calculate such diagrams as a function of specific cooling conditions, the steel composition and the parent austenite grain structure can be very useful in the the design of steels. After all, large steel components cannot be transformed isothermally into a homogeneous microstructure because the temperature distribution within the component will not be uniform. The fact that the surface regions will experience a greater cooling rate than those within the depths of the sample can sometimes be used to advantage. Large diameter concrete-reinforcing steel bars in their austenitic condition, when quenched into water form untempered martensite at the surface, but the residual heat within the core then diffuses and tempers the martensite, thus optimising properties. The vast majority of steel produced undergoes anisothermal transformation; the sort of information presented in Figure 13.8 is traditionally represented as continuous cooling transformation (CCT) diagrams, usually determined using a dilatometer.

In comparison with a TTT diagram, the first phases (e.g. α, P) to form from austenite will occur at a greater undercooling during continuous cooling transformation. What happens to subsequent transformations is more complex. Any changes in the composition of the austenite due to the preceding phases can suppress or elevate the rate of transformation (e.g. of α_b) depending kinetics depending on how the changes influence the stability of the residual austenite. Figure 13.13 shows schematically how the initiation of bainite is affected beyond the vertical line c because of the prior formation of α. Although bainite is depressed to lower temperatures by the prior formation of allotriomorphic ferrite as the cooling rate decreases, the temperature range over which bainite forms is reduced eventually. This is because very slow cooling rates give ample opportunity for transformation to be completed over a smaller temperature range as illustrated by the rising curve de. Because the ferrite and bainite domains are separated by a time gap, the continuity of constant volume fraction contours is interrupted, but they must still be plotted so that their loose ends are connected by a cooling curve as illustrated by ab. Naturally, the temperature at which martensite is initiated also depends on prior transformation.

The rate of transformation in a given steel with a known austenite grain size can be described with just one TTT diagram. However, a different CCT diagram is required for each cooling function, e.g. whether the cooling rate is constant or Newtonian. It is therefore necessary to plot the actual cooling curves used in the derivation of the CCT diagram. Each cooling curve must begin at the highest temperature where transformation becomes possible (usually the Ae_3 temperature). Each CCT diagram requires a specification of the chemical composition of the steel, the austenitisation conditions, the austenite grain size and the cooling conditions. The diagrams are therefore specific to particular processes and therefore lack the generality of TTT diagrams.

It may be possible to adapt TTT diagrams, for dealing with the many variables that determine

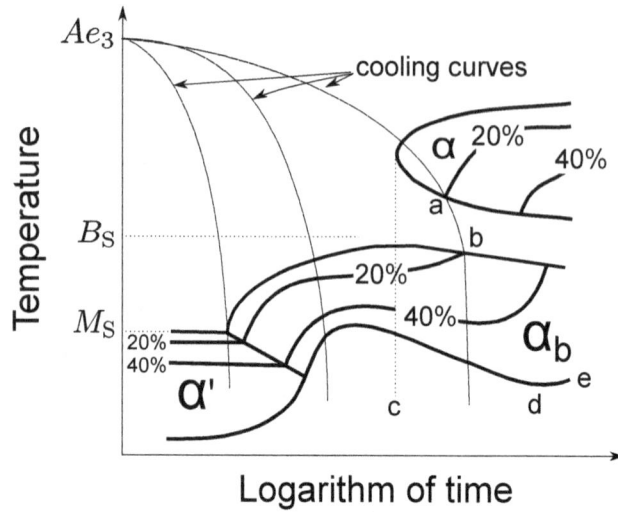

Figure 13.13 Schematic CCT diagram showing the terminology used in describing air-cooling from the austenitisation temperature (i.e. normalising) and furnace cooling.

continuous cooling transformation. In the Scheil *additive reaction rule* [52], a cooling curve is treated as a combination of a isothermal reaction steps. In Figure 13.14, a fraction $\xi = 0.05$ of transformation is achieved during continuous cooling when

$$\sum_i \frac{\Delta t_i}{t_i} = 1 \qquad (13.20)$$

with the summation beginning as soon as the parent phase cools below the equilibrium temperature.

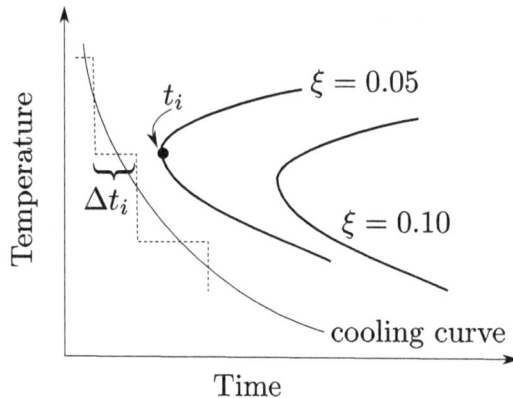

Figure 13.14 The Scheil method for converting between isothermal and anisothermal transformation data. The C-curves plotted represent isothermal transformation to two different volume fractions. The continuous cooling curve is divided into isothermal steps (dashed line).

The rule can be justified if the reaction rate depends solely on ζ and T. This is unlikely, but there are examples where the rule has been empirically applied to bainite with success [53].

REFERENCES

1. M. Hillert: 'On the theory of normal and abnormal grain growth', *Acta Metallurgica*, 1965, **13**, 227–238.

2. E. Nes, N. Ryum, and O. Hunderi: 'On the Zener drag', *Acta Metallurgica*, 1985, **33**, 11–22.

3. I. Andresen, and Ø. Grong: 'Analytical modelling of grain growth in metals and alloys in the presence of growing and dissolving precipitates-I. Normal grain growth', *Acta Metallurgica et Materalia*, 1995, **43**, 2673–1688.

4. H. Pous-Romero, I. Lonardelli, D. Cogswell, and H. K. D. H. Bhadeshia: 'Austenite grain growth in a nuclear pressure vessel steel', *Materials Science & Engineering A*, 2013, **567**, 72–79.

5. I. Andresen, and Ø. Grong: 'Analytical modelling of grain growth in metals and alloys in the presence of growing and dissolving precipitates', *Acta Metallurgica et Materalia*, 1995, **43**, 2689–2700.

6. T. Furu, K. Marthinsen, and E. Nes: 'Modelling recrystallisation', *Materials Science and Technology*, 1990, **6**, 1093–1102.

7. S. B. Singh, and H. K. D. H. Bhadeshia: 'Topology of grain deformation', *Materials Science and Technology*, 1998, **15**, 832–834.

8. Q. Zhu, C. M. Sellars, and H. K. D. H. Bhadeshia: 'Quantitative metallography of deformed grains', *Materials Science and Technology*, 2007, **23**, 757–766.

9. M. Tanaka, A. Kayama, Y. Ito, and R. Kato: 'Change in fractal dimension of the grain boundaries in pure Zn polycrystals during creep', *Journal of Material Science*, 1998, **33**, 5747–5757.

10. M. Umemoto, H. Ohtsuka, and I. Tamura: 'Prediction of ferrite grain size transformed from thermomechanically processed austenite', In: D. P. Dunne, and T. Chandra, eds. *Proceedings of International Conference on HSLA steels*. Sou, 1984:107–112.

11. R. Bengochea, B. Lopez, and I. Gutierrez: 'Microstructural evolution during the austenite to ferrite transformation from deformed austenite', *Metallurgical and Materials Transactions A*, 1998, **29**, 417–426.

12. M. F. Ashby: 'Work hardening of dispersion-hardened crystals', *Acta Metallurgica*, 1966, **14**, 1157–1178.

13. F. J. Humphreys: 'Local lattice rotations at second phase particles in deformed metals', *Acta Metallurgica*, 1979, **27**, 1801–1814.

14. W. B. Hutchinson: 'Recrystallisation textures in iron resulting from nucleation at grain boundaries', *Acta Metallurgica*, 1989, **37**, 1047–1056.

15. J. H. Beynon, and C. M. Sellars: 'Modelling microstructure and its effects during multipass hot rolling', *ISIJ international*, 1992, **32**, 359–367.

16. C. Zener, and J. H. Hollomon: 'Effect of strain rate upon plastic flow of steel', *Journal of Applied Physics*, 1944, **15**, 22–32.

17. C. Zener, and J. H. Hollomon: 'Plastic flow and rupture of metals', *Transactions of ASM*, 1944, **33**, 163–215.

18. C. M. Sellars: 'Modelling microstructural development during hot-rolling', *Materials Science and Technology*, 1990, **6**, 1072–1081.

19. D. Bombac, M. Peet, S. Zenitani, S. Kimura, T. Kurimura, and H. K. D. H. Bhadeshia: 'An integrated hot-rolling and microstructure model for dual-phase steels', *Modelling and Simulation in Materials Science and Engineering*, 2014, **22**, 045005.

20. I. Tamura, H. Sekine, T. Tanaka, and C. Ouchi: Thermomechanical processing of high-strength low-alloy steels: London, U.K.: Butterworth & Co., 1988.

21. T. Gladman: The Physical Metallurgy of Microalloyed Steels: IOM Communications, London, 1996.

22. C. M. Sellars, and J. A. Whiteman: 'Recrystallization and grain growth in hot rolling', *Metal Science*, 1979, **13**, 187–194.

23. C. M. Sellars: 'Computer modelling of hot working processes', *Materials Science and Technology*, 1985, **1**, 325–332.

24. E. J. Palmiere, C. M. Sellars, and S. V. Subramanian: 'Modelling of thermomechanical rolling', In: *Niobium 2001*. Warrendale, Pennsylvania, USA: TMS, 2001:501–526.

25. R. C. Hwang, Y. J. Chen, and H. C. Huang: 'Artificial intelligent analyzer for mechanical properties of rolled steel bar by using neural networks', *Expert Systems with Applications*, 2010, **37**, 3136–3139.

26. I. Mohanty, S. Sarkar, B. Jha, S. Das, and R. Kumar: 'Online mechanical property prediction system for hot rolled IF steel', *Ironmaking and Steelmaking*, 2014, **41**, 618–627.

27. M. Schlang, B. Feldkeller, B. Lang, T. Poppe, and T. Runkler: 'Neural computation in steel industry', In: *Control Conference (ECC), 1999 European*. Washington, D.C., USA: IEEE Computer Society, 1999:2922–2927.

28. S. B. Singh, H. K. D. H. Bhadeshia, D. J. C. MacKay, H. Carey, and I. Martin: 'Neural network analysis of steel plate processing', *Ironmaking and Steelmaking*, 1998, **25**, 355–365.

29. T. Yokota, C. Garcia-Mateo, and H. K. D. H. Bhadeshia: 'Formation of nanostructured steel by phase transformation', *Scripta Materialia*, 2004, **51**, 767–770.

30. W. A. Johnson, and R. F. Mehl: 'Reaction kinetics in processes of nucleation and growth', *TMS-AIMME*, 1939, **135**, 416–458.

31. M. Avrami: 'Kinetics of phase change 1', *Journal of Chemical Physics*, 1939, **7**, 1103–1112.

32. M. Avrami: 'Kinetics of phase change 2', *Journal of Chemical Physics*, 1940, **8**, 212–224.

33. M. Avrami: 'Kinetics of phase change 3', *Journal of Chemical Physics*, 1941, **9**, 177–184.

34. A. N. Kolmogorov: 'On statistical theory of metal crystallisation', *Izvestiya Akad. Nauk SSSR (Izvestia Academy of Science, USSR)*, 1937, **Ser. Math. 3**, 335–360.

35. J. D. Robson, and H. K. D. H. Bhadeshia: 'Modelling precipitation sequences in power plant steels: Part I, kinetic theory', *Materials Science and Technology*, 1997, **13**, 631–639.

36. S. Jones, and H. K. D. H. Bhadeshia: 'Kinetics of the simultaneous decomposition of austenite into several transformation products', *Acta Materialia*, 1997, **45**, 2911–2920.

37. T. Kasuya, K. Ichikawa, M. Fuji, and H. K. D. H. Bhadeshia: 'Real and extended volumes in simultaneous transformations', *Materials Science and Technology*, 1999, **15**, 471–473.

38. J. W. Cahn: 'The kinetics of grain boundary nucleated reactions', *Acta Metallurgica*, 1956, **4**, 449–459.

39. H. K. D. H. Bhadeshia, and J. W. Christian: 'The bainite transformation in steels', *Metallurgical & Materials Transactions A*, 1990, **21A**, 767–797.

40. J. W. Christian: Theory of Transformations in Metals and Alloys, Part I: 3 ed., Oxford, U. K.: Pergamon Press, 2003.

41. K. F. Kelton, and A. L. Greer: Nucleation in condensed matter: applications in materials and biology: Oxford, U. K.: Pergamon Press, 2010.

42. J. R. Bradley, J. M. Rigsbee, and H. I. Aaronson: 'Growth kinetics of grain boundary ferrite allot. in Fe-C alloys', *Metallurgical Transactions A*, 1977, **8**, 323–333.

43. R. L. Bodnar, and S. S. Hansen: 'Effects of austenite grain size and cooling rate on Widmanstätten ferrite formation in low alloy steels', *Metallurgical & Materials Transactions A*, 1994, **25A**, 665–675.

44. D. J. Abson, and R. J. Pargeter: 'Factors influencing the as-deposited strength, microstructure and toughness of manual metal arc welds suitable for C-Mn steel fabrications', *International Materials Reviews*, 1986, **31**, 141–196.

45. A. R. Mills, G. Thewlis, and J. A. Whiteman: 'Nature of inclusions in steel weld metals and their influence on the formation of acicular ferrite', *Materials Science and Technology*, 1987, **3**, 1051–1061.

46. H. Homma, S. Ohkita, S. Matsuda, and K. Yamamoto: 'Improvement of HAZ toughness in HSLA steel by introducing finely dispersed Ti-oxide', *Welding Journal, Research Supplement*, 1987, **66**, 301s–309s.

47. S. Jones, and H. K. D. H. Bhadeshia: 'Competitive formation of inter- and intragranularly nucleated ferrite', *Metallurgical & Materials Transactions A*, 1997, **28A**, 2005–2103.

48. A. Ali, and H. K. D. H. Bhadeshia: 'Aspects of the nucleation of Widmanstätten ferrite', *Materials Science and Technology*, 1990, **6**, 781–784.

49. J. R. Yang, and H. K. D. H. Bhadeshia: 'Thermodynamics of the acicular ferrite transformation in alloy-steel weld deposits', In: S. A. David, ed. *Advances in Welding Technology and Science*. Materials Park, Ohio, USA: ASM International, 1987:187–191.

50. C. Zener: 'Kinetics of the decomposition of austenite', *Transactions of the American Institute of Mining and Metallurgical Engineers*, 1946, **167**, 550–595.

51. H. K. D. H. Bhadeshia: 'Rationalisation of shear transformations in steels', *Acta Metallurgica*, 1981, **29**, 1117–1130.

52. E. Scheil: 'Anlaufzeit der austenitumwandlung', *Archiv für das Eisenhüttenwesen*, 1935, **12**, 565–567.

53. M. Umemoto, K. Horiuchi, and I. Tamura: 'Transformation kinetics of bainite during isothermal holding and continuous cooling', *Transaction of the Iron Steel Institute of Japan*, 1982, **22**, 854–861.

Notes

[1] This parabolic relationship arises because the driving force for grain growth decreases as $\overline{L}$ increases.

[2] If the specimen axes consist of the rolling direction, transverse direction and the direction normal to the planar sample, then the γ-fibre refers to the case where in a polycrystalline sample, the $\langle 111 \rangle$ directions tend to be parallel to the normal direction.

[3] The analysis presented is not quite correct because the relationship between S_V and $\overline{L}$ depends on the shape of the grains.

Indices

Author index

Subject index

For Product Safety Concerns and Information please contact our EU
representative GPSR@taylorandfrancis.com
Taylor & Francis Verlag GmbH, Kaufingerstraße 24, 80331 München, Germany